Studies in Systems, Decision and Control

Volume 427

The series "Studies in Systems, Decision and Control" (SSDC) covers both new developments and advances, as well as the state of the art, in the various areas of broadly perceived systems, decision making and control–quickly, up to date and with a high quality. The intent is to cover the theory, applications, and perspectives on the state of the art and future developments relevant to systems, decision making, control, complex processes and related areas, as embedded in the fields of engineering, computer science, physics, economics, social and life sciences, as well as the paradigms and methodologies behind them. The series contains monographs, textbooks, lecture notes and edited volumes in systems, decision making and control spanning the areas of Cyber-Physical Systems, Autonomous Systems, Sensor Networks, Control Systems, Energy Systems, Automotive Systems, Biological Systems, Vehicular Networking and Connected Vehicles, Aerospace Systems, Automation, Manufacturing, Smart Grids, Nonlinear Systems, Power Systems, Robotics, Social Systems, Economic Systems and other. Of particular value to both the contributors and the readership are the short publication timeframe and the world-wide distribution and exposure which enable both a wide and rapid dissemination of research output.

Indexed by SCOPUS, DBLP, WTI Frankfurt eG, zbMATH, SCImago.

All books published in the series are submitted for consideration in Web of Science.

More information about this series at https://link.springer.com/bookseries/13304

Nguyen Ngoc Thach · Vladik Kreinovich ·
Doan Thanh Ha · Nguyen Duc Trung
Editors

Financial Econometrics: Bayesian Analysis, Quantum Uncertainty, and Related Topics

Springer

Editors
Nguyen Ngoc Thach
Banking University of HCMC
Ho Chi Minh City, Vietnam

Doan Thanh Ha
Banking University of HCMC
Ho Chi Minh City, Vietnam

Vladik Kreinovich
Department of Computer Science
University of Texas at El Paso
El Paso, TX, USA

Nguyen Duc Trung
Banking University of HCMC
Ho Chi Minh City, Vietnam

ISSN 2198-4182 ISSN 2198-4190 (electronic)
Studies in Systems, Decision and Control
ISBN 978-3-030-98688-9 ISBN 978-3-030-98689-6 (eBook)
https://doi.org/10.1007/978-3-030-98689-6

This Springer imprint is published by the registered company Springer Nature Switzerland AG
The registered company address is: Gewerbestrasse 11, 6330 Cham, Switzerland

Preface

One of the main objectives of financial econometrics is to use finances to enhance economic development. The more information, the more knowledge we take into account, the more adequate methods we use to process this information and this knowledge, the better the results will be. In view of this, the main focus of this volume is:

- on Bayesian analysis—technique for taking prior knowledge into account, and
- on quantum uncertainty—application of well-tested sophisticated methods developed in modern physics for the analysis of economic phenomena.

Papers presented in this volume also cover related techniques, with economic and financial application being the unifying theme.

This volume shows what has been achieved, but even more important are remaining open problems. We hope that this volume will:

- inspire practitioners to learn how to apply state-of-the-art Bayesian, quantum, and related techniques to economic and financial problems, and
- inspire researchers to further improve the existing techniques and to come up with new techniques for studying economic and financial phenomena.

We want to thank all the authors for their contributions and all anonymous referees for their thorough analysis and helpful comments.

The publication of this volume is partly supported by the Banking University of Ho Chi Minh City, Vietnam. Our thanks to the leadership and staff of the Banking University, for providing crucial support. Our special thanks to Prof. Hung T. Nguyen for his valuable advice and constant support.

We would also like to thank Prof. Janusz Kacprzyk (Series Editor) and Dr. Thomas Ditzinger (Senior Editor, Engineering/Applied Sciences) for their support and cooperation for this publication.

Ho Chi Minh City, Vietnam — Nguyen Ngoc Thach
El Paso, TX, USA — Vladik Kreinovich
Ho Chi Minh City, Vietnam — Doan Thanh Ha
Ho Chi Minh City, Vietnam — Nguyen Duc Trung
December 2021

Contents

Correcting Interval-Valued Expert Estimates: Empirical Formulas Explained

Laura Adriana Berrout Ramos, Vladik Kreinovich, and Kittawit Autchariyapanitkul

Abstract Experts' estimates are approximate. To make decisions based on these estimates, we need to know how accurate these estimates are. Sometimes, experts themselves estimate the accuracy of their estimates—by providing the interval of possible values instead of a single number. In other cases, we can gauge the accuracy of the experts' estimates by asking several experts to estimates the same quantity and using the interval range of these values. In both situations, sometimes the interval is too narrow—e.g., if an expert is overconfident. Sometimes, the interval is too wide—if the expert is too cautious. In such situations, we correct these intervals, by making them, correspondingly, wider or narrower. Empirical studies show that people use specific formulas for such corrections. In this paper, we provide a theoretical explanation for these empirical formulas.

1 Formulation of the Problem

Expert estimates are needed. In economics—as in many other application areas—we often rely on expert estimates.

Expert estimates are approximate. Of course, expert estimates are approximate. So, to be able to use them effectively, we need to know how accurate they are.

L. A. Berrout Ramos (✉) · V. Kreinovich
University of Texas at El Paso, 500 W. University, El Paso, TX 79968, USA
e-mail: laberroutra@utep.edu

V. Kreinovich
e-mail: vladik@utep.edu

K. Autchariyapanitkul
Faculty of Economics, Maejo University, Chiang Mai, Thailand
e-mail: Kittawit_a@mju.ac.th

N. Ngoc Thach et al. (eds.), *Financial Econometrics: Bayesian Analysis, Quantum Uncertainty, and Related Topics*, Studies in Systems, Decision and Control 427,
https://doi.org/10.1007/978-3-030-98689-6_1

How do we usually gauge the accuracy of expert estimates. There are two main idea on how to gauge the accuracy of an expert estimate:

- the first is to ask the expert him/herself to gauge this accuracy, namely, to provide an *interval* of possible values instead of a single value;
- the second is to ask one or more other experts, and to consider the range formed by their values as the reasonable range of possible values for the corresponding quantity; for example, if one experts predicts the value 15, and the two others predict 10 and 20, then we take the interval [10, 20].

Need for correction. When an expert provides an interval of possible values:

- sometimes, the expert is too confident, and the interval provided by the expert is too narrow; in this case, a natural idea is to widen it;
- sometimes, the expert is too cautious, and the interval provided by the expert is too wide; in this case, a natural idea is to make it narrower.

Similarly, when we have an interval range formed by numerical estimates made by different experts:

- sometimes, the experts' estimates are too close to each other, so the resulting interval is too narrow; in this case, a natural idea is to widen it;
- sometimes, the experts' estimates are too far away from each other; for example, one expert's predictions are too optimistic, and another expert's predictions are too pessimistic; in this case, the resulting interval is too wide, so a natural idea is to make it narrower.

How do people correct the corresponding intervals? Empirical data shows the corrected version $[A, B]$ of the original interval $[a, b]$ usually follows the formula

$$[A, B] = \left[a \cdot \frac{1+\alpha}{2} + b \cdot \frac{1-\alpha}{2}, a \cdot \frac{1-\alpha}{2} + b \cdot \frac{1+\alpha}{2}\right] \tag{1}$$

for some value $\alpha > 0$;

- the values $\alpha < 1$ correspond to shrinking, while
- the values $\alpha > 1$ correspond to stretching of the interval;

see, e.g., Smithson (2015) and references therein.

Historical comment. The formula (1) was first proposed in Gajdos and Vergnaud (2013) for expert estimates of probabilities. In Smithson (2013), this formula was extended to general (not necessarily probabilistic) expert estimates.

Remaining problem. Why do people use this particular formula to correct interval-valued expert estimates?

In this paper, we provide a possible explanation for this formula.

2 Analysis of the Problem

What we look for. We are looking for a formula that describes the corrected interval $[A, B]$—i.e., that describes both endpoints A and B of this interval—in terms of the inputs interval $[a, b]$ – i.e., in terms of its endpoints a and b. In other words, we need to find algorithms $A(a, b)$ and $B(a, b)$ that, given the expert-provided values a and b, produced the corrected values A and B.

Let us analyze what are the natural properties of these algorithms.

The correction formula should not depend on the monetary unit. The same financial predictions can be described in different monetary units. For example, predictions related to Mexican economy can be made in Mexican pesos or in US dollars.

We are looking for general correction formulas, formulas that would be applicable to all possible interval-valued expert estimates. Suppose that we first applied this formula to the interval $[a, b]$ described in one monetary units. Then, in these units, the corrected interval takes the form

$$[A(a, b), B(a, b)]. \tag{2}$$

Another possibility is:

- to first translate into a different monetary unit,
- to make a correction there, and then
- to translate the result back into the original monetary unit.

If we select a different monetary unit which is λ times smaller than the original one, then all numerical values multiply by λ. In particular, in the new units, the original interval $[a, b]$ will take the form $[\lambda \cdot a, \lambda \cdot b]$. If we apply the same correction algorithm to this interval, we get—in the new units—the following corrected interval:

$$[A(\lambda \cdot a, \lambda \cdot b), B(\lambda \cdot a, \lambda \cdot b)]. \tag{3}$$

To describe the corrected interval (3) in the original units, we need to divide both its endpoints by λ. As a result, we get the corrected interval expressed in the original units:

$$\left[\frac{1}{\lambda} \cdot A(\lambda \cdot a, \lambda \cdot b), \frac{1}{\lambda} \cdot B(\lambda \cdot a, \lambda \cdot b)\right]. \tag{4}$$

As we have mentioned, it is reasonable to require that the corrected interval should be the same whether we use the original monetary units or different units, i.e., that we should have

$$[A(a, b), B(a, b)] = \left[\frac{1}{\lambda} \cdot A(\lambda \cdot a, \lambda \cdot b), \frac{1}{\lambda} \cdot B(\lambda \cdot a, \lambda \cdot b)\right]. \tag{5}$$

The two intervals are equal if their left endpoints are equal to each other and their right endpoints are equal to each other. So, the equality (5) means that:

$$A(a, b) = \frac{1}{\lambda} \cdot A(\lambda \cdot a, \lambda \cdot b) \tag{6}$$

and

$$B(a, b) = \frac{1}{\lambda} \cdot B(\lambda \cdot a, \lambda \cdot b). \tag{7}$$

Terminological comment. The properties (6)–(7) describing the fact that the transformation

$$[a, b] \mapsto [A(a, b), B(a, b)]$$

does not change if we change the measuring units (i.e., *re-scale* all the numerical values) is known as *scale-invariance.*

Shift-invariance. Suppose that the expected company's income consists of:

- the fixed amount f—e.g., determined by the current contracts—and
- some additional amount x that will depend on the relation between supply and demand.

Suppose that the expert predicts this additional amount to be somewhere in the interval $[a, b]$. This means that the overall company's income is predicted to be between $f + a$ and $f + b$, i.e., somewhere in the interval $[f + a, f + b]$.

If we believe that the expert estimate needs corrections, then we have two possible ways to perform this correction:

- we can apply the correction to the original interval $[a, b]$, resulting in the corrected interval estimate $[A(a, b), B(a, b)]$ for the additional income; in this case, the interval estimate for the overall income will be

$$[f + A(a, b), f + B(a, b)]; \tag{8}$$

- alternatively, we can apply the correction to the interval $[f + a, f + b]$ describing the overall income; in this case, the resulting corrected interval for the overall income will have the form

$$[A(f + a, f + b), B(f + a, f + b)]. \tag{9}$$

It is reasonable to require that the two methods should lead to the exact same interval estimate for the overall income:

$$[f + A(a, b), f + B(a, b)] = [A(f + a, f + b), B(f + a, f + b)], \tag{10}$$

i.e., equivalently,

$$f + A(a, b) = A(f + a, f + b) \quad (11)$$

and

$$f + B(a, b) = B(f + a, f + b). \quad (12)$$

Terminological comment. The properties (11)–(12) describing the fact that the transformation

$$[a, b] \mapsto [A(a, b), B(a, b)]$$

does not change if we shift all the inputs by the same amount f is known as *shift-invariance*.

Sign invariance. One of the possible expert predictions is, e.g., how much bank B_1 will owe a bank B_2 at a certain future date. This amount can be positive—meaning that the bank B_1 will owe some money to the bank B_2. This amount can also be negative—meaning that, according to the expert, at the given future date, the bank B_2 will owe money to the bank B_1.

Suppose that the expert estimates this amount by an interval $[a, b]$. This means that if we ask the same expert a different question: how much money will the bank B_2 owe to the bank B_1—this expert will provide the interval $[-b, -a]$, i.e., the set of all the values $-x$ when $x \in [a, b]$.

In this case, we also have two possible ways to perform this correction:

- we can apply the correction to the original interval $[a, b]$, resulting in the corrected interval estimate
$$[A(a, b), B(a, b)]; \quad (13)$$
- alternatively, we can apply the correction to the interval $[-b, -a]$ describing how much the bank B_2 will owe to the bank B_1, and get the corrected interval $[A(-b, -a), B(-b, -a)]$ for this amount; by changing the sign, we get an interval estimate of how much the bank B_1 will owe to the bank B_2:
$$[-B(-b, -a), -A(-b, -a)]. \quad (14)$$

It is reasonable to require that the two methods should lead to the exact same interval estimate for the overall amount:

$$[A(a, b), B(a, b)] = [-B(-b, -a), -A(-b, -a)], \quad (15)$$

i.e., equivalently,

$$A(a, b) = -B(-b, -a) \quad (16)$$

and

$$B(a, b) = -A(-b, -a). \quad (17)$$

Terminological comment. We will call the properties (16)–(17) describing the fact that the transformation

$$[a, b] \mapsto [A(a, b), B(a, b)]$$

does not change if change all the signs, *sign-invariance*.

3 Definitions and the Main Result

Definition 1 We say that a mapping $[a, b] \mapsto [A(a, b), B(a, b)]$ is:

- *scale-invariant* if it satisfies the properties (6) and (7) for all $a < b$ and for all $s\lambda > 0$;
- *scale-invariant* if it satisfies the properties (11) and (12) for all $a < b$ and f; and
- *sign-invariant* if it satisfies the properties (16) and (17) for all $a < b$.

Comment. One can easily check that the transformation (1) is scale-, shift-, and sign-invariant. It turns out that every scale-, shift-, and sign-invariant transformation has the form (1). This explains why people use such transformations to correct interval-valued expert estimates.

Proposition 1 *Every scale-, shift-, and sign-invariant transformation has the form (1).*

Proof Indeed, due to shift-invariance for $f = a$, we have

$$B(a, b) = a + B(0, b - a). \tag{18}$$

Due to scale-invariance for $\lambda = b - a$, we have

$$B(0, 1) = \frac{1}{b - a} \cdot B(0, b - a),$$

hence

$$B(0, b - a) = (b - a) \cdot B(0, 1). \tag{19}$$

Let us denote

$$\alpha \stackrel{\text{def}}{=} 2B(0, 1) - 1,$$

then

$$B(0, 1) = \frac{1 + \alpha}{2},$$

and the formula (19) takes the form

$$B(0, b - a) = (b - a) \cdot \frac{1 + \alpha}{2} = b \cdot \frac{1 + \alpha}{2} - a \cdot \frac{1 + \alpha}{2}. \tag{20}$$

Substituting the expression (20) into the formula (18), we get

$$B(a,b) = a + b \cdot \frac{1+\alpha}{2} - a \cdot \frac{1+\alpha}{2} = a \cdot \frac{1-\alpha}{2} + b \cdot \frac{1+\alpha}{2}.$$

This is exactly the expression for $B(a, b)$ corresponding to the formula (1).

Now, by using sign-invariance, namely, the formula (16), we conclude that

$$A(a,b) = -B(-b,-a) = -\left((-b) \cdot \frac{1-\alpha}{2} + (-a) \cdot \frac{1+\alpha}{2}\right) =$$

$$a \cdot \frac{1+\alpha}{2} + b \cdot \frac{1-\alpha}{2}.$$

This is also exactly the expression for $A(a, b)$ corresponding to the formula (1). Thus, the proposition is proven.

Comment. In addition to empirically confirming the formula (1), the paper Smithson (2015) also mentions an additional empirical fact that it finds difficult to explain: that if two experts provide interval estimates, people put trust into pairs of intervals that have comparable width.

From our viewpoint, this is easy to explain: if two experts, based on the same data, provide completely different estimates of how accurately we can make a prediction based on this data, then we do not trust any of these experts. This is one of the cases when the two supposed experts are inconsistent. A similar phenomenon happens when the two experts provide drastically different numerical estimates: in this case, we do not trust either of them.

Acknowledgements This work was supported in part by the National Science Foundation grants 1623190 (A Model of Change for Preparing a New Generation for Professional Practice in Computer Science), and HRD-1834620 and HRD-2034030 (CAHSI Includes), and by the AT&T Fellowship in Information Technology. It was also supported by the program of the development of the Scientific-Educational Mathematical Center of Volga Federal District No. 075-02-2020-1478, and by a grant from the Hungarian National Research, Development and Innovation Office (NRDI).

References

Gajdos, T., Vergnaud, J.-C.: Decisions with conflicting and imprecise information. Soc. Choice Welfare **41**, 427–452 (2013). https://doi.org/10.1007/s00355-012-0691-1

Smithson, M.: Conflict and ambiguity: preliminary models and empirical tests. In: Proceedings of the Eighth International Symposium on Imprecise Probability: Theories and Applications ISIPTA'2013, Compiègne, France, July 2–5, pp. 303–310 (2013)

Smithson, M.: Probability judgments under ambiguity and conflict. Front. Psychol. **6**, 674 (2015). https://doi.org/10.3389/fpsyg.2015.00674

On The Skill of Influential Predictions

William M. Briggs

Abstract The skill of a model is defined as its ability to predict better than a simpler or default model. These default models almost always exist. Models without skill should not be used. All predictions are based on models, tacit or formal. Predictions to be useful must therefore be skillful. Some predictions are influential in the sense that the observable predicted is caused, in whole or in part, by the prediction itself. Skill measures common in assessing non-influential predictions are developed here for use with influential models.

Keywords Explanation · Inference · Models · Prediction · Probability

1 Check Me Out

Unless a model can be deduced from causal premises, it needs to be verified against reality before it can be trusted to make decisions with. Causal premises are those that specify all causes, and their requisite conditions, that entail without uncertainty an observable's state. See Briggs (2016) for a description of causal and non-causal models.

These causal models are rare enough, and scarcely found outside physics, and only then in highly controlled situations. Outside these special circumstances, the vast majority of models in use, both formal and informal, are either approximations or are *ad hoc* to some extent (most are entirely *ad hoc*), and therefore need to be verified.

There is in some disciplines, meteorology being a prime example, a tradition and culture of model verification; the text Wilks (2006) is fundamental. Verification practices have inarguably led to vast improvements in weather forecasts, primarily because when a modeler in this field gets it wrong, he hears about it from everybody. Practicing doctors are similar in this, though they use a mixture of formal and highly

W. M. Briggs (✉)
Michigan, USA
e-mail: matt@wmbriggs.com

N. Ngoc Thach et al. (eds.), *Financial Econometrics: Bayesian Analysis, Quantum Uncertainty, and Related Topics*, Studies in Systems, Decision and Control 427,
https://doi.org/10.1007/978-3-030-98689-6_2

informal—which is to say, non-quantitative—models. Doctors, too, are provided constant feedback on their performance, but because of the (unavoidable) informality, not all take the lessons of model failure to heart, see Shapiro (1985). Just as with weathermen, there are good and bad doctors.

Almost every probability model is *ad hoc* to some extent. Regressions are the leading example of wholly *ad hoc* models. The use and abuse of these models need no extra comment here. There are, of course, myriad other kinds of probability models in use, having many names: AI, neural nets, machine learning, statistical, and on and on. A *probability model* is any model that makes a statement about an observable's state that contains uncertainty in the state or in the premises of the model, or both.

In some fields, like sociology, anything with "education" appended to it, and all the so-called "soft" sciences, models are hardly ever, and usually never, checked against independent reality; see Briggs (2016) for an extensive criticism. At best, these models are contrasted in some way with the observations (a.k.a. the "data") that were used to create the models. This process is circular, as is well known, and therefore cannot constitute an independent test.

Hypothesis testing, either of the frequentist or Bayesian flavor, also fails for the same reason; see Briggs and Nguyen (2019), Briggs (2019a, b). Testing should only be used in a predictive sense. It isn't needed otherwise because the observations themselves exist and can be queried. For example, a hypothesis test might say, "Women had higher anxiety scores than men ($p < 0.05$)". Theory insists this is meant as a statement about mysterious properties of the observables themselves, a statement which can never be verified in practice—probabilities cannot be observed. These statements aren't even needed, because we can look at the observations and (in this example) say with certainty which group had the higher scores.

The only test of any probability model is in independent tests of that model against reality. The concept is simplicity itself. Use the model to make predictions of observations never before seen or used in any way in creating the model, then use formal verification techniques, in association with the actual decisions made with the model, to test model goodness, Briggs (2019b).

Alas, though it is simple to explain, it is expensive and difficult to do. Whatever time and expense that went into gathering the observations to create a model have to be paid again, more or less, to do the verification. Academics in particular, in their desperate need to publish, haven't the patience. Most haven't the thick skin weathermen have developed. Criticism is rarely well taken. It is much simpler, and faster and more remunerative, to produce a hypothesis test, flaunt a wee p-value, and proclaim victory.

Nevertheless, there is some hope predictive approaches in probability modeling will increase—at least among those who have to make real-life decisions based on their models.

There are a great many verification techniques in use; good summaries are Murphy (1991), Gneiting and Raftery (2007), Gneiting et al. (2007). Here we use the concept of skill, which is a comparison between models, where one is an expert-based or touted model, and the other is a simple, usually the simplest, model given the situation. Our special interest in assessing the skill of models that exhibit influence; that is, models which cause, partially or wholly, their observables to take certain states.

2 Under The Influence

Weather predictions belong to a large class of *non-influential* models. That is, regardless what the forecast is, the weather will do what it will do because of causes external to the prediction itself. It is easy to rate the goodness of or compare non-influential models. Decide the relevant verification metric, and see which model did best.

A second class of predictions, far less studied, are *influential* models. This is when the prediction itself can cause, partially or fully, the observable to take a certain state. The observables in influential predictions will take a state, partially or wholly, be*cause* of the influence of the prediction maker.

One example is of the Fed Chairman predicting that interest rates will rise in the next term. Since the Fed is in charge of setting, i.e. causing, the rate, this is an excellent prediction, though not without some uncertainty because events may intrude at the last minute to change the rate decision.

The Chairman's model which led to that prediction, as informal as it is, *is* a model. And the prediction is still a prediction that can be assessed for goodness. A rival economist's model (perhaps a formal mathematical time series), ignorant of the Chairman's comments, can be compared in performance to the Chairman's. It is obvious, however, that this comparison is unfair because of the influence. To make it fair, we have to "subtract" the influence, if possible, leaving a bare prediction stripped of its causal powers.

This will not always be possible in a literal sense, as is clear enough. However, it may be possible to assess model skill by estimating the amount of influence, and subtracting it off on average or accounting for it in some other way.

It is not surprising models predicting human behavior are the most likely to be influential. Economic models fall into this class.

One study of the Central Bank of Japan found that the bank's economic forecasts affected economists' models, while the bank's forecasts were not affected by the economists' forecasts, Fujiwara (2005). That is, the bank's predictions caused the economists to change their models, but the economists' predictions did not cause the bank to change theirs.

Now this is an odd situation, because we can view the bank's prediction as not just about the economy, but also about what state the economists' predictions will take. That is, the bank's forecast can be taken as a prediction of the economists' forecasts. This situation, curious sounding as it is, can yield benefits if bets are being made on or with those economists' predictions.

In this case, the bank influenced the economists, but not vice versa. That process has, however, been seen to work the other way around. For instance, analysts' predictions have been found to influence managers' guidance in public companies, Zhou (2021).

One paper found "managers' willingness to misrepresent their forwardlooking information as a function of their incentives varies with the market's ability to detect misrepresentation", Rogers and Stocken (2005). In other words, managers' predictions can be influenced by what they believe (or what is) the market's expectations. That means the market's predictions are influential.

Others found "that firms with an influential CMO [Chief Marketing Officer] provide more accurate revenue forecasts than other firms", Koo and Lee (2017). This could be because a CMO has a better handle on the finances of companies, or because CMO's are in a position to steer (influence) those finances to match the predictions. Either way, the prediction is influential by our terms.

Influence has been found in stock market prices, too, as found in Loh and Stulz (2010). They confirmed that analysts' recommendations can change average stock-price reactions. Analyst forecasts also influence other analyst forecasts, Cohn and Juergens (2014).

Predictions are well known to be influential in certain fields. For instance, a forecast of poor performance of a product may delay or even stop a product release; see Ehrman and Shugan (1995). These authors highlight the problem of verification, arguing that only optimistic forecasts (those saying, for instance, products will do well) are eventually verified. Or that modelers can be discouraged from predicting poor performance, which shows the modelers' bosses are influential in changing predictions.

It is clear techniques for gauging model goodness in the presence of influence are needed. We begin with defining skill, and present skill for simple forecasts. We then expand the technique when influence is known to be present.

3 Simple Skill

It is less known outside fields like meteorology, but a model can look good in some verification measures, but still be a failure.

There is always a model behind any set of predictions. It may be formal or informal, rigorous or *ad hoc*, but it must be there. Predictions, if you like, are caused by models.

The most infamous example of an appealing set of predictions that were nevertheless failures was due to Sergeant John Finley in 1884, Finley (1884). He predicted whether tornadoes would hit in several areas in the US. Here are the predictions:

Table 1 are Finley's 2,803 tornado forecasts. He got 2,708 correct, for an accuracy of 96.6%. Which sounds mighty impressive. However, if he had said "no tornado" each time, he would have got $72 + 2680 = 2752$ correct, for an accuracy of 98.2%.

Table 1 Finley's tornado forecasts of 1884. There were 2,803 predictions, of which he got 2,708 correct, for an accuracy of 96.6%. However, if he had said no tornado each time, he would have got $72 + 2680 = 2752$, for an accuracy of 98.2%

Prediction	Observed	Number
1	1	28
1	0	72
0	1	23
0	0	2680

That "no tornado" prediction is itself a better model, at least as far as accuracy goes. Of course, for certain decisions it may be better to worry less about false positives than false negatives, and vice versa. However, a meteorologist making predictions for the public is not in a position to know how the predictions will be used by all. The costs of error and gains of accuracy or decisions dependent on the prediction are unknown to the modeler. Therefore model prediction accuracy is worthy goal to assess at least sometimes.

Ideally, any model should be tested against the actual costs and gains made using it. However, sometimes, as in this case, simple comparative models suggest themselves. It is cost free, and sensible, to predict "no tornado" over every time period, and highly accurate. Any expert-based model should be able to beat this trivial model in accuracy. If it can, it is said to have *skill*; if it cannot, it has no skill with respect to this simple model. It may have skill with respect to other models.

Simple inexpert models can be found with ease. Here is a another example. Regressions use normal distributions to quantify the uncertainty in the observable. The central parameter of the normal distribution is allowed to vary in accord with values in certain measures. A simple comparative model is just the normal distribution itself applied to the observable, its parameter unmodified by any other measure. If an expert regression can't "beat" this simple model, in some relevant verification metric, it should not be used.

The following is a summary of Briggs and Ruppert (2005), developing notation and concepts we use in the next section.

We are here interested here in simple predictions, and the skill in making them. A simple prediction is to bet (X) whether, or act as if, a thing (Y) occurs. Obviously, a perfect set of predictions is when $X = Y$ for all times a prediction is made.

Our focus is acts, times when we act as if Y will occur, or when we act as if Y will not occur. For instance, we carry an umbrella ($X = 1$) because we believe it will rain ($Y = 1$), or we do not carry ($X = 0$) because we believe it will be dry ($Y = 0$). A set of perfect predictions would have $X = 1$ when $Y = 1$ and $X = 0$ when $Y = 0$.

Without loss of generality, assume $\Pr(Y = 1|\mathrm{E}) < 1/2$, where E is the evidence assumed; i.e. the evidence which lets us deduce the numerical value of the probability of Y.

An obvious simple prediction, since Y is "rare", is to say Y will not happen. This naive prediction $W \equiv 0$ will be right, therefore, at least 50% of the time. To have *skill*, any forecast X must be able to "beat" W. Which is to say, skill is defined as when:

$$\Pr(Y = X|\mathrm{E}) > \Pr(Y = W|\mathrm{E}), \tag{1}$$

and where the inequality is strict.

Naturally, $\Pr(Y = W|\mathrm{E}) = \Pr(Y = 1, W = 1|\mathrm{E}) + \Pr(Y = 0, W = 0|\mathrm{E})$. But $\Pr(Y = 1, W = 1|\mathrm{E}) = 0$ since $W \equiv 0$. Thus $\Pr(Y = 0, W = 0|\mathrm{E}) = \Pr(Y = 0|\mathrm{E})$.

Now we also can write $\Pr(Y = 0|\mathrm{E}) = \Pr(Y = 0, X = 1|\mathrm{E}) + \Pr(Y = 0, X = 0|\mathrm{E})$, and $\Pr(Y = X|\mathrm{E}) = \Pr(Y = 1, X = 1|\mathrm{E}) + \Pr(Y = 0, X = 0|\mathrm{E})$. So skill comes down to

$$\begin{aligned}
\Pr(Y = 1, X = 1|\mathrm{E}) + \Pr(Y = 0, X = 0|\mathrm{E}) &> \Pr(Y = 0, X = 1|\mathrm{E}) + \Pr(Y = 0, X = 0|\mathrm{E}) \\
\Pr(Y = 1, X = 1|\mathrm{E}) &> \Pr(Y = 0, X = 1|\mathrm{E}) \\
\Pr(Y = 1|X = 1, \mathrm{E}) \Pr(X = 1|\mathrm{E}) &> \Pr(Y = 0|X = 1, \mathrm{E}) \Pr(X = 1|\mathrm{E}) \\
\Pr(Y = 1|X = 1, \mathrm{E}) &> \Pr(Y = 0|X = 1, \mathrm{E}). \qquad (2)
\end{aligned}$$

This assumes $\Pr(X = 1|\mathrm{E}) > 0$, which are the only cases of interest.

This is the basic skill requirement. It can be modified to account for asymmetric costs and gains. For Finely, it is clear Y (tornadoes), are rare. Using the ordinary estimates (we don't overly care here about estimation procedures), $\Pr(Y = 1|X = 1, \mathrm{E}) \approx 28/100 = 0.28$, and $\Pr(Y = 0|X = 1, \mathrm{E}) \approx 72/100 = 0.72$. The skill condition has been violated; skill has not been achieved.

Interpretation of skill is easy: the forecaster has to be right when he says Y will happen more often than he is wrong to have skill. It is easy to show the reverse is true when $\Pr(Y = 1|\mathrm{E}) \geq 1/2$.

This method can be expanded to when the prediction X is probabilistic Briggs and Zaretzki (2008) or when the loss and gains are not symmetric Briggs (2005). For the sake of ease in presenting the concept of influence, we investigate only simply skill here.

4 Simple Influence

One idea is to think of influence as a switch Z. When it is on ($Z = 1$), Y must take either 0 or 1 depending on whether $X = 0$ or $X = 1$ because of the influence: Y is *caused* to equal X regardless of any other cause. When the switch is off ($Z = 0$), Y is under the influence of all other causes except the influence. That is, Y will become equal to 1 or 0 depending on its "non-influential" causes. The prediction X then becomes genuine and non-influential.

Those cases which are influenced should not, of course, be counted in the skill calculation, because it takes no skill to make perfect predictions when you are under complete control of Y. Only those cases which are free of the influence prove predictive ability.

If we knew when $Z = 1$, this is easy. We just remove those cases from the collection of predictions and proceed with the usual skill calculation. But if we don't know when Z is active, only that it might be activity with a certain frequency or tendency, then we must estimate.

The skill condition can be modified with ease. Recall that skill is when

$$\Pr(Y = 1, X = 1|\text{E}) > \Pr(Y = 0|\text{E}),$$

assuming $\Pr(Y = 0|\text{E}) > 1/2$, which we continue to assume even under influence. That is, Y is not caused to become "common" and is still rare. (When Y is common, we can just switch labels in all the calculations.)

Since we want to exclude those cases when $Z = 1$, we calculate skill conditional only on $Z = 0$.

$$\Pr(Y = 1, X = 1|Z = 0, \text{E}) > \Pr(Y = 0|Z = 0, \text{E}). \tag{3}$$

Henceforth for ease of notation, we drop the conditioning on E, recognizing it is *always* there. And we write, e.g. $Y_0 \equiv Y = 1$ and so forth.

It's easy to see that (3) is, by similar calculations that led to (2),

$$\begin{aligned} \Pr(Y_1 X_1|Z_0) &> \Pr(Y_0 X_1|Z_0), \\ \Pr(Y_1 X_1|Z_0)\Pr(Z_0) &> \Pr(Y_0 X_1|Z_0)\Pr(Z_0). \end{aligned} \tag{4}$$

where it is assumed $\Pr(Z_0) > 0$, the only interesting case.

Now

$$\begin{aligned} \Pr(Y_1 X_1) &= \Pr(Y_1 X_1|Z_0)\Pr(Z_0) + \Pr(Y_1 X_1|Z_1)\Pr(Z_1), \\ \Pr(Y_1 X_1) &= \Pr(Y_1 X_1|Z_0)\Pr(Z_0) + \Pr(Z_1), \end{aligned} \tag{5}$$

because $\Pr(Y_1 X_1|Z_1) = 1$ by definition: $Z = 1$ demands $X = 1$, which causes $Y = 1$.

And

$$\begin{aligned} \Pr(Y_0 X_1) &= \Pr(Y_0 X_1|Z_0)\Pr(Z_0) + \Pr(Y_0 X_1|Z_1)\Pr(Z_1), \\ \Pr(Y_0 X_1) &= \Pr(Y_0 X_1|Z_0)\Pr(Z_0), \end{aligned} \tag{6}$$

because $\Pr(Y_0 X_1|Z_1) = 0$: again, $Z = 1$ demands $X = 1$, which causes $Y = 1$, so $Y = 0$ is impossible under influence.

We next substitute (5) and (6) into (4) and arrive at the final skill condition under influence:

$$\Pr(Y_1 X_1) - \Pr(Z_1) > \Pr(Y_0 X_1). \tag{7}$$

This is close to the original simple skill, and has a nice interpretation. We, more or less, subtract off those times when Y had no choice but to become $Y = 1$. The observed counts can in the obvious way serve as simple estimates of $\Pr(Y_1 X_1)$ and $\Pr(Y_0 X_1)$.

In the original calculation, skill exists when the number of times $Y = X = 1$ is larger than the number of times the forecast of $X = 1$ is "blown", i.e. when $Y = 0$. It is almost the same under influence, except we subtract the number of times $Z = 1$.

Of course, it might not be known *when* $Z = 1$, but we have some idea of its tendency, or an outside estimate (conditioned on E) of $\Pr(Z = 1|\text{E})$. Obviously, the larger this is, the harder it is to claim genuine skill. In the absence of any influence, the skill condition remains unchanged.

It turns out we can break down (7) as we did (2), which leads to an small bit of help.

$$\Pr(Y_1|X_1)\Pr(X_1) - \Pr(Z_1|X_1)\Pr(X_1) - \Pr(Z_1|X_0)\Pr(X_0) > \Pr(Y_0|X_1)\Pr(X_1),$$

$$\Pr(Y_1|X_1) - \Pr(Z_1|X_1) - \Pr(Z_1|X_0)\frac{\Pr(X_0)}{\Pr(X_1)} > \Pr(Y_0|X_1). \quad (8)$$

Equation (8) can be used if it is known or believed there is asymmetric influence. That the influencer prefers causing Y to equal 1 more (or less) than in causing Y to equal 0. If influence is symmetric in Y, as it were, then $\Pr(X_0)/\Pr(X_1) = 1$, which is equivalent to (7).

5 A Federal Case

The monthly effective Federal funds rate, or FEDFUNDS, from July 1954 to August 2021, is pictured in Fig. 1; data from the Fed (2021).

The absolute number of the rate is, of course, important for innumerable reasons, and knowing it ahead of time via a reliable predictive model can be invaluable. But often what is wanted is a simple prediction of whether the rate will fall, or whether it will rise or stay the same.

Models of this are plentiful. One such model, also common in meteorology, is *persistence*. The persistence model predicts whatever happened in the last period will also happen in the next. Today's high temperature will be tomorrow's; it rained today, it will rain tomorrow. There are excellent reasons to use persistence in weather, because the atmosphere is large and hard to change. In practice, the persistence model performs well, too.

Persistence is also clear in human behavior, and it is human behavior that causes the fund rate to be what it is. And it turns out (we see next) persistence is also a good model to predict the rate.

There are $n = 806$ observations of the rate; 325 times the rate decreased from its previous value, and 480 times it rose or stayed the same. In accord with treating

Fig. 1 The monthly effective Federal funds rate, or FEDFUNDS, from July 1954 to August 2021. A clear persistence is seen in the data; i.e. the tendency of the observations to stay the same

Table 2 Persistence forecast for the Fed fund rate. There were 804 predictions, of which 516 are correct, for an accuracy of 64.1%. A prediction which always said the Fed would rise or stay the same would have got 479 correct, for an accuracy of 59.6%. Persistence has skill

Prediction	Observed	Number
1	1	181
1	0	144
0	1	144
0	0	335

the observable as "rare", we will define "$Y = 1$" as the rate falling, and "$Y = 0$" as the rate increasing or remaining the same. This is backwards, in a way, but for the purposes of this paper, and the desire not to multiply notation, we'll use it. X is persistence and is defined similarly.

Table 2 shows the persistence forecast for the Fed fund rate. There were 804 predictions (we lose 2 observations accounting for the lags in Y and X), of which 516 are correct, for an accuracy of 64.1%. A prediction which always said the Fed would rise or remain the same would have got 479 correct, for an accuracy of 59.6%. The quick estimate for $\Pr(Y_1|X_1)$ is $181/(181 + 144) = 0.557$, which is greater than 0.5. Either look shows persistence has skill relative to the model of always predicting a rise or remain the same.

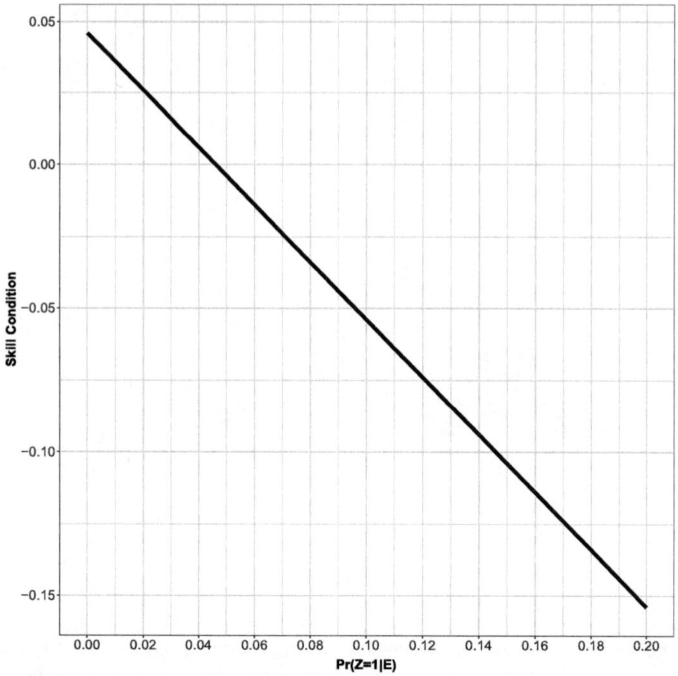

Fig. 2 The skill condition as given in (7) for various values of $\Pr(Z = 1|\text{E})$ from 0 to 0.2. Numbers greater than 0 indicate the persistence model is skillful, even with influence. Numbers less than 0 indicate the always-increasing or staying-the-same model is superior

There could be, and probably is, some influence in the persistence model. There is no outside estimate of $\Pr(Z = 1|\text{E})$ readily available. One could be had by scouring through old press reports and other similar material to gauge how firm the Fed was in announcing what they would do. This may even be available now, but I am unaware of it.

We don't need to know the exact amount of influence, and can calculate skill at various levels. This shown in Figs. 2 and 3.

There is skill in the persistence model when influence is symmetric, but only when it happens about 4.5% of the time or less. Influence does not have to be high to eliminate skill of the persistence model. If influence is open and known to be more than roughly one out of every twenty predictions, it's better to use the naive model (up or stay the same) than persistence.

The changes if influence is asymmetric. If $\Pr(Z = 1|X = 0, \text{E}) = 0$, which is when there is no influence for the for rising of remaining rates, there can be skill for persistence all the way to about $\Pr(Z = 1|X = 1, \text{E}) = 0.12$. If $\Pr(Z = 1|X = 1, \text{E}) = 0$, which is when there is no influence when there are falling rates, skill for the persistence model exists out to $\Pr(Z = 1|X = 1, \text{E}) = 0.075$.

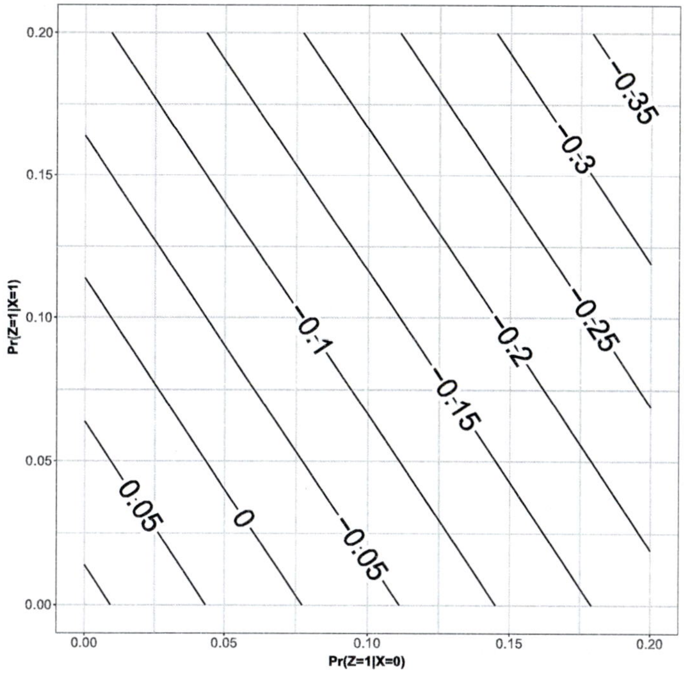

Fig. 3 The skill condition as given in (8) for various values of $\Pr(Z = 1|X = 1, \mathrm{E})$ and $\Pr(Z = 1|X = 0, \mathrm{E})$, both from from 0 to 0.2. Numbers greater than 0 indicate the persistence model is skillful, even with influence. Numbers less than 0 indicate the always-increasing or staying-the-same model is superior

6 Influencing The Influencers

This has been an exercise in expanding the most basic skill score for yes-no forecasts into conditions when influence is present for some, but not all, of the a set of predictions. Perhaps the Fed rate is a good example of this, given it is the Fed itself that causes the rate to take its value. Other examples exist.

An interesting one is polls. Political polls can, of course, be used to estimate "mood" or to give ratings. But they can also be used as projections, forecasts of how elections might turn out, for instance. The results of polls can "push" the results. Which is to say, the outcome would have been different had a public poll not been issued. Polls in this sense can be influential; see Renka (2010).

This is not the same as "push polls", which use deceitful wording to manipulate poll results, or even to create false memories; see e.g. Murphy et al. (2021). Push polls often occur in politics and can and are used in attempts to influence outcomes; Feld (2000). Skill scores and verification techniques that account for the influence are thus needed to rate and compare polls.

This paper only did the simplest possible skill test, but it can be expanded when instead of $X = 1$ or $X = 0$, X is given a probability. This has already been done for non-influential forecasts, as given in Briggs and Ruppert (2005). Beyond that, skill tests for non-binary outcomes also need to be developed.

References

Briggs, W.M., Nguyen, H.T., Trafimow, D., Kreinovich, V., Sriboonchitta, S. (eds.) Structural Changes and Their Econometric Modeling, pp. 3–17. Springer, New York (2019)

Briggs, W.M., Ruppert, D.: Biometrics **61**(3), 799 (2005)

Briggs, W.M., Zaretzki, R.A.: Biometrics **64**, 250 (2008). (with discussion)

Briggs, W.M.: Asian J. Bus. Econ. **1**, 37 (2019)

Briggs, W.M.: Monthly Weather Rev. **133**(11), 3393 (2005)

Briggs, W.M.: Uncertainty: The Soul of Probability. Modeling and Statistics, Springer, New York (2016)

Briggs, W.M.: Economics in beyond traditional probabilistic methods. In: Kreinovich, V., Thach, N., Trung, N., Thanh, D. (eds.) Economics, pp. 22–44. Springer, New York (2019)

Cohn, J.B., Juergens, J.L.: Quart. J. Financ. **04**(03), 1450017 (2014)

Ehrman, C.M., Shugan, S.M.: Market. Sci. **14**(2), 123 (1995)

Fed, S.L.: Effective federal funds rate (2021). https://fred.stlouisfed.org/series/FEDFUNDS

Feld, K.G.: Campaigns & Elections **21**(4), 62 (2000)

Finley, J.: Am. Meteorol. J. 85–88 (1884)

Fujiwara, I.: Econ. Lett. **89**(3), 255 (2005)

Gneiting, T., Raftery, A.E., Balabdaoui, F.: J. Royal Statist. Soc. Ser. B. Statist. Methodol. **69**, 243 (2007)

Gneiting, T., Raftery, A.E.: JASA **102**, 359 (2007)

Koo, D.S., Lee, D.: Account. Rev. **93**(4), 253 (2017)

Loh, R.K., Stulz, R.M.: Rev. Financ. Stud. **24**(2), 593 (2010)

Murphy, G., Lynch, L., Loftus, E., Egan, R.: Memory **29**(6), 693 (2021). PMID: 34080495

Murphy, A.H.: Monthly Weather Rev. **119**, 1590 (1991)

Renka, R.D.: Article published on http://cstl-cla.semo.edu/rdrenka/renka_papers/polls.htm (2010)

Rogers, J.L., Stocken, P.C.: Account. Rev. **80**(4), 1233 (2005)

Shapiro, A.R.: The Evaluation of Clinical Predictions: A Method and Initial Application, pp. 189–201. Springer New York, New York, NY (1985). https://doi.org/10.1007/978-1-4612-5108-8_10

Wilks, D.S.: Statistical Methods in the Atmospheric Sciences, 2nd edn. Academic Press, New York (2006)

Zhou, J.: J. Account. Audit. Financ. **36**(2), 405 (2021)

How to Find the Dependence Based on Measurements with Unknown Accuracy: Towards a Theoretical Justification for Midpoint and Convex-Combination Interval Techniques and Their Generalizations

Somsak Chanaim and Vladik Kreinovich

Abstract In practice, we often need to find regression parameters in situations when for some of the values, we have several results of measuring this same value. If we know the accuracy of each of these measurements, then we can use the usual statistical techniques to combine the measurement results into a single estimate for the corresponding value. In some cases, however, we do not know these accuracies, so what can we do? In this paper, we describe two natural approaches to solving this problem. In addition to describing general techniques, our results also provide a theoretical explanation for several semi-heuristic ideas proposed for solving an important particular case of this problem—the case when we deal with interval uncertainty.

1 Formulation of the Problem

General problem. In many practical situations:

- we know that the general form of the dependence of a quantity y on quantities $x_1, \ldots, x_n$, i.e., we know that $y = f(x_1, \ldots, x_n, c_1, \ldots, c_m)$ for some known function $f(x_1, \ldots, x_n, c_1, \ldots, c_m)$, but
- we do not know the values of the parameters $c_1, \ldots, c_m$; these values need to be determined empirically, from the known results of observations and measurements.

This general situation is known as *regression*.

S. Chanaim
International College of Digital Innovation, Chiang Mai University, Chiang Mai 50200, Thailand
e-mail: somsak_ch@cmu.ac.th

V. Kreinovich (✉)
University of Texas at El Paso, El Paso, TX 79968, USA
e-mail: vladik@utep.edu

N. Ngoc Thach et al. (eds.), *Financial Econometrics: Bayesian Analysis, Quantum Uncertainty, and Related Topics*, Studies in Systems, Decision and Control 427,
https://doi.org/10.1007/978-3-030-98689-6_3

A simple example. The simplest example if when $n = 1$, and y is simply proportional to x_1, with an unknown coefficient of proportionality c_1, so that $y = c_1 \cdot x_1$. In this case, we have $m = 1$ parameter c_i, and $f(x_1, c_1) = c_1 \cdot x_1$.

Econometric example. We may want to know the parameter β that describes, for a given stock, how the difference $r - r_f$ between the stock's rate of return r and the risk-free interest rate r_f depends on the difference $r_m - r_f$ between the overall market's rate of return r_m and the value r_f:

$$r - r_f = \beta \cdot (r_m - r_f).$$

General problem: usual case. Usually, we have several (K) cases $k = 1, \ldots, K$ in each of which we measure x_i and y, resulting in the values $x_1^{(k)}, \ldots, x_n^{(k)}$ and $y^{(k)}$. In this case, to find the values of the parameters c_i, a reasonable idea is to apply the Least Squares method (see, e.g., Sheskin (2011)), i.e., to find the values $c_1, \ldots, c_m$ of the parameters that minimize the expression

$$\sum_{k=1}^{K} \left(y^{(k)} - f\left(x_1^{(k)}, \ldots, x_n^{(k)}, c_1, \ldots, c_m\right)\right)^2. \tag{1}$$

Alternatively, we can minimize the sum of the absolute values of the differences $y^{(k)} - f\left(x_1^{(k)}, \ldots, x_n^{(k)}, c_1, \ldots, c_m\right)$, or any other appropriate objective function.

What if for each case, we have several measurement results? Sometimes, in each case k, we have several different measurement results of each of the variables:

- for each k and i, instead of a single measurement result $x_i^{(k)}$, we have several values $x_{i1}^{(k)}, \ldots, x_{iv_i}^{(k)}$ measured, in general, by several different measuring instruments, and
- for each k, instead of a single result $y^{(k)}$ of measuring y, we have several values $y_1^{(k)}, \ldots, y_v^{(k)}$ measured, in general, by several different measuring instruments.

In such situation, a natural idea is to do the following:

- first, for each k and for each i, we use all the results $x_{i1}^{(k)}, \ldots, x_{iv_i}^{(k)}$ of measuring x_i to come up with a single estimate $x_i^{(k)}$;
- then, for each k, we use all the results $y_1^{(k)}, \ldots, y_v^{(k)}$ of measuring y to come up with a single estimate $y^{(k)}$;
- then, we find the values of the parameters $c_1, \ldots, c_m$ that minimize the objective function (1)—or the corresponding alternative objective function.

To implement this idea, we need to be able to combine several estimates into a single one.

Econometric example. The stock price fluctuates during the day. The usual economic assumption is that:

- on any day, there is the fair price of the stock—the price that reflects its current value and its prospects;
- this fair price changes rarely—definitely rarely several times a day, it only changes based on the new information;
- on the other hand, the observed minute-by-minute price changes all the time, because it is obtained by adding some random fluctuations to the fair price.

In this example, we do not know the fair daily price of the stock x_i, but we can measure several characteristics that provide an approximate description of this fair price: the smallest daily price $x_{i1}^{(k)}$, the largest daily price $x_{i2}^{(k)}$, the closing price $x_{i3}^{(k)}$, the starting price $x_{i4}^{(k)}$, etc. If we limit ourselves to these four characteristics, then we have $v_i = 4$.

Instead of these four measurement results, we can use only two: the smallest daily price $x_{i1}^{(k)}$ and the largest daily price $x_{i2}^{(k)}$. In this case, what we know is an *interval* $\left[x_{i1}^{(k)}, x_{i2}^{(k)}\right]$ that contains the actual (unknown) fair price $x_i^{(k)}$ on day k.

Comment. There are other practical examples where, as a result of measurements, we get a lower bound $x_{i1}^{(k)}$ and an upper bound $x_{i2}^{(k)}$ for the desired quantity $x_i^{(k)}$, i.e., where, as a result of the measurements, we get an interval $\left[x_{i1}^{(k)}, x_{i2}^{(k)}\right]$ that contains the actual (unknown) value $x_i^{(k)}$.

We can naturally combine measurement results when we know the accuracy of each measurement. In many practical situations, we know the accuracy of different measuring instruments. For example:

- for each input i and for each instrument $j = 1, \ldots, v_i$ used to measure x_i, we know the corresponding standard deviation σ_{ij}, and
- for each instrument $j = 1, \ldots, v$ used to measure y, we know the corresponding standard deviation σ_j.

In this case, a natural idea for estimating $x_i^{(k)}$ is to use the least squares approach, i.e., to minimize the sum

$$\sum_{j=1}^{v_i} \frac{\left(x_i^{(k)} - x_{ij}^{(k)}\right)^2}{\sigma_{ij}^2}.$$

This minimization results in the estimate

$$x_i^{(k)} = \sum_{j=1}^{v_i} w_{ij} \cdot x_{ij}^{(k)}, \tag{2}$$

where

$$w_{ij} = \frac{\sigma_{ij}^{-2}}{\sum\limits_{j'=1}^{v_i} \sigma_{ij'}^{-2}}. \tag{3}$$

Similarly, a natural idea for estimating $y^{(k)}$ is to use the least squares approach, i.e., to minimize the sum

$$\sum_{j=1}^{v} \frac{\left(y^{(k)} - y_j^{(k)}\right)^2}{\sigma_j^2}.$$

This minimization results in the estimate

$$y^{(k)} = \sum_{j=1}^{v} w_j \cdot y_j^{(k)}, \tag{4}$$

where

$$w_j = \frac{\sigma_j^{-2}}{\sum\limits_{j'=1}^{v} \sigma_{j'}^{-2}}. \tag{5}$$

Comment. In both cases, the coefficients w add to 1: $\sum\limits_{j=1}^{v_i} w_{ij} = 1$ and $\sum\limits_{j=1}^{v} w_j = 1$.

Remaining problem. In some cases—e.g., in the econometric example—we do not know the corresponding accuracies. What shall we do?

This is a problem that we consider in this paper. Specifically, we describe two natural general solutions—and we explain how each of them is related to previously proposed methods. It turns out that this way, several previous proposed semi-empirical methods can be theoretically justified.

2 First Approach: Laplace's Indeterminacy Principle

Main idea. In its most general form, Laplace's Indeterminacy Principle states that if we have no reason to assume that one quantity is smaller or larger than the other one, then it is reasonable to assume that these two quantities are equal to each other; see, e.g., Jaynes and Bretthorst (2003).

Let us apply this idea to our problem. For each i, we have several unknown values σ_{ij}. Since we have no reason to believe that one of these values is larger, we conclude that all these values are equal to each other: $\sigma_{i1} = \sigma_{i2} = \ldots$ In this case, formula (3) leads to $w_{ij} = \frac{1}{v_i}$, and the estimate (2) becomes simply the arithmetic mean

$$x_i^{(k)} = \frac{1}{v_i} \cdot \sum_{j=1}^{v_i} x_{ij}^{(k)}. \tag{6}$$

Similarly, since we have no reason to believe that one of the values σ_j is larger, we conclude that all these values are equal to each other: $\sigma_1 = \sigma_2 = \ldots$ In this case, formula (5) leads to $w_j = \dfrac{1}{v}$, and the estimate (4) becomes the arithmetic mean

$$y^{(k)} = \frac{1}{v} \cdot \sum_{j=1}^{v} y_j^{(k)}. \tag{7}$$

Interval case. In the case when the two estimates are the two endpoints of the interval, formulas (6)-(7) result in a midpoint of this interval. Thus, in situations when we only know the intervals $\left[x_{i1}^{(k)}, x_{i2}^{(k)}\right]$ and $\left[y_1^{(k)}, y_2^{(k)}\right]$ containing the desired values x_i and y, this approach recommends applying the regression technique to midpoints $x_i^{(k)} = \dfrac{x_{i1}^{(k)} + x_{i2}^{(k)}}{2}$ and $y^{(k)} = \dfrac{y_1^{(k)} + y_2^{(k)}}{2}$ of these intervals.

Comment. The use of midpoints is exactly what was proposed in Billard and Diday (2002). Thus, our analysis provides a theoretical explanation for this semi-heuristic method.

3 Second Approach: Using the Known Dependence Between x_i and y

Alternative idea. We consider the case when do not know the measurement accuracies σ_{ij} and σ_j, so we cannot use these accuracies to find the coefficients w_{ij} and w_j. In other words, we do not know which linear combinations of the measurement results most adequately represent the actual values $x_i^{(k)}$ and $y^{(k)}$.

A natural idea is to take into account that the actual (unknown) values x_i and y should satisfy the formula $y = f(x_1, \ldots, x_n, c_1, \ldots, c_m)$. Thus, it is reasonable to select the coefficients w_{ij} and w_j for which the resulting linear combination $y^{(k)}$ is as close as possible to the value $f\left(x_1^{(k)}, \ldots, x_n^{(k)}, c_1, \ldots, c_m\right)$. To be more precise, we find the parameters $c_1, \ldots, c_m$ *and* the coefficients w_{ij} and w_j from the condition that the expression (1) (or any other selected objective function) attains its smallest possible value, where $x_i^{(k)}$ and $y^{(k)}$ are determined by the formulas (2) and (4).

In this case, the minimized objective function (1) takes the form

$$\sum_{k=1}^{K} \left(\sum_{j=1}^{v} w_j \cdot y_j^{(k)} - f\left(\sum_{j=1}^{v_1} w_{1j} \cdot x_{1j}^{(k)}, \ldots, \sum_{j=1}^{v_n} w_{nj} \cdot x_{nj}^{(k)}, c_1, \ldots, c_m \right) \right)^2.$$

Interval case. In the interval case, when we know the intervals $\left[x_{i1}^{(k)}, x_{i2}^{(k)}\right]$ and $\left[y_1^{(k)}, y_2^{(k)}\right]$, the idea is to select appropriate convex combinations $x_i^{(k)} = w_{i1} \cdot x_{i1}^{(k)} + (1 - w_{i1}) \cdot x_{i2}^{(k)}$ and $y^{(k)} = w_1 \cdot y_1^{(k)} + (1 - w_1) \cdot y_2^{(k)}$, i.e., coefficients for which the following expression is the smallest possible:

$$\sum_{k=1}^{K} \left(w_1 \cdot y_1^{(k)} + (1 - w_1) \cdot y_2^{(k)} - f\left(w_{11} \cdot x_{11}^{(k)} + (1 - w_{11}) \cdot x_{12}^{(k)}, \ldots, c_1, \ldots \right) \right)^2.$$

Comment. This idea of using convex combinations has indeed been proposed and successfully used; see, e.g., Chanaim et al. (2016, 2018). Thus, our analysis provides a theoretical explanation for this semi-heuristic idea as well.

Acknowledgements This work was supported in part by the National Science Foundation grants 1623190 (A Model of Change for Preparing a New Generation for Professional Practice in Computer Science), and HRD-1834620 and HRD-2034030 (CAHSI Includes). This work was also supported by the Center of Excellence in Econometrics, Faculty of Economics, Chiang Mai University, Thailand.

References

Billard, L., Diday, E.: Symbolic regression analysis. In: Jajuga, K., Sokolowski, A., Bock, H.H. (eds.), Classification, Clustering, and Data Analysis, pp. 281–288. Springer, Berlin, Heidelberg (2002). https://doi.org/10.1007/978-3-642-56181-8_31

Chanaim, S., Sriboonchitta, S., Rungruang, C.: A convex combination method for linear regression with interval data. In: Proceeding of the 5th International Symposium on Integrated Uncertainty in Knowledge Modelling and Decision Making IUKM'2016, pp. 469–480. Springer (2016)

Chanaim, S., Khiewngamdee, C., Sriboonchitta, S., Rungruang, C.: A convex combination method for quantile regression with interval data. In: Anh, L.H., Kreinovich, V., Thach, N.N. (eds.) Econometrics for Financial Applications, pp. 440–449. Springer, Cham, Switzeerland (2018). https://doi.org/10.1007/978-3-319-73150-6_35

Jaynes, E.T., Bretthorst, G.L.: Probability Theory: The Logic of Science. Cambridge University Press, Cambridge, UK (2003)

Sheskin, D.J.: Handbook of Parametric and Non-Parametric Statistical Procedures. Chapman & Hall/CRC, London, UK (2011)

An Alternative Extragradient Method for a Vector Quasi-Equilibrium Problem to a Vector Generalized Nash Equilibrium Problem

P. Dechboon, P. Kumam, and P. Chaipunya

Abstract In this paper, we consider a problem related to a generalized Nash equilibrium problem, known as, a quasi-equilibrium problem. Generally, such problem is discussed using a real valued bifunction. The so called vector quasi-equilibrium problem is the extension to the vector setting of the quasi-equilibrium problem. Therefore, we consider a sequence generated from a modified extragradient method, called an alternative extragradient method, for obtaining the convergence theorem to a solution of the vector quasi-equilibrium problem.

1 Introduction

Between 1928 and 1941, Morgenstern and Neumann (1953) proposed a general framework in view of social sciences which conflicts and cooperation of players are taken into account. This concept is the most widely used method of predicting the outcome of a strategic interaction. Later Nash (1951) introduced the concrete definition of the equilibrium of the situations in which there are no players who can take profit if they change their strategies (see Chin et al. (1974), Kreps (2017)). Let E be a Banach

P. Dechboon · P. Kumam
KMUTTFixed Point Research Laboratory, KMUTT Fixed Point Theory and Applications Research Group, SCL 802 Fixed Point Laboratory, Department of Mathematics, Faculty of Science, King Mongkut's University of Technology Thonburi (KMUTT), 126 Pracha Uthit Rd., Bang Mod, Thung Khru, Bangkok 10140, Thailand
e-mail: premyuda.e@mail.kmutt.ac.th

P. Kumam (✉)
Center of Excellence in Theoretical and Computational Science (TaCS-CoE), Science Laboratory Building, Faculty of Science, King Mongkut's University of Technology Thonburi (KMUTT), 126 Pracha Uthit Rd., Bang Mod, Thung Khru, Bangkok 10140, Thailand
e-mail: poom.kum@kmutt.ac.th

P. Chaipunya
Department of Mathematics, King Mongkut's University of Technology Thonburi (KMUTT), 126 Pracha Uthit Road, Bang Mod, Thung Khru, Bangkok 10140, Thailand
e-mail: parin.cha@mail.kmutt.ac.th

N. Ngoc Thach et al. (eds.), *Financial Econometrics: Bayesian Analysis, Quantum Uncertainty, and Related Topics*, Studies in Systems, Decision and Control 427,
https://doi.org/10.1007/978-3-030-98689-6_4

space with norm $\|\cdot\|$. Given $N = \{1, 2, \ldots, n\}$. Consider a convex set $X^i \subseteq E$ for all $i \in N$. Let us denote that $X = \bigoplus_{i \in N} X^i = X^1 \times X^2 \times X^3 \times \cdots \times X^n$ and $X^{-i} = \bigoplus_{j \in N, j \neq i} X^j = X^1 \times X^2 \times X^3 \times \cdots \times X^{i-1} \times X^{i+1} \times \cdots \times X^n$. That is, for any $x \in X$, $x = (x^1, x^2, \ldots, x^n)$, and, for any $x^{-i} = (x^1, x^2, \ldots, x^{i-1}, x^{i+1}, \ldots, x^n)$. Thus, $x = (x^{-i}, x^i)$. Recall that, for each $i \in N$, we define $f^i : X \to \mathbb{R}$ and $K^i : X^{-i} \rightrightarrows X^i$. The generalized Nash equilibrium problem (GNEP) is to find $x \in X$ such that, for each $i \in N$, $x^i \in K^i(x^{-i})$ and x^i is an optimal solution of

$$\left.\begin{array}{ll} \text{minimize} & f^i\left(x^{-i}, y^i\right) \\ \text{subject to} & y^i \in K^i(x^{-i}) \end{array}\right\} \tag{1}$$

In other words, GNEP is to find $x \in X$ such that, for all $i \in N$,

$$x^i \in K^i(x^{-i}) \text{ and } f^i(x) \leq f^i\left(x^{-i}, y^i\right)$$

for all $y^i \in K^i(x^{-i})$. (see Facchinei and Kanzow (2007)) Regarding the variational inequality problem, consider a continuous $F : E \to E^*$ and the problem consisting of finding a point $\overline{x} \in \mathcal{K} \subseteq E$ such that $\langle y - \overline{x}, F(\overline{x}) \rangle \geq 0$ for all $y \in \mathcal{K}$ where $\langle \cdot, \cdot \rangle : E \times E^* \to \mathbb{R}$ denotes the duality coupling, that is, $\langle z, x \rangle = z(x)$. Recall Kassay and Radulescu (2018) that the equilibrium problem of function $f : E \times E \to \mathbb{R}$ is to find an element $\overline{x} \in \mathcal{K} \subseteq E$ such that $f(\overline{x}, y) \geq 0$ for all $y \in \mathcal{K}$. Then the equilibrium problem of function f on $\mathcal{K}$ is equivalent to the variational inequality problem, when we define $f(\overline{x}, y) = \langle y - \overline{x}, F(\overline{x}) \rangle$. Whenever the constraint set $\mathcal{K}$ varies by an element $x \in \mathcal{K}$, i.e., we replace $\mathcal{K}$ by a set-valued mapping $K : \mathcal{K} \rightrightarrows \mathcal{K}$ such that, for any $x \in X$, $K(x)$ is a nonempty closed convex set of $\mathcal{K}$, it turns out that both variational inequality problems (VIP) and equilibrium problems (EP) become quasi-variational inequality problems (QVIP) and quasi-equilibrium problems (QEP), respectively. Moreover, it can be derived that if the loss function f^i is differentiable with some supplementary conditions then GNEP can be formulated as QVIP (see Zhang et al. (2010), Ye (2017), Kocvara and Outrata (1995), Han et al. (2012)). On the other hand, all functions do not need to be differentiable in general, therefore, GNEP reduces to the QEP. It should be noted that such conditions are known for QVIP but not for QEP (see Cotrina and Zuniga (2018), Aussel et al. (2018)). Generally, an equilibrium is the solutions of a Nash equilibrium problem if and only if it is the solutions of an equilibrium problem under some setting. Indeed, in case of finding the equilibrium satisfying GNEP and QEP, it can be proved in the same direction under some conditions (see Nasri and Sosa (2011)). As a matter of fact, the objective functions in GNEP and QEP are real-valued functions. It is precious to consider the vector-valued functions as extension problems (see Bonnel et al. (2005), Ansari and Yao (2003)). Hence, we analyze the vector quasi-equilibrium problem for solving vector generalized Nash equilibrium problem. Let us derive the vector generalized Nash equilibrium problem (VGNEP) and vector quasi-equilibrium problems (VQEP) as follows. Now, let $m \in \mathbb{N}$, we consider $f^i : X \to \mathbb{R}^m$ and $K^i : X^{-i} \rightrightarrows X^i$ for all $i \in N$. Let $C^i \subseteq \mathbb{R}^m$ be a closed convex and pointed cone with $\text{int}(C^i)$ for all $i \in N$. Normally, for any closed convex and pointed cone $C \subseteq \mathbb{R}^m$ with nonempty interior, a partial order on $\mathbb{R}^m$ induced by C, denoted by $\preccurlyeq_C$, is described by, for all $a, b \in \mathbb{R}^m$,

$$a \preccurlyeq_C b \text{ if and only if } b - a \in C,$$
$$a \prec_C b \text{ if and only if } b - a \in \text{int}(C).$$

Observe that $a \not\prec_C b$ if and only if $a - b \notin -\text{int}(C)$ for all $a, b \in \mathbb{R}^m$. A **vector generalized Nash equilibrium problem (VGNEP)** is to find $x \in X$ such that, for all $i \in N$, $x^i \in K^i(x^{-i})$ and

$$f^i\left(x^{-i}, y^i\right) - f^i(x) \in C^i \tag{2}$$

for all $y_i \in K^i(x^{-i})$. Sometimes this following equation holds instead of (2),

$$f^i\left(x^{-i}, y^i\right) - f^i(x) \notin -\text{int}(C^i).$$

Set $K(x) := \oplus_{i \in N} K^i(x^{-i})$ and $C = \bigoplus_{i \in N} C^i$. A **vector quasi-equilibrium problems (VQEP)** is to find $x \in K(x)$ such that

$$f(x, y) \notin -\text{int}(C) \tag{3}$$

for all $y \in K(x)$ where $f : X \times X \to \mathbb{R}^m$. Denote SVQEP the set of solution of VQEP, i.e., SVQEP $= \{x \in K(x) : f(x, y) \notin -\text{int}(C) \text{ for all } y \in K(x)\}$. Moreover, in 2000, Fu (2000) studied generalized vector quasi-equilibrium problems as multivalued vector equilibrium problems. Mathematically speaking, to study QEP and VQEP, many results are related to existence and convergence of the solutions of the problems. In view of the fact that schemes of quasi-equilibrium issues are utilized as tools to demonstrate of a solution of constrained Nash equilibrium problems, the existence of solutions of these problems is also examined under some assumptions (see Ansari et al. (2004), Ansari (2008), Ram and Lal (2020)). On other sides, many works develop iterative algorithms for solving and reaching the solutions of these problems with computational results (see Tran et al. (2008), Rouhani and Mohebbi (2020), Rouhani and Mohebbi (2020)). Recently, in 2019, Iusem and Mohebbi (2019) established some methods to vector equilibrium problem under some assumptions. Let $C \subseteq \mathbb{R}^m$ is a closed, convex, and pointed cone with $\text{int}(C) \neq \emptyset$, and $f : E \times E \to \mathbb{R}^m$ is a vector valued bifunction. The **vector equilibrium problem (VEP)** is to find $\overline{x} \in \mathcal{K}$ such that

$$f(\overline{x}, y) \notin -\text{int}(C)$$

for all $y \in \mathcal{K}$. Their method is related to backtracking procedure for determining the right step size which is sometimes called an Armijo-type search. Previously Chen et al. (2018) proposed an alternative extragradient projection method for quasi-equilibrium problems in 2018. First, choose $\vartheta, \mu \in (0, 1), x_0 \in X, k = 0$. For current iterate x_k, compute y_k by solving the following optimization problem:

$$\min_{y\in K(x_k)} \left\{ f(x_k, y) + \frac{1}{2}\|y - x_k\|^2 \right\}.$$

If $x_k = y_k$, then stop. Otherwise, let $z_k = (1-\eta_k)x_k + \eta_k y_k$, where $\eta_k = \mu^{m_k}$ with m_k being the smallest nonnegative integer such that

$$f\left(\left(1-\mu^m\right)x_k + \mu^m y_k, x_k\right) - f\left(\left(1-\mu^m\right)x_k + \mu^m y_k, y_k\right) \geq \vartheta \|x_k - y_k\|^2.$$

Compute $x_{k+1} = \prod_{H_k^1 \cap H_k^2 \cap X}(x_0)$ where

$$H_k^1 = \{x \in E : f(z_k, x) \leq 0\},$$
$$H_k^2 = \{x \in E : \langle x - x_k, x_0 - x_k\rangle \leq 0\}.$$

There are also many mathematicians who studied an alternating method (see Solodov and Svaiter (1999), Han et al. (2007), Kopecka and Reich (2012)).

Therefore, in this work, we consider a vector quasi-equilibrium problem which is an extension scalar quasi-equilibrium problem. A modified extragradient method motivated by an alternative extragradient projection method with Armijo-type search for this problem is proposed. A sequence generated from the algorithm weakly converges to a solution of the vector quasi-equilibrium problem.

The paper is organized as follows. Preliminaries, many tools, and some useful definitions are provided in the Sect. 2. Next, Sect. 3, we formulate a generalized Nash equilibrium problem to a quasi equilibrium problem. Finally, statement and discussion are contained in the Sect. 4 with the corollary.

2 Preliminaries

Let E be a real Banach space with norm $\|\cdot\|$. In this section, the boundary, interior and closure of a set $A \subseteq E$ are denoted by $\text{bd}(A)$, $\text{int}(A)$ and $\text{cl}(A)$. For given $A, B \subseteq E$ and $t \in \mathbb{R}$, the algebraic sum $A + B$ and the scalar multiplication tA are defined as follows:

$$A + B := \{a + b : a \in A, b \in B\};$$

$$tA := \{ta : a \in A\}.$$

In particular, we denote $A + \{y\}$ by $A + y$ (or $y + A$) and $(-1)A$ by $-A$ for $A \subseteq E$ and $y \in E$. We define $Ty := \{ty : t \in T\}$ for $y \in E$ and $T \subseteq \mathbb{R}$.

Let X be a nonempty set and $\preccurlyeq$ a binary relation on X. The relation $\preccurlyeq$ is said to be

(1) reflexive if for all $x \in X$, $x \preccurlyeq x$,
(2) irreflexive if for all $x \in X$, $x \not\preccurlyeq x$,
(3) transitive if for all $x, y, z \in X$, $x \preccurlyeq y$ and $y \preccurlyeq z$ imply $x \preccurlyeq z$,

(4) antisymmetric if for all $x, y \in X$, $x \preccurlyeq y$ and $y \preccurlyeq x$ imply $x = y$,
(5) complete if for all $x, y \in X$, $x \preccurlyeq y$ or $y \preccurlyeq x$ imply $x = y$.

The relation $\preccurlyeq$ is called

(1) a preorder if it is reflexive and transitive,
(2) a strict order if it is irreflexive and transitive,
(3) a partial order if it is reflexive, transitive, and antisymmetric,
(4) a total order if it is reflexive, transitive, antisymmetric, and complete.

Definition 1 A nonempty set $C \subseteq E$ is a **cone** if for every $x \in C$ and for every $\lambda \in \mathbb{R}_+$ we have $\lambda x \in C$.

Clearly, if C is a cone then $0 \in C$.

Definition 2 Let $C \subseteq E$ be a cone. The cone C is called

(a) **convex** if for all $x_1, x_2 \in C$ we have $x_1 + x_2 \in C$,
(b) **nontrivial** or **proper** if $C \neq \{0\}$ and $C \neq E$,
(c) **reproducing** if $C - C = X$,
(d) **pointed** if $C \cap (-C) = \{0\}$.

Let C be a convex cone in E. Then, $C + C = C$ holds, and int(C) and cl(C) are also convex cones. Since $0 \in C$, we define a preoder $\preccurlyeq_C$ on E induced by C as follows:

$$\text{for } y_1, y_2 \in E,\ y_1 \preccurlyeq_C y_2 \text{ if and only if } y_2 - y_1 \in C.$$

This preorder is compatible with the linear structure

$$\text{for } y_1, y_2, y_3 \in E, \text{ if } y_1 \preccurlyeq_C y_2 \text{ then } y_1 + y_3 \preccurlyeq_C y_2 + y_3,$$
$$\text{for } y_1, y_2 \in E,\ y_1 \preccurlyeq_C y_2 \text{ and } t > 0, \text{ if } y_1 \preccurlyeq_C y_2 \text{ then } ty_1 \preccurlyeq_C ty_2.$$

Note that if C is pointed then $\preccurlyeq_C$ is a partial order. As usual, we denote by $C^+ := \{z \in \mathbb{R}^m : \langle y, z\rangle \geq 0 \text{ for all } y \in C\}$ the **positive dual cone** of C.

For a real Banach space E with norm $\|\cdot\|$, the topological dual of E is denoted as E^*. Then the duality mapping $J : E \rightrightarrows E^*$ is defined as

$$J(x) = \left\{v \in E^* : \langle x, v\rangle = \|x\|^2 = \|v\|^2\right\}.$$

It is well-known that when E is smooth, the duality operator J is single valued. Let E be a smooth Banach space. We define $\phi : E \times E \to \mathbb{R}$ as

$$\phi(x, y) = \|x\|^2 - 2\langle x, J(y)\rangle + \|y\|^2.$$

This function can be seen as a distance-like function, better condition than the square of the metric distance, namely $\|x - y\|^2$. It is elementary that

$$0 \leq (\|x\| - \|y\|)^2 \leq \phi(x, y)$$

for all $x, y \in E$. In Hilbert spaces, where the duality mapping J is the identity operator, it holds that $\phi(x, y) = \|x - y\|^2$. Let $\mathcal{K}$ be a subset in E. We define $\prod_{\mathcal{K}} : E \to \mathcal{K}$ by taking as $\prod_{\mathcal{K}}(x)$ the unique $x_0 \in \mathcal{K}$ such that

$$\phi(x_0, x) = \inf\{\phi(z, x) : z \in \mathcal{K}\}.$$

Moreover, $\prod_{\mathcal{K}}$ is called the generalized projection on to $\mathcal{K}$. When E is a Hilbert space, $\prod_{\mathcal{K}}$ is the orthogonal projection on to $\mathcal{K}$.

Proposition 1 *Consider a smooth Banach space E, and a closed and convex set $\mathcal{K} \subseteq E$. Take $x \in E$, $x_0 = \prod_{\mathcal{K}}(x)$ if and only if*

$$\langle z - x_0, J(x) - J(x_0)\rangle \leq 0$$

for all $z \in \mathcal{K}$.

Definition 3 Given a set $\mathcal{K} \subseteq \mathbb{R}^n$, we associate the following set

$$\inf_w^C(\mathcal{K}) := \{y \in \mathrm{cl}\mathcal{K} : \nexists z \in \mathcal{K} \text{ such that } z \prec y\}.$$

Definition 4 Given $G : \mathcal{K} \to \mathbb{R}^n$ and $\mathcal{K} \subseteq E$, the point $a \in E$ is called **weakly efficient** if $a \in \mathcal{K}$ and $G(a) \in \inf_w^C(G(S))$.

We denote as $\mathrm{argmin}_w^C\{G(x) : x \in \mathcal{K}\}$ the set of weakly efficient points. We observe that

$$\mathrm{argmin}_w^C\{G(x) : x \in \mathcal{K}\} = \mathcal{K} \cap G^{-1}\left(\inf_w^C(G(S))\right).$$

Definition 5 A map $G : E \to \mathbb{R}^n$ is called **C-convex** whenever

$$G(tx + (1-t)y) \preccurlyeq_C tG(x) + (1-t)G(y),$$

for all $x, y \in E$ and $t \in [0, 1]$.

Definition 6 Let $h : E \times E \to \mathbb{R}^n$ is said to be

(i) **C—pseudomonotone** whenever $g(x, y) \in \mathbb{R}^n \setminus (-C)$ with $x, y \in E$, it holds that $g(y, x) \in -C\setminus\{0\}$,
(ii) **weakly C—pseudomonotone** whenever $g(x, y) \notin -\mathrm{int}(C)$ with $x, y \in E$, it holds that $g(y, x) \in -C$.

Definition 7 A Banach space E is said to be

(i) **strictly convex** if $\left\|\frac{x+y}{2}\right\| < 1$ for all $x, y \in E$ with $\|x\| = \|y\| = 1$ and $x \neq y$,
(ii) **uniformly convex** if for each $\epsilon \in (0, 2]$, there exists $\delta > 0$ such that for all $x, y \in E$ with $\|x\| = \|y\| = 1$ and $\|x - y\| \geq \epsilon$, it holds that $\left\|\frac{x+y}{2}\right\| < 1 - \delta$.

It is known that uniformly convex Banach space are reflexive and convex.

Definition 8 A Banach space E is said to be **smooth** if

$$\lim_{t\to 0}\frac{\|x+ty\|-\|x\|}{t}$$

exists for all $x, y \in \{z \in R : \|z\| = 1\}$. Moreover, it is said to be **uniformly smooth** if the limit is attained uniformly for all $x, y \in \{z \in R : \|z\| = 1\}$.

Theorem 1 *Let E be a reflexive Banach space. If $f : E \to \mathbb{R} \cup \{+\infty\}$ is proper, strongly convex, and lower-semicontinuous, then argmin(f) is a singleton.*

Lemma 1 *Let $f : E \to \mathbb{R} \cup \{+\infty\}$ be a function. Then,*

(i) f is weakly lower-semicontinuous if and only if for any $\{x_k\} \subseteq E$ with $x_k \rightharpoonup \overline{x}$ implies

$$\liminf_k f(x_k) \geq f(\overline{x}),$$

(ii) f is weakly upper-semicontinuous if and only if for any $\{x_k\} \subseteq E$ with $x_k \rightharpoonup \overline{x}$ implies

$$\limsup_k f(x_k) \leq f(\overline{x}).$$

Definition 9 A map $G : E \to \mathbb{R}^m \cup \{+\infty\}$ is called positively weakly lower-semicontinuous, if for every $z \in C^+$ with $x \mapsto \langle G(x), z\rangle$ is weakly lower-semicontinuous. Moreover, G is called positively weakly upper-semicontinuous, if $-G$ is positively weakly lower-semicontinuous.

Lemma 2 *If $G : E \to \mathbb{R}^m \cup \{+\infty\}$ is C-convex and positively upper-semicontinuous, then G is positively weakly upper-semicontinuous.*

Theorem 2 *If $S \subseteq E$ is a convex set and $G : S \to \mathbb{R}^m \cup \{+\infty\}$ is a C-convex proper map, then*

$$argmin_w^C\{G(x) : x \in S\} = \cup_{z\in C^+\setminus\{0\}} argmin_w^C\{\langle G(x), z\rangle : x \in S\}.$$

Proposition 2 *Let E be a smooth and uniformly convex Banach space. Take two sequences $\{x_k\}, \{y_k\} \subseteq E$. If $\lim_{k\to+\infty} \phi(x_k, y_k) = 0$ and either $\{x_k\}$ or $\{y_k\}$ is bounded, then $\lim_{k\to+\infty} \|x_k - y_k\| = 0$.*

Definition 10 Let X be a nonempty, closed and convex subset of a Banach space. A set-valued mapping $K : X \rightrightarrows X$ is said to be

(i) weak-to-strong upper-semicontinuous at $x \in X$ if for any sequence $(x_n) \subseteq X$ with $x_n \rightharpoonup \overline{x}$ and for any sequence $(y_n) \subseteq X$ with $y_n \in K(x_n)$ and $y_n \to \overline{y}$ then $\overline{y} \in K(\overline{x})$,

(ii) weak-to-strong lower-semicontinuous at $x \in X$ if for any sequence $(x_n) \subseteq X$ with $x_n \rightharpoonup \overline{x}$ and for any $\overline{y} \in K(\overline{x})$ then there exists a sequence $(y_n) \subseteq X$ such that $(y_n) \subseteq K(x_n)$ with $y_n \to \overline{y}$ for all $n \in \mathbb{N}$,

(iii) weak-to-strong semicontinuous at $x \in X$ if it is both weak-to-strong upper-semicontinuous and weak-to-strong lower-semicontinuous at x.

3 Formulation to a Quasi-Equilibrium Problem

Recall that, consider n player, let X^i be a nonempty set of pure strategies of player i for all $i = 1, 2, \ldots, n$. We gave a notation that $X := X^1 \times X^2 \times \ldots \times X^n$. Consider a function $f^i : X \to \mathbb{R}^{m_i}$ for some $m_i \in \mathbb{N}$ and for all $i = 1, 2, \ldots, n$ as a loss function of each player. Let C^i be a cone in $\mathbb{R}^{m_i}$. Moreover, we define $K^i : X^{-i} \rightrightarrows X^i$ be a set-valued mapping. In the n-person noncooperative game, the n-tuple $\overline{x} = (\overline{x}^1, \overline{x}^2, \ldots, \overline{x}^n) \in X$ is said to be a **vector generalized Nash equilibrium point** of $f^1, f^2, \ldots, f^n$ on X if for all $i = 1, 2, \ldots, n$, $\overline{x}^i \in K^i(\overline{x}^{-i})$ and

$$f^i(\overline{x}^1, \overline{x}^2, \ldots, \overline{x}^{i-1}, y^i, \overline{x}^{i+1}, \ldots, \overline{x}^n) - f^i(\overline{x}^1, \overline{x}^2, \ldots, \overline{x}^{i-1}, \overline{x}^i, \overline{x}^{i+1}, \ldots, \overline{x}^n) \notin -\text{int}(C^i)$$

for all $y^i \in K^i(\overline{x}^{-i})$.

Now, we set $K(x) := \oplus_{i=1}^n K^i(\overline{x}^{-i})$ and define $f : X \times X \to \mathbb{R}^{m_1+m_2+\ldots+m_n}$ by

$$f(x, y) = \oplus_{i=1}^n \left(f^i(\overline{x}^{-i}, y^i) - f^i(\overline{x})\right) \tag{4}$$

for all $x, y \in X$. Note that a **vector quasi-equilibrium problem** is to find $\overline{x} \in K(\overline{x})$ such that

$$f(x, y) \notin -\text{int}(C^i) \tag{5}$$

for all $y \in K(x)$.

Therefore, we can see that $\overline{x} = (\overline{x}^1, \overline{x}^2, \ldots, \overline{x}^n)$ is a vector generalized Nash equilibrium point for $f^1, f^2, \ldots, f^n$ if and only if it is a vector quasi-equilibrium point of f.

4 Convergence Results

Let E be a strictly convex, smooth and reflexive Banach space, $m \in \mathbb{N}$, and $C \subseteq \mathbb{R}^m$ be a closed convex and pointed cone with $\text{int}(C)$. Denote E^* the topological dual of E. The duality mapping $J : E \rightrightarrows E^*$. Given $N = \{1, 2, \ldots, n\}$. Consider a closed and convex set $X^i \subseteq E$ for all $i \in N$. Denote $X = \oplus_{i \in N} X^i$ and $X^{-i} = \oplus_{j \in N, j \neq i} X^j$. It would be noted that X and X^{-i} are closed and convex. Define $f^i : X \to \mathbb{R}^m$ and $K^i : X^{-i} \rightrightarrows X^i$ related to a Nash equilibrium problem. To formulate the problem to VQEP, we set $K(x) := \oplus_{i\in N} K^i(x^{-i})$ and $f : E \times E \to \mathbb{R}^m$. We assume the following assumptions in our work.

(A1) $f(x, \cdot)$ is C-convex for any fixed $x \in E$,
(A2) $f(\cdot, y)$ is positively weakly upper-semicontinuous for all $y \in E$,
(A3) $f(\cdot,\cdot) : E \times E \to \mathbb{R}^m$ is uniformly continuous on bounded sets,
(A4) $f(x, x) = 0$ for all $x \in E$,
(A5) f is weakly C-pseudomonotone on E with respect to SVQEP*, i.e., if $f(x, y) \notin \text{int}(C)$ then $f(y, x) \in -C$ for all $x \in$ SVQEP*, for all $y \in E$.
(A6) $K(x)$ is a nonempty closed convex subset of X for all $x \in X$,
(A7) K is weak-to-strong lower-semicontinuous on X,
(A8) $x \in K(x)$ for all $x \in X$,
(A9) SVQEP* $:= \left\{x \in \bigcap_{x\in X} K(x) \mid f(x, y) \notin -\text{int}(C) \text{ for all } y \in \bigcup_{x\in X} K(x)\right\}$ is nonempty,

Algorithm 1 (An Alternative Extragradient Method)
Step 1 Given constants $\delta \in (0, 1)$, $\hat{\beta}, \tilde{\beta}$, satisfying $0 < \hat{\beta} < \tilde{\beta}$, a sequence $\{\beta_k\} \subseteq [\hat{\beta}, \tilde{\beta}]$ and a sequence $\{e_k\} \subseteq \text{int}(C)$ such that $\|e_k\| = 1$. Take $x_0 \in X$. Set $k = 0$.
Step 2 Define $z_k \in \text{argmin}_w^C \left\{f(x_k, y) + \frac{1}{2\beta_k}\|y\|^2 e_k - \frac{1}{\beta_k}\langle y, J(x_k)\rangle e_k : y \in K(x_k)\right\}$.
Step 3 If $x_k = z_k$ stop. Otherwise, let

$$l(k) = \min\{l \geq 0 : -\beta_k f(y_l, x_k) + \beta_k f(y_l, z_k) + \frac{\delta}{2}\phi(z_k, x_k)e_k \notin \text{int}(C)\}$$

where $y_l = 2^{-l}z_k + (1 - 2^{-l})x_k$.
Step 4 Take $\alpha_k = 2^{-l(k)}$, and $y_k = \alpha_k z_k + (1 - \alpha_k)x_k$.
Step 5 Set

$$H_k^1 = \{x \in E : f(y_k, x) \in -C\}$$

and

$$H_k^2 = \{x \in E : \langle x - x_k, J(x_0) - J(x_k)\rangle e_k \in -C\},$$

compute $x_{k+1} = \prod_{H_k^1 \cap H_k^2 \cap X}(x_0)$.
Step 6 Set $k := k + 1$ and go to **Step 2**.

By the assumption (B5), we have $x_k \in K(x_k)$. Moreover $z_k \in K(x_k)$, hence, by (B2), $y_k \in K(x_k)$. Moreover, H_k^1 and H_k^2 are closed and convex.

Definition 11 A sequence $\{x_k\}$ is said to be an asymptotically solving sequence for VQEP if for each $y_k \in K(x_k)$, there exist a sequence $\{\varepsilon_k\} \subseteq \mathbb{R}^m$ and $k_0 \in \mathbb{N}$ such that $\|\varepsilon_k\| \to 0$ and for all $k \geq k_0$,

$$f(x_k, y_k) + \varepsilon_k \notin -\text{int}(C).$$

Lemma 3 *The sequence $\{z_k\}$ is well-defined.*

Proof Take any $c \in C^+\backslash\{0\}$. Because $e_k \in \text{int}(C)$, by the definition of C^+, we have that $\langle e_k, c\rangle > 0$. Define $\psi : E \to \mathbb{R} \cup \{+\infty\}$ as

$$\psi(y) = \langle f(x_k, y), c\rangle + \frac{1}{2\beta_k} \|y\|^2 \langle e_k, c\rangle - \frac{1}{\beta_k} \langle y, J(x_k)\rangle \langle e_k, c\rangle .$$

Since f is C-convex, for all $x, y \in E$ and $t \in [0, 1]$, we have that, $f(x_k, tx + (1-t)y) \preccurlyeq_C tf(x_k, x) + (1-t)f(x_k, y)$. That is, $tf(x_k, x) + (1-t)f(x_k, y) - f(x_k, tx + (1-t)y) \in C$. Since $c \in C^+\backslash\{0\}$, we get that

$$\langle tf(x_k, x) + (1-t)f(x_k, y) - f(x_k, tx + (1-t)y), c\rangle \geq 0.$$

Observe that

$$\begin{aligned} &\langle tf(x_k, x) + (1-t)f(x_k, y) - f(x_k, tx + (1-t)y), c\rangle \\ &= t\langle f(x_k, x), c\rangle + (1-t)\langle f(x_k, y), c\rangle - \langle f(x_k, tx + (1-t)y), c\rangle \end{aligned}$$

which implies that

$$t\langle f(x_k, x), c\rangle + (1-t)\langle f(x_k, y), c\rangle \geq \langle f(x_k, tx + (1-t)y), c\rangle .$$

Since E is strictly convex then so is $\|\cdot\|^2$. It implies that ψ is strongly convex. Moreover, we have ψ is proper and lower-semicontinuous. By Theorem 1 and Theorem 2, such minimizer satisfies

$$f(x_k, y) + \frac{1}{2\beta_k} \|y\|^2 e_k - \frac{1}{\beta_k} \langle y, J(x_k)\rangle$$

which can be taken as z_k. □

Lemma 4 *If $x_k \neq z_k$ then the half space H_k^1 separates the point x_k from the set SVQEP $= \{x \in K(x) : g(x, y) \notin -int(C)$ for all $y \in K(x)\}$. Moreover, SVQEP $\subseteq H_k^1 \cap X$ for all $k \geq 0$.*

Proof By C-convexity of $f(y_k, \cdot)$ and since $y_k = \alpha_k z_k + (1-\alpha_k)x_k$ then we have

$$(\alpha_k f(y_k, z_k) + (1-\alpha_k) f(y_k, x_k)) - f(y_k, y_k) \in C,$$

i.e. $\alpha_k f(y_k, z_k) + (1-\alpha_k) f(y_k, x_k) \in C$. Thus, $\beta_k f(y_k, z_k) + \left(\frac{\beta_k}{\alpha_k} - \beta_k\right) f(y_k, x_k) \in C$. That is, $\frac{\beta_k}{\alpha_k} f(y_k, x_k) + \beta_k f(y_k, z_k) - \beta_k f(y_k, x_k) \in C$. From the algorithm, we know that $-\beta_k f(y_k, x_k) + \beta_k f(y_k, z_k) + \frac{\delta}{2}\phi(z_k, x_k) e_k \notin \text{int}(C)$. We then simplify by addition and subtraction $\frac{\beta_k}{\alpha_k} f(y_k, x_k)$, it follows that

$$-\beta_k f(y_k, x_k) + \beta_k f(y_k, z_k) + \frac{\beta_k}{\alpha_k} f(y_k, x_k) - \frac{\beta_k}{\alpha_k} f(y_k, x_k) + \frac{\delta}{2}\phi(z_k, x_k) e_k \notin \text{int}(C).$$

Observe that $\frac{\delta}{2}\phi(z_k, x_k)\, e_k \in \text{int}(C)$ because $e_k \in \text{int}(C)$ and $\frac{\delta}{2}\phi(z_k, x_k) \geq 0$. It implies that $-\frac{\beta_k}{\alpha_k} f(y_k, x_k) \notin C$, i.e., $f(y_k, x_k) \notin -C$. Hence, $x_k \notin H_k^1$. By the assumption that SVQEP is nonempty, for any $x \in$ SVQEP, we have $f(x, y_k) \notin -\text{int}(C)$ because $y_k \in K(x_k)$. By weakly C-pseudomonotone property of f, we have that $f(y_k, x) \in -C$. Therefore, $x \in H_k^1$ for all $x \in$ SVQEP. That is $\text{bd}(H_k^1)$ separates x_k from SVQEP and SVQEP $\subseteq H_k^1 \cap X$ for all $k \geq 0$. □

Lemma 5 *The SVQEP* $\subseteq H_k^1 \cap H_k^2 \cap X$ *for all* $k \geq 0$.

Proof By Lemma 4, it is sufficient to prove that SVQEP $\subseteq H_k^2$ for all $k \geq 0$. By induction, if $k = 0$, it can be seen that SVQEP $\subseteq H_k^2$ because $H_0^2 = E$. Suppose that SVQEP $\subseteq H_k^2$ for some $k = l \geq 0$. Then

$$\text{SVQEP} \subseteq H_l^1 \cap H_l^2 \cap X. \tag{6}$$

Since SVQEP is nonempty, we can let $\overline{x} \in \text{SVQEP} \subseteq H_l^1 \cap H_l^2 \cap X$. By the fact that $x_{l+1} \in \prod_{H_l^1 \cap H_l^2 \cap X}(x_0)$ which $H_l^1 \cap H_l^2 \cap X$ is closed convex. Then, by Lemma 1, we obtain that

$$\langle \overline{x} - x_{l+1}, J(x_0) - J(x_{l+1}) \rangle \leq 0.$$

Since $e_{l+1} \in \text{int}(C)$ then

$$\langle \overline{x} - x_{l+1}, J(x_0) - J(x_{l+1}) \rangle\, e_{l+1} \in -C,$$

it implies that $\overline{x} \subseteq H_{l+1}^2$. Hence, SVQEP $\subseteq H_{l+1}^2$ □

Proposition 3 *Assume that f is C-convex. Take $x \in E$, $\beta \in \mathbb{R}^+$ and $e \in int(C)$ such that $\|e\| = 1$. If*

$$z \in argmin_w^C \left\{ f(x, y) + \frac{1}{2\beta}\|y\|^2 e - \frac{1}{\beta}\langle y, J(x)\rangle\, e : y \in K(x) \right\} \tag{7}$$

then there exists $c \in C^+ \backslash \{0\}$ such that

$$\langle y - z, J(x) - J(z)\rangle \langle e, c\rangle \leq \beta\left(\langle f(x, y), c\rangle - \langle f(x, z), c\rangle\right)$$

for all $y \in K(x)$.

Proof Let $N_{K(x)}(z)$ be a normal cone of $K(x)$ at $z \in K(x)$, i.e.,

$$N_{K(x)}(z) = \left\{ v \in E^* : \langle y - z, v\rangle \leq 0 \text{ for all } y \in K(x) \right\}.$$

Since z satisfies (7), by Theorem 2, there is $c \in C^+ \backslash \{0\}$ such that z satisfies the first order optimality condition given by

$$0 \in \partial \left\{ \langle f(x, \cdot), c\rangle + \frac{1}{2\beta}\|\cdot\|^2 \langle e, c\rangle - \frac{1}{\beta}\langle \cdot, J(x)\rangle \langle e, c\rangle \right\}(z) + N_{K(x)}(z).$$

Thus, we have that there are $w \in \partial \{\langle f(x, \cdot), c\rangle\}(z)$ and $w_x \in N_{K(x)}(z)$ such that

$$0 = w + \frac{\langle e, c\rangle}{\beta} J(z) - \frac{\langle e, c\rangle}{\beta} J(x) + w_x.$$

Therefore, since $w_x \in N_{K(x)}(z)$, we obtain that

$$\left\langle y - z, -w - \frac{\langle e, c\rangle}{\beta} J(z) + \frac{\langle e, c\rangle}{\beta} J(x)\right\rangle \leq 0,$$

so that, using the fact that $w \in \partial \{\langle f(x, \cdot), c\rangle\}(z)$,

$$\frac{\langle e, c\rangle}{\beta} \langle y - z, J(x) - J(z)\rangle \leq \langle y - z, w\rangle \leq \langle f(x, y), c\rangle - \langle f(x, z), c\rangle.$$

The proof is complete as desired. □

Corollary 1 *Assume that $\{x_k\}$ and $\{z_k\}$ are the sequences generated by Algorithm 1. Then there exists $\{c_k\} \subseteq C^+\backslash\{0\}$ such that*

$$\langle y - z_k, J(x_k) - J(z_k)\rangle \langle e_k, c_k\rangle \leq \beta_k \left(\langle f(x_k, y), c_k\rangle - \langle f(x_k, z_k), c_k\rangle\right)$$

for all $y \in K(x_k)$.

Proof By the proximal step in Algorithm 1, and Proposition 3, the proof can be easily seen. □

Proposition 4 *If the Algorithm 1 stops at k^{th} iteration, then x_k is a solution of VQEP.*

Proof By the assumption, we have $x_k = z_k$. By Corollary 1, it implies that there is $c_k \in C^+\backslash\{0\}$ such that

$$\langle f(x_k, y), c_k\rangle \geq 0$$

for all $y \in K(x_k)$. Since $c_k \in C^+\backslash\{0\}$, we obtain that $f(x_k, y) \notin -\text{int}(C)$ for all $y \in K(x_k)$. Hence x_k satisfies VQEP. □

Proposition 5 *The Armijo-type search for α_k is finite, i.e., the value $l(k)$ is well-defined. Moreover, the same holds for the sequence $\{x_k\}$.*

Proof By Mathematical induction, we assume that x_k is well-defined, and it can be proceed to establish that the same holds for x_{k+1}. Note that z_k is well-defined by Proposition 3. It is suffices to check that $l(k)$ is well-defined. Suppose to the contrary that

$$-\beta_k f(y_l, x_k) + \beta_k f(y_l, z_k) + \frac{\delta}{2}\phi(z_k, x_k)e_k \in \text{int}(C)$$

for all l. Because $c_k \in C^+ \setminus \{0\}$, therefore we have

$$\beta_k [\langle f(y_l, x_k), c_k\rangle - \langle f(y_l, z_k), c_k\rangle] \le \frac{\delta}{2}\phi(z_k, x_k)\langle e_k, c_k\rangle \tag{8}$$

for all l. Observe that the sequence $\{y_l\}$ is strongly convergent to x_k. By the uniform continuity on bounded set of f, talking limits in (8), we have

$$\beta_k [\langle f(x_k, x_k), c_k\rangle - \langle f(x_k, z_k), c_k\rangle] \le \frac{\delta}{2}\phi(z_k, x_k)\langle e_k, c_k\rangle$$

Since $x_k \in K(x_k)$, by Corollary 1, it follows that

$$\langle x_k - z_k, J(x_k) - J(z_k)\rangle \le \frac{\delta}{2}\phi(z_k, x_k).$$

By the definition of ϕ, then

$$\phi(z_k, x_k) + \phi(x_k, z_k) \le \delta\phi(z_k, x_k).$$

Since $\delta < 1$, it implies that $\phi(x_k, z_k) < 0$, which is a contradiction with the nonnegativity of ϕ. □

Proposition 6 *If the Algorithm 1 does not stop at iteration k^{th}, then $f(y_k, x_k) \notin -C$.*

Proof Suppose that $f(y_k, x_k) \in -C$. Note that

$$0 = f(y_k, y_k) \preccurlyeq_C \alpha_k f(y_k, z_k) + (1 - \alpha_k) f(y_k, x_k).$$

Because $-(1-\alpha_k) f(y_k, x_k) \in C$ and $\alpha_k f(y_k, z_k) + (1-\alpha_k) f(y_k, x_k) \in C$, and C is a convex cone, we can conclude that $f(y_k, z_k) \in C$. Therefore, we have

$$-\beta_k f(y_k, x_k) + \beta_k f(y_k, z_k) + \frac{\delta}{2}\phi(z_k, x_k)e_k \in \text{int}(C),$$

which is a contradiction because the assumption that $x_k \neq z_k$ and the fact that $e_k \in \text{int}(C)$. □

Proposition 7 *Assume that E is uniformly convex and smooth. If the Algorithm 1 does not stop at iteration k, then*

(i) the sequence $\{x_k\}$ is bounded,
(ii) the sequence $\{z_k\}$ is bounded,
(iii) the sequence $\{y_k\}$ is bounded.

Proof (i) Let $\overline{x} \in$ SVQEP. By Lemma 5, we have SVQEP $\subseteq H_k^1 \cap H_k^2 \cap X$. Since the algorithm does not have finite termination which means $x_k \notin H_k^1$, and $x_{k+1} = \prod_{H_k^1 \cap H_k^2 \cap X}(x_0)$ then

$$\langle \overline{x} - x_{k+1}, J(x_0) - J(x_{k+1}) \rangle \leq 0.$$

That is

$$\phi(x_{k+1}, x_0) + \phi(\overline{x}, x_{k+1}) - \phi(\overline{x}, x_0) \leq 0$$

which implies that $\phi(\overline{x}, x_{k+1}) \leq \phi(\overline{x}, x_0) - \phi(x_{k+1}, x_0) \leq \phi(\overline{x}, x_0)$. Hence, $\{\phi(\overline{x}, x_k)\}$ is bounded. By that fact that

$$0 \leq (\|\overline{x}\| - \|x_k\|)^2 \leq \phi(\overline{x}, x_k).$$

Therefore, $\{x_k\}$ is bounded. It would be noted that we consider the bounded sets in reflexive Banach spaces then $\{x_k\}$ has weak cluster points.

(ii) Since $x_k \in K(x_k)$ and $x_{k+1} = \prod_{H_k^1 \cap H_k^2 \cap X}(x_0)$, by Theorem 2, then there exists $c_k \in C^+ \setminus \{0\}$ such that

$$\begin{aligned}
&\beta_k \langle f(x_k, z_k), c_k \rangle + \tfrac{1}{2} \|z_k\|^2 \langle e_k, c_k \rangle - \langle z_k, J(x_k) \rangle \langle e_k, c_k \rangle \\
&\leq \beta_k \langle f(x_k, x_k), c_k \rangle + \tfrac{1}{2} \|x_k\|^2 \langle e_k, c_k \rangle - \langle x_k, J(x_k) \rangle \langle e_k, c_k \rangle \\
&= \tfrac{1}{2} \|x_k\|^2 \langle e_k, c_k \rangle - \langle x_k, J(x_k) \rangle \langle e_k, c_k \rangle \\
&= -\tfrac{1}{2} \|x_k\|^2 \langle e_k, c_k \rangle \\
&\leq 0.
\end{aligned} \tag{9}$$

That is,

$$\beta_k \langle f(x_k, z_k), c_k \rangle + \frac{1}{2} \|z_k\|^2 \langle e_k, c_k \rangle - \langle z_k, J(x_k) \rangle \langle e_k, c_k \rangle \leq 0.$$

It implies that

$$\begin{aligned}
\|z_k\|^2 \langle e_k, c_k \rangle &\leq 2 \langle z_k, J(x_k) \rangle \langle e_k, c_k \rangle - 2\beta_k \langle f(x_k, z_k), c_k \rangle \\
&\leq 2 \|z_k\| \|x_k\| \langle e_k, c_k \rangle - 2\beta_k \langle f(x_k, z_k), c_k \rangle .
\end{aligned} \tag{10}$$

We will now use the subdifferential by taking $u_k^* \in \partial \langle f(x_k, \cdot), c_k \rangle (x_k)$ and define $u_k = \frac{1}{\langle e_k, c_k \rangle} u_k^*$, by the definition of $\partial \langle f(x_k, \cdot), c_k \rangle (x_k)$, we obtain that

$$\langle z_k - x_k, u_k \rangle \langle e_k, c_k \rangle \leq \langle f(x_k, z_k), c_k \rangle - \langle f(x_k, x_k), c_k \rangle = \langle f(x_k, z_k), c_k \rangle . \tag{11}$$

Combining (10) and (11), and dividing by $\langle e_k, c_k \rangle$, we have

$$\|z_k\|^2 \leq 2 \|z_k\| \|x_k\| + 2\beta_k \langle x_k - z_k, u_k \rangle \leq 2 \|z_k\| \|x_k\| + 2\tilde{\beta} \|u_k\| (\|x_k\| + \|z_k\|) .$$

Considering separately the cases of $\|x_k\| \geq \|z_k\|$ and $\|x_k\| \leq \|z_k\|$, it follows that

$$\|z_k\| \leq 4\tilde{\beta} \|u_k\| + 2 \|x_k\| .$$

Since $u_k^* \in \partial \langle f(x_k, \cdot), c_k \rangle (x_k)$ for any $k \in \mathbb{N}$ and E is reflexive then $\partial \langle f(x_k, \cdot), c_k \rangle (x_k)$ is maximal monotone. Hence, it is bounded on bounded sets. By Proposition 7 (i), we have $\{x_k\}$ is bounded. Therefore, $\{z_k\}$ is bounded.

(iii) Since $y_k = \alpha_k z_k + (1 - \alpha_k) x_k$, by Proposition 7 (i), (ii), it immediately implies the result as desired. □

Lemma 6 $\lim_{k \to +\infty} \|x_{k+1} - x_k\| = 0$

Proof Since $x_{k+1} = \prod_{H_k^1 \cap H_k^2 \cap X}(x_0)$ then $x_{k+1} \in H_k^2$ where

$$H_k^2 = \{x \in E : \langle x - x_k, J(x_0) - J(x_k) \rangle e_k \in -C\}.$$

That is, $\langle x_{k+1} - x_k, J(x_0) - J(x_k) \rangle e_k \in -C$. Since $e_k \in \text{int}(C)$, we have

$$\langle x_{k+1} - x_k, J(x_0) - J(x_k) \rangle \leq 0.$$

Note that

$$\begin{aligned}\langle x_{k+1} - x_k, J(x_0) - J(x_k) \rangle &= \langle x_{k+1} - x_k, J(x_0) \rangle - \langle x_{k+1} - x_k, J(x_k) \rangle \\ &= \langle x_{k+1}, J(x_0) \rangle - \langle x_k, J(x_0) \rangle - \langle x_{k+1}, J(x_k) \rangle + \langle x_k, J(x_k) \rangle .\end{aligned}$$

Moreover, by the definition of ϕ, we obtain that

$$\begin{aligned}\langle x_{k+1}, J(x_0) \rangle &= \frac{-\phi(x_{k+1}, x_0) + \|x_{k+1}\|^2 + \|x_0\|^2}{2}, \\ \langle x_k, J(x_0) \rangle &= \frac{-\phi(x_k, x_0) + \|x_k\|^2 + \|x_0\|^2}{2}, \\ \langle x_{k+1}, J(x_k) \rangle &= \frac{-\phi(x_{k+1}, x_k) + \|x_{k+1}\|^2 + \|x_k\|^2}{2}, \\ \langle x_k, J(x_k) \rangle &= \frac{-\phi(x_k, x_k) + \|x_k\|^2 + \|x_k\|^2}{2}.\end{aligned}$$

It follows that

$$\phi(x_{k+1}, x_k) \leq \phi(x_{k+1}, x_0) - \phi(x_k, x_0) \tag{12}$$

because $\phi(x_k, x_k) = 0$. Since $0 \leq \phi(x, y)$ for all $x, y \in E$, we have that $0 \leq \phi(x_{k+1}, x_k) \leq \phi(x_{k+1}, x_0) - \phi(x_k, x_0)$ which implies that $\phi(x_k, x_0) \leq \phi(x_{k+1}, x_0)$. That is, $\phi(x_k, x_0)$ is nondecreasing. Also, we observe that $\phi(x_{k+1}, x_0)$ is bounded above because E is smooth,

$$\begin{aligned}\phi(x_{k+1}, x_0) &= \|x_{k+1}\|^2 - 2\langle x_{k+1}, J(x_k) \rangle + \|x_k\|^2 \\ &\leq \|x_{k+1}\|^2 + 2\|x_{k+1}\| \|J(x_k)\| + \|x_k\|^2 ,\end{aligned}$$

and $\{x_k\}$ is bounded by Proposition 7 (i). Then, we have that $\phi(x_k, x_0)$ is convergent. By (12), it turns out that $\lim_{k\to+\infty} \phi(x_{k+1}, x_k) = 0$. By Proposition 2 and Proposition 7 (i), we get that $\lim_{k\to+\infty} \|x_{k+1} - x_k\| = 0$. □

Lemma 7 *Assume that E is uniformly convex and smooth. If the Algorithm 1 does not stop at iteration k, then all limit points of* $\{f(y_k, x_k)\}$ *belong to* $-C$.

Proof Since $\{x_k\}$ and $\{z_k\}$ are bounded and $y_k = \alpha_k z_k + (1-\alpha_k)x_k$ where $\alpha_k \in [0, 1]$ then $\{y_k\}$ is bounded. Because f is uniformly continuous on bounded sets and, by Lemma 6, we have $\lim_{k\to+\infty} \|x_{k+1} - x_k\| = 0$, we can conclude that

$$\lim_{k\to+\infty} \|f(y_k, x_k) - f(y_k, x_{k+1})\| = 0.$$

Since the Algorithm 1 does not stop within finite iteration, then $f(y_k, x_k) \notin -C$. Observe that $x_{k+1} = \prod_{H_k^1 \cap H_k^2 \cap X}(x_0)$ then $x_{k+1} \in H_k^1$ which implies that $f(y_k, x_{k+1}) \in -C$. Since C is closed thus the limit points of $f(y_k, x_k)$ belong to $-C$. □

Proposition 8 *Assume that E is uniformly convex and uniformly smooth with the nonempty set SVEQP. If* $\{x_{k_i}\}$ *is a subsequence* $\{x_k\}$ *satisfying* $\lim_{i\to+\infty} \phi(z_{k_i}, x_{k_i}) = 0$, *then* $\{x_{k_i}\}$ *is an asymptotically solving sequence for VQEP.*

Proof By the assumption and Proposition 2, we get

$$\lim_{k\to+\infty} \|z_k - x_k\| = 0.$$

The uniform smoothness of E, implies uniformly norm-to-norm continuity of J on each bounded set of E. Therefore, we have

$$\lim_{k\to+\infty} \|J(z_k) - J(x_k)\| = 0. \tag{13}$$

We observe that

$$\left\|f(x_{k_i}, z_{k_i})\right\| = \left\|f(x_{k_i}, z_{k_i}) - f(x_{k_i}, x_{k_i})\right\|.$$

Since both $\{x_k\}$ and $\{z_k\}$ are bounded, we then obtain that

$$\lim_{i\to+\infty} f(x_{k_i}, z_{k_i}) = 0. \tag{14}$$

Take now any $y_k \in K(x_k)$, by Corollary 1, we have

$$\langle y_{k_i} - z_{k_i}, J(x_{k_i}) - J(z_{k_i})\rangle \langle e_{k_i}, c_{k_i}\rangle \le \beta_{k_i}\left[\langle f(x_{k_i}, y_{k_i}), c_{k_i}\rangle - \langle f(x_{k_i}, z_{k_i}), c_{k_i}\rangle\right],$$

which implies that

$$-\left\|y_{k_i} - z_{k_i}\right\| \left\|J(x_{k_i}) - J(z_{k_i})\right\| \langle e_{k_i}, c_{k_i}\rangle \le \beta_{k_i}\left[\langle f(x_{k_i}, y_{k_i}), c_{k_i}\rangle - \langle f(x_{k_i}, z_{k_i}), c_{k_i}\rangle\right],$$

i.e., $\|y_{k_i} - z_{k_i}\| \, \|J(x_{k_i}) - J(z_{k_i})\| \langle e_{k_i}, c_{k_i}\rangle + \beta_{k_i} [\langle f(x_{k_i}, y_{k_i}), c_{k_i}\rangle - \langle f(x_{k_i}, z_{k_i}), c_{k_i}\rangle] \geq 0$. Since $c_{k_i} \in C^+\backslash\{0\}$, we have

$$\frac{1}{\beta_{k_i}} \|y_{k_i} - z_{k_i}\| \, \|J(x_{k_i}) - J(z_{k_i})\| \, e_{k_i} + f(x_{k_i}, y_{k_i}) - f(x_{k_i}, z_{k_i}) \notin -\text{int}(C)$$

for all $y_k \in K(x_k)$. Define

$$\varepsilon_{k_i} := \frac{1}{\beta_{k_i}} \|y_{k_i} - z_{k_i}\| \, \|J(x_{k_i}) - J(z_{k_i})\| \, e_{k_i} - f(x_{k_i}, z_{k_i}).$$

By Proposition 7, (13), (14), it follows that $\lim_{i\to+\infty} \varepsilon_{k_i} = 0$. Therefore, $f(x_{k_i}, y_{k_i}) + \varepsilon_{k_i} \notin -\text{int}(C)$. Hence, $\{x_{k_i}\}$ is an asymptotically solving sequence for VQEP. □

Proposition 9 *Assume that E is uniformly convex and uniformly smooth with the nonempty set SVEQP. If a subsequence* $\{\alpha_{k_i}\}$ *of* $\{\alpha_k\}$ *converges to 0 then* $\lim_{i\to+\infty} \phi(z_{k_i}, x_{k_i}) = 0$.

Proof Suppose to the contrary that $\liminf_{i\to+\infty} \phi(z_{k_i}, x_{k_i}) \geq \eta > 0$ because $\phi(\cdot, \cdot)$ is nonnegative. Define $\hat{y}_i = 2\alpha_{k_i} z_{k_i} + (1 - 2\alpha_{k_i})x_{k_i}$. In other words, we have

$$\hat{y}_i - x_{k_i} = 2\alpha_{k_i} \left(z_{k_i} - x_{k_i}\right). \tag{15}$$

By the assumption and $l(k_i) > 1$ for large enough i. From the linesearch determining α_{k_i}, we have that $\hat{y}_i = y^{l(k_i)-1}$. Since $l(k_i)$ is the first integer satisfying the linesearch, for $l(k_i) - 1$, we obtain that

$$-\beta_{k_i} f(\hat{y}_i, x_{k_i}) + \beta_{k_i} f(\hat{y}_i, z_{k_i}) + \frac{\delta}{2}\phi(z_{k_i}, x_{k_i}) e_{k_i} \in \text{int}(C)$$

for large enough i where $\delta \in (0, 1)$. Since $\lim_{i\to+\infty} \alpha_{k_i} = 0$ and $\{z_{k_i} - x_{k_i}\}$ is bounded. By (15), it implies that $\lim_{i\to+\infty} \|\hat{y}_i - x_{k_i}\| = 0$. Because $f(\cdot, \cdot)$ is uniformly continuous on bounded sets, for large enough i, we have

$$\beta_{k_i} f(x_{k_i}, z_{k_i}) + \frac{1}{2}\phi(z_{k_i}, x_{k_i}) e_{k_i} \in \text{int}(C). \tag{16}$$

By Corollary 1, take $y = x_{k_i}$, we have

$$\begin{aligned} \frac{1}{2}\left[\phi(z_{k_i}, x_{k_i}) + \phi(x_{k_i}, z_{k_i})\right] \langle e_{k_i}, c_{k_i}\rangle &= \langle x_{k_i} - z_{k_i}, J(x_{k_i}) - J(z_{k_i})\rangle \langle e_{k_i}, c_{k_i}\rangle \\ &\leq \beta_{k_i} \left[\langle f(x_{k_i}, x_{k_i}), c_{k_i}\rangle - \langle f(x_{k_i}, z_{k_i}), c_{k_i}\rangle\right], \end{aligned}$$

which implies that

$$\frac{1}{2}\left[\phi(z_{k_i}, x_{k_i}) + \phi(x_{k_i}, z_{k_i})\right] \langle e_{k_i}, c_{k_i}\rangle + \beta_{k_i} \langle f(x_{k_i}, z_{k_i}), c_{k_i}\rangle \leq 0.$$

Since $c_{k_i} \in C^+\backslash\{0\}$, we have

$$\frac{1}{2}\phi(z_{k_i}, x_{k_i})e_{k_i} + \frac{1}{2}\phi(x_{k_i}, z_{k_i})e_{k_i} + \beta_{k_i} f(x_{k_i}, z_{k_i}) \notin \mathrm{int}(C).$$

Since $\phi(x_{k_i}, z_{k_i}) > 0$, it contradicts with (16). □

Proposition 10 *Assume that E is uniformly convex and uniformly smooth with the nonempty set SVEQP. The sequence $\{x_k\}$ is an asymptotically solving sequence for VQEP.*

Proof Assume that there exists a subsequence $\{\alpha_{k_i}\}$ of $\{\alpha_k\}$ which converges to 0. By Proposition 8 and Proposition 9, we can conclude the result. Otherwise, we assume that $\{\alpha_{k_i}\}$ is bounded away from 0, said $\alpha_{k_i} \geq \bar{\alpha} > 0$. Thus, we obtain that

$$-\beta_{k_i} f(y_{k_i}, x_{k_i}) + \beta_{k_i} f(y_{k_i}, z_{k_i}) + \frac{\delta}{2}\phi(z_{k_i}, x_{k_i})e_{k_i} \notin \mathrm{int}(C).$$

From the fact that $\alpha_{k_i} leq 1$ and since $y_{k_i} = \alpha_{k_i} z_{k_i} + (1 - \alpha_{k_i})x_{k_i}$, we have

$$\alpha_{k_i} f(y_{k_i}, z_{k_i}) + (1 - \alpha_{k_i}) f(y_{k_i}, x_{k_i}) \in C$$

due to C-convexity. Hence, we have

$$-\beta_{k_i} f(y_{k_i}, z_{k_i}) + \frac{-\beta_{k_i}(1 - \alpha_{k_i})}{\alpha_{k_i}} f(y_{k_i}, x_{k_i}) \in C.$$

It implies that

$$\frac{-\beta_{k_i}}{\alpha_{k_i}} f(y_{k_i}, x_{k_i}) + \frac{\delta}{2}\phi(z_{k_i}, x_{k_i})e_{k_i} \notin \mathrm{int}(C).$$

By Proposition 7, closeness, and convexity of the cone C, we obtain that

$$\lim_{i\to+\infty} \phi(z_{k_i}, x_{k_i}) = 0.$$

By Proposition 8, we can conclude that $\{x_{k_i}\}$ is an asymptotically solving sequence for VQEP. Since this result holds for every subsequence of $\{x_k\}$, then the same holds for the whole sequence $\{x_k\}$. □

Lemma 8 *Assume that E is uniformly convex and uniformly smooth with the nonempty set SVEQP. If $x_{k_n} \rightharpoonup \bar{x}$ and the sequence $\{x_k\}$ is an asymptotically solving sequence for VQEP then $\bar{x} \in$ SVQEP.*

Proof For each $y \in K(\bar{x})$, we suppose that there exists $c^* \in C^+\backslash\{0\}$ such that

$$0 \leq \limsup_{n\to+\infty} \langle f(x_{k_n}, y), c^* \rangle.$$

By positively weakly upper-semicontinuity of $f(\cdot, y)$, Lemmas 1 and 2, we have

$$\limsup_{n\to+\infty} \langle f(x_{k_n}, y), c^* \rangle \leq \langle f(\bar{x}, y), c^* \rangle.$$

Since $c^* \in C^+ \backslash \{0\}$, we obtain that $f(\bar{x}, y) \notin -\text{int}(C)$ for all $y \in K(\bar{x})$. In other words, $\bar{x}$ is a solution of VQEP. Otherwise, there exists $y \in K(\bar{x})$ such that for all $c^* \in C^+ \backslash \{0\}$, it holds that

$$\limsup_{n\to+\infty} \langle f(x_{k_n}, y), c^* \rangle < 0.$$

By weak-to-strong lower-semicontinuity of K, we have that there exists a sequence $y_n \in K(x_{k_n})$ such that $y_n \to y$. It turns out that

$$\limsup_{n\to+\infty} \langle f(x_{k_n}, y_n), c^* \rangle < 0. \tag{17}$$

Hence, each limit point of $f(x_{k_n}, y_n)$ belongs to $-\text{int}(C)$ because (17) holds for all $c^* \in C^+ \backslash \{0\}$. Since $-\text{int}(C)$ is open, there exists $n_0 \in \mathbb{N}$ such that $f(x_{k_n}, y_n) \in -\text{int}(C)$ for all $n \geq n_0$ which contradicts that $\{x_k\}$ is an asymptotically solving sequence for VQEP. □

Theorem 3 *Assume that E is uniformly convex and uniformly smooth with the nonempty set SVEQP. Then*

(i) the sequence $\{x_k\}$ has weak cluster points and all of them solves VQEP,
(ii) Assume that f has a unique solution, then the whole sequence $\{x_k\}$ is weakly convergent to a solution of VQEP.

Proof (i) As a result of Proposition 7, the sequence $\{x_k\}$ is bounded, it implies that $\{x_k\}$ has weak limit points. Let $v \in K(v)$ be a weak limit point of $\{x_k\}$, and a subsequence $\{x_{k_n}\}$ with $x_{k_n} \rightharpoonup v$. By Proposition 10, $\{x_{k_n}\}$ is an asymptotically solving sequence for VQEP. Now, by Lemma 8, we have that v is a solution to VQEP.

(ii) If VQEP has a unique solution, then the result is proved.

□

It should be noted that Hilbert space is uniformly convex space. If a vector-valued mapping $f : X \times X \to \mathbb{R}^m$ becomes a function, i.e. $m = 1$, then VQEP is reduced to be QEP. Therefore, we receive the following corollary.

Corollary 2 *Let $X \subseteq \mathbb{R}^n$ be a closed convex set with the nonempty set SQEP* $:= \{x \in \bigcap_{x\in X} K(x) \mid f(x, y) \geq 0$ *for all* $y \in \bigcup_{x\in X} K(x)\}$. *Assume that a bifunction $f : X \times X \to \mathbb{R}$ satisfies*

(i) $f(x, \cdot)$ *is convex for any fixed* $x \in X$,
(ii) f *is continuous on* $X \times X$,
(iii) $f(x, x) = 0$ *for all* $x \in X$,
(iv) f *is pseudomonotone on* X *with respect to SQEP*, i.e. if* $f(\overline{x}, y) \geq 0$ *then* $f(y, \overline{x}) \geq 0$ *for all* $\overline{x} \in$ *SQEP*, for all* $y \in X$.

Moreover, a multi-valued mapping $K : X \rightrightarrows \mathbb{R}^n$ *satisfies*

(i) K *is continuous on* X
(ii) $K(x)$ *is a nonempty closed convex subset of* X *for all* $x \in X$,
(iii) $x \in K(x)$ *for all* $x \in X$.

If the Algorithm 1 does not stop at iteration k, then

(i) each cluster point of $\{x_k\}$ *is a solution of the QEP,*
(ii) the sequence $\{x_k\}$ *converges to a solution* $\overline{x}$ *such that* $\overline{x} = \prod_{SQEP}(x_0)$.

Acknowledgements The first author would like to thank Science Achievement Scholarship of Thailand (SAST). The authors acknowledge the financial support provided by the Center of Excellence in Theoretical and Computational Science (TaCS-CoE), KMUTT. Finally, the authors would like to express our thanks to referees for their valuable comments and suggestions.

References

Ansari, Q.H.: Existence of solutions of systems of generalized implicit vector quasi-equilibrium problems. J. Math. Anal. Appl. **341**(2), 1271–1283 (2008). https://doi.org/10.1016/j.jmaa.2007.11.033

Ansari, Q.H., Yao, J.C.: On vector quasi-equilibrium problems. Nonconvex Optim. Appl. **68**, 1–18 (2003). https://doi.org/10.1007/978-1-4613-0239-1_1

Ansari, Q.H., Chan, W.K., Yang, X.Q.: The system of vector quasi-equilibrium problems with applications. J. Global Optim. **29**, 45–57 (2004). https://doi.org/10.1080/10556789508805617

Aussel, D., Dutta, J., Pandit, T.: About the links between equilibrium problems and variational inequalities. Indian Stat. Inst. Ser., 115–130 (2018). https://doi.org/10.1007/978-981-13-3059-9_6

Bonnel, H., Iusem, A.N., Svaiter, B.F.: Proximal methods in vector optimization. SIAM J. Optim. **15**(4), 953–970 (2005). https://doi.org/10.1137/S1052623403429093

Chen, H., Wang, Y., Xu, Y.: An alternative extragradient projection method for quasi-equilibrium problems. J. Inequal. Appl. **26** (2018). https://doi.org/10.1186/s13660-018-1619-9

Chin, H.H., Parthasarathy, T., Raghavan, T.E.S.: Structure of equilibria in N-person non-cooperative games. Int. J. Game Theory. **3**, 1–19 (1974). https://doi.org/10.1007/BF01766215

Cotrina, J., Zúñiga, J.: A note on quasi-equilibrium problem. Oper. Res. Lett. **46**(1), 138–140 (2018). https://doi.org/10.1016/j.orl.2017.12.002

Facchinei, F., Kanzow, C.: Generalized Nash equilibrium problems. 4OR. **5**, 173–210 (2007). https://doi.org/10.1007/s10288-007-0054-4

Fu, J.Y.: Generalized vector quasi-equilibrium problems. Math. Methods Oper. Res. **52**(1), 57–64 (2000). https://doi.org/10.1007/s001860000058

Han, D., Lo, H.K., Wang, Z.: A simple self-adaptive alternating direction method for linear variational inequality problems. Comput. Math. Appl. **53**(10), 1595–1604 (2007). https://doi.org/10.1016/j.camwa.2006.10.025

Han, D., Zhang, H., Qian, G., Xu, L.: An improved two-step method for solving generalized Nash equilibrium problems. Europ. J. Oper. Res. **216**(3), 613–623 (2012). https://doi.org/10.1023/B:JOGO.0000035018.46514.ca

Iusem, A.N., Mohebbi, V.: Extragradient methods for vector equilibrium problems in Banach spaces. Numer. Funct. Anal. Optim. **40**(9), 993–1022 (2019). https://doi.org/10.1080/01630563.2019.1578232

Kassay, G., Radulescu, V.: Equilibrium Problems and Applications. Academic Press, United States (2018)

Kočvara, M., Outrata, J.V.: On a class of quasi-variational inequalities. Optim. Methods Softw. **5**(4), 275–295 (1995). https://doi.org/10.1080/10556789508805617

Kopec?a, E., Reich, S.: A note on alternating projections in Hilbert space. J. Fixed Point Theory Appl. **12**, 41–47 (2012). https://doi.org/10.1007/s11784-013-0097-4

Kreps, V.: On maximal vector spaces of finite noncooperative games. Int. Game Theory Rev. **19**(2), 1–7 (2017). https://doi.org/10.1142/S0219198917500037

Morgenstern, O., Von Neumann, J.: Theory of Games and Economic Behavior. Princeton University Press (1953)

Nash, J.: Non-cooperative games. Ann. Math. Sec. Ser. **54**, 286–295 (1951). https://doi.org/10.2307/1969529

Nasri, M., Sosa, W.: Equilibrium problems and generalized Nash games. Optimization **60**(8–9), 1161–1170 (2011). https://doi.org/10.1080/02331934.2010.527341

Ram, T., Lal, P.: Existence results for generalized vector quasi-equilibrium problems. Malaya J. Mat. **8**(4), 1675–1680 (2020). https://doi.org/10.26637/mjm0804/0059

Rouhani, B.D., Mohebbi, V.: Extragradient methods for quasi-equilibrium problems in Banach spaces. J. Austr. Math. Soc. First View, 1–25 (2020). https://doi.org/10.1017/S1446788720000233

Rouhani, B.D., Mohebbi, V.: Proximal point method for quasi-equilibrium problems in Banach Spaces. Numer. Functi. Anal. Optim. **41**(9), 1007–1026 (2020). https://doi.org/10.1080/01630563.2019.1708392

Solodov, M.V., Svaiter, B.F.: A new projection method for variational inequality problems. SIAM J. Control Optim. **37**(3), 765–776 (1999). https://doi.org/10.1137/S0363012997317475

Tran, D.Q., Dung, M.L., Nguyen, V.H.: Extragradient algorithms extended to equilibrium problems. Optimization **57**(6), 749–776 (2008). https://doi.org/10.1080/02331930601122876

Ye, M.: A half-space projection method for solving generalized Nash equilibrium problems. Optimization **66**(7), 1119–1134 (2017). https://doi.org/10.1080/02331934.2017.1326045

Zhang, J., Qu, B., Xiu, N.: Some projection-like methods for the generalized Nash equilibria. Comput. Optim. Appl. **45**(1), 89–109 (2010). https://doi.org/10.1007/s10589-008-9173-x

Introduction to Rare-Event Predictive Modeling for Inferential Statisticians—A Hands-On Application in the Prediction of Breakthrough Patents

Daniel Hain and Roman Jurowetzki

Abstract Recent years have seen a substantial development of quantitative methods, mostly led by the computer science community with the goal to develop better machine learning application, mainly focused on predictive modeling. However, economic, management, and technology forecasting research has up to now been hesitant to apply predictive modeling techniques and workflows. In this paper, we introduce to a machine learning (ML) approach to quantitative analysis geared towards optimizing the predictive performance, contrasting it with standard practices inferential statistics which focus on producing good parameter estimates. We discuss the potential synergies between the two fields against the backdrop of this at first glance, target-incompatibility. We discuss fundamental concepts in predictive modeling, such as out-of-sample model validation, variable and model selection, generalization and hyperparameter tuning procedures. Providing a hands-on predictive modelling for an quantitative social science audience, while aiming at demystifying computer science jargon. We use the example of high-quality patent identification guiding the reader through various model classes and procedures for data preprocessing, modelling and validation. We start of with more familiar easy to interpret model classes (Logit and Elastic Nets), continues with less familiar non-parametric approaches (Classification Trees and Random Forest) and finally presents artificial neural network architectures, first a simple feed-forward and then a deep autoencoder geared towards anomaly detection. Instead of limiting ourselves to the introduction of standard ML techniques, we also present state-of-the-art yet approachable techniques from artificial neural networks and deep learning to predict rare phenomena of interest.

D. Hain (✉) · R. Jurowetzki
Aalborg University Business School, Aalborg 9220, Denmark
e-mail: dsh@business.aau.dk

R. Jurowetzki
e-mail: roman@business.aau.dk

N. Ngoc Thach et al. (eds.), *Financial Econometrics: Bayesian Analysis, Quantum Uncertainty, and Related Topics*, Studies in Systems, Decision and Control 427,
https://doi.org/10.1007/978-3-030-98689-6_5

1 Introduction

Recent years have seen a substantial development of machine learning (ML) methods, mostly led by the computer science community with the goal to increase the predictive power of statistical models. This progress ranges over a broad host of applications—from computer vision, speech recognition, synthesis, machine translation, only to name a few. Many models at the core of such ML applications resemble those used by statisticians in social science context, yet a paradigmatic difference persists. Social science researchers traditionally apply methods and workflows from inferential statistics, focusing on the identification and isolation of causal mechanisms from sample data which can be extrapolated to some larger population. In contrast, ML models are typically geared towards fitting algorithms that map some input data to an outcome of interest in order to perform predictions for cases there the outcome is not (yet) observed. While this has led to many practical applications in the natural sciences and business alike, the social sciences have until recently been hesitant to include ML in their portfolio of research methods. Yet, the recent availability of data with improvements in terms of quantity, quality and granularity Einav and Levin (2014a, b), led to various calls in the business studies (McAfee et al. 2012) and related communities for exploring potentials of ML methods for advancing their own research agenda.

This paper provides a condensed and practical introduction to ML for social science researchers which are commonly received training in methods and workflows from inferential statistics. We discuss the potential synergies and frictions between the two approaches, and elaborate on fundamental ML concepts in predictive modeling, such as generalization via out-of-sample model validation, variable and model selection, and hyperparameter tuning procedures. By doing so, we demystify computer science jargon common to ML, draw parallels and highlight commonalities as well as differences with popular methods and workflows in inferential statistics.

We illustrate the introduced concept by providing a hands-on example of high-quality patent identification.[1] We guide the reader through various model classes and procedures for data pre-processing, modelling and validation. After starting of with model classes familiar to applied inferential statisticians (Logit and Elastic Nets), we continues with nonparametric approaches less commonly applied in the social sciences (simple Classification Trees, Random Forest and Gradient Boosted Trees), gradually shifting the emphasis towards hyperparameter tuning and model

[1] Often, the challenge in adapting ML techniques for social science problems can be attributed to two issues: (1) Technical lock-ins and (2) Mental lock-ins against the backdrop of paradigmatic contrasts between research traditions. For instance, many ML techniques are initially demonstrated at a collection of—in the ML and Computer Science—well known *standard datasets* with specific properties. For an applied statistician particularly in social science, however, the classification of Netflix movie ratings or the reconstruction of handwritten digits form the MNIST data-set may appear remote or trivial. These two problems are addressed by contrasting ML techniques with inferential statistics approaches, while using the non-trivial example of patent quality prediction which should be easy to comprehend for scholars working in social science disciplines such as economics.

architecture engineering. Instead of limiting ourselves to the introduction of standard ML techniques, we also present state-of-the-art yet approachable techniques from artificial neural networks (ANN) and deep learning to predict rare phenomena of interest. Finally, we provide guidance on how to apply these techniques for social science research and point towards promising avenues of future research which could be enabled by the use of new data sources and estimation techniques.

2 An Introduction to Machine Learning and Predictive Modeling

2.1 Predictive Modeling, Machine Learning, and Social Science Research

Until recently, the social sciences where not overly eager to embrace and apply the methodological toolbox and procedural routines developed within the ML discipline. An apparent reason is given by inter-disciplinary boundaries and intra-disciplinary methodological "comfort zones" (Aguinis et al. 2009) as well as by path-dependencies, reinforced through the way how researchers are socialized during doctoral training (George et al. 2016). However, there also seems to be inherent—if not epistemological—tension between the social science inferential statistics and ML to data analysis, and how both could benefit from each other's insights is not obvious on first glance.

We here argue the ML community has not only developed "tricks" an social science statisticians in applied fields such as econometrics, sociometrics, or psychometrics might find extremely useful (Varian 2014), but also enable new research designs to tackle a host of questions currently hard to tackle with the traditional toolbox of statistical inference. We expect such methods to broadly diffuse within quantitative social science research, and suggest the upcoming liaison of inferential statistics and ML to shake up our current routines. Here, highly developed workflows and techniques for predictive modelling appear to be among the most obvious ones.

In Figs. 1 and 2, we depict two trade-offs that we find relevant to consider in a paradigmatic discussion of data science and econometric approaches. On the one hand, and as presented in Fig. 2, there is a general trade-off between the learning capacity of model classes and their interpretability. First, richer model classes are able to fit a complex functional form to a prediction problem, which improves their performance over simple (e.g. linear parametric) models if (and only if) the underlying real-world relationship to model is equally complex. It can be assumed that this is the case for many relevant outcomes of the interaction in complex social systems (e.g. economic growth, technology development, success of a start-up), and consequently, that richer models would have an edge for prediction tasks in such systems. With larger amounts of data available, such complex functional forms can be fitted more

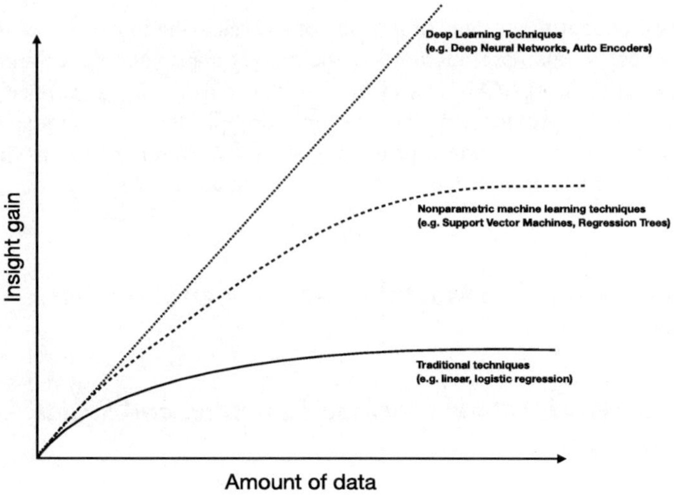

Fig. 1 Gain of insight versus data amount for different model classes

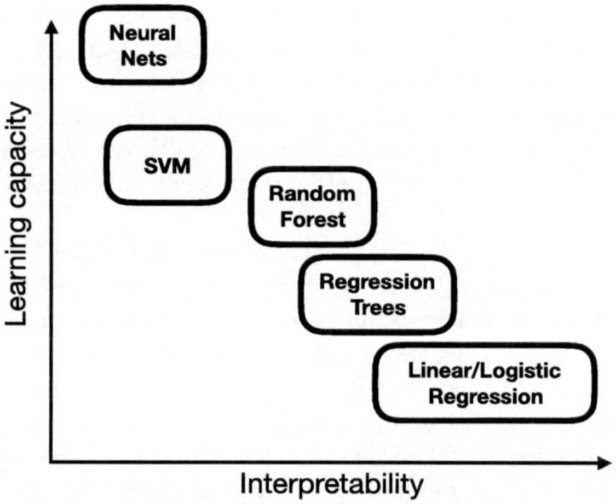

Fig. 2 Learning capacity versus interpretability of selected quantitative methods

accurately, resulting in a higher learning capability of richer model classes compared to simple models which tend to saturate early, as illustrated in Figs. 1 and 2.

The relationships between inputs and outputs captured by a linear regression model are easy to understand and interpret. As we move up and to the left in this chart, the learning capacity of the models increases. Considering the extreme case of deep neural networks, we find models that can capture interactions and nonlinear relations across large datasets, fitting in their complex functions between in- and outputs across the different layers with their multiple nodes. However, for the most part, it is fairly difficult if not impossible to understand the fitted functional relationship. This is not necessarily a problem for predictive modeling but of much use in cases where the aim is to find causal relationships between in- and outputs.

2.2 Contrasting Causal and Predictive Modeling

As applied econometricians, we are for the most part interested in producing good *parameter estimates.*[2] We construct models with unbiased estimates for some parameter β, capturing the relationship between a variable of interest x and an outcome y. Such models are supposed to be "structural", where we not merely aim to reveal correlations between x and y, but rather a causal effect of directionality $x \rightarrow y$, robust across a variety of observed as well as up to now unobserved settings. Therefore, we carefully draw from existing theories and empirical findings and apply logical reasoning to formulate hypotheses which articulate the expected direction of such causal effects. Typically, we do so by studying one or more bivariate relationships under *cetris paribus* conditions in a regression model, hand-curate with a set of supposedly causal variables of interest. The primary concern here is to minimize the standard errors ϵ of our β estimates, the difference between our predicted $hat(y)$ and the observed y, conditional to a certain level of x, *ceteris paribus*. We are less interested in the overall predictive power of our model (Usually measured by the models R^2), as long as it is in a tolerable range.[3]

However, we are usually worried about the various type of endogeneity issues inherent to social data which could bias our estimates of β. For instance, when our independent variable x can be suspected to have a bidirectional causal relationship with the outcome y, drawing a causal inference of our interpretation of β is obviously limited. To produce unbiased parameter estimates of arguably causal effects, we are indeed willing to sacrifice a fair share of our models' explanatory power.

A ML approach to statistical modeling is, however, fundamentally different. To a large extent driven by the needs of the private sector, data analysis here concentrates on producing trustworthy predictions of outcomes. Familiar examples are the recommender systems employed by companies such as Amazon and Netflix, which predict with "surprising" accuracy the types of books or movies one might find interesting. Likewise, insurance companies or credit providers use such predictive models to calculate individual "risk scores", indicating the likelihood that a particular person has an accident, turns sick, or defaults on their credit. Instances of such applications are numerous, but what most of them have in common is that: (i) they rely on a lot of data, in terms of the number of observations as well as possible predictors, and (ii) they are not overly concerned with the properties of parameter estimates, but very rigorous

[2] We here blatantly draw from stereotypical workflows inherent to the econometrics and ML discipline. We apologize for offending whoever does not fit neatly in one of these categories.

[3] At the point where our R^2 exceeds a threshold somewhere around 0.1, we commonly stop worrying about it.

in optimizing the overall prediction accuracy. The underlying socio-psychological forces which make their consumers enjoy a specific book are presumably only of minor interest for Amazon, as long as their recommender system suggests them books they ultimately buy.

2.3 The Predictive Modeling Workflow

2.3.1 General Idea

At its very core, in predictive modeling and for the most part the broader associated ML discipline, we seek for models and functions that do the best possible job in predicting some output variable y. This is done by considering some loss function $L(\hat{y}, y)$, such as the popular root-mean-square error (RMSE)[4] or the rate of missclassified observations, and then searching for a function $\hat{f}$ that minimizes our predicted loss $E_{y,x}[L(\hat{f}(x), y)]$.

To do so, the broader ML community has developed an enormeous set of techniques from traditional statistics but also computer science and other disciplines to tackle prediction problems of various nature. While some of those techniques are widely known and applied by econometricians and the broader research community engaged in causal modeling (e.g., linear and logistic regression) or lately started to receive attention (e.g., elastic nets, regression- and classification-trees, kernel regressions, and to some extend random forests), others are widely unknown and rarely applied (e.g., support vector machines, artificial neural networks).[5]

However, fundamental differences in general model building workflows and underlying philosophies makes the building as well as interpretation of (even familiar) predictive models with the "causal lense" of a trained econometrician prone to misunderstanding, misspecification, and misleading evaluation. Therefore, we in the following outlay some general principles of predictive modeling, before we in the following section illustrate them in an example.

First, in contrast to causal modeling, most predictive models have no a priori assumption regarding the direction of the effect, or any causal reason behind it. Therefore, predictive models exploit correlation rather that causation, and to predict rather than explain an outcome of interest. This provides quite some freedom in terms of which and how many variables (to introduce further ML jargon, henceforth called *features*) to select, if and how to transform them, and so forth. Since we are not interested in parameter estimates, we also do not have to worry so much about

---

[4] As the name already suggest, this simply expresses by how much our prediction is on average off: $RMSE = \sqrt{\frac{\sum_{i=1}^{n}(\hat{y}_i - y_i)^2}{n}}$.

[5] Interestingly, quite some techniques associated with identification strategies which are popular among econometricians, such as the use of instrumental variables, endogeneous selection models, fixed and random effect panel regressions, or vector autogregressions, are little known by the ML community.

asymptotic properties, assumptions, variance inflation, and all the other common problems in applied econometrics which could bias parameter estimates and associated standard errors. Since parameters are not of interest, there is also no urgent need to capture their effect properly, or have them at all. Indeed, many popular ML approaches are non-parametric and characterized by a flexible functional form to be fitted to whatever the data reveals. Equipped with such an arsenal, achieving a high explanatory power of a model appears quite easy, but every econometrician would doubt how well such a model generalizes. Therefore, without the limitations but also guarantees of causal modeling, predictive models are in need of other mechanisms to ensure their generalizability.

2.3.2 Out-of-Sample Validation

Again, as econometricians, we focus on parameter estimates, and we implicitly take their out-of-sample performance for granted. Once we set up a proper identification strategy that delivers unbiased estimates of a causal relationship between x and y, depending on the characteristics of the sample, this effect supposedly can be generalized on a larger population. Such an exercise is per se less prone to over-specification since the introduction of further variables with low predictive power or correlation with x tends to "water down" our effects of interest. Following a machine learning approach geared towards boosting the prediction of the model, the best way to test how a model predicts is to run it on data it was not fitted for. This can be done upfront dividing your data in a training sample, which you use to fit the model, and a test (or hold-out) sample, which we set aside and exclusively use to evaluate the models final prediction. This should only be done once, because a forth and back between tweaked training and repeated evaluation on the test sample otherwise has the effect of an indirect overfitting of the model.

Therefore, it is common in the training data also set a validation sample aside to first test the performance of different model configurations out-of-sample. Consequently, we aim at minimizing the *out-of-sample* instead of the *within sample* loss function. Since such a procedure is sensitive to potential outliers in the training or test sample, it is good practice to not validate your model on one single test-sample, but instead perform a *k-fold cross-validation*, where the loss function is computed as the average loss of k (commonly 5 or 10)) separate test samples.[6] Finally, the best performing configuration is used to fit this model on the whole training sample. The final performance of this model is in a last step then evaluated by its prediction on the test sample, to which the model up to now has not been exposed to, neither direct nor indirect. This procedure is illustrated in Fig. 3.

[6] Such k-fold cross-validations can be conveniently done in R with the `caret`, and in Python with the `scikit-learn` package.

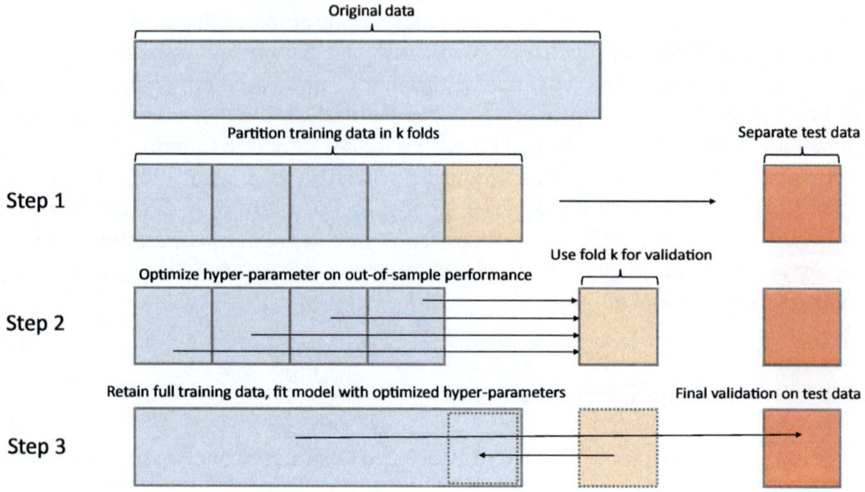

Fig. 3 Intuition behind K-fold crossvalidation

While out-of-sample performance is a standard model validation procedure in machine learning, it has yet not gained popularity among econometricians.[7] As a discipline originating from a comparably "small data" universe, it appears counter-intuitive for most cases to "throw away" a big chunk of data. However, the size of data-sources available for mainstream economic analysis, such as register data, has increased to a level, where sample size cannot be taken anymore as an excuse for not considering such a goodness-of-fit test, which delivers much more realistic measures of a model's explanatory power. What econometricians like to do to minimize unobserved heterogeneity and thereby improve parameter estimates is to include a battery of categorical control variables (or in panel models, fixed effects) for individuals, sectors, countries, *et cetera*. It is needless to say that this indeed improves parameter estimates in the presence of omitted variables but typically leads to terrible out-of-sample prediction.

2.3.3 Regularization and Hyperparameter Tuning

Turning back to our problem of out-of-sample prediction, now that we have a good way of measuring it, the question remains how to optimize it. As a general rule, the higher the complexity of a model, the better it tends to perform within-sample,

[7] However, one instantly recognizes the similarity to the nowadays common practice among econometricians to bootstrap standard errors by computing them over different subsets of data. The difference here is that we commonly use this procedure, (i) to get more robust parameter estimates instead of evaluating the model's overall goodness-of-fit, and (ii) we compute them on subsets of the same data the model as fitted on.

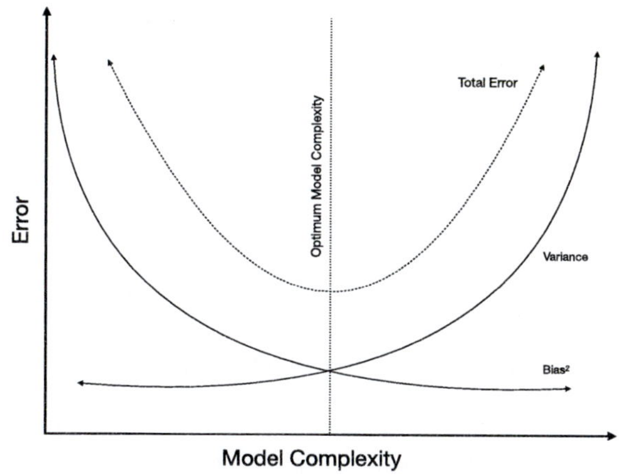

Fig. 4 In- versus out-of-sample loss relationship

but also to loose predictive power when performing out-of-sample prediction. Since finding the right level of complexity is a crucial, researchers in machine learning have put a lot of effort in developing "regularization" techniques which penalize model complexity. In addition to various complexity restrictions, many ML techniques have additional options, called *hyperparameter*, which influence their process and the resulting prediction. The search for optimal tuning parameters (in machine learning jargon called *regularization*, or hyperparameter tuning)[8] is at the heart of machine learning research efforts, somewhat its secret sauce. The idea in it's most basic form can be described by the following equation, as expressed by Mullainathan and Spiess (2017) (Fig. 4).

$$minimize \underbrace{\sum_{i=1}^{n} L(f(x_i), y_i)}_{in-sample\ loss},\ over \overbrace{f \in F}^{function\ class}\ subject\ to \underbrace{R(f) \leq c.}_{complexity\ restriction} \tag{1}$$

Basically, we here aim at minimizing the in-sample loss of a prediction algorithm of some functional class subject to some complexity restriction, with the final aim to minimize the expected out-of-sample loss. Depending on the technique applied, this can be done by either selecting the functions features x_i (as we discussed before in variable selection), the functional form and class f, the complexity restrictions c, or other hyperparameters that influence the models internal processes. This process of

[8] For exhaustive surveys on regularization approaches in machine learning particularly focused on high-dimensional data, consider Wainwright (2014), Pillonetto et al. (2014).

model tuning in practice often is a mixture of internal estimation from the training data, expert intuition, and best practice, as well as trial-and-error. Depending on the complexity of the problem, this can be a quite tedious and lengthy process.

The type of regularizations and model tuning techniques one might apply varies, depending on the properties of the sample, the functional form, and the desired output. For parametric approaches such as OLS and logistic regressions, regularization is primarily centered around feature selection and parameter weighting. Many model tuning techniques are iterative, such as model *boosting*, an iterative technique involving the linear combination of prediction of residuals, where initially misclassified observations are given increased weight in the next iteration.

Bootstrapping, the repeated estimation of random subsamples, is in ML used primarily to adjust the parameter estimates by weighting them across subsamples (which is then called *bagging*).[9] Finally, *ensemble* techniques use the weighted combination of predictions done by independent models to determine the final classification.

3 An Application on Patent Data

In this section, we will illustrate the formerly discussed methods, techniques, and concepts at the example of PATSTAT patent data in order to develop a predictive model of high-impact (breakthrough) patents. In addition, we will "translate" necessary vocabulary differences between ML and econometrics jargon, and point to useful packages in R and Python, the current *de factor* standards for statistical programming in ML and also increasingly popular among applied econometricians. Here, our task is to predict a dichotomous outcome variable. In ML jargon, this is the simplest form of a *classification problem*, where the available classes are $0 =$ no and $1 =$ yes. As econometricians, probably our intuition would lead us to apply a linear probability (LPM) or some form of a logistic regression model. While such models are indeed very useful to deliver parameter estimates, if our goal is pure prediction, there exist much richer model classes, as we demonstrate in the following.

3.1 Data and Context

3.1.1 Context

Patent data has long been used as a widely accessible measure of inventive and innovative activity. Besides its use as an indicator of inventive activity, previous research shows that patents are a valid indicator for the output, value and utility of

[9] Bootstrapping is a technique most applied econometricians are well-acquainted with, yes used for a slightly different purpose. In econometrics, bootstrapping represents a powerful way to circumvent problems arising out of selection bias and other sampling issues, where the regression on several subsamples is used to adjust the standard errors of the estimates.

inventions (Trajtenberg et al. 1997), innovations (Hagedoorn and Schakenraad 1993), and resulting economic performance on firm level (Ernst 2001). These signals are also useful and recognized by investors (Hirschey and Richardson 2004), making them more likely to provide firms with external capital (Hall and Harhoff 2012). Yet, it has widely been recognized that the technological as well as economic significance of patents varies broadly (Basberg 1987). Consequently, the *ex-ante* identification of potential high value and impact is of high relevance for firms, investors, and policy makers alike. Besides guiding the allocation of investments and internal resources, it might enable "nowcasting" and "placecasting" of the quality of inventive and innovative activity (consider Andrews et al. 2017; Fazio et al. 2016; Guzman and Stern 2015, 2017 for an application in entrepreneurship). However, by definition patents of abnormally high value and impact are rare in nature. Together with the broad availability of structured patent data via providers such as PATSTAT and the OECD, this makes the presented setting an useful and informative case for a predictive modeling exercise.

3.2 Data

For this exercise, we draw from the patent database provided by PATSTAT. To keep the data volume moderate and the content somewhat homogeneous, we here limit ourselves to patents granted at the USTPO and EPO in the 1980–2016 period, leading to a number of roughly 6.5 million patents. While this number appears already large compared to other datasets commonly used by applied econometricians in the field of entrepreneurship and innovation studies, according to ML standards it can still be considered as small, both in terms of the number of observation as well as available variables. While offering reasonable analytic depth, such amounts of data can still be conveniently processed with standard in-memory workflows on personal computers.[10]

We classify high impact patents following (Ahuja and Lampert 2001) as the patents within a given cohort receiving the most citations by other patents within the following 5 year window. Originally, such "breakthrough patents" are defined as the ones in the top 1% of the distribution. For this exercise, we also create another outcome, indicating the patent to be in the top 50% of the distribution, indicating above-average successful but not necessarily breakthrough patents.

For the sake of simplicity and reconstructability, we in the following models mostly use features either directly contained in PATSTAT, or easily derived from it. In detail, we create a selection of *ex-ante* patent novelty and quality indicators,[11]

[10] This is often not the case for typical ML problems, drawing from large numbers of observations and/or a large set of variables. Here, distributed or cloud-based workflows become necessary. We discuss the arising challenges elsewhere (e.g., Hain and Jurowetzki 2020).

[11] For a recent and exhaustive review on patent quality measures, including all used in this exercise, consider Squicciarini et al. (2013).

summarized in Table 1. The only data external to PATSTAT data we used are the temporal technological similarity indicator proposed by Hain (2022).

Notice that the breakthrough features are calculated based on the distribution of patents that receive citations. Since a growing number of patents never get cited, the percentage of patents that fall within the top-n% appears less than expected. Notice also that we abstain of including a lot of categorical features, which would traditionally be included in a causal econometric exercise, for example dummies for the patent's application year, the inventor and applicant firm. Again, since we aim at building a predictive model that fits well on new data. Obviously, such features would lead to overfitting, and reduce its performance when predicting up to now unobserved firms, inventors, and years. We include dummy features for the technological field, though, which is a somewhat more static classification. However, since the importance of technological fields also change over time, such a model would be in need of retraining as time passes by and the importance of technological fields shift.

3.3 First Data Exploration

The ML toolbox around predictive modeling is rich and diverse, and the variety of available techniques in many cases can be tuned along a set of parameters, depending on the structure of data and problem at hand. Therefore, numerical as well as visual data inspection becomes an integral part of the model building an tuning process.

First, for our set of features to be useful for a classification problem, it is useful (and for the autoencoder model introduced later necessary) they indeed display a different distribution conditional to the classes we aim at predicting. In Fig. 5, we plot this conditional distribution for all candidate features for our outcome classes of `breakthrough` and `breakthrough50`, where we indeed observe such differences.

3.4 Preprocessing

Again, as an remainder, for a predictive modeling exercise we are per se not in need of producing causal, robust, or even interpretative parameter estimates. Consequently, we enjoy a higher degree of flexibility in the way how we select, construct, and transform model features. First, moderate amounts of missing feature values are commonly imputed, while observations with missing outcomes are dropped.

Second, various kind of "feature scaling" techniques are usually performed in the preprocessing stage, where feature values are normalized in order to increase accuracy as well as computational efficiency of predictive models. The selection of appropriate scaling techniques again depends on the properties of the data and model at hand, where popular approaches are minMax rescaling ($x' = \frac{x-\bar{x}}{\sigma}$), mean normal-

Table 1 Descriptive statistics: USTPO patents 2010–2015

Feature	N	Mean	St. Dev.	Min	Max	Description
Breakthrough	6,571,931	0.006	0.074	0	1	Top 1%-cited patent in annual cohort
Breakthrough50	6,571,931	0.175	0.380	0	1	Top 50%-cited patent in annual cohort
sim.past	6,571,931	0.088	0.185	0	1	Technological similarity to past (Hain 2022)
sim.present	6,571,931	0.153	0.262	0	1	Technological similarity to present (Hain 2022)
many_field	6,571,931	0.398	0.489	0	1	Multiple IPC classes (Lerner 1994)
patent_scope	6,571,931	1.854	1.162	1	31	Number of IpC classes
family_size	6,571,931	4.251	3.906	1	57	Size of the patent family (Harhoff et al. 2003)
bwd_cits	6,571,931	15.150	25.640	0	4,756	Backward citations (Harhoff et al. 2003)
npl_cits	6,571,931	3.328	12.690	0	1,592	NPL backward citations (Narin et al. 1997)
claims_bwd	6,571,931	1.673	3.378	0	405	Backward claims
Originality	6,571,931	0.707	0.248	0	1	*originality* index (Trajtenberg et al. 1997)
Radicalness	6,571,931	0.382	0.288	0	1	*radicalness* index (Shane 2001)
nb_applicants	6,571,931	1.849	1.705	0	77	Number of applicants
nb_inventors	6,571,931	2.666	1.925	0	99	Number of Inventors
patent_scope.diff	6,571,931	0.008	1.091	−2.806	29.130	Δ patent–cohort scope
bwd_cits.diff	6,571,931	0.222	24.560	−42.050	4,732.000	Δ patent–cohort backwards citations
npl_cits.diff	6,571,931	0.112	12.100	−30.510	1,579.000	Δ patent–cohort nlp citations
family_size.diff	6,571,931	0.031	3.536	−11.090	50.090	Δ patent–cohort family size
originality.diff	6,571,931	−0.029	0.242	−0.911	0.431	Δ patent–cohort originality
radicalness.diff	6,571,931	−0.018	0.277	−0.751	0.808	Δ patent–cohort radicalness
sim.past.diff	6,571,931	−0.000	0.180	−0.247	0.990	Δ patent–cohort similarity to past
sim.present.diff	6,571,931	0.000	0.260	−0.315	0.942	Δ patent–cohort similarity present

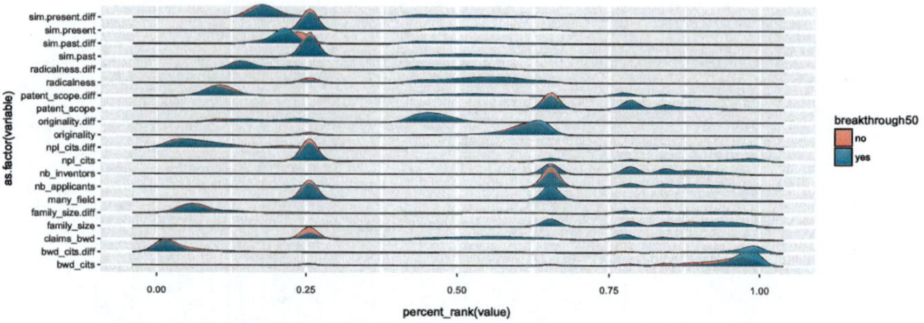

(a) Conditional to *breakthrough50* ($\geq$ 50% forward citations in cohort)

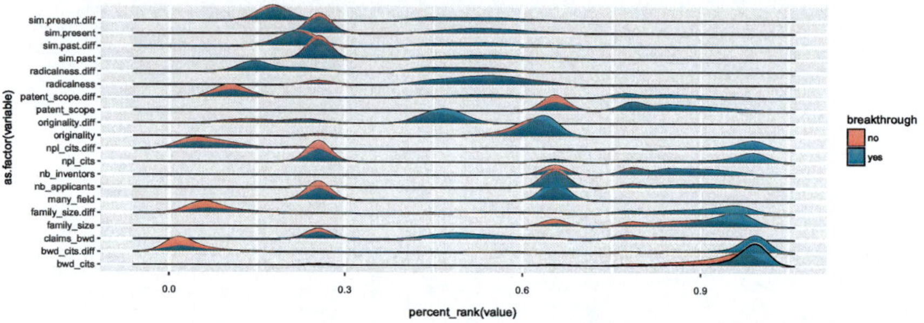

(b) Conditional to *breakthrough01* ($\geq$ 99% forward citations in cohort)

Fig. 5 Conditional distribution of predictors

ization ($x' = \frac{x - \text{mean}(x)}{\max(x) - \min(x)}$), standardization ($x' = \frac{x - \bar{x}}{\sigma}$), dimensionality reduction of the feature space with a principal component analysis (PCA), and binary recoding to “one-hot-encodings”. In this case, we normalize all continuous features to $\mu = 0,\ \sigma = 1$, and categorical features to one-hot-encoding.

Before we do so, we split our data in the test sample we will use for the model and hyperparameter tuning phase (75%), and a test sample, which we will only use once for the final evaluation of the model (25%). It is important to do this split before the preprocessing, since otherwise a common feature scaling with the test sample could contaminate our training sample.

3.5 *Model Setup and Tuning*

After exploring and preprocessing our data, we now select and tune a set of models to predict high impact patents, where we start with the outcome *breakthrough3*, indicating a patent to be among the 50% most cited patents in a given year cohort.

This classification problem calls for a class of models able to predict categorical outcomes. While the space of candidate models is vast, we limit ourselves to the demonstration of popular and commonly well performing model classes, namely the traditional logit, elastic nets, boosted classification trees, and random forests. Most of these models include tunable hyperparameter, leading to varying model performance. Given the data at hand, we aim at identifying the best combination of hyperparameter values for every model before we evaluate their final performance and select the best model for our classification problem.

We do so via a hyperparameter "grid search" and repeated 5-fold crossvalidation, For every hyperparameter, we define a sequence of possible values. In case of multiple hyperparameters, we create a tune grid, a matrix containing a cell for every unique combination of hyperparameter values. Then we perform the following steps[12]:

1. Partition the training data into 5 equally sized folds.
2. Fit a model with a specific hyperparameter combination separate on fold 1–4, evaluate its performance by predicting the outcome of fold 5.
3. Repeat the process up to now 5 times.
4. Calculate the average performance of a hyperparameter combination.
5. Repeat the process up to now for every hyperparameter combination.
6. Select the hyperparameter combination with the best average model performance.
7. Fit the final model with optimal hyperparameters on the full training data.

It is easy to see that this exhaustive tuning process results in a large amount of models to run, of which some might be quite computationally intensive. To reduce the time spent on tuning, we here separate hyperparameter tuning and fitting the final model, where the tuning is done on only a subset of 10% of the training data, and only the fit of the final model on the full training data.

3.5.1 Logit

The class of logit regressions for binary outcomes is well known and applied in econometrics and ML alike, and will serve as a baseline for more complex models to follow. In its relatively simple and rigid functional form, there are no tunable parameters. Taking this functional form as given, then minimizing the out-of-sample loss function $L(\hat{y}, y)$ becomes a question of (i) how many variables, and (ii) which variables to include. Such problems of *variable selection* are well known to econometricians, which use them mainly for the selection of control variables, including stepwise regressions (one-by-one adding control variables with the highest impact on our $\bar{R}^2$), partial least squares (PLS), different information criterion (e.g., Aikon:

[12] While the described process appears rather tedious by hand, specialized ML packages such as `caret` in R provide efficient workflows to automatize the creation of folds as well as hyperparamether grid search.

AIC, Bayesian: BIC), to only name a few.[13] While in reality there might be incentives to create a sparse model, we for this exercise use all variables available.

3.5.2 Elastic Nets

We proceed a second parametric approach, a class of estimators for penalized linear regression models that lately also became popular among econometricians, *elastic nets*. Generally, the functional form is identical to a generalized linear model, with a small addition. Our β parameters are weighted by an adittional parameter λ, which penalizes the coefficient by its contribution to the models loss in the form of:

$$\lambda \sum_{p=1}^{P} [(1-\alpha)|\beta_p| + \alpha|\beta_p|^2] \tag{2}$$

Of this general formulation, we know two popular cases. When $\alpha = 1$, we are left with the quadratic term, leading to a *ridge regression*. If $\alpha = 0$, we are left with $|\beta_i|$, turning it to a lately among econometricians very popular "Least Absolute Shrinkage and Selection Operator" (LASSO) regression. Obviously, when $\lambda = 0$, the whole term vanishes, and we are again left with a generalized linear model[14] Consequently, the model has two tunable parameters, α and λ, over which we perform a grid search., illustrated in Fig. 6.

While for low α values the model performance turns out to be somewhat insensitive to changes in λ, with increasing α values, λ 1 leads to sharply decreasing model performance. With a slight margin, the pest performing hyperparameter configuration resembles a LASSO ($= 1, \alpha = 0$).

3.5.3 Classification Tree

Next, this time following a non-parametric approach, we fit a *classification and regression trees* (CART, in business application also known as *decision trees*).[15] The rich class of classification trees is characterized by a flexible functional form able to fit complex relationships between predictors and outcomes, yet is can be illustrated in an accessible way. They appear to show their benefits over traditional logistic regression approaches mostly in settings where we have a large sample size (Perlich

[13] For an exhaustive overview on model and variable selection algorithms consider Castle et al. (2009).

[14] For an exhaustive discussion on the use of LASSO, consider Belloni et al. (2014). Elastic nets are integrated, among others, in the R package `Glmnet`, and Python's`scikit-learn`.

[15] There are quite many packages dealing with different implementations of regression trees in common data science environments, such as `rpart, tree, party` for R, and again the machine learning allrounder `scikit-learn` in Python. For a more exhaustive introduction to CART models, consider Strobl et al. (2009).

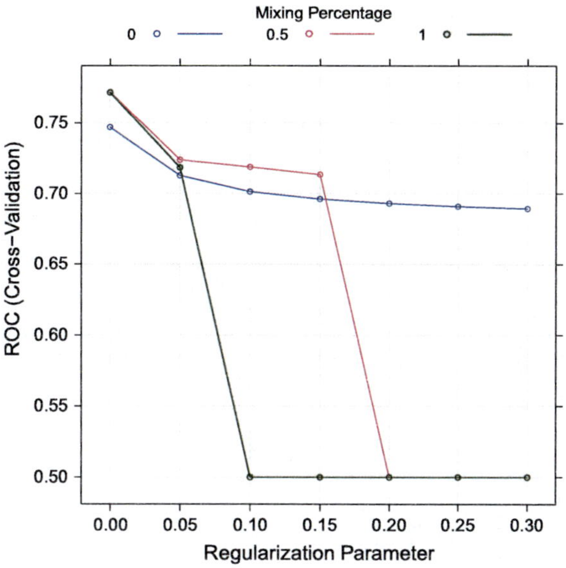

Fig. 6 Hyper-parameter tuning elastic nets

et al. 2003), and where the underlying relationships are really non-linear (Friedman and Popescu 2008). The general idea behind this approach is to step-wise identify feature explaining the highest variance of outcomes. This can be done in various ways, but in principle you aim to at every step use some criterion to identify the most influential feature X of the model (e.g., the lowest p value), and then another criterion (e.g., lowest χ^2 value) to determine a cutoff value of this feature. Then, the sample is split according to this cutoff. This is repeated for every subsample, leading to a tree-like decision structure, which eventually ends at a terminal node (a *leaf*), which in the optimal case contains only or mostly observation of one class. While simple and powerful, classification trees are prone to overfitting, when left to grow unconstrained, since this procedure can be repeated until every observation ends up in an own leaf, characterized by an unique configuration of features. In practice, a tree's complexity can be constrained with a number of potential hyperparameters, including a limit the maximum depth, or criteria if a split is accepted or the node becomes a terminal one (e.g., certain $p-value$, certain improvement in the predictive performance, or a minimum number of observations falling in a split).

In this exercise, we fit a classification tree via a "Recursive Partitioning" implemented in the `rpart` package in R, cf. Therneau et al. (1997). The resulting tree structure can be inspected in Fig. 7.

Here we are able to restrict the complexity via a hyperparameter α. This parameter represents the complexity costs of every split, and allows further splits only if it leads to an decrease in model loss below this threshold. Figure 8 plots the result of the hyperparameter tuning of α.

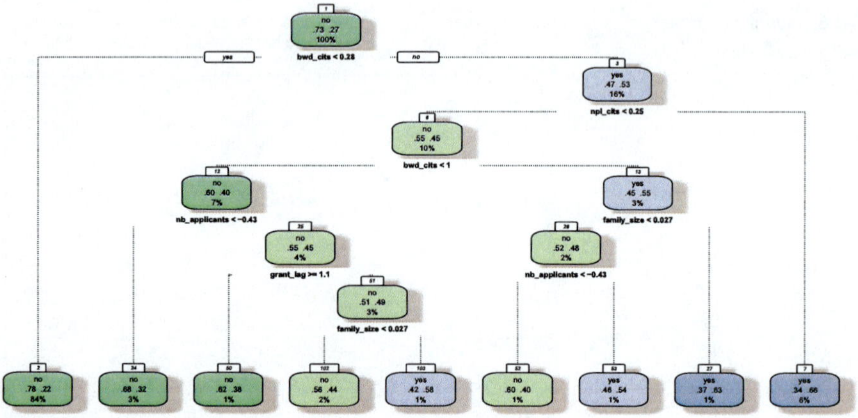

Fig. 7 Structure of the decision tree

Fig. 8 Hyper-parameter tuning classification tree

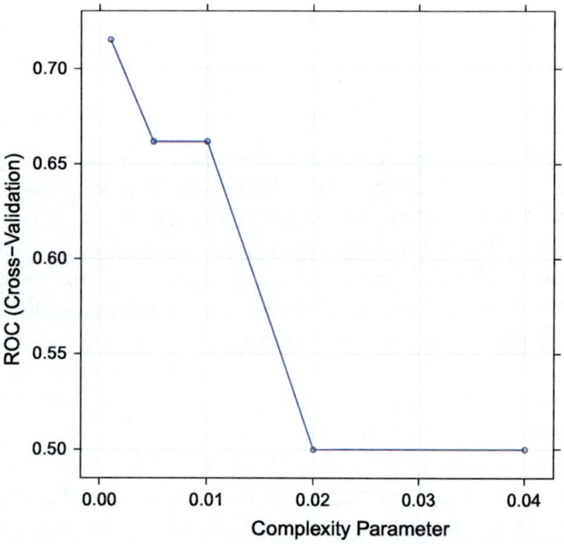

We directly see that in this case, increasing complexity costs lead to decreasing model performance. Such results are somewhat typical for large datasets, where high complexity costs prevent the tree to fully exploit the richness of information. Therefore, we settle for a minimal α of 0.001.

3.5.4 Random Forest

Finally, we fit another class of models which has gained popularity in the last decade, and proven to be a powerful and versatile prediction technique which performs well in almost every setting, a random forest. As a continuation of tree-based classification methods, random forests aim at reducing overfitting by introducing randomness via bootstrapping, boosting, and ensemble techniques. The idea here is to create an "ensemble of classification trees", all grown out of a different bootstrap sample. These trees are typically not pruned or otherwise restricted in complexity, but instead, a random selection of features is chosen to determine the split at the next decision nodes.[16] Having grown a "forest of trees", every tree performs a prediction, and the final model prediction is formed by a "majority vote" of all trees. The idea is close to the Monte-Carlo approach, assuming a large population of weak predictions injected by randomness leads to overall stronger results than one strong and potentially overfitted prediction. However, the robustness of this model class comes with a price. First, the large amount of models to be fitted is computationally rather intensive, which becomes painfully visible when working with large datasets. Further, the predictions made by a random forest are more opaque than the ones provided by the other model classes used in this example. While the logit and elastic net delivers easily interpretable parameter estimates and the classification tree provides a relatively intuitive graphical representation of the classification process, there exists no way to represent the functional form and internal process of a classification carried out by a random forest in a way suitable for human annotation.

In this case, we draw from a number of tunable hyperparameters. First, we tune the number of randomly selected features which are available candidates for every split on a range $[1, k - 1]$, where lower values introduce a higher level of randomness to every split. Our second hyperparameter is the minimal number of observations which have to fall in every split, where lower numbers increase the potential precision of splits, but also the risk of overfitting. Finally, we also use the general splitrule as an hyperparameter, where the choice is between (i) a traditional split according to a the optimization of the gini coefficient of the distribution of classes in every split, and (ii) according to the "Extremely randomized trees" (ExtraTree) procedure by Geurts et al. (2006), where adittional randomness is introduced to the selection of splitpoints.

In Fig. 9 we see that number of randomly selected features per split of roughly half (22) of all available features in all cases maximizes model performance. Same goes for a high minimal number of observations (100) per split. Finally, the ExtraTree procedure first underperforms at a minimal amount of randomly selected features, but outperforms the traditional gini-based splitrule when the number of available features increases. Such results are typical for large samples, where a high amount of injected randomness tends to make model predictions more robust.

[16] Indeed, it is worth mentioning here that many model tuning techniques are based on the idea that adding randomness to the prediction process—somewhat counter-intuitively—increases the robustness and out-of-sample prediction performance of the model.

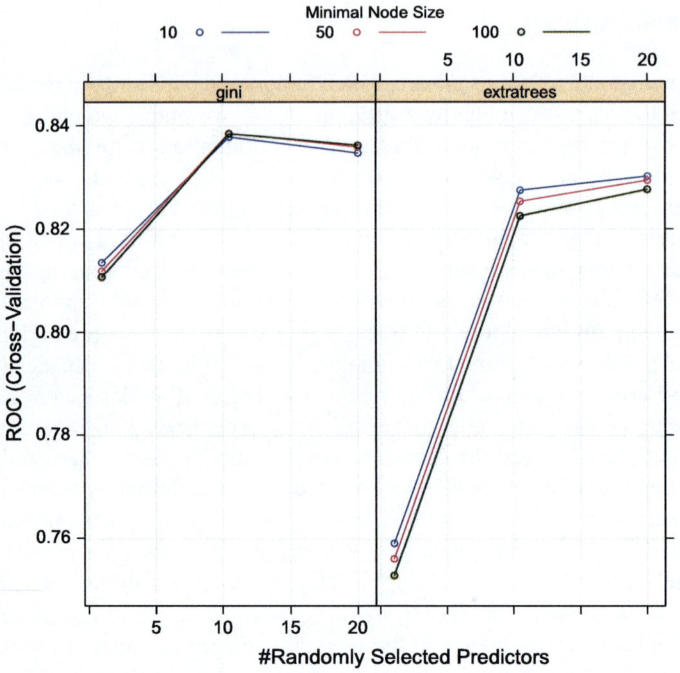

Fig. 9 Hyper-parameter tuning random forest

3.5.5 Explainability

In the exercise above we demonstrate that richer model classes with a flexible functional form indeed enable us to better capture complex non-linear relationships and enable us to tackle hard prediction problems more efficiently that traditional methods and techniques from causal modeling, which are usually applied by econometricians. First, even in parametric approaches, feature effects in predictive modeling are explicitly non-causal.[17] This holds true for most machine learning approaches and represents a danger for econometricians using them blindly. Again, while an adequately tuned machine learning model may deliver very accurate estimates, it is misleading to believe that a model designed and optimized for predicting $\hat{y}$ per se also produces β's with the statistical properties we usually associate with them in econometric models.

Second, with increasing model complexity, the prediction process becomes more opaque, and the isolated (non-causal) effect of features on the outcome becomes harder to capture. While the feature effect of the logit and elastic net can be interpreted in a familiar and straightforward way as a traditional regression table, already

[17] Just to give an example, Mullainathan and Spiess (2017) demonstrate how a LASSO might select very different features in every fold.

in classification tree models (see Fig. 10) we do not get constant *ceteris paribus* parameter estimates. However, the simple tree structure still provides some insights into the process that leads to the prediction. The predictions of a forest consisting of thousands of such trees in a random forest obviously cannot be interpreted anymore in a meaningful way.

Some model classes have developed own metrics of variable impact as we depict for our example in Fig. 10. However, they are not for all model classes available, and sometimes hard to compare across models. In this cases, the most straightforward war to get an intuition of feature importance across models is to calculate the correlation between the features and predicted outcome, as we did in figure. Again, this gives us some intuition on the relative influence of a feature, but tells us nothing about any local prediction process.

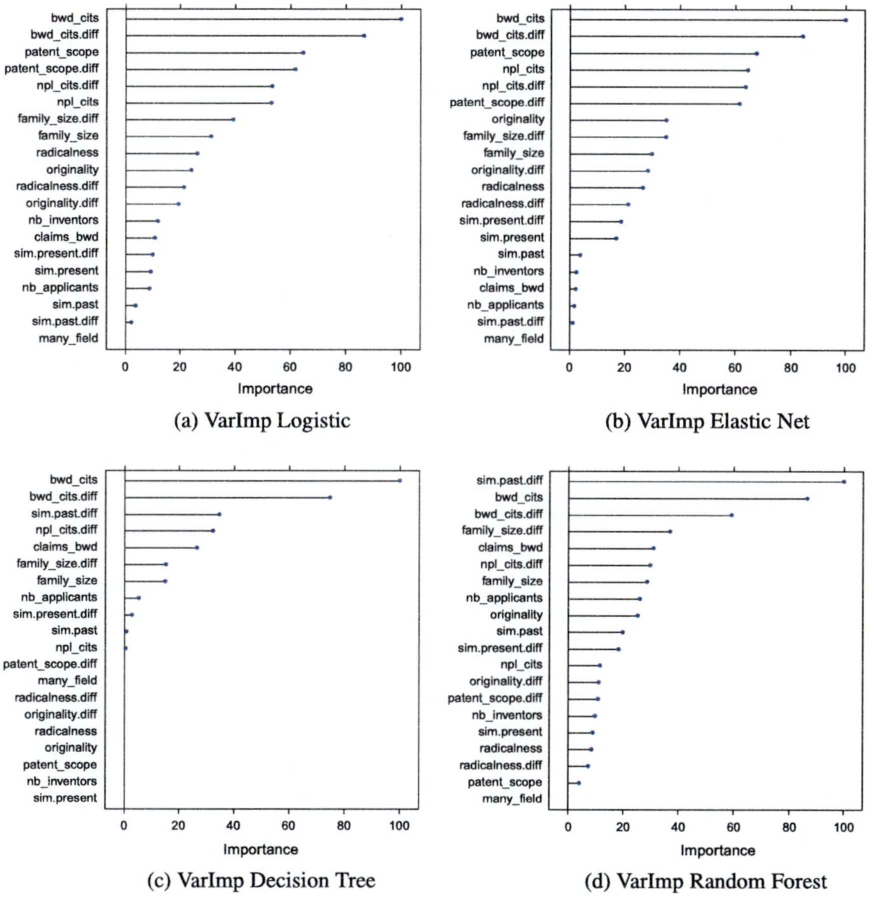

(a) VarImp Logistic (b) VarImp Elastic Net

(c) VarImp Decision Tree (d) VarImp Random Forest

Fig. 10 Variable importance of final models

Table 2 Final model evaluation with test sample

Names	Logit	ElasticNet	ClassTree	RandForest
Accuracy	0.758	0.758	0.754	0.77
Kappa	0.228	0.226	0.217	0.299
Sensitivity	0.222	0.219	0.219	0.304
Specificity	0.959	0.96	0.953	0.944
AUC	0.73	0.73	0.607	0.758

From investigating the relative variable importance, we gain a set of insights. First, we see quite some difference in relative variable importance across models. While backward citations across all models appears to be a moderate or good predictor, technology fields are assigned as highly predictive in the elastic net, but way less in the random forest, that ranks all other features higher. Furthermore, the extend to which the models draw from the available features differs. While the elastic net draws more evenly from a large set of features, in the classification tree only 8 are integrated. That again also reminds us that features importance, albeit informative, cannot be interpreted as a causal effect.

Discussion on local explainability.

3.5.6 Final Evaluation

After identifying the optimal hyperparameters for every model class, we now fit the final prediction models on the whole training data accordingly. As a final step, we evaluate the performance of the models by investigating its performance on a holdout sample, consisting of 25% of the original data, which was from the start set aside and never inspected, or used for any model fitting. Figure 10 displays the results of the final model's prediction on the holdout sample by providing a confusion matrix as well as the ROC curve to the corresponding models, while Table 2 provides a summary of standard evaluation metrics of predictive models.

We see that the logit and elastic net in this case leads to almost identical results across a variety of performance measures. Surprisingly, the classification tree in this case is the worst performing model, despite it's more complex functional form. However, earlier we saw that the classification tree takes only a small subset of variables into account, hinting at a to restrictive choice of the complexity parameter α during the hyperparameter tuning phase. This is also indicated by the kinked ROC curve in Fig. 11, indicating that after exploiting the effect of a few parameters the model has little to contribute. Across all measures, the random forest, as expected, performs best. Here we see how randomization and averaging over a large number of repetition indeed overcomes many issues of the classification tree. While the overall accuracy (ratio of correctly classified to all observations) only increases modestly, the low occurrence of positive outcomes makes this measure only partially informative

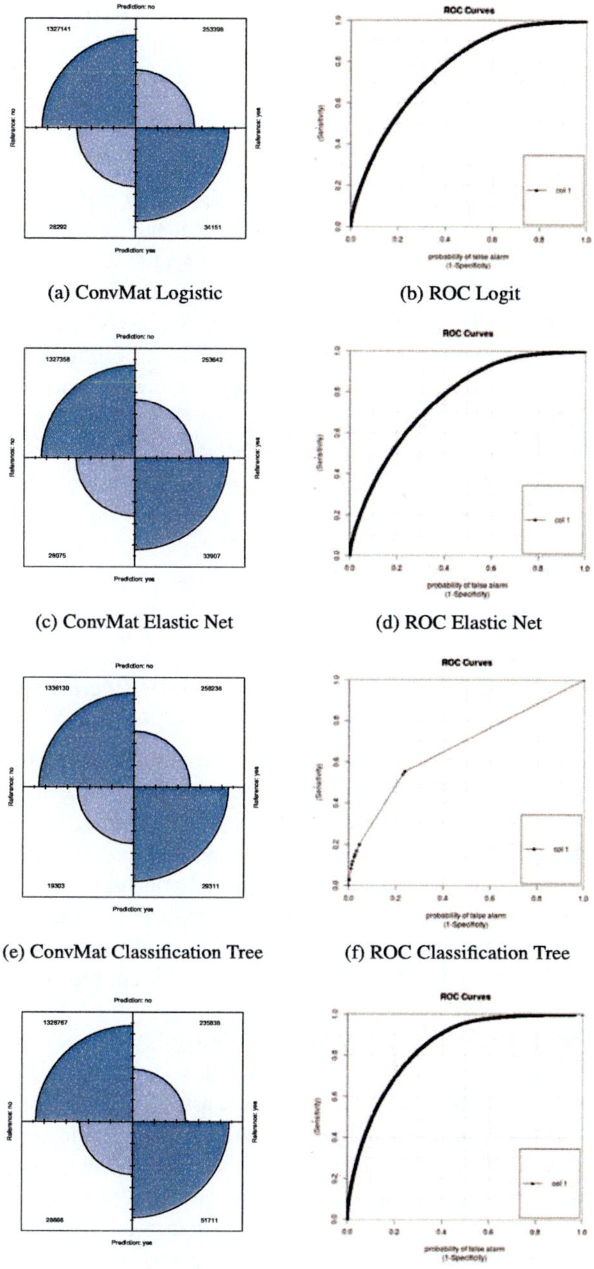

(a) ConvMat Logistic

(b) ROC Logit

(c) ConvMat Elastic Net

(d) ROC Elastic Net

(e) ConvMat Classification Tree

(f) ROC Classification Tree

(g) ConvMat Random Forest

(h) ROC Random Forest

Fig. 11 ROC curves of final models

in this case. However, the sensitivity (ratio correctly classified positives to all positive observation), increases quite visibly compared to other models. This also highlights issues of traditional methods often occurring when facing unbalanced outcome states.

Explain performance measures

Up to now we demonstrated a predictive modelling workflow using traditional machine learning techniques, which performs reasonably well for the exercise at hand. While the prediction of successful patents can be seen as analytically non-trivial, we are able to achieve an acceptable level of accuracy by only using features that can be computed directly from the PATSTAT data with little effort. With a more complex feature generation procedure (e.g., matching with further datasources to include inventor or applicant characteristics), this performance could likely be improved further. However, be reminded that up to now we only fitted models to predict *breakthrough3*, the outcome indicating that the patent is in the top 50% of received citations within its cohort (ca. 37% of cases). We repeated this exercise for the outcome *breakthrough* (top 1%, 0.6% of cases), where we get less optimistic results. For our initial target outcome, the rare *breakthrough*, all models are of no help whatsoever and predict 100% non-breakthroughs, except of the random forest, which here predicts marginally better than a coin toss.

3.6 Outlook: Deep Learning and Rare Event Prediction

Up to now, we demonstrated how various more traditional model classes can be applied in a predictictive modelling context at the example of a classification problem with approximately 15% of positive cases of interest. But what if we were interested in the "breakthrough" patents? The really rare events that receive large amounts of citations? This is a situation where more traditional approaches fail. In such cases the dataset is heavily unbalanced. Most supervised machine learning approaches are sensitive to such scenarios and a common approach has been to undersample the data. More recently deep learning approaches to such problems have become popular, due to their efficiency, ability to handle large amount of data in a sequential manner (data can be streamed line by line) and flexibility that allows to work with different types of inputs (cross-sectional, sequential, text, images etc.). In the following we will start by repeating the exercise from the previous section using a relatively simple neural network architecture. Then, we will apply the same architecture to the breakthrough-identification problem. Lastly, we will use a deep autoencoder and an anomaly detection approach to identify the most rare breakthrough patents. Before that, we will, however, provide a short, more general introduction to artificial neural networks and deep learning.

3.6.1 Introduction to Artificial Neural Networks and Deep Learning

Regression trees might still be familiar to some applied statisticians and quantitative researchers. Now we would like to introduce to another class of models which due to current breakthroughs delivered unprecedented prediction performance on large high-dimensional data and thus enjoys a lot of popularity: Neural networks. Connecting to the former narrative, neural networks represent *regression trees on steroids*, which are flexible enough to—given enough data—fit every functional form and thereby theoretically can produce optimal predictions to every well-defined problem.

While early ideas about artificial neural networks (ANNs) were already developed in the 1950s and 60s by among others Frank Rosenblatt 1958 and the formal logic of neural calculation described by McCulloch and Pitts (1943), it took several decades for this type of biology-inspired models to see a renaissance in the recent few years.[18]

This revival can be attributed to three reasons: (i) New training techniques, (ii) the availability of large training datasets, and (iii) hardware development, particularly the identification of graphical processing units (GPUs)—normally used, as the name suggests, for complex graphics rendering tasks in PCs and video game consoles—as extremely well suited for modeling neural networks (LeCun et al. 2015).

To understand the neural network approach to modeling, it is essential to get a basic grasp of two main concepts. First, the logic behind the functioning of single neurons,[19] and second the architecture and sequential processes happening within ANNs.

A single neuron receives the inputs $x_1,x_2,...,x_n$ with weights $w_1,w_2,...,w_n$ that are passed to it through "synapses" from previous layer neurons (i.e. the input layer). Given these inputs the neuron will "fire" and produce an output y passing it on to the next layer, which can be a hidden layer or the output layer. In praxis, the inputs can be equated to standardized or normalized independent variables in a regression function. The weights play a crucial role, as they decide about the strength with which signals are passed along in the network. As the network learns, the initially randomly assigned weights are continuously adjusted. As the neuron receives the inputs, it first calculates a weighted sum of $w_i x_i$ and then applies an activation function ϕ.

$$\phi\left(\sum_{i=1}^{m} w_i x_i\right) \tag{3}$$

[18] It has to be stressed that even though neural networks are indeed inspired by the most basic concept of how a brain works, they are by no means mysterious artificial brains. The analogy goes as far as the abstraction that a couple of neurons that are interconnected in some architecture. The neuron is represented as some sigmoid function (somewhat like a logistic regression) which decides based on the inputs received if it should get activated and send a signal to connected neurons, which might again trigger their activation. Having that said, calling a neural network an artificial brain is somewhat like calling a paper-plane an artificial bird.

[19] for the sake of simplicity here we will not distinguish between the simple perceptron model, sigmoid neurons or the recently more commonly used rectified linear neurons (Glorot et al. 2011).

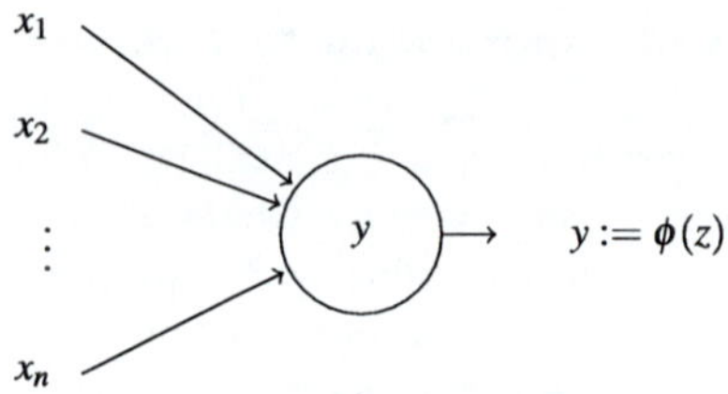

Fig. 12 Illustration of a neuron

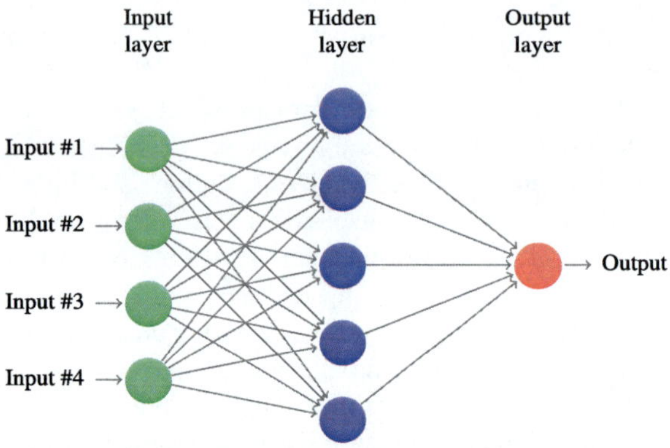

Fig. 13 Illustration of a neural network

Depending on the activation function the signal is passed on or not (Fig. 12).

Figure 13 represents an artificial neural network with three layers: One input layer with four neurons, one fully connected hidden layer with five neurons and one output layer with a single neuron. As the model is trained for each observation inputs are passed on from the input layer into the neurons of the hidden layer and processed as described above. This step is repeated and an output value $\hat{y}$ is calculated. This process is called forward propagation. Comparing this value with the actual value y (i.e. our dependent variable for the particular observation) allows to calculate a *cost function* e.g. $C = \frac{1}{2}(\hat{y} - y)^2$. From here on *backpropagation*[20] is used to update the weights. The network is trained as these processes are repeated for all observations in the dataset.

Artificial neural networks have many interesting properties that let them stand out from more traditional models and make them appealing when approaching complex pattern discovery tasks, confronting nonlinearity but most importantly dealing with large amounts of data in terms of the number of observations and the number of inputs. These properties, coupled with the recent developments in hardware and data availability, led to a rapid spread and development of artificial nets in the 2010s. Today, a variety of architectures has evolved and is used for a large number of complex tasks such as speech recognition (Recurrent neural networks: RNNs and Long Short Term Memory: LSTMs), computer vision (CNNs and Capsule Networks, proposed

[20] This complex algorithm adjusts simultaneously all weight in the network, considering the individual contribution of the neuron to the error.

in late October 2017) and as backbones in artificial intelligence applications. They are used not only because they can approach challenges where other classes of models struggle technically but rather due to their performance.

Despite the numerous advantages of artificial neural nets they are yet rarely seen applied in non-technical research fields. Here CNNs may be so far the most often used type, where its properties were employed to generate estimates from large image datasets (e.g. Gebru et al. 2017). The simplest architecture of a CNN puts several convolutional, and pooling layers in front of an ANN. This allows transforming images, which are technically two-dimensional matrices, into long vectors, while preserving the information that describes the characteristic features of the image.

The predictive performance of neural nets stands in stark contrast to the explainability of these models, meaning that a trained neural net is more or less a black box, which produces great predictions but does not allow to make causal inference. In addition, this leads to asking: What is the reason the model produced this or that prediction. This becomes particularly important when such models are deployed for instance in diagnostics or other fields to support decision making. There are several attempts to address this problem under the heading of "explainable AI" (e.g. Ribeiro et al. 2016).

3.6.2 Application: Simple Feed Forward Neural Network for Predicting Successful and Breakthrough Patents

Before approaching a complex neural architecture in the following section, here we use a relatively simple feed forward neural network to first tackle the problem, presented in earlier sections. The network is composed of 3 dense layers of 22, 20, 15 neurons respectively and an output layer. In addition two dropout layers with 0.3 and 0.2 rates have been added for regularization. We used gridsearch and k-fold cross-validation to determine optimal hypermarameter settings. Such a setup represents a relatively standard architecture and no further variation has been explored regarding adding and removing of further neurons or layers.

3.6.3 Application: Prediction of Breakthrough Patents as Anomaly Detection with Stacked Autoencoder

An autoencoder is a neural network architecture that is used to learn a representation (encoding) for a set of data. Autoencoders are referred to as "self-suprevised" models, as their inputs are usually identical to their outputs (Hinton and Salakhutdinov 2006).

$$f_{W,b}(x) \approx x \tag{4}$$

The typical use case is dimensionality reduction: Training such models to reconstruct some inputs from a low-dimensional representation in the latent hidden layer

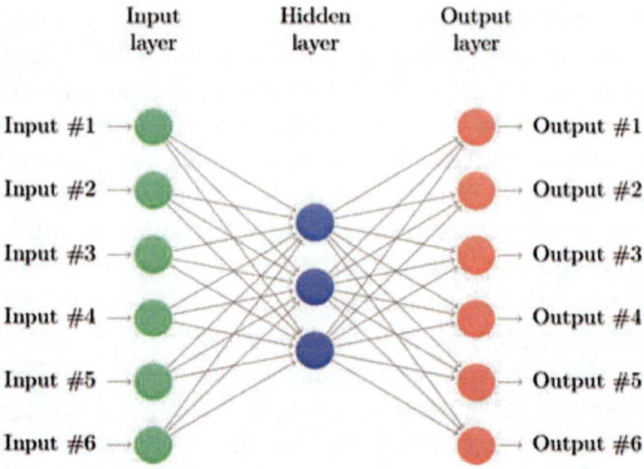

Fig. 14 Illustration of an autoencoder architecture

as shown in Fig. 14 makes them very powerful general dimensionality reduction tools. In comparison to techniques such as PCA, autoencoders capture all kinds of non-linearities and interactions. The encoder part of such a model can be used separately to create some low-dimensional representations of inputs that consequently can be for instance clustered or used to identify similar inputs, which is used in state-of-the-art recommender systems (Sedhain et al. 2015). In combination with recurrent or convolutional layers inputs can be sequences or images, respectively (Tian et al. 2014). Sequence to Sequence (Seq2seq) models with recurrent layers on both ends are increasingly used for neural machine translation and contributed to great improvements in the field (Sutskever et al. 2014). Such models are trained by using phrases in the source language as inputs and the same in the target language as outputs. The hidden layer of such models is called "thought" vector, as it incodes a universal representation meaning representation of the input that the model can translate into another language. Another recent application of autoencoders has been anomaly detection (Sakurada and Yairi 2014). The idea is very simple: Given the vast amount of "normal" cases, it is easy to train the autoencoder to "reconstruct" the *status quo*. This information will constitute all what the autoencoder is exposed to, and therefore we would expect that non-normal inputs would confuse the autoencoder. Two approaches to capture such confusion are proposed[21]: We could now (a) extract the latent layer and cluster it, reduce dimensionality with for example PCA and measure the distance of the known outliers to the centeroids of our the clusters (Sakurada and Yairi 2014). (b) An even easier approach is to look at the behaviour of the autoencoder more directly, particularly through the reconstruction error—e.g. Mean Square Error or Eucledian Distance, Eq. 5 and 6 respectively (An and Cho 2015).

$$L(x, x') = \| x - x' \|^2 \tag{5}$$

[21] For some overview on other methods using similar logic, consider: Wang (2005), Zhou and Lang (2003), Shyu et al. (2003).

$$L(x, x') = \| x - x' \|_2 \tag{6}$$

In a nutshell: An autoencoder that is really good at reconstructing "boring normality" should experience considerable difficulties when trying to reconstruct anomaly events.

The latter is the approach that we take to detect the the 0.6% breakthrough patents in our dataset that more traditional models would most likely oversee. We use the same data as above, however removing the categorical variable `technology field`. All other variables are normalised (Fig. 16).

We train a stacked autoencoder model using Tensorflow-Keras in Python.[22] Neural network models, and particularly autoencoders, are difficult to train, and while some guidelines on hyperparameter optimization exist, one has to experiment with different settings. A systematic grid-search has not been carried out due to the vast number of parameters and exponentially scaling combinations of those. In fact, recently services have been found that support researchers and developers when performing such experiments (e.g. comet.ml). There are a number of architectural hyperparameters that can be adjusted: The architecture of the encoder (number of layers and their respective size), regularization, activation functions, optimization, and loss functions. In addition, during training, we have to decide on batch size, shuffling, and the number of epochs.

Our input data has 11 variables and we decide not to have more nodes than this number in any layer. Thereby we do not have to address challenges of overcomplete layers that can lead to the model simply passing the inputs without any learning happening (Vincent et al. 2010). Figure 15 summarizes the relatively simple network architecture that achieved good performance: The first two dense layers comprise the encoder, while the last two decode the data. In the encoded latent layer, the data is compressed down to 4 dimensions. We use a combination of hyperbolic tangent (tanh, *dense_5*, *dense_7*) and rectified linear unit (ReLU, *dense_6*, *dense_8*) activation functions. Again, this combination is not fully theory based but has shown good performance with autoencoder models more generally. For the sake of simplicity, we started with a mean-squared-error as our loss function/reconstruction error. However, experiments with various loss functions (Kullback-Leibler divergence, Poisson, Cosine Proximity), found Cosine Proximity loss and the Adam optimizer to deliver the best results. To prevent overfitting—in our case most likely leading to high reconstruction errors for any inputs previously unseen by the model—we introduce activity regularization in the first dense layer. We use L2 regularization, which is a method to penalize weights with large magnitudes, that are an indicator for overfitting (Ng 2004). The model gets "smoothed out", which should make it more alert to anomalies.

[22] Variational autoencoers are a slightly more modern and interesting take on this class of models which also performed well in our experiments. Following the KISS principle, we decided to use the more traditional and simpler autoencoder architecture that is easier to explain and performed almost equally well.

```
Layer (type)                 Output Shape              Param #
=================================================================
input_2 (InputLayer)         (None, 11)                0
_________________________________________________________________
dense_5 (Dense)              (None, 9)                 108
_________________________________________________________________
dense_6 (Dense)              (None, 4)                 40
_________________________________________________________________
dense_7 (Dense)              (None, 4)                 20
_________________________________________________________________
dense_8 (Dense)              (None, 11)                55
=================================================================
Total params: 223
Trainable params: 223
Non-trainable params: 0
_________________________________________________________________
```

Fig. 15 Autoencoder model architecture summary

Fig. 16 ROC-AUC for the stacked autoencoder

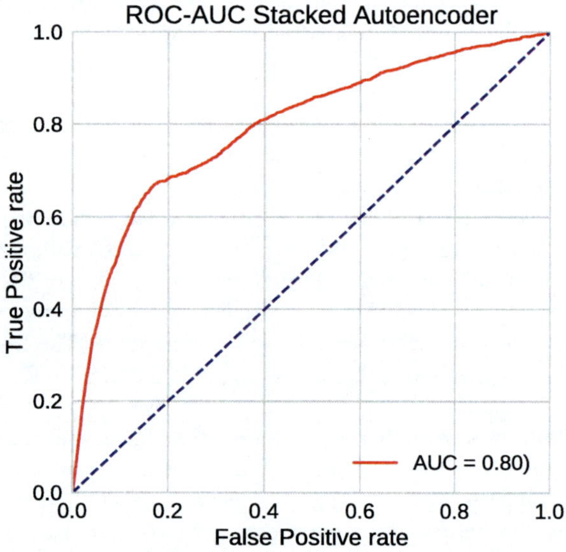

The data is divided into 80% for training and 20% for validation. All anomalies are removed from the training set, leaving 1,715,843 observations. The network is trained in batches of 512 observations which are reshuffled in each epoch. We found that training the relatively small network for 10 epochs has been enough (we experimented with values up to 100 epochs). The training converged fast with no major accuracy gains being achieved after 2 epochs. Training time for one epoch on the Google Colaboratory GPU Engine averaged at 27 s. Given the very small model (only 223 trainable parameters—weights and biases), the performance bottleneck was actually not the neural net computation itself and computation times dropped to 16 s on a 4 Core CPU machine, showing that in our case a GPU infrastructure was not necessary.

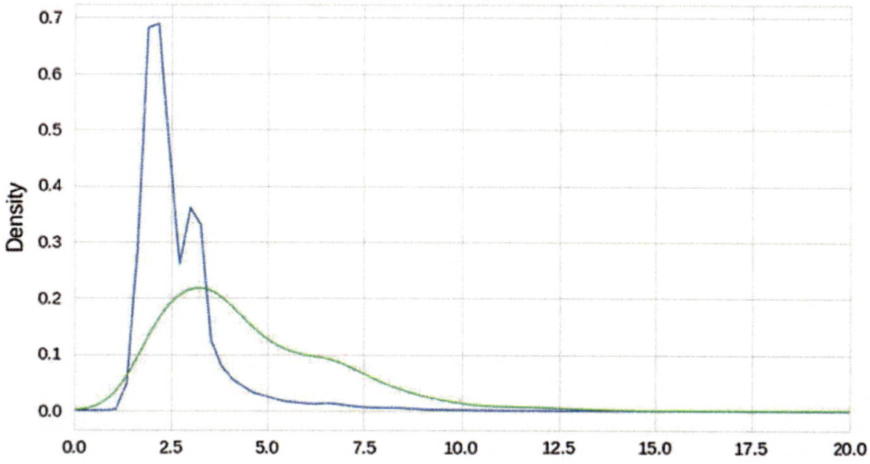

Fig. 17 Density distribution of the reconstruction error, breakthrough patents in green

Fig. 18 Confusion matrix, error [3.8, 11]

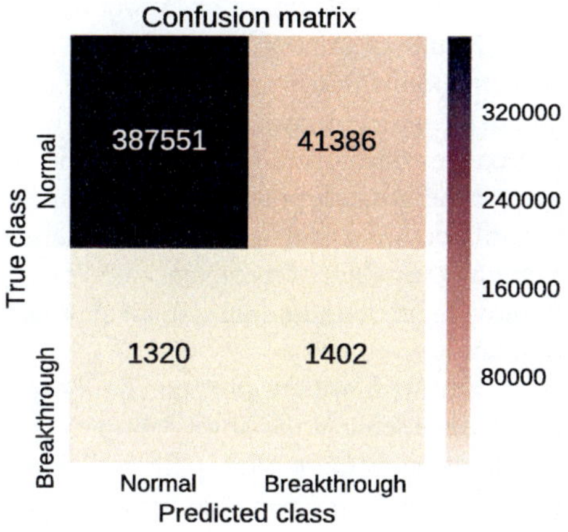

Once trained, the autoencoder can be used to reconstruct the test set of 431.659 observations, 2722 of which, approx 0.6% are our cases of interest. Here, all observations (normal and anomalies) are fed to the model and the reconstruction error is calculated as the Euclidean Distance, where we can clearly observe in Fig. 17 the difference between the classes.

The error-term can be further used to calculate the ROC and AUC indicators. We achieve an intermediate AUC value of 0.8, which is not excellent but for this application reasonably good. We refer to this value as intermediate because the calculated error term functions as an ex-post parameter that can be arbitrarily set depending on the application. We can, for instance, decide to classify all estimations with an error between 3.8 and 11 to be considered breakthrough patents (Fig. 18).

This would leave us with 1402 correctly identified breakthrough patents, a ROC-AUC value of 0.71. While far from an excellent result, this is a good result for this application. Not only is this a proof of concept but it also shows that breakthrough patents expose some significantly distinctive patterns.

4 Conclusion, Ways Forward, and Avenues for Future Research

In this paper, we introduced the readership to the main idea behind predictive modeling, which somewhat stands in stark contrast to common intuition, workflows and data analysis routines of a trained econometricians developing causal models. Particularly, we elaborate on central concepts such as the establishment of generalizability via out-of-sample validation techniques. As a "bonus", we finally to a deep learning based workflow for anomaly detection, that can be used in cases when extremely rare observations need to be identified—the proverbial needle in the haystack problem.

Predictive modeling offers a promising methodology that can be used in diverse settings and is gaining relevance in a big data world. In this exercises our analysis stopped with the predicted results and their validation. We argue that scholars within the wider innovation studies and entrepreneurship community can adapt many approaches developed in ML, and in the following point towards some areas where we see the greatest potentials. These are (i) the generation of quality indicators to quantify complex and up to now often unmeasurable theoretical concepts, (ii) understanding the nature of rare events, (iii) exploratory phenomenon spotting, and (iv) the improvement of traditional statistical models particularly with explicit out-of sample evaluation.

To start with, we see great potential in employing predictions from a ML architecture as independent variables for up to now unobservable qualitative measures in a traditional regression setting. Such combinations of prediction and causal inference techniques offer the potential for granular and timely analysis of phenomena which currently cannot, or only to a limited extent, be addressed using traditional techniques and data sources. The "Startup Cartography Project" at the MIT (Andrews et al. 2017; Fazio et al. 2016; Guzman and Stern 2015, 2017) provides a good example of such efforts. Coining it "nowcasting" and "placecasting", the project uses large amounts of business registration records and predictive analytics to estimate entrepreneurial quality for a substantial portion of registered firms in the US (about 80%) over 27 years. When engaging in such "predictions in the service of estimation" (Mullainathan and Spiess 2017), it is not hard to see how these predictions of start-up quality (and similar quality indicators) might serve as dependent or independent variables in many interesting hypothesis-testing settings.

Related, one task our traditional econometrics toolkit has performed particularly bad, is the explanation but also prediction of extremely *rare events*, as we demonstrated in our empirical exercise. However, being able to explain impactful low probability events (also coined as black swans, cf. Taleb 2010) such as which start-up is going to be the next gazelle, which technology our patent is going to be the futures "next big thing", when does the next financial crisis hit or firm defaults (cf. e.g. van der Vegt et al. 2015), and so forth, would certainly be of enormous interest for research, policy, and business alike.

Along that line, a predictive model can be deployed in a more exploratory way for "phenomena spotting", to test if "there is something in the data". In our example, the anomaly detection part indicated that for over half of the breakthrough patents, there seem to exist some latent divergent patterns that the model picks up and that may be worth exploring for potential causality.

Lastly, the practice of out-of-sample testing might help to improve the external validity of our models by explicitly testing how good our model performs in terms of parameter estimates and overall prediction (Athey and Imbens 2017).

We hope that this paper is also a contribution to initiating and shaping a dialogue between the disciplines. While the ML community may be ahead within predictive modelling of big data, applied econometricians have developed a deep understanding of causality, which is currently mostly lacking among ML practitioners. Social sciences have also a much richer tradition thinking about epistemology and—ideally—sampling, research design, ethics and arguably context-awareness. Such an interaction may be rather helpful when discussing current issues of algorithmic bias and explainable AI.

References

Aguinis, H., Pierce, C.A., Bosco, F.A., Muslin, I.S.: First decade of organizational research methods: trends in design, measurement, and data-analysis topics. Organ. Res. Methods **12**(1), 69–112 (2009)

Ahuja, G., Lampert, C.: Entrepreneurship in the large corporation: a longitudinal study of how established firms create breakthrough inventions. Strateg. Manag. J. **22**(6–7), 521–543 (2001)

An, J., Cho, S.: Variational autoencoder based anomaly detection using reconstruction probability. Rep, SNU Data Mining Center, Tech (2015)

Andrews, R.J., Fazio, C., Guzman, J., Stern, S.: The startup cartography project: A map of entrepreneurial quality and quantity in the united states across time and location. MIT Working Paper (2017)

Athey, S., Imbens, G.W.: The state of applied econometrics: causality and policy evaluation. J. Econ. Perspect. **31**(2), 3–32 (2017)

Basberg, B.L.: Patents and the measurement of technological change: a survey of the literature. Res. Policy **16**(2–4), 131–141 (1987)

Belloni, A., Chernozhukov, V., Hansen, C.: Inference on treatment effects after selection among high-dimensional controls. Rev. Econ. Stud. **81**(2), 608–650 (2014)

Castle, J.L., Qin, X., Reed, W.R., et al.: How to pick the best regression equation: A review and comparison of model selection algorithms. Working Paper No. 13/2009, Department of Economics and Finance, University of Canterbury (2009)

Einav, L., Levin, J.: The data revolution and economic analysis. Innov. Policy Econ. **14**(1), 1–24 (2014)

Einav, L., Levin, J.: Economics in the age of big data. Science **346**(6210), 1243089 (2014)

Ernst, H.: Patent applications and subsequent changes of performance: evidence from time-series cross-section analyses on the firm level. Res. Policy **30**(1), 143–157 (2001)
Fazio, C., Guzman, J., Murray, F., and Stern, S.: A new view of the skew: quantitative assessment of the quality of American entrepreneurship. MIT Innovation Initiative Paper (2016)
Friedman, J.H., Popescu, B.E.: Predictive learning via rule ensembles. Ann. Appl. Stat. 916–954 (2008)
Gebru, T., Krause, J., Wang, Y., Chen, D., Deng, J., Aiden, E.L., Fei-Fei, L.: Using deep learning and google street view to estimate the demographic makeup of neighborhoods across the united states. Proc. Natl. Acad. Sci. 201700035 (2017)
George, G., Osinga, E.C., Lavie, D., Scott, B.A.: From the editors: big data and data science methods for management research. Acad. Manag. J. **59**(5), 1493–1507 (2016)
Geurts, P., Ernst, D., Wehenkel, L.: Extremely randomized trees. Mach. Learn. **63**(1), 3–42 (2006)
Glorot, X., Bordes, A., and Bengio, Y.: Domain adaptation for large-scale sentiment classification: A deep learning approach. In: Proceedings of the 28th International Conference on Machine Learning (ICML-11), pp. 513–520 (2011)
Guzman, J., Stern, S.: Where is silicon valley? Science **347**(6222), 606–609 (2015)
Guzman, J., Stern, S.: Nowcasting and placecasting entrepreneurial quality and performance. In: Haltiwanger, J., Hurst, E., Miranda, J., Schoar, A. (eds.) Measuring Entrepreneurial Businesses: Current Knowledge and Challenges, Chapter 2. University of Chicago Press (2017)
Hagedoorn, J., Schakenraad, J.: A comparison of private and subsidized r&d partnerships in the European information technology industry. JCMS J. Common Market Stud. **31**(3), 373–390 (1993)
Hain, D.S., Jurowetzki, R.: The potentials of machine learning and big data in entrepreneurship research-the liaison of econometrics and data science. In: Cowling, M., Saridakis, G. (eds.) Handbook of Quantitative Research Methods in Entrepreneurship. Edward Elgar Publishing (2020)
Hain, D.S., Jurowetzki, R., Buchmann, T., Wolf, P.: A text-embedding-based approach to measuring patent-to-patent technological similarity. Technol. Forecast. Soc. Change **177**, 121559 (2022). https://doi.org/10.1016/j.techfore.2022.121559
Hall, B.H., Harhoff, D.: Recent research on the economics of patents. Annu. Rev. Econ. **4**(1), 541–565 (2012)
Harhoff, D., Scherer, F.M., Vopel, K.: Citations, family size, opposition and the value of patent rights. Res. Policy **32**(8), 1343–1363 (2003)
Hinton, G.E., Salakhutdinov, R.R.: Reducing the dimensionality of data with neural networks. Science **313**(5786), 504–507 (2006)
Hirschey, M., Richardson, V.J.: Are scientific indicators of patent quality useful to investors? J. Empir. Financ. **11**(1), 91–107 (2004)
LeCun, Y., Bengio, Y., Hinton, G.: Deep learning. Nature **521**(7553), 436 (2015)
Lerner, J.: The importance of patent scope: an empirical analysis. RAND J. Econ. 319–333 (1994)
McAfee, A., Brynjolfsson, E., Davenport, T.H., et al.: Big data: the management revolution. Harv. Bus. Rev. **90**(10), 60–68 (2012)
McCulloch, W.S., Pitts, W.: A logical calculus of the ideas immanent in nervous activity. Bull. Math. Biophys. **5**(4), 115–133 (1943)
Mullainathan, S., Spiess, J.: Machine learning: an applied econometric approach. J. Econ. Perspect. **31**(2), 87–106 (2017)
Narin, F., Hamilton, K.S., Olivastro, D.: The increasing linkage between US technology and public science. Res. Policy **26**(3), 317–330 (1997). https://doi.org/10.1016/S0048-7333(97)00013-9
Ng, A.Y.: Feature selection, l 1 vs. l 2 regularization, and rotational invariance. In: Proceedings of the Twenty-first International Conference on Machine Learning, pp. 78. ACM (2004)
Perlich, C., Provost, F., Simonoff, J.S.: Tree induction vs. logistic regression: a learning-curve analysis. J. Mach. Learn. Res. **4**(Jun), 211–255 (2003)
Pillonetto, G., Dinuzzo, F., Chen, T., De Nicolao, G., Ljung, L.: Kernel methods in system identification, machine learning and function estimation: a survey. Automatica **50**(3), 657–682 (2014). cited By 115

Ribeiro, M.T., Singh, S., Guestrin, C.: Why should i trust you?: Explaining the predictions of any classifier. In: Proceedings of the 22nd ACM SIGKDD International Conference on Knowledge Discovery and Data Mining, pp. 1135–1144. ACM (2016)
Rosenblatt, F.: The perceptron: a probabilistic model for information storage and organization in the brain. Psychol. Rev. **65**(6), 386 (1958)
Sakurada, M., Yairi, T.: Anomaly detection using autoencoders with nonlinear dimensionality reduction. In: Proceedings of the MLSDA 2014 2nd Workshop on Machine Learning for Sensory Data Analysis, pp. 4. ACM (2014)
Sedhain, S., Menon, A.K., Sanner, S., Xie, L.: Autorec: Autoencoders meet collaborative filtering. In: Proceedings of the 24th International Conference on World Wide Web, pp. 111–112. ACM (2015)
Shane, S.: Technological opportunities and new firm creation. Manag. Sci. **47**(2), 205–220 (2001)
Shyu, M.-L., Chen, S.-C., Sarinnapakorn, K., Chang, L.: A novel anomaly detection scheme based on principal component classifier. Technical report, Miami University Coral Gables FL Department of Electrical and Computer Engineering (2003)
Squicciarini, M., Dernis, H., Criscuolo, C.: Measuring patent quality, Indicators of technological and economic value (2013)
Strobl, C., Malley, J., Tutz, G.: An introduction to recursive partitioning: rationale, application, and characteristics of classification and regression trees, bagging, and random forests. Psychol. Methods **14**(4), 323 (2009)
Sutskever, I., Vinyals, O., Le, Q.V.: Sequence to sequence learning with neural networks. In: Advances in Neural Information Processing Systems, pp. 3104–3112 (2014)
Taleb, N.: The black swan: The impact of the highly improbable. Random House Trade Paperbacks (2010)
Therneau, T.M., Atkinson, E.J., et al.: An introduction to recursive partitioning using the rpart routines (1997)
Tian, F., Gao, B., Cui, Q., Chen, E., Liu, T.-Y.: Learning deep representations for graph clustering. In: AAAI, pp. 1293–1299 (2014)
Trajtenberg, M., Henderson, R., Jaffe, A.: University versus corporate patents: a window on the basicness of invention. Econ. Innov. New Technol. **5**(1), 19–50 (1997)
van der Vegt, G.S., Essens, P., Wahlström, M., George, G.: Managing risk and resilience. Acad. Manag. J. **58**(4), 971–980 (2015)
Varian, H.R.: Big data: New tricks for econometrics. J. Econ. Perspect. **28**(2), 3–27 (2014)
Vincent, P., Larochelle, H., Lajoie, I., Bengio, Y., Manzagol, P.-A.: Stacked denoising autoencoders: learning useful representations in a deep network with a local denoising criterion. J. Mach. Learn. Res. **11**(Dec), 3371–3408 (2010)
Wainwright, M.: Structured regularizers for high-dimensional problems: Statistical and computational issues. Ann. Rev. Stat. Appl. **1**, 233–253 (2014). cited By 24
Wang, Y.: A multinomial logistic regression modeling approach for anomaly intrusion detection. Comput. Secur. **24**(8), 662–674 (2005)
Zhou, M., Lang, S.-D.: Mining frequency content of network traffic for intrusion detection. In: Proceedings of the IASTED International Conference on Communication, Network, and Information Security, pp. 101–107 (2003)

Logical Aspects of Quantum Structures

J. Harding and Z. Wang

Abstract We survey several problems related to logical aspects of quantum structures. In particular, we consider problems related to completions, decidability and axiomatizability, and embedding problems. The historical development is described, as well as recent progress and some suggested paths forward.

1 Introduction

This note takes an overview of several problems related to logical aspects of quantum structures. The quantum structures we consider are motivated by the ortholattice $\mathscr{P}(\mathscr{H})$ of projection operators of a Hilbert space $\mathscr{H}$. These include orthomodular lattices and orthomodular posets on the more general end of the spectrum, as well as more specialized structures such as projection lattices of finite-dimensional Hilbert spaces and ortholattices of projections of von Neumann algebras.

The problems we consider are largely based on our personal experience and interests, and represent only a fragment of the the subject. Some have a long history, including the completion problem and word problem for orthomodular lattices. The intension is to provide a survey of old results, more recent developments, and some

J. Harding: Partially supported by US Army grant W911NF-21-1-0247. Z. Wang: Partially supported by NSF grants FRG-1664351, CCF 2006463, and DOD MURI grant ARO W911NF-19-S-0008.

J. Harding (✉)
New Mexico State University, Las Cruces, NM 88003, USA
e-mail: hardingj@nmsu.edu

Z. Wang
Microsoft Station Q and Department of Mathematics, University of California at Santa Barbara, Santa Barbara, CA 93106, USA
e-mail: zhenghwa@microsoft.com

N. Ngoc Thach et al. (eds.), *Financial Econometrics: Bayesian Analysis, Quantum Uncertainty, and Related Topics*, Studies in Systems, Decision and Control 427,
https://doi.org/10.1007/978-3-030-98689-6_6

open problems. There are a few novel contributions here, but the intent is to provide easy access to areas we feel are deserving of further attention. For further reading on some of the topics discussed here, and related topics, the reader may consult Dunn et al. (2013), Fritz (2021) and Herrmann (2010).

The second section provides a brief background and our perspective on quantum structures. The third section discusses completions, the fourth section deals with matters related to decidability and axiomatizability, and the fifth section discusses embedding problems.

2 Background

In their 1936 paper Birkhoff and von Neumann (1936), it was noted that the closed subspaces $\mathscr{C}(\mathscr{H})$ of a Hilbert space $\mathscr{H}$ form a lattice with an additional unary operation $\perp$ where $A^{\perp}$ is the closed subspace of vectors orthogonal to those in A. They proposed that this lattice serve as a type of non-distributive "logic" for a calculus of propositions involving quantum mechanical events. Husimi (1937) noted that $\mathscr{C}(\mathscr{H})$ satisfies the identity $A \subseteq B \Rightarrow B = A \vee (A^{\perp} \wedge B)$ now known as the *orthomodular law*. This led to the study of orthomodular lattices (see Kalmbach 1983).

Definition 1 An *ortholattice* (abbrev.: OL) $(L, \wedge, \vee, 0, 1, ')$ is a bounded lattice with a period two order-inverting unary operation $'$ so that $x \wedge x' = 0$ and $x \vee x' = 1$. An *orthomodular lattice* (abbrev.: OML) is an OL satisfying $x \leq y \Rightarrow y = x \vee (x' \wedge y)$.

There were other reasons for the interest of Birkhoff and von Neumann in the lattice $\mathscr{C}(\mathscr{H})$. Birkhoff (1935) and Menger (1936) had recently developed the lattice-theoretic view of projective geometry. This spurred von Neumann's development of his "continuous geometry" von Neumann (1960) which occurred in parallel to his work with (Murray and Von Neumann 1936, 1937, 1943; Von Neumann 1940) on "rings of operators", subalgebras of the algebra $\mathscr{B}(\mathscr{H})$ of bounded operators on $\mathscr{H}$ that are closed under the weak operator topology. In modern terms, these rings of operators are known as *von Neumann algebras*. Due to the bijective correspondence between closed subspaces of $\mathscr{H}$ and self-adjoint projection operators of $\mathscr{H}$, the OML $\mathscr{C}(\mathscr{H})$ is isomorphic to the OML $\mathscr{P}(\mathscr{H})$ of self-adjoint projection operators of $\mathscr{H}$. In fact, for any von Neumann algebra $\mathscr{A}$, its self-adjoint projections $\mathscr{P}(\mathscr{A})$ form an OML that comes remarkably close to determining the structure of $\mathscr{A}$ (see Dye (1955) and Hamhalter (2003)).

It is difficult to piece together the full motivation of von Neumann during this period of amazing activity. There are his published papers and notes in von Neumann (1962), accounts of Halperin of various issues in places such as his foreword to von Neumann (1960), and work of Redei and Stoltzner (2001). It is fair to say that von Neumann had a mix of logical, geometric, as well as probabilistic and measure-theoretic motivations that were never completely implemented due to the onset of the second world war.

The view that the lattice $\mathscr{C}(\mathscr{H})$, or its incarnation as projections $\mathscr{P}(\mathscr{H})$, plays a key role in quantum theory seems to have been completely born out by subsequent events. Gleason showed that the states of a quantum system modeled by $\mathscr{H}$ correspond to σ-additive measures on $\mathscr{P}(\mathscr{H})$; Ulhorn's formulation of Wigner's theorem characterizes projective classes of unitary and anti-unitary operators on $\mathscr{H}$ as automorphisms of $\mathscr{P}(\mathscr{H})$; and of course the spectral theorem describes self-adjoint operators on $\mathscr{H}$ as σ-additive homomorphisms from the Borel sets of the reals to $\mathscr{P}(\mathscr{H})$. See Dvurečenskij (1983) and Uhlhorn (1963).

In a different vein, Mackey (1963) took the task of motivating the structure $\mathscr{P}(\mathscr{H})$ from simple physically meaningful assumptions. He began with abstract sets $\mathscr{O}$ of observables, and $\mathscr{S}$ of states, and used $\mathscr{B}$ for the Borel subsets of the reals. He assumes a function $p : \mathscr{O} \times \mathscr{S} \times \mathscr{B} \to [0, 1]$ where $p(A, \alpha, E)$ is the probability that a measurement of observable A when the system is in state α yields a result in the Borel set E. He defines the set $\mathscr{Q}$ of *questions* to be those observables A that take only two values 0,1, i.e. with $p(A, \alpha, \{0, 1\}) = 1$ for each state α. From minimal assumptions, he shows that $\mathscr{Q}$ has the structure of what we now call an orthomodular poset (see below). He then makes the quantum leap to require that the orthomodular poset associated to a quantum system is the projection lattice $\mathscr{P}(\mathscr{H})$ of a Hilbert space.

Definition 2 An *orthocomplemented poset* (abbrev.: OP) $(P, \leq, ', 0, 1)$ is a bounded poset with a period two order-inverting operation $'$ so that $x \wedge x' = 0$ and $x \vee x' = 1$. For elements $x, y \in P$, we say that x, y are *orthogonal* and write $x \perp y$, if $x \leq y'$.

In this definition we use the meet and join symbols to indicate that the elements have a meet or join, and to also express to what this meet or join is equal.

Definition 3 An *orthomodular poset* (abbrev.: OMP) is an OP where every pair of orthogonal elements have a join, and satisfies $x \leq y \Rightarrow y = x \vee (x \vee y')'$. An OMP is called σ-complete if every countable set of pairwise orthogonal elements has a join.

The area of study that uses the lens of projection lattices, or more generally OMLs and OMPs, to motivate and study quantum foundations is often known as "quantum logic". Varadarajan (1985) uses "geometric quantum theory" for this study. There is also a notion of "quantum logic" much more closely aligned to traditional algebraic logic, based on the idea of replacing the Boolean algebras used in classical logic with some particular type of quantum structure such as projection lattices $\mathscr{P}(\mathscr{H})$, or general OMLs. There are different flavors of this, see for example Weaver (2001).

3 The Completion Problem

We group our first set of problems under the banner of "the completion problem". This consists of several different problems with an obviously similar theme. While they appear quite similar, there may turn out to be substant differences in detail.

Problem 1 Can every OML be embedded into a complete OML?

Problem 2 Can every OML/OMP be embedded into a σ-complete OML/OMP?

In considering an OMP or OML as a model for the events of a quantum system, it is natural to consider σ-completeness. This is directly interpreted as providing an event comprised from a countable collection of mutually exclusive events, and makes analysis using conventional techniques from probability theory tractable. Completeness implies σ-completeness, so serves the same purpose, but it has less physical motivation. Indeed, it is difficult to motivate even the existence of binary joins and meets of non-compatible events. But completeness is used in logical applications where it provides a means to treat quantifiers, so is also of interest for this reason. It is known that there are OMPs that cannot be embedded into an OML (see Sect. 5), so these questions may have different content. We begin with the following result of MacLaren (1964).

Theorem 1 *The MacNeille completion of an* OL *naturally forms an* OL.

Several core results about completions are consequences of deep early results from the study of OMLs. For the first, recall that the orthogonal of a subspace S of an inner product space V is given by $S^{\perp} = \{v : s \cdot v = 0 \text{ for all } s \in S\}$, and that S is biorthogonal if $S = S^{\perp\perp}$. Amemiya and Araki (1966) provided the following influential result viewed in part as justification of the use of Hilbert space in quantum mechanics.

Theorem 2 *The* OL *of biorthogonal subspaces of an inner product space is orthomodular iff the inner product space is complete, i.e. is a Hilbert space.*

Corollary 1 *The MacNeille completion of an* OML *need not be an* OML.

Proof Let L be the OL of finite and co-finite dimensional subspaces of an incomplete inner product space V. This is a modular OL, hence an OML. Since L is join and meet dense in the OL $\overline{L}$ of biorthogonal subspaces of V and $\overline{L}$ is complete, we have that $\overline{L}$ is the MacNeille completion of L.

Remark 1 The OML L of Corollary 1 can be embedded into a complete OML. Embed V into a complete inner product space $\overline{V}$. Then L embeds into the subalgebra of finite or co-finite dimensional subspaces of $\overline{V}$, and this is a subalgebra of the OML of all closed subspaces of $\overline{V}$.

A deep result of Kaplansky (1955) settles negatively the situation for modular ortholattices (abbrev.: MOLs). We recall that every MOL is an OML but not conversely. For space, we will not provide the details of continuous geometries, the reader can see von Neumann (1960). The crucial fact we use is that a continuous geometry L has a dimension function $d : L \to [0, 1]$ that satisfies among other conditions $d(x \vee y) = d(x) + d(y) - d(x \wedge y)$.

Theorem 3 *A complete, directly irreducible* MOL *is a continuous geometry.*

Corollary 2 *There is a* MOL *that cannot be embedded into a complete* MOL.

Proof Let L be the OL of all subspaces A of a Hilbert space where A or $A^{\perp}$ is finite-dimensional. It is well-known that lattice operations with such subspaces are obtained via intersection and sum of subspaces, so this is a MOL. The atoms of L are 1-dimensional subspaces. It is easily seen that any two atoms have a common complement, i.e. are perspective. So there is an infinite set of pairwise orthogonal pairwise perspective elements in L. But a continuous geometry cannot have such a set of elements since they would all have the same non-zero dimension because of being perspective.

So MOLs do not admit completions, and for OMLs the MacNeille completion does not always remain within OML. To gain a better understanding of the situation for OMLs, we can further limit expectations.

Definition 4 An order embedding of posets $\varphi : P \to Q$ is *regular* if it preserves all existing joins and meets.

The following result is found in Harding (1998), and is based on a result of Palko (1993). The proof extends in an obvious way to other situations. For instance, it shows that there is an OMP that cannot be regularly embedded into a σ-complete OMP.

Theorem 4 *A regular completion of an* OML *factors as a pair of regular embeddings through the MacNeille completion.*

Corollary 3 *There is no regular completion for* OMLs.

Remark 2 The examples discussed so far have all involved the OML of subspaces of a Hilbert space. There is an alternate source of examples of interest for the completion problem. A construction of Kalmbach (1983) builds from a bounded poset P an OMP $K(P)$. In the case that P is a lattice, $K(P)$ is an OML. This construction works by *gluing* the free Boolean extensions of chains of P. The forgetful functor $\mathsf{U} : \mathsf{Omp} \to \mathsf{Pos}$ takes the category of orthomodular posets and maps that preserve orthocomplementation and finite orthogonal joins to the category of bounded posets and order-preserving maps. Harding (2004) showed that the Kalmbach construction provides an adjoint $\mathsf{K} : \mathsf{Pos} \to \mathsf{Omp}$ to the forgetful functor and Jenča (2015) showed that effect algebras are the Eilenberg-Moore category over this monad.

There is a simple condition on a lattice L equivalent to the MacNeille completion of $K(L)$ being an OML. This provides a rich source of relatively transparent examples. For any lattice completion C of L, we have that the MacNeille completion of $K(C)$ is an OML, and this provides a completion of $K(L)$. Again, as in Remark 1, a completion of an OML is obtained by first completing some underlying structure.

We turn our attention to a method of completing OLs that is far from regular in that it destroys all existing joins except those that are essentially finite. This is termed the canonical completion. The canonical completion has a long history originating with completing Boolean algebras with operators. Here, the canonical completion of a Boolean algebra B is realized as the embedding into the power set of the Stone space of B. For its application to OLs, see Gehrke and Harding (2001) and Harding (1998).

Definition 5 A canonical completion of an OL L is an embedding of L into a complete OL L^σ such that every element of L^σ is a join of meets of elements of L and for each $S, T \subseteq L$, if $\bigwedge S \leq \bigvee T$, then there are finite $S' \subseteq S$ and $T' \subseteq T$ with $\bigwedge S' \leq \bigvee T'$.

Theorem 5 *Every* OL *has a canonical completion, and this is unique up to unique commuting isomorphism.*

Canonical completions are better at preserving algebraic properties than MacNeille completions. In Gehrke et al. (2006) it is shown that any variety that is closed under MacNeille completions is closed under canonical completions. However, canonical completions do not provide the answers we seek here.

Theorem 6 *The canonical completion of an* MOL *need not be an* MOL *and the canonical completion of an* OML *need not be an* OML.

Proof The first statement follows from Corollary 2. The second is established in Harding (1998) where necessary and sufficient conditions are given for the canonical completion of $K(L)$ to be an OML.

We have occasion now to discuss matters related to states. We recall that a *state* on an OMP P is a map $\sigma : P \to [0, 1]$ that preserves bounds and is *finitely additive* meaning that $x \perp y$ implies $\sigma(x \vee y) = \sigma(x) + \sigma(y)$. *Pure states* are ones that cannot be obtained as a non-trivial convex combination of others.

Remark 3 The proof of Theorem 6 gives more. Each $K(L)$ has a full set of 2-valued states, i.e. states taking only values 0,1, hence is what is known as a *concrete* OML. The concrete OMLs form a variety, and the above results show that this variety is not closed under MacNeille completions or canonical completions.

Having given a number of negative results, we mention a direction that produces strong positive results in a physically motivated setting. The first result in this area was by Bugajska and Bugajski (1973). Here we follow a sequel to this result by Guz (1978) which we reformulate below.

Theorem 7 *Let P be an* OMP *and $\mathscr{S}$ be its set of pure states. Suppose P satisfies*

1. *For each non-zero $x \in P$ there is $\sigma \in \mathscr{S}$ with $\sigma(x) = 1$*
2. *If $x \not\leq y$ then there is $\sigma \in \mathscr{S}$ with $\sigma(x) = 1$ and $\sigma(y) \neq 1$*
3. *For each $\sigma \in \mathscr{S}$ there is $x \in P$ with $\sigma(x) = 1$ and $\sigma'(x) \neq 1$ for all $\sigma \neq \sigma' \in \mathscr{S}$.*

Then P is atomistic and its MacNeille completion is an OML.

Remark 4 Theorem 7 is intended to provide simple physical assumptions on the events of a quantum system that ensure it can be embedded into a complete OML. The assumptions here have physical interpretation. The first says that each event is certain in some pure state. The second says that $x \leq y$ iff whenever x is certain in some state, then so also is y. Guz describes the third as saying that for any pure state, there is an event that can be used to test for it. The line of reasoning is certainly of interest, but the axioms chosen are not without issue. The third axiom in particular is quite strong. For instance, the axioms given will not hold in a non-atomic Boolean algebra. It is of interest to see if less restrictive conditions on the set of states provides similar results.

For the remainder of this section, we discuss some possible directions for approaches to completion problems. These are speculative, and may not turn out to be of use. But they seem worthy of further consideration. We begin with the following result of Bruns and Harding (1998) called the Boolean amalgamation property.

Theorem 8 *Let L_1, L_2 be* OML*s that intersect in a common Boolean subalgebra. Then there is an* OML *L containing L_1, L_2 as subalgebras.*

Remark 5 As with any lattice, an OML L is complete iff each chain in L has a join. Since a chain in an OML is contained in a *block* (a maximal Boolean subalgebra), an OML is complete iff all of its blocks are complete. The Boolean amalgamation property lets us complete a given block B of an OML L. Taking a completion C of B, we have that L and C intersect in common Boolean subalgebra B, so there is an OML M containing L and C as subalgebras. In effect, B has been completed within M.

One could hope to iterate this process to obtain a σ-completion of L. Unfortunately, such an iterative approach requires that we preserve joins completed at an early stage, and the Boolean amalgamation described in Theorem 8 does not do this. Perhaps there is a modification of the proof of Theorem 8 that does allow this. On the other hand, Theorem 8 has implications in terms of constructing an example of an OML that cannot be completed to an OML. Roughly, one particular join can always be inserted.

Our next topic, due to Rump (2018), is a translation of orthomodular lattices into structures belonging to classical algebra. In Foulis et al. (1998) Foulis, Greechie and Bennett associated to an OML L its *unigroup*. This unigroup $G(L)$ is a partially ordered abelian group with strong order unit u and map μ_L from L into the interval $[0, u]$ of G. The map $\mu_L : L \to G(L)$ is universal among abelian group-valued measures, meaning that for any abelian group-valued measure $\mu : L \to G$ there is a group homomorphism $f_\mu : G(L) \to G$ with $f \circ \mu_L = \mu$. However, μ_L need not be an embedding. Rump repairs this defect by extending to the setting of non-commutative groups.

Definition 6 A *right ℓ-group* is a group G equipped with a lattice structure $\leq$ such that $a \leq b \Rightarrow ac \leq bc$. An element u is a *strong order unit* if $b \leq c \Leftrightarrow ub \leq uc$ for all $b, c \in G$ and each $a \in G$ satisfies $a \leq u^n$ for some natural number n. A strong order unit u is called *singular* if $u^{-1} \leq xy \Rightarrow yx = x \wedge y$.

Rump called a right ℓ-group with singular strong order unit an *orthomodular group*. Using a construction that extends in Foulis (1960) construction of a Baer *-semigroup from an OML, he provided a method to associate to any OML L an orthomodular group $G_R(L)$, called the structure group of L. We denote this $G_R(L)$ to distinguish it from the unigroup of L. However, it has the same universal property for group-valued measures as the unigroup has for abelian group-valued measures. Moreover, he shows the following (Rump 2018, Theorem 4.10).

Theorem 9 *If G is an orthomodular group with singular strong order unit s, then the interval $[s^{-1}, 1]$ is an* OML *with structure group G. Conversely, each* OML *arises this way.*

As is usual with ordered groups, we say an orthomodular group is complete iff every bounded subset has a least upper bound. This property is often known as Dedekind complete. Rump showed that an OML is complete iff its structure group is complete. This leads to the question of whether techniques from the theory of ℓ-groups can be used to find a completion for OMLs. Of course, the results of Foulis on Baer *-semigroups allowed a similar algebraic path for many decades without result, but there has been much progress in the study of ℓ-groups.

4 Decidability and Axiomatizability

Given a class of algebras $\mathscr{K}$, several logical questions arise. One can ask if there is an algorithm to decide the equational theory $\mathsf{Eq}(\mathscr{K})$ or first order theory $\mathsf{Th}(\mathscr{K})$, and if there is a finite set of equations, respectively first order formulas, that axiomatize these theories. When $\mathscr{K}$ consists of a single algebra A we write $\mathsf{Eq(A)}$ and $\mathsf{Th(A)}$. We begin with some standard terminology.

Definition 7 A variety V has *solvable free word problem* if there is an algorithm to determine if an equation holds in V.

Since V has the same equational theory as its free algebra $F_V(\omega)$ over countably many generators, solving the free word problem for V amounts to giving an algorithm to decide if an equation holds in its countably generated free algebra. Before discussing the situation for specific varieties, we describe a general technique.

Definition 8 A *partial subalgebra* of an algebra A is a subset $S \subseteq A$ equipped with partial operations being the restriction of the operations of A to those tuples in S where the result of the operation in A belongs to S. A variety V has the *finite embedding property* if each finite partial subalgebra of an algebra in V is a partial subalgebra of a finite algebra in V.

There are a number of connections between partial algebras and word problems, see for example Evans (1978) and Kharlampovich and Sapir (1995). The following is obtained via a back and forth argument of checking for a proof of an equation and looking for a counterexample among finite algebras.

Theorem 10 *If a variety V is finitely axiomatized and has the finite embedding property, then it has solvable free word problem.*

We begin with two varieties at the opposite ends of the spectrum of quantum structures, the variety BA of Boolean algebras and OL. As is often the case where there is a great deal of structure, or relatively little structure, we have solvable free word problems.

Theorem 11 *The free word problems in* BA *and in* OL *are solvable.*

In both cases the result follows easily from Theorem 10. The variety BA has the finite embedding property since every finitely generated BA is finite. The variety OL has the finite embedding property since MacLaren (1964) showed that the MacNeille completion of an orthocomplemented poset is an OL and MacNeille completions preserve all existing joins and meets. In both cases we can give a much more tractable algorithm to decide the free word problem than that provided by Theorem 10. Since BA is generated by the 2-element BA we need only check validity of an equation in 2, essentially the method of truth tables. For OL Bruns (1976) gave an explicit algorithm based on Whitman's algorithm for free lattices which is found in Freese et al. (1995).

Problem 3 Is the free word problem for the variety OML solvable?

This problem has received considerable attention over the years, but without a great deal of progress. Trying to establish the finite embedding property for OMLs was a motivation behind some of the work of Bruns and Greechie on commutators, and the idea of Kalmbach's attempt at a solution to the free word problem in Kalmbach (1986) that unfortunately seems to have a gap. There are other hopes to solve the free word problem for OML that proceed by finding some other free word problem whose solution would yield that of OML. This is the direction of the recent work of Fussner and John (2021) involving ortholattices where a derived operation is residuated.

We turn briefly to a more general discussion of free algebras in BA, OL, and OML. The following collects a number of known results. We recall, that for a cardinal κ, that MO_κ is the MOL of height 2 with a bottom, a top, and κ pairs of incomparable orthocomplemtary elements in the middle.

Theorem 12 *In* OL *the free algebra on 2 generators contains the free one on countably many generators as a subalgebra. In* OML *the free algebra on 2 generators is $MO_2 \times 2^4$ and the free algebra on 3 generators contains the free one on countably many generators as a subalgebra.*

The results about free OLs are due to Bruns (1976). The result about the free OML on 2 generators is due to Beran and is found in Kalmbach (1983). It has a significant impact on the study of OMLs since it makes calculations involving 2 variables tractable. The result about the free OML on 3 generators containing the countably generated free one is due to Harding (2002). The situation for OL is similar to that of lattices with very similar algorithms providing a solution to the free word problem. While there is an extensive literature on properties of free lattices (see Freese et al. (1995)), relatively little is known about the structure of free OLs.

Problem 4 Obtain a better understanding of the structure of free OLs and free OMLs. In particular, what are their finite subalgebras? In a free OL, if b is a complement of a, are $a' \vee b$ and $a' \wedge b$ also complements of a? Can a free OML contain an uncountable Boolean subalgebra?

The situation for MOLs also leaves a great deal open. The free MOL on 2 generators is also $MO_2 \times 2^4$ since this is modular and is free on 2 generators in OML. The free MOL on 3 generators is infinite. We are not aware of whether it contains the free one on countably many generators as a subalgebra. Roddy has provided a finitely presented MOL with unsolvable word problem in Roddy (1989). Bruns (1983) shows that every finite subdirectly irreducible MOL is in the variety generated by MO_ω, hence satisfies the 2-distributive law. Thus MOL is not generated by its finite members and so cannot have the finite embedding property. This leaves the following open problem.

Problem 5 Does MOL have solvable free word problem?

Dunn, Moss, and Wang in their "Third life of quantum logic" Dunn et al. (2013) pointed to the value of studying free word problems for the algebras most tightly tied to quantum computing. They used $\mathsf{QL}(\mathbb{C}^n)$ for the equational theory of the MOL of closed subspaces $\mathscr{C}(\mathbb{C}^n)$, and they called this the *quantum logic* of $\mathbb{C}^n$. We extend this practice and use $\mathsf{QL}(\mathscr{R})$ for the equational theory of the projection lattice of an arbitrary type II_1 factor $\mathscr{R}$, and $\mathsf{QL}(\mathsf{CG}(\mathbb{C}))$ for the equational theory of the orthocomplemented continuous geometry $CG(\mathbb{C})$ constructed by von Neumann (1960) via a metric completion of a limit of subspace lattices. We summarize results obtained in Dunn et al. (2005), Harding (2013) and Herrmann (2010) below. In particular, the note of Herrmann (2019) provides an excellent description.

Theorem 13 $\mathsf{QL}(\mathbb{C}) \supset \mathsf{QL}(\mathbb{C}^2) \supset \cdots \supset \bigcap\{\mathsf{QL}(\mathbb{C}^n) : n \geq 1\} = \mathsf{QL}(\mathsf{CG}(\mathbb{C})) = \mathsf{QL}(\mathscr{R})$ *for each type II_1 factor $\mathscr{R}$. Each of these containments is strict. Each of these equational theories is decidable, and the first order theory of each* MOL $\mathscr{C}(\mathbb{C}^n)$ *for $n \geq 1$ is decidable.*

The containments among the $\mathsf{QL}(\mathbb{C}^n)$ and $\bigcap\{\mathsf{QL}(\mathbb{C}^n) : n \geq 1\}$ are trivial. That they are strict follows from the fact given in Huhn (1972) that $\mathscr{C}(\mathbb{C}^k)$ is n-distributive iff $k \leq n$. The two equalities are established in Herrmann (2010). Decidability of the first order theory of each $\mathscr{C}(\mathbb{C}^n)$ is given in Dunn et al. (2005) by translating formulas of the MOL to formulas about $\mathbb{C}$, and using a theorem of Tarski (1948) on the decidability of the first order theory of $\mathbb{C}$. This has as a consequence the decidability of the equational theories $\mathsf{QL}(\mathbb{C}^n)$ for $n \geq 1$. The decidability of the equational theories in the remaining cases was established in Herrmann (2010), and independently for $\mathsf{QL}(\mathsf{CG}(\mathbb{C}))$ in Harding (2013).

Definition 9 A *quasi-equation* is a formula $\forall x_1 \ldots \forall x_n (s_1 = t_1 \,\&\, \cdots \,\&\, s_k = t_k \rightarrow s = t)$ where $s_1 = t_1, \ldots, s_k = t_k$ and $s = t$ are equations. The *uniform word problem* for a variety V asks if there is an algorithm that determines which quasi-equations are valid in V.

The uniform word problem asks if there is a single algorithm that decides when two words are equal for a finitely presented algebra in V. Of course, each equation is equivalent to a quasi-equation, so it is (much) more difficult to have a positive solution to the uniform word problem than to have a positive solution to the free word problem.

Theorem 14 MOL *has unsolvable uniform word problem, as does the variety generated by the projection lattice of any type II_1 factor, and the variety generated by $CG(\mathbb{C})$.*

The first statement was shown by Roddy, who gave a finitely presented MOL with unsolvable word problem. The second statement is in Herrmann (2010).

Proposition 1 *The first order theory of $\mathscr{C}(\mathbb{C}^n)$ is finitely axiomatizable iff $n = 1$. The first order theory of the closed subspaces $\mathscr{C}(\mathscr{H})$ of an infinite-dimensional Hilbert space is not finitely axiomatizable.*

Since $\mathscr{C}(\mathbb{C})$ is the 2-element Boolean algebra, the case $n = 1$ is trivial. For $n \geq 3$ one can recover the field $\mathbb{C}$ from the MOL $\mathscr{C}(\mathbb{C}^n)$ by the standard lattice-theoretic treatment of the usual techniques from projective geometry (see e.g. Crawley and Dilworth 1973; Faure and Frölicher 2000). Since this process is first order, a finite axiomatization of the first order theory of $\mathscr{C}(\mathbb{C}^n)$ would give a finite axiomatization of the first order theory of the field $\mathbb{C}$. This is not possible since any sentence true in $\mathbb{C}$ is true in algebraically closed fields of sufficiently large prime characteristic (see e.g. Marker 1996, p. 2). For further discussion of the case $n = 3$, see Herrmann (2019).

For the case $n = 2$, note that $\mathscr{C}(\mathbb{C}^2)$ is $\mathrm{MO}_{\mathfrak{c}}$ where $\mathfrak{c}$ is the continuum. There is a non-principle ultraproduct of the MO_n for $n \in \mathbb{N}$ of cardinality $\mathfrak{c}$ (see Frayne et al. (1962)), and by Łoś's Theorem this ultrapower must be $\mathrm{MO}_{\mathfrak{c}}$. If $\mathsf{Th}(\mathrm{MO}_{\mathfrak{c}})$ can be finitely axiomatized, then it can be axiomatized by a single sentence φ. But then each $\mathrm{MO}_n \models \neg\varphi$ and by Łoś's theorem, the ultraproduct $\mathrm{MO}_{\mathfrak{c}} \models \neg\varphi$, an impossibility.

If $\mathscr{H}$ is infinite-dimensional, then $\mathscr{C}(\mathscr{H})$ has an element p of height 3, and $[0, p]$ is isomorphic to $\mathscr{C}(\mathbb{C}^3)$. The result follows from the previous ones.

Remark 6 At this point, there are many questions related to this line of investigation. Several are raised in Herrmann (2019). We resist the temptation to formulate more problems related to the current material, but we will post a further problem raised in Dunn et al. (2013).

Problem 6 Is the equational theory of the OML $\mathscr{C}(\mathscr{H})$ decidable?

This problem is one way to address what has been a primary issue since the early days of quantum logic, understanding more deeply the OL $\mathscr{C}(\mathscr{H})$. Birkhoff and von

Neumann (1936) knew that $\mathscr{C}(\mathscr{H})$ did not belong to MOL. Husimi (1937) formulated the orthomodular law that separated the equational theory of $\mathscr{C}(\mathscr{H})$ from that of OL. Day introduced the "ortho-Arguesian law", an equation in six variables related to the Arguesian condition of projective geometry. He showed that this equation is valid in $\mathscr{C}(\mathscr{H})$ and not in OML. Many refinements of this ortho-Arguesian identity have been found (Megill and Mladen Pavičić 2010; Megill 2014) providing other equations valid in $\mathscr{C}(\mathscr{H})$ and not in OML. A further source of equations valid in $\mathscr{C}(\mathscr{H})$ and not in OML is provided by the fact that $\mathscr{C}(\mathscr{H})$ has an ample supply of well-behaved states (Godowski and Greechie 1981; Mayet 2006, 2007).

In a different direction, Fritz (2021) used results of Slofstra (2020) from combinatorial group theory to establish the following.

Theorem 15 *For an infinite-dimensional Hilbert space $\mathscr{H}$, the uniform word problem for the variety generated by $\mathscr{C}(\mathscr{H})$ is unsolvable.*

The result shown is actually quite a bit more specific than this, showing that there is no decision procedure for quasi-equations of a very specific form. These quasi-equations encode when certain configurations can be embedded into $\mathscr{C}(\mathscr{H})$, and a discussion of them naturally leads to the topic of our next section.

5 Embedding Problems

An embedding of one OML into another is a one-one OL homomorphism. Our first problem is formulated in an open-ended fashion, but accurately reflects the intent.

Problem 7 Increase our understanding of which OMLs can be embedded into $\mathscr{C}(\mathscr{H})$ for some Hilbert space $\mathscr{H}$.

Note that each $\mathscr{C}(\mathbb{C}^n)$ for $n \geq 1$ embeds into $\mathscr{C}(\mathscr{H})$ for $\mathscr{H}$ an infinite-dimensional separable Hilbert space. So our interest primarily lies in embeddings into $\mathscr{C}(\mathscr{H})$ when $\mathscr{H}$ is infinite-dimensional.

The discussion at the end of Sect. 4 gives several equations that are valid in $\mathscr{C}(\mathscr{H})$ but not in all OMLs. These include the ortho-Arguesian law and its variants, and also equations holding in all OMLs with a sufficient supply of certain types of states. Failure of any such equation in an OML implies that it cannot be embedded into $\mathscr{C}(\mathscr{H})$. Also, $\mathscr{C}(\mathscr{H})$ has a strongly order determining set of states, meaning that $a \leq b$ iff each finitely additive state s with $s(a) = 1$ has $s(b) = 1$. Any OML without a strongly order determining set of states cannot be embedded into $\mathscr{C}(\mathscr{H})$.

For the other side of the question, it seems that relatively little is known about methods to determine that a given OML can be embedded into $\mathscr{C}(\mathscr{H})$. The projections of a von Neumann algebra are an OML that can be embedded into $\mathscr{C}(\mathscr{H})$, so in a sense this provides a source of examples. There is also an example Bruns and Roddy (1992) of a MOL with interesting properties constructed as a subalgebra of $\mathscr{C}(\mathscr{H})$. This example is constructed by carefully choosing bases of infinite-dimensional subspaces of $\mathscr{C}(\ell^2)$ and using delicate arguments. Aside from relatively simple cases that can easily be seen to embed in $\mathbb{C}(\mathbb{C}^n)$ for some n, we are aware of few positive results in this direction. To illustrate the situation, consider the following.

Definition 10 The diagram below at left is the OML constructed as the *horizontal sum* of the Boolean algebras 2^3 and 2^2 and is written $2^3 \oplus 2^2$. The diagram in the middle consists of n copies of an 8-element Boolean algebra glued together at an atom, coatom, and 0,1 as shown. This is called an *n-element chain* because of its appearance when viewed as a Greechie diagram. The diagram at right is obtained from the one in the middle by identifying the two copies of a and a'. This is called an *n-loop*. This is an OMP when $n \geq 4$ and an OML when $n \geq 5$.

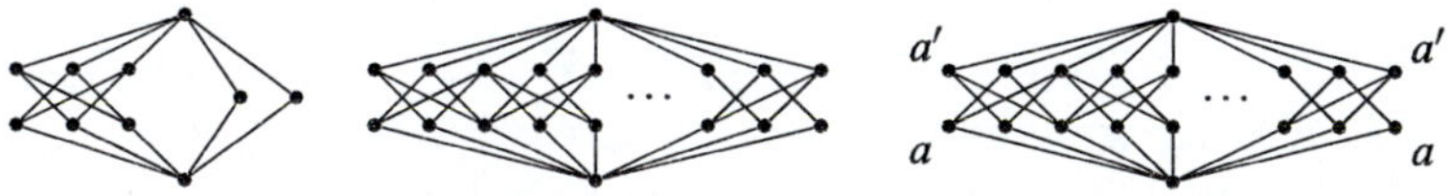

We are not aware if it is known when an n-chain or n-loop can be embedded as an OML into $\mathscr{C}(\mathscr{H})$. It would be desirable to have technology sufficient to answer such basic questions, even if it winds up being undecidable when a finite OML can be embedded into $\mathscr{C}(\mathscr{H})$. For the OML $2^3 \oplus 2^2$, it is remarked in Greechie (1969) that Ramsey had shown that it could be embedded into $\mathscr{C}(\mathscr{H})$, but the result is unpublished and we know of no proof in print. We add this below.

Proposition 2 *There is an embedding of the* OML $2^3 \oplus 2^2$ *into* $\mathscr{C}(\mathscr{H})$ *where* $\mathscr{H}$ *is a separable Hilbert space.*

Proof Let $\mathscr{H} = L^2(\mathbb{R})$ be the square integrable complex functions on $\mathbb{R}$ modulo equivalence a.e. and let $\mathscr{F}$ be the Fourier transform and $\mathscr{F}^{-1}$ the inverse Fourier transform. Let A, B, C be the closed subspaces of all functions vanishing a.e. on $(-\infty, -1)$, $(-1, 1)$, and $(1, \infty)$ respectively; and let D, E be the closed subspaces of all functions whose Fourier transforms vanish a.e. on $(-\infty, 0)$ and $(0, \infty)$ respectively. Then A, B, C are the atoms of an 8-element Boolean subalgebra of $\mathscr{C}(\mathscr{H})$, and D, E are the atoms of a 4-element Boolean subalgebra of $\mathscr{C}(\mathscr{H})$. To establish the result, it is sufficient to show that any of A', B', C' intersect with D, E trivially.

Suppose $f \in D$. Since the Fourier transform $\hat{f}$ of f vanishes a.e. on the negative reals, by Titchmarsh's theorem (Titchmarch 1948, Theorem 95) there is a holomorphic function F defined on the upper half-plane so that $f(x) = \lim_{z \to x} F(z)$ a.e. If f belongs to one of A', B', C', then it is zero a.e. on a set of positive measure, so by the Luzin-Privolov theorem (Lusin and Priwaloff 1925) is zero a.e. Thus D intersects each of A', B', C' trivially. The argument for $f \in E$ follows since for $g(x) = f(-x)$ we have $\hat{g}(\xi) = \hat{f}(-\xi)$.

Remark 7 The proof of Proposition 2 shows more. It shows that for B the Boolean algebra of Lebesgue measurable subsets of $\mathbb{R}$ modulo sets of measure zero, that $B \oplus 2^2$ is a subalgebra of $\mathscr{C}(\mathscr{H})$. Thus $2^n \oplus 2^2$ is a subalgebra of $\mathscr{C}(\mathscr{H})$ for each natural number n. One might hope that more is true, that for B as described, that $B \oplus B$ is a subalgebra of $\mathscr{C}(\mathscr{H})$. This may or may not be the case, but there is a difficulty in extending the proof in the obvious way. In Kargaev (1982) it is shown that there is a set $E \subseteq \mathbb{R}$ of positive finite measure so that the Fourier transform of its characteristic function vanishes on an interval.

This line of investigation illustrates the difference between having a OML embedding of an OML L into $\mathscr{C}(\mathscr{H})$ and having an OMP embedding of L into $\mathscr{C}(\mathscr{H})$. Indeed, any horizontal sum of Boolean algebras is *concrete* since it has a full set of 2-valued states, and so can easily be embedded as an OMP into a Boolean algebra. So it is trivial that $2^3 \oplus 2^2$ has an OMP embedding into $\mathscr{C}(\mathscr{H})$, as well as many other easy facts. This leads us to our next problem.

Problem 8 When can an OMP be embedded into $\mathscr{C}(\mathscr{H})$ for some Hilbert space $\mathscr{H}$?

While in many cases it is easy to embed an OMP into $\mathscr{C}(\mathscr{H})$, not all finite OMPs can be embedded into $\mathscr{C}(\mathscr{H})$ since there are finite OMPs without any states. The most notable work on determining which OMPs can be embedded into $\mathscr{C}(\mathscr{H})$ is from Fritz (2021) using work of Slofstra (2020) as we mentioned at the end of Sect. 4. We describe this in more detail.

Definition 11 Let M be an $m \times n$ matrix with coefficients in $\mathbb{Z}_2$ and b be a column vector of length m with coefficients in $\mathbb{Z}_2$. A *quantum solution* to a linear equation $Mx = b$ over $\mathbb{Z}_2$ is a sequence $A_1, \ldots, A_n$ of self-adjoint bounded operators of a Hilbert space $\mathscr{H}$ such that

1. $A_i^2 = 1$ for each $i \leq n$,
2. A_i and A_j commute if x_i and x_j both appear in some equation,
3. For each each equation $x_{k_1} + \cdots + x_{k_r} = b_r$ we have $A_{k_1} \cdots A_{k_r} = (-1)^{b_r} 1$.

In Slofstra (2020) it was shown that it is undecidable whether a given linear equation over $\mathbb{Z}_2$ has a quantum solution. Fritz (2021) translated this into a form that begins to resemble the problem of embedding an OMP into $\mathscr{C}(\mathscr{H})$ as we now explain.

Definition 12 A *hypergraph* is a set V of *vertices* and a collection $E \subseteq \mathrm{Pow}(V)$ of subsets of V called *edges* such that each vertex lies in at least one edge. A *quantum representation* of a hypergraph is a mapping ρ from the set of vertices to the projection operators of a Hilbert space $\mathscr{H}$ with $\dim(\mathscr{H}) > 0$ such that for any edge E we have $\sum_{v \in E} \rho(v) = 1$.

Remark 8 Note that the condition $\sum_{v \in E} \rho(v) = 1$ implies that $\rho(v)$ is orthogonal to $\rho(w)$ for $v \neq w$ belonging to a common edge. It is however useful to note that being orthogonal does not mean being distinct since 0 is orthogonal to itself.

Example 1 A hypergraph with 7 vertices and 3 edges is shown below.

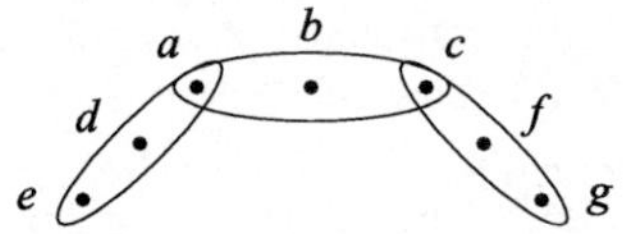

There are many quantum representations of this hypergraph. Let i, j, k be the standard basis vectors of $\mathbb{C}^3$. Using P_v for the projection onto the one-dimensional subspace spanned by the vector v, set

$$\rho(a) = P_i, \rho(b) = P_j, \rho(c) = P_k, \rho(d) = P_{j+k}, \rho(e) = P_{j-k}, \rho(f) = P_{i+j}, \rho(g) = P_{i-j}.$$

Another representation is obtained by setting

$$\rho(a) = P_i, \rho(b) = P_j, \rho(c) = P_k, \rho(d) = P_j, \rho(e) = P_k, \rho(f) = P_j, \rho(g) = P_i.$$

Of course, this second representation does not embed the vertex set into the projections, but this is not required. A further representation can be found even in $\mathbb{C}^n$ for any $n \geq 1$. Here we use 0 and 1 for projections onto the zero subspace and the whole space. Set $\rho(a) = \rho(c) = \rho(e) = \rho(g) = 0$ and $\rho(b) = \rho(d) = \rho(f) = 1$.

A principle contribution of Fritz (2021) is to provide a translation between quantum solutions of linear equations over $\mathbb{Z}_2$ and quantum representations of hypergraphs. The key result is the following (Fritz 2021, Lemma 10).

Theorem 16 *There is an algorithm to compute, for every linear system $Mx = b$, a finite hypergraph so that the quantum solutions of the linear system are in bijective correspondence with the representations of the hypergraph.*

The first idea behind the translation is that symmetries of a Hilbert space $\mathscr{H}$, i.e. bounded self-adjoint operators A with $A^2 = 1$, are in bijective correspondence with projections of $\mathscr{H}$. A symmetry A yields the projection $\frac{1}{2}(1 + A)$ and a projection P gives the symmetry $2P - 1$. In fact, a symmetry A has spectral decomposition $\frac{1}{2}(1 + A) - \frac{1}{2}(1 - A)$ so corresponds to an experiment with 2 outcomes. Commutivity of symmetries corresponds to that of their associated projections. So some aspects of the translation are relatively straightforward, but a clever construction is needed to match all the requirements. Combining this with Slofstra's result gives the following.

Corollary 4 *It is undecidable whether a finite hypergraph has a quantum representation.*

For a given hypergraph, it is a simple matter to encode the conditions for it to have a quantum representation as a conjunction of equations. For example, if v, w appear in a common edge then we require the equation $\rho(v) = \rho(v) \wedge \rho(w)'$ giving their orthogonality, and so forth. Then taking the conjunction of these equations to imply $0 = 1$ gives a quasi-equation that is valid in $\mathscr{C}(\mathscr{H})$ for all $\mathscr{H}$ of dimension greater than 0 iff the hypergraph has no quantum representation. This yields the following result that was stated at the end of the previous section.

Theorem 17 *For an infinite-dimensional Hilbert space $\mathscr{H}$, the uniform word problem for the variety generated by $\mathscr{C}(\mathscr{H})$ is unsolvable.*

Remark 9 There is a connection between OMPs and hypergraphs. For simplicity, we restrict discussion to finite OMPs, but many things hold in a wider setting. There are several ways to attach a hypergraph to a finite OMP P. One way takes all elements of P as vertices and uses as edges all pairwise orthogonal subsets of P with join 1. Another way of doing this, an extension of "Greechie diagrams" described in Kalmbach (1983) takes as vertices the atoms of P and takes as edges those pairwise orthogonal sets of atoms with join 1. The hypergraph shown in Example 1 is the 3-chain of Definition 10 considered as an OMP.

One easily gains a feeling that the work on quantum solutions of linear systems and representations of hypergraphs has implications for the problem of embedding finite OMPs into $\mathcal{C}(\mathcal{H})$. But the situation is not so clear. Is it decidable whether members of the very special class of hypergraphs that arise from finite OMPs are representable? On the other hand, representability of the hypergraph of an OMP does not imply that it is embedable, embedability is equivalent to the existence of an injective representation. Never-the-less, this seems an area worthy of further study.

Problem 9 When can an OMP be embedded into an OML?

To reiterate the setting, an embedding $f : P \to L$ of an OMP P into an OML L is a one-one map that preserves orthocomplementation and finite orthogonal joins. This implies that it preserves bounds and order. The following is given in Harding (1996).

Theorem 18 *The* OMP *with Greechie diagram below cannot be embedded into an* OML.

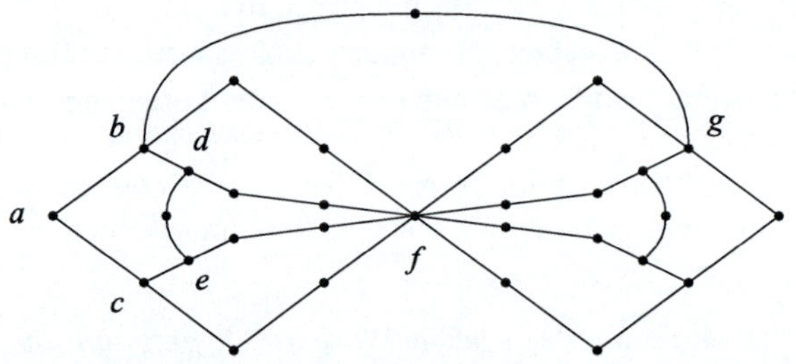

Note that this Greechie diagram has all edges with 3 elements.

Proof Suppose that this OMP is embedded into an OML L. We recall that the extended Foulis-Holland theorem Greechie (1977) says that if S is a subset of an OML and for any 3-element subset of S there is one element that commutes with the other two, then the sublattice generated by S is distributive. Using this and writing $+$ for join and (suppressed) multiplication for meet in L we have

$$
\begin{aligned}
f &\leq (a+b)(a+c)(b+d)(c+e) \\
&= [(a+b)(a+c)]\,[(b+a)(b+d)]\,[(c+a)(c+e)] \\
&= (a+bc)(b+ad)(c+ae) \\
&= [(a+bc)(b+ad)]\,[(a+bc)(c+ae)] \\
&= [ab+bc+ad+abcd]\,[ac+bc+ae+abce] \\
&= (bc+ad)(bc+ae) \\
&= bc+ade \\
&= bc
\end{aligned}
$$

From this, we have $f \leq b$, and by symmetry, $f \leq g$, and as b, g are orthogonal, $f = 0$.

We move to what at first seems an unrelated topic.

Definition 13 For a set X, an ordered pair (α, α') of equivalence relations on X is a *factor pair* if $\alpha \cap \alpha'$ is the diagonal relation Δ and the relational product $\alpha \circ \alpha'$ is the universal relation $\nabla = X^2$. Let Fact (X) be the set of all factor pairs of X.

The motivation behind factor pairs is that they encode the direct product decompositions of a set. Indeed, factor pairs are exactly the kernels of the projection operators associated to a direct product decomposition $X \simeq Y \times Z$. The following was established in Harding (1996).

Theorem 19 *Fact* (X) *is an* OMP *with* $0 = (\Delta, \nabla)$, $1 = (\nabla, \Delta)$; *orthocomplementation given by* $(\alpha, \alpha')^{\perp} = (\alpha', \alpha)$; *and* $(\alpha, \alpha') \leq (\beta, \beta')$ *iff* $\alpha \subseteq \beta$, $\beta' \subseteq \alpha'$, *and all equivalence relations involved permute.*

As was first observed by Chin and Tarski (1951), under special circumstances, a fragment of distributivity holds among equivalence relations (see also Harding 1996, Lemma 7.2). Using this, one can establish the following, where an embedding of one OMP into another is a one-one map that preserves orthocomplementation and finite orthogonal joins.

Theorem 20 *The* OMP *from Theorem 18 cannot be embedded into an* OMP *Fact* (X) *for any set* X.

The resemblance between Theorems 18 and 20 goes beyond their statements. Their proofs are nearly identical, using the fragment of distributivity that holds in relation algebras in place of the generalized Foulis-Holland theorem that provides a fragment of distributivity in OMLs. There are other finite OMPs that cannot be embedded into OMLs, and in each case, the proof of non-embedability into an OML transfers transparently to a proof of non-embeddability into an OMP Fact (X). This raises a number of issues, such as whether Fact (X) can be embedded into an OML, whether every OML can be embedded as an OMP into some Fact (X), and whether there are further fragments of distributivity in a relation algebra to mirror the situation with the generalized Foulis-Holland theorem.

References

Adams, D.H.: The completion by cuts of an orthocomplemented modular lattice. Bull. Austral. Math. Soc. **1**, 279–280 (1969)

Amemiya, I., Araki, H.: A remark on Prion's paper, publ. Res. Inst. Math. Ser. A **2**, 423–427 (1966)

Birkhoff, G.: Combinatorial relations in projective geometries. Ann. Math. **36**, 743–748 (1935)

Birkhoff, G.: Lattice Theory, Third edition. American Mathematical Society Colloquium Publications, Vol. XXV American Mathematical Society, Providence, R.I. (1967)

Birkhoff, G., von Neumann, J.: The logic of quantum mechanics. Ann. Math. (2) **37**(4), 823–843 (1936)

Bruns, G.: Free ortholattices. Canad. J. Math. **5**, 977–985 (1976)

Bruns, G.: Varieties of modular ortholattices. Houston J. Math. **9**(1), 1–7 (1983)

Bruns, G., Harding, J.: Amalgamation of ortholattices. Order **14**, 193–209 (1998)

Bruns, G., Roddy, M.: A finitely generated modular ortholattice. Canad. J. Math. **35**(1), 29–33 (1992)

Bugajska, K., Bugajski, S.: The lattice structure of quantum logics. Ann. Inst. Henri Poincaré **19**, 333–340 (1973)

Chin, L.H., Tarski, A.: Distributive and modular laws in relation algebras. University of California Publications in Mathematics N.S. 1, pp. 341-383 (1951)

Crawley, P., Dilworth, R.P.: Algebraic Theory of Lattices. Prentice-Hall (1973)

Dunn, J.M., Hagge, T.J., Moss, L.S., Wang, Z.: Quantum logic as motivated by quantum computing. J. Symbol. Logic **70**(2), 353–369 (2005)

Dunn, J.M., Moss, L.S., Wang, Z.: Editors introduction: the third life of quantum logic: quantum logic inspired by quantum computing. J. Phil. Logic **42**(3), 443–459 (2013)

Dvurečenskij, A.: Gleason's Theorem and its Applications. Mathematics and its Applications, vol. 60. Springer (1983)

Dye, H.A.: On the geometry of projections in certain operator algebras. Ann. Math. **61**, 73–89 (1955)

Evans, T.: Word problems. Bull. Amer. Math. Soc. **84**, 789–802 (1978)

Faure, C., Frölicher, A.: Modern Projective Geometry. Springer-Science+Business Media, Dordrecht (2000)

Foulis, D.J.: Baer *-semigroups. Proc. Amer. Math. Soc. **11**, 648–654 (1960)

Foulis, D.J., Greechie, R.J., Bennett, M.K.: A transition to unigroups. Int. J. Theoret. Phys. **37**(1), 45–63 (1998)

Frayne, T., Morel, A.C., Scott, D.: Reduced direct products. Fund. Math. **51**, 195–228 (1962)

Freese, R., Ježek, J., Nation, J.B.: Free Lattices, Mathematical Surveys and Monographs. AMS, Providence, RI (1995)

Fritz, T.: Quantum logic is undecidable. Arch. Math. Logic **60**(3–4), 329–341 (2021)

Fussner, W., St. John, G.: Negative translations of orthomodular lattices and their logic. arxiv.org/abs/2106.03656 (2021)

Gehrke, M., Harding, J.: Bounded lattice expansions. J. Algebra **238**(1), 345–371 (2001)

Gehrke, M., Harding, J., Venema, Y.: MacNeille completions and canonical extensions. Trans. Amer. Math. Soc. **358**(2), 573–590 (2006)

Godowski, R., Greechie, R.J.: A non-standard quantum logic with a strong set of states. In: Beltrametti, E., van Fraassen, B.C. (eds.) Current Issues in Quantum Logic. Plenum Press, New York (1981)

Godowski, R., Greechie, R.J.: Some equations related to states on orthomodular lattices. Demonstratio Math. **17**, 241–250 (1984)

Greechie, R.J.: An orthomodular poset with a full set of states not embeddable in Hilbert space. Carribean J. Sci. Math. 1, 15–26 (1969)

Greechie, R.J.: On generating distributive sublattices of orthomodular lattices. Proc. Am. Math. Soc. **67**(1), 17–22 (1977)

Guz, W.: On the lattice structure of quantum logics. Ann. Inst. Henri Poincaré Sec. A **28**(1), 1–7 (1978)
Hamhalter, J.: Quantum Measure Theory, Fundamental Theories of Physics. Kluwer Academic Publishers Group, Dordrecht (2003)
Harding, J.: Decompositions in quantum logic. Trans. AMS **348**, 1839–1862 (1996)
Harding, J.: Canonical completions of lattices and ortholattices. Tatra Mt. Math. Publ. **15**, 85–96 (1998)
Harding, J.: The free orthomodular lattice on countably many generators is a subalgebra of the free orthomodular lattice on three generators. Algebra Univ. **48**(2), 171–182 (2002)
Harding, J.: Remarks on concrete orthomodular lattices. Int. J. Theor. Phys. **43**(10), 2149–2168 (2004)
Harding, J.: Decidability of the equational theory of the continuous geometry CG(F). J. Phil. Logic **42**(3), 461–465 (2013)
Herrmann, C.: On the equational theory of projection lattices of finite von-Neumann factors. J. Symbolic Logic **75**(2), 1102–1110 (2010)
Herrmann, C.: A note on the "Third life of quantum logic" (2019). arXiv:1908.02639
Huhn, A.P.: Schwach distributive Verbände I. Acta Scientiarum Math. **33**, 297–305 (1972)
Husimi, K.: Studies on the foundations of quantum mechanics I. Proc. Physico-Math. Soc. Jpn. **9**, 766–78 (1937)
Jónsson, B.: Modular lattices and Desargues theorem. Math. Scand. **2**, 295–314 (1954)
Jenča, G.: Effect algebras are the Eilenberg-Moore category for the Kalmbach Monad. Order **32**(3), 439–448 (2015)
Kalmbach, G.: Orthomodular Lattices. London Mathematical Society Monographs, Academic Press, London (1983)
Kalmbach, G.: The free orthomodular word problem is solvable. Bull. Aust. Math. Soc. **34**(2), 219–223 (1986)
Kaplansky, I.: Any orthocomplemented complete modular lattice is a continuous geometry. Ann. Math. **61**(3), 524–541 (1955)
Kargaev, P.P.: The Fourier transform of the characteristic function of a set that is vanishing on the interval. (Russian) Mat. Sb. (N.S.) 117(159), 3, 397–411, 432 (1982)
Kharlampovich, O.G., Sapir, M.V.: Algorithmic problems in varieties. Int. J. Algebra Comput. **5**(4–5), 379–602 (1995)
Lusin, N.N., Priwaloff, I.I.: Sur l'unicité et la multiplicit des fonctions analytiques. Ann. Sci. Ecole Norm. Sup. **42**(3), 143–191 (1925)
MacLaren, M.D.: Atomic orthocomplemented lattices. Pacific J. Math. **14**, 597–612 (1964)
Mackey, G.: Mathematical Foundations of Quantum Mechanics. Benjamin, W. A (1963)
Marker, D.: Introduction to the model theory of fields. Lecture Notes Logic **5**, 1–37 (1996)
Mayet, R.: Equations holding in Hilbert lattices. Int. J. Theoret. Phys. **45**(7), 1257–1287 (2006)
Mayet, R.: Equations and Hilbert lattices, Handbook of quantum logic and quantum structures, pp. 525–554. Elsevier Science BV, Amsterdam (2007)
Megill, N.D., Pavičić, M.: Hilbert lattice equations. Ann. Henri Poincaré 10(7), 1335–1358 (2010)
Megill, N.: Quantum logic explorer home page (2014). us.metamath.org/qlegif/mmql.html
Menger, K.: New foundations of projective and affine geometry. Ann. Math. **37**, 456–482 (1936)
Murray, F., Von Neumann, J.: On rings of operators. Ann. Math. 371(2), 116–229 (1936)
Murray, F., von Neumann, J.: On rings of operators. II. Trans. Amer. Math. Soc. **41**(2), 208–248 (1937)
Murray, F., von Neumann, J.: On rings of operators. IV. Ann. Math. **44**(2), 716–808 (1943),
von Neumann, J.: On rings of operators. III. Ann. Math. **41**(2), 94–161 (1940)
von Neumann, J.: Continuous geometry. Foreword by Israel Halperin, Princeton Mathematical Series, No. 25 Princeton University Press, Princeton, NJ (1960)
von Neumann, J.: Collected Works. In: Taub, A.H. (eds), vol. I–IV. Pergamon Press (1962)
Palko, V.: Embeddings of orthoposets into orthocomplete posets. Tatra Mt. Math. Publ. **3**, 7–12 (1993)

Redei , M., Stoltzner, M.: John von Neumann and the Foundations of Quantum Physics. Vienna Circle Institute Yearbook (2001)
Roddy, M.S.: On the word problem for orthocomplemented modular lattices. Can. J. Math. **41**, 961–1004 (1989)
Rump, W.: Von Neumann algebras, L-algebras, Baer *-monoids, and Garside groups. Forum. Math. **30**(3), 973–995 (2018)
Slofstra, W.: Tsirelson's problem and an embedding theorem for groups arising from non-local games. J. Amer. Math. Soc. 33(1), 1–56 (2020). arXiv:1606.03140
Tarski, A.: A Decision Method for Elementary Algebra and Geometry. RAND Corporation, California, Santa Monica (1948)
Titchmarch, E.C.: Introduction to the Theory of Fourier Integrals, 2nd edn. Clarendon Press (1948)
Uhlhorn, U.: Representation of symmetry transformations in quantum mechanics. Ark. Fys. **23**, 307–340 (1963)
Varadarajan, V.S.: Geometry of Quantum Theory, 2nd edn. Springer (1985)
Weaver, N.: Mathematical Quantization. Studies in Advanced Mathematics. Chapman & Hall/CRC, Boca Raton, FL (2001)

Distributions on an Interval as a Scale-Invariant Combination of Scale-Invariant Functions: Theoretical Explanation of Empirical Marchenko-Pastur-Type Distributions

Vladik Kreinovich, Kevin Alvarez, and Chon Van Le

Abstract In many practical situations, we know the lower and upper bounds $\underline{x}$ and $\overline{x}$ on possible values of a quantity x. In such situations, the probability distribution of this quantity is also located on the corresponding interval $[\underline{x}, \overline{x}]$. In many such cases, the empirical probability distribution has the form

$$\rho(x) = \text{const} \cdot (x - \underline{x})^{\alpha_-} \cdot (\overline{x} - x)^{\alpha_+} \cdot x^{\alpha}.$$

In the particular case $\alpha_- = \alpha_+ = 0.5$ and $\alpha = -1$, we get the Marchenko-Pastur distribution that describes the distribution of the eigenvalues of a random matrix. However, in some cases, the empirical distribution corresponds to different values of α_-, α_+, and α. In this paper, we show that by using the general idea of scale-invariance, we can provide a theoretical explanation for the ubiquity of such Marchenko-Pastur-type distributions.

1 Formulation of the Problem

Some distributions are located on an interval. For many physical and economic quantities x:

- there are is a lower bound $\underline{x}$ on its possible values and
- there is an upper bound $\overline{x}$ on its possible values.

V. Kreinovich (✉) · K. Alvarez
University of Texas at El Paso, 500 W University Ave., Paso, TX 79968, USA
e-mail: vladik@utep.edu

K. Alvarez
e-mail: kalvarez9@miners.utep.edu

C. V. Le
International University, Ho-Chi-Minh City, Vietnam
e-mail: lvchon@hcmiu.edu.vn

N. Ngoc Thach et al. (eds.), *Financial Econometrics: Bayesian Analysis, Quantum Uncertainty, and Related Topics*, Studies in Systems, Decision and Control 427,
https://doi.org/10.1007/978-3-030-98689-6_7

This means that all possible values on the quantity x are located on the interval $[\underline{x}, \overline{x}]$. In particular, this means that the probability distribution of this quantity is located on this interval.

Empirical fact. In economics, as shown, e.g., in Hieu et al. (2018), many such distributions have the form

$$\rho(x) = \text{const} \cdot (x - \underline{x})^{\alpha_-} \cdot (\overline{x} - x)^{\alpha_+} \cdot x^{\alpha}. \quad (1)$$

In particular, for $\alpha_- = \alpha_+ = 0.5$ and $\alpha = -1$, we get the Marchenko-Pastur distribution – the distribution of eigenvalues of a random matrix Marchenko and Pastur (1967). For example, this is how the eigenvalues of the cross-correlation matrix of different stocks are distributed Hieu et al. (2018).

However, in other cases, we have distributions of type (1) with different values of α_-, α_+, and α.

A natural question. How can we explain the ubiquity of such Marchenko-Pastur-type distributions?

What we do in this paper. In this paper, we use the idea of scale invariance to provide a theoretical explanation for this empirical family of distributions.

2 Analysis of the Problem and the Resulting Explanation

Scale-invariance: a brief reminder. The numerical value of a quantity depends on the choice of a measuring unit. For example, we can describe the price of a financial instrument in Euros, in US Dollars, in Japanese Yen, or in any other currency. The instrument is the same in all cases, but for different currencies, we will get different numerical representations of the same price.

In many situations, there is no reason to select this or that measuring unit, the choice of the unit is just a matter of convention. In such situations, it makes sense to require that the formula $y = f(x)$ describing the dependence between quantities x and y should not change if we replace the original measuring unit for x. If we replace the original measuring unit for x by a new unit which is λ times smaller, then all the numerical values of this quantity will be multiplied by λ:

$$x \to x' = \lambda \cdot x.$$

Of course, for the formula $y = f(x)$ to be valid in the new units, we need to appropriately change the unit for y. For example, the formula $y = x^2$ that describes how the area of a square depends on its size does not depend on the choice of units, but if we replace, e.g., meters with centimeters, we need to also replace square meters with square centimeters.

In general, scale-invariance of a function $f(x)$ takes the following form: *For every* $\lambda > 0$, *there exists a value* $\mu(\lambda) > 0$ *for which* $y = f(x)$ *implies* $y' = f(x')$, *where* $x' \stackrel{\text{def}}{=} \lambda \cdot x$ *and* $y' \stackrel{\text{def}}{=} \mu(\lambda) \cdot y$.

Which functions are scale-invariant? Substituting the expressions $x'=\lambda \cdot x$ and $y'=\mu(\lambda) \cdot y$ into the formula $y' = f(x')$, we get $\mu(\lambda) \cdot y = f(\lambda \cdot x)$. Here, $y = f(x)$, so we get

$$f(\lambda \cdot x) = \mu(\lambda) \cdot f(x). \tag{2}$$

It is known that every continuous (and even every measurable) function $f(x)$ that satisfies the Eq. (2) for all x and λ has the form

$$f(x) = c \cdot x^a, \tag{3}$$

for some constants c and a; see, e.g., Aczél and Dhombres (2008).

Starting point can also be different. For many quantities—e.g., for time—we can also select different starting points. If we replace the original starting point with a new starting point which is x_0 units before, then all the numerical values x of this quantity are replaced by new values:

$$x' = x + x_0.$$

If we have a scale-invariant dependence $f(x) = c \cdot (x')^a$ in the new scale, then, in the old scale, this dependence takes the form $y = c \cdot (x + x_0)^a$.

What are natural starting points for functions located on an interval $[\underline{x}, \overline{x}]$. If we know that a quantity x is always located on an interval $[\underline{x}, \overline{x}]$, then we have two natural starting points: $\underline{x}$ and $\overline{x}$. Thus, in addition to the original scale-invariant functions $f(x) = c \cdot x^a$, we also get functions

$$f(x) = c_- \cdot (x - \underline{x})^{a_-} \tag{4}$$

and

$$f(x) = c_+ \cdot (\overline{x} - x)^{a_+}. \tag{5}$$

Since we need a single function, we need to combine these functions.

How can we use scale-invariance to combine different functions? We want to combine several functions $y_1 = f_1(x)$, …, $y_n = f_n(x)$ into a single quantity $y = F(y_1, \ldots, y_n)$. In view of the above, it makes sense to do it in scale-invariant way. In other words, we want to find a function $F(y_1, \ldots, y_n)$ that has the following property: *For every combination of possible values* $\lambda_1 > 0$, …, $\lambda_n > 0$, *there should exist a value* $\mu(\lambda_1, \ldots, \lambda_n)$ *for which* $y = F(y_1, \ldots, y_n)$ *implies that* $y' = F(y'_1, \ldots, y'_n)$, *where* $y'_1 \stackrel{\text{def}}{=} \lambda_1 \cdot y_1$, …, $y'_n \stackrel{\text{def}}{=} \lambda_n \cdot y_n$, *and* $y' \stackrel{\text{def}}{=} \mu(\lambda_1, \ldots, \lambda_n) \cdot y$.

Which combination operations are scale-invariant? Substituting the expressions $y_i'=\lambda_i \cdot y_i$ and $y'=\mu(\lambda_1,\ldots,\lambda_n)\cdot y$ into the formula $y' = F(y_1',\ldots,y_n')$, we get

$$\mu(\lambda_1,\ldots,\lambda_n)\cdot y = F(\lambda_1\cdot y_1,\ldots,\lambda_n\cdot y_n).$$

Here, $y = F(y_1,\ldots,y_n)$, so we get

$$F(\lambda_1\cdot y_1,\ldots,\lambda_n\cdot y_n) = \mu(\lambda_1,\ldots,\lambda_n)\cdot F(y_1,\ldots,y_n). \tag{6}$$

It is known that every continuous (and even every measurable) function $F(y_1,\ldots,y_n)$ that satisfies the Eq. (6) for all $y_1,\ldots,y_n$ and $\lambda_1,\ldots,\lambda_n$ has the form

$$y = F(y_1,\ldots,y_n) = C\cdot y_1^{a_1}\cdot\ldots\cdot y_n^{a_n}, \tag{7}$$

for some constants $C, a_1,\ldots,a_n$; see, e.g., Aczél and Dhombres (2008).

Resulting expression. In our case, we combine three expression:

- the expression for y_1 described by the formula (4),
- the expression for y_2 described by the formula (5),
- the expression for y_3 described by the formula (3).

Substituting $n = 3$ and the expressions (4), (5), and (3) for y_i into the formula (7), we conclude that

$$y = c_0\cdot(x-\underline{x})^{\alpha_-}\cdot(\overline{x}-x)^{\alpha_+}\cdot x^{\alpha}, \tag{8}$$

where we denoted $c_0 \stackrel{\text{def}}{=} C\cdot c^{a_1}\cdot c_-^{a_-}\cdot c_+^{a_+}$, $\alpha_- \stackrel{\text{def}}{=} a_2\cdot a_-$, $\alpha_+ \stackrel{\text{def}}{=} a_3\cdot a_+$, and $\alpha \stackrel{\text{def}}{=} a_1\cdot a$.

For the case when y is the probability density, this is exactly the desired formula (1). So, we have indeed explained the empirical formula (1) by using scale-invariance.

Comment. Our explanation is more general that explaining the empirical dependence (1) of the distribution located on an interval.

It also explains, e.g., why in many cases, Bernstein polynomials, i.e., sums of monomials of the type $(x-\underline{x})^{a_-}\cdot(\overline{x}-x)^{a_+}$, provide a good approximation to functions located on an interval; see, e.g., Lorentz (2012).

Acknowledgements This work was supported in part by the National Science Foundation grants 1623190 (A Model of Change for Preparing a New Generation for Professional Practice in Computer Science), and HRD-1834620 and HRD-2034030 (CAHSI Includes). It was also supported by the program of the development of the Scientific-Educational Mathematical Center of Volga Federal District No. 075-02-2020-1478.

References

Aczél, J., Dhombres, J.: Functional Equations in Several Variables. Cambridge University Press, Cambridge (2008)

Lorentz, G.G.: Bernstein Polynomials. American Mathematical Society, Providence, Rhode Island (2012)

Marchenko, V.A., Pastur, L.A.: Distribution of eigenvalues for some sets of random matrices. Mathematics of the USSR-Sbornik **1**(4), 457–483 (1967)

Nguyen, H.T., Tran, P.N.U., Nguyen, Q.: An analysis of eigenvectors of a stock market cross-correlation matrix. In: Anh, L.H., Kreinovich, V., Thach, N.N. (eds.), Econometrics for Financial Applications, pp. 504–513. Springer Verlag, Cham, Switzerland (2018)

A Panorama of Advances in Econometric Analysis

Hung T. Nguyen

Abstract The purpose of this paper is to spell out current interests in advanced statistical methods for statistics in general, and econometrics in particular, in the spirit of credible and efficient statistics. By pointing out these advances, the hope is to make researchers aware of them, so that future research will be credible and significant. The panorama of recent advances covered in this paper consists of Bayesian assett allocation (Black-Litterman model), copula theory, partial identification of statistical models, the role of random set theory in set estimation, the optimal transport methods, and the quantum modeling in economics.

Keywords Black-litterman · Copulas · Multivariate quantile regression · Optimal transport · Partial identification · Quantum modeling · Random sets

1 Introduction

Recent advances (in the literature) in uncertainty modeling and statistical methods in econometrics, and statistics in general, seem to focus upon quantum probability (and its calculus), partial identification (of population parameters), and Optimal Transport theory. These topics seem not be quite familiar with the empirical research community. This paper aims at, not only making applied researchers aware of these topics, but mainly emphasizing their rationale and usefulness in making empirical research more credible and efficient.

With respect to partial identification of parameters, we emphasize the crucial role played by random set theory where a theory of statistics of random sets needs to be further developed.

The topic of optimal transport, which as reformulated by Kantorovich for economics, finally surfaces again in economics (!) and deserves full attention of statisti-

H. T. Nguyen (✉)
New Mexico State University (USA), Las Cruces, NM, USA
e-mail: hunguyen@nmsu.edu

Chiang Mai University, Chiang Mai, Thailand

N. Ngoc Thach et al. (eds.), *Financial Econometrics: Bayesian Analysis, Quantum Uncertainty, and Related Topics*, Studies in Systems, Decision and Control 427,
https://doi.org/10.1007/978-3-030-98689-6_8

cians and econometricians. Its amazing contributions to econometric analysis should be welcome.

While the question whether uncertainty so far modelled by standard Kolmogorov probability theory is sufficient and adequate has been raised for some time, with various proposals, it seems finally quantum uncertainty takes a central position. As such, it is about time to make empirical researchers aware of this more general probability theory to apply to econometrics.

2 A Reminder of Bayesian Asset Allocation (Black-Litterman Model)

Without "taking side"(!) and evoking the current objection to the use of p-values to carry out frequentist testing of statistical hypotheses, we start out by reminding researchers of the state-of-the-art of portfolio theory where the Bayesian statistics seems to be the most advanced approach.

In general, it could be said that the Bayesian approach to decision analysis provides two basic additional ingredients. In one hand, prior information is useful when we lack information to make decisions under uncertainty, such as in the formulation of games of incomplete information. On the other hand, when we have prior information, we should use it to improve our decision analysis, in the sense of making statistics more efficient, such as in the problem of portfolio selection. We are going to elaborate on the latter issue.

The story of portfolio selection so far is this. Having a set of n risky assets a_j, $j = 1, 2, ..., n$ in mind, we seek a weight (proportion) vector $w = (w_1, w_2, ..., w_n)$, $w_j \geq 0$, $\sum_{j=1}^n w_j = 1$, to form an investment portfolio.

Let R_j be the return (at some fixed time in the future) of asset a_j, and $R = \sum_{j=1}^n w_j R_j$ be the return of the portfolio. Assuming that returns have finite second moments, the expected return of the portfolio is $ER = \sum_{j=1}^n w_j E(R_j)$. Each asset is characterized by (μ_j, σ_j), where $\mu_j = E(R_j)$, $\sigma_j^2 = Var(R_j)$.

If we let $\mu = (\mu_1, \mu_2, ..., \mu_n)'$, and Σ being the covariance matrix of assets, then each portfolio is characterized by (μ, Σ), whose risk is

$$\sigma^2 = \sigma^2(R) = \sigma^2(\sum_{j=1}^n w_j R_j) = \sum_{j=1}^n \sum_{k=1}^n w_j w_k Cov(R_j, R_k)$$

Since we are looking for choices of weights to obtain a "good" portfolio, we need to agree on "what is a good portfolio?". Well, from common sense, a good portfolio is could be one with high expected return μ and low risk σ. Can we get such a portfolio? To answer this question, we need to look at the map $f(.) : H$ (a hyperplane) $\subseteq \mathbb{R}^n \to \mathbb{R}^2$, namely $f(w) = (\mu(w), \sigma(w))$ to see if there is some relation between μ and σ to explore. Since mean and variance of a random variable are unrelated, we could treat them as such, and the best we can do is to balance (trade-offs) them appropriately.

The Markowitz theory of portfolio selection is based on the following. First, with respect to unsystematic risk, minimizing $w'\Sigma w$ is the same as maximizing $-w'\Sigma w$. Secondly, consider risk averse investors (whose utility functions are concave). For a concave quadratic utility function $u(x) = x - \frac{\lambda}{2}x^2$, the "rational" investors will seek to $\max_w E(u(R))$ which is equivalent (via Taylor expansion) to $E(R) - \frac{\lambda}{2}Var(R)$, so that the integrated optimization problem is

$$\max_w [w'\mu - \frac{\lambda}{2}w'\Sigma w] \quad \text{subject to} \quad \sum_{j=1}^{n} w_j = 1$$

The closed form solution of the above integrated optimization problem is $w = \frac{1}{\lambda}\Sigma^{-1}\mu$ with appropriate calibration of λ (say, as the historical Sharpe ratio/ average market return over volatility).

The theory is so compact and beautiful, but how about applications? Well, in applications, we do not know the "input" "ingredient" (μ, Σ) to obtain the solution!

Relying on the availability of historial data, we replace (μ, Σ) by their estimates. It turns out that with the sample estimates, the optimization machine is badly behaved, producing not good portfolio to use. It is a statistical problem. For example, the estimation of the mean vector of the return μ, when there are $n > 2$ risky assets, the sample mean is inadmissible (a James-Stein's result). Also the sample mean is not a robust estimator. More importantly, does the past "represent" thr future?

Now, while the CAPM is called an asset pricing model, it does contribute to Markowitz's optimal portfolio theory. Needless to say, CAPM suggested the inclusion of a risk-free asset into the collection of risky assets, and change our view on risk, namely from usystematic risk to systematic risk.

Let μ_f be the risk-free rate, and μ_m be the market mean of return, then the CAPM formula is, for any $1 \leq j \leq n$:

$$\mu_j - \mu_f = \beta_j(\mu_m - \mu_f)$$

where $\beta_j = \frac{Cov(R_j, R_m)}{Var(R_m)}$.

The main point of CAPM is this. If the market is efficient, then instead of sample means, take equilibrium vector of expected returns, denoted as μ_{eq}, also called "implied expected returns" which is derived as follows. The equilibrium weights are computed from Markowitz's solution

$$w = \frac{1}{\lambda}\Sigma^{-1}\mu$$

By *reverse optimization* (or reverse engineering), we get

$$\mu_{eq} = \lambda \Sigma w_m$$

where we can calibrate as follows. Multiply the above by w_m', we get

$$w'_m \mu_{eq} = \lambda w'_m \Sigma w_m \Longrightarrow \lambda = \frac{w'_m \mu_{eq}}{w'_m \Sigma w_m} = \frac{\mu_m}{\sigma_m^2}$$

(in practice, $\lambda = 2.5$, average risk aversion in the market).

The Back-Litterman (BL) model Black and Litterman (1991) is really a statistical estimation of μ to make the Markowitz's optimization better behaved, by estimating μ by a Bayesian method.

Rather than estimating the excess mean return μ of the portfolio only from historical data, the BL regards μ as influenced by three factors: the implied returns, the market noise, and the subjective views (opinions) of experts about the values of future asset returns.

Using CAPM as a starting point, the BL proceeds to establish a conditional distribution $f(v|\mu)$ as a "model" (in Bayesian statistics language), where V is an observable random variable called a *view*, then a prior distribution $f(\mu)$, from which, Bayes' rule of updating gives the posterior distribution $f(\mu|v)$, resulting, in square loss function, the posterior mean as an admissible estimator for μ.

Note that, this new input to Markowitz's optimization machine is admisible, it also has the flavor of a *shrinkage estimator*.

The BL model provides new "ingredients" as input to Markowitz's optimization machine to make it better behaved!

Roughly speaking, under the standard global assumption that the excess return vector $X = (R_1 - R_f, ..., R_n - R_f)'$ is $N_n(\mu, \Sigma)$, the BL proceeds to supply an estimate for μ using the Bayesian setting. The "data" consists of the equilibrium CAPM, the investor's subjective views on CAPM data, together with uncertainties involved. The "model" is the conditional density $f(v|\mu)$, where the prior is a density $f(\mu)$, producing the posterior density of μ given the views $f(\mu|v)$. With respect to square loss function, the admissible Bayes estimator of μ is the posterior mean which is used as input to Markowitz's optimization procedure. See e.g. Meucci (2003) for details.

3 Subcopulas for Discrete Data

Upfront, the purpose of this Section is to emphasize the use of subcopulas, in modeling of multivariate discrete distributions and their dependence structure, rather than copulas, to avoid not only the problem of identification, but also the characterization of dependence structures.

By now (i.e., since the early 1990's), Sklar's theory of copulas seems "familiar" with statisticians and econometricians, despite the fact that Sklar's work appeared back in 1959. Although unlike univariate distribution theory, modeling of multivariate distributions is difficult, and mostly relying upon conditional distributions, it took almost 30 years to finally recognize that copulas are useful, not only for modeling multivariate distributions, but more importantly, for modeling general dependence structure among univariate variables. See, e.g. Durante and Sempi (2016).

As we will see in Sect. 6 of this paper, it is interesting to note that while copula theory was motivated by Frechet's problem in 1951, there was no mention of *optimal transport theory* as reformulated by Kantorovich in 1941, let alone of the original work of Monge in 1781 !

The theory of copulas is very short! It consists simply with one theorem and a corollary. The so-called "copula technology" is now widely applied in both statistics and econometrics, consisting of parametric models of copulas for *continuous* random variables, together, of course, with computational algorithms. This can be called "the golden standard of copulas"!

However, there are many important applications where data are discrete, such as in actuarial science, and the "state-of-the-art" of copulas for discrete data is not satisfactory. It is so since, instead of focusing on subcopulas, researchers tend to use copulas to study discrete variables as well. The credibility of such analyses is in question, in view of the problem of *partial identification* that we will discuss in the next section, as an advanced statistical philosophy. The best criticisms of the current approach to discrete data using copulas are in Faugeras (2017) where the main message is: "to warn the practicioners who would like to naively extend the copula paradigm to discrete data, especially by using parametric copula models to making inference".

Here we will lay out the mathematics pointing to the seemingly obvious fact that when joint distributions are not continuous, the credible method for their studies should be *subcopulas*, and not copulas.

Sklar's theorem (in the case of two random variables) is this. Let $H(x, y)$ be the joint distribution function of the (real-valued) random vector (X, Y), then there is a unique subcopula S, defined on $\mathscr{R}(F) \times \mathscr{R}(G) \subseteq [0, 1]^2$ (ranges of the marginal distributions F, G of X, Y, respectively) such that $H(x, y) = S(F(x), G(y)$. In particular, if $H(x, y)$ is conrinuous, then S is a (unique) copula.

Remark. The unique subcopula

$$S(u, v) = H(F^{-1}(u), G^{-1}(v)), \quad (u, v) \in \mathscr{R}(F) \times \mathscr{R}(G) \supseteq \{0, 1\}$$

is characterized by the following axioms: for $(u, v) \in \mathscr{R}(F) \times \mathscr{R}(G)$,

(i) $S(u, 0) = S(0, v) = 0$
(ii) $S(u, 1) = u$, $S(1, v) = v$
(iii) If $(u, v) \leq (u', v')$ (componentwise), then $S(u', v') - S(u', v) - S(u, v') + S(u, v) \geq 0$

Note that a copula is a special case of a subcopula. Thus, subcopula is a more fundamental notion than copula. Although, a subcopula can be extended to be a copula, but the extension is not unique. This last fact is the cause of the non "point-identification" issue.

In the literature, the focus is on the "continuous case", i.e., for continuous random variables (F, G are continuous functions, so that $\mathscr{R}(F) \times \mathscr{R}(G) = [0, 1]^2$), since then the situation is very simple to proceed. For example, the (unknown) joint distribution $H(.,.)$ of (X, Y) can be correctly modeled by $C(F(x), G(y))$ by a unique copula C, when F, G are known, so that, as a "tradition", it suffices to consider *parametric* copula models (!), and using copulas to "capture" the dependence between X and Y without paying attention to the marginals (saying that it is "the" copula which describes the dependence structure, and not the marginals).

It is well known that in the discrete case, not only that the dependence of X and Y does depend on the marginals, see Heckman (1979), but also modeling their joint discrete distribution by a copula (let alone a parametric copula) is not appropriate. These basic things alone render the "popular" practice of using copula models, together with all statistical inference based on copulas, incredible. There is no compelling reason to use copulas when data are discrete. Thus, the message is this. We should treat discrete data differently than continuous ones as far as copula-like modeling is concerned.

In applications, we are mainly concerned with two issues:

(a) Modeling a joint distribution of two unknown marginals based on observed data, especially for discrete data. Usually, the marginals F, G could be paametric in nature, such as Poisson distributions with unknown means. In such a case, we wish to provide a subcopula S to couple with F, G to form the true joint distribution $H(x, y) = S(F_\alpha(x), G_\beta(y))$. The problem is how? Estimating S, α, β will be a semiparametric estimation problem. Modeling the subcopula S parametrically, by the restriction of a parametric copula model (for "ease" of estimation !) could lead to incredible results. A nonparametric estimation of a subcopula should be a good research problem to render empirical results more credible.
(b) Investigating the dependence structure of variables. As stated above, the dependence structure of discrete variables is much more delicate than for continuous variables, as it could be dependent on the marginals. This issue is also a research problem to be considered.

4 Partial Identification of Population Parameters

Perhaps, the notion of *partial identification* of population parameters (together with its associated philosophy/ credible statistics), initiated in Manski (2003), Manski (2007)) is a revolution in the way we used to perform inference with observed data and maintained assumptions.

It is a revolution in at least two important aspects. The first one is about credibility of empirical research, namely unjustified assumptions could lead to results which are not believable. The second one is about the nature of population parameters: Instead of being finitely or infinitely dimensional (e.g., elements of some $\mathbb{R}^d$, of of some functional space $\mathcal{L}^2(\mathbb{R}^d, \mathcal{B}(\mathbb{R}^d), dx)$), the parameters are *sets* (of some parameter space). As such, for carrying out estimation and inference, we need to call upon the theory of *random sets* (but still sitting inside the general theory of probability, since random sets are simply random elements).

While the second point will be elaborated in the next Section, we devote this Section to the notion of partial identification.

In empirical research, we have data $X_1, X_2, ..., X_n$ which are observations for some variable X of interest. When we view X as a random variable, we say that the data are generated by the distribution F_o of X, and call F_o the data generating process (DGP). Having only the data, we do not know F_o (of course, except in simulations!). In fact, F_o can be "parametric" (i.e., known except some finitely dimensional quantity) or nonparametric, otherwise.

Usually, we are interested in estimating some quantity of interest (for various purposes) such as the mean of X. The data alone are not sufficient to use to accomplish our task.

Indeed, the mean $EX = \int_{\mathbb{R}} x dF_o(x)$ can be finite or infinite. Thus, to estimate the mean, we need to "assume" (or justify) that $EX < \infty$, so that, for example, we can use the strong law of large numbers to estimate EX consistently. The point is this. We need not only data but also a "model". Of course, the delicate thing will be "how to propose a model for our observed data?"

A parametric model is a family of distributions indexed by some finitely dimensional parameter, such as $\mathcal{P} = \{N(\theta^2, 1), \theta \in \Theta = \mathbb{R}\}$, and we think that F_o is in it, i.e., there is θ_o such that $F_o = N(\theta_o^2, 1)$ (of course, mispecification often occurs!). A parameter is a quantity determined by the "true" distribution F_o of the DGP X. For example, $EX = \lambda(F_o) = \int_{\mathbb{R}} x dF_o(x)$. Note that, in general, a population parameter is a function of the distribution generating our observed data. As such, a parameter can be of finite or infinite dimensions, such as the distribution function (or probability measure) of the DGP itself.

Now suppose that our model is $\{N(\theta^2, 1), \theta \in \Theta = \mathbb{R}\}$, and we wish to estimate the parameter $\lambda(F_o) = EX$. A parameter is said to be estimable if it can be estimated consistently, i.e., there is a sequence of statistics converging to it (in some probabilistic fashion) if the sample size $n \to \infty$ (i.e., with infinite data, we can find it). But we only have a finite set of data? The consistency of an estimator tells us that we can be confident that our estimator is a good approximation of the true parameter, and if we want to quantify its uncertainty, we can construct confidence regions for it.

All the above are "elementary" in a first course in Statistics. What we forgot to tell students is this. There is something very important we need to examine before doing estimation. Instead, we assume that when we face an estimation problem, it can be done, and the problem is how to get a consistent estimator (together with its desirable properties). The question "Can we actually estimate the parameter of interest from data and maintained assumptions?" or more generally "What can we learn about the parameter?" was not asked. In other words, "identification" is always assumed. By identification, in fact *point identification* (as a parameter value is a "point" in the parameter space), we mean the possibility to "identify" the parameter, from data and assumed model, i.e., with infinite data we can recover it (via consistency). Point identification is a necessary condition for estimability of a parameter.

Just for an example, consider the model $\mathcal{P} = \{N(\theta^2, 1), \theta \in \Theta = \mathbb{R}\}$, with $E_\theta(X) = \theta^2$. Two different $\theta' \neq \theta''$ corresponds to the same distribution generating our observed data (we say that they produce *"observationally equivalent"* data).

The population parameter θ in this model is not point identified. The reason is that the map $\theta \in \mathbb{R} \to P_\theta = N(\theta^2, 1)$ is not injective. Thus, let's say that a parameter θ in a model $\{P_\theta : \theta \in \Theta\}$ is point identifiable if $\theta \in \Theta \to P_\theta$ is injective (if θ_o is the true parameter, then for any $\theta \in \Theta \backslash \{\theta_o\}$, $P_\theta \neq P_{\theta_o}$).

More generally, a function of the model parameter, $\lambda(\theta)$, e.g., a linear regression coefficient, is point identified if it is uniquely determined from the observable population (its distribution) that generates the data. Specifically, $\lambda(\theta)$ is point identified if the set $\{\lambda(\theta) : P_\theta = P_{\theta_o}\}$ is a singleton.

Point identification is a precursor to estimation since logically, we need to know whether or not a parameter is "estimable" before trying to estimate it! If the parameter is not identified, then we cannot hope to find a consistent estimator for it.

In traditional analysis, when we can construct a consistent estimator for $\lambda(\theta)$, then by construction, we have proved that $\lambda(\theta)$ is point identified. For example, if our model is the collection of all distributions with finite means, then the law of large numbers, say, for i.i.d. data, indicates that the population mean is point identified.

How to prove that a parameter is point identified? A parameter is point identified if it can be determined from our data and maintained assumptions. For example, in a linear regression $Y = \beta X + \varepsilon$, with our assumption $E(\varepsilon X) = 0$, then $YX = \beta X^2 + \varepsilon X$, so that $E(YX) = \beta E(X^2)$, and hence $\beta = \frac{E(YX)}{E(X^2)}$ which can be estimated from observed data.

As we will see shortly, there are many reasons why a parameter is not point identified. But, first let's take a closer look at how, traditionally, statisticians "handle" or behave when they face a non point identification problem.

Essentially, statisticians proceed to add more assumptions to render the problem point identified.

Well, so far so good!? But are there any problems with this traditional approach?

Yes, there are problems, but it is not easy to see them! When conductiong theoretical research, statisticians or econometricians, with good mathematical training, are free to impose additional assumptions to obtain point identification (in fact, this is always assumed!), and from which, impose more "necessary" assumptions on the considered model to reach consistency of estimators. From a theoretical (or mathematical) point of view, there is nothing wrong with it. What is "wrong" is this. Theoretical results (based on lots of assumtions) might not be applicable in real-world problems. We can "assume" things in a theoretical research, but we cannot assume things in an applied problem, since if we do so, our asumptions are poorly justified, leading to results which are not credible. Remember empirical research must be credible. The buzz word is credibility of statistical analyses. Note that "efficiency" (of statistical methods) is another buzz word to be kept in mind.

Although, it has been said so many times in the literature, it does not ring the bell within the "community" of "applied people" who use statistical toolkits to do empirical research in their fields. By trusting theoretical results, applied people just pick up tools they need from the available toolkit without realising that these tools are heavily constructed from their "domains of applicability". Empirical results coming out from available toolkit are they credible? Results are credible or not have something to do with assumptions they used. Roughly speaking, a result is not credible if the asumptions they used to get it are not justifiable.

As applied people rely on "theories" provided to them by theoreticians, it is not their fault to arrive at incredible conclusions affecting real economic phenomena! It is the "duty" of theoreticians to provide an update toolkit for practicioners. This Section is precisely about the revealation of a concept which changes the way we do statistics, at the benefit of applied people.

Here is a "concrete" example illustrating the tradition of standard (classical) statistics. Whether or not the additional statistical assumptions are justfied should be examined with care, the purpose of evoking James Heckman's correction (See Heckman (1979)) is simply to illustrate a situation where *the observed data are not sufficient to point identify parameters of interest*, and, from a "traditional point of view", statisticians will "consider" additional assumptions or information to make a parameter of interest point identifiable (by "construction", i.e., by providing a consistent estimator for it).

This is about a familiar situation in studying labor economics where observed data manifest the so-called "sample selection bias".

Suppose we wish to investigate the question "Does X_1 affect Y_1?" by considering the linear model, with "standard" maintained assumptions to make the regression coefficient point identified from a random sample

$$Y_1 = \beta_1 X_1 + u_1$$

However, our intended (i.i.d.) random sample $(Y_{1j}, X_{1j} : j = 1, 2, ..., n)$ cannot be completely observed, so that we have only available a subsample of it. In other words, some data are missing, and we have to use the available observed data (a non random sample). But then, β_1 is no longer point identified.

Facing a situation such as this, statisticians think about what additional information is needed to make β_1 point identified. Now, while data could be missing for various reasons (e.g., Nonresponse, dropout, researchers' mistakes, self selection), here is a possible reason. Each individual j has another factor (e.g., utility) Y_{2j} of the form

$$Y_{2j} = \beta_2 X_{2j} + u_{2j}$$

and the data Y_{1j} is observed (available) only when $Y_{2j} \geq 0$ (which is observable). This is referred to as a sample selection rule.

Conditional on available sample,

$$E(Y_{1j}|X_{1j}, \text{ sample selection rule}) = \beta_1 X_{1j} + E(u_1|X_{1j}, \text{ sample selection rule})$$

where

$$E(u_1|X_{1j}\text{sample selection rule}) = E(u_{1j}|X_{1j}, Y_{2j} \geq 0) = E(u_{1j}|X_{1j}, u_{2j} \geq -\beta_2 X_{2j})$$

In other words, the subsample regression is

$$E(Y_{1j}|X_{1j}, Y_{2j} \geq 0) = \beta_1 X_{1j} + E(u_{1j}|X_{1j}, u_{2j} \geq -\beta_2 X_{2j})$$

which depends also on X_{2j}.

Thus, if we use

$$Y_1 = \beta_1 X_1 + u_1$$

with selected subsample, we are omitting the term $E(u_{1j}|X_{1j}, u_{2j} \geq -\beta_2 X_{2j})$ which is a random variable, playing the role of another (relevant) regressor.

Therefore, the missing data regression takes the form of the problem of misspecification of model which is caused by omitting variables. *The implication is that omitting relevant regressors leads to inconsistency of OLS as well as their efficiency.* The *inconsistency of OLS* in a linear model is caused by the fact that regressors are correlated with the error

How to fix it? well, if we do not omit the regressor $E(u_{1j}|X_{1j}, u_{2j} \geq -\beta_2 X_{2j})$, then things will be fine. Usually, as in the case considered by Heckman, even if we recognize the problem, we still cannot retain this second regressor since it is not observable. The novel idea of Heckman is this. Estimate the second regressor (by using probit analysis), then use the correct model (with two regressors) for OLS. For details, see Hartigan (1987).

Remark. The Heckman's two-step estimation procedure in linear regression model with missing data is somewhat similar the use of Lagrange multipliers to transform an optimization under constraint to an optimation without constraint to apply standard calculus.

It turns out that, from Manski's works, Manski (2003),Manski (2007), when a parameter is not point identified (from available data and maintained assumptions) and additional assumptions to make it point identified are not justified, or no such assumptions are possible, we are not in an impossible case !. In other words, identification is not an all or nothing classification. There is something in between. Specifically, the *identified set* $\Theta_I = \{\theta \in \Theta : P_\theta = P_{\theta_o}\}$, while not being a singleton, could be a proper subset of Θ, a situation we refere to as *partial identification*, and we can "learn" about it, i.e., estimating it.

As such, our focus is shifted from point parameter to *set parameter*, and hence from point estimator to set estimator.

Here are some simple examples of partial identification and their statistical analyses.

(a) Let X, Y be real-valued random variables with distribution functions F, G, respectively. Consider the case where F, G are known (as in the case of an optimal transport problem, see Sect. 6, or can be consistently estimated from observed data). Let's the population be the random vector (X, Y) whose (true) joint distribution H_o has F, G as marginals. In orther words, our statistical model (parameter space) is the set of all joint distribution functions H, such that $H(x, \infty) = F(x)$ and $H(\infty, y) = G(y)$.

Then the parameter H is not point identifiable, since its identified set is the Frechet's class

$$\Theta_I = \{H : \max\{F(x) + G(y) - 1, 0\} \leq H(x, y) \leq \min\{F(x), G(y)\}\}$$

which is not a singleton.

Note that this Θ_I is sharp since the bounds $\max\{F(x) + G(y) - 1, 0\}, \min\{F(x), G(y)\}\}$ are elements of the identified set.

(b) A standard example of partial identification of a parameter when data are missing.

Suppose we wish to predict the outcome of a future election involving two candidates A and B in a physical population. We send out questionaires to a random sample of individuals in that polulation, and receive back only a sub sample of it from individuals who did response, i.e., there are missing data (nonresponses). Can we estimate (consistently), say, the chance that candidate A will win the election?

Let X be a binary random variable, defined on $(\Omega, \mathcal{A}, P)$ (the source of uncertainty), and taking values in $\{0, 1\}$, and we are interested in $P(X = 1)$, the (true) proportion of individuals who will vote for A. The missing of data in a random sample $X_1, X_2, ..., X_n$, drawn from X, can be described by an (binary) indicator Y: If $Y_i = 1$, we observe X_i, if $Y_i = 0$, we do not observe X_i. Of course, we observe Y since its values $0, 1$ correspond to nonresponse and response (to our questionaires), respectively. In this problem, the parameter $P(X = 1)$ is the mean θ (lying in the parameter space $\Theta = [0, 1]$) of our statistical population X, but we only have a (non-random) subsample of data from it. If there are no missing data, then θ is point identified, since it is so "by construction" of a consistent estimator, namely the sample mean, in view of the strong law of large numbers. But, unfortunately, we have missing data. So the first question is "Is $\theta = EX$ point identifiable?" Well, assuming that X and Y are defined on the same probability space $(\Omega, \mathcal{A}, P)$ (otherwise, consider a coupling of them), we have

$$\{\omega \in \Omega : X(\omega) = 1\} = \{\omega \in \Omega : X(\omega) = 1\} \cap \Omega =$$

$$\{\omega \in \Omega : X(\omega) = 1\} \cap \{\{\omega : Y(\omega) = 1\} \cup \{\omega : Y(\omega) = 0\}\} =$$

so that

$$\theta = P(X = 1) = P(X = 1, Y = 1) + P(X = 1, Y = 0) =$$

$$P(X = 1|Y = 1)P(Y = 1) + P(X = 1|Y = 0)P(Y = 0)$$

in which the data contain no information about $q = P(X = 1|Y = 0) \in [0, 1]$.

Thus θ is of the form $P(X = 1|Y = 1)P(Y = 1) + qP(Y = 0)$ for $q \in [0, 1]$, i.e., θ is only partially identified by the identified set $\Theta_I =$

$$\{\theta \in [0, 1] : \theta = P(X = 1|Y = 1)P(Y = 1) + qP(Y = 0) \text{ for some } q \in [0, 1]\}$$

Clearly, the above is similar for the (finite) mean of an arbitrary population X with missing data indicating by some missing data indicator (binary) Y.

Remark. The goal of an empirical research, such as the above, is not only to understand a phenomenon, but also to make prediction and decision. For a follow-up statistical inference of the above partial identification problem, see e.g. Manski (2007).

5 Random Sets for Estimating Set Parameters

Upfront, among other scientific investigations, random sets and their statistics, formulated within general probability theory, are necessary tools to study partially identified parameters in econometrics.

It is quite surprising that while general random elements were somewhat popular in both theoretical and applied research, the case of random sets seems "hidden". Specifically, statistics with random sets seems "rare" in the literature. Perhaps the phenomenon is familiar: Until applications call for, a theoretical theory remains hidden. This is particular true with the theory of copulas. Thus, upfront, although we might not be at the level of "Nothing is so powerful as an idea whose time has come" (Victor Hugo), we are glad to declare that "The time has come for using random set theory in applications", thanks to partial identification in econometrics.

In standard statistical analysis, we do run sometimes into random sets (although not at a "theoretical" level), i.e., sets obtained at random, e.g., confidence regions for parameters, not for estimating a set parameter, but for quantifying the ambiguity of a point parameter, although the probem of computing the expected value of the size of a random region (say, a confidence region) did require the concept of a random set. On the other hand, we did face the problem of estimating sets, not as set parameters per se, but as an intermediary step toward estimating a point parameter.

In sampling from a finite population (experimental designs), a sampling design is a finite random set S characterized by its *covering function*, where S is a random element $S : (\Omega, \mathcal{A}, P) \to 2^U$, where U is a finite set (a finite population), and its covering function $\pi(.) : U \to [0, 1]$, $\pi(u) = P(S \ni u)$.

Formally, as a random element in general Kolmogorov probability theory, the "distribution function" of a finite random set S is $F(.) : 2^U \to [0, 1]$, $F(A) = P(S \subseteq A)$ (with the set inclusion $\subseteq$ among subsets of U replacing the order relation $\leq$ on $\mathbb{R}$). By Mobius inversion, the probability density of S is $f(.) : 2^U \to \mathbb{R}^+$,

$$f(A) = P(S = A) = \sum_{B \subseteq A} (-1)^{|A \setminus B|} F(B)$$

where $|B|$ is the number of elements in the subset B.

A suitable generalization of a random vector X, i.e., a random element taking values in $\mathbb{R}^d$, is the notion of a *random closed set*, as developed by G. Matheron in 1975, Matheron (1975) (an introduction to it is Molchanov and Molinari (2018)).

A random closed set X on $\mathbb{R}^d$ is a random element, defined on a probability space $(\Omega, \mathscr{A}, P)$, with values in the measurable space $(\mathscr{F}(\mathbb{R}^d), \sigma(\mathscr{F}))$, such that $X^{-1}(\sigma(\mathscr{F})) \subseteq \mathscr{A}$, where $\mathscr{F}(\mathbb{R}^d)$ is the set of closed subsets of $\mathbb{R}^d$, and $\sigma(\mathscr{F})$ is the Borel $\sigma-$field with respect to the hit-or-miss topology of $\mathscr{F}(\mathbb{R}^d)$.

By Choquet's theorem, the probability measure (law) $P_X = PX^{-1}$ of the random set X on $\sigma(\mathscr{F})$ is characterized by its *capacity functional* T_X (the counterpart of distribution function of a random vector), namely $T_X(K) = P\{\omega \in \Omega : X(\omega) \cap K \neq \emptyset\}$, for K compact, which can be extended to all $B \in \mathscr{B}(\mathbb{R}^d)$, generalizing P_X when the random set X is singleton, i.e., a random vector.

For example, if $\varphi(.) : \mathbb{R}^d \to [0, 1]$, upper semi-continuous, then $T(K) = \sup_{x \in K} \varphi(x)$ is a capacity functional.

As we will see, when applying Optimal Transport theory (next Sect. 6), we need "coupling", not only of random vectors, but also of random sets. For such general couplings, the concept of probability measures is needed, not just distribution functions. Thus, if we let P_X the probability measure of the random set X on $(\mathscr{F}(\mathbb{R}^d), \sigma(\mathscr{F}))$, corresponding to the capacity functional T_X, then by a coupling, say, of (X, Y), where Y is a random vector with distribution F_Y, or equivalently, probability measure P_Y on $(\mathbb{R}^d, \mathscr{B}(\mathbb{R}^d))$, we mean a joint probability measure π on $\sigma(\mathscr{F}) \otimes \mathscr{B}(\mathbb{R}^d)$ with marginal probability measures P_X, P_Y. This framework extends Copula framework from distribution functions (of random vectors) to probability measures (of random sets).

Here is a situation where estimating of set parameters leads to a nonparametric estimation of a probability density function $f(.)$ of a real-valued random variable X, as proposed by Hartigan (1987).

Let X be a random vector taking values in $\mathbb{R}^d$ with probability density $f(.) : \mathbb{R}^d \to \mathbb{R}^+$. For $\alpha > 0$, the $\alpha-$level set of f is the set $L_\alpha(f) = \{x \in \mathbb{R}^d : f(x) \geq \alpha\}$. Since,

$$f(x) = \int_0^{+\infty} A_\alpha(x) d\alpha$$

where $A_\alpha(.)$ denotes the indicator function of the set $L_\alpha(f)$, i.e., $A_\alpha(x) = 1_{L_\alpha(f)}(x)$, a set estimator $L_{\alpha,n}$ of $L_\alpha(f)$ could be used to estimate $f(.)$ by plug in

$$f_n(x) = \int_0^{+\infty} A_{\alpha,n}(x) d\alpha$$

where $A_{\alpha,n}(x) = 1_{L_{\alpha,n}(f)}(x)$.

How to estimate the set $L_\alpha(f)$? The following "maximum excess mass" principle, the counterpart of maximum likelihood principle, could be used.

Let $F(.)$ be the distribution function of X, $dF(x)$ the associated probability measure on $\mathscr{B}(\mathbb{R}^d)$, and $\mu(dx)$ the Lebesgue measure on $\mathscr{B}(\mathbb{R}^d)$. For each $\alpha > 0$, the signed measure $\mathscr{E}_\alpha(dx) = (dF - \alpha\mu)(dx)$ on $\mathscr{B}(\mathbb{R}^d)$ is called the excess measure at level α (its name comes from the fact that $(dF - \alpha\mu)(L_\alpha(f))$ is the excess mass of $L_\alpha(f)$).

Now, writing, for any $A \in \mathscr{B}(\mathbb{R}^d)$, as $A = (A \cap L_\alpha(f)) \cup (A \cap L_\alpha^c(f))$ we see that $\mathscr{E}_\alpha(A) \leq \mathscr{E}_\alpha(L_\alpha(f))$ so that $L_\alpha(f) = \arg\max_{A \in \mathscr{B}(\mathbb{R}^d)} \mathscr{E}_\alpha(A)$. Therefore, as in extrumum estimation, an estimator of $L_\alpha(f)$ can be proposed as iys empirical counterpart, namely $L_{\alpha,n}(f) = \arg\max_{A \in \mathscr{B}(\mathbb{R}^d)} \mathscr{E}_{\alpha,n}(A)$ where $\mathscr{E}_{\alpha,n}(A) = (dF_n - \alpha\mu)(A)$, $F_n(.)$ being the empirical distribution of $F(.)$.

Of course, consistency of the set estimator $L_{\alpha,n}(f)$ needs to be established (with additional assumptions) using random set theory. It is interesting to note that, why Hartigan's maximum excess mass principle is the counterpart of the maximum likelihood princeple, the optimization is not! It is an optimization of a *set function* $\mathscr{E}_\alpha(.)$ where a variational calculus is needed. See this issue in Nguyen (2006).

Now as we often face partial identified parameters in econometrics, as spelled out in Molchanov and Molinari (2018) (see also Nguyen (2021)), the theory of random sets appears as an additional necessary tool for statistical estimation and inference for set parameters. More specifically, the theory of random sets provides the foundations for statistics of random sets, including coarse data, generalizing statistics of random vectors. If "analogy is everything", then the extension from point statistics to set statistics is somewhat similar to the extention from Kolmogorov probability to *quantum probability* (see Sect. 7).

6 Optimal Transport for Econometrics

As stated in Sect. 3, it is surprizing that A. Sklar did not mention the old theory of optimal transport in his work on copula theory given that, while the goals of two theories are different, the internal setting is somewhat similar. A connection between optimal transport and copula could be beneficial to both theories, just like the connection between Potential Theory and Markov Processes.

This Section is about other advances in econometric analyses arising from Optimal Transport (OT) theory whose reformulation by Kantorovich 1941 was precisely for economics (allocation of resources).

By now, especially with the specific text "Optimal Transport Methods in Economics" Galichon (2016), econometricians should be aware of new tools, not only for analyzing new economic applications, but also for developing new tools. This Section is very specific: It is written for researchers who are not yet aware of OT theory, let alone its usefulness in economics. The purpose is obvious: encouraging researchers to study in depth OT applications, such as partial identification, random set theory, multivariate quantile regression.

The original optimal transport problem of Monge is reformulated by Kantotovich, with economic applications in mind, as follows (see, e.g., Nguyen (2021)).

Let μ, ν be two probability measures on $\mathscr{X}, \mathscr{Y}$ (e.g., $\subseteq \mathbb{R}^d$), respectively, and $c(.,.) : \mathscr{X} \times \mathscr{Y} \rightarrow [0, \infty)$, a "cost" function. Denoting by $\mathscr{M}(\mu, \nu)$ the set of joint probability measures on $\mathscr{X} \times \mathscr{Y}$ admitting μ, ν as marginal measures, i.e., for $\pi \in \mathscr{M}(\mu, \nu)$, we have $\pi(A \times \mathscr{Y}) = \mu(A), \pi(\mathscr{X} \times B) = \nu(B)$.

Remark. Each $\pi \in \mathscr{M}(\mu, \nu)$ generalizes the notion of a transport map, i.e., a map $T : \mathscr{X} \rightarrow \mathscr{Y}$, such that $\mu T^{-1} = \nu$, since for a such T, for $I \times T : \mathscr{X} \rightarrow \mathscr{X} \times \mathscr{Y}$, $(I \times T)(x) = (x, T(x))$, we have

$$\mu \circ (I \times T)^{-1}(A \times B) = \mu(A \cap T^{-1}(B))$$

in particular,

$$\mu \circ (I \times T)^{-1}(A \times \mathscr{Y}) = \mu(A \cap T^{-1}(\mathscr{Y}) = \mu(A \cap \mathscr{X}) = \mu(A)$$

and

$$\mu \circ (I \times T)^{-1}(\mathscr{X} \times B) = \mu(\mathscr{X} \cap T^{-1}(B) = \mu(T^{-1}(B)) = \nu(B)$$

In other words, $\mu \circ (I \times T)^{-1} \in \mathscr{M}(\mu, \nu)$. Elements of $\mathscr{M}(\mu, \nu)$ are called transport plans. For the case of $\mathbb{R}^d$ where μ, ν have associated distribution functions F, G, respectively, the joint measure $\pi \in \mathscr{M}(\mu, \nu)$ has the joint distribution function H belonging to the well known Frechet's class. Equivalently, $H(x, y) = S(F(x), G(y))$, where $S(.,.) : \mathscr{R}(F) \times \mathscr{R}(G) \rightarrow [0, 1]$ is a (uinique) subcopula; it is a copula if both F, G are continuous, where $\mathscr{R}(F)$ stands for the range of the function F.

An optimal transport plan is the joint probability measure $\pi^* \in \mathscr{M}(\mu, \nu)$ minimizing the objective function $\int_{\mathscr{X} \times \mathscr{Y}} c(x, y) d\pi(x, y)$ over $\pi \in \mathscr{M}(\mu, \nu)$. For example, with $c(x, y) = ||x - y||$, the euclidean distance on $\mathbb{R}^d$,

$$\pi^* = \arg \min_{\pi \in \mathscr{M}(\mu,\nu)} \int_{\mathscr{X} \times \mathscr{Y}} ||x - y|| d\pi(x, y)$$

Remark. A notion of (*optimal transport*) distance between two (arbitrary) probability measures μ, ν on $\mathscr{B}(\mathbb{R}^d)$, arising from optimal transport idea, suitable for Machine Learning, Image Processing and Data Science, is the Wasserstein distance defined as follows. For $p \geq 1$, the quantity

$$\mathscr{W}_p(\mu, \nu) = [\inf_{\pi \in \mathscr{M}(\mu,\nu)} \int_{\mathscr{X} \times \mathscr{Y}} ||x - y||^p d\pi(x, y)]^{\frac{1}{p}}$$

quantifies how much μ, ν are close to each other, and satisfies the axioms of a distance.

Note that the connection with copulas seems obvious by using Wasserstein distance between copulas to model dependence structures.

While there are applications of Optimal Transport in many fields (see e.g., Villani (2003)), we elaborate here only some main advanced tools, derived from it, for econometrics, as spelled out in Faugeras (2017) to call attention of empirical researchers.

Basically, Optimal Transport (OT) theory can contribute to econometric analyses when an economic issue can be formulated as an OT problem, since then we can take advantage, at least of two main things from OT theory: Existence of solutions, and computation via linear programming with available softwares.

As stated in Sect. 4, partial identification plays an essential role in credible statistical analysis. The crucial problem for a partially identified parameter in an econometric model is how to determine its identified set (for estimation)?

Partial identified parameters appear when models or data are incomplete such as interval data (coarse data, i.e., low quality data), missing data (censored, sample selection), or data with measurement errors.

Consider the case of coarse data, e.g., interval valued observations, say in a linear regression Y (income) on X (tax), where Y is only observed to be in a random set S (e.g., a random interval $[Y_*, Y^*]$), i.e., Y is an a.s. selection of S ($P(Y \in S) = 1$).

Suppose a parametric model for S is $S_\theta, \theta \in \Theta$, so that Y is an a.s. selection of S_θ. We are interested in estimating the true θ_o from data consisting of precise observations on X, and coarse data on Y. Clearly, θ_o is only partially identified, so that the first task is how to determine its identified set Θ_I?

Let P_X be the law of X on $\mathbb{R}^d$, and T_θ the capacity functional of the random set S_θ on $\mathcal{F}(\mathbb{R}^d)$. It should be clearly that $\theta \in \Theta_I$ if and only if X is an a.s. selector of S_θ. More specifically, $\theta \in \Theta_I$ if and only if (X, S_θ) is a "coupling" (i.e., X, S_θ are distributed as P_X, T_θ, respectively) such that X is an a.s. selector of S_θ.

Fortunately, the above condition (a characterization of selectionable distributions) is equivalent to $P_X(.) \leq T_\theta(.)$ on $\mathcal{B}(\mathbb{R}^d)$ see Artstein (1983), Norberg (1992), or Nguyen (2006).

For a "practical" determination of Θ_I, it turns out that we need OT, as spelled out in Galichon (2016): The above characterization of Θ_I is equivalent to an OT problem from which Θ_I can be actually obtained. See Galichon (2016) for details.

We turn now to another important contribution of OT to statistics in general, namely the extension of univariate quantiles to *multivariate quantiles*, and hence multivariate quantile regression.

For some time now, unlike the straightforward generalization of univariate mean regression to multivariate mean regression, various attempts to extend the notion of univariate quantiles to multivariate quantiles, let alone multivariate quantile regression, were not satisfactory. The difficulty is due to the fact that there is no natural total order on $\mathbb{R}^d$ for $d > 1$. But such approaches are "direct" approaches! In mathematics, we run into extension problems often, such as from ordinary sets to fuzzy sets, from Kolmogorov probability to noncommutative (quantum) probability, and we did that "indirectly": When we cannot extend a notion directly, we do it indirectly, i.e., looking for another equivalent form of that notion which can be extended.

Upfront, just like a complex number $z = x + iy$ that can be written in polar coordinates as $z = re^{i\theta}$, a random variable Y with distribution F can be "factored" as $Y = F^{-1}(U)$, where U is a random variable, uniformly distributed on [0, 1], called a *polar factorization* of Y. It is this polar factorization which is the appropriate equivalent representation for univariate quantile function to be extended to higher dimensions, as $Y = \nabla\varphi(U)$, when Y is a random vector in $\mathbb{R}^d$, $d \geq 2$, U is uniformly distributed on $[0, 1]^d$, and $\nabla\varphi$ is the gradient of a (unique) convex function $\varphi : [0, 1]^d \to \mathbb{R}$. Specifically, the vector quantile of a multivariate distribution function F is the gradient of a convex function, and its justification is within Optimal Transport Theory. If X is a $d-$dimensional random vector, then the *conditional vector quantile* of Y given $X = x$ is the multivariate quantile of the random vector $Y|X = x$.

While all details can be found in Galichon (2016), we will elaborate a bit on it.

In order to extend the notion of quantiles of an one-dimensional distribution function F (of a real-valued random variable X), we need to look closely at its quantile map $F^{-1}(.) : [0, 1] \to \mathbb{R}$, defined as

$$F^{-1}(u) = \inf\{x \in \mathbb{R} : F(x) \geq u\}$$

It is impossible to generalize this one dimensional quantile function to $\mathbb{R}^d$, with $d > 1$, since there is no (natural) total order relation of $\mathbb{R}^d$, if we try to generalize this function so defined. In other words, we cannot "directly" generalize this concept. We could try to generalize it "indirectly"?

Remember, how Kantorovich generalized Monge's OT formulation? For example, how to generalize a permutation σ (a pure assignment) on $(1, 2, ..., n\}$ to transport plan?

We cannot do it "directly", so we search for an "equivalent representation" of σ, i.e., looking for some indirect way. An equivalent representation (an one-to-one map) of a permutation is a permutation matrix to be generalized.

We could do the same thing to generalze quantiles. Perhaps, the difficulty is to find a "canonical" equivalent representation for the quantile map $F^{-1}(.)$ which could be extended.

Perhaps, it was so since an equivalent respresentation of $F^{-1}(.)$ is somewhat "hidden"!

Although we all know that $F^{-1}(.)$ is basic for **simulations** because if U is a random variable, uniformly distributed on [0, 1], then the random variable $F^{-1}(U) = F^{-1} \circ U$ has $F(.)$ as its distribution.

Note again that, while the polar factorization of a random variable is used for simulations, it is somewhat hidden (latent) in quantile regression analysis (not needed).

Thus, a characteristic of $F^{-1}(.) : [0, 1] \to \mathbb{R}$ is that it transports the uniform distribution $\mathcal{U}$ on [0, 1] to dF on $\mathbb{R}$, in the "language" of OT, in other words, the quantile function $F^{-1}(.)$ is a transport map in OT theory. Is it an equivalent representation for quantiles? Not obviously!

Any way, what seems to be missing is that the probability space $([0, 1], \mathcal{U})$ is hidden in the "background": When we define $F^{-1}(.)$, we did not (in fact, need) mention it at all. Only it surface after, for simulations.

It is hidden, but it's there! in the language OT, we need to involve the "background" $([0,1], \mathscr{U})$ to describe $F^{-1}(.)$ as a transport map.

So let say this. The quantile function $F^{-1}(.) : [0,1] \to \mathbb{R}$ is a transport map pushing $\mathscr{U}$ forward to dF.

If this is an equivalent representation of $F^{-1}(.)$ in the context of OT, then we hope to be able to say this.

Let $X : \Omega \to \mathbb{R}^d$ with multivariate distribution function $F(.) : \mathbb{R}^d \to [0,1]$. Then the quantile map of F is $Q_F(.) : [0,1]^d \to \mathbb{R}^d$, defined as the transport map pushing forward the uniform probability on $[0,1]^d$ to dF on $\mathbb{R}^d$.

We proceed now to justify the above definition of multivariate (vector) quantiles, to specify it, to give meaning to it, to provide examples, to define *conditional multivariate quantiles, and multivariate quantile regression.*

If we look closerly at the notion of (univariate) quantile function F^{-1} of a random variable X with distribution function F, then we realize something fundamental in Monte Carlo (simulation), namely $F^{-1}(U) = X$, for a random variable U, uniformly distributed on $[0, 1]$.

The upshot is this. Rather than "look" at the very definition of $F^{-1}(.)$, we could "look" at $F^{-1}(.)$ as a map from $[0, 1]$ to $\mathbb{R}$, having the property that $F^{-1}(U) = X$.

Specifically, consider $(\mathscr{X}, \mu) = ([0,1], u)$, where u is the uniform probability measure on $[0, 1]$, and $(\mathscr{Y}, \nu) = (\mathbb{R}, dF)$. Then we realize that $F^{-1}(.) : \mathscr{X} \to \mathscr{Y}$, such that $F^{-1}(X) = Y$. a.s., where $X \sim u$, and $Y \sim dF$, so that F^{-1} is a transport map (of Monge !).

However, in order to say that $F^{-1}(.)$ is characterized by such an OT map, we need to show that it is the only transport map in this OT formulation.

Next, for extending this to the multivariate case, we need to show that in the extended OT formulation, namely $(\mathscr{X}, \mu) = ([0,1]^n, u_n)$, where u_n is the uniform probability measure on the unit cube $[0,1]^n$, and $(\mathscr{Y}, \nu) = (\mathbb{R}^n, dF_n)$, where $F_n(.)$ is the multivariate distribution function on $\mathbb{R}^n$, there is a unique tramsport map.

If it is so, then the unique transport map between $([0,1]^n, u_n)$ and $(\mathbb{R}^n, dF_n)$ *can be used as the multivariate quantile function of the distribution* F_n.

It turns out that we do have a theoretical result confirming the above! See Galichon (2016).

Remark. For $n = 1$, consider $(\mathscr{X}, \mu) = ([0,1], u)$, noting that the uniform measure u is continuous, and $(\mathscr{Y}, \nu) = (\mathbb{R}, dF)$. The quantile function $F^{-1}(.) : [0,1] \to \mathbb{R}$, which is non decreasing, and transporting u to dF, because $F^{-1}(U) \sim dF$. The quantile function F^{-1} is nondecreasing and hence is the derivative of a convex function.

Definition of a multivariate quantile function. The $d-$quantile function of a multivariate distribution function $F(.) : \mathbb{R}^d \to [0,1]$ is the gradient $\nabla\varphi : [0,1]^d \to \mathbb{R}^d$, of some convex function $\varphi : [0,1]^d \to \mathbb{R}$, such that $\nabla\varphi(U) \sim dF$, where U is the uniform random vector on $[0,1]^d$.

The above map $\nabla\varphi$ (Brenier map) is the map between dU (uniform probability measure on the unit cube $[0,1]^d$) and dF.

In one dimension, $\nabla\varphi$ is $F^{-1}(.) : [0,1] \to \mathbb{R}$ (a nondecreasing function, such that $F^{-1}(U) \sim F$).

Conditional multivariate quantiles. Let X, Y be random vectors on $\mathbb{R}^d$, $\mathbb{R}^k$ with distribution F, G, respectively. Then the conditional multivariate quantile function of $Y|X = x$ is the Brenier map between dU on $[0, 1]^d$ and the conditional probability measure of $Y|X = x$, i.e., the multivariate quantile of the conditional distribution.

Specifically, the conditional quantile function of $Y|X = x$ is $\nabla\varphi_x$ where $\varphi_x(.)$ is a convex function on $[0, 1]^d$ with $Y = \nabla\varphi_X(U)$.

Note that there are many attempts to define multivariate quantiles in the literature, but as R. Koenker said, this approach based on OT seems the best! mainly because it captures two basic properties of the univariate quantile function $F^{-1}(.) : [0, 1] \to \mathbb{R}$ (as a kind of "inverse" of F, with a precise meaning, e.g., of median), namely $F^{-1}(.)$ is a monotone (nondecreasing) function, and $F^{-1}(U) = Y$ (where $Y \sim dF$). This is so because, as the gradient of a convex function, $\nabla\varphi$ is the natural generalization of monotonicity in one dimension case, and $Y = \nabla\varphi_X(U)$ when X is a covariate.

Linear multivariate quantile regression. Let $Q_{Y|X}(u|x)$ be the conditional (multivariate) quantile of $Y|X = x$ at level $u \in [0, 1]^d$. A linear model for it is

$$Q_{Y|X}(u|x) = \beta_o(u)^T g(x)$$

so that we have the representation

$$Y = \beta_o(U)^T g(X)$$

with $U|X \sim$ uniform $[0, 1]^d$, $\beta(u)$ is $k \times d$ matrix ($X \in \mathbb{R}^k$).

This formulation leads to a linear programming to computing $\beta(u)$ both for population and sample settings.

7 Quantum Modeling of Economic Phenomena

Just like a century ago when quantum mechanics appeared to improve Newtonian mechanics, recently, quantum probability setting started entering social sciences, including economics. The recent literature argues the necessity to develop *quantum econometrics* and *quantum computational economics.* Of course, by quantum economics we mean economics where standard probability (for modeling uncertainty) is replaced by quantum probability. The rationale of using quantum probability (a generalization of standard probability) in social sciences is that it models more faithfully human decision-making under uncertainty that we observe.

Essentially, as we can guess, the movitation for considering quantum economics is rooted in the usual uncertainty we employ so far in modeling of economic phenomena. It is all about the type of uncertainty we use to model economic phenomena, as uncertain (dynamical) systems, and from which to make decisions. Since there are human agents involved, behavioral economic analyses are much more delicate than physical systems. Thus, the question is "Is Kolmogorov probability theory sufficient or adequate to model the uncertainty in social or economic phenomena?". The answer seems to be "We need to enlarge Kolmogorov probability to quantum probability".

We choose to elaborate a bit on the notion of quantum probability here because it is the core of the literature on quantum economics. For more details on quantum probability calculus, see e.g., Parthasarathy (1992).For a flavor of quantum computing, say, in financial econometrics, see Milek (2020), Norberg (1992), Woerner and Egger (2019).

Roughly speaking, quantum probability is a noncommutative extension of Kolmogorov probability. The following simple setting will suffice to digest the idea.

The source of uncertainty in standard probability theory is a probability space $(\Omega, \mathscr{A}, P)$. In the simplest case, the sample space Ω is a finite set, say, $\Omega = \{1, 2, ..., n\}$, the $\sigma-$field $\mathscr{A}$ is the power set 2^{Ω}, and the probability measure P is determined by its density function $f : \Omega \to [0, 1]$, $f(j) = P(\{j\})$, $\sum_{j=1}^{n} f(j) = 1$.

A real-valued random variable $X(.) : \Omega \to \mathbb{R}$, in this simple setting, is identified as the vector $(X(1), X(2), ..., X(n))'$, and an even $A \subseteq \Omega$ is idenfied as its indication function $1_A(.) : \Omega \to \{0, 1\}$. Thus, everything in this setting is identified as an element of $\mathbb{R}^n$, including the probability density function $f(.)$.

Now a vector $X = (X(1), X(2), ..., X(n))' \in \mathbb{R}^n$ can be further idenitifed as an $n \times n$ diagonal matrix

$$[X] = \begin{bmatrix} X(1) & 0 & . & . & 0 \\ 0 & X(2) & 0 & . & 0 \\ . & 0 & . & . & . \\ . & . & 0 & . & . \\ 0 & . & . & . & X(n) \end{bmatrix}$$

In particular, the event A is identified as $[1_A]$, and $f(.)$ as $[f]$.

Now, a $n \times n$ matrix with real entries acts as a linear operator on $\mathbb{R}^n$. Thus, in terms of operators, a random variable is a symmetric (self-adjoint) operator, an event is a projection operator, and a probability density is a positive operator with unit trace.

This equivalent representation of elements of probability theory is suitable for extension to quantum probability as follows.

First, the sample space $\Omega = \{1, 2, ..., n\}$ is replaced by $\mathbb{R}^n$ (a real, finitely dimensional, separable Hilbert space), an event is an arbitrary projection operatot, i.e., a linear map $A : \mathbb{R}^n \to \mathbb{R}^n$ such that $A = A^2 = A^*$ (transpose), the density function T is an arbitrary positive operator, i.e., $\langle Tx, x\rangle \geq 0$, for all $x \in \mathbb{R}^n$, with unit trace (sum of the diagonal terms), and a random variable X, called an observable, is an arbitrary (not necessarily diagonal) symmetric matrix. In view of the non commutativity of arbitrary matrix multiplication, quantum events are non commutative in general.

Moving a step further, we replace $\mathbb{R}^n$ by a complex, infinitely dimensional, separable Hilbert space H, we extend all operators above to self adjoint ones (complex symmetric), so that the set $\mathscr{P}$ of all projections on H plays the role of quantum events, the set $\mathscr{S}$ of all self adjoint operators are quantum observables, and a positive operator ρ with unit trace is a quantum density matrix. In summary, a *quantum probability space* is a triple $(H, \mathscr{P}, \rho)$.

For quatum probability calculus, let's pursue a little more in a finite quantum probability space $(\mathbb{R}^n, \mathscr{P}, \rho)$ to be concrete. When a random variable $X : \Omega \to \mathbb{R}$ is represented by the matrix (observable) $[X]$, its possible values are on the diagonal of $[X]$, i.e., the range of X is $\sigma([X])$, the spectrum of the operator $[X]$. For $A \subseteq \Omega$, $\Pr(A)$ is taken to be $P([1_A]) = \sum_{j\in A} f(j) = tr([f][1_A])$. More generally, $EX = tr([f][X])$ exhibiting the important fact that the concept of trace (of a matrix/operator) replaces integration, a fact which is essential when considering infinitely quantum spaces. Next, the spectral measure of a random variable X, represented by the observable $[X]$ is the projection-valued "measure" $\zeta_{[X]} : \mathscr{B}(\mathbb{R}) \to \mathscr{P}(\mathbb{R}^n)$, $\zeta_{[X]}(B) = \sum_{X(j)\in B} \pi_{X(j)}$, where $\pi_{X(j)}$ is the (orthogonal) projection on the space spanned by $X(j)$.The quanum probability of the event $(X \in B)$, $B \in \mathscr{B}(\mathbb{R})$ is $P(X \in B) = \sum_{X(j)\in B} f(j) = tr([f]\zeta_{[X]}(B))$. For quantum stochastic calculus, see e.g., Parthasarathy (1992).

References

Artstein, Z.: Distribution of random sets and random selections. Israel J. Math. **46**, 313–324 (1983)

Black, F., Litterman, R.: Global asset allocation with equities, bonds, and currencies. Fixed Income Res. **2**, 15–28 (1991)

Durante, F., Sempi, C.: Principles of Copula Theory. Chapman & Hall/CRC Press (2016)

Faugeras, O.P.: Inference for copula modeling of discrete data: a cautionary tale and some facts. Dep. Model. **5**(1), 121–132 (2017)

Galichon, A.: Optimal Transport Methods in Economics. Princeton University Press, Princeton (2016)

Hartigan, J.A.: Estimation of a convex density contour in two dimensions. J. Am. Statist. Assoc. **82**, 267–270 (1987)

Heckman, J.: Sample selection bias as a specification error. Econometrica **47**, 153–161 (1979)

Manski, C.F. : Partial Identification of Probability Distributions. Springer (2003)

Manski, C.F.: Identification for Prediction and Decision. Harvard University Press (2007)

Marshall, A.W.: Copulas, marginals, and joint distributions. In: Distributions with Fixed Marginals and Related Topics. IMS Lecture Notes, vol. 28, pp. 213–222 (1996)

Matheron, G.: Random Set and Integral Geometry. Wiley (1975)

Meucci, A.: Risk and Asset Allocation. Springer (2003)

Milek, J.: Quantum Implementation of Risk Analysis-Relevant Copulas (2020). arXiv: 2020.07389v2

Molchanov, I., Molinari, F.: Random Sets in Econometrics. Cambridge University Press (2018)

Nguyen, H.T.: An Introduction to Random Sets. Chapman & Hall/CRC Press (2006)

Nguyen, H.T.: On random sets for inference in statistics and econometrics. In: Prediction and Causality in Econometrics and Related Topics, Springer Series "Studies in Systems, Decision, and Control", to appear 2021

Norberg, T.: On the existence of ordered couplings of random sets with applications. Israel J. Math. **77**, 241–264 (1992)

Orus, R., Mulgel, S., Lizaso, E.: Quantum computing for finance: overview and prospects (2018). arXiv:1807.03890v1

Parthasarathy, K.R.: An Introduction to Quantum Stochastic Calculus. Springer (1992)

Villani, C. : Topics in optimal transportation. Am. Math. Soc. (2003)

Woerner, S., Egger, D.J.: Quantum risk analysis. njp Quantum Inf. (2019). https://doi.org/10.1038/s41534-019-0130-6

What's Wrong with How We Teach Estimation and Inference in Econometrics? And What Should We Do About It?

Mark E. Schaffer

Abstract The widespread use of "Null hypothesis significance testing" and *p*-values in empirical work has come in for widespread criticism from many directions in recent years. Nearly all this literature and commentary has, understandably, focused on practice: how researchers use, and abuse, these methods and tools, and what they should do instead. Surprisingly, relatively little attention has been devoted to what to do about how we teach econometrics and applied statistics more generally. I suggest that it is possible to teach students how to practice frequentist statistics sensibly if the core concepts they are taught at the start are "coverage" and interval estimation. I suggest various tools that can be used to convey these concepts.

1 Introduction

The widespread use of "Null hypothesis significance testing" and *p*-values in empirical work has come in for widespread criticism from many directions in recent years. Nearly all this literature and commentary has, understandably, focused on practice: how researchers use, and abuse, these methods and tools, and what they should do instead.

Surprisingly, relatively little attention has been devoted to what to do about how we teach econometrics and applied statistics more generally. It is surprising because widespread poor practice by applied researchers is a clear reflection of what researchers were taught when they were first learning basic statistics. It will be difficult to make progress in changing empirical practice unless we also change how we teach econometrics and statistics.

Invited paper for the Fifth International Econometric Conference of Vietnam, Ho-Chi-Minh City, Vietnam, 10–12 January 2022. All errors are my own.

M. E. Schaffer (✉)
Heriot-Watt University, Edinburgh, UK
e-mail: M.E.Schaffer@hw.ac.uk

N. Ngoc Thach et al. (eds.), *Financial Econometrics: Bayesian Analysis, Quantum Uncertainty, and Related Topics*, Studies in Systems, Decision and Control 427,
https://doi.org/10.1007/978-3-030-98689-6_9

In this paper I suggest that it is possible to teach students how to practice frequentist statistics sensibly if the core concepts they are taught at the start are "coverage" and interval estimation. I suggest various tools that can be used to convey these concepts.

2 What's Wrong?

2.1 The NHST Problem

A researcher estimates

$$y_i = \beta x_i + \varepsilon_i \tag{1}$$

usually with a constant and some "controls", and then tests the null hypothesis $H_0 : \beta = \bar{\beta}$ based on the estimated $\hat{\beta}$ and its standard error. The most common choice of $\bar{\beta}$ is zero, i.e., the researcher is testing whether or not the variable of interest x_i has any causal or predictive role. If the p-value is less than 5%, the researcher declares victory: β is "statistically significant"—shorthand for "statistically significantly different from zero"—and it's time to write it up and send it off to a journal.

The use of this practice of "null hypothesis significance testing" (NHST) and p-values has attracted an enormous amount of attention in the applied statistics literature in recent years. To cite just two examples, the American Statistical Association released a "Statement on Statistical Significance and P-Values" in 2016 (Wasserstein and Lazar (2016)), with six principles including "Scientific conclusions and business or policy decisions should not be based only on whether a p-value passes a specific threshold." And in 2019 Amrhein et al. (2019) published a paper in *Nature*, cosigned by over 800 researchers (including myself), enjoining researchers to "retire statistical significance" in favor of more nuanced interpretation. Most of the economics discipline largely ignored this debate until fairly recently, despite the earier efforts of some, notably McCloskey and Ziliak (see e.g. McCloskey and Ziliak (1996)).

The good news is that the economics discipline now is starting to take this issue seriously; for example, the summer 2021 issue of the *Journal of Economic Perspectives* contains a symposium on statistical significance, with contributions from Kasy (2021), Miguel (2021), and one of that year's Nobel Laureates, Imbens (2021).

Most of the wider debate has been about the problems of misuse of statistical significance, p-values, NHST etc. in the practice of research. But the problem is rife in teaching, as a casual skim of econometrics (or indeed applied statistics textbooks in general) will reveal. (NB: *Mea culpa*. Looking at my old teaching materials makes for uncomfortable reading in places.) It is not hard to make the case that the problem starts at home, in the classroom. If we are to make progress in improving general practice in and understanding applied econometrics and statistics, we have to start with our students.

2.2 The Teaching Problem

In most textbooks with a frequentist orientation, the building blocks of inference are assembled piece by piece, starting with probability and distributions, then moving on to point estimates, sampling distributions, and then hypothesis tests. Point estimates and hypothesis tests are typically presented either explicitly or implicitly as the end product, i.e., the objective of the empirical research.

The number one entry on what should be the economists' list of problems with hypothesis tests of a parameter: it rarely helps to answer any question of economic interest. As economists, we almost always want to know the answers to "How big is the effect?" and "How precisely is it estimated?" Testing $H_0 : \beta = \bar{\beta}$ answer neither of these questions.

Say the researcher rejects the null and concludes that β is likely nonzero.

- What if $\hat{\beta}$ is extremely small but extremely precisely estimated?
- What if $\hat{\beta}$ is very large but the standard error is also huge?

It is obvious the researcher should treat these two cases very differently: the researcher should treat the first case as a solid finding that the effect of x is negligible, and the second case as effectively inconclusive because β is estimated so imprecisely, unless the precision is enough to perhaps say something about the sign of the effect if not its magnitude. Stopping with "statistical significance" is clearly not enough.

The same sort of reasoning applies if the researcher fails to reject the null of a zero effect. If the standard error of $\hat{\beta}$ is extremely small, the researcher has a solid result—no effect of x. But if the standard error is very large, there is little to report—the finding is consistent with a possible large effect of either sign, or no effect at all.

Here is a modified example of an end-of-chapter exercise from a leading textbook. (I should note that the textbook is actually very good. These are not empty words; I use this very textbook myself for teaching undergraduates.)

> A researcher wishes to investigate a possible gender wage gap in a firm. The researcher has a sample of 1,000 employees, roughly equally divided between men and women. The individuals in the sample were randomly selected from the firm's payroll and are employeed in the same occupation. [Data and estimation detail follows.]
>
> 1. What do the findings of the researcher suggest about the firm's gender wage gap? Does estimated difference in wages between men and women amount to statistically significant evidence?
> 2. Do the findings of the research suggest that the firm is guilty of wage discrimination? Discuss.

What's wrong with this? The students are asked to report the estimated wage gap (the point estimate), which is good. But reporting just whether we can reject a zero gap or not is clearly not enough.

- What if we can reject that the gender wage gap is zero but we can't reject that it's either \$10/year or \$25k/year?

- What if we can't reject that the gender wage gap is zero but we also can't reject that it's \$25k/year?
- What if ...
- The key unasked question—and therefore unanswered by students when they do this end-of-chapter exercise: *How precisely estimated is the gap?*

3 Interval Estimation

What should be done instead? In sum: interval estimation should be the key teaching outcome.

Teaching interval estimation rather than point estimation as the key estimand automatically emphasises uncertainty. An interval estimate has the key ingredients to understanding an econometric estimate: it conveys the estimated magnitude of the quantity of interest, and at the same time it conveys information about how precisely that magnitude is estimated.

If students remember only one thing from their basic econometrics or statistics training, and it's to look for and how to interpret an interval estimate, they are well prepared: if an interval estimate is reported, they know how to interpret it; if an interval estimate is not reported but the ingredients for it are (a point estimate and standard errors, say), they know how to construct it; and if neither the interval nor the ingredients are reported, they they know to ask "but how precise is this estimate?"

More generally, the key estimand for students should be $[\hat{\beta}_{LL}, \hat{\beta}_{UL}]$; *not* $\hat{\beta}_{OLS}$

I focus on frequentist inference for the purposes of this paper, which means here confidence intervals. But a similar argument can be constructed for a introductory Bayesian course.

Using the gender wage gap example and frequentist confidence intervals, students should automatically summarize the findings like these examples:

- "Based on this sample, and using a 95% confidence interval, we estimate the firm's gender wage gap to be [17%, 19%]." ... -or-
- "Based on this sample, and using a 95% confidence interval, we estimate the firm's gender wage gap to be [1%, 35%]."

It's easy to see, and to teach, the difference between these two results: the first estimate is obviously more precise than the second, and the metric is easy to understand. Compare the traditional NHST approach that also shows up as an early exercise in many textbooks: "At the 5% significance level, we can reject the null hypothesis that there is no discrimination. The p-value is $xxx\%$." What are students to make of this?

3.1 Coverage and Confidence Intervals

The key concept behind confidence intervals is *coverage*. Teaching this concept is easier than it sounds, and much easier than teaching p-values. But coverage has its subtleties. The two trickiest aspects of coverage to convey are that the coverage is property of the *procedure*, and that the parameter to be estimated is constant and it is the *interval* generated by the procedure that is *random*.

Two definitions of coverage:

Definition: The coverage probability (or just coverage) of an estimation procedure for a parameter β is the probability that the estimated interval will contain the true β.

Definition: The coverage probability of an estimation procedure for a parameter β is the frequency with which the intervals from the procedure contain the true β in repeated samples.

The advantage of the first definition is that it makes clear that the interval is what is random. The advantage of the second definition is that it is explicitly frequentist—the concept of "in repeated samples" comes up any time we teach frequentist statistics—and that it is the property of the procedure and not any particular interval that it generates. In particular, using the second definition it should be easy for the student to see that it is the interval is that is random (random draw of the data, use it to calculate an interval, a function of random variables is itself random, etc.).

The last conceptual ingredient is "nominal coverage". This can be taught as the objective of econometric theory: can theory deliver a procedure whose nominal coverage—the coverage probability specified by the applied econometrician—is equal to actual coverage? And of course the answer is "yes, if certain assumptions are satisfied".

And now we can start with students by defining a 95% confidence interval in a simple and intuitive way:

Definition: A 95% confidence interval procedure is a procedure that enables us to construct intervals with 95% coverage.

And under certain assumptions to be made explicit, nominal coverage of the procedure equals actual coverage: if we aim to construct 95% confidence intervals, and these assumptions are satisfied, then our intervals will have actual coverage of 95%.

3.2 Teaching Coverage

Teaching this is easier than it sounds, because there are good analogies available. Here is one: "Mystery Pin-the-Ring-on-the-Donkey".

Pin-the-Tail-on-the-Donkey is a well-known children's party game. In Pin-the-Tail-on-the-Donkey, a large poster of a donkey is put on a wall. The donkey is missing its tail. The child playing is given a tail with a pin or something sticky so that the tail can be attached to the donkey in the appropriate place [sic]. The catch

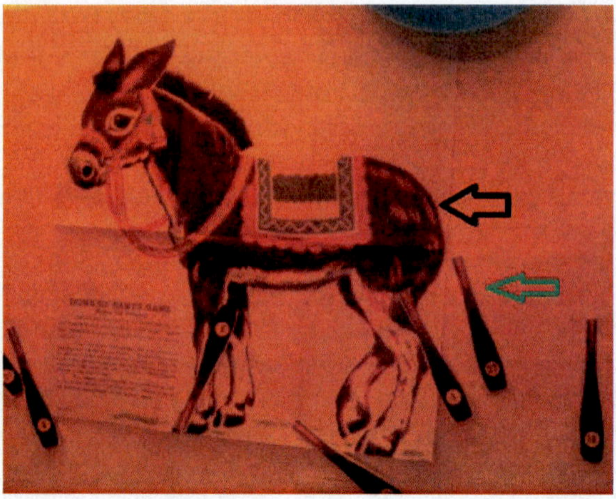

Fig. 1 Pin-the-Tail-on-the-Donkey. The black arrow indicates where the tail belongs; the green arrow points to the tail placed by the winner

is that the child is blindfolded and then spun around several times so that they are disoriented. They are then pointed towards to donkey poster and told to try to put the tail as close as possible to where the tail belongs. The other children at the party can yell clues and suggestions to the blindfolded child: "Higher!" "To your right!" "Lower!" And so on.

The picture here, from Wikimedia Commons, shows the aftermath of a play of the game. The target point—where the tail belongs—is indicated by the black arrow. The winning player's tail is indicated by the green arrow.[1]

Pin-the-Tail-on-the-Donkey is a common children's game in some countries but not in others; but it is a simple game aimed at very young children, so explaining this to students, perhaps with a short video of children playing, should be very easy.

"Mystery Pin-the-Ring-on-the-Donkey" differs from conventional Pin-the-Tail-on-the-Donkey in two key respects (see Fig. 1):

- In Pin-the-**Ring**-on-the-Donkey you're blindfolded as usual, but instead of a tail with a pin, you're trying to place a ring on the poster where the donkey's tail goes. If the ring contains the point where the tail goes, you get a point; if it doesn't, you get nothing.
- In **Mystery** Pin-the-Ring-on-the-Donkey, you never find out the result of any particular game. The blindfold never comes off, or, equivalently, the donkey is removed from the poster before the bindfold comes off.

The features of this game share a near-complete range of characteristics with frequentist confidence intervals:

- The ring is our confidence interval.

---

[1] Source: Wikimedia Commons. https://en.wikipedia.org/wiki/File:Pin_the_Tail_On_the_Donkey-example.jpg. Licensed under the Creative Commons Attribution 2.0 Generic license. The photograph has been trimmed slightly and the two arrows added; it is otherwise unchanged.

- The placement of the ring is random; each time we play the game, the ring will end up in a different place.
- But any one time we play the game, it will either contain, or not contain, the point where the tail belongs. The ring is no longer random after the end of play.
- The size of the ring corresponds to the confidence level. Want to win more often? Use a larger ring. (Set the confidence level higher and the confidence interval will be wider.)
- The clues yelled out by our friends are datapoints. Very few clues (a small dataset) and we are likely to do poorly; more clues and we will do better.
- The frequentist properties of interval construction are analogous to playing the game repeatedly. We know that if we play the game over and over, 95% of the time we will score a point, just as 95% of the time our confidence interval will contain the true β. But in any one play of the game we might or might not score.
- Our skills at understanding the clues and acting on them is analogous to econometric theory and efficient estimation, but with an important difference. In the game, the only way to find out the actual coverage of our ring-placing skills is to play the game repeatedly and then take off the blindfold. In econometrics, the magic of mathematical statistics tells us that under certain assumptions, we know already what the coverage of our procedure will be; and we can check this using simulated data and Monte Carlos.
- The other key difference with the conventional children's game is that we never get to take off the blindfold (or, equivalently, the donkey is removed and only then the blindfold comes off). In "Mystery Pin-the-Ring-on-the-Donkey" we never find out if we won, and in real-word applied econometrics, we never find out if our interval really did contain the true β. (But see below—sometimes we find out if it *didn't*.)

This can be done live in the classroom with a real Pin-the-Tail-on-the-Donkey kit, and will be engaging and entertaining.[2]

Note, by the way, that "Mystery Pin-the-Ring-on-the-Donkey" makes it easy to teach two-dimensional confidence sets as well. The winner needs to be correct in both dimensions, vertically and horizontally, in order to get a point; and a confidence set has 95% coverage if, in repeated samples, 95% of confidence sets contain the pair (β_1, β_2).

4 Confidence Intervals and Realized Confidence Intervals

A subtle point that is easy for students and practitioners to miss is that the coverage properties of confidence intervals apply to the procedure we use to construct them and not to the intervals they actually calculate. The good news here is that this subtlety can be missed and the user can still employ confidence intervals sensibly. Students

[2] Contrast this with the more traditional live classroom approach of using a computer simulation to illustrate the coverage properties of confidence intervals. While this is instructive and can be helpful, it is very dry and on its own is likely to go over heads of some students.

know that [17%, 19%] is a precise estimate of the gender wage gap in a company, and [1%, 35%] is a noisy estimate, even if they forget or don't understand the subtle point that the coverage property applies to the procedure rather than to a particular realized interval.

But it's clearly important for students in advanced courses, and for practitioners, that this is understood. We can still teach this point, and teach it an accessible way, by extending the game analogy above. We can do this by focusing on the possibility that realized confidence intervals can be empty, or more generally can include values of the parameter β that are literally impossible, without violating the coverage properties of the estimation procedure. This is not an empty thought experiment—such estimation procedures are actually in common use in econometrics, as I will discuss below.

We extend the "Mystery Pin-the-Ring-on-the-Donkey" by introducing a formal set of rules—call it "*Olympic* Mystery Pin-the-Ring-on-the-Donkey".

In the modern javelin competition, the javelin has to land in a well-defined "sector" that extends outwards from the point where the javelin is launched. If the thrown javelin lands outside this sector, it is an illegal throw and does not count.

Similarly, in "Olympic Mystery Pin-the-Ring-on-the-Donkey", the ring, when placed, has to contain part of the poster on which the donkey appears.[3] If the ring is placed outside the poster, that play is disqualified. Crucially, the player is told after the play *whether or not the ring was placed on the poster*, i.e.,whether or not it was a legal play. Or, equivalently, the donkey is removed from the poster and then the blindfold comes off, so the player can see whether the ring was placed legally or not.

This is a simple but effective way to convey that the coverage property of confidence intervals applies only to the procedure. Prior to any play of "Olympic Mystery Pin-the-Ring-on-the-Donkey", a player may have a coverage probability of 95%, i.e., a 95% change of scoring a point. But after the donkey is removed from the poster and the player removes their blindfold, they can see whether it was a legal play or not. If it was not a legal play, then the ring covers an area where it is literally impossible for the donkey's tail to belong. Yet the ex ante coverage probability of the procedure was (and remains, for future plays) 95%. Similarly, a realized confidence interval can contain values that are literally impossible for β to take, and yet the procedure can have nominal coverage equal to average coverage.

This is not an empty academic point. Indeed, understanding this point is essential to a proper understanding of a now-common set of empirical methods that extend the instrumental variables (IV) setting to the case where the parameters are not assumed to be identified. In these procedures, where we denote the set of excluded instruments by $\mathbf{z}_i'$, the rank condition $E(\mathbf{z_i}'x_i) \neq \mathbf{0}$ is dropped. The earliest and best known of these weak-instrument-robust methods for constructing confidence intervals is based on the Anderson-Rubin (1949) test. The AR test of the null $H_0 : \beta = \bar{\beta}$ is performed by defining a new variable $\tilde{y}_i = y_i - \bar{\beta}$ and estimating the following auxiliary regression using OLS:

$$\tilde{y}_i = \mathbf{z_i}'\Theta + u_i \tag{2}$$

[3] Note that in the photograph of an actual play of Pin-the-Tail-on-the-Donkey above, some of the pinned tails have indeed missed the "playing area".

The null hypothesis to test becomes $H_0 : \Theta = \mathbf{0}$. Rejection is interpreted as a rejection of the original null $H_0 : \beta = \bar{\beta}$ or of some other aspect of the model (in particular, a failure of the orthogonality condition $E(\mathbf{z_i}'x_i) = \mathbf{0}$ can also lead to rejecting the new null). A confidence interval is constructed by assembling the set of all $\bar{\beta}$s that cannot be rejected at some α significance level.

The intuition behind this approach is that if $\beta = \bar{\beta}$, the new auxiliary variable $\tilde{y}_i$ should now be orthogonal to the instruments $\mathbf{z}_i$, since the only route by which the instruments can be correlated with y_i is via x_i and β. The appeal of this approach is that it dispenses with the rank condition entirely. Outright failure of the rank condition (underidentification) or even just near-failure (weak identification) is fatal to consistent estimation by IV, GMM or their relatives. In weak-instrument-robust inference, the rank condition never appears. Rather, as instruments get weaker ($E(\mathbf{z_i}'x_i)$ gets smaller), the confidence interval for β gets wider. This is exactly as it should be and is a very appealing feature of the procedure: the weakness of instruments is built into the precision of the estimates.

It's easy to see, though, that in the overidentified case where there is more than one instrument, one can always construct an ex post AR confidence interval that is empty, just by choosing a large-enough α. A test for the joint null $H_0 : \Theta = 0$) will in practice never have a p-value of 1. See, for example, the graph below for the case of 2 excluded instruments and a heteroskedasticity-robust weak-inference-robust AR confidence interval for a wage equation (the coefficient is the return to education). Note that the vertical axis the is rejection probability, defined as $r = 1 - p$. The AR 95% confidence interval is $[-0.02, 0.14]$. The minimum p-value across the full range of possible $\bar{\beta}$s is 0.80, so a $1 - 0.80 = 20\%$ AR confidence interval will be empty.

Now consider the same estimation, but where we have replaced one of the excluded instruments with a variable that is deliberately "bad", i.e., correlated with u_i so that the orthogonality condition fails. The AR 95% confidence interval is now [0.0, 0.13]. But this apparently more precise estimation is an illusion. The minimum p-value across the full range of possible $\bar{\beta}$s is 0.20, so a $1 - 0.20 = 80\%$ AR confidence interval will be empty. In other words,the apparent precision of the estimate is actually indicative of *misspecification*. The confidence interval is narrow because the data are hard to reconcile with both the hypothesized $\bar{\beta}$s and the orthogonality condition $E(\mathbf{z_i}'x_i) = \mathbf{0}$. The empty 80% confidence interval is actually very informative here.

The narrow lesson here is that in this category of problem, just the confidence interval is not enough; the research should examine the p-value function graphically, as we did above. The broader lesson is that understanding the difference between confidence intervals and realized confidence intervals can be essential for some now-standard procedures (see Figs. 2 and 3).

For more on realized confidence intervals and how to interpret them, see Müller and Norets (2016). They show how to use the concept of "bet-proofness": if you observe a realized $(1 - \alpha)$ confidence interval, you cannot make money betting that the true β isn't in the interval. Realized confidence intervals are "bet proof" for

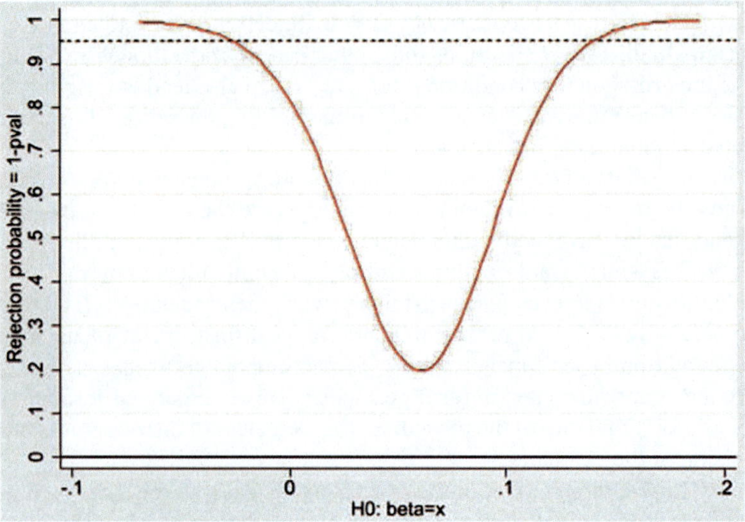

Fig. 2 Anderson-Rubin realized confidence interval

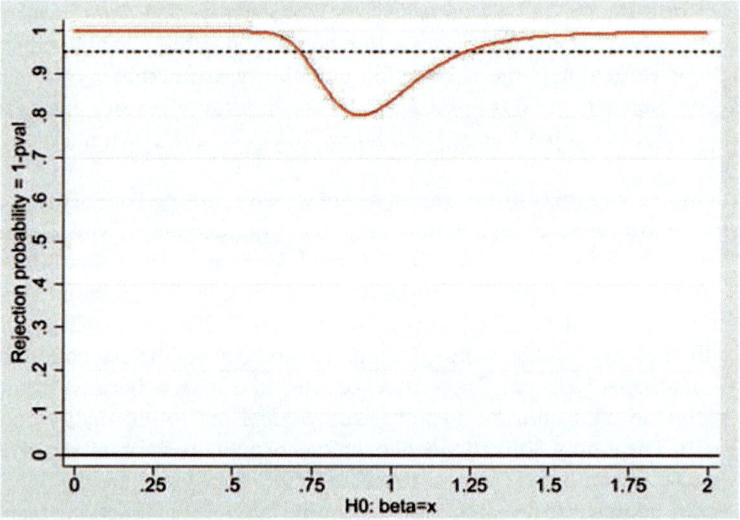

Fig. 3 Anderson-Rubin realized confidence interval + "bad" IV

standard problems, where "standard problems" means the observer doesn't see, e.g., empty intervals or impossible values (or illegal plays of the ring in our Olympic game). They also discuss the connection between frequentist realized confidence intervals and Bayesian credible intervals.

5 Teaching the Formalities: Confidence Intervals and Hypothesis Tests

We need to teach how we know frequentist CIs have these coverage properties, even if many of the students will forget. For most students—after all, most will not move into professions where they need to use formal statistical tools—this not a problem if they forget the formal definitions and derivations but remember the coverage interpretation. But good students will remember, as will those who go on to later, more advanced study. So we need to teach Neyman-Pearson hypothesis testing and, in fact, p-values.

Denote by $t(\bar{\beta})$ the test statistic t as a function of the hypothesized $\bar{\beta}$. Write $p(t(\bar{\beta}))$ to mean the p-value function, i.e., the probability of observing a test statistic at least as extreme as t, assuming that the model is correct and the null hypothesis $H_0 : \beta = \bar{\beta}$ is true. We then define a confidence interval in the usual way:

Definition: A $(1 - \alpha)$ confidence interval C for β is the set of all hypothesized $\bar{\beta}$ such that we will fail to reject $H_0 : \beta = \bar{\beta}$ at the α significance level: $C = \bar{\beta} : p(t(\bar{\beta})) \leq \alpha$.

How should we teach this definition when it depends on such a tricky and misunderstood concept, namely the p-value?

5.1 *Teaching Hypothesis Tests and p-Values*

The main answer to this question is to teach the p-value as an intermediate input that figure on the way to the construction of the final product, namely the interval construction procedure. If we take this route to discussing and interpreting p-values, it is hopefully clear to students that a small p-value is not an end in itself; rather, it is a tool used in the interval construction procedure. This is as it should be: p-values have their values.

But the definition of p-value is famously difficult to teach and interpret. Demoting the p-value to an intermediate input is helpful, but of course it would be better if our students (and those who become practitioners) had a good intuitive grasp of what it actually means. Plus, as I discuss below, there are cases in econometrics where the p-value does have a use as a final output.

Thanks to Greenland (2019), a new interpretation of p-values is now available, one that lends itself well to teaching. The idea is simple: convert p-values to the Shannon information s-value or "surprisal value":

$$s \equiv -log_2(p) \tag{3}$$

The surprisal value s is the information content of the p-value in bits. It is surprisingly [sic] easy to explain.

Consider repeated flips of a fair coin. How surprised would you be if you observed s heads in a row? This is exactly how you should respond if your test statistic has p as its p-value. Applying estimation procedures to a dataset to obtain a test statistic corresponds to flipping a coin. The assumption of a fair coin corresponds to the assumption that the model and null hypothesis are correct, and hence the estimation procedure yields tests that are correctly sized. If you are very surprised by the number of heads you flip in a row, you may conclude that something is fishy about the coin, just as a small p-value can lead the researcher to question their assumptions about the null hypothesis and/or the model.

An important feature of the s-value transformation is that it puts the probability of obtaining an extreme test statistic into a metric that is easy to explain and understand. Note that $-\log_2(0.05) \approx 4.3$. Is 4 heads in a row surprising? Yes, it's surprising ... but not *very* surprising. This is an excellent way to get across to students (and practitioners) that the traditional 5% significance level is not very tough, and that a 95% confidence level yields interval estimates that are rather loose.

Having demonstrated the role of the p-value as an intermediate input in the construction of confidence intervals, we can also present other uses of the concept, at the same time stressing that these uses apply when the standard frequentist tool of confidence intervals is not enough or has its limitations. Here are three examples.

There are times when a sharp null is a sensible way to formulate a hypothesis. Spiegelhalter (2019)'s excellent *The Art of Statistics* motivates the first part of the book, covering frequentist inference and hypothesis testing, with just such a sharp null: was Harold Shipman, Britain's most prolific serial killer, guilty or not guilty? Shipman was a physician who was found guilty of having killed over 200 of his elderly patients by lethal injection over the period 1975-98. The accumulated evidence pointed strongly towards "guilty"; or, using Greenland's surprisal s, if we assumed he was innocent and we considered this evidence, we would have been as surprised as if we had flipped a very, very large number of heads in a row.

Sharp nulls like this are rare in economic questions when the objective is to estimate the value of a parameter; for the reasons discussed above, it is much more natural to use an interval estimator. Imbens (2021) comes out strongly against misuse of statistical significance and p-values, but, somewhat surprisingly, suggests that "cases commonly arise in economics" where "if a researcher is legitimately interested in assessing a null hypothesis versus an alternative hypothesis". I think this is less common than he suggests. Most of the examples he cites—"sheepskin effects" of diplomas in estimating the return to education, the returns to scale, the scale of discrimination in the labour market and some others—are much better served with an answer that includes "how precisely estimated is it?"

But such sharp nulls do occasionally arise in economics, and probably the best illustration is one that Imbens cites and that will also be familiar to many students: the Efficient Markets Hypothesis (EMH). In its simplest form, the EMH implies that in an efficient market, asset prices should incorporate all available information. A simple version of a test of the EMH is a test of predictability; does a variable incorporating past information predict price movements? In this case, summarizing the result of a test of predictability using the p-value of the null that the coefficient is zero is not unreasonable: traders can make large sums from even ϵ-deviations.

A second example arises with nested tests of functional form. Hayashi (2000)'s textbook *Econometrics* provides a good illustration: Nerlove (1963)'s famous study of returns to scale in the electricity supply industry, and its extension in the study by Christensen and Greene (1976). The latter study used a general translog cost function; it is possible to recast Nerlove's preferred specification as an intermediate case between a basic Cobb-Douglas specification and a full translog specification, where all but one of the higher-order translog terms is restricted to zero. The translog parameters are not easy to interpret (the factor elasticities are readily interpretable, but are complicated functions of these parameters that also depend on the levels of factor inputs), and constructing interval estimates for them in order to examine the reasonableness of Nerlove's restrictions is impractical. Here a sharp null is reasonable: can we reject the null of Nerlove's restrictions on the full translog specification? (The answer, by the way, is "no" at any traditional significance level.)

A third example comes from misspecification testing, where the model specification rather than a specific parameter is to be tested. I have considered this at length in a separate paper with Austin Nichols (Nichols and Schaffer (2021)). Here I will repeat just one example from that paper, namely testing for underidentification in an IV model. A test of underidentification is a test of the rank condition: $H_0 : E(\mathbf{z}_i' x_i))$ is full column rank. If the rank condition fails, the IV estimate is inconsistent. This is a sharp null where a failure to reject is informative and useful, and the researcher should consider, e.g., weak-instrument-robust methods such as those discussed earlier.

6 Conclusion

The problems with NHST and the misuse of p-values are still widespread in empirical social science, including in economics. Progress is being made, and we should be optimistic: the very fact that there is widespread recognition in the research and academic community that there is indeed a problem is good news indeed. But progress will be slow until how we teach our students changes. We should teach our students interval estimation as the key learning outcome, and coverage as the key (frequentist) concept. I have suggested some tools for how to do this in an accessible and easy-to-understand way.

References

Amrhein, V., Greenland, S., McShane, B.: Scientists rise up against statistical significance. Nature, 305–307 (2019)

Christensen, L.R., Greene, W.: Economies of scale in U.S. electric power generation. J. Polit. Econ. **84**, 655–76 (1976)

Greenland, S.: Valid P-values behave exactly as they should: some misleading criticisms of P-values and their resolution with S-values. Am. Stat. **73**, 106–114 (2019)

Hayashi, F.: Econometrics. Princeton University Press, Princeton, NJ [u.a.] (2000)

Imbens, G.W.: Statistical significance, p-values, and the reporting of uncertainty. J. Econ. Perspect. **35**, 157–74 (2021)

Kasy, M.: Of forking paths and tied hands: selective publication of findings, and what economists should do about it. J. Econ. Perspect. **35**, 175–92 (2021)

McCloskey, D.N., Ziliak, S.T.: The standard error of regressions. J. Econ. Lit., 97–114 (1996)

Miguel, E.: Evidence on research transparency in economics. J. Econ. Perspect. **35**, 193–214 (2021)

Müller, U.K., Norets, A.: Credibility of confidence sets in nonstandard econometric problems. Econometrica **84**, 2183–2213 (2016)

Nerlove, M.: Returns to scale in electricity supply. In: Christ, C. (ed.) Measurement in Economics: Studies in Mathematical Economics and Econometrics in Memory of Yehuda Grunfeld. Stanford University Press

Nichols, A., Schaffer, M.: Practical steps to improve specification testing. In: Ngoc Thach, N., Ha, D., Trung, N., Kreinovich, V. (eds.) Prediction and Causality in Econometrics and Related Topics. Springer, Studies in Computational Intelligence, 4th International Econometric Conference of Vietnam 2021; Conference date: 11-01-2021 Through 13-01-2021, pp. 75–88 (2021)

Spiegelhalter, D.J.: The Art of Statistics: How to Learn from Data, 1st ed. Basic Books, New York (2019)

Wasserstein, R.L., Lazar, N.A.: The ASA statement on p-values: context, process, and purpose. Am. Stat. **70**, 129–133 (2016)

Time Series Forecasting Using a Markov Switching Vector Autoregressive Model with Stochastic Search Variable Selection Method

Katsuhiro Sugita

Abstract This paper investigates forecasting performance using a Markov switching vector autoregressive (MSVAR) model with stochastic search variable selection (SSVS) method. An MSVAR model has been widely used for empirical macroeconomics. However, an MSVAR model usually fits in-sample better but forecasts poorly relative to a linear VAR model, and typically has a large number of parameters, leading to over-parameterization problem. The use of SSVS prior is expected to mitigate this over-parameterization problem by setting insignificant parameters to be zero in an automatic fashion. In recursive forecasting exercises of empirical study and Monte Carlo simulation, I find that implementing SSVS to unrestricted VAR or MSVAR model typically improve forecasting performance. However, the results show that the MSVAR model with the SSVS prior does not always provide superior forecasting performance relative to the linear VAR model with SSVS prior, and that the improvement is not significantly large enough to alleviate over-parametrization problem of an MSVAR model.

1 Introduction

This paper evaluates forecasting performance of a Markov switching vector autoregressive (MSVAR) model with Bayesian stochastic search variable selection prior. Since the pioneering work of Sims (1980), VAR models have been widely used to analyze and forecast macroeconomics. VAR models tend to contain a large number of parameters and thus the problem of over-parametrization may arise. To overcome this problem, Bayesian methods have been popular among researchers, see Kadiyala and

K. Sugita (✉)
Faculty of Global and Regional Studies, University of the Ryukyus, Nishihara, Senbaru, Okinawa 903-0213, Japan
e-mail: ksugita@grs.u-ryukyu.ac.jp

N. Ngoc Thach et al. (eds.), *Financial Econometrics: Bayesian Analysis, Quantum Uncertainty, and Related Topics*, Studies in Systems, Decision and Control 427,
https://doi.org/10.1007/978-3-030-98689-6_10

Karlsson (1997) and Andersson and Karlsson (2007) in detail. Koop and Korobilis (2010) provide an excellent survey of the development of Bayesian VAR models. For nonlinear VAR models, an MSVAR model, that is a VAR model with Hamilton (1989)'s state-dependent Markov switching, has been utilized by many empirical researchers to capture nonlinearities and asymmetry. An MSVAR model allows the regime shift in any parameter in the model to explore the regime-dependent dynamic relationships among several macroeconomic variables, see Krolzig (1997), Krolzig et al. (2002), Clements and Krolzig (2002), Beine et al. (2003), Fallahi (2011), Balcilar et al. (2017), Droumaguet et al. (2017), Akgul et al. (2015) among others.

Engel (1994), Dacco and Satchell (1999), Clements and Krolzig (1998) and Nikolsko-Rzhevskyy and Prodan (2012) show that a good in-sample fit by Markov switching models will not necessarily lead to a good forecasting performance. As Terasvirta, Timo and Anderson (1992) note, out-of-sample by non-linear models is superior only if the non-linear features also characterize the later period. Aside from this non-linear property for out-of-sample, VAR models have over-parameterization problem, and thus MSVAR models have much serious over-parametrization problem. In case of a two-regime MSVAR model that allows all coefficients and covariances to shift with regime, there are just twice more parameters in the MSVAR model than in linear VAR. This over-parameterization problem of MSVAR models leads to imprecise inference and thus deteriorates the forecast performance. Furthermore, selecting an appropriate number of lags p in the MSVAR is often problematic. A lag length that varies over regimes can be accommodated by defining p as the maximum number of lags across the regimes. Thus, the lag order is the same across regimes, and there is a possibility that for one regime the number of lags is appropriate though for other regimes it is too long and thus causes the over-parameterization problem.

To remedy the over-parameterization problem of VAR model, George and McCulloch (1993) and George and McCulloch (1997) propose a Bayesian stochastic search variable selection (SSVS) method. The SSVS method uses a hierarchical prior where each of the parameters in the model is generated in a Markov chain Monte Carlo (MCMC) from one of two normal distributions with a zero mean—one with a large variance and the other with a small variance. A normal prior distribution with a zero mean and a large variance implies a relatively non-informative, and thus the parameter is non-zero and the corresponding variable is included in the model, while the other normal prior distribution with a zero mean and a small variance implies that the parameters is virtually zero and the corresponding variable is excluded from the model. George et al. (2008) investigate numerical simulations and show that implementing the SSVS method in VAR can be effective at both selecting a satisfactory model and improving forecast performance based on the 1-step ahead mean squared error of forecast error and Kullback-Liebler divergence.

Jochmann et al. (2010) investigate forecasting performance using VAR models with the SSVS prior subject to structural breaks. Their study involves difficulty in forecasting of the model that allows for non-recursive regime shifts. In a recursive forecasting exercise, they find slight improvement over the standard VAR, although most of these improvements are due to the use of SSVS rather than to allowing structural breaks. VAR with structural breaks allows for uncertainty about whether a break occurs at time τ, the period in which the forecast is being made. To account for this, two sets of the estimates of the VAR coefficients and error covariance matrix should be provided - one is assuming a break does not occur at time τ with probability of no break at time τ, and the other is assuming that a break has occurred in the period of forecasting with probability of break at time τ. Since the set of parameters for new regime is unknown at time τ, they use the prior for the model of new regime. Unlike the VAR model subject to breaks by Jochmann et al. (2010), an MSVAR model allows to use the set of estimated parameters in VAR of all regimes which occurred in the past prior to time τ.

In this paper, the SSVS hierarchical prior is applied to an MSVAR model to study whether multistep-ahead forecasting performance is improved over the unrestricted MSVAR or linear VAR models. In Sect. 3 the empirical study is illustrated, using three variables of US macroeconomic data: the unemployment rate, the interest rate and the inflation rate, running from 1953:Q1 to 2020:Q1. In order to compare the forecasting performance, I consider four models—the unrestricted VAR, the restricted VAR with the SSVS prior, the unrestricted MSVAR and the restricted MSVAR with the SSVS prior. In Sect. 4 Monte Carlo simulation is presented using the data-generating process based on the US macroeconomic data. Both the out-of-sample recursive forecasting and the Monte Carlo simulation show that SSVS offers an improvement in forecast performance and helps to alleviate the problem of the over-parametrization of the VAR or MSVAR models. However, the MSVAR model with the SSVS prior does not always provide superior forecasting performance relative to the linear VAR model with SSVS prior, and that the improvement is not significantly large enough to alleviate over-parametrization problem of MSVAR models. All computation in this paper are performed using code written by the author with Ox v8.02 for Linux (Doornik 2013).

2 MSVAR Model with Stochastic Search Variable Selection

2.1 Vector Autoregression with the SSVS Prior

George and McCulloch (1993) and George and McCulloch (1997) propose the SSVS method, and George et al. (2008) apply the SSVS approach to a VAR model. A VAR model typically contains a very large number of parameters to estimate relative to the number of the observations. This over-parameterization problem leads to imprecision of inference and thus worsens forecasting performance. To overcome this problem, George et al. (2008) implement the SSVS method in a VAR. SSVS is a Bayesian MCMC method to take restrictions on the parameters of the model by using a hierarchical prior on the parameters.

Let y_t be an $n \times 1$ vector of observations at time t, then a VAR model with p lag is written as

$$y'_t = \mu + \sum_{i=1}^{p} y'_{t-i}\Phi_i + \varepsilon'_t \tag{1}$$

for $t = 1, \ldots, T$, where μ is a $1 \times n$ vector of an intercept term; Φ_i are $n \times n$ matrices of coefficients for $i = 1, \ldots, p$; ε_t are $n \times 1$ independent $N_n(0, \Sigma)$ errors; and the covariance matrix Σ is an $n \times n$ positive definite matrix. Let x_t be a $(1 + np) \times 1$ vector with $x_t = \left(1, y'_{t-1}, \ldots, y'_{t-p}\right)'$, then the VAR model (1) can be rewritten in matrix form as

$$Y = X\Phi + \varepsilon \tag{2}$$

where the $T \times n$ matrix Y is defined as $Y = (y_1, \ldots, y_T)'$; the $T \times (1 + np)$ matrix X is defined as $X = (x_1, \ldots, x_T)'$; the $(1 + np) \times n$ matrix Φ is defined as $\Phi = (\mu', \Phi'_1, \ldots, \Phi'_p)'$; and the ε is a $T \times n$ matrix with $\varepsilon = (\varepsilon_1, \ldots, \varepsilon_T)'$.

SSVS defines the prior for the VAR coefficient Φ not as a whole but as all of the elements in Φ. Let $\phi = vec(\Phi)$ and m be the number of restricted elements in Φ, then the prior for each element, ϕ_j, $j = 1, 2, \ldots, m$, is a hierarchical prior with mixture of two normal distributions conditional on an unknown dummy variable γ_j that takes zero or one:

$$\phi_j|\gamma_j \sim (1 - \gamma_j)N(0, \tau_{0,j}^2) + \gamma_j N(0, \tau_{1,j}^2) \tag{3}$$

where $\tau_{0,j}^2$ is small enough to be near zero and $\tau_{1,j}^2$ is large enough to be non-informative. This implies that if $\gamma_j = 0$, then the element ϕ_j is restricted to be zero, and thus the prior for ϕ_j is virtually zero, while if $\gamma_j = 1$, then the element ϕ_j is unrestricted, and thus the prior is relatively non-informative. With regard to priors on γ_j, SSVS assumes independent Bernoulli $p_i \in (0, 1)$ random variables:

$$\begin{aligned} P(\gamma_j = 1) &= p_j \\ P(\gamma_j = 0) &= 1 - p_j \end{aligned} \tag{4}$$

where p_j is the prior hyperparameter that is set to be 0.5 for a natural default choice. Let $\gamma = (\gamma_1, \ldots, \gamma_m)$, then the prior for ϕ in (3) can be written as:

$$\phi|\gamma \sim N(0, DD) \tag{5}$$

where D is a diagonal matrix as $D = diag(d_1, \ldots, d_m)$; where

$$d_j = \begin{cases} \tau_{0j} & \text{if} \quad \gamma_j = 0 \\ \tau_{1j} & \text{if} \quad \gamma_j = 1 \end{cases} \tag{6}$$

George and McCulloch (1997) and George et al. (2008) recommend to use a default semiautomatic approach that sets $\tau_{kj} = c_k\hat{\sigma}_{\phi_j}$ for $k = 0, 1$, where $\hat{\sigma}_{\phi_j}$ is the least squares estimates of the standard error of ϕ_j, which is the coefficients in an unrestricted VAR. The pre-selected constants c_0 and c_1 must be $c_0 < c_1$. George et al. (2008) and Jochmann et al. (2010) set $c_0 = 0.1$ and $c_1 = 10$, however these numbers should be adjusted by researcher to obtain an optimal forecasting performance. George and McCulloch (1993) recommend these numbers to be such that $\tau_{k1}^2/\tau_{k0}^2 \leq 10000$, otherwise the MCMC would be very slow to converge if τ_{k1}/τ_{k0} is chosen extremely large.

George et al. (2008) also consider the restrictions on the covariances in Σ in VAR. The error covariance matrix can be decomposed as $\Sigma^{-1} = \Psi\Psi'$, where Ψ is upper-triangular of the Choleski decomposition of Σ. The prior for the diagonal elements of Ψ, $\psi_{11}, \psi_{22}, \ldots, \psi_{nn}$, is assumed

$$\psi_{ii}^2 \sim G(a_i, b_i) \tag{7}$$

where $G(a_i, b_i)$ denotes the gamma distribution with mean a_i/b_i and variance a_i/b_i^2. For the elements above the diagonal, SSVS considers restrictions on the off-diagonal elements of Ψ, and the hierarchical prior for each element of ψ_{ij} takes the same mixture of two normal distributions as ϕ_j so that

$$\psi_{ij}|\omega_{ij} \sim (1 - \omega_{ij})N(0, \kappa_{0ij}^2) + \omega_{ij}N(0, \kappa_{1ij}^2) \tag{8}$$

where κ_{0ij}^2 is small enough to be near zero and κ_{1ij}^2 is large enough to be non-informative. With regard to priors on ω_{ij}, SSVS assumes independent Bernoulli as:

$$\begin{aligned} P(\omega_{ij} = 1) &= q_{ij} \\ P(\omega_{ij} = 0) &= 1 - q_{ij} \end{aligned} \tag{9}$$

where q_{ij} is equal to 0.5 for a natural default choice. Let define $\eta_j = (\psi_{1j}, \psi_{2j}, \ldots, \psi_{j-1,j})'$ for $j = 2, \ldots, n$, and $\eta = \left(\psi_{12}, \psi_{13}, \psi_{23}, \ldots, \psi_{n-1,n}\right)' = \left(\eta_2', \ldots, \eta_n'\right)'$, then priors on η is assumed as:

$$\eta_j|\omega_j \sim N(0, D_j D_j) \tag{10}$$

where $\omega_j = \left(\omega_{1j}, \ldots, \omega_{j-1,j}\right)'$ is a vector of dummy variables which are assumed to be independent D_j in (10) is defined as $D_j = diag\left(h_{1j}, \ldots, h_{j-1,j}\right)$; where

$$h_{ij} = \begin{cases} \kappa_{0ij} & \text{if } \omega_{ij} = 0 \\ \kappa_{1ij} & \text{if } \omega_{ij} = 1 \end{cases}. \tag{11}$$

The choice of κ_{kij} for $k = 0, 1$ can be determined by using a semi-automatic default approach that is similar considerations for setting τ_{kj}, that is, we set $\kappa_{kij} = c_k\hat{\sigma}_{\psi_{ij}}$ with values of $c_0 < c_1$, where $\hat{\sigma}_{\psi_{ij}}$ is an estimate of the standard error associated with off-diagonal element of Ψ.

This summarizes the SSVS hierarchical prior for VAR model. George et al. (2008) and Jochmann et al. (2010) consider three patterns of SSVS—restrictions only for Ψ and for Φ separately, and then for both. In this paper, only stochastic search for both of Ψ and Φ together is considered. Note that, if the unknown indicator parameters is set to equal to 1, that is $\gamma_j = \omega_{ij} = 1$ for all j and i, then the SSVS-VAR is just unrestricted VAR. I use this prior for the unrestricted VAR in my empirical and simulation sections.

With above SSVS priors, George et al. (2008) derive the conditional posterior distribution for each parameter. Let s_{ij} be the elements of $S = (Y - X\Phi)'(Y - X\Phi)$, $s_j = \left(s_{1j}, \ldots, s_{j-1,j}\right)'$, and S_j be the upper left $j \times j$ block of S, then the likelihood function is

$$\begin{aligned}\mathscr{L}(Y|\Phi, \Sigma) &\propto |\Sigma|^{-T/2} \exp\left[-\frac{1}{2}(Y - X\Phi)\,\Sigma^{-1}\,(Y - X\Phi)'\right] \\ &\propto \prod_{i=1}^{n} \psi_{ii}^{T} \exp\left[-\frac{1}{2}\left\{\sum_{i=1}^{n} \psi_{ii}^{2} \upsilon_i + \sum_{j=2}^{n} \left(\eta_j + \psi_{jj} S_{j-1}^{-1} s_j\right)' S_{j-1}\right.\right. \\ &\qquad \left.\left.\left(\eta_j + \psi_{jj} S_{j-1}^{-1} s_j\right)\right\}\right] \end{aligned} \tag{12}$$

where $\upsilon_1 = s_{11}$ and $\upsilon_i = |S_i|/|S_{i-1}|$ for $i = 2, \ldots, n$.

With the likelihood function (12) and priors of (3)–(5), (7)–(10), George et al. (2008) derive the conditional posterior distributions as follows. For the VAR coefficients Φ, the conditional posterior is given as:

$$\Phi|\gamma, \eta, \psi; Y \sim N_m(\mu, \Xi) \tag{13}$$

where

$$\begin{aligned}\Xi &= \left[(\Psi\Psi') \otimes (X'X) + (DD)^{-1}\right],^{-1} \\ \mu &= \Xi\left[\left\{(\Psi\Psi)' \otimes (X'X)\right\} \hat{\psi}_M\right], \\ \hat{\psi}_M &= vec(\hat{\Psi}_M) = vec\left[\left(X'X\right)^{-1} X'Y\right].\end{aligned}$$

For the conditional posterior of γ, let define $\gamma_{(-i)} = (\gamma_1, \ldots, \gamma_{i-1}, \gamma_{i+1}, \ldots, , \gamma_m)$, then we have

$$\gamma_j|\phi, \gamma_{j-1}, \eta, \psi; Y \sim \textit{Bernoulli}\ \left(u_{j1}/(u_{j1} + u_{j2})\right) \tag{14}$$

where

$$u_{j1} = \frac{1}{\tau_{0j}} \exp\left(-\frac{\phi_j^2}{2\tau_{0j}^2}\right) p_i,$$

$$u_{j2} = \frac{1}{\tau_{1j}} \exp\left(-\frac{\phi_j^2}{2\tau_{1j}^2}\right)(1 - p_i).$$

The conditional posterior distributions of $\psi_{11}^2, \psi_{22}^2, \ldots, \psi_{nn}^2$ are independent and gamma distributions as:

$$\psi_{jj}^2 | \phi, \omega; Y \sim G\left(a_j + \frac{T}{2}, b_j\right) \tag{15}$$

where

$$b_j = \begin{cases} b_1 + \frac{s_{11}}{2} & \text{if} \quad j = 1 \\ b_j + \frac{1}{2}\left\{s_{jj} - s_j' \left[V_{j-1} + \left(D_j D_j\right)^{-1}\right]^{-1} s_j\right\} & \text{if} \quad j = 2, \ldots, n \end{cases}$$

The conditional posterior distributions of $\eta_2, \ldots, \eta_n$ are independent and given as;

$$\eta_j | \phi, \omega, \psi; Y \sim N_{j-1}(\mu_j, \Delta_j) \tag{16}$$

where

$$\Delta_j = \left[S_{j-1} + \left(D_j D_j\right)^{-1}\right]^{-1},$$

$$\mu_j = -\psi_{jj} \Delta_j s_j.$$

Finally, the conditional posterior distribution of ω_{ij} for $j = 2, \ldots, n$ and $i = 1, \ldots, j-1$ is derived as:

$$\omega_{ij} | \phi, \psi, \omega_k, k \neq j; Y \sim \textit{Bernoulli}\ \left(u_{ij1}/(u_{ij1} + u_{ij2})\right) \tag{17}$$

where

$$u_{ij1} = \frac{1}{\kappa_{1ij}} \exp\left(-\frac{\psi_{ij}^2}{2\kappa_{1ij}^2}\right) q_{ij},$$

$$u_{ij2} = \frac{1}{\kappa_{0ij}} \exp\left(-\frac{\psi_{ij}^2}{2\kappa_{0ij}^2}\right)(1 - q_{ij}).$$

The MCMC stochastic search algorithm is obtained by drawing sequentially the above conditional distributions (13)–(17).

2.2 Markov Switching VAR Model with SSVS Prior

MSVAR model allows regime shifts in parameters of (1):

$$y_t' = \mu(s_t) + \sum_{i=1}^{p} y_{t-i}' \Phi_i(s_t) + \varepsilon'_t \tag{18}$$

where $\varepsilon_t \sim N(0, \Sigma(s_t))$. Here all coefficients in VAR and covariance terms are subject to an unobservable discrete state variable $s_t = 1, ..., R$, that evolves according to an R-state, first-order Markov switching process with the transition probabilities, $\Pr(s_t = i|s_{t-1} = j) = q_{ij}, i, j = 1, \ldots R$. This MSVAR model in (18) can be written in matrix form as:

$$Y = \sum_{m=1}^{R} X(s_t = m)\Phi(s_t = m) + E \tag{19}$$

where $Y = (y_1, \ldots, y_T)'$ is a $T \times n$ matrix, and $\Phi(s_t) = \left(\mu(s_t)', \Phi_1(s_t)', ..., \Phi_p(s_t)'\right)'$ is a $(1 + np) \times n$ matrix, and $E = \left(\varepsilon_1', ..., \varepsilon_T'\right)'$ is a $T \times n$ matrix. $X(s_t)$ in (19) is a $T \times (1 + np)$ matrix containing an intercept and lagged dependent variables $x_t = \left(1, y_{t-1}', y_{t-2}', ..., y_{t-p}'\right)'$ such as

$$X(s_t) = (\iota_1(s_t)x_1, \iota_2(s_t)x_2, ..., \iota_T(s_t)x_T)'$$
$$= \begin{bmatrix} \iota_1(s_t) & \iota_0(s_t)y_{-1} & \cdots & \iota_{-p}(s_t)y_{-p} \\ \iota_2(s_t) & \iota_1(s_t)y_0 & \cdots & \iota_{-p+1}(s_t)y_{-p+1} \\ \vdots & \vdots & \ddots & \vdots \\ \iota_T(s_t) & \iota_{T-1}(s_t)y_{T-1} & \cdots & \iota_{T-p}(s_t)y_{T-p} \end{bmatrix}$$

where $\iota_t(s_t)$ in $X(s_t)$ is an indicator variable that equals to 1 if regime s_t is m, and 0 otherwise.

In Markov switching model the regime is generated by process of an ergodic Markov chain with a finite number of states $s_t = 1, \ldots, R$ defined by the transition probability matrix P

$$P = \begin{bmatrix} \Pr(s_t = 1|s_{t-1} = 1) & \Pr(s_t = 1|s_{t-1} = 2) & \cdots & \Pr(s_t = 1|s_{t-1} = R) \\ \Pr(s_t = 2|s_{t-1} = 1) & \Pr(s_t = 2|s_{t-1} = 2) & \cdots & \Pr(s_t = 2|s_{t-1} = R) \\ \vdots & \vdots & \ddots & \vdots \\ \Pr(s_t = R|s_{t-1} = 1) & \Pr(s_t = R|s_{t-1} = 2) & \cdots & \Pr(s_t = R|s_{t-1} = R) \end{bmatrix} \tag{20}$$

where the last row of the i-th column of the transition probability is $\Pr(s_t = R|s_{t-1} = i) = 1 - \sum_{j=1}^{R-1} \Pr(s_t = j|s_{t-1} = i)$ for $i = 1, \ldots, R - 1$ so that sum of each column in (20) is 1.

To draw the regime variable, I employ the multi-move Gibbs sampling method proposed by Carter and Kohn (1994) and Shephard (1994). To generate the transition probabilities I follow Albert and Chib (1993). After generating the regime variables, parameters of the VAR in each regime are drawn using the SSVS method described in the previous subsection.

3 Empirical Analysis

In this section, I present empirical results using US macroeconomic variables that include three variables—the unemployment rate, the interest rate and the inflation rate. A VAR model that uses these variables has been analyzed by Cogley and Sargent (2005), Primiceri (2005), Koop et al. (2009), and Jochmann et al. (2010), among many others. The data are quarterly, from 1953:Q1 to 2020:Q1 with sample size 268. The unemployment rate is measured by the civilian unemployment rate. The inflation rate is the 400 times the difference of the log of CPI, which is the GDP chain-type price index. The interest rate is the 3-month treasury bill. I obtained these data from the Federal Reserve Bank of St. Louise.[1] These data are plotted in Fig. 1.

The choice of the number of lags in a VAR affects efficiency in estimation and thus forecasting performances. Cogley and Sargent (2005) and Primiceri (2005) work with VAR(2) to analyze US macroeconomy with the three variables. Jochmann et al. (2010) use VAR(4) for their SSVS-VAR model with reason that the SSVS can find zero restrictions on the parameters of longer lags even if the true lag length is <4. In this paper, the number of lags in all models is set to be four. For Markov switching models, the number of regime is set to be two[2].

3.1 In-Sample Results

I begin by briefly discussing the performance of SSVS using the full sample. I focus on two models: the unrestricted MSVAR (hereafter MSVAR) model and the restricted MSVAR model with the SSVS prior (hereafter S-MSVAR). The prior for the intercept term is not restricted and assigned a normal with a zero mean and a variance of 10 for relatively non-informative prior. For the SSVS priors, the hyperparameters are set at $p_i = 0.5$ in (4), $q_i = 0.5$ in (9), $a = b = 0.01$ in (7), $c_0 = 0.1$ and $c_1 = 20$ for τ_{kj} and κ_{kij} in (6) and (11). These prior hyperparameters are also used in the out-sample forecasting and Monte Carlo simulation in the next section. MCMC is run with 20,000 draws after 5,000 burn-in. SSVS is the joint restriction search for both Φ and Ψ. A lag order p is set to be 4.

[1] https://fred.stlouisfed.org.

[2] I also consider the case of the three-regime MSVAR model. However, I found the most evidence in favor of the two-regimes MSVAR.

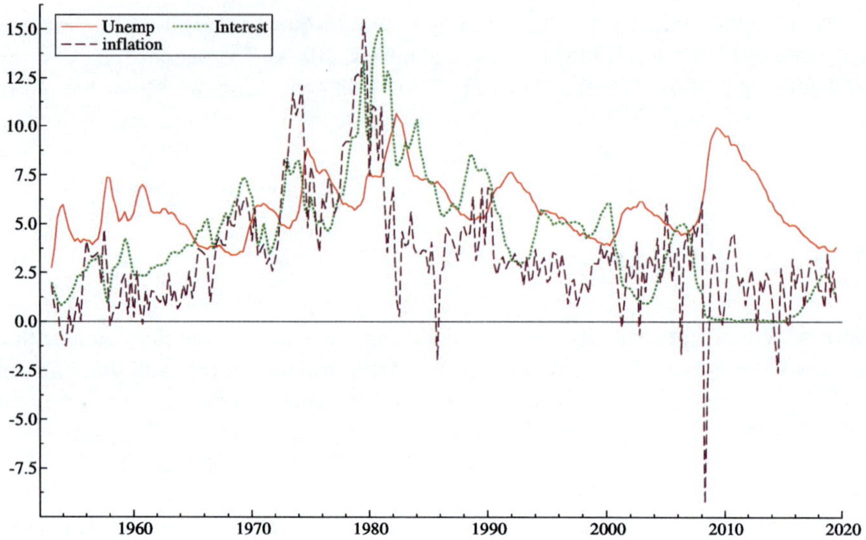

Fig. 1 US data

Tables 1 and 2 present the posterior results of two-regime MSVAR and S-MSVAR models. The columns labelled MSVAR present the posterior results of the unrestricted MSVAR with $\gamma_j = \omega_{ij} = 1$ for all j and i in (4) and (9), while the columns labelled S-MSVAR present the posterior results of the S-MSVAR model. The tables show a set of point estimates, standard deviations, and the posterior probability of inclusion of each parameter (the posterior probability that $\gamma_j = 1$ for the VAR coefficients for each regime and the posterior probability that $\omega_{ij} = 1$ for the off-diagonal elements of Ψ for each regime). Here, I summarize a few noteworthy aspects of the results given in Tables 1 and 2. In this discussion I will refer to a parameter as being included if its posterior inclusion probability Pr(inc.) is >0.5.

Implementing the SSVS prior on MSVAR model results in the exclusion of many coefficients. The S-MSVAR model includes 15 of the 36 possible VAR coefficients and 2 of the 3 possible off-diagonal elements of Ψ for regime 1, while only 7 of the 36 possible VAR coefficients and 1 of the 3 possible off-diagonal elements of Ψ for regime 2. This indicates that SSVS is an effective for ensuring parsimony in over parameterized VAR with Markov regime switching. The diagonal elements of Ψ are much smaller in regime 1 than those in regime 2, indicating that the regime 1 is high volatility and the regime 2 is low volatility since Ψ is defined as $\Sigma^{-1} = \Psi\Psi'$ so that smaller diagonal elements indicate larger error variances.

The S-MSVAR model always includes the own lags for the three variables of both regimes. The unemployment rate equation includes only its own lags in both regimes and thus is not affected by any other variables. The interest rate equation includes its own lags and the lags of the unemployment rate in regime 1, however, does not include the lag of the unemployment rate in regime 2, suggesting that the interest

Table 1 Results for MSVAR and S-MSVAR

	Regime 1			Regime 2		
	MSVAR	S-MSVAR		MSVAR	S-MSVAR	
	Mean (s.d.)	Mean (s.d.)	Pr(inc.)	Mean(s.d.)	Mean (s.d.)	Pr(inc.)
Dep Var: unemp						
Const	0.588 (0.118)	0.459 (0.107)	—	0.072 (0.047)	0.054 (0.044)	—
Unemp(−1)	1.556 (0.072)	1.543 (0.060)	1.000	1.143 (0.074)	1.140 (0.059)	1.000
Unemp(−2)	−0.738 (0.132)	−0.650 (0.103)	1.000	−0.072 (0.112)	-0.031 (0.072)	0.445
Unemp(−3)	−0.197 (0.133)	−0.099 (0.125)	0.594	−0.031 (0.099)	-0.070 (0.082)	0.591
Unemp(−4)	0.258 (0.073)	0.112 (0.088)	0.754	−0.129 (0.059)	−0.063 (0.056)	0.684
Interest(−1)	−0.052 (0.035)	0.002 (0.007)	0.016	−0.054 (0.043)	−0.001 (0.007)	0.010
Interest(−2)	0.125 (0.049)	0.008 (0.012)	0.017	0.054 (0.054)	0.004 (0.009)	0.009
Iinterest(−3)	−0.156 (0.046)	0.001 (0.012)	0.012	0.034 (0.036)	0.003 (0.008)	0.009
Interest(−4)	0.102 (0.034)	0.006 (0.011)	0.060	−0.030 (0.026)	0.000 (0.006)	0.013
Inflaton(−1)	−0.019 (0.012)	−0.002 (0.006)	0.055	0.001 (0.008)	0.000 (0.004)	0.010
Inflation(−2)	−0.023 (0.012)	−0.003 (0.005)	0.027	0.001 (0.008)	-0.001 (0.004)	0.008
Inflation(−3)	0.042 (0.016)	0.006 (0.010)	0.127	−0.005 (0.007)	−0.001 (0.003)	0.013
Inflation(−4)	0.026 (0.015)	0.017 (0.017)	0.469	0.007 (0.006)	0.001 (0.002)	0.015
Dep Var: interest						
Const	−0.522 (0.269)	−0.330 (0.265)	—	0.134 (0.089)	0.109 (0.083)	—
Unemp(−1)	−0.546 (0.185)	−0.358 (0.136)	0.997	−0.220 (0.117)	−0.016 (0.037)	0.360
Unemp(−2)	0.474 (0.331)	0.085 (0.208)	0.249	0.386 (0.172)	0.004 (0.035)	0.060
Unemp(−3)	0.129 (0.341)	0.241 (0.228)	0.588	−0.390 (0.161)	-0.005 (0.022)	0.023
Unemp(−4)	0.081 (0.194)	0.120 (0.170)	0.389	0.201 (0.093)	0.002 (0.016)	0.033
Interest(−1)	1.121 (0.090)	1.081 (0.068)	1.000	1.398 (0.096)	1.469 (0.082)	1.000
Interest(−2)	−0.812 (0.114)	−0.686 (0.092)	1.000	−0.384 (0.118)	−0.468 (0.095)	1.000
Interest(−3)	0.843 (0.116)	0.641 (0.111)	1.000	−0.084 (0.068)	−0.009 (0.029)	0.206
Interest(−4)	−0.328 (0.085)	−0.180 (0.088)	0.896	0.048 (0.055)	−0.004 (0.021)	0.168
Inflaton(−1)	−0.011 (0.027)	−0.005 (0.015)	0.009	0.023 (0.014)	0.011 (0.009)	0.020
Inflation(−2)	0.108 (0.029)	0.068 (0.022)	0.096	0.028 (0.014)	0.018 (0.010)	0.013
inflation(−3)	0.026 (0.040)	0.045 (0.025)	0.032	−0.008 (0.012)	−0.004 (0.008)	0.006
Inflation(−4)	0.011 (0.037)	0.010 (0.018)	0.016	0.002 (0.010)	0.005 (0.007)	0.009

(continued)

Table 1 (continued)

	Regime 1			Regime 2		
	MSVAR	S-MSVAR		MSVAR	S-MSVAR	
	Mean (s.d.)	Mean (s.d.)	Pr(inc.)	Mean(s.d.)	Mean (s.d.)	Pr(inc.)
Dep Var: inflation						
Const	1.096 (0.668)	1.462 (0.612)	—	0.624 (0.441)	0.729 (0.401)	—
Unemp(−1)	−1.716 (0.469)	−1.596 (0.560)	0.996	−0.050 (0.554)	0.021 (0.084)	0.164
Unemp(−2)	2.141 (0.846)	1.240 (0.635)	0.893	0.877 (0.876)	0.024 (0.090)	0.023
Unemp(−3)	−0.955 (0.876)	0.032 (0.209)	0.066	−0.435 (0.838)	0.004 (0.095)	0.014
Unemp(−4)	0.349 (0.485)	0.039 (0.124)	0.042	−0.358 (0.470)	−0.028 (0.080)	0.016
Interest(−1)	0.577 (0.215)	0.377 (0.251)	0.858	0.991 (0.338)	0.155 (0.101)	0.856
Interest(−2)	−0.234 (0.290)	−0.123 (0.220)	0.292	−1.087 (0.433)	−0.002 (0.053)	0.038
Interest(−3)	0.066 (0.296)	−0.081 (0.159)	0.217	0.410 (0.386)	0.007 (0.050)	0.022
Interest(−4)	−0.341 (0.218)	−0.086 (0.148)	0.239	−0.145 (0.275)	0.007 (0.043)	0.019
Inflaton(−1)	0.432 (0.075)	0.388 (0.070)	0.981	0.324 (0.082)	0.447 (0.086)	0.999
Inflation(−2)	−0.050 (0.074)	−0.006 (0.027)	0.071	0.055 (0.069)	0.014 (0.041)	0.126
Inflation(−3)	0.610 (0.218)	0.564 (0.080)	1.000	0.082 (0.063)	0.018 (0.045)	0.159
Inflation(−4)	−0.085 (0.100)	−0.017 (0.051)	0.117	0.087 (0.050)	0.016 (0.038)	0.163

Table 2 Results for MSVAR and S-MSVAR cont

	Regime 1			Regime 2		
	MSVAR	S-MSVAR		MSVAR	S-MSVAR	
	Mean (s.d.)	Mean (s.d.)	Pr(inc.)	Mean(s.d.)	Mean (s.d.)	Pr(inc.)
Ψ-matrix						
Unemp, unemp	5.043 (0.265)	4.665 (0.227)	—	9.130 (0.563)	9.338 (0.605)	—
Interest, interest	2.203 (0.106)	2.156 (0.103)	—	5.585 (0.388)	6.057 (0.536)	—
Inflation, inflation	0.816 (0.044)	0.784 (0.042)	—	1.078 (0.072)	1.092 (0.081)	—
Unemp, interest	2.379 (0.351)	2.405 (0.334)	1.000	2.086 (0.668)	1.620 (1.121)	0.741
Unemp, inflation	−0.368 (0.381)	−0.019 (0.106)	0.036	−0.714 (0.707)	−0.013 (0.140)	0.021
Interest, inflation	−0.658 (0.157)	−0.559 (0.142)	0.994	−0.867 (0.406)	−0.113 (0.323)	0.124
	MSVAR	S-MSVAR				
	Mean (s.d.)	Mean (s.d.)				
$\Pr(s_t = 1 \mid s_{t-1} = 1)$	0.906 (0.025)	0.870 (0.038)				
$\Pr(s_t = 2 \mid s_{t-1} = 2)$	0.863 (0.041)	0.898 (0.027)				

rate is affected by the past unemployment rate only in regime 1. The inflation rate equation in regime 1 includes its own lags and the lags of the unemployment rate and the interest rate. However the inflation rate equation in regime 2 includes the first lag of the inflation rate and the interest rate but not the lags of the unemployment rate. Thus, in both regimes the inflation rate is affected by the interest rate as expected.

With regard to Ψ, in regime 1 the off-diagonal element of Ψ relating to the correlation between the errors in the unemployment and interest rate equations and between the errors in the interest and inflation rate equations are included. However, in regime 2, which is low volatility regime, only off-diagonal element of Ψ relating to the correlation between the errors in the unemployment and interest rate equations is included.

Compared with the unrestricted MSVAR model and the restricted S-MSAVR model, the S-MSVAR model generally estimates much smaller values (in absolute values) for the parameters in the case of lower probability of inclusion. For example, the MSVAR model estimates the coefficient of the second lag of the interest rate in the inflation equation in regime 2 as -1.087 while the S-MSVAR model estimates the parameter as -0.002 with the inclusion probability 0.038. Thus, SSVS contributes to shrinkage virtually to zero for some coefficients with lower probability of inclusion.

Figure 2 presents the posterior expectation of the regime variables $\Pr(s_t = 2|I(t))$ where $I(t)$ is the information set at the time t. The *non-borrowed reserves operating procedure* between 1979 and 1982 is detected as high volatility regime, when the Federal Reserve moved from interest rate targeting to money growth targeting and allowed the interest rate to fluctuate freely. A regime shift occurs around in 2005 that leads to Lehman's crash in 2008.

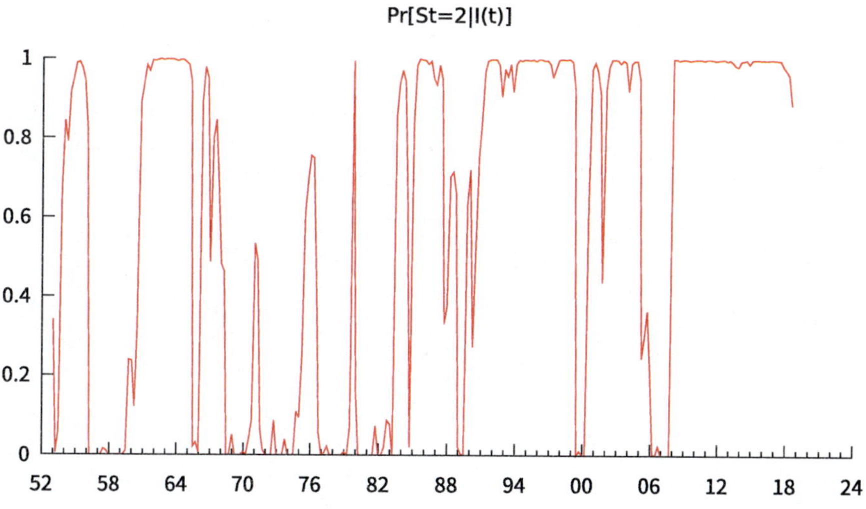

Fig. 2 Posterior expectation of the regime variables

3.2 Out-of-Sample Iterated Forecasts

In this subsection, I first discuss methods for evaluating the forecasting performances, then present results from a recursive forecasting exercise. For multi-step ahead forecasting, there are two methods to compute forecasting values of the time series outside the estimation sample - iterated forecasting method and direct forecasting method. Chevillon and Hendry (2005) evaluate the asymptotic and finite-sample properties of the direct forecasting method, and show that the direct method is more efficient asymptotically, more precise in finite samples and more robust against model misspecification, than the iterated method. However, Marcellino et al. (2006) evaluate a large-scale empirical comparison of iterated and direct forecasts using U.S. macroeconomic time series data, and find that iterated forecasts tend to have smaller MSFEs than direct forecasts, contrary to the theoretical preference of direct forecasts. In this paper, the iterated forecast method is adopted.

Let θ denote all of the parameters of the model and $\theta^{(m)}$ denote the MCMC replication (after the burn-in replications) for $m = 1, \dots M$, then standard MCMC theory implies that

$$\begin{aligned}\hat{p}(\hat{y}_{t+h+1}|I(t)) &= \frac{1}{M}\sum_{m=1}^{M} p(\hat{y}_{t+h+1}|I(t), \theta^{(m)}) \\ &\to p(\hat{y}_{t+h+1}|I(t)) \text{ as } M \to \infty.\end{aligned}$$

Since the predicted regime probabilities are given by

$$\Pr(s_{t+1} = j|I(t)) = \sum_{i=1}^{R} \Pr(s_{t+1} = j|s_t = i)\Pr(s_t = i|I(t))$$

where $\Pr(s_{t+1} = j|s_t = i)$ is the transition probabilities and $\Pr(s_t = i|I(t))$ is the filtered regime probabilities for $i, j = 1, ..., R$. The one-step ahead forecast $E(\hat{y}'_{t+1|t})$ of the MS-VAR model is obtained by estimating the parameters in (18) as:

$$\begin{aligned}E(\hat{y}'_{t+1|t}) &= \frac{1}{M}\sum_{m=1}^{M}\sum_{i=1}^{R}\sum_{j=1}^{R}\Big[\hat{\Pr}^{(m)}(s_{t+1} = j|s_t = i)\,\hat{\Pr}^{(m)}(s_t = i|I(t)) \\ &\quad\times\Big\{\hat{\mu}^{(m)}(s_t = j) + \sum_{k=1}^{p} \hat{y}'_{t+1-k}\hat{\Phi}_k^{(m)}(s_t = j)\Big\}\Big] \\ &= \frac{1}{M}\sum_{m=1}^{M} \hat{S}_t^{(m)}\hat{P}^{(m)}\hat{Z}_t^{(m)} \qquad (21)\end{aligned}$$

where $\hat{S}_t^{(m)} = \Big(\hat{\Pr}^{(m)}(s_t = 1|I(t)), \hat{\Pr}^{(m)}(s_t = 2|I(t)), \cdots, \hat{\Pr}^{(m)}(s_t = R|I(t))\Big)$, $\hat{P}$ is the estimated transition probability matrix defined in (20), and

$$\hat{Z}_t^{(m)} = \begin{bmatrix} \hat{\mu}^{(m)}(s_t = 1) + \sum_{k=1}^{p} \hat{y}'_{t+1-k} \hat{\Phi}_k^{(m)}(s_t = 1) \\ \hat{\mu}^{(m)}(s_t = 2) + \sum_{k=1}^{p} \hat{y}'_{t+1-k} \hat{\Phi}_k^{(m)}(s_t = 2) \\ \vdots \\ \hat{\mu}^{(m)}(s_t = R) + \sum_{k=1}^{p} \hat{y}'_{t+1-k} \hat{\Phi}_k^{(m)}(s_t = R) \end{bmatrix}$$

with $\hat{y}'_{t+1-k} = y'_{t+j-k}$ for $k < h$ and $\hat{y}'_{t+1-k}$ otherwise. To make forecasting further than one-period ahead into $h+1$-period, iterated forecasts for the multi-period forecasts are obtained recursively as $\hat{y}'_{t+h+1|t} = \frac{1}{M} \sum_{m=1}^{M} \hat{S}_t^{(m)} \hat{P}^{(m)} \hat{Z}_{t+h}^{(m)}$ for $h = 0, 1, \ldots$.

To evaluate the forecasting performances among several different models in a recursive out-of-sample prediction exercise, the mean squared forecast error (MSFE) and the mean absolute forecast error (MAFE) are widely used. Let $y'_{\tau+h+1}$ is a vector of observations at time $\tau + h + 1$ for $\tau = \tau_0, ..., T - h$. Then, $\hat{\Phi}(s_t) = \left(\hat{\mu}(s_t)', \hat{\Phi}_1(s_t), \ldots, \hat{\Phi}_p(s_t)\right)'$ is estimated, using information up to τ to forecast values $\hat{y}_{\tau+h+1}$ starting from $\tau = \tau_0$ up to $\tau = T - h$, and calculate the MSFE/MAFE defined as:

$$\text{MSFE} = \frac{1}{T - h - \tau_0 + 1} \sum_{\tau=\tau_0}^{T-h} \left[y_{\tau+h+1} - E(\hat{y}_{\tau+h+1} | \hat{\Phi}, I(\tau)) \right]^2. \quad (22)$$

and

$$\text{MAFE} = \frac{1}{T - h - \tau_0 + 1} \sum_{\tau=\tau_0}^{T-h} \left| y_{\tau+h+1} - E(\hat{y}_{\tau+h+1} | \hat{\Phi}, I(\tau)) \right|. \quad (23)$$

The empirical forecast comparison is based on recursive rolling forecasts. The recursive exercise involves using data up to time τ to forecast at time $\tau + h + 1$ for $h = 0, 1, ..., 11$. These are one-, two-, ... , twelve-step-ahead forecasts respectively. A sequence of 1–12-step ahead forecasts are generated at the starting date τ_0. Then the forecast origin is rolled forward one period, and another sequence of 1 to 12-step ahead forecasts is generated. The procedure is then repeated until $(T - \tau_0) \times$ 1-step forecasts, down to $(T - \tau_0 - 11) \times$ 12-step forecasts, where T equals the number of observation less the number of lags. This enables MSFEs and MAFEs to be calculated for each forecast horizon, and thus for long forecast horizons the smaller number of forecasts is available. The results of the forecast performance depend upon the choice of the starting date τ_0. The earlier the starting date is set, the smaller number of observations are available for VAR model in each regime, leading to poor accuracy of estimation. In this empirical exercise, the starting date τ_0 is set to be 180 so that the MSFE and MAFE statistics are computed over the forecast sample 1998:Q1–2020:Q1. The number of lag is set to be 4 and all prior parameters are set to be the same as those of the in-sample analysis in the previous subsection. MCMC is run with 20,000 draws after 5000 burn-in. SSVS is joint restriction search for both Φ and Ψ.

Table 3 presents MSFEs and the standard deviations in parentheses and Table 4 presents MAFEs and the standard deviations in parentheses for the three variables, twelve forecast horizons, using four models—the linear unrestricted VAR (VAR), the restricted linear VAR with SSVS prior (S-VAR), the unrestricted two-regime MSVAR (MSVAR), and the restricted two-regime MSVAR with the SSVS prior (S-MSVAR) models. For each variable, I computed the average of MSFEs/MAFEs and standard deviations over all forecast horizons $h = 0, 1, \ldots, 11$.

Overall, the results are mixed. Using the SSVS prior improves the forecasting performance for all three variables, compared with unrestricted models. For the unemployment rate, the S-MSVAR model leads to improvement in forecast performance with lower MSFEs and MAFEs than other models for most of the forecast horizons. For the interest rate and the inflation rate, however, the S-VAR model produces the lowest MSFEs and MAFEs. It seems that the nonlinear models do not improve forecasting performance relative to linear S-VAR model.

Comparison between VAR and MSVAR shows that allowing regime shifts does not improve or rather deteriorate in the forecast performance in this data set except for the unemployment rate. For the interest rate and the inflation rate, adding a Markov switching nonlinearity to a VAR model deteriorates in forecast performance for all forecasting horizons. There are several research that find this deterioration in the forecast performance by using Markov regime switching, see Engel (1994), Dacco and Satchell (1999) and Clements and Krolzig (1998) among others. Nikolsko-Rzhevskyy and Prodan (2012) and Jochmann et al. (2010) show that model with high dimensional spaces such as MSVAR fit well in-sample but often do not forecast well.

Comparison between MSVAR and S-MSVAR shows that implementing the SSVS prior on MSVAR improves the forecasting performance. Comparison between VAR and S-VAR tells the same story that the SSVS prior on the linear VAR improves the forecasting performance.

In terms of standard deviation for each MSFE/MAFE, adding the SSVS prior to VAR or MSVAR typically decreases standard deviation and thus provides more stable forecasting, and adding the nonlinearity of Markov switching tends to increase standard deviation and thus provides more unstable forecasting. For the three variables, standard deviation for each MSFE/MAFE are higher according to the magnitude for the value of MSFE/MAFE, and thus it becomes higher for longer forecasting horizons for the unemployment rate and the interest rate. For the inflation rate, the standard deviations of the MSFEs/MAFEs is remarkably high for all forecast horizons, suggesting that inflation rate is difficult to predict. The MSFEs/MAFEs show that S-VAR or S-MSVAR predicts better but not significantly with high standard deviations.

In short, models with the SSVS prior predict better against non-SSVS unrestricted models slightly, though adding nonlinearity does not improve the forecasting performance except for the unemployment rate.

Table 3 MSFEs for out-of-sample of US data

h	MSFE (s.d.)			
	VAR	S-VAR	MSVAR	S-MSVAR
For unemp				
0	0.078 (0.244)	0.060 (0.168)	0.117 (0.521)	0.067 (0.224)
1	0.318 (0.936)	0.254 (0.744)	0.444 (1.772)	0.278 (0.931)
2	0.743 (1.979)	0.601 (1.611)	0.877 (2.952)	0.638 (1.973)
3	1.310 (3.332)	1.079 (2.751)	1.418 (4.014)	1.114 (3.180)
4	1.908 (4.547)	1.609 (3.814)	1.962 (4.924)	1.632 (4.299)
5	2.420 (5.545)	2.108 (4.687)	2.356 (5.276)	2.103 (5.141)
6	2.845 (6.146)	2.559 (5.408)	2.666 (5.759)	2.519 (5.816)
7	3.188 (6.628)	2.947 (6.003)	2.896 (6.237)	2.895 (6.414)
8	3.440 (6.893)	3.244 (6.391)	3.034 (6.842)	3.089 (6.794)
9	3.634 (7.139)	3.473 (6.747)	3.172 (7.434)	3.249 (7.140)
10	3.753 (7.098)	3.612 (6.804)	3.289 (7.615)	3.362 (7.350)
11	3.819 (6.887)	3.687 (6.678)	3.367 (7.678)	3.415 (7.406)
Average	2.288 (4.774)	2.103 (4.317)	2.133 (5.085)	2.030 (4.722)
For interest				
0	0.231 (0.404)	0.235 (0.484)	0.332 (1.185)	0.221 (0.569)
1	0.860 (1.310)	0.905 (1.570)	0.890 (1.833)	0.852 (1.728)
2	1.498 (1.955)	1.532 (2.225)	1.902 (4.131)	1.405 (2.100)
3	2.305 (2.860)	2.208 (3.008)	3.298 (7.002)	2.135 (3.178)
4	3.464 (4.019)	3.137 (3.872)	4.357 (6.721)	3.245 (4.520)
5	4.664 (5.209)	4.118 (4.776)	5.597 (8.012)	4.353 (5.728)
6	5.745 (6.078)	5.018 (5.447)	7.031 (9.559)	5.364 (6.776)
7	6.832 (6.904)	5.887 (6.025)	8.148 (10.18)	6.422 (7.618)
8	7.908 (7.799)	6.755 (6.702)	9.087 (10.73)	7.592 (8.662)
9	8.867 (8.541)	7.572 (7.329)	10.18 (11.24)	8.707 (9.576)
10	9.779 (9.205)	8.358 (7.925)	11.28 (11.53)	9.822 (10.29)
11	10.66 (9.824)	9.124 (8.461)	12.19 (11.77)	11.04 (11.05)
Average	5.235 (5.342)	4.571 (4.819)	6.191 (7.834)	5.096 (5.983)
For inflation				
0	5.726 (21.72)	5.578 (21.32)	6.233 (23.40)	5.555 (22.34)
1	7.571 (25.02)	7.396 (24.45)	8.450 (29.91)	7.818 (26.60)
2	7.595 (22.81)	7.386 (22.86)	8.598 (26.35)	7.748 (24.24)
3	6.312 (20.08)	6.506 (21.76)	6.664 (21.68)	6.690 (21.70)
4	6.097 (19.19)	5.877 (19.21)	6.528 (19.49)	6.179 (20.01)
5	6.488 (19.78)	6.077 (18.97)	6.511 (20.73)	6.604 (20.79)
6	7.096 (17.50)	6.698 (18.23)	7.105 (20.02)	7.721 (21.18)
7	7.505 (15.86)	6.967 (16.29)	7.970 (16.31)	8.036 (18.18)
8	7.998 (20.31)	7.082 (20.06)	8.452 (22.24)	8.364 (22.62)
9	7.785 (20.27)	6.833 (20.45)	8.086 (22.74)	8.103 (22.28)
10	7.778 (21.38)	6.766 (21.49)	7.943 (22.39)	7.761 (21.57)
11	8.276 (23.67)	6.972 (22.65)	8.316 (23.27)	8.250 (23.96)
Average	7.186 (20.63)	6.678 (20.64)	7.575 (22.38)	7.403 (22.12)

Table 4 MAFEs for out-of-sample of US data

h	MAFE (s.d.)			
	VAR	S-VAR	MSVAR	S-MSVAR
For unemp				
0	0.188 (0.207)	0.167 (0.179)	0.187 (0.287)	0.165 (0.198)
1	0.357 (0.436)	0.329 (0.382)	0.351 (0.567)	0.304 (0.430)
2	0.549 (0.664)	0.508 (0.586)	0.518 (0.780)	0.448 (0.661)
3	0.724 (0.887)	0.683 (0.783)	0.672 (0.983)	0.583 (0.880)
4	0.893 (1.054)	0.846 (0.945)	0.816 (1.138)	0.720 (1.055)
5	1.018 (1.177)	0.977 (1.074)	0.936 (1.217)	0.843 (1.180)
6	1.114 (1.267)	1.082 (1.178)	1.020 (1.275)	0.938 (1.280)
7	1.190 (1.332)	1.159 (1.267)	1.081 (1.314)	1.018 (1.363)
8	1.245 (1.375)	1.212 (1.333)	1.108 (1.344)	1.083 (1.420)
9	1.308 (1.382)	1.262 (1.371)	1.133 (1.374)	1.159 (1.451)
10	1.349 (1.391)	1.301 (1.385)	1.153 (1.400)	1.224 (1.471)
11	1.387 (1.377)	1.337 (1.378)	1.178 (1.407)	1.277 (1.478)
Average	0.943 (1.046)	0.905 (0.988)	0.846 (1.091)	0.814 (1.072)
For interest				
0	0.361 (0.317)	0.346 (0.339)	0.358 (0.452)	0.311 (0.353)
1	0.710 (0.597)	0.706 (0.638)	0.690 (0.643)	0.654 (0.652)
2	0.965 (0.752)	0.957 (0.785)	1.020 (0.928)	0.897 (0.775)
3	1.199 (0.932)	1.168 (0.919)	1.374 (1.187)	1.130 (0.926)
4	1.487 (1.119)	1.408 (1.074)	1.640 (1.292)	1.407 (1.125)
5	1.745 (1.272)	1.641 (1.194)	1.873 (1.446)	1.655 (1.271)
6	1.971 (1.364)	1.841 (1.276)	2.163 (1.535)	1.857 (1.384)
7	2.160 (1.471)	2.017 (1.349)	2.382 (1.573)	2.058 (1.478)
8	2.329 (1.575)	2.169 (1.433)	2.544 (1.617)	2.260 (1.577)
9	2.490 (1.633)	2.304 (1.505)	2.719 (1.669)	2.446 (1.651)
10	2.636 (1.683)	2.424 (1.575)	2.892 (1.708)	2.623 (1.715)
11	2.761 (1.744)	2.538 (1.638)	3.025 (1.743)	2.796 (1.794)
Average	1.735 (1.205)	1.627 (1.144)	1.890 (1.316)	1.674 (1.225)
For inflation				
0	1.508 (1.858)	1.483 (1.838)	1.534 (1.962)	1.483 (1.732)
1	1.729 (2.141)	1.696 (2.126)	1.738 (2.339)	1.722 (2.103)
2	1.784 (2.101)	1.757 (2.074)	1.857 (2.269)	1.735 (2.077)
3	1.693 (1.857)	1.652 (1.943)	1.709 (1.935)	1.679 (1.868)
4	1.639 (1.847)	1.563 (1.853)	1.670 (1.934)	1.578 (1.821)
5	1.780 (1.822)	1.196 (1.789)	1.754 (1.853)	1.665 (1.858)
6	1.937 (1.829)	1.794 (1.866)	1.832 (1.936)	1.800 (2.016)
7	2.007 (1.865)	1.836 (1.897)	2.010 (1.982)	1.885 (2.017)
8	2.022 (1.977)	1.819 (1.942)	1.944 (2.162)	1.950 (2.036)
9	1.972 (1.974)	1.764 (1.929)	1.901 (2.115)	1.909 (1.911)
10	1.966 (1.978)	1.750 (1.924)	2.016 (1.970)	1.931 (1.908)
11	2.019 (2.049)	1.770 (1.959)	2.087 (1.990)	2.002 (1.959)
Average	1.838 (1.941)	1.715 (1.929)	1.838 (2.037)	1.778 (1.942)

4 Monte Carlo Simulation

In this section, simulated numerical example is used to illustrate the forecast performance of the S-MSVAR model in comparison with other models. The data-generating process (DGP) is taken from the in-sample estimation by the S-MSVAR model presented in Tables 1 and 2. Thus, the DGP is intended to mimic the behavior of the three variables of US macroeconomics - the unemployment rate, the interest rate and the inflation rate, containing the nonlinearity. The simulation is conducted 100 times with size of 268, and for each sample the four models—VAR, S-VAR, MSVAR and S-MSVAR—are estimated to calculate MSFEs/MAFEs with forecast horizons of 1-step ahead to 12-step ahead with $h = 0, 1, \ldots, 11$. The prior parameters are the same as used in the previous empirical studies. For each τ, a Markov chain is simulated 20,000 draws after 5000 burn-in. The starting date τ_0 is set to be 180 for each sample. A lag order is set to be four.

Tables 5 and 6 present the average MSFEs/MAFEs over 100 sample and the standard deviations in parentheses by the Monte Carlo simulations. Overall, the results suggest that the S-MSVAR model performs better than other models for most of the forecast horizons for the three variables. Implementing the SSVS restrictions to MSVAR model improves the forecasting performance. However, implementing the SSVS restrictions to linear VAR does not improve forecast except for the unemployment rate. Compared S-VAR with VAR, the SSVS prior deteriorates forecast for the interest rate and inflation rate. For the inflation rate, MSVAR provides the largest MSFEs/MAFEs.

The standard deviations of MSFEs/MAFEs in Tables 5 and 6 suggest that nonlinear Markov switching models typically provides unstable forecasting in terms of accuracy. The standard deviations of the MSFEs/MAFEs for the three variables are substantially high and become higher for longer forecasting horizons. For this simulation, I find that S-MSVAR predicts better than other models but improvement in forecasting performance is modest with relatively high standard deviations.

Figure 3 presents the forecast density error for selected forecast horizon 1-step, 4-step, 8-step and 12-step ahead corresponding $h = 0, 3, 7$ and 11. The first row is for the VAR, the second is for the S-VAR, the third is for the MSVAR and the fourth is for the S-MSVAR model. All these densities show that forecast errors tend to be more dispersed at longer forecast horizon, and the variances of the third variable (which represents the inflation rate) are larger than those of other variables, suggesting the inflation rate is difficult to predict.

This simulation suggests that adding the SSVS prior on nonlinear MSVAR model improves forecasting performance and helps to alleviate the over-parametrization problem of VAR or MSVAR models modestly.

Table 5 Monte Carlo results: MSFEs

h	MSFE (s.d.)			
	VAR	S-VAR	MSVAR	S-MSVAR
For unemp				
0	0.029 (0.053)	0.028 (0.052)	0.027 (0.050)	0.025 (0.046)
1	0.098 (0.193)	0.092 (0.187)	0.085 (0.174)	0.079 (0.165)
2	0.200 (0.393)	0.187 (0.380)	0.173 (0.361)	0.161 (0.343)
3	0.321 (0.614)	0.300 (0.597)	0.286 (0.615)	0.266 (0.587)
4	0.439 (0.776)	0.409 (0.752)	0.402 (0.830)	0.375 (0.796)
5	0.547 (0.887)	0.510 (0.868)	0.512 (1.007)	0.479 (0.977)
6	0.641 (0.959)	0.597 (0.937)	0.612 (1.142)	0.575 (1.113)
7	0.727 (1.002)	0.677 (0.980)	0.700 (1.231)	0.660 (1.201)
8	0.806 (1.045)	0.750 (1.024)	0.783 (1.327)	0.741 (1.293)
9	0.873 (1.083)	0.815 (1.062)	0.857 (1.431)	0.814 (1.394)
10	0.928 (1.112)	0.868 (1.094)	0.913 (1.513)	0.874 (1.478)
11	0.975 (1.139)	0.914 (1.125)	0.956 (1.542)	0.919 (1.510)
Average	0.549 (0.771)	0.512 (0.755)	0.526 (0.935)	0.497 (0.909)
For interest				
0	0.139 (0.286)	0.144 (0.305)	0.126 (0.299)	0.129 (0.303)
1	0.452 (0.824)	0.471 (0.873)	0.406 (0.855)	0.414 (0.874)
2	0.771 (1.262)	0.807 (1.311)	0.716 (1.376)	0.722 (1.368)
3	1.129 (1.675)	1.177 (1.717)	1.080 (1.889)	1.085 (1.877)
4	1.550 (2.091)	1.595 (2.122)	1.503 (2.478)	1.497 (2.431)
5	1.992 (2.518)	2.024 (2.537)	1.929 (3.004)	1.916 (2.932)
6	2.412 (2.939)	2.436 (2.948)	2.341 (3.406)	2.317 (3.336)
7	2.862 (3.410)	2.884 (3.417)	2.778 (3.809)	2.748 (3.752)
8	3.326 (3.891)	3.343 (3.888)	3.217 (4.155)	3.176 (4.109)
9	3.819 (4.412)	3.828 (4.400)	3.672 (4.605)	3.621 (4.564)
10	4.343 (4.967)	4.336 (4.937)	4.148 (5.163)	4.083 (5.130)
11	4.885 (5.535)	4.860 (5.477)	4.635 (5.769)	4.554 (5.726)
Average	2.307 (2.818)	2.326 (2.828)	2.213 (3.067)	2.189 (3.033)
For inflation				
0	0.798 (1.395)	0.791 (1.405)	0.761 (1.398)	0.731 (1.356)
1	0.965 (1.662)	0.965 (1.685)	0.946 (1.750)	0.900 (1.666)
2	1.056 (1.780)	1.053 (1.757)	1.047 (1.894)	1.005 (1.802)
3	1.210 (2.067)	1.210 (2.061)	1.199 (2.185)	1.157 (2.113)
4	1.294 (2.413)	1.312 (2.334)	1.297 (2.599)	1.254 (2.505)
5	1.370 (2.514)	1.388 (2.450)	1.360 (2.564)	1.325 (2.579)
6	1.466 (2.619)	1.475 (2.579)	1.459 (2.727)	1.417 (2.731)
7	1.521 (2.695)	1.522 (2.657)	1.552 (2.927)	1.499 (2.876)
8	1.531 (2.671)	1.530 (2.636)	1.592 (2.891)	1.538 (2.833)
9	1.563 (2.679)	1.561 (2.669)	1.619 (2.949)	1.553 (2.876)
10	1.581 (2.697)	1.589 (2.684)	1.659 (3.056)	1.583 (2.986)
11	1.644 (2.758)	1.653 (2.755)	1.691 (3.122)	1.630 (3.051)
Average	1.333 (2.329)	1.337 (2.306)	1.348 (2.505)	1.299 (2.448)

Table 6 Monte Carlo results: MAFEs

h	MSFE (s.d.)			
	VAR	S-VAR	MSVAR	S-MSVAR
For unemp				
0	0.131 (0.111)	0.126 (0.109)	0.125 (0.106)	0.121 (0.102)
1	0.231 (0.211)	0.222 (0.206)	0.215 (0.198)	0.207 (0.190)
2	0.332 (0.300)	0.318 (0.293)	0.303 (0.285)	0.293 (0.274)
3	0.426 (0.374)	0.407 (0.366)	0.386 (0.371)	0.371 (0.358)
4	0.509 (0.424)	0.486 (0.415)	0.457 (0.440)	0.438 (0.427)
5	0.579 (0.460)	0.554 (0.450)	0.519 (0.493)	0.497 (0.482)
6	0.640 (0.482)	0.613 (0.471)	0.573 (0.534)	0.550 (0.522)
7	0.694 (0.496)	0.664 (0.486)	0.621 (0.561)	0.597 (0.550)
8	0.740 (0.509)	0.707 (0.501)	0.664 (0.585)	0.641 (0.574)
9	0.778 (0.518)	0.745 (0.510)	0.702 (0.604)	0.679 (0.594)
10	0.808 (0.524)	0.774 (0.519)	0.729 (0.617)	0.710 (0.608)
11	0.832 (0.532)	0.800 (0.524)	0.753 (0.624)	0.736 (0.614)
Average	0.558 (0.412)	0.535 (0.404)	0.504 (0.451)	0.487 (0.441)
For interest				
0	0.272 (0.254)	0.275 (0.252)	0.246 (0.257)	0.248 (0.259)
1	0.514 (0.433)	0.522 (0.446)	0.463 (0.438)	0.474 (0.446)
2	0.686 (0.548)	0.705 (0.557)	0.626 (0.570)	0.627 (0.574)
3	0.847 (0.642)	0.868 (0.652)	0.787 (0.678)	0.787 (0.682)
4	1.015 (0.722)	1.031 (0.730)	0.942 (0.784)	0.940 (0.783)
5	1.169 (0.791)	1.178 (0.798)	1.087 (0.864)	1.085 (0.859)
6	1.294 (0.858)	1.308 (0.852)	1.225 (0.916)	1.223 (0.906)
7	1.417 (0.924)	1.430 (0.917)	1.357 (0.968)	1.353 (0.958)
8	1.533 (0.987)	1.544 (0.980)	1.478 (1.016)	1.473 (1.003)
9	1.645 (1.055)	1.653 (1.048)	1.599 (1.056)	1.587 (1.050)
10	1.755 (1.124)	1.758 (1.116)	1.705 (1.113)	1.689 (1.110)
11	1.863 (1.189)	1.861 (1.182)	1.809 (1.167)	1.789 (1.164)
Average	1.168 (0.794)	1.178 (0.795)	1.110 (0.819)	1.105 (0.816)
For inflation				
0	0.680 (0.580)	0.677 (0.577)	0.659 (0.572)	0.642 (0.565)
1	0.757 (0.626)	0.753 (0.630)	0.737 (0.635)	0.719 (0.619)
2	0.793 (0.654)	0.793 (0.651)	0.770 (0.674)	0.761 (0.653)
3	0.841 (0.709)	0.850 (0.699)	0.817 (0.729)	0.807 (0.711)
4	0.859 (0.745)	0.877 (0.737)	0.839 (0.769)	0.829 (0.752)
5	0.887 (0.763)	0.903 (0.756)	0.870 (0.776)	0.854 (0.772)
6	0.925 (0.782)	0.935 (0.776)	0.914 (0.789)	0.891 (0.789)
7	0.940 (0.799)	0.948 (0.790)	0.939 (0.819)	0.919 (0.809)
8	0.942 (0.802)	0.949 (0.794)	0.950 (0.830)	0.934 (0.816)
9	0.964 (0.797)	0.965 (0.793)	0.959 (0.836)	0.940 (0.818)
10	0.974 (0.795)	0.980 (0.793)	0.976 (0.841)	0.948 (0.828)
11	0.997 (0.806)	0.999 (0.809)	0.986 (0.848)	0.963 (0.839)
Average	0.880 (0.738)	0.886 (0.734)	0.868 (0.760)	0.851 (0.848)

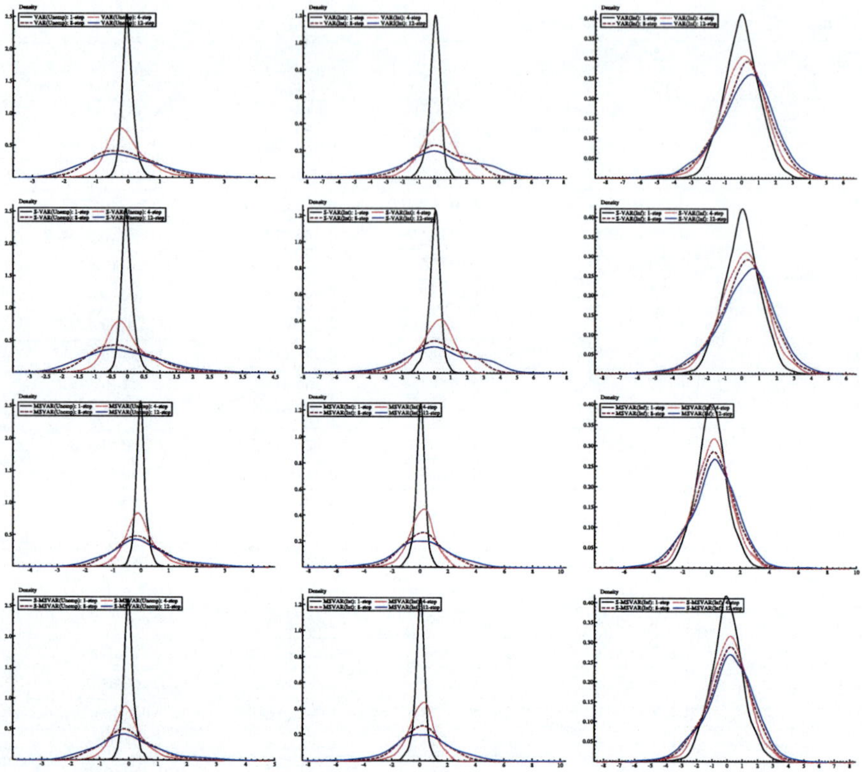

Fig. 3 Monte Carlo. Forecast error density

5 Conclusion

In this paper, I have evaluated the forecast performance of MSVAR model with the SSVS prior using US macroeconomic variables—the unemployment rate, the interest rate and the inflation rate. VAR models and especially MSVAR models have a large number of parameters with many of them being insignificant, leading to over-parameterization problems, overfitting in-sample with inaccuracy in out-of-sample forecasting. It is also difficult to decide the appropriate (parsimonious but good in fit) lag length in MSVAR models with the same lag length in all regimes. To mitigate these problems, I implement the SSVS method to a MSVAR model to improve forecasting performance in an application of a standard trivariate macroeconomic variables.

In the recursive forecasting exercise, the use of SSVS in a linear VAR without regime switching provides improvements relative to a standard unrestricted VAR. The use of SSVS in a nonlinear MSVAR brings slight improvements relative to a standard unrestricted MSVAR, but not relative to VAR with the SSVS prior for all variables. For the interest rate and the inflation rate, linear VAR with the SSVS prior is superior in forecasting. From the Monte Carlo study, I find that MSVAR model with the SSVS prior is superior to other models with slight improvements due to the DGP containing nonlinearity. However, the standard deviations of MSFEs/MAFEs for MSVAR with/without the SSVS prior are too large to conclude that the SSVS method does overcome over-parametrization problems of MSVAR model.

Acknowledgements This work was supported by JSPS KAKENHI Grant Number 20K01591.

References

Akgul, I., Bildirici, M., Ozdemir, S.: Evaluating the nonlinear linkage between gold prices and stock market index using Markov-switching Bayesian VAR models. Procedia—Soc. Behav. Sci. **210**, 408–415 (2015)

Albert, J.H., Chib, S.: Bayes inference via Gibbs sampling of autoregressive time series subject to Markov mean and variance shifts. J. Bus. Econ. Stat. **11**, 1–15 (1993)

Andersson, M.K. Karlsson, S.: Bayesian forecast combination for VAR models. In: Advances in Econometrics, pp. 501–524 (2007)

Balcilar, M., van Eyden, R., Uwilingiye, J., Gupta, R.: The impact of oil price on South African GDP growth: a Bayesian Markov switching-VAR analysis. Afr. Dev. Rev. **29**, 319–336 (2017)

Beine, M., Candelon, B., Sekkat, K.: EMU membership and business cycle phases in Europe: Markov-switching VAR analysis. J. Econ. Integr. **18**, 214–242 (2003)

Carter, C.K., Kohn, R.: On Gibbs sampling for state space models. Biometrika **81**, 541–553 (1994)

Chevillon, G., Hendry, D.F.: Non-parametric direct multi-step estimation for forecasting economic processes. Int. J. Forecast. **21**, 201–218 (2005)

Clements, M.P., Krolzig, H.-M.: A comparison of the forecast performance of Markov-switching and threshold autoregressive models of US GNP. Economet. J. **1**, 47–75 (1998)

Clements, M.P., Krolzig, H.M.: Can oil shocks explain asymmetries in the US business cycle? Emp. Econ. **27**, 185–204 (2002)

Cogley, T., Sargent, T.J.: Drifts and volatilities: monetary policies and outcomes in the post WWII US. Rev. Econ. Dyn. **8**, 262–302 (2005)

Dacco, R., Satchell, S.: Why do regime-switching models forecast so badly? J. Forecast. **18**, 1–16 (1999)

Doornik, J.A.: Object-Oriented Matrix Programming using Ox. Timberlake Consultants Press, London (2013)

Droumaguet, M., Warne, A., Woźniak, T.: Granger causality and regime inference in Markov switching VAR models with Bayesian methods. J. Appl. Economet. **32**, 802–818 (2017)

Engel, C.: Can the Markov switching model forecast exchange rates? J. Int. Econ. **36**, 151–165 (1994)

Fallahi, F.: Causal relationship between energy consumption (EC) and GDP: a Markov-switching (MS) causality. Energy **36**, 4165–4170 (2011)

George, E.I., McCulloch, R.E.: Variable selection via Gibbs sampling (1993)

George, E.I., McCulloch, R.E.: Approaches for Bayesian variable selection. Statistica Sinica **7**, 339–373 (1997)

George, E.I., Sun, D., Ni, S.: Bayesian stochastic search for VAR model restrictions. J. Economet. **142**, 553–580 (2008)

Hamilton, J.D.: A new approach to the economic analysis of nonstationary time series and the business cycle. Economet.: J. Economet. Soc. 357–384 (1989)

Jochmann, M., Koop, G., Strachan, R.W.: Bayesian forecasting using stochastic search variable selection in a VAR subject to breaks. Int. J. Forecast. **26**, 326–347 (2010)
Kadiyala, K.R., Karlsson, S.: Numerical methods for estimation and inference in Bayesian VAR-models. J. Appl. Economet. **12**, 99–132 (1997)
Koop, G., Korobilis, D.: Bayesian Multivariate Time Series Methods for Empirical Macroeconomics, vol. 3 (2010)
Koop, G., Leon-Gonzalez, R., Strachan, R.W.: On the evolution of the monetary policy transmission mechanism. J. Econ. Dyn. Control **33**, 997–1017 (2009)
Krolzig, H.M.: Markov-Switching Vector Autoregressions. Springer, New York (1997)
Krolzig, H.-M., Marcellino, M., Mizon, G.E.: A Markov-switching vector equilibrium correction model of the UK labour market. In: Advances in Markov-Switching Models, pp. 91–112. Springer (2002)
Marcellino, M., Stock, J.H., Watson, M.W.: A comparison of direct and iterated multistep AR methods for forecasting macroeconomic time series. J. Economet. **135**, 499–526 (2006)
Nikolsko-Rzhevskyy, A., Prodan, R.: Markov switching and exchange rate predictability. Int. J. Forecast. **28**, 353–365 (2012)
Primiceri, G.E.: Time varying structural vector autoregressions and monetary policy. Rev. Econ. Stud. **72**, 821–852 (2005)
Shephard, N.: Partial non-Gaussian state space. Biometrika **81**, 115–131 (1994)
Sims, C.A.: Macroeconomics and reality. Econometrica **48**, 1–48 (1980)
Terasvirta, T., Anderson, H.M.: Characterizing nonlinearities in business cycles using smooth transition autoregressive modelse. J. Appl. Economet. **7**, S119–S136 (1992)

Estimating the Correlation Coefficients in a Multivariate Skew Normal Population Using the a Priori Procedure (APP)

Cong Wang, Tonghui Wang, David Trafimow, and Tingting Tong

Abstract This paper expands the a priori procedure (APP) to enable researchers to determine appropriate sample sizes for using sample correlation coefficients to estimate corresponding population correlation coefficients. The underlying assumption is that the population follows a skew normal distribution, which is more general than the typical assumption of a normal distribution. Furthermore, we work out the statistical implications and provide links to free and user-friendly programs.

Keywords A priori procedure · Sample sizes · Correlation coefficients · Skew normal distribution

1 Introduction

The correlation coefficient, often used in investing and finance, indexes the strength of the relationship between two variables, or measures the degree to which two securities move in relation to each other. It ranges between -1 (perfect negative relationship) and 1 (perfect positive relationship), with 0 indicating no relationship whatsoever. The most common type of correlation coefficient is the Pearson correlation, denoted by r, that measures the strength and direction of the linear relationship between two variables, but should not be used for nonlinear relationships.

C. Wang
University of Nebraska Omaha, Omaha, Nebraska 68182, USA
e-mail: congwang@unomaha.edu

T. Wang (✉) · D. Trafimow · T. Tong
New Mexico State University, Las Cruces, New Mexico 88003, USA
e-mail: twang@nmsu.edu

D. Trafimow
e-mail: dtrafimo@nmsu.edu

T. Tong
e-mail: tongtt@nmsu.edu

N. Ngoc Thach et al. (eds.), *Financial Econometrics: Bayesian Analysis, Quantum Uncertainty, and Related Topics*, Studies in Systems, Decision and Control 427,
https://doi.org/10.1007/978-3-030-98689-6_11

Several researchers have reported their findings regarding the correlational studies, see Olsson (1979) and Schisterman et al. (2003). Robust Estimation of the Correlation Coefficient was studied by Shevlyakov and Smirnov (2011), and Sari et al. (2017) worked on the sample size for estimation of the Pearson correlation coefficient in cherry tomato tests.

In contrast to other work concerned with establishing sample sizes to meet power specifications, the a priori procedure (APP) is concerned with how well the sample statistic estimates the corresponding population parameter. The researcher renders a desired precision level, and level of confidence to meet the precision specification, and an APP equation provides the necessary sample size to meet the precision and confidence specifications. Although there have been many important APP advances (see Wang et al. (2019), Wang et al. (2019), Wang et al. (2019), and Wang et al. (2020)), there is currently no way to determine the necessary sample size to meet precision and confidence specifications for correlation coefficients based on skew normal distributions. Our goal is to propose APP equations to enable researchers to determine the minimum sample size they need to obtain to meet specifications for precision and confidence for using the sample-based correlation coefficient estimator to estimate the population correlation coefficient ρ_{ij} for $i,\ j = 1,\ \ldots,\ p\ (i \neq j)$ from a skew normal population. To set up the present argument, it is necessary first to review the family of skew normal distribution. As a generalization of the normal distribution, Azzalini (1985) formalized the skew normal distribution for modeling asymmetric data and provided the probability density function of the skew normal distribution as follows:

$$f_Z(z) = 2\phi(z)\Phi(\gamma z), \tag{1}$$

where $\phi(\cdot)$ and $\Phi(\cdot)$ are the density and cumulative distribution function of the standard normal distribution, respectively. γ is called the skewness parameter. For its sampling distributions, see Azzalini and Capitanio (1999), Wang et al. (2019) and Zhu et al. (2019).

The paper is organized as follows. Properties of skew normal distributions are reviewed briefly in Sect. 2. In Sect. 3, the distribution of the sample correlation coefficient and its properties are discussed. Then the a priori procedure is set up in Sect. 4. To illustrate our results, the numerical results as well as the simulation works are listed in Sect. 5.

2 Some Properties of Skew Normal Distribution

First we introduce some notations that will used in this paper. Let $M_{n\times m}$ be the set of all $n \times m$ matrices over $\Re$ so that $M_{n\times 1} = \Re^n$. For any nonnegative definite matrix $T \in M_{n\times n}$, let T' and $T^+ \in M_{n\times n}$ be the transpose and Moore-Penrose inverse of T, respectively, and let $T^{-\frac{1}{2}}$ and $T^{\frac{1}{2}}$ be symmetric such that $T^{-\frac{1}{2}}T^{-\frac{1}{2}} = T^+$ and $T^{\frac{1}{2}}T^{\frac{1}{2}} = T$. Also, $I_n \in M_{n\times n}$ is the identity matrix, $\mathbf{1}_n \in \Re^n$ is the column vector with entries of 1's, and $\mathbf{e_i} \in \Re^\mathbf{n}$ is the column vector with 1 in the ith position and 0's otherwise. For $A = (a_{ij}) \in M_{m\times n}$ and $B = (b_{k\ell}) \in M_{p\times q}$ the Kronecker product of A and B, denoted by $A \otimes B$, is defined by $A \otimes B = (a_{ij}B) = (a_{ij}b_{k\ell}) \in M_{np\times m\ell}$.

The univariate skew normal distribution Azzalini (1985) was studied as an extension of the normal distribution to accommodate asymmetry. This new class of distribution shared similar properties with the normal distribution. The skew normal distribution denoted as SN (given in Eq. (1) in Sect. 1) introduced the skewness or shape parameter to model skewness. A location-scale extension of the skew normal distribution has been defined as follows.

Definition 1 (Azzalini (1985)) If Z is a continuous random variable with density function 1, then the variable

$$Y = \xi + \omega Z \quad (\xi \in \Re, \quad \omega > 0)$$

will be called a skew normal (SN) variable with location parameter ξ, scale parameter ω, and skewness parameter γ, which is denoted by $SN(\xi,\ \omega^2,\ \gamma)$. Its density function at $y \in \Re$ is

$$f_Y(y) = 2\phi(y,\ \xi,\ \omega^2)\Phi\left(\gamma\,\frac{y-\xi}{\omega}\right). \tag{2}$$

The univariate skew normal distribution defined above was extended to its multivariate case.

Definition 2 (Azzalini and Dalla Valle 1996) A random vector $\mathbf{X}$ is said to have an n-dimensional multivariate skew normal distribution with vector of location parameters $\boldsymbol{\mu} = (\mu_1,\ \mu_2,\ \ldots,\ \mu_n)' \in \Re^n$, scale parameter of nonnegative definite $\Sigma \in M_{n\times n}$, and the vector of skewness (shape) parameters $\boldsymbol{\gamma} = (\gamma_1, \gamma_2, \ldots, \gamma_n)' \in \Re^n$, denoted as $\boldsymbol{X} \sim SN_n(\boldsymbol{\mu},\ \boldsymbol{\Sigma},\ \boldsymbol{\gamma})$, if its density function (pdf) is

$$f_{\mathbf{X}}(\mathbf{x}) = 2\phi_n(\mathbf{x};\ \boldsymbol{\mu},\ \Sigma)\Phi\left(\boldsymbol{\gamma}'\Sigma^{-1/2}(\mathbf{x}-\boldsymbol{\mu})\right), \tag{3}$$

where $\phi_n(\mathbf{x};\ \boldsymbol{\mu},\ \Sigma)$ is the density of the n-dimensional multivariate normal distribution $N_n(\boldsymbol{\mu},\ \Sigma)$ with mean vector $\boldsymbol{\mu}$ and covariance matrix Σ, and $\Phi(z)$ is the cumulative distribution function (cdf) of the standard normal random variable.

Lemma 1 *Suppose that $\mathbf{X} \sim SN_n(\boldsymbol{\mu},\ \Sigma,\ \boldsymbol{\gamma})$. Then for any matrix $A \in M_{n\times k}$ with full column rank, $A'\mathbf{X} \sim SN_k(A'\boldsymbol{\mu},\ \Sigma_*,\ \boldsymbol{\gamma}_*)$, where*

$$\boldsymbol{\gamma}_* = \frac{\Sigma_*^{-1/2}A'\Sigma^{\frac{1}{2}}\boldsymbol{\gamma}}{\sqrt{1+\boldsymbol{\gamma}'(I_n-P)\boldsymbol{\gamma}}}, \quad \Sigma_* = A'\Sigma A, \quad \textit{and} \quad P = \Sigma^{\frac{1}{2}}A\Sigma_*^{-1}A'\Sigma^{\frac{1}{2}}.$$

For more details, see Wang et al. (2019). Also if we let $A = \frac{1}{n}\mathbf{1}_n$ and $A = \mathbf{e}_i$, respectively, then it is easy to obtain the following result and its proof is given in Wang et al. (2016).

Corollary 1 *Suppose that* $\mathbf{X} = (X_1, X_2, ..., X_n)' \sim SN_n(\boldsymbol{\mu}, \Sigma, \boldsymbol{\alpha})$ *with* $\boldsymbol{\mu} = \xi \mathbf{1_n}$, $\Sigma = \omega^2 I_n$ *and* $\boldsymbol{\gamma} = \gamma \mathbf{1_n}$, *where* $\xi, \lambda \in \Re$ *and* $\omega > 0$. *Then the following results hold.*

(a) The sample mean $\bar{X} = \sum_{i=1}^{n} X_i/n$ *has a skew normal distribution:*

$$\bar{X} \sim SN\left(\xi, \frac{\omega^2}{n}, \sqrt{n}\gamma\right),$$

(b) Random variables $X_1, \cdots, X_n$ *are identically distributed:*

$$X_i \sim SN\left(\xi, \omega^2, \gamma_*\right) \qquad i = 1, \dots, n,$$

where $\gamma_* = \frac{\gamma}{\sqrt{1+(n-1)\gamma^2}}$.

3 Distribution of the Sample Correlation Coefficient in the Skew Normal Population

In the normal case, the following results hold for setting up the sample correlation coefficients.

Lemma 2 *Let* $\mathbf{X}_1, \dots, \mathbf{X}_N$ *be a random sample of size* N *from* $N_p(\boldsymbol{\mu}, \Sigma)$ *and* $\bar{\mathbf{X}}$ *be its sample mean. Then* $\bar{\mathbf{X}} \sim N_p(\boldsymbol{\mu}, \Sigma/N)$ *and is independent of* $\hat{\Sigma}$, *the maximum likelihood estimator of* Σ. *Also* $N\hat{\Sigma}$ *is distributed as* $\sum_{\alpha=1}^{N-1} \mathbf{Z}_\alpha \mathbf{Z}_\alpha' \sim W_p(N-1, \Sigma)$, *the Wishart distribution with* $N-1$ *degrees of freedom and covariance matrix* Σ, *where* $\mathbf{Z}_\alpha$ *is distributed according to* $N_p(\mathbf{0}, \Sigma)$, $\alpha = 1, \dots, N-1$, *and* $\mathbf{Z}_\alpha$*'s are independent.*

Now we consider that $\mathbf{X}_\alpha = (X_{1\alpha}, \dots, X_{p\alpha})'$ for $\alpha = 1, \dots, N$ and $\mathbf{X} = (\mathbf{X}_1', \dots, \mathbf{X}_N')' \in M_{N\times p}$ has the joint distribution

$$\mathbf{X} \sim SN_{N\times p}(\mathbf{1}_N \otimes \boldsymbol{\xi}', I_N \otimes \Sigma, \mathbf{1}_N \otimes \boldsymbol{\gamma}'), \tag{4}$$

where $\boldsymbol{\xi} = (\xi_1, \dots, \xi_p)'$, $\boldsymbol{\gamma} = (\gamma_1, ..., \gamma_p)'$, and

$$\Sigma = \begin{pmatrix} \sigma_1^2 & \sigma_{12} & \cdots & \sigma_{1p} \\ \sigma_{21} & \sigma_2^2 & \cdots & \sigma_{2p} \\ \vdots & \vdots & \ddots & \vdots \\ \sigma_{p1} & \sigma_{p2} & \cdots & \sigma_p^2 \end{pmatrix} \quad \text{with} \quad \sigma_{ij} = \rho_{ij}\sigma_i\sigma_j.$$

Then we have the following result.

Theorem 1 *Suppose that the assumption given in (4) holds, and let*

$$A = \sum_{\alpha=1}^{N} (\mathbf{X}_\alpha - \bar{\mathbf{X}})(\mathbf{X}_\alpha - \bar{\mathbf{X}})'.$$

Then A is distributed as $\sum_{\alpha=1}^{n} \mathbf{Z}_\alpha \mathbf{Z}_\alpha' \sim W_p(n, \Sigma)$, where $\mathbf{Z}_\alpha \sim N_p(\mathbf{0}, \Sigma)$, $\alpha = 1, \ldots, n$, $n = N - 1$, and $\mathbf{Z}_\alpha$'s are independent.

Proof Let $B = \left(b_{\alpha\beta}\right)$ be an orthogonal matrix in $M_{N\times N}$ with last row $\mathbf{1}_N'/N$ and consider the linear transformations of $\mathbf{X}_\alpha$'s:

$$\mathbf{Y}_\beta = \sum_{\alpha=1}^{N} b_{\alpha\beta}\mathbf{X}_\alpha, \qquad \beta = 1, \ldots, N-1,$$

and

$$\mathbf{Y}_N = \frac{1}{\sqrt{N}}\mathbf{1}_N'\mathbf{X} = \frac{1}{\sqrt{N}}\left(\sum_{\alpha=1}^{N} X_{1\alpha}, \ \ldots, \sum_{\alpha=1}^{n} X_{p\alpha}\right) = \sqrt{N}\bar{\mathbf{X}}'.$$

Note that

$$\begin{aligned}
\sum_{\alpha=1}^{N} \mathbf{Y}_\alpha \mathbf{Y}_\alpha' &= \sum_{\alpha=1}^{N} \left(\sum_{\beta=1}^{N} b_{\alpha\beta}\mathbf{X}_\beta\right)\left(\sum_{\tau=1}^{N} b_{\alpha\tau}\mathbf{X}_\tau\right)' \\
&= \sum_{\beta=1}^{N}\sum_{\tau=1}^{N}\left(\sum_{\alpha=1}^{N} b_{\alpha\beta}b_{\alpha\tau}\right)\mathbf{X}_\beta\mathbf{X}_\tau' \\
&= \sum_{\beta=1}^{N} \mathbf{X}_\beta\mathbf{X}_\beta'.
\end{aligned}$$

Thus

$$A = \sum_{\alpha=1}^{N} \mathbf{X}_\alpha\mathbf{X}_\alpha' - N\bar{\mathbf{X}}\bar{\mathbf{X}}' = \sum_{\beta=1}^{n} \mathbf{Y}_\beta\mathbf{Y}_\beta',$$

where $n = N - 1$. Since $\mathbf{Y}_\alpha = B_\alpha'\mathbf{X} = (I_p \otimes B_\gamma')vec(\mathbf{X})$, by Lemma 1, $\mathbf{Y}_\alpha \sim SN_p(\boldsymbol{\mu}_*, \Sigma_*, \boldsymbol{\alpha}_*)$, with

$$\boldsymbol{\mu}_* = (I_p \otimes B_\alpha')(\boldsymbol{\xi} \otimes \mathbf{1}_N) = \mathbf{0},$$

$$\Sigma_* = (I_p \otimes B_\alpha')(\Sigma \otimes I_N)(I_p \otimes B_\alpha')' = \Sigma$$

and

$$\boldsymbol{\gamma}_* = \frac{\Sigma^{-1/2}(I_p \otimes B_\gamma')(\Sigma \otimes I_N)^{1/2}(\boldsymbol{\gamma} \otimes \mathbf{1}_N)}{\sqrt{1 + (\boldsymbol{\gamma} \otimes \mathbf{1}_N)'(I_{Np} - P)(\boldsymbol{\gamma} \otimes \mathbf{1}_N)}} = \mathbf{0},$$

that is, $\mathbf{Y}_\alpha \sim N_p(\mathbf{0},\ \Sigma)$ so that the desired result follows. □

For estimating the correlation coefficient, ρ_{ij}, between X_i and X_j (two components of $\mathbf{X}$), under the normal assumption, we know that the corresponding sample correlation coefficient, r_{ij}, is the maximum likelihood estimator of ρ_{ij}. We need to find the distribution of r_{ij}. For the convenience, we will find the distribution of r_{12}, the same theory holds for each r_{ij}. Since $r \equiv r_{12}$ depends only on X_1 and X_2 so that we just consider the bivariate normal distribution of X_1 and X_2. Let $(X_{1\alpha}, X_{2\alpha})'$, $\alpha = 1, \ldots, N$ be a random sample from $N_2(\boldsymbol{\mu}, \Sigma)$. Then we have the following Lemma and its proof is given in Anderson (2022), applying for our bivariate skew normal case.

Lemma 3 *Suppose that* $\mathbf{V_i} = (Z_{i1}, \ldots, Z_{in})'$, $i = 1, 2$ *are two random vectors and* $A = (a_{ij}) \in M_{2\times 2}$ *is given in Theorem 1. For given* $\mathbf{V_1} = \mathbf{v_1}$, *let*

$$b = \frac{\mathbf{v_1}'\mathbf{V_2}}{\mathbf{v_1}'\mathbf{v_1}} \quad \text{and} \quad U = (\mathbf{V_2} - b\mathbf{v_1})'(\mathbf{V_2} - b\mathbf{v_1}).$$

Then, given $\mathbf{V_1} = \mathbf{v_1}$,

(i) the conditional distribution of b is normal with mean $\rho\sigma_2/\sigma_1$ *and variance* $\sigma^2/c^2 \equiv \sigma_2^2(1-\rho^2)/c^2$, *where* $c^2 = \mathbf{v_1}'\mathbf{v_1}$;
(ii) the conditional distribution of $\frac{U}{\sigma^2} \sim \chi^2_{n-1}$, *the chi-square distribution with degrees of the freedom* $n-1$;
(iii) b and U are independent; and
(iv) the distribution of

$$T = \frac{\sqrt{n-1}r}{\sqrt{1-r^2}},$$

is t-distribution with degrees of the freedom $n-1$ *when* $\rho = 0$, *where* $r = \frac{a_{12}}{\sqrt{a_{11}a_{22}}}$.

Now we suppose that $X_1, \ldots, X_N$ be random variables from a bivariate skew normal population and their joint distribution is

$$SN_{N\times 2}(\mathbf{1}_N \otimes \boldsymbol{\xi}',\ I_N \otimes \Sigma, \mathbf{1}_N \otimes \boldsymbol{\gamma}'), \tag{5}$$

where

$$\boldsymbol{\xi} = \begin{pmatrix}\xi_1\\ \xi_2\end{pmatrix}, \quad \Sigma = \begin{pmatrix}\sigma_1^2 & \rho\sigma_1\sigma_2\\ \rho\sigma_1\sigma_2 & \sigma_2^2\end{pmatrix}, \quad \boldsymbol{\gamma} = \begin{pmatrix}\gamma_1\\ \gamma_2\end{pmatrix}.$$

In the following, we focus on the distribution of T for the case where $\rho \neq 0$ under the bivariate skew normal setting. From Theorem 1, we know that the distribution of

T is free of the skewness vector $\boldsymbol{\gamma}$ so that it is the same under the bivariate normal case. Also by the invariance property of r under linear transformations of $Z_{1\alpha}$'s and $Z_{2\alpha}$'s, without loss of generality, we may assume that $\sigma_1 = \sigma_2 = 1$ in deriving the density of T.

Theorem 2 *Under the above assumption given in Lemma 3, we have the following results.*

(a) The probability density function (pdf) of T, which is defined in (iv) of Lemma 3 is

$$f_T(t) = a \int_0^\infty \int_0^\infty (xy)^{\frac{N-3}{2}} \exp\left\{-\frac{1}{2}\left[x + y + \left(t\sqrt{\frac{x}{N-2}} - \sqrt{y}\lambda\right)^2\right]\right\} dydx, \tag{6}$$

where $a = (4\pi\sqrt{N-2}\Gamma(N-2))^{-1}$.

(b) The pdf of $S = r/\sqrt{1-r^2}$ is

$$f_S(s) = \frac{N-2}{\pi} \int_0^{\frac{\pi}{2}} \frac{(\sin 2\theta)^{N-2}}{[1 + (s\cos\theta - \lambda\sin\theta)^2]^{N-1}} d\theta. \tag{7}$$

Proof (a) From Lemma 3 and the proof of Theorem 1, we obtain that $b \sim N\left(\rho, \frac{1-\rho^2}{c^2}\right)$. Then the conditional distribution of T given $Y \equiv \mathbf{V}_1'\mathbf{V}_1 = c^2$ is of non-central t distribution with non-central parameter $c\lambda$ where $\lambda = \frac{\rho}{\sqrt{1-\rho^2}}$. Therefore, the density function of T is

$$f_T(t) = \int_0^\infty f_{T|y}(t) f_Y(y) dy,$$

where $T|y \sim t(\sqrt{y}\lambda)$ and $Y \sim \chi^2_{N-1}$. Note that the pdf of $T|y$ is given by

$$f_{T|y}(t) = b \int_0^\infty x^{(N-3)/2} \exp\left(-\frac{1}{2}\left[x + \left(t\sqrt{\frac{x}{N-2}} - \lambda\right)^2\right]\right) dx,$$

where $b = \left(2^{(N-1)/2}\Gamma[(N-2)/2]\sqrt{(N-2)\pi}\right)^{-1}$, and the pdf of Y is

$$g(y) = (2^{(N-1)/2}\Gamma[(N-1)/2])^{-1} y^{(N-1)/2-1} e^{-y/2}.$$

Therefore, the pdf of T, in the integral form, is

$$f_T(t) = a \int_0^\infty \int_0^\infty (xy)^{(N-3)/2} \exp\left\{-\frac{1}{2}\left[x + y + (t\sqrt{x} - \sqrt{y}\lambda)^2\right]\right\} dydx,$$

where

$$a = \left(2^{N-1}\Gamma[(N-1)/2]\Gamma[(N-2)/2]\sqrt{(N-2)\pi}\right)^{-1}.$$

Note that $\Gamma[(N-1)/2]\Gamma[(N-2)/2] = 2^{3-N}\Gamma(N-2)\sqrt{\pi}$ so that (a) follows.

(b) Let $S = r/\sqrt{1-r^2}$. Then $T = \sqrt{N-2}S$, and the density function of S is

$$f_S(s) = \frac{1}{4\pi\Gamma(N-2)}\int_0^\infty\int_0^\infty (xy)^{(N-3)/2}\exp\left\{-\frac{1}{2}\left[x+y+(s\sqrt{x}-\sqrt{y}\lambda)^2\right]\right\}dydx.$$

Now let $x = u^2$ and $y = v^2$, and we consider substitution $u = w\cos\theta$ and $v = w\sin\theta$, where $w > 0$ and $\theta \in [0, \pi/2]$. Then

$$f_S(s) = \frac{1}{2^{N-2}\pi\Gamma(N-2)}\int_0^{\pi/2}\int_0^\infty [\sin 2\theta]^{N-2} w^{2N-3}\exp\left(-d^2w^2/2\right)dwd\theta, \tag{8}$$

where $d^2 = 1 + (s\cos\theta - \lambda\sin\theta)^2$. By substituting dw by v and the fact that the integral of the pdf of a chi-distribution over the interval $[0, \infty)$ is 1,

$$f_S(s) = \frac{2^{N-2}\Gamma(N-1)}{2^{N-2}\pi\Gamma(N-2)}\int_0^{\pi/2}\frac{(\sin 2\theta)^{N-2}}{d^{2N-2}}d\theta,$$

so that (b) follows after simplification. □

Remark 1 For the case where $\rho \neq 0$, densities of the sample correlation coefficient are given by Fisher (1915) in both the summation and differentiation forms, and by Hotelling (1953) using the hypergeometric function. For more details, see Anderson (2022).

Corollary 2 *The mean and variance of S are*

$$E(S) = \frac{N-2}{N-3}\lambda, \quad Var(S) = \left[\frac{N-1}{N-4} - \left(\frac{N-2}{N-3}\right)^2\right]\lambda^2 + \frac{1}{N-4},$$

where $\lambda = \frac{\rho}{\sqrt{1-\rho^2}}$.

Let $Z = \frac{S-\lambda}{\sigma_S}$. By Eq. (7), the pdf of Z is

$$f_Z(z) = \frac{(N-2)\sigma_s}{\pi}\int_0^{\frac{\pi}{2}}\frac{(\sin 2\theta)^{N-2}}{(1+[(\sigma_s z+\lambda)\cos\theta - \lambda\sin\theta]^2)^{N-1}}d\theta$$

The following graphs show that density curves of S (Fig. 1) and Z (Fig. 2) for different values of N and λ.

Remark 2 The distribution of the sample correlation coefficient was given in Anderson (2022) and the pdf of $S = \frac{r}{\sqrt{1-r^2}}$ is

$$h_0(s) = \frac{\Gamma(n/2)}{\Gamma(\frac{n-1}{2})\sqrt{\pi}}(1+s^2)^{-n/2} \tag{9}$$

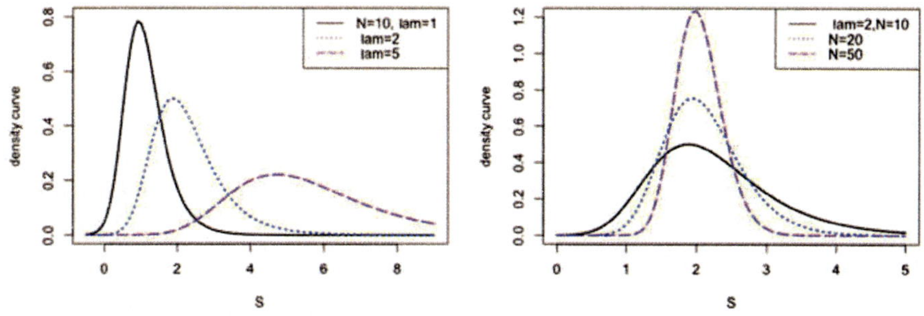

Fig. 1 Density curves of S for different λ (left) and N(right)

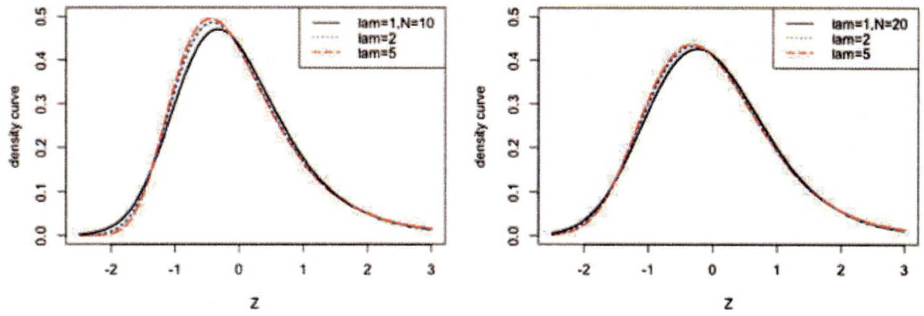

Fig. 2 Density curves of Z for different λ and N

for $\rho = 0$, and

$$h_1(s) = \frac{2^{n-2}(1-\rho^2)^{n/2}(1+s^2)^{-n/2}}{(n-2)!\pi} \sum_{i=0}^{\infty} \frac{(2\rho s/\sqrt{1+s^2})^i}{i!} \Gamma^2\left[\frac{1}{2}(n+i)\right] \quad (10)$$

for $\rho \neq 0$, where $n = N - 1$.

The following graphs show that density curves of $S = \frac{r}{\sqrt{1-r^2}}$ given in Remark (2) and Eq. (7), are the same as those given in Anderson (2022) when $\rho = 0$ and $\rho = 0.5$ (Fig. 3).

Recall that we have the following result on the approximate distribution of sample correlation coefficient.

Lemma 4 *If r is the sample correlation coefficient of a sample of $N(= n + 1)$ from a normal distribution with correlation ρ, then $\sqrt{n}(r - \rho)/(1 - \rho^2)$ has the limiting distribution $N(0, 1)$.*

Then we can obtain the limiting pdf of S, which is given below.

Proposition 1 *Suppose that r has the distribution given in Lemma 4, then $S = \frac{r}{\sqrt{1-r^2}}$ has density function*

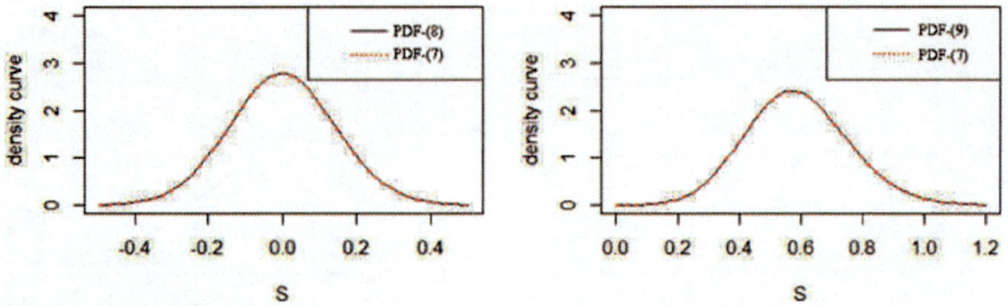

Fig. 3 Density curves of S given by Eqs. (9) and (7) for $N = 50$, $\rho = 0$ (left), and Eqs. (10) and (7) for $\rho = 0.5$ (right)

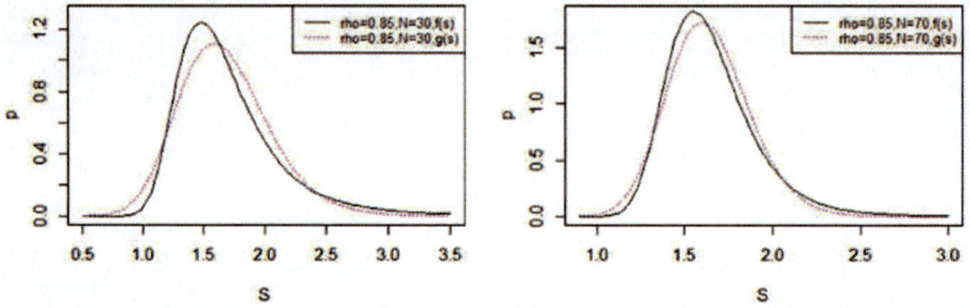

Fig. 4 Density curves of S for obtained by Eqs. (7) and (11) for given $\rho = 0.8$ and $N = 30$

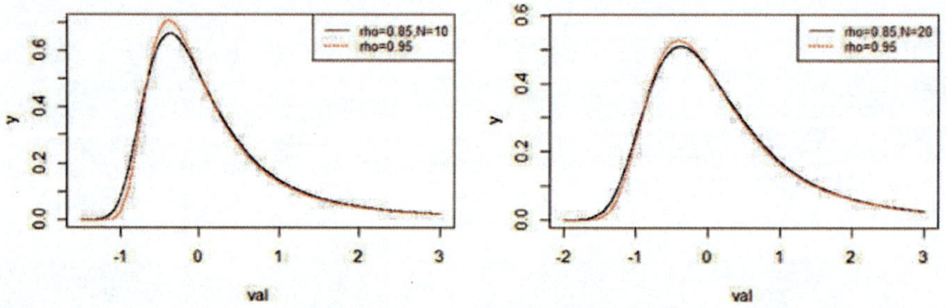

Fig. 5 Density curves of Z obtained by Eq. (11) for given $\rho = 0.85, 0.95$ and $N = 10, 20$

$$g_s(s) = \frac{\sqrt{n}}{\sqrt{2\pi}(1-\rho^2)(1+s^2)^{3/2}} \exp\left[-\frac{\left(\frac{s}{\sqrt{1+s^2}} - \rho\right)^2}{2(1-\rho^2)^2/n}\right]. \tag{11}$$

The following graphs (Fig. 4) compared the density curves of S obtained by Eqs. (7) and (11) for given ρ and N.

For density curves of Z with respect to the density function (11), they are showed in the Fig. 5.

4 The APP for the Correlation Coefficient in Skew Normal Distribution

In order to determine the required sample size N to be $c--100\%$ confident for the given sampling precision on estimating ρ, we consider the density of S given in Theorem 2.

Theorem 3 *Let c be the confidence level and f be the precision which are specified such that the error associated with estimator S is $E = f\sigma_T$. More specifically,*

$$P\left[f_1\sigma_T \leq S - \lambda \leq f_2\sigma_T\right] = c. \tag{12}$$

Here, f_1 and f_2 are left and right precision which is satisfied $max\{|f_1|, f_2\} \leq f$, and σ_T is the standard deviation of T given in Corollary 2, respectively. Then the required sample size $N(N > 3)$ is obtained by

$$\int_L^U f(z)dz = c \tag{13}$$

such that the length of the confidence interval $U - L$ is the minimized, where $f(z)$ is the pdf of Z which is assumed to be free of ρ, and

$$L = \sqrt{N-2}f_1, \qquad U = \sqrt{N-2}f_2.$$

Proof Let $Z = \frac{S-\lambda}{\sigma_S}$. Then the density function of Z can be obtained by Eq. (7) and

$$f(z) = \frac{(N-2)\sigma_S}{\pi}\int_0^{\frac{\pi}{2}} \frac{(\sin(2\theta))^{N-2}}{[1 + ((\sigma_S z + \lambda)\cos\theta - \lambda\sin\theta)^2]^{N-1}}d\theta.$$

By Corollary 2, Equation (12) can be simplified to the integral Equation (13) by dividing σ_s, where $L = \sqrt{N-2}f_1$ and $U = \sqrt{N-2}f_2$. □

Remark 3 The computer program for finding the required sample size N based on Eq. (13), the distribution of S, is listed below.

https://applinks.shinyapps.io/ncoefficientncst/

The input data values are ρ_0 specified correlation from the previous data (use 0, 0.5, or 0.95 if no previous information about it), the confidence level c, and the precision f. The output value is the required sample size N. You may see the values of N don't change much for different input ρ_0 values, which can be explained by Fig. 2.

By Theorem 3, we can construct the $c \times 100\%$ confidence region for λ, $P(f_1\sigma_T + \lambda \leq S \leq f_1\sigma_T + \lambda) = c$, by Eq. (12).

Corollary 3 *Under the assumptions in Theorem 3, the necessary sample size N with respect to the density function given by Eq. (11) in Section 3 can be from solving the integration equation*

$$\int_{L_*}^{U_*} g(z)dz = c \tag{14}$$

such that $U_* - L_*$ *is the shortest. where* $g(z)$ *is the pdf of Z given in Eq. (11) which is assumed to be free of* ρ*, and*

$$L_* = \sqrt{N-2}f_{1*}, \qquad U_* = \sqrt{N-2}f_{2*}$$

so that $max\{|f_{1*}|, f_{2*}\} \leq f$.

The proof is similar as that of Theorem 3.

Remark 4 The computer program that can be used to find the required sample size based on Eq. (14) for different research goals is listed below.

https://applinks.shinyapps.io/ncoeffaprxn/

The input variables are (i) a specified correlation ρ_0 from the previous data (otherwise use the default values of $\rho_0 = 0.5$ *or* 0.95) and (ii) the confidence level c and precision f. However, from Fig. 5, it shows the graphs don't change significantly for different values of ρ_0. The output value is the required sample size N.

Although the programs mentioned in Remark 3 and Remark 4, include an entry for the size of the correlation coefficient, ρ_0, that entry makes little difference for reasons illustrated by Figs. 2 and 5. Also for the convenience, researchers can use either one in Remark 3 or the one Remark 4. But the second one requires few more observations than the first one.

5 Simulation Results

In this section, we provide some results of the necessary sample size for specific precision and confidences by Eqs. (13) and (14) given in Theorem 3 and Corollary 3, respectively. Here we consider $c = 0.95,\ 0.9$ and $f = 0.1,\ 0.15,\ 0.2,\ 0.25,\ 0.3$, the corresponding necessary sample sizes are given in Table 1 and Table 2 for given $\rho_0 = 0,\ 0.5,\ 0.95$.

It shows that the necessary sample size obtained from Theorem 3 is a bit smaller than that from Corollary 3, which is reasonable since the approximate normal distribution was used for obtaining the density function of Z in Corollary 3. Moreover, both tables indicate that the information provided for ρ_0 from previous data doesn't significantly affect the required sample sizes, which can also be obtained from Figs. 2 and 5.

Using the Monte Carlo simulations, we obtain relative frequency for different values of ρ_0, which are given Tables 3 and 4. All results are illustrated with a number

Table 1 The value of sample size N for under different precision f and ρ_0 for the given $c = 0.95,\ 0.9$

f	c	N		
		$\rho_0 = 0$	$\rho_0 = 0.5$	$\rho_0 = 0.95$
0.1	0.95	379	380	379
	0.9	265	265	265
0.15	0.95	166	166	165
	0.9	115	115	114
0.2	0.95	100	100	100
	0.9	71	71	71
0.25	0.95	66	65	65
	0.9	47	47	46
0.3	0.95	47	47	46
	0.9	34	34	33

Table 2 The value of sample size N for different f and ρ_0 for the given confidence $c = 0.95,\ 0.9$

f	c	N		
		$\rho_0 = 0$	$\rho_0 = 0.5$	$\rho_0 = 0.95$
0.1	0.95	380	382	385
	0.9	265	266	268
0.15	0.95	167	178	171
	0.9	115	115	118
0.2	0.95	199	100	104
	0.9	69	70	72
0.25	0.95	64	65	69
	0.9	45	46	47
0.3	0.95	46	47	49
	0.9	32	32	35

of simulation runs $M = 10000$. From both tables we can see that coverage rates on 90% and 95% confidence intervals for $f = 0.1,\ 0.15,\ 0.2,\ 0.3,\ \rho_0 = 0,\ 0.5$ and 0.95 corresponding to Tables 1 and 2, indicate that our proposed APPs work well.

6 Conclusion Remarks

Our goal was to derive ways to perform the APP for estimating correlation coefficients under multivariate skew normal distributions. The present mathematics provide those derivations and show that the distribution of sample correlations is free of skewness parameters for the multivariate skew normal populations. In turn, computer

Table 3 The coverage rate for confidence intervals with confidence $c = 0.95,\ 0.9$, different precision f and ρ_0

f	c	Coverage rate		
		$\rho_0 = 0$	$\rho_0 = 0.5$	$\rho_0 = 0.95$
0.1	0.95	0.9469	0.9447	0.9428
	0.9	0.8918	0.8924	0.8917
0.15	0.95	0.9414	0.9444	0.9457
	0.9	0.8956	0.8972	0.8914
0.2	0.95	0.9440	0.9497	0.9560
	0.9	0.8909	0.8951	0.8973
0.25	0.95	0.9462	0.9457	0.9464
	0.9	0.9090	0.9023	0.9074
0.3	0.95	0.9455	0.9499	0.9507
	0.9	0.8989	0.9041	0.9096

Table 4 The coverage rate for confidence intervals of confidence $c = 0.95,\ 0.9$, different precision f and ρ_0

f	c	Coverage rate		
		$\rho_0 = 0$	$\rho_0 = 0.5$	$\rho_0 = 0.95$
0.1	0.95	0.9469	0.9454	0.9428
	0.9	0.8918	0.8932	0.8917
0.15	0.95	0.9414	0.9443	0.9457
	0.9	0.8956	0.8937	0.8914
0.2	0.95	0.9440	0.9523	0.9560
	0.9	0.8909	0.8946	0.8973
0.25	0.95	0.9462	0.9477	0.9464
	0.9	0.9090	0.9053	0.9074
0.3	0.95	0.9455	0.9501	0.9507
	0.9	0.8989	0.8999	0.9096

simulations support the mathematical derivations. We also provide links to free and user friendly programs to facilitate researchers performing the APP to determine sample sizes to meet their specifications for precision and confidence. An advantage of the programs is that even researchers who are unsophisticated in mathematics nevertheless can avail themselves of APP advantages.

References

Azzalini, A.: A class of distributions which includes the normal ones. Scandinavian J. Stat. 171–178 (1985)

Azzalini, A., Dalla Valle, A.: The multivariate skew-normal distribution. Biometrica **83**(4), 715–726 (1996)

Azzalini, A., Capitanio, A.: Statistical application of the multivariate skew normal distribution. J. Roy. Statist. Soc. B **83**, 579–602 (1999)

Anderson, T. W.: An Introduction to Multivariate Statistical Analysis–3rd ed. Wiley series in probability and mathematical statistics, ISBN 0-471-36091-0

Olsson, U.: Maximum likelihood estimation of polychoric correlation coefficient. PSYCHOMETRIKA **44**(4) (1979)

Schisterman, E.F., Moysich, K.B., England, L.J. , Rao, M.: Estimation of the correlation coefficient using the Bayesian Approach and its applications for epidemiological research. BMC Med. Res. Methodol. 3:5 (2003)

Shevlyakov, G., Smirnov, P.: Robust estimation of the correlation coefficient: an attempt of survey. Aust. J. Stat. **40**(1 & 2), 147–156 (2011)

Sari, B.G., Lúcio, A.D., Santana, C.S., Krysczun, D.K., Tischler, A.L., Drebes, L.: Sample size for estimation of the Pearson correlation coefficient in cherry tomato tests. Ciência Rural, Santa Maria, v.47: 10, e20170116 (2017)

Wang, C., Wang, T., Trafimow, T, Zhang, X.: Necessary sample size for estimating the scale parameter with specified closeness and confidence. Int. J. Intell. Technol. Appl. Stat. **12**(1), 17–29 (2019). https://doi.org/10.6148/IJITAS.201903_12(1).0002

Wang, C., Wang, T., Trafimow, D., Myuz, H.: Desired sample size for estimating the skewness parameter under skew normal settings. In: Kreinovich, V., Sriboonchitta, S. (eds.) Structural changes and their economic modeling. Springer-Verlag, Switzerland, pp. 152–162 (2019)

Wang, C., Wang, T., Trafimow, D., Chen, J.: Extending a priori procedure to two independent samples under skew normal setting. Asian J. Econ. Bank. **03**(02), 29-40 (2019)

Wang, C., Wang, T., Trafimow, D., Li, H., Hu, L., Rodriguez, A.: Extending a priori procedure (APP) to address correlation coefficients. To appear In: Trung, N., Thach, N., Kreinovich, V. (eds.) Data science for financial econometrics. Springer-Verlag (2020)

Wang, C., Wang, T., Trafimow, D., Myuz, H.A.: Necessary sample size for specified closeness and confidence of matched data under the skew normal setting. Commun. Stat. Simul. Comput. (2019). https://doi.org/10.1080/03610918.2019.1661473

Wang, Z., Wang, C., Wang, T.: Estimation of location parameter on the skew normal setting with known coefficient of variation and skewness. Int. J. Intell. Technol. Appl. Stat. **9**(3), 45–63 (2016)

Zhu, X., Li, B., Wang, T., Gupta, A.: Sampling distributions of skew normal populations associated with closed skew normal distributions. Random Oper. Stoch. Equ. **2019** (2019). https://doi.org/10.1515/rose-2018-2007

Comparison of Entropy Measures in Panel Quantile Regression and Applications to Economic Growth Analysis

Woraphon Yamaka, Wilawan Srichaikul, and Paravee Maneejuk

Abstract The three entropy measures (Shannon, Tsallis, and Renyi entropy) are used as the objective of the entropy functions in Generalized Maximum Entropy(GME) to estimate the unknown parameters in the panel quantile regression model. This study applies these estimators to the macroeconomic dataset. The results show that Tsallis entropy is the most appropriate measure to describe the effect of macroeconomic variables on economic growth in G20 countries as it provides the lowest mean squared error (MSE) and root mean squared error (RMSE). The results also show that the Shannon entropy GME estimates a bit different from Tsallis and Renyi in terms of the magnitudes of the estimates, particularly in the extreme quantiles(10th and 90th)

Keywords Generalized maximum entropy · Panel data · Quantile regression · Renyi · Tsallis

1 Introduction

One of the challenges of using the Generalized Maximum Entropy(GME) approach (Golan et al. 1996) as an estimator is to choose an appropriate entropy measure which reflects the uncertainty that we have about the occurrence of a collection of events (Golan and Perloff 2002). In the literature, the Shannon entropy measure has been mainly used as the objective function of the GME estimator in Panel regression models (Tibprasorn et al. 2017; Lee and Cheon 2014). But as will be shown later,

W. Yamaka (✉) · W. Srichaikul · P. Maneejuk
Center of Excellence in Econometrics, Faculty of Economics, Chiang Mai University, Chiang Mai 50200, Thailand
e-mail: woraphon.econ@gmail.com

W. Srichaikul
e-mail: srichaikul.w@gmail.com

P. Maneejuk
e-mail: mparavee@gmail.com

N. Ngoc Thach et al. (eds.), *Financial Econometrics: Bayesian Analysis, Quantum Uncertainty, and Related Topics*, Studies in Systems, Decision and Control 427,
https://doi.org/10.1007/978-3-030-98689-6_12

different entropy measure may reflect different uncertainty structure, hence it is not reasonable to apply Shannon entropy (Shannon 1948) to all datasets and model estimations. Therefore, in this study, we nest the GME estimator into two more general classes of estimators, namely Renyi and Tsallis (1970, 1988) entropy measures and then the performance of these two GME estimators are compared to the Shannon entropy under the Panel quantile regression models (Canay 2011; Srichaikul et al. 2019). Although, the GME estimation for panel quantile regression has already been introduced by Srichaikul et al. (2018), the GME still relies on the Shannon's entropy measure. In this study, we replace the Shannon entropy measure with either the Renyi or Tsallis in the objective function. By carefully selecting the entropy measure in the GME, we can estimate efficiently the Panel quantile regression model which is flexible enough to reflect all possible uncertainty in the model. In addition, several advantages are obtained from the GME. Firstly, it can solve underdetermined or ill-posed problems for the observed data, or regressions when small data set is applied. Secondly, it can solve the multicollinearity problem (Pipitpojanakarn et al. 2017).

To examine the performance of Renyi and Tsallis GME estimations for panel quantile regression, we apply these methods to the real dataset of macroeconomic panel data. Although studies on the macroeconomic variables and growth nexus have grown in number significantly recently, researchers have not paid much attention to identifying if the impacts of macroeconomic variables vary across the different conditional distributions of economic growth. Our concern is the potentially unreliable results of the conventional linear model which is usually employed in previous studies. Only estimating how on average the macroeconomic factors affect economic growth will miss considering some important causes or extreme events such as natural calamity, war, energy shock, and financial crisis. Furthermore, the change in macroeconomic variables, on average, may affect economic growth in one country but not the others which have different economic structure. In other words, it is important to examine the impact of the macroeconomic variables on economic growth at different points of the conditional function of economic growth. To address this issue, we apply the panel quantile regression model to investigate the factors affecting economic growth in G20 countries.

Economic growth means the increase of the production of economic goods and services, compared from one period to another, which can be measured and understood in terms of national income, aggregated output or GDP, and total expenditure. It is vital and instrumental for improving living standard, enlarging job opportunities hence lowering unemployment rate, raising tax revenues, reducing certain government spending e.g. on crime control and pollutions, and finally attracting new and heavier investment from both domestic and international sources. Thus, many countries are focusing on sustainable economic growth to guarantee the overall economic development. There are many factors contributing to the actualization and the maintenance of sustainable economic growth of a nation. The measurement of genuine economic growth of a country is complex as there are so many determinants of growth contributing simultaneously to the attainment of elevated gross domestic product.

The main contribution of this paper is that we develop a GME for estimating fixed effects-panel quantile regression. Two more entropy measures, namely Renyi and

Tsallis are considered in our estimation. Our approach is then applied to investigate the impact of macroeconomic variables on economic growth of G20 countries.

The remainder of the paper is organized as follows. Section 2 briefly explains the methodology used in this study, Sect. 3 provides data description and model specification. In Sect. 4, we present the comparison results. Section 5 gives conclusions.

2 Methodology

2.1 Panel Quantile Regression Models

Panel quantile regression is one of the important tools for estimating the relationships between independent variables and different conditional quantiles of dependent variable. The model is based on panel data, consisting of both cross-sectional and time series data. Like in the general panel regression context, the appropriate estimation method for this model should address the two error components. Generally, we can express the model as

$$y_{it} = X'_{it}\beta^{\tau} + \alpha_i^{\tau} + u_{it}^{\tau} \quad (1)$$

where $i = 1, ..., N$ is the G20 countries, $t = 1, ..., T$ is the time index, y_{it} is $NT \times 1$ economic growth variable, X_{it} is $NT \times K$ macroeconomic variables, β^{τ} is $K \times 1$ coefficient parameters at given quantile level τ. u_{it}^{τ} is a common stochastic error term at given quantile level τ and is usually not correlated with X_{it}. The country-specific effect error component α_i^{τ} varies across G20 countries but is constant over time. Thus, α_i^{τ} are assumed to be fixed parameters to be estimated in the model expressed in Eq. (1). This is called the fixed effects panel data model which usually gives consistent estimates for parameters. The advantages of the fixed effects model are that: (a) it can allow unobserved individual effect to be correlated with explanatory variable X_{it} and (b) there is no need to specify the conditional density for unknown individual effect.

2.2 Entropy Approach

The maximum entropy estimator is introduced to estimate unknown parameters in Eq. (1). We can maximize the entropy of the model with respected to the structure of Eq. (1). In this study,three entropys measures, Shannon, Renyi, and Tsallis are considered and their functions can be expressed as.

Shannon entropy measure (Shannon 1948)

$$H^S(\mathbf{p}) = -\sum_{k=1}^{K} p_k \log p_k, \tag{2}$$

Renyi entropy measure (Renyi 1970)

$$H^R(\mathbf{p}) = \frac{1}{1-q} \log \sum_{k=1}^{K} p_k^q, \tag{3}$$

Tsallis entropy measure (Tsallis 1988)

$$H^T(\mathbf{p}) = \frac{1}{1-q} \left(\sum_{k=1}^{K} p_k^q - 1 \right), \tag{4}$$

where p_k is a probability with value form [0,1] and $\sum_{k=1}^{K} p_k = 1$. Renyi and Tsallis entropy measures are indexed by a single parameter q, which we restrict to be strictly positive: $q > 0$. If $q = 1$, they become Shannon entropy. In this study, we assume $q = 2$ (see, Golan and Perloff 2002) to simplify the estimation, thus

$$H^R(p) = -\log \sum_{k=1}^{K} p_k^2, \tag{5}$$

$$H^T = -\sum_{k=1}^{K} p_k(p_k - 1). \tag{6}$$

We then generalize the maximum entropy to the inverse problem of the Panel quantile regression. In this estimation, the estimated parameters cannot obtained directly. Thus, we considered these unknown parameters as expectations of random variables with M support value for each estimated parameter value (k), $Z = [z_1, ..., z_K]$ where $z_k = [\underline{z}_{k1}, ..., \bar{z}_{km}]$ for all $k = 1, ...K$. Note that $\underline{z}_{k1}$ and $\bar{z}_{km}$ denote the lower bound and upper, respectively, of each support z_k. Thus, we can express parameter β_k^τ as.

$$\beta_k^\tau = \sum_m p_{km}^\tau z_{km}^\tau, \tag{7}$$

where p_{km}^τ are the M dimensional estimated probability defined on the set z_{km}^τ at given quantile τ. Likewise, the error u_{it}^τ is also constructed as the expectation. Therefore, each u_{it}^τ is assumed to be a random vector with finite and discrete random variable with M support value, $v_i^\tau = [v_{t1}^\tau, ..., v_{tM}^\tau]$. Let w_{itm}^τ be an M dimensional proper probability weights defined on the set v_i^τ such that

$$u_{it}^{\tau} = \rho_{\tau}\left(\sum_{m} w_{itm}^{\tau} v_{itm}^{\tau}\right) \tag{8}$$

where $\rho_{\tau}(L) = L(\tau - I(L < 0))$ is the check function. Finally, the country specific error α_i^{τ} can be written as

$$\alpha_i^{\tau} = \sum_{m} f_{im}^{\tau} g_{im}^{\tau}, \tag{9}$$

where f_{im}^{τ} and g_{im}^{τ} are the M dimensional estimated probability distribution and support value of α_i^{τ}, respectively. Note that the support vectors,$v_{itm}^{\tau}$and f_{im}^{τ}, z_{km}^{τ}, are convex set that is symmetric around zero with $2 \leq M < \infty$. The objective of our panel quantile regression model is to estimate the unknown parameters by rewriting the entropy function as

$$H(\mathbf{p}, \mathbf{g}, \mathbf{w} \,|\, \tau) = \{H(\mathbf{p} \,|\, \tau) + H(\mathbf{g} \,|\, \tau) + H(\mathbf{w} \,|\, \tau)\}\ , \tag{10}$$

subject to the constraint $y_{it} = X_{it}\beta^{\tau} + \alpha_i^{\tau} + u_{it}^{\tau}$. Then, we need to reparametrize β^{τ}, α_i^{τ}, and u_{it}^{τ} in our constraints as in the following

$$y_{it} = \sum_{k}\sum_{m} p_{km}^{\tau} z_{km}^{\tau} X_{kit} + \sum_{m} f_{im}^{\tau} g_{im}^{\tau} + \rho_{\tau}\left(\sum_{m} w_{itm}^{\tau} v_{itm}^{\tau}\right), \tag{11}$$

2.3 *Generalized Entropy Estimation*

2.3.1 The Generalized Maximum Renyi Entropy Estimation

We can construct our Generalized Maximum Renyi Entropy estimator as

$$\begin{aligned} H^{R}(\mathbf{p}, \mathbf{g}, \mathbf{w} \,|\, \tau) &= \underset{\mathbf{p},\mathbf{g},\mathbf{w}}{\arg\max} \left\{H^{R}(\mathbf{p} \,|\, \tau) + H^{R}(\mathbf{g} \,|\, \tau) + H^{R}(\mathbf{w} \,|\, \tau)\right\} \\ &\equiv -\sum_{k}\sum_{m} \ln p_{k}^{2,\tau} - \sum_{i}\sum_{m} \ln g_{im}^{2,\tau} - \sum_{i}\sum_{t}\sum_{m} \ln w_{itm}^{2,\tau} \end{aligned} \tag{12}$$

subject to

$$y_{it} = \sum_{k}\sum_{m} p_{km}^{\tau} z_{km}^{\tau} X_{kit} + \sum_{m} f_{im}^{\tau} g_{im}^{\tau} + \rho_{\tau}\left(\sum_{m} w_{itm}^{\tau} v_{itm}^{\tau}\right), \tag{13}$$

$$\sum_{m} p_{km}^{\tau} = 1, \sum_{m} g_{im}^{\tau} = 1, \sum_{m} w_{itm}^{\tau} = 1 \tag{14}$$

Then, the Lagrange equation is

$$L^R|\tau = H^R(\mathbf{p},\mathbf{f},\mathbf{w}|\tau) + \lambda'_1\left(y_{it} - \sum_k\sum_m p^\tau_{km}z^\tau_{km}X_{kit} - \sum_m f^\tau_{im}g^\tau_{im} - \rho_\tau\left(\sum_m w^\tau_{itm}v^\tau_{itm}\right)\right)$$
$$+\lambda'_2(1-\sum_m p^\tau_{km}) + \lambda'_3(1-\sum_m g^\tau_{im}) + \lambda'_4(1-\sum_m w^\tau_{itm}), \tag{15}$$

2.3.2 The Generalized Maximum Tsallis Entropy Estimation

We can construct our Generalized Maximum Tsallis Entropy estimator as

$$H^T(\mathbf{p},\mathbf{g},\mathbf{w}|\tau) = \underset{\mathbf{p},\mathbf{g},\mathbf{w}}{\arg\max}\left\{H^T(\mathbf{p}|\tau) + H^T(\mathbf{g}|\tau) + H^T(\mathbf{w}|\tau)\right\} \equiv$$
$$-\sum_k\sum_m p^\tau_{km}(p^\tau_{km}-1) - \sum_i\sum_m g^\tau_{im}(g^\tau_{im}-1) - \sum_i\sum_t\sum_m w^\tau_{itm}(w^\tau_{itm}-1) \tag{16}$$

subject to

$$y_{it} = \sum_k\sum_m p^\tau_{km}z^\tau_{km}X_{kit} + \sum_m f^\tau_{im}g^\tau_{im} + \rho_\tau\left(\sum_m w^\tau_{itm}v^\tau_{itm}\right), \tag{17}$$

$$\sum_m p^\tau_{km} = 1, \sum_m g^\tau_{im} = 1, \sum_m w^\tau_{itm} = 1 \tag{18}$$

Then, the Lagrange equation is

$$L^T|\tau = H^T(\mathbf{p},\mathbf{f},\mathbf{w}|\tau) + \lambda'_1\left(y_{it} - \sum_k\sum_m p^\tau_{km}z^\tau_{km}X_{kit} - \sum_m f^\tau_{im}g^\tau_{im} - \rho_\tau\left(\sum_m w^\tau_{itm}v^\tau_{itm}\right)\right)$$
$$+\lambda'_2(1-\sum_m p^\tau_{km}) + \lambda'_3(1-\sum_m g^\tau_{im}) + \lambda'_4(1-\sum_m w^\tau_{itm}), \tag{19}$$

2.3.3 The Generalized Maximum Shannon Entropy Estimation

We can construct our Generalized Maximum Shannon Entropy (GME) estimator as

$$H^S(\mathbf{p},\mathbf{g},\mathbf{w}|\tau) = \underset{p,g,w}{\arg\max}\left\{H^S(\mathbf{p}|\tau) + H^S(\mathbf{g}|\tau) + H^S(\mathbf{w}|\tau)\right\} \equiv$$
$$-\sum_k\sum_m p^\tau_{km}\log p^\tau_{km} - \sum_i\sum_m g^\tau_{im}\log g^\tau_{im} - \sum_i\sum_t\sum_m w^\tau_{itm}\log w^\tau_{itm} \tag{20}$$

subject to

$$y_{it} = \sum_k \sum_m p_{km}^{\tau} z_{km}^{\tau} X_{kit} + \sum_m f_{im}^{\tau} g_{im}^{\tau} + \rho_{\tau} \left(\sum_m w_{itm}^{\tau} v_{itm}^{\tau} \right), \tag{21}$$

$$\sum_m p_{km}^{\tau} = 1, \sum_m g_{im}^{\tau} = 1, \sum_m w_{itm}^{\tau} = 1 \tag{22}$$

Then, the Lagrange equation is

$$\begin{aligned} L^S|\tau &= H^S(\mathbf{p}, \mathbf{f}, \mathbf{w}|\tau) + \\ &\lambda'_1 \left(y_{it} - \sum_k \sum_m p_{km}^{\tau} z_{km}^{\tau} X_{kit} - \sum_m f_{im}^{\tau} g_{im}^{\tau} - \rho_{\tau} \left(\sum_m w_{itm}^{\tau} v_{itm}^{\tau} \right) \right) \\ &+ \lambda'_2 (1 - \sum_m p_{km}^{\tau}) + \lambda'_3 (1 - \sum_m g_{im}^{\tau}) + \lambda'_4 (1 - \sum_m w_{itm}^{\tau}), \end{aligned} \tag{23}$$

We can do the first difference with respect to **p**,**f**and **w**to obtain the optimal $\widehat{\mathbf{p}}$, $\widehat{\mathbf{f}}$ and $\widehat{\mathbf{w}}$. Then, substituting the solutions of $\widehat{\mathbf{p}}$, $\widehat{\mathbf{f}}$ and $\widehat{\mathbf{w}}$in Eqs. (7–9) to get the estimated β^{τ}, $\alpha_i^{\tau}$and u_{it}^{τ}.

3 Data Analysis

Several recent theoretical and empirical papers have raised the issue of whether macroeconomic factors affect economic growth. For instance, Garrison and Lee (1995) attempted to examine the impact of macroeconomic variables on economic growth during the period 1960–1987. They found a weak negative effect of marginal tax rates on growth, but the foreign trade is likely to have strong impact on growth. Antwi et al. (2013) confirmed the relationship between real GDP per capita (economic growth) and many macroeconomic factors such as physical capital, labour force, foreign direct investment, foreign aid, inflation and government expenditure. Mbulawa (2015) suggested that foreign direct investment and inflation had a positive effect on economic growth. Sarker (2016) investigated the effect of urban population on economic growth of the South Asian countries and found that the growth of urban population had a significant impact on economic growth in the long run. Yalcinkaya et al. (2018) showed that international tourism revenues have a positive and statistically significant effect on economic growth. Most recently, Srichaikul et al. (2019) studied the impacts of macroeconomic variables, namely foreign direct investment, population, and real effective exchange rate on the economic growth of China, Japan, and South Korea. They found that all the considered variables significantly affected economic development with FDI apparently having both negative and positive effects in the GDP while population being positive in the GDP.

According to above literature, the impact of macroeconomic variables and economic growth in G20 countries over the period 2000–2016 is replicated and used as the example dataset for our suggested estimations. We would like to note that the G20 countries in this study are consist of Argentina, Australia, Brazil, Canada, China, France, Germany, India, Indonesia, Italy, Japan, the Republic of Korea, Mexico, Russia, Saudi Arabia, South Africa, Turkey, the United Kingdom, and the United States of America. The Group of Twenty (G20) is the premier forum for international cooperation on the most important issues of the global economic and financial agenda. The objectives of the G20 refer to (i) Policy coordination between its members to achieve global economic stability, sustainable growth; (ii) Promoting financial regulations that reduce risks and prevent future financial crises; (iii) Modernizing international financial architecture. The G20 brings together finance ministers and central bank governors from 19 countries, which is represented by the President of the European Council and by Head of the European Central Bank (2013).

All data are transformed into natural log before estimating the model. In this example, we consider the following model

$$GDP_{it} = \beta_{0\tau} + \beta_{1\tau} UR_{it} + \beta_{2\tau} TR_{it} + \beta_{3\tau} FDI_{it} + \beta_{4\tau} PM2.5_{it} + u_{it}, \quad (24)$$

where GDP is Gross Domestic Product, PPP (constant 2011 US$), UR is Urban population (% of total population), TR is International tourism receipts (current US$). FDI is Foreign Direct Investment and PM 2.5 is Fine Particulate Matter (micrograms per cubic meter).

The summary of the descriptive statistics is illustrated in Table 1. We conduct the Minimum Bayes Factor (MBF) to make the statistical inference (Maneejuk and Yamaka 2021). The MBF is interpreted following Goodman (1999) about labelled intervals. As a result, an MBF between 1–1/3 is considered weak evidence for H_1, 1/3–1/10 moderate evidence, 1/10–1/30 substantial evidence, 1/30–1/100 strong evidence, 1/100–1/300 very strong evidence, and <1/300 decisive evidence. The result shows that UR has the highest standard deviation, while GDP has the lowest. The distributions of all the variables are skewed, and the kurtosis values show that the data series distributions with longer tails are more concentrated than the normal distribution. The Jarque-Bera statistic indicates that all series are not normally distributed because it rejects the null hypothesis. Furthermore, Table 1 shows the results of the panel unit root test according to Levin et al. (2002) which confirm that GDP, UR, TR, FDI and PM 2.5 are in favor of H_1 with decisive evidence. Therefore, these variables are stationary.

4 Result

Tables 2, 3 and 4 present the results of the panel quantile regression estimation, which report the 10th , 20th, 30th , 40th, 50th, 60th, 70th, 80th and 90th percentiles of the conditional concentrations distribution. The lower quantiles, such as the 10th, 20th, 30th, 40th, and 50th, refer to the countries with lower economic growth. The higher quantiles, such as the 60th, 70th, 80th, 90th refer to the countries with higher

Table 1 Descriptive statistics

Variable	GDP	UR	TR	FDI	PM 2.5
Mean	10.0961	74.3894	16.7705	2.0129	2.9409
Median	10.3905	79.3940	16.7202	1.7963	2.6847
Maximum	10.8856	91.6270	18.2517	10.0274	5.2358
Minimum	8.3904	30.9300	15.4566	−0.0138	1.7717
Std. Dev.	0.6339	14.6489	0.8534	1.3248	0.7869
Skewness	−0.8120	−1.4324	0.0736	1.9051	0.9830
Kurtosis	2.7757	4.5639	1.7619	11.7029	3.4775
Jarque-Bera	14.8937 (0.0000)	59.0372 (0.0000)	8.6151 (0.0000)	500.1755 (0.0000)	22.6816 (0.0000)
Unit root test					
None	4.75695 (0.0000)	73.1129 (0.0000)	10.8743 (0.0000)	−3.95395 (0.0000)	2.27867 (0.0000)
Intercept	−6.77885 (0.0000)	3.73986 (0.0000)	7.74530 (0.0000)	−26.3838 (0.0000)	−4.57775 (0.0000)
Intercept and trend	−10.8825 (0.0000)	−22.2026 (0.0000)	−8.19095 (0.0000)	−3.95395 (0.0000)	−8.38663 (0.0000)

Note ( ) is Minimum Bayes factor

Table 2 Results of maximum Tsallis entropy quantile regression for panel data

Tsallis Entropy									
Variable	10th	20th	3th	40th	50th	60th	70th	80th	90th
UR	0.0234 [0.0046] (0.0000)	0.0368 [0.0037] (0.0000)	0.0248 [0.0045] (0.0000)	0.0240 [0.0045] (0.0000)	0.0152 [0.0050] (0.0000)	0.0218 [0.0045] (0.0000)	0.0246 [0.0045] (0.0000)	0.0233 [0.0046] (0.0000)	0.0247 [0.0046] (0.0000)
TR	0.0335 [0.0227] (0.3365)	0.0566 [0.0181] (0.0075)	0.0454 [0.0223] (0.1258)	0.0522 [0.0219] (0.0583)	0.0337 [0.0247] (0.3942)	0.0503 [0.0223] (0.0785)	0.0448 [0.0223] (0.1329)	0.0320 [0.0228] (0.3734)	0.0251 [0.0221] (0.5246)
FDI	−0.0077 [0.0037] (0.1146)	−0.0100 [0.0030] (0.0038)	−0.0069 [0.0037] (0.1757)	−0.0056 [0.0036] (0.2982)	−0.0034 [0.0041] (0.7090)	−0.0052 [0.0037] (0.3724)	−0.0069 [0.0037] (0.1757)	−0.0077 [0.0038] (0.1283)	−0.0082 [0.0038] (0.0974)
PM2.5	−0.1113 [0.0638] (0.2183)	0.0112 [0.0511] (0.9762)	−0.1586 [0.0627] (0.0407)	−0.1401 [0.0617] (0.0759)	−0.0771 [0.0696] (0.5414)	−0.1101 [0.0626] (0.2129)	−0.1548 [0.0627] (0.0474)	−0.1123 [0.0641] (0.2155)	−0.1134 [0.0641] (0.2091)
MSE	0.0016	0.001	0.0015	0.0015	0.0019	0.0015	0.0015	0.0016	0.0016
RMSE	0.0398	0.0319	0.0391	0.0385	0.0434	0.0391	0.0391	0.0400	0.0400
Average									
MSE	0.00152								
RMSE	0.03896								

Note [ ] is standard error, and ( ) is Minimum Bayes factor

Table 3 Results of maximum Renyi entropy quantile regression for panel data

Renyi Entropy

Variable	10th	20th	3th	40th	50th	60th	70th	80th	90th
UR	0.0234 [0.0046] (0.0000)	0.0236 [0.0046] (0.0000)	0.0224 [0.0046] (0.0000)	0.0229 [0.0045] (0.0000)	0.0167 [0.0050] (0.0000)	0.0333 [0.0037] (0.0000)	0.0216 [0.0046] (0.0000)	0.0236 [0.0046] (0.0000)	0.0234 [0.0046] (0.0000)
TR	0.0335 [0.0227] (0.3365)	0.0328 [0.0226] (0.3488)	0.0399 [0.0226] (0.2104)	0.0527 [0.0219] (0.0552)	0.0333 [0.0243] (0.3910)	0.0842 [0.0181] (0.0000)	0.0420 [0.0227] (0.1805)	0.0328 [0.0226] (0.3488)	0.0335 [0.0227] (0.3365)
FDI	−0.0078 [0.0037] (0.6065)	−0.0077 [0.0037] (0.1146)	−0.0063 [0.0037] (0.2346)	−0.0054 [0.0036] (0.3246)	−0.0042 [0.0040] (0.5762)	−0.0074 [0.0030] (0.0477)	−0.0063 [0.0038] (0.2530)	−0.0077 [0.0037] (0.1146)	−0.0078 [0.0037] (0.1083)
PM2.5	−0.1113 [0.0638] (0.2183)	−0.1118 [0.0637] (0.2143)	−0.1143 [0.0637] (0.1999)	−0.1139 [0.0615] (0.1799)	−0.0780 [0.0683] (0.5209)	−0.0790 [0.0510] (0.3012)	−0.1132 [0.0640] (0.2092)	−0.1118 [0.0637] (0.2143)	−0.1113 [0.0638] (0.2183)
MSE	0.0016	0.0016	0.0016	0.0015	0.0018	0.0010	0.0016	0.0016	0.0016
RMSE	0.0398	0.0397	0.0397	0.0384	0.0426	0.0318	0.0399	0.0397	0.0398
Average									
MSE	0.00153								
RMSE	0.03905								

Note [ ] is standard error, and ( ) is Minimum Bayes factor

Table 4 Results of maximum Shannon entropy quantile regression for panel data

Shannon Entropy

Variable	10th	20th	3th	40th	50th	60th	70th	80th	90th
UR	0.0234 [0.0046] (0.0000)	0.0236 [0.0046] (0.0000)	0.0216 [0.0046] (0.0000)	0.0218 [0.0045] (0.0000)	0.0168 [0.0049] (0.0000)	0.0218 [0.0045] (0.0000)	0.0219 [0.0046] (0.0000)	0.0233 [0.0046] (0.0000)	0.0234 [0.0046] (0.0000)
TR	0.0335 [0.0227] (0.3365)	0.0328 [0.0226] (0.3488)	0.0420 [0.0227] (0.0227)	0.0503 [0.0223] (0.0785)	0.0376 [0.0241] (0.2960)	0.0503 [0.0223] (0.0785)	0.0520 [0.0227] (0.0725)	0.0322 [0.0228] (0.3688)	0.0335 [0.0227] (0.3365)
FDI	−0.0078 [0.0037] (0.1083)	−0.0077 [0.0037] (0.1146)	−0.0063 [0.0038] (0.2530)	−0.0052 [0.0037] (0.3724)	−0.0062 [0.0040] (0.3008)	−0.0052 [0.0037] (0.3724)	−0.0067 [0.0037] (0.1940)	−0.0077 [0.0038] (0.1283)	−0.0077 [0.0037] (0.1283)
PM2.5	−0.1113 [0.0638] (0.2183)	−0.1118 [0.0637] (0.2173)	−0.1132 [0.0640] (0.2092)	−0.1102 [0.0626] (0.2123)	−0.0841 [0.0678] (0.4633)	0.1102 [0.0626] (0.2123)	−0.1486 [0.0638] (0.0663)	−0.1122 [0.0641] (0.2161)	−0.1113 [0.0638] (0.2183)
MSE	0.0016	0.0016	0.0016	0.0015	0.0018	0.0015	0.0016	0.0016	0.0016
RMSE	0.0398	0.0397	0.0399	0.0391	0.0423	0.0391	0.0398	0.0400	0.0398
Average									
MSE	0.00159								
RMSE	0.03993								

Note [ ] is standard error, and ( ) is Minimum Bayes factor economic growth. Prior to interpreting the results, we compare the performance of three entropy measures, Tsallis, Renyi and Shannon, which are presented in Tables 2, 3 and 4, respectively. The comparison is based on the average Mean Squared Error (MSE) and the average Root Mean Squared Error (RMSE) of quanitles 10th–90th. Thus, we can compute MSE and RMSE as in the followings:

$$RMSE = \sum_{j=1}^{9}\left(\sqrt{\left(\sum_{i=1}^{N}\sum_{t=1}^{T}(\widehat{u_{it}^{\tau=j}})^2\right)/(N+T)}\right)/9 \quad (25)$$

$$MSE = \sum_{j=1}^{9}\left(\left(\sum_{i=1}^{N}\sum_{t=1}^{T}(\widehat{u_{it}^{\tau=j}})^2\right)/(N+T)\right)/9 \quad (26)$$

The comparison result shows that Tsallis entropy is the most appropriate model to describe the effect of macroeconomic variables on economic growth in G20 countries as it provides the lowest MSE and RMSE. The results also show that the Shannon entropy GME estimates a bit different from Tsallis and Renyi in terms of the magnitudes of the estimates, particularly in the extreme quantiles(10th and 90th). But both the Tsallis and Renyi estimates are near. However, our results show that all estimations provide the same sign of parameter estimates in all quantile levels. This indicates the robustness of GME estimation.

From the economic perspective, the results show that the impact of UR on economic growth in G20 countries has the same direction in all quantiles. The coefficient of UR conveys highly decisive evidence for $H_1 : \beta \neq 0$ and is positive at lower and upper quantiles (10th, 20th, 30th, 40th, 60th, 70th, 80th and 90th quantile), except the 50th quantile which has very strong evidence for $H_1 : \beta \neq 0$. At the 20th quantile, the coefficient value of UR is the highest among all quantiles. This result indicates an asymmetric impact of UR on the economic growth in G20 countries.

Furthermore, the coefficient of TR is positive at lower and higher quantiles (the 20th, 30th, 40th, 60th and 70th quantile). At the 20th quantile, the coefficient value of TR is the highest among all quantiles and conveys very strong evidence for $H_1 : \beta \neq 0$ on economic growth in G20 countries. For 30th and 70th quantiles, the coefficient of TR conveys moderate evidence for $H_1 : \beta \neq 0$. Likewise, at the 40th and 60th quantiles, TR has coefficient values indicating strong evidence for $H_1 : \beta \neq 0$ but weak evidence for H_1 at the 10th, 50th, 80th and 90th quantiles. Besides, our results show that the coefficient of TR indicates a positive impact on economic growth for all quantiles, meaning that a higher International tourism receipts level is associated with an increase in economic growth in both low and high economic growth countries.

However in the case of FDI, its coefficient provides a very strong evidence for $H_1 : \beta \neq 0$ at the 20th quantile. For the 10th, 30th, 40th, 70th, and 80th quantiles, the coefficient of FDI conveys moderate evidence for$H_1 : \beta \neq 0$. Likewise, at the 90th quantile, FDI has coefficient value with substantial evidence for $H_1 : \beta \neq 0$ but weak evidence for H_1 at the 50th and 60th quantile levels. Furthermore, our results show that the coefficient of FDI indicates a negative impact on economic growth for all quantiles, meaning that a higher foreign direct investment inflows level is associated with a decrease in economic growth in both low and high economic growth countries.

Finally, the coefficient of PM 2.5 provides substantial evidence for $H_1 : \beta \neq 0$at the 30th, 40th and 70th quantiles. For the 10th, 60th, 80th and 90th quantiles, the coefficient of PM 2.5 conveys moderate evidence for$H_1 : \beta \neq 0$. Likewise, at the 20th and 50th quantile, PM 2.5 has coefficient values with weak evidence for$H_1 : \beta \neq 0$. In addition, our results show that the coefficient of PM 2.5 indicates a negative impact on economic growth, meaning that a higher fine particulate matter level is associated with a decrease in economic growth in both low and high economic growth countries.

5 Conclusion

In this study, we develop generalized maximum Shannon, Tsallis and Renyi entropy measures as an estimator for the panel quantile regression approach. The benefit of this estimator is that it can solve the limited data and multicollinearity problems in the regression context. We then use this model to consider an unobserved individual heterogeneity and distributional heterogeneity. By using the G20 countries datasets, the results show that Tsallis entropy outperforms the other two entropy measures in terms of the lowest mean squared error (MSE) and root mean squared error (RMSE). Therefore, the Generalized Maximum Tsallis entropy is conducted to be the estimator of our model for explaining the effect of macroeconomic variables on economic growth in G20 countries.

The Tsallis entropy is likely to perform better than other twos, the estimated parameters are not much different. The important point is that the Tsallis entropy does not depend on the logarithm like the Shannon entropy. Or finding is consisting to the wotk of Tabass and Borzadaran (2017) who revealed the higher performance of Tsallis entropy in linear regression model.

In the future study, we suggest to consider different number support points and parameter supports in order to improve the parameter estimation of GME method.

Acknowledgements The authors would like to thank the Dr. Laxmi Worachai for her helpful comments on an earlier version of the paper. The authors are also grateful for the financial support offered by Center of Excellence in Econometrics, Chiang Mai University, Thailand.

References

Antwi, S., Mills, E.F.A., Zhao, X.: Impact of macroeconomic factors on economic growth in Ghana: a cointegration analysis. Int. J. Acad. Res. Account. Finance Manag. Sci. **3**(1), 35–45 (2013)

Canay, I.A.: A simple approach to quantile regression for panel data. Economet. J. **14**(3), 368–386 (2011)

Garrison, C.B., Lee, F.Y.: The effect of macroeconomic variables on economic growth rates: a cross-country study. J. Macroecon. **17**(2), 303–317 (1995)

Goodman, S.N.: Toward evidence-based medical statistics. 1: The P value fallacy. Annals Int Med **130**(12), 995-1004 (1999)
Golan, A., Judge, G., Miller, D.: Maximum Entropy Econometrics: Robust Estimation with Limited Data. JohnWiley Sons, New York (1996)
Golan, A., Perloff, J.M.: Comparison of maximum entropy and higher-order entropy estimators. J. Economet. **107**(1–2), 195–211 (2002)
Held, L., Ott, M.: On p-values and Bayes factors. Annual Rev. Stat. Appl. **5**, 393–419 (2018)
Lee, J.J., Cheon, S.Y.: Dual generalized maximum entropy estimation for panel data regression models. Commun. Stat. Appl. Methods **21**(5), 395–409 (2014)
Levin, A., Lin, C.F., Chu, C.S.J.: Unit root tests in panel data: asymptotic and finite-sample properties. J. Economet. **108**(1), 1–24 (2002)
Maneejuk, P., Yamaka, W.: Significance test for linear regression: how to test without P-values? J. Appl. Stat. **48**(5), 827–845 (2021)
Mbulawa, S.: Effect of macroeconomic variables on economic growth in Botswana (2015)
Pipitpojanakarn, V., Maneejuk, P., Yamaka, W., Sriboonchitta, S.: How does economic growth affect the well-being in Asia? Int. J. Econ. Res. **14**(6), 231–244 (2017)
Renyi, A.: Probability Theory North-Holland Amsterdam (1970)
Sarker, S., Khan, A., Mannan, M.M.: Urban population and economic growth: South Asia perspective. Eur. J. Govern. Econ. **5**(1), 64–75 (2016)
Shannon, C.E.: A mathematical theory of communication. Bell Syst. Techn. J. **27**(3), 379–423 (1948)
Srichaikul, W., Yamaka, W., Sriboonchitta, S.: The impacts of macroeconomic variables on economic growth: evidence from China, Japan, and South Korea. In: International Conference of the Thailand Econometrics Society, pp. 552–562. Springer, Cham (2019)
Srichaikul, W., Yamaka, W., Maneejuk, P., Sriboonchitta, S.: Comparison of entropy measures in generalized maximum entropy estimation. J. Phys. Conf. Series **1053**(1), 012021. IOP Publishing (2018)
Tas, N., Onder, E., Hepsen, A.: Determinants of economic growth in G20 Countries: a panel data approach. In: International Conference on Applied Business and Economics (ICABE-13), pp. 572–582 (2013)
Tibprasorn, P., Maneejuk, P., Sriboochitta, S.: Generalized information theoretical approach to panel regression kink model. Thai J. Mathemat. 133–145 (2017)
Tabass, M.S., Borzadaran, G.M.: The generalized maximum Tsallis entropy estimators and applications to the Portland cement dataset. Commun. Stat. Simul. Comput. **46**(4), 3284–3293 (2017)
Tsallis, C.: Possible generalization of Boltzmann-Gibbs statistics. J. Stat. Phys. **52**(1–2), 479–487 (1988)
Yalcinkaya, O., Dastan, M., Karabulut, K.: The effects of international tourism receipts on economic growth: evidence from the first 20 highest income earning countries from tourism in the world (1996–2016). Montenegrin J. Econ. **14**(3), 055–071 (2018)

Quantum Uncertainty in Decision Theory

Vyacheslav I. Yukalov

Abstract An approach is presented treating decision theory as a probabilistic theory based on quantum techniques. Accurate definitions are given and thorough analysis is accomplished for the quantum probabilities describing the choice between separate alternatives, sequential alternatives characterizing conditional quantum probabilities, and behavioral quantum probabilities taking into account rational-irrational duality of decision making. The comparison between quantum and classical probabilities is explained. The analysis demonstrates that quantum probabilities serve as an essentially more powerful tool of characterizing various decision-making situations including the influence of psychological behavioral effects.

1 Introduction

Decision theory is an interdisciplinary topic widely used in various applications, such as economics, statistics, finances, data analysis, psychology, biology, politics, social sciences, philosophy, in computer and artificial intelligence studies. For example, one has to decide choosing between several portfolios of assets. The basis of decision theory is expected utility theory (Neumann and Morgenstern 1953) that is a deterministic normative theory prescribing that with probability one the optimal choice is associated with the largest expected utility. There exist as well stochastic decision theories (Berger 1985; Raiffa and Schlaifer 2000) which nevertheless are based on a deterministic approach decorated by superimposed randomness of utility.

However the overwhelming majority of empirical studies demonstrates that the choice of decision makers is not deterministic but in principle probabilistic. It is never happens that among a given group of people all without exception would make the

V. I. Yukalov (✉)
Bogolubov Laboratory of Theoretical Physics, Joint Institute for Nuclear Research, Dubna 141980, Russia
e-mail: yukalov@theor.jinr.ru

Instituto de Fisica de São Carlos, Universidade de São Paulo, CP 369, São Carlos, São Paulo 13560-970, Brazil

N. Ngoc Thach et al. (eds.), *Financial Econometrics: Bayesian Analysis, Quantum Uncertainty, and Related Topics*, Studies in Systems, Decision and Control 427,
https://doi.org/10.1007/978-3-030-98689-6_13

identical choice prescribed by the standard deterministic utility theory. There always exist fractions of subjects preferring different alternatives. That is, there always exists a distribution of decisions over the set of the given alternatives. Moreover, as has been recently summarized by Woodford (2020), psychological and neurological studies persuasively demonstrate that even a single decision maker in a single choice acts probabilistically because of the noisy functioning of the brain. This tells us that the correct description of decision making has to be probabilistic.

Two types of probabilities are known, classical and quantum. The natural question is: Which of these two types is more appropriate for describing human decision making? To convincely answer this question, it is necessary to thoroughly compare what situations in decision making can be treated by classical and quantum probabilities. That of them having the wider region of applicability would be more general, hence preferable, although the other one also can be employed in its domain of validity.

The basics of classical probability theory are well known (Kolmogorov 2018). Quantum probability, as applied to physical measurements, also has been described in several books (Neumann 1955). Here we give the exposition of quantum probability in the language of decision theory, keeping in mind a decision maker performing a choice between several alternatives. We start from the notion of the probability of a single choice, then consider sequential choices among alternatives and compare these with classical probabilities.

It is important to stress that quantum probabilities can be generalized to the form that allows us to take into account behavioral effects and the dual nature of human decision making, comprising its rational cognitive side as well as irrational emotional feelings. This generalization also is described.

2 Probability of Separate Alternatives

A decision problem is assumed to consist in choosing one of the alternatives from the set $\{A_n : n = 1, 2, \ldots, N_A\}$ of alternatives. Each alternative is associated with a vector $|A_n\rangle$ from a Hilbert space $\mathcal{H}_A$. Here and in what follows, we employ the Dirac bracket notation (Dirac 1958). It is possible to accept that the set of the vectors $|A_n\rangle$ forms a basis of $\mathcal{H}_A$, although, for generality, we can accept any basis whose span composes the space $\mathcal{H}_A$. The vectors $|A_n\rangle$ are orthonormalized, such that

$$\langle A_m \mid A_n\rangle = \delta_{mn} \qquad (n = 1, 2, \ldots, N_A). \tag{1}$$

Since we are planning to develop a probabilistic approach, we need a statistical object in order to introduce later quantum probability. The statistics of the problem at time t is represented by a statistical operator $\hat{\rho}(t)$, which is a semipositive trace-class operator on $\mathcal{H}_A$, briefly called the state. The pair $\{\mathcal{H}_A, \hat{\rho}(t)\}$ forms a *quantum statistical ensemble*. The evolution of the state in time, from an initial state $\rho(0)$, is given by the relation

$$\hat{\rho}(t) = \hat{U}(t, 0)\ \hat{\rho}(0)\ \hat{U}^+(t, 0) \tag{2}$$

through the action of a unitary evolution operator $\hat{U}(t, 0)$.

The set of alternatives is usually associated with the eigenvectors of operators of local observables acting on the Hilbert space $\mathcal{H}_A$. The collection of these operators $\{\hat{A}\}$ composes a von Neumann algebra $\mathcal{A}(\mathcal{H}_A)$ which is a norm-closed algebra of bounded self-adjoint operators, containing the identity operator, on the Hilbert space $\mathcal{H}_A$. The triple

$$\{\mathcal{H}_A,\ \hat{\rho}(t),\ \mathcal{A}(\mathcal{H}_A)\}$$

is termed a *quantum statistical system.*

To each alternative, one puts into correspondence a projector

$$\hat{P}(A_n) = |\ A_n\rangle\langle A_n\ |. \tag{3}$$

In quantum measurements, there can occur the so-called degenerate states with projectors given by a sum of projectors. However in decision theory the degenerate states need to be specified, otherwise they do not have much sense. Therefore in decision theory we shall deal with the standard projectors (3). The collection of these projectors constitutes a projector-valued measure

$$\mathcal{P}(A) = \{\hat{P}(A_n) :\ n = 1, 2, \ldots, N_A\}. \tag{4}$$

And the triple

$$\{\mathcal{H}_A,\ \hat{\rho}(t),\ \mathcal{P}(A)\} \tag{5}$$

is a *quantum probability space.*

The probability of choosing at the moment of time t an alternative A_n is

$$p(A_n, t) = \text{Tr}\ \hat{\rho}(t)\ \hat{P}(A_n), \tag{6}$$

with the trace over $\mathcal{H}_A$. The probability is normalized, so that

$$\sum_n p(A_n, t) = 1\ , \qquad 0 \leq p(A_n, t) \leq 1. \tag{7}$$

Unfolding the probability (6), we get

$$p(A_n, t) = \langle A_n \mid \hat{\rho}(t) \mid A_n\rangle. \tag{8}$$

Note that this result does not depend on the used basis, if one takes into account that for any basis $\{|j\rangle\}$, one has

$$\sum_j |\ j\rangle\langle j\ | = \hat{1}.$$

Similarly, for any problem represented by an operator $\hat{B} \in \mathcal{A}(\mathcal{H}_A)$, one has a set of vectors $\{|B_k\rangle\}$ that are ortonormalized,

$$\langle B_k \mid B_p \rangle = \delta_{kp} \qquad (k = 1, 2, \ldots, N_B) \tag{9}$$

and define the corresponding projectors

$$\hat{P}(B_k) = \mid B_k\rangle\langle B_k \mid . \tag{10}$$

The family of these projectors composes the operator valued measure

$$\mathcal{P}(B) = \{\hat{P}(B_k) : \; k = 1, 2, \ldots, N_B\}. \tag{11}$$

The corresponding quantum probability space is

$$\{\mathcal{H}_A, \; \hat{\rho}(t), \; \mathcal{P}(B)\}. \tag{12}$$

The probability of choosing at time t an alternative B_k is

$$p(B_k, t) = \text{Tr}\; \hat{\rho}(t)\; \hat{P}(B_k) \tag{13}$$

satisfying the normalization condition

$$\sum_k p(B_k, t) = 1 \; , \qquad 0 \;\leq\; p(B_k, t) \;\leq\; 1. \tag{14}$$

3 Probability of Sequential Choices

Considering sequential choices, it is necessary to accurately take into account the temporal evolution of probabilities (Yukalov 2020). The probability of an alternative A_n, defined in Eq. (6), is the a priori expected probability of choosing this alternative. Suppose at the moment t_0 a decision has been made and the alternative A_n has actually been chosen. Hence a priori form (6) is valid only before this time up to the time $t_0 - 0$, when

$$p(A_n, t_0 - 0) = \text{Tr}\; \hat{\rho}(t_0 - 0)\; \hat{P}(A_n). \tag{15}$$

If at the time t_0 the alternative A_n has certainly been chosen, this means that the a posteriori probability of choosing A_n becomes

$$p(A_n, t_0 + 0) = 1. \tag{16}$$

One tells that at the moment t_0 the state reduction has happened,

$$\hat{\rho}(t_0 - 0) \longmapsto \hat{\rho}(A_n, t_0 + 0), \tag{17}$$

leading to the probability update

$$p(A_n, t_0 - 0) \longmapsto p(A_n, t_0 + 0). \tag{18}$$

Condition (16) implies

$$p(A_n, t_0 + 0) = \text{Tr } \hat{\rho}(A_n, t_0 + 0) \ \hat{P}(A_n) = 1. \tag{19}$$

The solution to this equation, describing the state reduction (17), reads as

$$\hat{\rho}(A_n, t_0 + 0) = \frac{\hat{P}(A_n)\hat{\rho}(t_0 - 0)\hat{P}(A_n)}{\text{Tr}\hat{\rho}(t_0 - 0)\hat{P}(A_n)}, \tag{20}$$

which is the von Neumann-Lüders state (Neumann 1955; Lüders 1951). This form serves as a new initial condition for the state evolution

$$\hat{\rho}(A_n, t) = \hat{U}(t, t_0) \ \hat{\rho}(A_n, t_0 + 0) \ \hat{U}^+(t, t_0) \tag{21}$$

after the time t_0.

Aiming at making a choice among the set of alternatives $\{B_k\}$, we deal with the quantum probability space

$$\{\mathcal{H}_A, \ \hat{\rho}(A_n, t), \ \mathcal{P}(B)\}. \tag{22}$$

The related probability plays the role of the *quantum conditional probability*

$$p(B_k, t | A_n, t_0) = \text{Tr } \hat{\rho}(A_n, t) \ \hat{P}(B_k) \tag{23}$$

of choosing an alternative B_k at a time t, after the alternative A_n has certainly been chosen at the time t_0.

Introducing the notation

$$p(B_k, t, A_n, t_0) \equiv \text{Tr } \hat{U}(t, t_0) \ \hat{P}(A_n) \ \hat{\rho}(t_0 - 0) \ \hat{P}(A_n) \ \hat{U}^+(t, t_0) \ \hat{P}(B_k) \tag{24}$$

allows us to represent the conditional probability (23) in the form

$$p(B_k, t | A_n, t_0) = \frac{p(B_k, t, A_n, t_0)}{p(A_n, t_0 - 0)}. \tag{25}$$

This relation is similar to the classical relation between conditional and joint probabilities, which justifies the admissibility of naming the probability (24) *quantum joint probability* of choosing an alternative A_n at the time t_0 and an alternative B_k at the time $t > t_0$. The difference with the classical probabilities is that now the choices are made at different times, but not simultaneously.

An important particular case is when the second choice is made immediately after the first one (Goodman et al. 1991). Then the evolution operator reduces to the identity,

$$\hat{U}(t_0 + 0, t_0) = \hat{1}. \tag{26}$$

This simplifies the conditional probability

$$p(B_k, t_0 + 0|A_n, t_0) = \text{Tr } \hat{\rho}(A_n, t_0 + 0) \hat{P}(B_k) \tag{27}$$

and the joint probability

$$p(B_k, t_0 + 0, A_n, t_0) = \text{Tr } \hat{P}(A_n) \hat{\rho}(t_0 - 0) \hat{P}(A_n) \hat{P}(B_k). \tag{28}$$

The latter is analogous to the Wigner probability (Wigner 1932). The relation (25) now reads as

$$p(B_k, t_0 + 0|A_n, t_0) = \frac{p(B_k, t_0 + 0, A_n, t_0)}{p(A_n, t_0 - 0)}. \tag{29}$$

Accomplishing the trace operation in the joint probability (28) yields

$$p(B_k, t_0 + 0, A_n, t_0) = |\langle B_k \mid A_n \rangle|^2 p(A_n, t_0 - 0). \tag{30}$$

Hence the conditional probability (29) becomes

$$p(B_k, t_0 + 0|A_n, t_0) = |\langle B_k \mid A_n \rangle|^2. \tag{31}$$

4 Symmetry Properties of Probabilities

It is important to study the symmetry properties of the quantum probabilities when the choice order reverses, that is, first one chooses an alternative B_k at the time t_0 and then one considers the probability of choosing an alternative A_n at the time t. The symmetry properties should be compared with those of classical probabilities. Not to confuse the latter with the quantum probabilities, denoted by the letter p, we shall denote the classical probabilities by the letter f. Thus the classical conditional probability of two events, A_n and B_k, is

$$f(B_k|A_n) = \frac{f(B_k A_n)}{f(A_n)}, \tag{32}$$

where $f(B_k A_n)$ is the classical joint probability, which is symmetric:

$$f(A_n B_k) = f(B_k A_n), \tag{33}$$

while the conditional probability is not,

$$f(A_n|B_k) \neq f(B_k|A_n). \tag{34}$$

For the quantum probabilities with the reversed order, acting as in the previous section, we obtain the conditional probability

$$p(A_n, t|B_k, t_0) = \text{Tr}\; \hat{\rho}(B_k, t)\; \hat{P}(A_n), \tag{35}$$

with the state

$$\hat{\rho}(B_k, t) = \hat{U}(t, t_0)\; \hat{\rho}(B_k, t_0 + 0)\; \hat{U}^+(t, t_0), \tag{36}$$

where

$$\hat{\rho}(B_k, t_0 + 0) = \frac{\hat{P}(B_k)\hat{\rho}(t_0 - 0)\hat{P}(B_k)}{\text{Tr}\hat{\rho}(t_0 - 0)\hat{P}(B_k)}. \tag{37}$$

Introducing the notation of the joint probability

$$p(A_n, t, B_k, t_0) \equiv \text{Tr}\; \hat{U}(t, t_0)\; \hat{P}(B_k)\hat{\rho}(t_0 - 0)\; \hat{P}(B_k)\; \hat{U}^+(t, t_0)\; \hat{P}(A_n), \tag{38}$$

results in the relation

$$p(A_n, t|B_k, t_0) = \frac{p(A_n, t, B_k, t_0)}{p(B_k, t_0 - 0)}. \tag{39}$$

For different times t and t_0, neither conditional nor joint quantum probabilities are symmetric:

$$p(A_n, t|B_k, t_0) \neq p(B_k, t|A_n, t_0),$$

$$p(A_n, t, B_k, t_0) \neq p(B_k, t, A_n, t_0) \qquad (t > t_0). \tag{40}$$

The quantum and classical probabilities satisfy the same normalization conditions, such as for the conditional probability

$$\sum_k p(B_k, t|A_n, t_0) = \sum_n p(A_n, t|B_k, t_0) = 1 \tag{41}$$

and for the joint probability

$$\sum_k p(B_k, t, A_n, t_0) = p(A_n, t_0 - 0),$$

$$\sum_n p(A_n, t, B_k, t_0) = p(B_k, t_0 - 0). \tag{42}$$

Then the normalization condition follows:

$$\sum_{nk} p(B_k, t, A_n, t_0) = \sum_{nk} p(A_n, t, B_k, t_0) = 1, \tag{43}$$

from which

$$\sum_{nk} [\, p(B_k, t, A_n, t_0) - p(A_n, t, B_k, t_0)\,] = 0. \tag{44}$$

In the case when in the second choice at the time $t_0 + 0$ one estimates the probability of an alternative A_n immediately after the first choice at the time t_0 has resulted in an alternative B_k, similarly to the previous section, we find the joint probability

$$p(A_n, t_0 + 0, B_k, t_0) = |\langle A_n \mid B_k \rangle|^2 \, p(B_k, t_0 - 0) \tag{45}$$

and the conditional probability

$$p(A_n, t_0 + 0 | B_k, t_0) = |\langle A_n \mid B_k \rangle|^2. \tag{46}$$

Therefore the conditional probability is symmetric, while the joint probability is not:

$$p(A_n, t_0 + 0 | B_k, t_0) = p(B_k, t_0 + 0 | A_n, t_0),$$

$$p(A_n, t_0 + 0, B_k, t_0) \neq p(B_k, t_0 + 0, A_n, t_0) \qquad (t = t_0 + 0), \tag{47}$$

which is contrary to the classical case (33) and (34).

If the projectors $\hat{P}(A_n)$ and $\hat{P}(B_k)$ commute, then the joint probability (28) becomes symmetric:

$$p(B_k, t_0 + 0, A_n, t_0) = \mathrm{Tr}\; \hat{\rho}(t_0 - 0)\; \hat{P}(B_k)\hat{P}(A_n) = p(A_n, t_0 + 0, B_k, t_0), \tag{48}$$

provided that $[\hat{P}(A_n), \hat{P}(B_k)] = 0$. Taking into account the form of the joint probability (45), we get the equality

$$p(B_k, t_0 - 0) = p(A_n, t_0 - 0) \qquad \left(\left[\, \hat{P}(A_n),\; \hat{P}(B_k) \,\right] = 0 \right). \tag{49}$$

Thus in that case both the joint and conditional probabilities are symmetric, which contradicts the asymmetry of the classical conditional probability.

If the repeated choice is made among the same set of alternatives, say $\{A_n\}$, that is when $B_k = A_k$, we obtain

$$p(A_k, t_0 + 0 | A_n, t_0) = \delta_{nk}. \tag{50}$$

This equation represents the principle of the *choice reproducibility*, according to which, when the choice, among the same set of alternatives, is made twice, immediately one after another, the second choice reproduces the first one. This sounds reasonable for decision making. Really, when a decision maker accomplishes a choice immediately after another one, there is no time for deliberation, hence this decision maker just should repeat the previous choice (Yukalov 2021).

5 Duality in Decision Making

Human decision making is known to be of dual nature, including the rational (slow, cognitive, conscious, objective) evaluation of alternatives and their irrational (fast, emotional, subconscious, subjective) appreciation (Sun 2002; Paivio 2007; Stanovich 2011; Kahneman 2011). This feature of decision making that can be called *rational-irrational duality*, or *cognition-emotion duality*, or *objective-subjective duality*, can be effectively described in the language of quantum theory that also possesses a dual nature comprising the so-called particle-wave duality. To take into account the dual nature of decision making, the quantum decision theory has been advanced (Yukalov and Sornette 2008, 2009a, b, c, 2010, 2011). In the frame of this theory, quantum probability, taking account of emotional behavioral effects, becomes behavioral probability (Yukalov and Sornette 2017). Below, we briefly delineate quantum decision theory following the recent papers (Yukalov 2020, 2021).

The space of alternatives $\mathcal{H}_A$ is composed of the state vectors characterizing the rational representation of these alternatives whose probabilities can be rationally and objectively evaluated. Since there also exist subjective emotional feelings, for taking them into account, the space of the state vectors has to be extended by including the *subject space*

$$\mathcal{H}_S = \text{span}\,\{|\,\alpha\,\rangle\} \tag{51}$$

formed by the vector representations $|\alpha\rangle$ of all admissible elementary feelings. These vectors $|\alpha\rangle$ form an orthonormal basis,

$$\langle\,\alpha\,|\,\beta\,\rangle = \delta_{\alpha\beta}.$$

Thus, the total *decision space* is the tensor product

$$\mathcal{H} = \mathcal{H}_A \bigotimes \mathcal{H}_S. \tag{52}$$

The statistical state $\hat{\rho}(t)$ now acts on the decision space (52) where it evolves as

$$\hat{\rho}(t) = \hat{U}(t,0)\;\hat{\rho}(0)\;\hat{U}^+(t,0). \tag{53}$$

Respectively, the quantum statistical ensemble is

$$\left\{\mathcal{H} = \mathcal{H}_A \bigotimes \mathcal{H}_S,\ \hat{\rho}(t)\right\}. \tag{54}$$

Each alternative A_n is accompanied by a related set of emotions x_n that is represented in the subject space by an emotion vector $|x_n\rangle \in \mathcal{H}_S$, which can be written as an expansion

$$|\ x_n\ \rangle = \sum_\alpha b_{n\alpha}\ |\ \alpha\ \rangle. \tag{55}$$

Strictly speaking, emotions are contextual and are subject to variations, which means that the coefficients $b_{n\alpha}$ can vary and, generally, fluctuate with time depending on the state of a decision maker and the corresponding surrounding.

The emotion vectors can be normalized,

$$\langle\ x_n\ |\ x_n\ \rangle = \sum_\alpha |\ b_{n\alpha}\ |^2 = 1, \tag{56}$$

but they are not necessarily orthogonal, so that

$$\langle\ x_m\ |\ x_n\ \rangle = \sum_\alpha b^*_{m\alpha} b_{n\alpha} \tag{57}$$

is not compulsorily a Kronecker delta.

An emotion operator

$$\hat{P}(x_n) = |\ x_n\ \rangle\ \langle x_n\ | \tag{58}$$

is idempotent,

$$[\ \hat{P}(x_n)\ ]^2 = \hat{P}(x_n), \tag{59}$$

but different operators are not orthogonal, since

$$\hat{P}(x_m)\ \hat{P}(x_n) = \langle\ x_m\ |\ x_n\ \rangle |\ x_m\ \rangle\langle\ x_n\ |. \tag{60}$$

The emotion operators of elementary feelings $|\alpha\rangle$, forming a complete orthonormal basis in the space (51), sum to one

$$\sum_\alpha \hat{P}(\alpha) = \sum_\alpha |\ \alpha\ \rangle\langle\ \alpha\ | = \hat{1},$$

but the emotion operators (58) do not necessarily sum to one, giving

$$\sum_n \hat{P}(x_n) = \sum_n \sum_{\alpha\beta} b_{n\alpha} b^*_{n\beta} |\ \alpha\ \rangle\langle\ \beta\ |,$$

from where

$$\langle \alpha \mid \sum_n \hat{P}(x_n) \mid \beta \rangle = \sum_n b_{n\alpha} b^*_{n\beta}.$$

The emotion vectors $|x_n\rangle$ do not form a basis, hence the emotion operators (58) do not have to sum to one. The projector (58) projects onto the subspace of feelings associated with the alternative A_n.

The pair of an alternative A_n and the set of the related emotions x_n composes a prospect $A_n x_n$ whose representation in the decision space (52) is given by the vector

$$\mid A_n x_n \rangle = \mid A_n \rangle \bigotimes \mid x_n \rangle = \sum_\alpha b_{n\alpha} \mid A_n \alpha \rangle. \tag{61}$$

These vectors are orthonormalized,

$$\langle x_m A_m \mid A_n x_n \rangle = \delta_{mn}.$$

The prospect projector

$$\hat{P}(A_n x_n) = \mid A_n x_n \rangle\langle x_n A_n \mid = \hat{P}(A_n) \bigotimes \hat{P}(x_n) \tag{62}$$

is idempotent,

$$[\ \hat{P}(A_n x_n)\ ]^2 = \hat{P}(A_n x_n). \tag{63}$$

The projectors of different prospects are orthogonal,

$$\hat{P}(A_m x_m)\ \hat{P}(A_n x_n) = \delta_{mn} \hat{P}(A_n x_n) \tag{64}$$

and commute with each other,

$$[\ \hat{P}(A_m x_m),\ \hat{P}(A_n x_n)\ ] = 0.$$

The vectors $|A_n \alpha\rangle$ generate a complete basis in the decision space (52), because of which

$$\sum_{n\alpha} \hat{P}(A_n \alpha) = \sum_{n\alpha} \mid A_n \alpha \rangle\langle \alpha A_n \mid = \hat{1}. \tag{65}$$

But the prospect projectors (62) on subspaces do not necessarily sum to one,

$$\sum_n \hat{P}(A_n x_n) = \sum_n \sum_{\alpha\beta} b_{n\alpha} b^*_{n\beta} \mid A_n \alpha \rangle\langle \beta A_n \mid,$$

as far as

$$\langle \alpha A_m \mid \sum_n \hat{P}(A_n x_n) \mid A_n \beta \rangle = \delta_{mn} b^*_{n\alpha} b_{n\beta}.$$

However, it is admissible to require that the prospect projectors would sum to one on average, so that

$$\mathrm{Tr}\, \hat{\rho}(t) \sum_n \hat{P}(A_n x_n) = 1, \tag{66}$$

which is equivalent to the condition

$$\sum_n \sum_{\alpha\beta} b^*_{n\alpha} b_{n\beta} \, \langle \, \alpha A_n \mid \hat{\rho}(t) \mid A_n \beta \, \rangle = 1. \tag{67}$$

The trace operation in Eq. (66) and below is over the total decision space (52). The projection-valued measure on the space (52) is

$$\mathcal{P}(Ax) = \{\hat{P}(A_n x_n) : \; n = 1, 2, \ldots, N_A\}, \tag{68}$$

so that the quantum probability space is

$$\{\mathcal{H}, \; \hat{\rho}(t), \; \mathcal{P}(Ax)\}. \tag{69}$$

The prospect probability reads as

$$p(A_n x_n, t) = \mathrm{Tr}\, \hat{\rho}(t) \, \hat{P}(A_n x_n) \tag{70}$$

and satisfies the normalization conditions

$$\sum_n p(A_n x_n, t) = 1 \, , \qquad 0 \, \leq \, p(A_n x_n, t) \, \leq \, 1. \tag{71}$$

In expression (70), it is possible to separate the diagonal part

$$f(A_n x_n, t) \equiv \sum_\alpha \mid b_{n\alpha} \mid^2 \langle \, \alpha A_n \mid \hat{\rho}(t) \mid A_n \alpha \, \rangle \tag{72}$$

and the nondiagonal part

$$q(A_n x_n, t) \equiv \sum_{\alpha \neq \beta} b^*_{n\alpha} b_{n\beta} \langle \, \alpha A_n \mid \hat{\rho}(t) \mid A_n \beta \, \rangle. \tag{73}$$

The diagonal part has the meaning of the rational fraction of the total probability (70), because of which it is called the *rational fraction* and is assumed to satisfy the normalization condition

$$\sum_n f(A_n x_n, t) = 1 \, , \qquad 0 \, \leq \, f(A_n x_n, t) \, \leq \, 1. \tag{74}$$

The rational fraction satisfies the standard properties of classical probabilities.

The nondiagonal part is caused by the quantum interference of emotions and fulfills the conditions

$$\sum_n q(A_n x_n, t) = 0 \; , \qquad -1 \; \leq \; q(A_n x_n, t) \; \leq \; 1 . \tag{75}$$

As far as emotions describe the quality of alternatives, the quantum term (75) can be called the *quality factor*. Being due to quantum interference, it also can be named the *quantum factor*. And since the quality of alternatives characterizes their attractiveness, the term (75) can be called the *attraction factor*.

Thus the prospect probability (70) reads as the sum of the rational fraction and the quality factor:

$$p(A_n x_n, t) = f(A_n x_n, t) + q(A_n x_n, t). \tag{76}$$

6 Conditional Behavioral Probability

Sequential choices in quantum decision theory can be treated by analogy with Sect. 3. The a priori probability at any time $t < t_0$ is defined in Eq. (70), provided no explicit choice has been done before the time t_0. Just until this time, the a priori probability of an alternative A_n is

$$p(A_n x_n, t_0 - 0) = \text{Tr}\; \hat{\rho}(t_0 - 0)\; \hat{P}(A_n x_n). \tag{77}$$

If at the moment of time t_0 a choice has been made and an alternative A_n is certainly chosen, then the a posteriori probability becomes

$$p(A_n x_n, t_0 + 0) = 1. \tag{78}$$

This implies the reduction of the probability

$$p(A_n x_n, t_0 - 0) \longmapsto p(A_n x_n, t_0 + 0) \tag{79}$$

and the related state reduction

$$\hat{\rho}(t_0 - 0) \longmapsto \hat{\rho}(A_n x_n, t_0 + 0). \tag{80}$$

Equation (78), asserting that

$$\text{Tr}\; \hat{\rho}(A_n x_n, t_0 + 0)\; \hat{P}(A_n x_n) = 1, \tag{81}$$

possesses the solution

$$\hat{\rho}(A_n x_n, t_0 + 0) = \frac{\hat{P}(A_n x_n)\hat{\rho}(t_0 - 0)\hat{P}(A_n x_n)}{\mathrm{Tr}\hat{\rho}(t_0 - 0)\hat{P}(A_n x_n)}. \tag{82}$$

The state (82) serves as an initial condition for the new dynamics prescribed by the equation

$$\hat{\rho}(A_n x_n, t) = \hat{U}(t, t_0)\; \hat{\rho}(A_n x_n, t_0 + 0)\; \hat{U}^+(t, t_0). \tag{83}$$

The a priori probability of choosing an alternative B_k at any time $t > t_0$, after the alternative A_n has certainly been chosen, is the conditional probability

$$p(B_k x_k, t | A_n x_n, t_0) = \mathrm{Tr}\; \hat{\rho}(A_n x_n, t)\; \hat{P}(B_k x_k). \tag{84}$$

Introducing the joint behavioral probability

$$p(B_k x_k, t, A_n x_n, t_0) \equiv \mathrm{Tr}\; \hat{U}(t, t_0)\; \hat{P}(A_n x_n)\; \hat{\rho}(t_0 - 0)\; \hat{P}(A_n x_n)\; \hat{U}^+(t, t_0)\; \hat{P}(B_k x_k), \tag{85}$$

allows us to represent the conditional probability in the form

$$p(B_k x_k, t | A_n x_n, t_0) = \frac{p(B_k x_k, t, A_n x_n, t_0)}{p(A_n x_n, t_0 - 0)}. \tag{86}$$

If the probability of choosing an alternative B_k is evaluated immediately after t_0, then we need to consider the conditional probability

$$p(B_k x_k, t_0 + 0 | A_n x_n, t_0) = \mathrm{Tr}\; \hat{\rho}(A_n x_n, t_0 + 0)\; \hat{P}(B_k x_k) \tag{87}$$

and the joint probability

$$p(B_k x_k, t_0 + 0, A_n x_n, t_0) = \mathrm{Tr}\; \hat{P}(A_n x_n)\; \hat{\rho}(t_0 - 0)\; \hat{P}(A_n x_n)\; \hat{P}(B_k x_k). \tag{88}$$

As a result, the conditional probability (87) becomes

$$p(B_k x_k, t_0 + 0 | A_n x_n, t_0) = \frac{p(B_k x_k, t_0 + 0, A_n x_n, t_0)}{p(A_n x_n, t_0 - 0)}. \tag{89}$$

For the joint probability, we obtain

$$p(B_k x_k, t_0 + 0, A_n x_n, t_0) = |\; \langle\; x_k B_k \mid A_n x_n \;\rangle\; |^2\; p(A_n x_n, t_0 - 0). \tag{90}$$

Therefore the conditional behavioral probability is

$$p(B_k x_k, t_0 + 0 | A_n x_n, t_0) = |\; \langle\; x_k B_k \mid A_n x_n \;\rangle\; |^2. \tag{91}$$

7 Symmetry of Behavioral Probabilities

Considering the symmetry properties of behavioral probabilities, it is useful to remember that, strictly speaking, emotions are contextual and can vary in time. In an approximate picture, it is possible to assume that emotions are mainly associated with the corresponding alternatives and are approximately the same at all times. Then the symmetry properties of the probabilities can be studied with respect to the interchange of the order of the prospects $A_n x_n$ and $B_k x_k$. Keeping in mind this kind of the order interchange, we can conclude that the symmetry properties of behavioral probabilities are similar to the order symmetry of the quantum probabilities examined in Sect. 4.

Thus for any time $t > t_0$, both the conditional and the joint behavioral probabilities are not order symmetric with respect to the prospect interchange,

$$p(A_n x_n, t | B_k x_k, t_0) \neq p(B_k x_k, t | A_n x_n, t_0) \qquad (t > t_0), \tag{92}$$

and

$$p(A_n x_n, t, B_k x_k, t_0) \neq p(B_k x_k, t, A_n x_n, t_0) \qquad (t > t_0), \tag{93}$$

which is analogous to Eq. (40).

When the second decision is being made at the time $t = t_0 + 0$ immediately after the first choice has been accomplished at the time t_0, the conditional behavioral probability is order symmetric,

$$p(A_n x_n, t_0 + 0 | B_k x_k, t_0) = p(B_k x_k, t_0 + 0 | A_n x_n, t_0), \tag{94}$$

but the joint probability, generally, is not order symmetric,

$$p(A_n x_n, t_0 + 0, B_k x_k, t_0) \neq p(B_k x_k, t_0 + 0, A_n x_n, t_0), \tag{95}$$

which is similar to property (47).

If one makes the immediate sequential choices, and in addition the prospect projectors commute with each other, so that

$$[\ \hat{P}(A_n x_n),\ \ \hat{P}(B_k x_k)\ ] = 0,$$

then the joint behavioral probability becomes order symmetric,

$$p(A_n x_n, t_0 + 0, B_k x_k, t_0) = p(B_k x_k, t_0 + 0, A_n x_n, t_0). \tag{96}$$

This property is in agreement with Eq. (48).

Recall that the conditional probability (93), because of its form (91), is symmetric in any case, whether the prospect operators commute or not.

As is seen, the symmetry properties of the quantum probabilities are in variance with the properties (33) and (34) of the classical probabilities, according to which the joint classical probability is order symmetric, while the conditional classical probability is not order symmetric.

The absence of the order symmetry in classical conditional probability is evident from the definition (32). Empirical investigations (Boyer-Kassem et al. 2016a, b) also show that the conditional probability is not order symmetric. However, as is seen from equality (94), the quantum conditional probability is order symmetric. Does this mean that the quantum conditional probability cannot be applied to the realistic human behavior?

To answer this question, it is necessary to concretize the realistic process of making decisions. In reality, any decision is not a momentary action, but it takes some finite time. The modern point of view accepted in neurobiology and psychology is that the cognition process, through which decisions are generated, involves three stages: the process of stimulus encoding through which the internal representation is generated, followed by the evaluation of the stimulus signal and then by decoding of the internal representation to draw a conclusion about the stimulus that can be consciously reported (Woodford 2020; Libert 2006). It has been experimentally demonstrated that awareness of a sensory event does not appear until the delay time up to 0.5 s after the initial response of the sensory cortex to the arrival of the fastest projection to the cerebral cortex (Libert 2006; Teichert et al. 2016). About the same time is necessary for the process of the internal representation decoding. So, the delay time of about 1 s is the minimal time for the simplest physiological processes involved in decision making. Sometimes the evaluation of the stimulus signal constitutes the total response time, necessary for formulating a decision, of about 10 s (Hochman et al. 2010). In any case, the delay time of order 1 s seems to be the minimal period of time required for formulating a decision. This assumes that in order to consider a sequential choice as following immediately after the first one, as is necessary for the quantum conditional probability (89) or (91), the second decision has to follow in about 1 s after the first choice. However, to formulate the second task needs time, as well as some time is required for the understanding the second choice problem. This process demands several minutes.

In this way, the typical situation in the sequential choices is when the temporal interval between the decisions is of the order of minutes, which is much longer than the time of 1 s necessary for taking a decision. Therefore the second choice cannot be treated as following immediately after the first one, hence the form of the conditional probability (91) is not applicable to such a situation. For that case, one has to use expression (86) which is not order symmetric, in agreement with the inequality (92) and empirical observations.

Thus the decisions can be considered as following immediately one after the other provided the temporal interval between them is of the order of 1 s. Such a short interval between subsequent measurements could be realized in quantum experiments, but it is not realizable in human decision making, where the interval between subsequent decisions is usually much longer than 1 s. Hence the form of the conditional probability (91), that one often calls the Lüders probability, is not applicable to human problems, but expression (86), valid for a finite time interval between decisions, has to be employed. The latter is not order symmetric similarly to the classical conditional probability.

Concluding, quantum probabilities, whose definition takes into account dynamical processes of taking decisions, are more general than simple classical probabilities (32), hence can be applied to a larger class of realistic human decision problems.

Acknowledgements The author is grateful to J. Harding and H. Nguyen for fruitful discussions and to E. P. Yukalova for useful advice.

References

Berger, J.O.: Statistical Decision Theory and Bayesian Analysis. Springer, New York (1985)

Boyer-Kassem, T., Duchêne, S., Guerci, E.: Testing quantum-like models of judgment for question order effect. Math. Soc. Sci. **80**, 33–46 (2016)

Boyer-Kassem, T., Duchêne, S., Guerci, E.: Quantum-like models cannot account for the conjunction fallacy. Theory Decision **81**, 479–510 (2016)

Dirac, P.A.M.: The Principles of Quantum Mechanics. Clarendon, Oxford (1958)

Goodman, I.R., Nguyen, H.T., Walker, E.A.: Conditional Inference and Logic for Intelligent Systems. North-Holland, Amsterdam (1991)

Hochman, G., Ayal, S., Glöckner, A.: Physiological arousal in processing recognition information: Ignoring or integrating cognitive cues? Judgment Decision Making **5**, 285–299 (2010)

Kahneman, D.: Thinking. Fast and Slow, Farrar, Straus and Giroux, New York (2011)

Kolmogorov, A.N.: Foundations of the Theory of Probability. Dover, New York (2018)

Libert, B.: Reflections on the interaction of the mind and brain. Progr. Neurobiol. **78**, 322–326 (2006)

Lüders, G.: Concerning the state change due to the measurement process. Annalen der Physik (Berlin) **8**, 322–328 (1951)

Paivio, A.: Mind and Its Evolution: A Dual Coding Theoretical Approach. Lawrence Erlbaum Associates, Mahwah (2007)

Raiffa, H., Schlaifer, R.: Applied Statistical Decision Theory. Wiley, New York (2000)

Stanovich, K.E.: Rationality and the Reflective Mind. Oxford University Press, New York (2011)

Sun, R.: Duality of the Mind. Lawrence Erlbaum Associates, Mahwah (2002)

Teichert, T., Grinband, J., Ferrera, V.: The importance of decision onset. J. Neurophysiol. **115**, 643–661 (2016)

von Neumann, J.: Mathematical Foundations of Quantum Mechanics. Princeton University, Princeton (1955)

von Neumann, J., Morgenstern, O.: Theory of Games and Economic Behavior. Princeton University, Princeton (1953)

Wigner, E.: On the quantum correction for thermodynamic equilibrium. Phys. Rev. **40**, 749–759 (1932)

Woodford, M.: Modeling imprecision in perception, valuation, and choice. Annual Rev. Econ. **12**, 579–601 (2020)

Yukalov, V.I.: Tossing quantum coins and dice. Laser Phys. **31**, 055201 (2021)

Yukalov, V.I.: Evolutionary proceses in quantum decision theory. Entropy **22**, 681 (2020)

Yukalov, V.I., Sornette, D.: Quantum decision theory as quantum theory of measurement. Phys. Lett. A **372**, 6867–6871 (2008)

Yukalov, V.I., Sornette, D.: Scheme of thinking quantum systems. Laser Phys. Lett. **6**, 833–839 (2009)

Yukalov, V.I., Sornette, D.: Physics of risk and uncertainty in quantum decision making. Eur. Phys. J. B **71**, 533–548 (2009)
Yukalov, V.I., Sornette, D.: Processing information in quantum decision theory. Entropy **11**, 1073–1120 (2009)
Yukalov, V.I., Sornette, D.: Mathematical structure of quantum decision theory. Adv. Complex Syst. **13**, 659–698 (2010)
Yukalov, V.I., Sornette, D.: Decision theory with prospect interference and entanglement. Theory Decision **70**, 283–328 (2011)
Yukalov, V.I., Sornette, D.: Quantum probabilities as behavioral probabilities. Entropy **19**, 112–30 (2017)

Does Internal Control Affect Bank Profitability in Vietnam? A Bayesian Approach

Pham Hai Nam, Nguyen Ngoc Thach, Ngo Van Tuan, Nguyen Minh Nhat, and Pham Thi Hong Nhung

Abstract This article studies the impact of internal control on the profitability of Vietnamese commercial banks. The influence of internal control on the profitability of commercial banks has been carried out by some researchers around the world. However, the research on the relationship between internal control and profitability of Vietnamese commercial banks is still limited. The article mentions the basic contents of internal control according to the Committee of Sponsoring Organization (COSO 2013) and, at the same time, determines the impact of internal control on the profitability of Vietnamese commercial banks. The research sample includes 30 Vietnamese commercial banks in the period 2007–2018, data is collected from audited financial statements and annual reports of banks, and macro data is collected from the General Statistics Office. The study uses return on total assets (ROA) and return on equity (ROE) to represent the profitability of commercial banks. By using the Gibbs sampling within the Bayesian approach, the research results show that there are five components of the COSO framework (2013) that affect the profitability of Vietnamese commercial banks, including: control environment, risk assessment, control activities, information and communication, monitoring activities. In addition, the study also shows that factors belonging to banking characteristics and macroeconomic factors affect the profitability of Vietnamese commercial banks: bank size, number of years of operation, inflation, GDP growth. The results of the study are the

P. H. Nam (✉) · N. Ngoc Thach · N. Van Tuan · N. M. Nhat
Banking University HCMC, Ho Chi Minh City, Vietnam
e-mail: namph@buh.edu.vn

N. Ngoc Thach
e-mail: thachnn@buh.edu.vn

N. Van Tuan
e-mail: tuannv@buh.edu.vn

N. M. Nhat
e-mail: nhatnm@buh.edu.vn

P. T. H. Nhung
Ho Chi Minh City College of Economics, Ho Chi Minh City, Vietnam

N. Ngoc Thach et al. (eds.), *Financial Econometrics: Bayesian Analysis, Quantum Uncertainty, and Related Topics*, Studies in Systems, Decision and Control 427,
https://doi.org/10.1007/978-3-030-98689-6_14

basis for making some recommendations to strengthen the internal control system of Vietnamese commercial banks to achieve more sustainable profitability.

Keywords COSO · Commercial bank · Internal control · Profitability

1 Introduction

In recent years, especially after the 2008 financial crisis, Vietnamese commercial banks have experienced great difficulties and fluctuations (Batten and Vo 2019), leading to mergers between banks, or becoming 100% state-owned banks. In the opposite direction, many banks have grown in both size and quality, becoming banks of great stature and influence in the country and reaching out to foreign markets. That shows that the development is uneven among Vietnamese banks; there is an obvious divergence among banks.

The banking sector is a unique one handling many transactions daily and providing financial banking services to the entire economy. The influence of the banking industry on the economy is enormous, dominating the activities of businesses, organizations and individuals to an increasing degree, contributing to the economic prosperity (Pham 2019). Therefore, the stable development of banks plays a vital role in the development of businesses in particular and the whole economy in general. To do that, banks need to establish a closed system of internal control to ensure targets are met, achieve good long-term profitability, maintain a reliable management system and financial statements (Nguyen and Duong 2015). This system can ensure banks comply with laws, internal processes and procedures, helping banks reduce the risk of reputational damage or unexpected losses for the bank. An ineffective internal control system is one of the severe problems banks face, leading to fraudulent transactions that cause significant financial losses to banks, even in some cases, can cause the bank to fail. In Vietnam, despite the restructuring of the banking system and renovation of the banking administration system towards modernity, in line with international practices and standards (Government 2012), banks still face great risks due to ineffective internal control systems. Therefore, one of the crucial issues raised is that banks need to establish an effective internal control system, which helps control risks in banking operations, and ensures the achievement of objectives in terms of profitability, growth, and going concerned (Arad and Jamshedy-Navid 2010).

The COSO framework with five components is applied (COSO 2013) to evaluate the effectiveness of the internal control system. These five components: Control environment, risk assessment, control activities, information and communication, monitoring activities. The COSO framework is also the framework used by many researchers to evaluate the effectiveness of the internal control system. Some studies assess the effectiveness of the internal control system on credit risk, while others focus on assessing the impact of internal control on operational efficiency or profitability. Typical studies evaluating the impact of internal control on the profitability of commercial banks can be mentioned as the study of Hamour et al. (2021), Hanoon et al. (2021), Channar and Khan (2015), Umar and Dikko (2018), Koutoupis and Malisiovas (2019). These studies all apply traditional methods to accomplish the research objectives. Due to the application of many different traditional methods, the results of these studies show the different influences of internal control on the profitability of commercial banks. In Vietnam, when evaluating the effectiveness of the internal control system on financial performance, most of the studies focus on enterprises such as Chu (2016), Ngo (2004) and Vu (2016). Meanwhile, when assessing the impact of internal control on the performance of commercial banks, according to the authors' summary, there are currently only studies by Hoang (2020) or Nguyen and Duong (2015). However, these studies only build a theoretical framework to examine the impact of internal control on the performance of Vietnamese commercial banks (Hoang, 2020) or use primary data sets (Nguyen and Duong 2015). Moreover, all of these studies apply the traditional approach, which has many disadvantages and causes controversy about the research results (Anh et al. 2018; Hung et al. 2019; Kreinovich et al. 2019). To overcome that problem, the authors apply Bayesian method, which is a new and more modern approach compared to previous studies. Therefore, this study will fill the gap of previous studies. The research results will provide some policy suggestions for commercial banks, policymaker, to ensure the healthy operation of banks and compliance with internal control processes and procedures. At the same time, the research will help to increase the suitability of the internal control system in the bank and enhance the effectiveness of the governance apparatus, complying with the requirements of the State Bank of Vietnam's internal control, help Vietnamese commercial banks achieve objectives in business activities, bringing financial efficiency and better risk control.

The remainder of this paper is structured as follows: Sect. 2 presents the theoretical background and related studies, Sect. 3 presents research method and models, Sect. 4 reports the research results and discussion, Sect. 5 concludes and suggests policy.

2 Literature Review

2.1 Agency Theory

The theory to study the impact of internal control on the financial performance of Vietnamese commercial banks is Agency Theory. This theory was developed by Jensen and Meckling (1976). Agency Theory is an assumption intended to explain the financial relationship between the principals and the agents. An agency relationship is a contract whereby one or more persons (the principals) assign the agents to perform a service on behalf of the principals, with the responsibility and authority to perform the work independently of the principals (Jensen and Meckling 1976). Agency Theory emphasizes problems arising from different goals and desires between the principals and the agents. This situation can occur because the principal may not fully understand the agents' actions or simply do not have complete information, such as the company's board of directors wishing to expand business operations to another market. This decision can reduce the company's profits in the short term but can be highly profitable in the long term. However, the company's shareholders, i.e. the principals, may not fully understand that and still want high profits in the short term. In addition, according to the Agency Theory, the agents can take advantage of their position to seek their interests, going against the general interests and interests of the principal. Therefore, Agency Theory also explains why it is necessary to establish an internal control system in each enterprise. An effective internal control system will help managers prevent and timely detect weaknesses and risks that may arise, with the highest aim of maximizing business value.

2.2 *COSO's Internal Control Framework*

According to COSO (1992), internal control consists of five components: The control environment, the entity's risk assessment process, system information and communication, control activities, and control monitoring. COSO (1992) argues that every organization, whether private or public, large or small, faces risks from external and internal sources, and therefore, those risks must be evaluated. Thus, the board of managers needs to take the necessary actions to prevent these risks. But sometimes, the board of managers cannot avoid the risk that occurs. In these situations, the board of managers needs to determine whether to accept the risk, reduce it to an acceptable level, or avoid it. Next, the COSO (2004) framework for internal control defines internal control as "a process performed by an entity's board of directors, board of managers and other personnel designed to provide reasonable assurance about the achievement of operational, reporting and compliance objectives". Internal control includes controls designed to provide reasonable assurance that the company's financial statements are reliable and are prepared under GAAP (Generally Accepted Accounting Principles). To update the novelty due to many changes in the world economy, COSO (2013) released a new report, according to which the internal control framework consists of five components: Control environment, risk assessment, control activities, information and communication, monitoring activities. Compared with the previous framework, the new framework does not change the core definition of internal control, including five elements and criteria for evaluating the effectiveness of internal control. However, it emphasizes the critical role of board oversight and corporate governance. In addition, the new framework covers both internal financial reporting and external non-financial reporting. It requires internal control to be transformed from a financial reporting-oriented control system to a comprehensive control system.

2.3 Basel Internal Control Framework in the Banking Sector

The Basel Committee on Banking Supervision (1998) stated that significant losses in banking activities mainly stem from the failure of banks to maintain an effective internal control system to prevent or early detection of signs of risk, thereby minimizing possible damage to the bank. According to the Basel Committee (1998), an effective internal control system is an important component in the management of a bank's operations. It is the foundation for a safe and sound banking operation. Therefore, on September 22, 1998, the Basel Committee released an internal control template specifically applied to the banking sector to guide banks in building an effective governance system. Through the analysis of the content of the Basel 1998 report, it can be seen that the Basel 1998 report does not provide new theories but only applies the basic theories of the internal control framework issued by COSO in 1992—the framework for internal control was first issued in the world in a comprehensive, systematic, recognized and widely used manner in many countries, in the banking sector.

After more than 20 years since the COSO committee issued the COSO 1992 report, the business environment has changed significantly. Therefore, in 2013, the COSO committee released a new internal control framework. However, up to now, the Basel Committee (1998) has not issued a new internal control framework specifically applied to the banking sector. Therefore, it is necessary to study based on the Basel 1998 report approach and inherit the updated points on internal control of the COSO 2013 report to approach the principles of setting up internal control according to practice into the banking sector to perfect the management system of the bank, minimize risks that may arise, ensure safe and healthy banking activities, and generate sustainable income and add value for banks.

In Vietnam, the Accounting Law (2015) states that internal control is the establishment and organization of internal implementation of the internal mechanisms, policies, processes and regulations under the provisions of the law to ensure the timely prevention, detection and handling of risks and to meet the set requirements.

According to Auditing Standard 35 (2012), internal control is the process which is designed, implemented and maintained by the board of managers and other individuals within the entity to provide reasonable assurance about the achieve the entity's objectives in ensuring the reliability of financial statements, ensuring operational efficiency and effectiveness, and in compliance with relevant laws and regulations.

In the banking sector, the State Bank of Vietnam (SBV) has paid special attention to the safety of the banking system and has issued many regulations related to internal control. Circular No. 44/2011/TT-NHNN dated December 29, 2011 stipulates internal control and audit of credit institutions and foreign bank branches. However, the disadvantage of this Circular is that it is still general, not clear and specific, and does not meet the vital role of the internal control system in the operations of banks. Therefore, to overcome the shortcomings of Circular No. 44/2011/TT-NHNN, and to meet the requirements of international integration increasingly more profound in the banking sector, the SBV continued to update the contents of internal control when it issued Circular No. 13/2018/TT-NHNN dated May 18, 2018 regulating the internal control system of commercial banks and foreign bank branches. The new point of this Circular is that it clearly and specifically regulates the commercial banks' internal control system according to international standards.

Thus, when comparing the framework of COSO (2013), Vietnam Auditing Standard No. 35, Law on Accounting (2015), Circular No. 13/2018/TT-NHNN, the view on internal control is similar. Therefore, the authors agree with COSO (2013) on internal control because this is widely applied in commercial banks and related empirical studies. On that basis, the authors build research models in this article.

2.4 Empirical Studies

Hamour et al. (2021) studied the impact of internal control on the performance of commercial banks in Jordan. Using primary and secondary data sets, the research results show a positive impact of internal control on the performance of Jordanian commercial banks. The study provides a unique contribution to other studies of the impact internal control systems based on the COSO components control and their impact on the financial performance, as well as the moderator effect (board independence) on this relationship in the Jordanian banks. The results of the study support the agency's theory in which the level of financial performance satisfactory to shareholders will reduce the agency's problem and costs, and thus a new scientific shareholder path added to accounting literature that explains the relationships and impacts between COSO.

The study by Hanoon et al. (2021) was conducted to assess the impact of internal control on the performance performance of commercial banks in Iraq. The five components of COSO include control environment, control activity, risk assessment, information & communication, and monitoring. The fundamental theory used in the study is Agency theory. In this study, the authors used Structural Equation Modeling (SEM), data were collected from primary and secondary sources. The elements of internal control that have a positive impact on bank performance are control environment, control activity, risk assessment, information & communication, and monitoring. In which, control activity is the factor that has the strongest impact on ROA and ROE. On that basis, the authors suggest that Iraqi banks should focus on the improvement of Internal Control Components to enhance financial performance.

Koutoupis and Malisiovas (2019) studied the impact of internal control on the profitability of 210 largest commercial banks in the US using the fixed effects model (FEM), during the period from 2013 to 2017. In this study, the authors use NIM (Net interest margin) to indicate the profitability of commercial banks. The research results show four components of internal control that affect the profitability of commercial banks in the US, namely control environment, control activities, information and communication, and monitoring activities.

Channar and Khan (2015) examined the effectiveness of the internal control system on the financial performance of six commercial banks in Pakistan. The authors use the indicators ROA, ROE, net profit to cost to represent financial performance, along with primary and secondary data sets in this study. The results show that the effectiveness of the internal control system has an impact on financial performance in Pakistani commercial banks.

Umar and Dikko (2018) studied the impact of internal control on the performance of commercial banks in Nigeria. The research results show that four components of the internal control system affect the performance of banks, namely control environment, control activities, monitoring activities and risk assessment.

Akwaa-Sekyi and Gené (2017) explored the impact of internal control on the credit risk of 91 commercial banks in Europe from 2008 to 2014. By GMM method, the research results show that a component of internal control that has a significant impact on credit risk is the risk assessment. Besides, the control variables that affect credit risk include return on risky assets, institutional ownership, bank size, inflation, interest rates and GDP growth.

In Vietnam, according to the authors' review, only researches by Hoang (2020), Nguyen and Duong (2015) is relevant to this study. However, in these studies, the authors have only developed a theoretical framework to examine the impact of internal control on the performance of Vietnamese commercial banks. The review of previous studies show that there has not been any research that has been fully studied, both theoretically and experimentally, regarding the influence of internal control on the profitability of commercial banks. Moreover, all of these studies apply the traditional approach, which has many disadvantages and causes controversy about the research results. Therefore, based on applying the COSO framework (2013) and relevant domestic and foreign studies, this study uses the quantitative method of Bayes and Gibbs sampling to study the impact of internal control on the profitability of Vietnamese commercial banks.

2.5 Research Hypothesis

2.5.1 Size of the Board of Directors (Control Environment)

The Board of Directors (BOD) is the place to set the development direction and policies of the bank, and is responsible for building the internal control system of the bank. Nodeh et al. (2015) demonstrated that board independence and board size have a positive effect on the profitability of Malaysian banks. In the same opinion, Isik and Ince (2016) proved that the larger the board size, the more profitable Turkish banks are. Finally, Agoraki et al. (2010), Pathan and Faff (2013) argue that smaller boards are more effective when studying US banks. For Vietnamese commercial banks, a large board size can help make decision-making more efficient due to the diverse opinions of board members. Therefore, the author expects the board size to have a positive impact on the profitability of Vietnamese commercial banks.

Hypothesis H1: Board size has a positive relationship with the profitability of commercial banks.

2.5.2 Management Experience (Risk Assessment)

With a background in banking and finance, BOD members can better prevent, detect and handle risks and have better cost control. Since then, the profitability of commercial banks has been higher. Board members of Vietnamese commercial banks need to have solid banking and finance knowledge to make accurate and reasonable decisions, thereby reducing risks and increasing profitability. Therefore, the authors put forward the following hypothesis:

Hypothesis H2: Management experience has a positive impact on the profitability of commercial banks.

2.5.3 Compliance with Credit Limit (Control Activities)

Banks that comply with credit limits will tend to be more cautious and stricter in their operations, which may reduce profitability in the short term, but the profitability will be better in the long term because less provision is required for credit risk. Thus, ensuring credit limits will help commercial banks achieve higher profitability. Therefore, the authors propose the following hypothesis:

Hypothesis H3: Compliance with credit limit has a positive impact on the profitability of commercial banks.

2.5.4 Reliability of Financial Statements (Information and Communication)

Timeliness, which reflects reliability, acts as a pillar of the relevancy of financial reporting information (Ohaka and Akani 2017). Timeliness of financial statements is the way to verify the level of transparency and quality of the reporting (Adebayo and Adebiyi 2016). Therefore, it is understandable that this fact easily affects the bank's reputation and positively impacts the bank's profitability. Therefore, the reliability of financial statements will have a positive impact on the profitability of commercial banks. On that basis, the authors propose the hypothesis:

Hypothesis H4: Reliability of financial statements has a positive effect on the profitability of commercial banks.

2.5.5 Audit Report Quality (Monitoring Activity)

Auditing firms, especially those operating globally, build their reputation on the quality of their audit reports. Therefore, banks audited by reputable auditing firms will have better performance quality and achieve higher profitability. As a result, the quality of the audit report will have a positive impact on the profitability of commercial banks. Thus, the authors propose the hypothesis:

Hypothesis H5: Audit report quality has a positive impact on the profitability of commercial banks.

2.5.6 Bank Size

Larger banks often have higher risk and lower efficiency, weaker internal control system due to operating in many fields, increasing operational complexity. However, the larger the bank size, the more closely the bank is forced to operate, and there is an advantage of scale (Schildbach 2017). For Vietnamese commercial banks, increasing size will lead to increased market share, increased credibility with customers and diversification of loan and investment portfolios, leading to higher profits. Therefore, the authors propose the hypothesis:

Hypothesis H6: Bank size and the profitability have a positive relationship.

2.5.7 Bank Age

Young banks often do not have an effective internal control system, and therefore, are more likely to report operational weakness. In contrast, banks with a long history of operation have the ability to improve the internal control system in a better and better direction. The longer a bank has been in business, the more profitable it is likely to be. Therefore, the authors propose the hypothesis:

Hypothesis H7: Bank age has a positive impact on the profitability of commercial banks.

2.5.8 Inflation

A moderate increase in annual inflation can promote sustainable and stable economic growth, increase consumption and investment demand, and thereby has a positive impact on the profitability of commercial banks. However, high inflation can help banks impose high lending rates, but there is a potential risk because high loan interest rates will burden borrowers. Therefore, non-performing loans increase and debt repayment ability will decrease, which has a negative impact on profitability of commercial banks. As a result, inflation has a negative impact on the profitability of commercial banks. On that basis, the authors propose the hypothesis:

Hypothesis H8: Inflation has a negative impact on the profitability of commercial banks.

2.5.9 GDP Growth

High GDP growth is expected to have a positive impact on the profitability of commercial banks because the solid economic development will boost consumption and investment demand, stimulate more debt. Thus, GDP growth will have a positive effect on the profitability of commercial banks. Therefore, the authors propose the hypothesis:

Hypothesis H9: GDP growth has a positive impact on the profitability of commercial banks.

3 Research Method and Model

Based on the framework of the COSO (2013) and empirical studies of Koutoupis and Malisiovas (2019), Akwaa-Sekyi and Gené (2017) and some other studies as well, the authors propose regression models as follows:

$$\begin{aligned}\text{ROA} = {} & \propto_0 + \propto_1 \text{BOD} + \propto_2 \text{FBB} + \propto_3 \text{LOAN} + \propto_4 \text{FINS} + \propto_5 \text{MONITOR} \\ & + \propto_6 \text{SIZE} + \propto_7 \text{AGE} + \propto_8 \text{INFLAT} + \propto_9 \text{GGDP} + \varepsilon\end{aligned}$$

$$\begin{aligned}\text{ROE} = {} & \beta_0 + \beta_1\text{BOD} + \beta_2\text{FBB} + \beta_3\text{LOAN} + \beta_4\text{FINS} + \beta_5\text{MONITOR} \\ & + \beta_6\text{SIZE} + \beta_7\text{AGE} + \beta_8\text{INFLAT} + \beta_9\text{GGDP} + u\end{aligned}$$

where $\propto_0$, β_0 are constant terms; $\propto$ i (i = 1, 9); β_j (j = 1, 9) are coefficients of variables; ε, u are error terms.

The descriptions of the variables are provided in Table 1.

3.1 Data Collection

We collected financial data from audited financial statements. For non-financial data, we collected from annual reports of 30 Vietnamese commercial banks in 2007–2018, combined with macro data collected from the General Statistics Office during the same period.

Table 1 Variable definitions

Variables		Notation	Previous studies	Proxy variables	Expected results
Dependent variable	Bank profitability	Return on assets (ROA)	Hoang (2020), Channar and Khan (2015), Umar and Dikko (2018)	Profit after tax/Total assets	
		Return on equity (ROE)	Channar and Khan (2015), Umar and Dikko (2018)	Profit after tax/Equity	
Independent variables	Control environment	Board Size (BOD)	Zhang et al. (2007), Koutoupis and Malisiovas (2019), Akwaa-Sekyi and Gené (2016, 2017)	Number of board members	+
	Risk assessment	Board Expertise in Finance (FBB)	Akwaa-Sekyi and Gené (2016, 2017)	Number of board members with banking or finance background as the ratio of total board members	+
	Control activities	Compliance with credit limits (LOAN)	Koutoupis and Malisiovas (2019)	Loans to total assets ratio	+
	Information and communication systems	Reliability of financial statements (FINS)	Koutoupis and Malisiovas (2019), Zhang et al. (2007)	The number of days between the year-end and the publication of the financial statements	+
	Monitoring	Audit quality (MONITOR)	Zhang et al. (2007), Tang et al. (2014)	Use of top four auditing firms	+

(continued)

Table 1 (continued)

Variables		Notation	Previous studies	Proxy variables	Expected results
		Bank size (SIZE)	Akwaa-Sekyi and Gené (2016, 2017), Zhang et al. (2007), Tang et al. (2014), Koutoupis and Malisiovas (2019)	Logarithm of bank total assets	+
		Bank age (AGE)	Tang et al. (2014), Koutoupis and Malisiovas (2019), Akwaa-Sekyi and Gené (2016, 2017)	Number of years in business	+
	Macroeconomic factors	Inflation (INFLAT)	Sufian (2011), Alexiou and Sofoklis (2009)	The General Statistics Office	−
		GDP growth rate (GGDP)	Kohlscheen et al. (2018)		−

3.2 Methodology

In recent years, scientists are increasingly realizing the disadvantages of the frequency statistical method (Frequentist) because this method leads to many false conclusions in scientific research (Nguyen 2011; Anh et al. 2018; Hung et al. 2019). Scientific conclusions in regular frequency statistics based on the dataset without regard to the known information (Nguyen 2019). In frequency statistics, the parameters of the population are treated as constants that are fixed but unknown. But for time series data, these parameters will change, so it is no longer appropriate to assume the parameters are constant. In Bayesian statistics, the parameters are assumed as random variable. From conditional probability:

$$p(A|B) = \frac{p(A, B)}{p(B)}$$

we have Bayes' theorem:

$$p(A|B) = \frac{p(B|A)p(A)}{p(B)}$$

where A, B are random vectors.

Since most previous research was performed using a frequency approach, prior information is not available. However, with a sample of 30 banks in Vietnam and a period of 12 years, the number of observations is relatively large, so prior information does not affect the posterior distribution too much. The prior distribution is chosen as the normal distribution for the regression coefficients of the observed variables because the normal distribution is common and is consistent with the distribution of the financial data sets used in the study; Igamma distribution for the variances because the financial data have many outliers and are not stable. Referring to Nguyen (2020), we propose prior information for the research model as follows:

$$\alpha \sim \text{N}(0;\ 100)$$

$$\sigma^2 \sim \text{Invgamma}(0.01;\ 0.01)$$

4 Results and Discussion

The dependent variables used in the research model are ROA and ROE. ROA is an indicator that measures the ability of commercial banks to manage and use financial resources to generate profits. ROA shows how much profit each dollar of the asset can generate, most clearly reflecting a bank's management's ability to use financial and investment resources to generate profits (Kosmidou 2008). ROE reflects management efficiency from equity. That is, it measures the efficiency of equity investment. Total assets are taken from the financial statements and profit after tax is taken from the income statement.

The results of Tables 2 and 3 show that the Monte Carlo standard errors (MCSE) of the parameters are very small. Also, the model's acceptance rate is 1, higher than the required level of 0.15. Therefore, the Metropolis–Hastings algorithm is efficient (Roberts and Rosenthal 2001).

To ensure that Bayesian inference based on Markov Chain Monte Carlo (MCMC) sample simulation is reasonable, the authors use the convergence diagnostic of MCMC chains for variables ROA and ROE. When MCMC chains converge, Bayesian inference is robust. Testing the convergence of MCMC chains is done through trace plots, histograms, and autocorrelation.

The test results from Figs. 1 and 2 show that the histograms run fast through the distribution, the autocorrelation plots are acceptable, the shape of the histograms simulates the shape of the probability of distributions. From this, we can conclude that the Bayesian inference is robust.

The results from Tables 2 and 3 show that Board size (Control environment) has a positive effect on ROA and ROE, similar to the research hypothesis and studies of Nodeh et al. (2015), Isik and Ince (2016). This result can be explained that the

Table 2 Summary of regression results for the dependent variable ROA

Variables	Mean coefficient	Std.dev	MCSE	Median
BOD	0.0002	0.0004	0.0000	0.0002
FBB	–0.0039	0.0046	0.0000	–0.0039
LOAN	–0.0037	0.0059	0.0000	–0.0038
FINS	–0.0001	0.0000	0.0000	0.0000
MONITOR	0.0028	0.0018	0.0000	0.0028
SIZE	–0.0006	0.0008	0.0001	–0.0006
AGE	0.0001	0.0000	0.0000	0.0002
INFLAT	0.0389	0.0121	0.0001	0.0389
GGDP	0.0578	0.0861	0.0008	0.0584
_cons	0.0263	0.0253	0.0002	0.0263
var	0.0001	0.0000	0.0000	0.0000

Source Authors' calculation

Table 3 Summary of regression results for the dependent variable ROE

Variables	Mean coefficient	Std.dev	MCSE	Median
BOD	0.0025	0.0021	0.0000	0.0025
FBB	–0.0220	0.0241	0.0002	–0.0219
LOAN	–0.0455	0.0304	0.0003	–0.0455
FINS	–0.0005	0.0001	0.0000	–0.0005
MONITOR	0.0135	0.0096	0.0000	0.0134
SIZE	0.0212	0.0041	0.0000	0.0213
AGE	–0.0001	0.0003	0.0000	–0.0001
INFLAT	0.3992	0.0613	0.0006	0.3986
GGDP	1.3876	0.4484	0.0084	1.3875
_cons	–0.6555	0.1322	0.0013	–0.6571
var	0.0028	0.0002	0.0000	0.0028

Source Authors' calculation

larger the board size, the higher the diversity, thereby reducing risks and increasing the profitability of commercial banks.

Management experience (Risk Assessment) has a negative effect on ROA and ROE and is contrary to the research hypothesis. This result can be explained that the board of directors with a background in finance-banking tend to come up with more risky business strategies to seek higher profits. However, in the long term, riskier activities can result in higher NPLs, requiring banks to make more provisions for credit losses and leading to lower profitability.

Compliance with credit limit (Control activities) has a positive effect on bank profitability, which means that the more banks lend, the lower the profitability

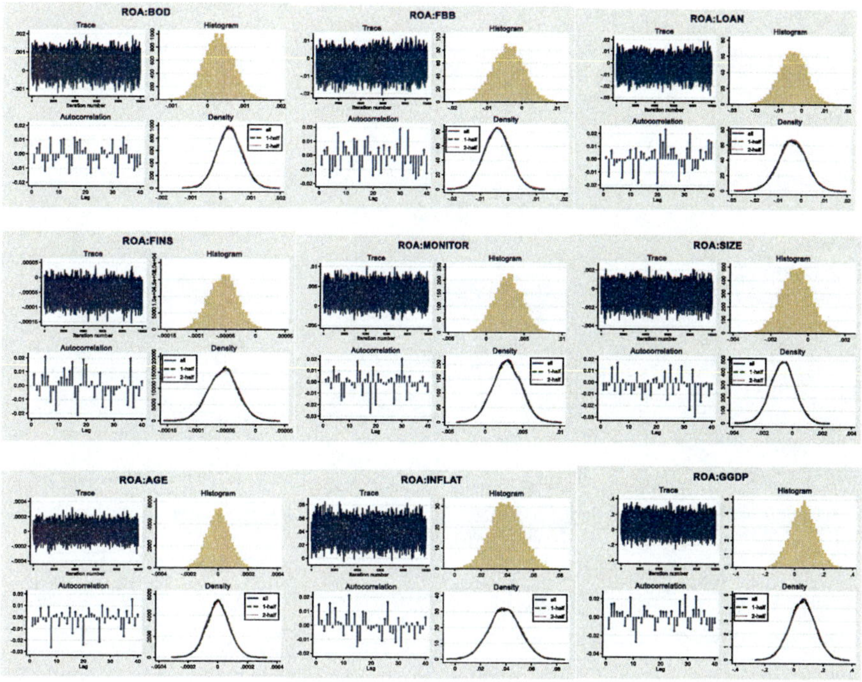

Fig. 1 Graphical diagnostics for MCMC convergence of ROA

ratio. Stricter lending by banks to ensure credit limit can lead to better profitability. This result is consistent with the research hypothesis but contrary to the studies of Koutoupis and Malisiovas (2019). When banks better control risks, strictly comply with credit-granting processes and procedures, and carefully select customers, the profitability of commercial banks will increase.

The fewer days from the end of the financial year to the signing of the audit report, the better the profitability of commercial banks. In other words, the reliability of financial statements (Information and communication) has a positive impact on the profitability of commercial banks, consistent with the hypothesis but contrary to the research results of Koutoupis and Malisiovas (2019). Banks provide quick and timely financial statements that reflect the stability of the bank's operations, improve the bank's reputation and position, and help the bank achieve higher profitability.

Audit report quality (Monitoring activities) positively impacts on bank profitability, consistent with the research hypothesis. When banks are audited by the Big Four (the top four auditing firms), those banks are forced to closely examine the evaluation of their loan portfolios and their venture capital activities (Hodgdon and Porter 2017). Moreover, the contribution of Big Four audit firms in improving the quality of internal control is significant, and the cooperation of the audit committee with the quality system of internal control is more noticeable when the audit takes place from a Big Four audit firm (Khlif and Samaha 2016).

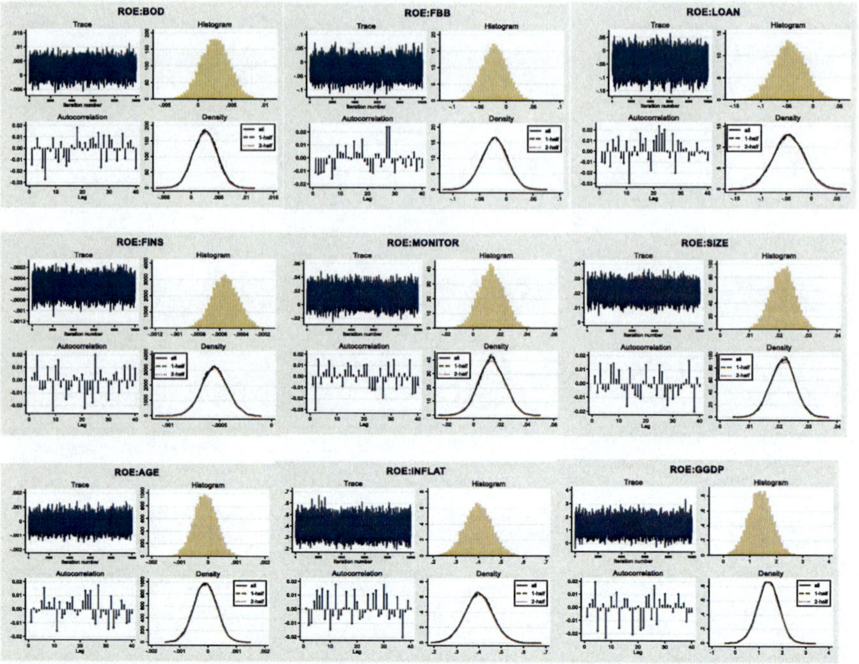

Fig. 2 Graphical diagnostics for MCMC convergence of ROE

In addition, control variables such as bank size and the number of years of operation have different effects on ROA and ROE. For macroeconomic factors, inflation and GDP growth all have a positive impact on the profitability of commercial banks. When the economy has a high growth rate, the demand for loans increases, the repayment ability is guaranteed when the customer's business is favourable. As a result, the bank achieves higher profitability.

Thus, it can be seen that an effective internal control system is an important mechanism to manage commercial banks reasonably and effectively. The results also show that there are four components of COSO (2013) that have a positive impact on the profitability of Vietnamese commercial banks. Specifically, there are four factors that positively affect bank profitability: Control environment, control activities, information and communication, and monitoring activities. In contrast, risk assessment has a negative impact on the profitability of Vietnamese commercial banks.

5 Conclusion and Policy Implications

5.1 Conclusion

The study applies the framework of COSO (2013) and related studies to build a research model. By the regression of Bayesian method and Gibbs sampling, the research results show that there are five components of COSO (2013) that affect the profitability of Vietnamese commercial banks in the period 2007–2018. These components: Control environment, risk assessment, control activities, information and communication, monitoring activities. In addition, factors belonging to bank characteristics and macroeconomic factors affect the profitability of Vietnamese commercial banks: Bank size, number of years of operation, inflation, GDP growth.

5.2 Policy Implications

The research results provide evidence showing the impact of internal control on the profitability of Vietnamese commercial banks. Based on those results, some recommendations proposed to help Vietnamese commercial banks improve the effectiveness of the internal control system, contributing to improving business efficiency as well as profitability, specifically as follows:

First, the research results show that the board of directors members (Risk Assessment) with background knowledge in finance-banking tend to act riskier. Therefore, banks need to review their portfolios and lend in a safer way instead of maximizing profits, and reviewing the risk assessment process, ensuring long-term safety goals, compatible with the State Bank of Vietnam's orientation in the bank's business activities.

Second, review the process of control activities. The more banks lend, the lower the profit margin. Therefore, developing strategies such as credit process, risk appraisal process, etc., in a simpler but stricter direction is necessary. Raise the awareness of compliance with banking procedures of employees and management levels.

Third, it is necessary to strengthen the reliability of financial statements, build a more effective and efficient information and communication system, and ensure prompt and transparent information both internally and externally. Upgrading the internal management system towards international standards is crucial. Especially when the information and communication system works effectively, the disclosure of information to the outside, such as financial statements, will not be delayed, increasing profitability for the bank.

Fourth, the results also show that monitoring activities play an important role in affecting the profitability of commercial banks. Therefore, banks need to be audited by leading auditing firms. These firms will help banks perfect their internal control systems according to international standards, meet the goals of the international economic integration of Vietnamese commercial banks, and take advantage of the deep integration of the Vietnamese economy into the world economies and global economic unions such as joining the Free Trade Agreement between Vietnam and the European Union (EVFTA), joining the Comprehensive and Progressive Agreement for Trans-Pacific Partnership (CPTPP).

References

Adebayo, P.A., Adebiyi, W.K.: Effect of firm characteristics on the timeliness of corporate financial reporting: evidence from Nigerian deposit money banks. Int. J. Econ. Commer. Manag. **4**(3), 369–381 (2016)

Agoraki, M., Delis, M., Staikouras, P.: The effect of board size and composition on bank efficiency. Int. J. Bank. Account. Finance **2**(4), 357–386 (2010)

Akwaa-Sekyi, E.K., Gené, J.M.: Effect of internal controls on credit risk among listed Spanish banks. Intang. Cap. **12**(1), 357–389 (2016)

Akwaa-Sekyi, E.K., Gené, J.M.: Internal controls and credit risk relationship among banks in Europe. Intang. Cap. **13**(1), 25–50 (2017)

Alexiou, C., Sofoklis, V.: Determinants of bank profitability: evidence from the Greek banking sector. Econ. Ann. **182**, 93–118 (2009)

Anh, L.H., Le, S.D., Kreinovich, V., Thach, N.N.: Econometrics for financial applications. Springer, Cham (2018). https://doi.org/10.1007/978-3-319-73150-6(2018)

Arad, H., Jamshedy-Navid, B.: A clear look at Internal Controls : Theory and concepts. Social Science Research Network, (March) (2010). Retrieved from http://ssrn.com/abstract=1342048

Basel Committee on Banking Supervision. Framework for Internal Control Systems in Banking Organization (1998). https://www.bis.org/publ/bcbs40.pdf. Accessed 5 Feb 2019

Batten, J., Vo, X.V.: Determinants of bank profitability—evidence from Vietnam. Emerg. Mark. Financ. Trade **55**(1), 1–12 (2019)

Committee of Sponsoring Organisations of the Treadway commission (COSO): Internal Control-Integrated Framework. AICPA, New York (1992)

Channar, J.A., Khan, M.: Internal control effectiveness & its relationship with financial performance. J. Bus. Stud. **11**(2), 92–117 (2015)

Chu, T.T.T.: Organization of internal control costs with improving financial efficiency in small and medium enterprises in Vietnam. Ph.D. thesis in economics, National Economics University, Hanoi (2016)

Committee of Sponsoring Organisations of the Treadway commission (COSO): Internal Control-Intergrated Framework, Committee of Sponsoring Orgarnisations of the Treadway Commission. Coopers and Librand, New York (2004)

Committee of Sponsoring Organisations of the Treadway commission (COSO): Internal Control-Integrated Framework. AICPA, New York (2013)

Jensen, M., Meckling, W.: Theory of the firm: managerial behavior, agency costs, and ownership structure. J. Financ. Econ. **3**(4), 305–360 (1976)

Hanoon, R.N., Khalid, A.A., Rapani, A., Aljajawy, T., Al-Waeli, A.: The impact of internal control components on the financial performance, in the Iraqi banking sector. J. Contemp. Issues Bus. Gov. **27**(3), 2517–2529 (2021)

Hamour, A.M., Massadeh, D., Bshayreh, M.: The impact of the COSO control components on the financial performance in the Jordanian banks and the moderating effect of board independence. J. Sustain. Financ. Invest. (2021). https://doi.org/10.1080/20430795.2021.1886553

Hung T. Nguyen, Trung N.D., Thach N.N.: Beyond traditional probabilistic methods in econometrics. In: Kreinovich, V., Thach, N., Trung, N., Van Thanh, D. (eds.) Beyond Traditional Probabilistic Methods in Economics. ECONVN 2019. Studies in Computational Intelligence, vol. 809. Springer, Cham (2019). https://doi.org/10.1007/978-3-030-04200-4_1

Hoang, T. H.: Research model of the impact of internal control on operational efficiency and risk at Vietnamese commercial banks. J. Ind. Trade **12** (2020)

Hodgdon, C., Porter, R.: Auditor choice and the consistency of bank accounting: are some auditors stricter than others when assessing the value of a bank's loan portfolio? J. Account. Financ. **17**(3), 40–54 (2017)

Isik, O., Ince, A.R.: Board size, board composition and performance: an investigation on Turkish banks. Int. Bus. Res. **9**(2), 74–84 (2016)

Khlif, H., Samaha, K.: Audit committee activity and internal control quality in Egypt: does external auditor's size matter? Manag. Audit. J. **31**(3), 269–289 (2016)

Kohlscheen, E., Murcia, A., Contreras, J.: Determinants of bank profitability in emerging markets. BIS Working Paper No. 686 (2018)

Koutoupis, A., Malisiovas, T.: The Effects of Internal Control Systems on Risk, Profitability and Compliance of the US Banking Sector: A Quantitative Approach (2019). SSRN: https://ssrn.com/abstract=3435626

Kosmidou, K.: The determinants of banks' profits in Greece during the period of EU financial integration. Manag. Financ. **34**(3), 146–159 (2008)

Kreinovich, V., Thach, N.N., Trung, N.D., Thanh, D.V. (eds.): Beyond Traditional Probabilistic Methods in Economics. Cham, Springer. https://doi.org/10.1007/978-3-030-04200-4 (2019)

Nodeh, F., Anuar, M., Ramakrishnan, S., Raftnia, A.: The effect of board structure on bank's financial performance by moderating firm size. Mediterr. J. Soc. Sci. **7**(1), 258–263 (2015)

Ngo, T.T.: Building internal control system with strengthening financial management at Vietnam Post and Telecommunications Corporation. National Economics University, Hanoi, Ministry-Level Scientific Research Project (2004)

Nguyen, T., Duong, N. H.: A theoretical model studying the impact of internal control on performance and risks of Vietnam commercial banks. In: International Conference on Accounting. ICOA, Da Nang (2015)

Nguyen, N.T.: How to explain when the ES is lower than one? A Bayesian nonlinear mixed-effects approach. J. Risk Financ. Manag. **13**(2), 1–17 (2020)

Nguyen, V.T.: An introduction to bayes. J. Med. Rev. **63**, 26–34 (2011)

Nguyen, N.T.: A Bayesian approach in the predictions of The US's gross domestic product. Asian J. Econ. Bank. **163**, 5–19 (2019)

Ohaka, J., Akani, F.: Timeliness and relevance of financial reporting in Nigerian quoted firms. Manag. Organ. Stud. **4**(2), 55–62 (2017)

Pathan, S., Faff, R.: Does board structure in banks really affect their performance? J. Bank. Finance **37**(5), 1573–1589 (2013)

Pham, T.T.: Positioning Vietnam's commercial banking system in the community of CPTPP countries. J. Monet. Financ. Mark. **2019**(13/2019) (2019)

Roberts, G.O., Rosenthal, J.S.: Optimal scaling for various Metropolis-Hastings algorithms. Stat. Sci. **16**, 351–367 (2001)

Schildbach, J.: Large or small? How to measure bank size. DBR.online (April 25) (2017). https://www.dbresearch.com/PROD/RPS_EN

Sufian, F.: Profitability of the Korean banking sector: panel evidence on banking-specific and macroeconomic determinants. J. Econ. Manag. **7**, 43–72 (2011)

Tang, D., Tian, F., Yan, H.: Internal control quality and credit default swap spreads. Account. Horiz. **29**(3), 603–629 (2014)

The Ministry of Finance of Vietnam. Auditing Standard No. 315 (2012)

Umar, H., Dikko, M.U.: The effect of internal control on performance of commercial banks in Nigeria. Int. J. Manag. Res. Rev. **8**(6), 13–32 (2018)

Vu, T.P.: Impact of internal control on the performance of Vietnam Electricity Corporation. Ph.D. thesis in economics, University of Economics, Ho Chi Minh City (2016)

Zhang, Y., Zhou, J., Zhou, N.: Audit committee quality, auditor independence, and internal control weaknesses. J. Account. Public Policy **26**(3), 300–327 (2007)

Determinants of Labor Productivity of Small and Medium-Sized Enterprises (SMEs) with Bayesian Approach: A Case Study in the Trade and Service Sector in Vietnam

Huong Thi Thanh Tran

Abstract In every economy, SMEs always play an important role because they can create jobs for the majority of employees. Improving the productivity of SMEs will play a key role in enhancing the labor productivity of the whole economy. To evaluate the impact of factors on labor productivity of SMEs, this study used the Bayesian Model Averaging (BMA) with data set of 14,768 micro, small and medium-sized enterprises in the trade and service sector obtained from the 2017 Enterprise Survey conducted by the General Statistics Office of Vietnam. The research results found 5 factors that had a positive influence on the labor productivity of SMEs in the trade and service sector of Vietnam, including capital productivity, average income per employee, equipped fixed assets per employee, investment capital-to-labor ratio, and profit.

Keywords Labor productivity · Small and medium-sized enterprises · Bayesian model averaging

1 Introduction

In the context of increasingly deep and wide international integration, labor productivity is one of the important factors contributing to improving the competitiveness of each enterprise in particular and each country in general. Vietnam is a country with low per capita income; therefore, if the traditional competitive advantages such as low cost of raw materials, cheap labor, etc. gradually disappear, the problem of low labor productivity will be a major barrier to foreign direct investment attraction and international integration. Consequently, increasing labor productivity of the whole economy in general and the enterprises in the economy, in particular, is one of the important factors contributing to improving competitiveness for Vietnam in the current period.

H. T. T. Tran (✉)
Faculty of Accounting and Auditing, Banking Academy Vietnam, Hanoi, Vietnam
e-mail: huongttt76@hvnh.edu.vn

N. Ngoc Thach et al. (eds.), *Financial Econometrics: Bayesian Analysis, Quantum Uncertainty, and Related Topics*, Studies in Systems, Decision and Control 427,
https://doi.org/10.1007/978-3-030-98689-6_15

Both theoretical and empirical researches have shown the extremely important role of SMEs in socio-economic development in countries around the world, especially in the developing economies in the context of globalization like Vietnam (Mansour et al. 2018; Agussalim et al. 2019; Al-Haddad et al. 2019). Because SMEs are small in size, they are easy to be adjusted and changed to suit the needs of the economy. Therefore, SMEs are a factor in creating the dynamism of the economy. However, due to the small scale of capital as well as many other limitations, SMEs often face a lot of difficulties in the process of improving and enhancing labor productivity. Thus, finding the factors promoting SMEs' labor productivity growth becomes extremely imperative for the economic growth of developing countries in general and Vietnam in particular in the short term and long-term.

In addition, most of the researches on determinants of labor productivity of small and medium-sized enterprises by Pham and Nguyen (2017), Cin et al. (2017), Calza et al. (2019), etc. have used estimation methods based on P-value. However, some recent researches have shown that traditional estimation methods have three serious problems when using p-value to evaluate research results. First, the p-value has not accurately measured the probability of the hypothesis when no regard to the probability that the sample was selected. Second, mostly, the logic in hypothesis construction is valuable for binary logics, and little valuable for multivalued logics. Meanwhile, because the majority of economic events happen in accordance with multivalued logic, the p-value will be less accurate. And the final problem is that when the p-value is small, the rejection of the null hypothesis H_0 is not completely convincing in both theoretical and practical manners (Nguyen 2019). Halsey (2019) also showed that the P-value provided quite limited information and was easily misunderstood. According to Halsey (2019), there were four statistical approaches that could extend or replace the P-value. One of the simple approaches was to use the Akaike information criterion (AIC) to replace the P-value. "The AIC approach encourages you to think hard about alternative models and thus hypotheses, in contrast to p-value interpretation that encourages rejecting the null when p is small, and supporting the alternative hypothesis by default. More broadly, the AIC paradigm involves dropping hypotheses judged implausible, refining remaining hypotheses and adding new hypotheses" (Halsey 2019, p. 6). Halsey (2019) also noted that "a key limitation of the AIC is that it provides a relative, not absolute, test of model quality. It is easy to fall into the trap of assuming that the best model is also a good model for your data" (p. 6). Another approach is to use the Bayesian Model Averaging (BMA) method. The BMA method is based on the statistical principle of the Bayesian statistics, and each model has a prior probability, along with the actual data, we can find out the independent variables affecting the dependent variables. Each BMA model reports the regression coefficient of the independent variable, the coefficient of determination R^2, the Bayesian Information Criterion (BIC) value, and the posterior probability. Like AIC, the model with the lowest BIC value is deemed to be the most optimal model.

To evaluate the impact of factors on labor productivity of SMEs, this study used the Bayesian Model Averaging (BMA) with data set of 14,768 micro, small and medium-sized enterprises in the trade and service sector obtained from the 2017 Enterprise

Survey conducted by the General Statistics Office of Vietnam. The research results found 5 factors that had a positive influence on the labor productivity of SMEs in the trade and service sector of Vietnam, including capital productivity (*CPR*), average income per employee (*AIE*), equipped fixed assets per employee (*EAE*), investment capital-to-labor ratio (*IC*), and profit (*PRO*).

The structure of this research paper, in addition to the introduction, includes literature reviews, research model and data source, results, discussion, and conclusion.

2 Literature Reviews

2.1 Productivity

Productivity is defined as a ratio between the output volume and the input volume (OECD 2001). In reality, there are two main options for measuring productivity, consisting of partial factor productivity (PFP) and total factor productivity (TFP). PFP is defined as the ratio between the output and a specific factor of input (capital or labor). Depending on the selected factor of input, labor productivity or capital productivity can be measured. TFP is defined as the ratio of an index of produced output to an index of aggregate inputs. Two main methods are chosen to measure TFP: the non-parametric approach (TFP index, inclusive data analysis) and the parametric approach (estimation of production functions, analysis of random variables). Estimation of production functions is performed via some available techniques such as the OLS estimation by Olley and Pakes (1996) and the approach of Levinsohn and Petrin (2003).

2.2 Factors Affecting Labor Productivity of SMEs

Labor quality is one of the important factors determining the growth of labor productivity. The more science and technology develops, the more modern the machinery and equipment are, the more relevant professional qualifications of employees are required. Employees with high education level will be able to quickly absorb scientific and technological advances into production, which creates the highest working efficiency. The positive impact indicates that enterprises with a high proportion of employees with high education level will have high labor productivity (Firouzi et al. 2010). Investment in human capital has an influence on business activities of enterprises via output, productivity, profit, and competitiveness (Black and Lynch 1997; Honig 2001; Blundell et al. 1999; Barron et al. 1989; Blackemore and Hoffman 1988). The research by Lynch and Sandra (1995) shows that there is a positive relationship between employees' years of schooling and productivity and the impact of training is highly dependent on training programs. Increasing productivity reflects an improvement in input efficiency. Therefore, the same level of inputs can result in a higher level of output and lower production costs. The positive relationship between human capital and productivity is greatly affected by salary. When employees receive a higher salary, they will be encouraged to work harder and contribute to higher productivity. Employees with higher education levels and training skills tend to receive higher salary, and they are more likely to contribute to career development, and further human capital accumulation (Blundell et al. 1999; Montague 1986), thereby contributing to labor productivity growth. Therefore, to achieve this stimulating effect, enterprises need a lot of highly educated employees. The research by Naoki (2011) also shows a positive impact of labor quality on labor productivity. Nguyen and Nguyen (2016) conducted a survey of 2000 enterprises in 7 economic sectors within 3 groups: low technology, medium technology, and high technology of Vietnam. The survey results show that if employees are highly skilled, productivity will be high. According to Cin et al. (2017), expenditures on employee education and training have a positive impact on the labor productivity of SMEs in Korea.

Labor productivity and salary have a reciprocal and positive relationship. Salary is an important factor contributing to motivating employees to improve labor productivity. In other words, for employees, salary is the main income; therefore, they must increase labor productivity to increase their salary. The research by Firouz et al. (2010), and the one by Nguyen (2016) showed that salary had a positive impact on labor productivity. Tran (2019) indicated that the average monthly income per employee of enterprise had a positive impact on Vietnam's labor productivity growth.. Pham and Nguyen (2017) used the spatial cross-regressive method with a dataset of 1943 small and medium-sized enterprises in 9 provinces and cities of Vietnam in order to investigate the determinants of labor productivity. The research results showed that the average income per employee was an important determinant that had the most impact on labor productivity of enterprises in all fields, and was the main motivation leading to the difference in labor productivity in different fields. Basically, the average income per employee (AIE) is more comprehensive than the average salary per employee because AIE measures the value that the employee receives, including salary, welfares, and benefits of the employee.

To improve labor productivity, enterprises themselves must have a source of capital large enough to invest in purchasing modern machinery, equipment, and technological lines. Limited capital size and outdated production technology are the factors that hinder labor productivity growth. The use of investment capital with high or low efficiency also has a significant impact on the growth rate of facilities of each sector and the whole economy, thereby affecting the growth of labor productivity. Papadogonas and Voulgaris (2005), Naoki (2011) stated that capital increase would boost the growth of labor productivity. Naoki (2011) used array data on Japanese enterprises in the period of 1977–2008 to demonstrate the impact of capital on labor productivity. According to Papadogonas and Voulgaris (2005), the capital-to-labor ratio and the average fixed asset per employee are the decisive factors to labor productivity, which implies that enterprises should invest in human capital for higher productivity. Rahmas (2009) conducted a survey of 574 enterprises (264 manufacturing enterprises, 310 service enterprises) of Malaysia in

2001–2002 to evaluate the human capital's impact on output and labor productivity of enterprises. Research results showed that the capital-to-labor ratio was the decisive factor in the labor productivity of manufacturing enterprises. However, for service enterprises, the number of trained employees reduced the rate of increase in labor productivity. This arose from the lack of employees that enterprises must face when sending their employees to join the training course. Research results of Cin et al. (2017) also confirmed that capital intensity had a positive relationship with labor productivity. Increasing capital intensity would increase the labor productivity of small and medium manufacturing enterprises of Korea. According to Nguyen (2016), investment capital had a positive impact on the labor productivity of enterprises in Vietnam. Nguyen and Nguyen (2016) indicated that the enterprises implementing R&D projects and having investment capital for R&D activities had higher labor productivity than the ones that did not pay attention to this investment activity.

Applying new and modern science and technology is one of the important factors contributing to increasing labor productivity. Scientific and technological advances not only speed up the development of production and business establishments but also are a decisive factor to improve the efficiency of resource exploitation and use, thereby increasing labor productivity. Application of scientific and technological advances will require improving the skills of employees, at the same time improving the level of organization, arrangement, and management, thereby improving labor productivity. To improve productivity, quality, and efficiency of production and business, enterprises need to increase capital investment in science and technology development. Using data from the SMEs survey in Italy from 1995 to 2003 by Hall et al. (2009) showed that research intensity, R&D costs had a positive impact on labor productivity. Cin et al. (2017) also indicated that the positive impact of Government R&D subsidies on R&D costs increased the labor productivity of SMEs. Through using two sets of data (surveying 192 SMEs accounting for 51% of 376 manufacturing enterprises, and balance sheet of 192 interviewed enterprises in the period of 1998–2004). Antonioli et al. (2010) showed that innovation activities were the important motivation of labor productivity growth. Innovation activities ranked based on the impact on labor productivity include training innovation, technological innovation, organizational innovation, innovation of information and communication technology (ICT). Baumann and Kritikos (2016) analyzed the relationship between R&D and innovation to labor productivity in micro-enterprises (with fewer than 10 employees) and SMEs (in the industry of Germany) with a particular focus on micro- enterprises with fewer than 10 employees. The research results showed that there was a positive impact of innovation and R&D on labor productivity, and generally, there was not much difference in the impact of innovation and R&D on labor productivity between small enterprises and large ones. Cin et al. (2017) surveyed the impact of R&D policies on promoting SMEs' operational efficiency in manufacturing enterprises in Korea. The research results proved that the positive impact of public R&D subsidies and enterprises' R&D expenditure increased labor productivity of SMEs of Korea. Government subsidies indirectly increased labor productivity by stimulating the private sector to invest in R&D, thereby increasing labor productivity in SMEs. Nguyen and Nguyen (2016) stated that the enterprises implementing R&D projects and having investment capital for the development of research activities had higher labor productivity than the ones that did not pay attention to this investment activity; The more information technology enterprises applied, the higher productivity was. Calza et al. (2018), using data from the SMEs survey in 10 provinces of Vietnam in 2013, 2014, 2015, studied the factors promoting labor productivity of SMEs in Vietnam. The research results showed that technological innovation (for product and process), improvement of management method and business organization had a positive impact on the labor productivity of SMEs.

3 Research Model and Data Source

3.1 Research Model

To quantify the impact of factors on labor productivity of SMEs in Vietnam, based on the literature reviews and collected data sources, the author built the model:

$$LPR_i = \alpha_0 + \alpha_1 CPR_i + \alpha_2 AIE_i + \alpha_3 EAE_i + \alpha_4 IC_i + \alpha_5 PRO_i + u_i$$

where: i is an enterprise; α is the regression coefficient; u_i is the random error; *LPR* is the labor productivity; *CPR* is the capital productivity, *AIE* is the average income per employee; *EAE* is the equipped fixed assets per employee; *IC* is the investment capital-to-labor ratio; PRO is the profit (Table 1).

To choose the best model, the author selected the Bayesian approach with the Bayesian Model Averaging (BMA) method. The BMA method is built on the principles of Bayesian statistics. The BMA method is based on the Bayesian Information Criterion (BIC) to select an optimal model. This is a new method that overcomes the problem of redundancy of variables (variables have no real impact) in the regression model. According to the BMA method, each model has a prior probability, together with the actual data collected, we will find the best variable to explain the variation of the dependent variable. Each BMA model will report the regression coefficients of independent variables (prognostic variables), coefficients of determination (R^2), BIC values and posterior probability. The most optimal model is the one with the smallest BIC value and the independent variables must be statistically significant. BIC is determined by the following formula:

$$\mathrm{BIC} = \mathrm{k} * \ln(\mathrm{n}) - 2 * \mathrm{LogLikelihood}$$

in which: n is the number of observations; k is the number of the estimated coefficients; LogLikelihood is determined by ln(Likelihood).

The Likelihood function measures the fitness of a statistical model to a sample of data for given values of unknown parameters. It is formed from the joint probability

Table 1 Measurement of variables used in the model

Name of variable	Abbreviation	Unit	Measurement
Labor productivity	LPR	Million VND/ person	Gross output/average number of employees per year
Capital productivity	CPR	Million VND/ Million VND	Gross output/average investment capital per year
Average income per employee	AIE	Million VND/ person	Total expenses paid to employees/average number of employees per year
Equipped fixed assets per employee	EAE	Million VND/ person	Average fixed assets per year/average number of employees per year
Investment capital-to-labor ratio	IC	Million VND/ person	Average investment capital per year/average number of employees per year
Profit	PRO	Million VND	Enterprise's profit

distribution of the sample, but only viewed and used as a function of parameters, thus treating the random variables as fixed at the observed values. The procedure for obtaining the arguments of the maximum of the likelihood function is called the maximum likelihood estimation, usually calculated by using the natural logarithm of the likelihood, known as the log-likelihood function.

For linear regression, with the Gauss assumption, $-2 *$ (loglikelihood) is proportional to n $*$ log(SSE/n). Therefore, the smaller the sum of squared errors (SSE) explained by the model is, the larger the BIC is.

3.2 Data Source

Data source to calculate the indicators, including labor productivity (*LPR*), capital productivity (*CPR*), average income per employee (*AIE*), equipped fixed assets per employee (*EAE*), profit (*PRO*), investment capital-to-labor ratio (*IC*) in SMEs was collected from the Enterprise Survey conducted by the General Statistics Office of Vietnam in 2017.

In Vietnam, the criteria for determining SMEs were issued following Article 6, Decree No. 39/2018/ND-CP dated March 11, 2018, of the Government of Vietnam. Accordingly, SMEs are classified by size, including micro-enterprises, small enterprises, and medium enterprises. The determination of SMEs is based on the field of economic activity and 3 groups of criteria on labor, capital, and revenue. On the basis of the criteria for determining SMEs, SMEs in the trade and service sector can be divided into three groups as follows: (1) Micro enterprises in the trade and service sector are the enterprises that meet the following criteria: the average number of employees participating in social insurance per year does not exceed 10 people and the total annual revenue does not exceed 10 billion VND or the total capital does not exceed 3 billion VND; (2) Small enterprises in the trade and service sector are the enterprises that satisfies the following criteria: the average number of employees participating in social insurance per year does not exceed 50 people and the total annual revenue does not exceed 100 billion VND or the total capital is not more than 50 billion VND, but they are not micro enterprises as prescribed by laws; (3) Medium enterprises in the trade and service sector are the enterprises satisfying the following criteria: the average number of employees participating in social insurance per year does not exceed 100 people and the total annual revenue is not more than 300 billion VND or the total capital is not more than 100 billion VND, but they are not micro enterprises or small enterprises in accordance with the provisions of Vietnam laws.

4 Results and Discussion

4.1 Results

To quantify the factors' impact on labor productivity of SMEs in the trade and service sector, the author used the data of 14,768 micro, small, and medium-sized enterprises collected from the Enterprise Survey conducted by the General Statistics Office of Vietnam in 2017.

Descriptive statistics of the variables used in the research model are specifically shown in Table 2.

To select the optimal model in analyzing the impact of factors on labor productivity of SMEs in the trade and service sector, the author used the BMA method. The results of Table 3 showed that with the BMA method, we could obtain two optimal models (Model 1 and Model 2) in assessing the determinants' impact on the labor productivity of SMEs in the trade and service sector of Vietnam. Model 1 and model

Table 2 Descriptive statistics

Variable	Obs	Mean	Std. Dev	Min	Max
LPR	14,768	139.9808	341.9807	0.0368421	16,398.63
CPR	14,768	0.2809282	0.4909045	9.62e-06	18.56647
AIE	14,768	19.2653	31.40288	−23.625	3000
EAE	14,768	301.6254	1907.728	0.0052632	159,193.5
IC	14,768	999.8474	1876.565	0.0631579	47,947.71
PRO	14,768	108.0603	3039.249	−108,126	264,075

Source Author's data processing

Table 3 Selection of the optimal model by the BMA method

	p! = 0	EV	SD	Model 1	Model 2
Intercept	100.0	−4.304151	3.484071	−4.609e + 00	−3.975e + 00
CPR	100.0	191.835482	5.152551	1.919e + 02	1.918e + 02
AIE	100.0	1.139758	0.079876	1.142e + 00	1.137e + 00
EAE	51.9	0.002151	0.002278	4.144e-03	...
IC	100.0	0.066091	0.001403	6.573e-02	6.648e-02
PRO	100.0	0.015786	0.000817	1.578e-02	1.579e-02
nVar				5	4
r2				0.228	0.227
BIC				−3.773e + 03	−3.773e + 03
post prob				0.519	0.481

2 models were selected
Best 2 models (cumulative posterior probability = 1)
Source Author's data processing

2 had the same low BIC (−3.773e + 03); however, Model 1 had a higher posterior probability (0.519), Model 2 had a lower posterior probability (0.481); therefore, the author selected Model 1 to assess the determinants' impact on the labor productivity of SMEs in the trade and service sector of Vietnam. The results of model 1 indicated 5 factors, including capital productivity (*CPR*), average income per employee (*AIE*), equipped fixed assets per employee (*EAE*), investment capital-to-labor ratio (*IC*), profit (*PRO*), had a positive impact on the labor productivity of SMEs in the trade and service sector with the posterior probability of 51.9%. In which, the probability of occurrence of the variables *CPR, AIE, IC,* and *PRO* in the models studying the factors' impact on labor productivity was 100% and that of the *EAE* variable was 51.9%.

The data in Table 4 showed that the variable having the highest explanation proportion of the variation in labor productivity of SMEs in the trade and service sector was the investment capital-to-labor ratio (*IC*), accounting for 11.57%; followed by the capital productivity (*CPR*), accounting for 6.57%; the profit (*PRO*) accounting for 2.28%; the average income per employee (*AIE*) accounting for 2.04%; and the

Table 4 Partial correlation coefficient

Variable	R^2
CPR	0.06573988
AIE	0.02035899
EAE	0.00338678
IC	0.11568124
PRO	0.02280607

Proportion of variance explained by model: 22.8%
Source Author's data processing

equipped fixed assets per employee (*EAE*) had the lowest explanation proportion of the variation in labor productivity of SMEs in the trade and service sector, accounting for 0.34%.

4.2 Discussion

The majority of Vietnamese enterprises are SMEs that have limited capital resources, limited technological investment capacity, weak experience in production management, and poor competitiveness. In fact, the size of Vietnamese enterprises is too small; the number of micro, small and medium-sized enterprises accounts for about 98% of the total enterprises of the whole economy. SMEs in Vietnam have a very low level of science, technology, and innovation; most SMEs have been using the technology which is 2–3 generations more outdated than the average level of the world. According to the Global Competitiveness Report of the World Economic Forum, Vietnam ranked 67th out of 141 economies in 2019 while the component index of innovation capacity was only ranked 76th. It is showed that Vietnam needs to continue to create a favorable environment, new institutions, and policies for enterprises to promote the process of technology enhancement and innovation.

The limited quality of human resources is also one of the reasons hindering the growth of labor productivity of Vietnamese enterprises. This is clearly shown in the low proportion of trained employees, inadequate training structure, lack of highly-skilled workers, and large gap between vocational education and labor market demand. According to the report of the General Statistics Office of Vietnam, the rate of trained employees aged 15 and over who have worked in the economy has gradually increased over the years (it was 21.8% in 2017, 22.2% in 2018, and 24.1% in 2020). However, this rate is still quite low compared to one of other countries in the same region. This is a major barrier to improving labor productivity. Next time, SMEs should focus on training employees, building business strategies in parallel with human resource development strategies. Besides, unemployment among young employees and the mismatch between job position and training qualification is quite common. A large number of employees have not been trained in industrial labor discipline; they also lack teamwork skills and knowledge as well as the ability to cooperate and bear risks and hesitate to promote initiatives. This is a major barrier to improving the labor productivity of Vietnamese enterprises.

Low salary is also one of the reasons that reduce employees' motivation, thereby decreasing labor productivity. According to the data of the General Statistics Office of Vietnam, from 2017 to 2020, the average monthly income per employee had no significant change. If in 2017 the average monthly income per employee was 5.41 million VND/person, there was only a slight increase in 2020 (5.5 million VND/person).

In the near future, to improve the labor productivity of SMEs in Vietnam, it is necessary to continue to maintain policies to effectively use investment capital;

continue to invest in purchasing equipment and fixed assets for production modernization; increase the level of capital investment per employee and manage enterprises effectively to increase the operating profit of enterprises. Especially, SMEs in the trade and service sector need to improve salary policies and efficiently manage the expenses paid to employees.

5 Conclusion

SMEs development plays an extremely important role in the economic development of Vietnam. To promote the development of SMEs, in the coming time, Vietnam needs to focus on several policies to improve labor productivity in SMEs, specifically:

Firstly, SMEs themselves must promote their advantages and improve their competitiveness. SMEs of Vietnam have a lot of advantages in business, market scope, and customers to engage in domestic, regional and international trading activities. Especially, because Vietnam is a member of the ASEAN Economic Community, utilizing the inherent advantages of SMEs will promote the strong economic development of Vietnam in Southeast Asia.

Secondly, in the context of economic openness and extensive international integration, to improve their competitiveness, SMEs need to know how to make the most of their potentials in terms of capital, human resources, market, culture, and business experience. In particular, SMEs need transparency in their activities and financial statements to improve the quality of corporate governance and risk management.

Thirdly, in the context of trade liberalization and the Fourth Industrial Revolution, competition is getting more fierce; therefore, enterprises need technological innovation, apply innovation and creativity in product design, manufacturing process, process management, etc. However, in fact, a lot of SMEs in Vietnam can not afford to take advantage of opportunities for digital transformation and application of science and technology into manufacturing and business. Especially, in the context of the economy seriously affected by the Covid-19 pandemic, solutions of science and technology, innovation, and digital transformation play an important role in helping SMEs overcome the difficulties caused by the pandemic. Science and technology-based innovation, and administrative regulation reform are considered as the key to helping Vietnamese enterprises "survive" in the current context.

The Government of Vietnam needs to promulgate and implement breakthrough solutions, create favorable conditions for economic organizations and enterprises to access and apply technology, gradually improve their innovation capacity in their production and business activities. Enterprises should be placed at the center of innovation policies, at the same time, it is essential to create a policy environment to support and improve technology capacity for enterprises to increase labor productivity with new products and high technology. Implement preferential policies on land, tax incentives, credit, etc.… for enterprises using modern and high technology lines; encourage enterprises to invest in innovation and creativity, apply science and technology to manufacturing, deeply participate in the global value chain to increase productivity, especially have the policies to support the early application of automation, and investment in automation technology and digital infrastructure in order to actively take advantage of the opportunities of the Fourth Industrial Revolution.

Fourthly, reform the salary and wage policies in accordance with the market principle to match salary increase with labor productivity growth. Take more care of the material and spiritual life of employees in enterprises, especially employees in industrial zones and SMEs.

Fifthly, human resources are the most important resource that decides the growth of labor productivity in particular and the sustainable development of enterprises in general. To improve the quality of human resources and create a competitive advantage, next time, SMEs should concentrate on training the skills for employees to increase the number of highly-skilled employees; Renew training methods and programs with the combination of theory and practice, vocational training in the direction of modernization, etc.... Facilitate employees to study such as funding and time support, etc.... In addition, enterprises should also encourage employees to self-explore, learn and improve their skills and knowledge through online training courses, as well as create a learning environment for employees right in enterprises; Applying technology to human resource management can avoid the errors caused by human, even save time, and increase management efficiency.

References

Agussalim, M.: Paulus Insap Santosa, Darmini Roza, Rina Asmeri: the role of small and medium enterprises (SMES)) in Improving the local economic growth. Int. J. Civ. Eng. Technol. **10**(3), 2954–2963 (2019)

Al-Haddad, L., Sial, M.S., Ali, I., Alam, R., Khuong, N.V., Khanh, T.H.T.: The role of small and medium enterprises (SMEs) in employment generation and economic growth: a study of marble industry in emerging economy. Int. J. Financ. Res. **10**(6), 174–187 (2019)

Antonioli, D., Mazzanti, M., Pini, P.: Productivity, innovation strategies and industrial relations in SMEs. In: Empirical Evidence for a Local Production System in Northern Italy, pp. 453–482. Published online: 04 Aug 2010. https://doi.org/10.1080/02692171.2010.483790

Barron, J.M., Black, D.A., Loewenstein, M.A.: Job Matching and on-the-job training. J. Labour Econ. **1**, 1–19 (1989)

Baumann, J., Kritikos, A.S.: The Link between R&D, innovation and productivity: are micro firms diferent. Res. Policy **45**, 1263–1274 (2016)

Black, S., Lynch, L.: How to compete: the impact of workplace practices and information technology on productivity. National Bureau of Economic Research Working Paper, No. 6120. Cambridge, Mass (1997)

Blakemore, A., Hoffman, D.: Seniority rules and productivity: an empirical test. Arizona State University, Sept (1988)

Blundell, R., Dearden, L., Meghir, C., Sianesi, B.: Human capital investment: the returns from education and training to the individual, the firm and the economy. Inst. Fisc. Stud. **20**(1), 1–23 (1999)

Calza, E., Goedhuys, M., Trifković, N.: Drivers of productivity in Vietnamese SMEs: the role of management standards and innovation. Ecocomics Innov. New Technol. **28**(1), 23–44 (2019). https://doi.org/10.1080/10438599.2018.1423765

Calza, E., Goedhuys, M., Trifković, N.: Drivers of productivity in Vietnamese SMEs: the role of management standards and innovation. In: Economics of Innovation and New Technology, pp. 23–44. Published online: 05 Feb 2018. https://doi.org/10.1080/10438599.2018.1423765

Cin, B.C., Kim, Y.J., Vonortas, N.S.: The impact of public R&D subsidy on small firm productivity: evidence from Korean SMEs. Small Bus. Econ. **48**, 345–360 (2017). https://doi.org/10.1007/s11187-016-9786-x

Firouzi, F., Sakineh, S., Mehin Aslaninia, N.: Determinants of Labour Productivity in Iran's Manufacturing Firms: Emphasizing on Labour Education and Training. University of Tabriz (2010)

Government of Vietnam: Decree No. 39/2018/ND-CP detailing a number of laws to support small and medium—sized enterprises (2018)

Hai, N.T., Le Hoa, N.T.: Situation and factors affecting labor productivity of Vietnam through the survey of industrial production enterprises (2016). http://www.khcnnamdinh.vn/index.php/news/804?fx=7

Hall, B.H., Lotti, F., Mairesse, J.: Innovation and productivity in SMEs: empirical evidence for Italy. Small Bus. Econ. **33**, 13–33 (2009). https://doi.org/10.1007/s11187-009-9184-8

Halsey, L.G.: The reign of the p-value is over: what alternative analyses could we employ to fill the power vacuum? Biol. Lett. **15**, 20190174 (2019). https://doi.org/10.1098/rsbl.2019.0174

Honig, B.: Human capital and structural upheaval: a study of manufacturing firms in the west bank. J. Bus. Ventur. **16**(6), 575–594 (2001)

Ismail, R.: The impact of human capital attainment on output and labor productivity of Malay firms. J. Int. Manag. Stud. **4**(1), 221–230 (2009)

Levinsohn, J., Petrin, A.: Estimating production functions using inputs to control for unobservables. Rev. Econ. Stud. **70**, 317–341 (2003). (CrossRef)

Lynch, L.M., Black, S.E.: Beyond the incidence of training: evidence from a national employers survey. National Bureau of Economic Research Working Paper No. 5231. Cambridge, Mass (1995)

Mansour, T.G.I, Eleshmawiy, K.H., Abdelazez, M.A., Abd El-Ghani, S.S.: The role of small and medium enterprises in economic development, case of Egypt. Int. J. Bus. Manag. Technol. **2**(5), 165–173 (2018)

Montague, L.: Training: an investment in human capital. Retail Distrib. Manag. **14**(2), 13–17 (1986)

Ngoc, P.T.B., Van Phuoc, N.H.: Small and medium enterprises' labor productivity in vietnam: a firm—level investigation. In: VEAM 2017 Conference. http://veam.org/wp-content/uploads/2017/12/131.-Pham-Thi-Bich-Ngoc.pdf (2017)

Nguyen, H.T.: How to test without P-value. Thail. Stat. **17**(2), i–x (2019)

Olley, G.S., Pakes, A.: The dynamics of productivity in the telecommunications equipment industry. Econometrica **1996**(64), 1263–1297 (1996). (CrossRef)

Organization for Economic Co-operation and Development: Measuring Productivity: Measurement of Aggregate and Industry-Level Productivity Growth. OECD Publishing, Paris, France (2001)

Papadogonas, T., Voulgaris, F.: Labor productivity growth in Greek manufacturing firms. Oper. Res. Int. Journal **5**(3), 459–472 (2005)

Shinada, N.: Qualily of labor, capital, and productivity growth in Japan: effects of employee age, seniority, and capital vintage. RIETI Discuss. Pap. Ser. **11**(E), 036 (2011)

Thanh, H.T.T.: The impact of factor on labor productivity growth in Vietnam. International Conference: Economics and Statistical Methods—Application in Economics and Finance, Information and Communucation Publishin House, January 2019, pp. 215–236. University of Ecconomics Hochiminh, Vietnam, ISBN 978-604-80-3675-1 (2019)

Van Dong N.: Factors affecting labor productivity and the problems posed. J. Financ. (II), (2016). https://tapchitaichinh.vn/nghien-cuu--trao-doi/trao-doi-binh-luan/cac-yeu-to-tac-dong-den-nang-suat-lao-dong-va-nhung-van-de-dat-ra-116664.html

MEM or/and LogARMA: Which Model for Realized Volatility?

Stanislav Anatolyev

Abstract There coexist two popular autoregressive conditional density model classes for series of positive financial variables such as realized volatility. One is a class of multiplicative error models (MEM), where the conditional mean is modelled autoregressively, while the specified shape of conditional distribution imposes evolution on higher order moments. The other class contains LogARMA models—ARMA models for logarithms of the original series, with a possibly time varying conditional distribution imposed on top of it. For MEM models, generating forecasts is straightforward, while for LogARMA models, additional numerical integration may be required. We compare small and big models from the two classes, along with their combinations, in terms of in-sample fit and out-of-sample predictability, using real data on realized volatility. The forecast combination weights show that both model classes are able to generate competitive forecasts, but the class of LogARMA models appears more reliable in forecasting than the class of MEM models.

1 Introduction

When the notion and analysis of realized volatility came to play in Andersen et al. (2003), the leading idea of how to model it was the use of *LogARMA models*, i.e., the class of ARMA models applied to the logarithm of the original series. Andersen and coauthors (Andersen et al. 2003) employed a LogARMA-spirited VAR-RV model for logarithmic volatility, somewhat motivated by the fact that the realized volatility is approximately log-normally distributed (see also Andersen et al. (2001)). Later, the LogARMA model took various fancy forms, such as the heterogeneous autoregressive (HAR) model (Corsi 2009) and its extensions.

However, realized volatility is a positive series. There is another option for modeling the dynamics of positive series. The story started developing from the work of Engle (Engle and Russell 1998) who proposed a class of autoregressive con-

S. Anatolyev (✉)
CERGE-EI, Politickych vězňů 7, 11121 Prague 1, Czech Republic
e-mail: stanislav.anatolyev@cerge-ei.cz

N. Ngoc Thach et al. (eds.), *Financial Econometrics: Bayesian Analysis, Quantum Uncertainty, and Related Topics*, Studies in Systems, Decision and Control 427,
https://doi.org/10.1007/978-3-030-98689-6_16

ditional duration (ACD) models for intradaily trade durations, a positive variable. Later, Engle (2002) renamed the class as *multiplicative error models (MEM)* noting that these models are convenient to use for any serially correlated stationary positive series. Various more generalized MEM-based models for realized volatility in levels were introduced into empirical work (e.g., Engle and Gallo (2006), Hautsch (2011)).

Nowadays, both approaches to modeling realized volatility coexist. Engle (Engle 2002) made a quick comparison of attractive and unpleasant features of MEM with LogARMA models, while Allen and coauthors (Allen et al. 2008) showed how the two model classes overlap (see also Sect. 2.3). Below we list in comparison four separate aspects related to both classes, the first two important for modeling decisions, and the other two less critical.

1. The MEM class is targeted primarily to model the conditional mean of the series, and hence is natural and sufficient for modeling the dynamics for the purposes of forecasting. At the same time, a model for the conditional mean of a log-transformed variable such as LogARMA needs to model the whole conditional distribution if one wants to eventually forecast the volatility in levels. This may be cumbersome to do, except under conditional homoskedasticity and/or conditional normality—the situations, when the conditional mean can be easily translated to the conditional mean of an exponent.
2. The simplest MEM model (so called exponential MEM) describing the dynamics of the conditional mean of the series automatically describes its higher order conditional moments, at least when the multiplicative innovation's conditional distribution is time-independent. The class of LogARMA models, in contrast, allows independent modeling higher order moments, at least its conditional variance by, for instance, an ARCH-type evolution, even when the standardized innovation's conditional distribution is time-independent. Thus, all else equal, the dynamics of conditional distribution is more flexible within the LogARMA framework.
3. If the support of the variable modelled contains values close to zero, construction of an LogARMA model may be problematic because of taking a logarithm of very small values, and zero values are totally prohibitive. At the same time, the MEM model successfully adapts for zeros in the support of the conditional distribution even in a logarithmic version that we use here. Moreover, Hautsch and coauthors (Hautsch et al. 2014) show how to take care of a probability mass at zero if zero values have a non-zero probability of occurrence.
4. Extensions of a scalar heteroskedastic LogARMA to a multivariate framework are familiar: the mean equation extends from ARMA to VARMA, and the variance equation from GARCH to multivariate GARCH. Extensions of a scalar MEM to a multivariate MEM are trickier and require the copula machinery (see, for example, Engle and Gallo (2006)).

The first two critical aspects suggest a trade-off between flexibility of modeling and complexity of two types: the usual one—degree of parameterization, and another one—producing forecasts. In this paper, we play one model class off against the other, using real data on realized volatility, to try to determine the 'optimal' modeling strategy. We specify two members of each class of models—one is a 'small' model and the other a 'big' model. Both use plausible specifications of volatility dynamics, but differ in specifications of the conditional distribution. The 'small model' uses a baseline conditional distribution that, in particular, allows quasi-maximum likelihood estimation, while the 'big model' uses a sophisticated conditional distribution with additional shape parameters—not the fanciest that one can find in the literature but one that is likely to be exploited by a practitioner. For the MEM class, these are the exponential and Burr distributions, respectively, while for the LogARMA class, these are the normal and skewed Student distributions. Both sophisticated distributions—Burr and skewed Student—possess two additional shape parameters.

We compare the in-sample and out-of-sample performance of the resulting four models, paying special attention to forecasting quality. Towards this end, we compare forecasts produced by the four models together with two types of model combinations, and construct the model confidence sets (MCS, Hansen et al. (2011)) of the best performing models. To do the numerical evaluation, we use a popular data-set of realized stock market volatility on ten stocks. We find that in terms of in-sample fit, the 'small' MEM model does not fit well compared to the other three, while among these three, usually two of three models tend to stand out, depending on the stock. In terms of forecasting quality, the four models perform quite similarly, and usually multiple models can be deemed the best in terms of MCS. Overall, the class of LogARMA models seems to be more reliable in forecasting than the class of MEM models, and small models tend to dominate big models from the same class. While model averaging using in-sample quality of fit does not tend to improve forecasting performance above the performance of best individual best models, model averaging using out-of-sample quality of fit is able, sometimes, to slightly improve forecasts.

The article is organized as follows. Section 2 describes the models, together with estimation and forecasting methods. Section 3 contains empirical results. Section 4 concludes.

2 Models

Denote the realized volatility by rv_t. We will compare four individual ('pure') models and two combinations of those. The individual models are: two MEM models based on the conditional exponential ('small') and Burr ('big') distributions and a linear dynamics in logs for the conditional mean of rv_t, and two LogARMA models based on the conditional normal ('small') and skewed Student ('big') distributions and an ARMA-EGARCH dynamics for logs of rv_t. The two model combinations are based on individual model performance: one on the in-sample performance as judged by the smoothed Takeuchi information criterion, and the other on the out-of-sample performance as judged by the forecasting quality in a validation subsample. For simplicity, all dynamic models have orders (1,1).

2.1 MEM

In both MEM models, $rv_t = \mu_t \varepsilon_t$, where μ_t is the conditional mean of rv_t, and ε_t has a positive distribution with conditional mean unity. The dynamics of μ_t is logarithmic:

$$\log \mu_t = \omega + \alpha \log rv_{t-1} + \beta \log \mu_{t-1}.$$

A small MEM is represented by the *ExpMEM* model

$$\varepsilon_t | I_{t-1} \sim \mathscr{E},$$

where $\mathscr{E}$ denotes standard exponential distribution having the density

$$f_{\mathscr{E}}(\varepsilon) = \exp(-\varepsilon).$$

A big MEM is represented by the *BurrMEM* model[1]

$$\varepsilon_t | I_{t-1} \sim \mathscr{B}(\zeta, \varrho),$$

where $\mathscr{B}(\zeta, \varrho)$ denotes Burr distribution with mean unity and shape parameters ζ and ϱ, thus having the density

$$f_{\mathscr{B}}(\varepsilon) = \frac{\zeta}{\chi^{\zeta}} \varepsilon^{\zeta-1} \left(1 + \varrho \left(\frac{\varepsilon}{\chi}\right)^{\zeta}\right)^{-1-\varrho^{-1}}, \tag{1}$$

where

$$\chi = \frac{\Gamma(1+\varrho^{-1})\varrho^{1+\zeta^{-1}}}{\Gamma(1+\zeta^{-1})\Gamma(\varrho^{-1}-\zeta^{-1})}, \quad \zeta, \varrho > 0.$$

For both ExpMEM and BurrMEM, the one-step forecast of realized volatility then has the following form:

$$\widehat{rv}_{t+1} = \hat{\mu}_{t+1} = \exp\left(\hat{\omega} + \hat{\alpha} \log rv_t + \hat{\beta} \log \hat{\mu}_t\right).$$

2.2 LogARMA

In both LogARMA models, $\log rv_t$ follows an ARMA dynamics

[1] This distribution was proposed in Grammig and Maurer (2000) for ACD models for trade durations in order to account for non-monotonicity of the conditional hazard function. An alternative flexible distribution is normalized generalized gamma (see Hautsch (2011)). We utilize Burr because it exhibits much higher stability than the generalized gamma in experiments with real data.

$$\log rv_t = \mu + \phi \log rv_{t-1} + e_t + \theta e_{t-1},$$

with semi-strong white noise innovations $e_t = \sigma_t \eta_t$, whose conditional variance ('volatility of volatility') follows EGARCH dynamics:

$$\log \sigma_t^2 = \omega + \alpha |\eta_{t-1}| + \gamma \eta_{t-1} + \beta \log \sigma_{t-1}^2,$$

where η_t is standardized innovation with conditional mean zero and conditional variance unity.

A small LogARMA is represented by the *NormLogARMA* model

$$\eta_t | I_{t-1} \sim \mathcal{N},$$

where $\mathcal{N}$ denotes standard normal distribution having the density

$$f_{\mathcal{N}}(\eta) = \frac{1}{\sqrt{2\pi}} \exp\left(-\frac{\eta^2}{2}\right).$$

A big LogARMA is represented by the *SkStLogARMA* model[2]

$$\eta_t | I_{t-1} \sim \mathcal{S}\mathcal{S}(\lambda, \nu).$$

where $\mathcal{S}\mathcal{S}(\lambda, \nu)$ denotes standardized (to have zero mean and unit variance) skewed Student distribution (Hansen 1994) with shape parameters λ (responsible for asymmetry) and ν (degrees of freedom, responsible for tail thickness), thus having the density

$$f_{\mathcal{S}\mathcal{S}}(\eta, \lambda, \nu) = c_0 c_1 \left(1 + \frac{\xi^2}{\nu - 2}\right)^{-(\nu+1)/2}, \tag{2}$$

where $\xi = (c_1\eta + c_2)/(1 - \lambda)$ if $\eta < -c_2/c_1$ and $\xi = (c_1\eta + c_2)/(1 + \lambda)$ otherwise, $c_0 = \Gamma((\nu + 1)/2)/\Gamma(\nu/2)/\sqrt{\pi(\nu - 2)}$, $c_2 = 4c_0\lambda(\nu - 2)/(\nu - 1)$, and $c_1 = \sqrt{1 + 3\lambda^2 - c_2^2}$.

For both LogARMA models, the one-step forecast of logarithmic realized volatility is given by

$$\widehat{\log rv}_{t+1} = \hat{\mu} + \hat{\phi} \log rv_t + \hat{\theta} e_t,$$

while the volatility prediction is

$$\hat{\sigma}_{t+1}^2 = \exp\left(\hat{\omega} + \hat{\alpha}|\hat{\eta}_t| + \hat{\gamma}\hat{\eta}_t + \hat{\beta} \log \hat{\sigma}_t^2\right).$$

[2] An alternative flexible distribution is normalized skewed generalized error distribution (SGED) (see, e.g., Anatolyev and Petukhov (2016)). We utilize skewed Student because it exhibits much higher stability than the SGED in experiments with real data.

For the NormLogARMA model, because of conditional normality, these forecasts are translated into the forecast for realized volatility via

$$\widehat{rv}_{t+1} = \hat{E}[rv_{t+1}|I_t] = \exp\left(\widehat{\log rv}_{t+1} + \frac{1}{2}\hat{\sigma}^2_{t+1}\right).$$

For the SkStLogARMA model, the conditional expectation $E[rv_{t+1}|I_t]$ does not have a closed form. Therefore, we form the forecasts using

$$\widehat{rv}_{t+1} = \exp\left(\widehat{\log rv}_{t+1}\right)\widehat{E\left[\exp\left(\sigma_{t+1}\eta_{t+1}\right)|I_t\right]},$$

where the integral $\widehat{E\left[\exp\left(\sigma_{t+1}\eta_{t+1}\right)|I_t\right]} = \int_{-\infty}^{+\infty}\exp\left(\hat{\sigma}_{t+1}\eta\right) f_{\mathcal{S}\mathcal{S}}(\eta, \hat{\lambda}, \hat{\nu})d\eta$ is computed using the Gauss-Chebychev quadrature (see, e.g., Judd (1998)):

$$\int_a^b g(x)dx \approx \frac{\pi(b-a)}{2n}\sum_{i=1}^{n}(1-x_i^2)^{\frac{1}{2}} g\left(a + \frac{(x_i+1)(b-a)}{2}\right),$$

where $x_i = \cos\left((2i-1)\pi/(2n)\right)$, $i = 1, \ldots, n$. We set $n = 100$, $a = -8$, and $b = 8$; these values deliver sufficient computational precision.

2.3 *Reconciliation of MEM and LogARMA*

Let us reconcile the dynamics of realized volatility in the two model classes. From the MEM multiplicative structure, it follows that

$$\log rv_t = \log \mu_t + \log \varepsilon_t = (1-\beta L)^{-1}(\omega + \alpha \log rv_{t-1}) + \log \varepsilon_t,$$

or

$$\log rv_t = \omega + (\alpha+\beta)\log rv_{t-1} + \log \varepsilon_t - \beta \log \varepsilon_{t-1},$$

which has a homoskedastic ARMA(1,1) for $\log rv_t$. On the other hand, in the LogARMA model class, $\log rv_t$ follows a heteroskedastic ARMA dynamics

$$\log rv_t = \mu + \phi \log rv_{t-1} + e_t + \theta e_{t-1},$$

with innovations e_t, whose conditional variance follows EGARCH dynamics. Hence, as viewed from the perspective of the mean logarithmic volatility dynamics, the MEM and LogARMA models are equally flexible, but are different in the flexibility of the volatility-of-volatility dynamics; LogARMA is more flexible in this respect.

However, even if the heteroskedasticity was shut down in LogARMA, the models would still not be equivalent in conditional distributional features, as the exponential/Burr distribution of ε_t does not correspond to normal/skewed Student distribution of e_t. Thus, neither model is a special case of the other.

2.4 Model Averaging

Along with the four pure models, we use their combinations and corresponding combined forecasts with weights based on in-sample and out-of-sample performance. In both cases, predictions are formed as a linear combination of predictions from individual models:

$$\widehat{rv}_{t+1} = \sum_{i=1}^{M} w_i \widehat{rv}_{t+1,i},$$

where M is number of pure models (4 in our case), and w_i and $\widehat{rv}_{t+1,i}$, $i = 1, \ldots, M$, are the model weights and individual forecasts, respectively.

The first model averaging combination based on an in-sample smoothed information criterion (Buckland et al. 1997), which is a convenient tool to track relative in-sample fit of several models. Denote by $f_{t-1}(rv_t|\theta)$ the conditional density of realized volatility at period t, where θ is a vector of all parameters in a given model, and let $\ell_n(\hat{\theta}) = \sum_t \log f_{t-1}(rv_t|\hat{\theta})$ be the loglikelihood function. The 5th set of predictions is produced by model averaging using smoothed Takeuchi information criterion (STICMA). The Takeuchi information criterion (TIC, Takeuchi (1976)) is a more general version of the familiar Akaike information criterion (AIC) that acknowledges misspecification of the conditional density, which is important in our setup:

$$TIC = -2\ell_n(\hat{\theta}) + 2\text{tr}(\hat{J}^{-1}\hat{I}),$$

where $\hat{J}$ and $\hat{I}$ are empirical analogs of $J = -E\left[\partial^2 \log f_{t-1}(rv_t|\theta)/\partial\theta\partial\theta'\right]$ and $I = E\left[\partial \log f_{t-1}(rv_t|\theta)/\partial\theta\, \partial \log f_{t-1}(rv_t|\theta)/\partial\theta'\right]$, the ingredients of the asymptotic variance of the quasi-ML estimator. When the given model is correctly specified, $J = I$ and $\text{tr}\left(J^{-1}I\right) = K$, and TIC reduces to AIC, $AIC = -2\ell_n(\hat{\theta}) + 2K$, where $K = \dim(\theta)$ is a total number of parameters in the model under consideration. Implementationwise, we compute $\hat{J}$ and $\hat{I}$ by using numerical derivatives (see, e.g., Judd (1998)). For the MEM models, formulas for the TICs are straightforward, but for LogARMA models formulated for log-transformed variables, the TIC that contains densities of observables should be adjusted, according to the relation between densities of original and transformed variables: if $z = \exp(x)$, then $f_Z(z) = f(\log z)/z$, and so $E[\log f_Z(z)] = E[\log f(x) - x]$. The STICMA weights w_i are given by

$$w_i = \frac{\exp(-TIC_i/2)}{\sum_{j=1}^{M} \exp(-TIC_j/2)}, \quad i = 1, \ldots, M.$$

The second model averaging combination, which produces the 6th set of predictions, is based on the out-of-sample performance of individual models. To this end, we adapt the jackknife model averaging (JMA) machinery (Hansen and Racine 2012) to the nonlinear modeling setup. Here, the weights w_i, $i = 1, \ldots, M$, are determined by minimizing the cross-validation (CV) criterion, which is computed from all M models' forecast errors on the validation subsample; for exact formulation, see equations (5) and (6) in Hansen and Racine (2012). The quadratic optimization problem subject to the constraint that all the M weights are non-negative and sum to unity, is a nice easily implementable problem even when numerically solved repeatedly in a rolling window.

3 Empirical Evaluation

3.1 Data

We perform the estimation and forecasting exercises using the popular elsewhere data-set of realized stock market volatility from (Noureldin et al. 2012). This data-set contains daily realized volatilities on 10 stocks: BAC, JPM, IBM, MSFT, XOM, AA, AXP, DD, GE, KO, and covers the period from February 1, 2001 to December 31, 2009. Because towards the end of the sample volatilities exhibit turbulence, we cut it at May 31, 2007. This leaves 1589 observations, of which the last 589 we use for forecast evaluation, and estimate the models in a 1000-observations window. The length of the validation subsample is set at 100 observations. The ten volatility series are depicted in Fig. 1.

3.2 Model Estimation

Box 1 shows estimation results for BAC, as a typical example, in the first estimation window. For the two MEM models, there is a stark difference in the degree of in-sample fit between the small and big models. The large gap between the corresponding loglikelihood values and information criteria is delivered by the two shape parameters of the Burr density sharply different from those implied by the standard exponential density. The parameters between the two conditional mean equations are very similar and well identified.

In contrast, the differences in the degree of fit between the two LogARMA models are modest at most, even though the density shape parameters of the skewed Student distribution statistically significantly differ from those implied by the standard normal density. The two conditional mean equations are very similar and pretty well identified, while some estimates in the variance equation differ quite a lot and have big standard errors. Evidently, the dynamics of 'volatility of volatility' is quite hard to identify.

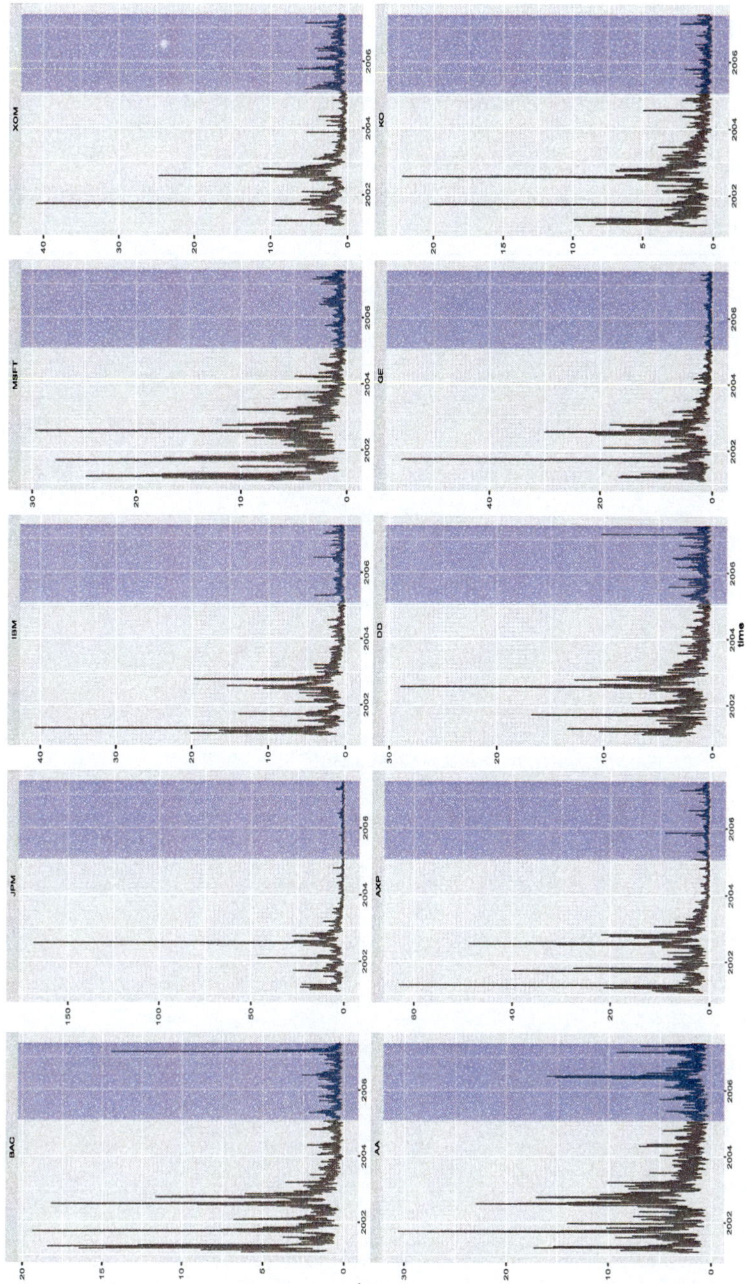

Fig. 1 Data on realized volatilities (shaded is out-of-sample period)

3.3 Model Average Weights

Figures 2 and 3 depict, correspondingly, the values of STICMA and JMA weights in a rolling window for all the stocks. When the weights are formed from the in-sample performance (Fig. 2), the small MEM model (ExpMEM) always has zero weights as a result of its having a much smaller in-sample quality than the big MEM model (BurrMEM).

ExpMEM model

$$\varepsilon_t | I_{t-1} \sim \mathscr{E}$$

$$\log \mu_t = \underset{(0.0096)}{0.0420} + \underset{(0.193)}{0.363} \log rv_{t-1} + \underset{(0.215)}{0.615} \log \mu_{t-1}$$

$$LL = -1165.7, \; TIC = 2333.8$$

BurrMEM model

$$\varepsilon_t | I_{t-1} \sim \mathscr{B}\left(\underset{(0.159)}{3.081}, \underset{(0.198)}{1.277}\right)$$

$$\log \mu_t = \underset{(0.0117)}{0.0438} + \underset{(0.140)}{0.371} \log rv_{t-1} + \underset{(0.150)}{0.611} \log \mu_{t-1}$$

$$LL = -675.0, \; TIC = 1361.2$$

NormLogARMA model

$$\log rv_t = \underset{(0.00569)}{-0.00267} + \underset{(0.009)}{0.977} \log rv_{t-i} + e_t - \underset{(0.040)}{0.611} e_{t-1}$$

$$\log \sigma_t^2 = \underset{(2.375)}{-0.155} + \underset{(0.275)}{0.071} |\eta_{t-1}| - \underset{(0.021)}{0.013} \eta_{t-1} + \underset{(0.979)}{0.938} \log \sigma_{t-1}^2$$

$$\eta_t | I_{t-1} \sim \mathscr{N}$$

$$LL = -625.6, \; TIC = 1383.4$$

SkStLogARMA model

$$\log rv_t = \underset{(0.00591)}{-0.00077} + \underset{(0.009)}{0.981} \log rv_{t-1} + e_t - \underset{(0.033)}{0.611} e_{t-1}$$

$$\log \sigma_t^2 = -\underset{(10.231)}{0.809} + \underset{(0.179)}{0.112} |\eta_{t-1}| - \underset{(0.046)}{0.000} \eta_{t-1} + \underset{(5.810)}{0.545} \log \sigma_{t-1}^2$$

$$\eta_t | I_{t-1} \sim \mathscr{S}\mathscr{S}\left(\underset{(0.0370)}{0.0927}, \underset{(0.102)}{1.546}\right)$$

$$LL = -615.3, \; TIC = 1361.4$$

Box 1: Results of estimation of the four models for BAC in the first window.

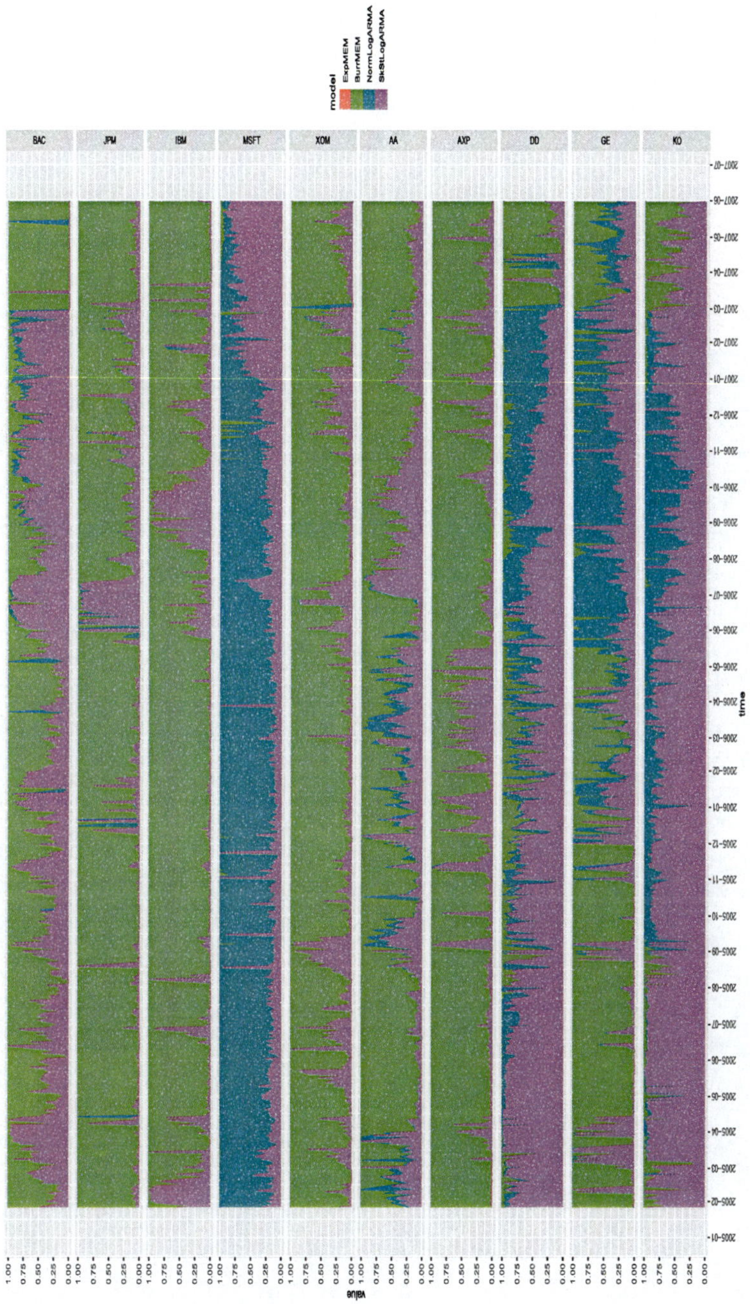

Fig. 2 Values of STICMA weights in rolling window

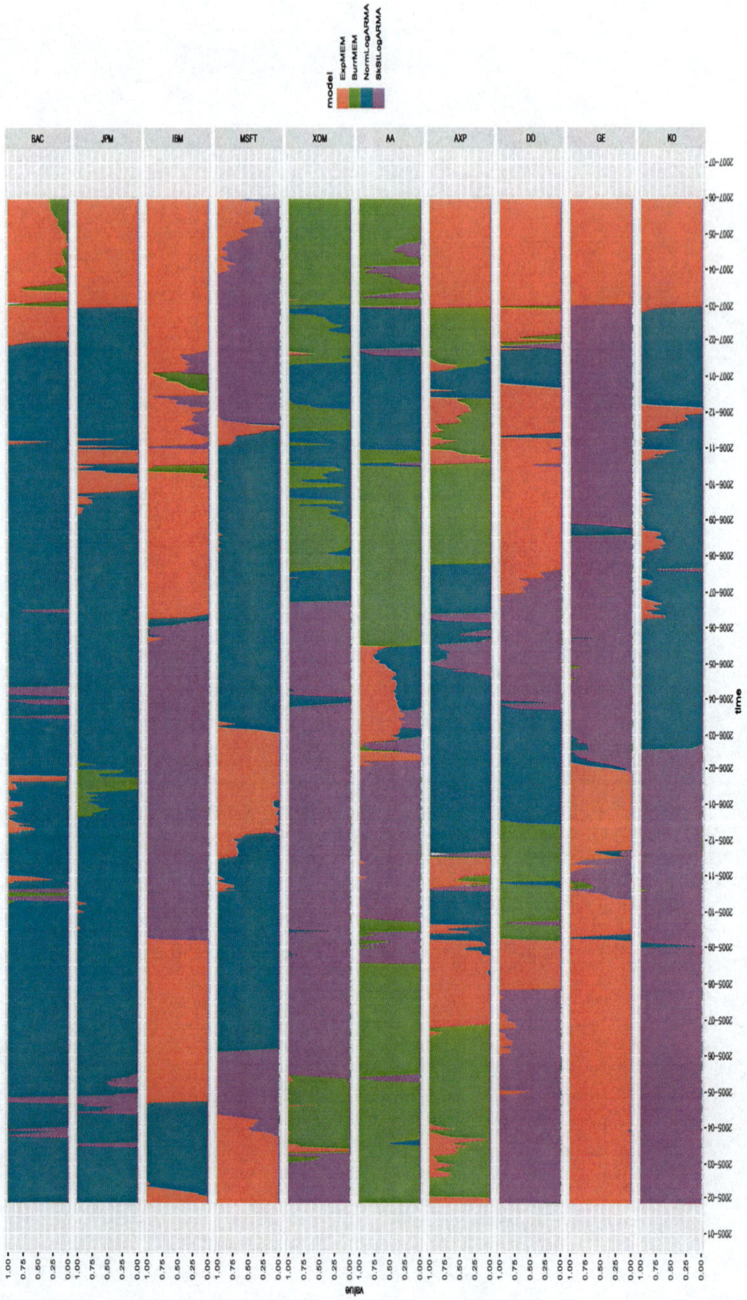

Fig. 3 Values of JMA weights in rolling window

The big MEM and both LogARMA models perform in-sample on par with each other, though the parity depends on the asset under consideration. For most stocks, two of the three models stand out—for example, for MSFT, only the two LogARMAs have nontrivial weights, while, for example, for IBM, it is the two big models that perform better. For other stocks, for example, for GE, all the three models are balanced. The balance between the non-trivially weighted models usually varies in time in clusters, with weights for one model being able to take values close to zero at times and values near unity at other times. Table 1 presents the average, together with standard deviations, STICMA weights from different models. Any of the models other than ExpMEM is able to dominate on average, although BurrMEM seems to be doing it most often.

When the weights are formed from the out-of-sample performance (Fig. 3), the role of the small MEM model (ExpMEM) ceases to be trivial, and for some stocks in some periods this models dominates. The other three models are keeping up, so for each of the four models there are combinations of stocks and periods, albeit short, when this model dominates the other three. There are much fewer instances when two models are on par with each other, than when the weights are driven by in-sample performance. Table 2 presents the average, together with standard deviations, JMA weights from different models. Each of the four models is able to exhibit average weights of 50% or higher for some of the stocks, and at the same time average weights close to zero for others. The standard deviations also point at the instability of weights as functions of a position of the validation subsample.

Table 1 Average (and standard deviations of) STICMA weights from different models

	BAC	JPM	IBM	MSFT	XOM	AA	AXP	DD	GE	KO
ExpMEM	0.00 (0.00)	0.00 (0.00)	0.00 (0.00)	0.00 (0.00)	0.00 (0.00)	0.00 (0.00)	0.00 (0.00)	0.00 (0.00)	0.00 (0.00)	0.00 (0.00)
BurrMEM	0.56 (0.33)	0.87 (0.24)	0.86 (0.26)	0.02 (0.10)	0.86 (0.19)	0.70 (0.26)	0.85 (0.23)	0.22 (0.29)	0.51 (0.43)	0.10 (0.20)
NormLogARMA	0.03 (0.11)	0.02 (0.11)	0.00 (0.02)	0.72 (0.26)	0.00 (0.04)	0.06 (0.12)	0.00 (0.00)	0.25 (0.26)	0.26 (0.31)	0.21 (0.24)
SkStLogARMA	0.41 (0.31)	0.12 (0.22)	0.14 (0.26)	0.26 (0.24)	0.13 (0.19)	0.24 (0.21)	0.15 (0.23)	0.53 (0.29)	0.23 (0.25)	0.69 (0.28)

Table 2 Average (and standard deviations of) JMA weights from different models

	BAC	JPM	IBM	MSFT	XOM	AA	AXP	DD	GE	KO
ExpMEM	0.15 (0.33)	0.16 (0.36)	0.56 (0.48)	0.24 (0.39)	0.01 (0.08)	0.07 (0.20)	0.29 (0.42)	0.40 (0.48)	0.50 (0.49)	0.15 (0.33)
BurrMEM	0.02 (0.09)	0.03 (0.14)	0.01 (0.09)	0.00 (0.00)	0.33 (0.45)	0.54 (0.49)	0.35 (0.45)	0.12 (0.32)	0.00 (0.03)	0.00 (0.02)
NormLogARMA	0.80 (0.39)	0.79 (0.39)	0.10 (0.29)	0.50 (0.49)	0.15 (0.33)	0.17 (0.35)	0.32 (0.46)	0.15 (0.36)	0.02 (0.13)	0.40 (0.47)
SkStLogARMA	0.03 (0.17)	0.02 (0.13)	0.33 (0.46)	0.26 (0.42)	0.51 (0.50)	0.22 (0.39)	0.04 (0.17)	0.33 (0.46)	0.47 (0.49)	0.45 (0.50)

3.4 Forecasting Performance

Now we turn to comparing the forecasting performance of individual models and their model averages. Table 3 contains out-of-sample average squared errors from different models for all the stocks, and Table 4 reports model confidence sets (MCS) for the 25% confidence level, which is a conventional level in volatility analysis (see Laurent et al. (2012)). The MCS machinery (Hansen et al. 2011) allows statistically correct multiple hypotheses testing, and the MCS is a subset of models from the pool of all predictive models under consideration that are statistically insignificanly different by their forecasting performance. The null hypothesis states that all the models inside the MCS perform equally well, while any model outside of the MCS performs worse; for more details, see Hansen et al. (2011).

One can immediately see that the four 'pure' models are very similar in forecasting performance, and for most stocks multiple models can be deemed the best. Often, it is two or three models out of the four, but it may be also all four. The model that always belongs to the MCS at the 25% level is the NormLogARMA model, i.e. small LogARMA. This happens despite the estimation noise in barely identifiable volatility-of-volatility dynamics. Also, recall that an additional advantage of this particular LogARMA model is that forecasts can be computed without numerical integration. The SkStLogARMA model, i.e. big LogARMA, is a bit less likely to be among the best, so the MCS contains this model for 7 out of 10 stocks. The ExpMEM model, i.e. small MEM, is much less likely to be among the best, and only for half of the stocks is this model contained in the MCS. Finally, the BurrMEM model, i.e. big MEM, enters the MCS only for 3 out of 10 stocks. Overall, the class of LogARMA

Table 3 Average out-of-sample losses from different models

	BAC	JPM	IBM	MSFT	XOM	AA	AXP	DD	GE	KO
ExpMEM	0.4364	0.2097	0.1538	0.1537	0.4946	1.4307	0.3422	0.4118	0.0748	0.0570
BurrMEM	0.4376	0.2090	0.1541	0.1555	0.4926	1.4329	0.3423	0.4136	0.0758	0.0571
NormLogARMA	0.4372	0.2070	0.1536	0.1525	0.4922	1.4318	0.3423	0.4136	0.0752	0.0565
SkStLogARMA	0.4378	0.2076	0.1532	0.1527	0.4907	1.4349	0.3424	0.4133	0.0754	0.0564
STICMA	0.4377	0.2084	0.1540	0.1525	0.4928	1.4325	0.3421	0.4136	0.0756	0.0564
JMA	0.4376	0.2069	0.1534	0.1527	0.4932	1.4332	0.3426	0.4121	0.0750	0.0563

Table 4 Predictive model confidence sets for 25% confidence level

	BAC	JPM	IBM	MSFT	XOM	AA	AXP	DD	GE	KO
ExpMEM	*					*	*	*	*	
BurrMEM					*	*	*	*		
NormLogARMA	*	*	*	*	*	*	*	*	*	*
SkStLogARMA		*	*		*	*	*	*	*	*
STICMA				*	*	*	*	*		*
JMA	*	*	*		*	*	*	*	*	*

models seems to be more reliable in forecasting than the class of MEM models. There is a slight tendency of small models to dominate big models from the same classes.

The two bottom lines in Tables 3 and 4 show the figures for the model averaging forecasts that are based on STICMA and JMA. The model average combinations do sometimes, though not always, improve the forecasting performance relative to individual models, confirming the common wisdom. It is also intuitive that model averaging based on out-of-sample performance fairs better than model averaging based on in-sample criteria. In fact, the JMA forecasts entered the 25% MCS for almost all stocks—9 out of 10, in contrast to STICMA, which is contained in the MCS only for 6 stocks.

4 Concluding Remarks

We have run a mini-competition among several models from two popular model classes—MEM and LogARMA—for realized volatility of ten liquid stocks, paying main attention to the forecasting quality. Overall, the class of LogARMA models seems to be more reliable in forecasting than the class of MEM models, while small models tend to dominate big models from the same class. The small LogARMA model and the model average based on forecasting performance have more chances to yield best predictions than the other individual models or the information criterion based model average, although this tendency is unstable through time and across stocks. For some stocks, the difference across all the models seems to be immaterial. The decision about which model class to select and how complex a model within the class to use does not seem empirically that big of a deal.

Acknowledgements The author thanks organizers and audiences of the 41st International Symposium on Forecasting (virtual, June 2021), 7th International Conference on Time Series and Forecasting (Gran Canaria, Spain, July 2021) and 5th International Conference on Financial Econometrics (hybrid, Ho Chi Minh City, Vietnam, January 2022). This research was supported by grant 20-28055S from the Czech Science Foundation. I also thank Filip Staněk for excellent research assistance.

References

Allen, D., Chan, F., McAleer, M., Peiris, S.: Finite sample properties of the QMLE- for the LogACD model: application to Australian stocks. J. Econometr. **147**, 163–185 (2008)

Anatolyev, S., Petukhov, A.: Uncovering the skewness news impact curve. J. Financ. Econometr. **14**(4), 746–771 (2016)

Andersen, T.G., Bollerslev, T., Diebold, F.X., Labys, P.: The distribution of realized exchange rate volatility. J. Amer. Stat. Assoc. **96**, 42–55 (2001)

Andersen, T.G., Bollerslev, T., Diebold, F.X., Labys, P.: Modeling and forecasting realized volatility. Econometrica **71**(2), 579–625 (2003)

Buckland, S.T., Burnham, K.P., Augustin, N.H.: Model selection: an integral part of inference. Biometrics **53**, 603–618 (1997)
Corsi, F.: A simple approximate long-memory model of realized volatility. J. Financ. Econometr. **7**(2), 174–196 (2009)
Engle, R.F.: New frontiers for ARCH models. J. Appl. Econometr. **17**(5), 425–446 (2002)
Engle, R.F., Russell, J.R.: Autoregressive conditional duration: a new model for irregularly spaced transaction data. Econometrica **66**(5), 1127–1162 (1998)
Engle, R.F., Gallo, G.M.: A multiple indicators model for volatility using intra-daily data. J. Econometr. **131**(1–2), 3–27 (2006)
Grammig, J., Maurer, K.-O.: Non-monotonic hazard functions and the autoregressive conditional duration model. Econometri. J. **3**, 16–38 (2000)
Hansen, B.E.: Autoregressive conditional density estimation. Int. Econ. Rev. **35**(3), 705–730 (1994)
Hansen, B.E., Racine, J.: Jackknife model averaging. J. Econ. **167**, 38–46 (2012)
Hansen, P.R., Lunde, A., Nason, J.M.: The model confidence set. Econometrica **79**(2), 453–497 (2011)
Hautsch, N.: Econometrics of Financial High-Frequency Data. Springer Science & Business Media, Berlin/Heidelberg, Germany (2011)
Hautsch, N., Malec, P., Schienle, M.: Capturing the zero: a new class of zero-augmented distributions and multiplicative error processes. J. Financ. Econometr. **12**(1), 89–121 (2014)
Judd, K.: Numerical Methods in Economics. MIT Press, Cambridge, MA (1998)
Laurent, S., Rombouts, J.V.K., Violante, F.: On the forecasting accuracy of multivariate GARCH models. J. Appl. Econometr. **27**(6), 934–955 (2012)
Noureldin, D., Shephard, N., Sheppard, K.: Multivariate high-frequency-based volatility (HEAVY) models. J. Appl. Econometr. **27**(6), 907–933 (2012)
Takeuchi, K.: Distributions of information statistics and criteria for adequacy of models (in Japanese). Suri-Kagaku (Math. Sci.) **153**, 12–18 (1976)

Crime and the Shadow Economy: Evidence from BRICS Countries

Nguyen Ngoc Thach, Duong Tien Ha My, Pham Xuan Thu, and Nguyen Van Diep

Abstract The paper investigates the impact of crime on the dimension of the shadow economy in the BRICS countries. To achieve this research objective, we collect data from the Numbeo database, World Development Indicators, Heritage Foundation, and Worldwide Governance Indicators for 2001–2017. The Bayesian linear regression strategy is employed to discover the determinants of the informal sector. We apply the normal prior suggested by (Block et al., J. Fam. Bus. Strat. 2:232–245, 2011). Also, the posterior distributions of all parameters in the model are generated through the Markov Chain Monte Carlo (MCMC) approach and Gibbs sampling. The results indicate that crime has a reversed U-shaped association with the shadow economy size in the BRICS countries between 2001–2017. Specifically, if the crime value does not exceed the turning point, this factor can positively impact the shadow economy. However, if the crime value is above the turning point, the relationship may change. Moreover, we find that gross domestic product (GDP) growth, the rule of law, corruption control, and government effectiveness adversely affect the shadow economy. By contrast, trade openness and foreign direct investment (FDI) have beneficial effects on the shadow economy. Meanwhile, government spending can positively impact the shadow economy once the expenditure is below a certain level.

N. N. Thach
Banking University of Ho Chi Minh City, 39 Ham Nghi, Nguyen Thai Binh Ward, District 1, Ho Chi Minh City, Vietnam
e-mail: thachnn@buh.edu.vn

D. T. H. My · N. Van Diep (✉)
Ho Chi Minh City Open University, 35-37 Ho Hao Hon, Co Giang Ward, District 1, Ho Chi Minh City, Vietnam
e-mail: diep.nv@ou.edu.vn

D. T. H. My
e-mail: my.dth@ou.edu.vn

P. X. Thu
College of Foreign Economic Relations, 287 Phan Dinh Phung, Ward 15, Phu Nhuan District, Ho Chi Minh City, Vietnam

N. Ngoc Thach et al. (eds.), *Financial Econometrics: Bayesian Analysis, Quantum Uncertainty, and Related Topics*, Studies in Systems, Decision and Control 427,
https://doi.org/10.1007/978-3-030-98689-6_17

Keywords Bayesian · Crime rate · Shadow economy · BRICS

1 Introduction

Shadow economy is a worldwide economic phenomenon. Therefore, many researchers have tried to discover the driving forces that encourage individuals to go underground. Findings from previous studies indicate that various factors can determine the dimension of the shadow economy. Specifically, some authors analyze the impact of economic indicators such as government expenditures, FDI, GDP growth, and trade openness on the shadow economy (Esaku 2021; Fugazza and Fiess 2010; Goel and Saunoris 2014; Jamalmanesh, Meidani, and Mashhadi 2014; Luong et al. 2020; Nguyen et al. 2019a, b; Nguyen and Thach 2019).

Meanwhile, other authors concentrate on the relationship between the social and psychological dimensions and this informal sector (Achim et al. 2018; Lee 2013; Torgler and Schneider 2009; Williams and Horodnic 2015). For instance, Torgler and Schneider (2009) argue that a rise in governance and institutional quality and social norms (tax morale) can reduce individuals' incentives to go illegal. Likewise, Lee (2013) claims that social capital is adversely connected to the dimension of the informal economy in numerous countries. Regarding this issue, Arrow (1973, p. 303) suggests that under some particular cases, "the economic agent should forgo profit or other benefits to himself in order to achieve some social goal, especially to avoid a disservice to other individuals".

Besides the above factors, a growing literature examines the role of the crime. Monticelli (2014, p. 1) supposes that criminology and economics "are irretrievably intertwined". The author also states that this relationship is probably not simple and can involve a wide range of motives and interests. Therefore, although there have been different studies on the shadow economy, the informal sector and crime relationship is still mixed. For instance, Habibullah and Eng (2011) reveal that criminal activities influence the Malaysian underground economy just in the armed robbery. Meanwhile, Gottschalk and Gunnesdal (2018) provide profound thoughts on the white-collar criminals in the informal economy.

The current study, applying the Bayesian approach, aims to supplement the literature by investigating the impact of criminal activities on the dimension of informality in the BRICS economies in 2001–2017. The BRICS represent the five countries of Brazil, Russia, India, China, and South Africa. They together contribute significantly to the global GDP (Li 2019). Between 2001 and 2017, the average dimension of the informal economy in the BRICS is approximately 26.50% (Medina and Schneider 2019). A study of the factors affecting this informal economy helps propose appropriate recommendations, whereby governments can consider measures to control underground activities better.

The paper is constructed as follows. In Sect. 2, we describe the literature review and the research hypotheses. Section 3 discusses the research model, methodology, data sources. Section 4 highlights our main empirical findings, while Sect. 5 is about MCMC diagnostics. Finally, Sect. 6 presents the conclusion.

2 Literature Review and Research Hypotheses

2.1 Theoretical Considerations

Criminals activities exist in different societies and can be studied by criminology. On the other side, economics examines the allocation of scarce resources. Monticelli (2014) argues that these two disciplines are "irretrievably intertwined" (Monticelli 2014, p. 1). Furthermore, the author also indicates that the connection between crime and the economy might be not simple. Indeed, they can be related to various factors such as individuals and group motives and stakeholders interests. Fleisher (1963) confirmed a direct connection between the crime level and the unemployment rate. Becker (1968) shows that criminals behave rationally, and they engage in illegal activities for profit. Studies by Becker and Stigler (1974) and Akerlof and Dickens (1982) suggest that low wages in the public sector cause corruption, and fair pay has been proposed to solve this problem. In addition, Witte (1996) also shows a link between drugs and organized crime gangs when drugs are traded in the shadow economy. Alternatively stated, the rise of the informal economy can create strong encouragement to lure laborers and other resources to leave the formal economy. In addition, over 65% of income earned in the underground economy is spent and strongly encourages the rise of the formal economy (Schneider 2000). Since crime and informal economic activities are a reality for all countries, governments try to control such activities using sanctions, prosecution, or education. These results show that the connection between crime and the shadow economy needs to be considered with caution.

Before analyzing such a relationship, it is essential to consider the definition of the shadow economy. So far, there have been numerous studies on the shadow economy. However, there is no unified definition of the shadow economy. Dell'Anno (2007) suggests that this type of economy involves informal production, underground production, and illegal production. Alternatively, Schneider (2010) argues that the shadow economy is related to production activities basis on the market and are intentionally hidden from public agencies because of specific reasons. Meanwhile, Medina and Schneider (2019) show that the shadow economy or the informal economy involves "all economic activities which are hidden from official authorities for monetary, regulatory, and institutional reasons" (Medina and Schneider 2019, p. 4). We employ the definition and data on the shadow economy size introduced by Medina and Schneider (2019) in the research model.

Regarding the crime, Naylor (1995) contends that criminal activities no longer only involve mere redistribution of wealth, such as burglary or ransom kidnapping. In fact, the number of economically motivated crimes is increasing. They are corporate crimes concerned with the production and dispensation of goods and services rather than just wealth redistribution. The author emphasizes that the boundary between the unambiguous criminal and the "informal" components of the underground economy has been equally blurred in modern days. Meanwhile, some other authors suggest that the white-collar crime plays a role in the informal economy. However, it should be noted that "white-collar crime is no specific type of crime, it is only a crime committed by a specific type of person" (Gottschalk and Gunnesdal 2018, p. 1).

2.2 *The Determinants of the Size of the Shadow Economy Size*

There have been numerous studies exploring the driving forces of the shadow economy. Notably, Petersen et al. (2010) suppose that evasion of social security contributions can discourage individuals from returning to the formal labor market. Instead, abusers become involved in the informal economy. Habibullah and Eng (2011) examine the linkage between crime and the shadow economy size in Malaysia over 1973–2003. The authors reveal that criminal activities result in the underground economy in Malaysia in case of armed robbery.

Meanwhile, Katsios (2019) claims that in a country where regulatory and law enforcement capacities are low, social exclusion is high, and/or tax considerations are involved, some economic activities may be unlisted and informal. The author also adds that economic activities that are informal/unlisted create difficulties for law enforcement authorities when they explore tax violations and proceeds of crime. Overall, different studies suggest that the linkage between crime and the informal economy might be positive. Based on theoretical considerations and empirical work, our main research hypothesis is:

Hypothesis 1 Crime is positively related to the extent of the informal economy.

In addition to the main explanatory variable, we also include other factors, namely GDP growth, the rule of law, government effectiveness, trade openness, FDI, government expenditure, and control of corruption in the model. Different studies have shown that these factors can impact the dimension of the shadow economy (Buehn and Schneider 2012; Dreher and Schneider 2010; Esaku 2021; Fugazza and Fiess 2010; Malaczewska 2013; Goel and Saunoris 2014; Jamalmanesh et al. 2014; Luong et al. 2020). For instance, Jamalmanesh et al. (2014) analyzed data from 37 Asian economies in the 2000–2007 period. The authors conclude that an increase in GDP growth can reduce the informal sector in these countries. Similarly, both the rule of law and government effectiveness are adversely associated with the dimension of the informal economy. Luong et al. (2020) investigate a sample of 18 transition countries between 2002–2015 to uncover the antecedents of the shadow economy. They show that economic growth and the rule of law harm the informal economy size.

Other determinants of the size of the shadow economy can be trade openness and FDI. Specifically, Johnson et al. (1997) state that there exists a significant linkage between liberalization and the unofficial sector. Meanwhile, Fugazza and Fiess (2010) argue that the connection between informality and trade liberalization has not gained the attention it might deserve. The authors add that macro-founded data support a positive relationship between trade liberalization the informal sector, while micro-founded data may not reinforce this view. Regarding the effect of FDI, Nikopour et al. (2009) examine a sample of 145 economies at five data points, namely 1999/2000, 2001/2002, 2002/2003, 2003/2004, and 2004/2005. The authors claim that an amplify in FDI can decline the shadow economy. Nevertheless, Luong et al. (2020) find that the linkage between FDI and the informal economy in transition countries is not statistically significant.

Finally, government expenditure and control of corruption are variables included in our models. Goel and Saunoris (2014) uncover the determinants of the underground economy using a sample of 162 economies for 1990–2011. The authors indicate that other things equal, increased military spending leads to a decrement in the shadow economy. However, the linkage between non-military government spending and the informal economy is statistically insignificant. By contrast, Malaczewska (2013) discovers that a rise in useful government spending can promote the shadow economy under certain circumstances.

Meanwhile, corruption is often considered as a driver of underground economic activities. For instance, Dreher and Schneider (2010) report that corruption and the shadow economy are supplements in low-income countries. Buehn and Schneider (2012) investigate a sample of 51 nations worldwide for the 2000–2005 period and reveal a positive linkage between corruption control and the dimension of the underground economy. Likewise, Esaku (2021) analyzes time series data between 1984–2008 to uncover the effect of corruption on informality. The author concludes that corruption positively impacts the shadow economy in Uganda.

Based on previous studies, we formulate the following research hypotheses:

Hypothesis 2 GDP growth is negatively associated with the size of the shadow economy.

Hypothesis 3 Rule of law is negatively associated with the size of the shadow economy.

Hypothesis 4 Government effectiveness is negatively associated with the size of the shadow economy.

Hypothesis 5 Trade openness is positively related to the dimension of the informal economy.

Hypothesis 6 FDI is negatively related to the dimension of the informal economy.

Hypothesis 7 Government expenditure is positively related to the dimension of the informal economy.

Hypothesis 8 Corruption control is positively related to the dimension of the informal economy.

3 Model, Methodology and Data

3.1 Model

To analyze the effect of crime on the shadow economy size of the BRICS countries, we use an econometric model of the following form:

$$SE = \alpha_0 + \beta_1 CI + \gamma_i Z + \varepsilon \quad (1)$$

To test the existence of a reversed U-shaped connection between crime and the shadow economy, the econometric representation in the regression approach is given by:

$$SE = \alpha_0 + \beta_1 CI + \beta_2 CIsquare + \gamma_i Z + \varepsilon \quad (2)$$

where SE = Shadow economy, CI = Crime index, CIsquare = the square of CI, Z is a vector of other determinants of the informal economy suggested by the literature (including GDPg = GDP growth, TO = Trade openness, FDI = Foreign direct investment, GovE = Government expenditure, GovEsquare = the square of GovE, RL = Rule of law, CC = Corruption Control, GE = Government effectiveness), and ε is error term.

3.2 Methodology

In this paper, we use a Bayesian approach. Bayesian statistics rely on Bayes' rule (Bayes 1763):

$$\Pr(\theta|\mathrm{y}) = \frac{\Pr(\mathrm{y}|\theta)\Pr(\theta)}{\Pr(\mathrm{y})} \quad (3)$$

where y is the data, and θ are unknown parameters. $\Pr(y|\theta)$ is the likelihood function, which is the probability of y given θ. $\Pr(\theta)$ is the prior distribution of θ. $\Pr(y)$ is the marginal distribution of y, and $\Pr(\theta|y)$ is the posterior distribution, which is the probability of θ given y.

When testing a hypothesis, the process of Bayesian analysis is as follows. First, we presume a normally distributed prior with a mean of zero for all parameters. Various researchers have used a priori normal distribution in previous empirical analyses (Block et al. 2011; Nguyen and Duong 2021; Nguyen et al. 2019a, b; Oanh et al. 2022; Thach 2021). A normal distributed prior will not tilt the outcomes of the Bayesian investigation about our hypotheses in either a negative or a positive direction (Block et al. 2011). Next, we presume normal distributions of the parameters coming from our models for the respective likelihood functions of the coefficients. In a third step, a simulation approach is used to update prior beliefs. We use MCMC techniques and Gibbs sampling to turn up at the corresponding univariate distributions of the parameters. The posterior distribution delivers a density function of the parameters.

3.3 Data

We use a panel data of the BRICS economies from 2001 to 2017. The dependent variable is a measure of the informal economy collected from Medina and Schneider (2019). The main explanatory variable, the crime index, comes from the online database of Numbeo. Control variables are from World Development Indicators (WDI), Worldwide Governance Indicators (WGI), and the Heritage Foundation. Table 1 describes the definitions and data sources of variables in the model.

Our dependent variable is crime, as measured by the crime index. The crime index, published by Numbeo, estimates the overall crime level in a given country. The Numbeo organization uses the scale [0, 100] for values since it is easier for users to read. Crime levels are classified into different groups: very low for crime levels below 20; low for crime degrees between 20 and 40; moderate for crime degrees between 40 and 60; high for crime degrees between 60 and 80; and very high for crime levels above 80.

Table 1 Variable definitions and data sources

Variables	Definition	Source
SE (shadow economy)	Size of informal economy as a percentage of GDP	Medina and Schneider (2019)
CI (crime index)	Ranges from 0 (very low crime rate) to 100 (very high crime rate)	Numbeo database
GDPg (GDP growth)	GDP growth (annual %)	WDI
TO (trade openness)	Trade as a percentage of GDP	WDI
FDI (foreign direct investment)	FDI inflows as a percentage of GDP	WDI
GovE (government expenditure)	Government expenditures as a percentage of GDP	Heritage Foundation
RL (rule of law)	Scopes from −2.5 (weak) to 2.5 (strong)	WGI
CC (corruption control)	Scopes from −2.5 (highly corrupt) to 2.5 (very clean)	WGI
GE (government effectiveness)	Scopes from −2.5 (weak) to 2.5 (strong)	WGI

Source Authors' synthesis

We employ an estimate of the size of the shadow economy (SE) developed by Medina and Schneider (2019). Specifically, they apply the multiple indicator-multiple causes (MIMIC) approach to estimate the dimension of the informal economy for more than 150 countries from 1991 to 2017. The MIMIC approach uses data from multiple causes and multiple indicators to determine the magnitude of the latent (unobservable) informal economy variable (Rocque et al. 2019; Schneider and Enste 2013).

We also include other variables that have been shown to affect the dimension of the shadow economy. First, GDP growth (GDPg) is measured by the growth of GDP. Second, trade openness (TO) is defined as the ratio of total trade to GDP. Third, FDI represents the FDI inflows comprising capital supplied by foreign direct investors to a foreign affiliate or capital received by foreign direct investors from a foreign affiliate (Giroud and Ivarsson 2020). All these variables come from the WDI. Fourth, government expenditure (GovE), including transfers and consumption, accounts for the entire score. This control variable comes from the Heritage Foundation. Fifth, the rule of law (RL) measures society's achievement in developing an environment where equitable and predictable rules form the foundation of social and economic interactions. Sixth, corruption control (CC), measure the perception of the level of corruption (see Kaufmann et al. 2004). Finally, government effectiveness (GE) encompasses the "perceptions of the quality of public services, the quality of the civil service and the degree of its independence from political pressures, the quality of policy formulation and implementation, and the credibility of the government's commitment to such policies" (Duho et al. 2020). These control variables come from the WGI.

Table 2 shows the shadow economy's descriptive statistics and the primary independent variable (crime index) for the BRICS countries. Among these economies, the dimension of the informal economy in China is the smallest, i.e., 12.91% of GDP, while Russia has the biggest dimension of the informal economy (37.71% of GDP). For crime rate, China is an economy with the lowest crime index (34.87). By contrast, South Africa has the highest crime index (77.65). Thus, China is a low crime rate country, India and Russia are moderate crime rates countries, and Brazil and South Africa are high crime rates countries.

Table 2 Descriptive statistics of the focus variables

Variables	Brazil	Russia	India	China	South Africa	BRICS
SE (shadow economy)	34.62	37.71	21.83	12.91	25.42	26.50
CI (crime index)	66.61	50.05	44.76	34.87	77.65	54.79

Source Authors' estimations

4 Empirical Results

Bayesian regression is not a point estimate but rather the whole distribution function of the effects of explanatory variables included in the corresponding regression (the posterior distributions). Therefore, Bayesian analysis allows announcing probable and unlikely parameter values of the respective coefficients. Bayesian results provide a credible interval to describe an effect that encloses the parameter with a specific probability (Block et al. 2011). First, if the probability that the effect of a parameter has a positive (or negative) effect is 90% or more, we classify the effect as extreme. Second, if the probability of an effect of a parameter is between 80 and 89%, we rank the effect as very strong. Third, if the probability of an effect of a parameter is between 70 and 79%, we classify the effect as strong. Fourth, if the probability of an effect of a parameter is between 60 and 69%, we rank the effect as moderate. Finally, if the probability of an effect of a parameter between 50 and 59%, we categorize the effect as anecdotal.

We use 12,500 draws from the posterior distribution for the simulation and discard the first 2,500 draws. Table 3 provides results for mean coefficients and the 95%-credible intervals for all parameters, aiming to test the hypotheses and the validity of our findings.

The results indicate that the mean parameters of the crime variable (CI) in models 1 and 2 are 0.2633 and 0.9952, respectively. Moreover, the probabilities of the CI coefficient in both models are above 99%. In other words, we find extreme evidence for the positive linkage between crime and the extent of the informal economy in the BRICS. This finding supports hypothesis H1 described in Sect. 2.

However, in model 2, the crime square (CIsquare) has a mean coefficient of −0.0090, while the negative probability impact of the variable is also above 99%. These numbers reveal a reversed U-shaped linkage connecting the crime and the extent of the informal economy in the BRICS countries. Specifically, the marginal effect of crime on the informal economy is positive as long as this informality's size does not exceed the turning point, i.e., 55.13. Once the crime value is higher than 55.13, the relationship becomes negative. We suppose that this can be because when the number of crimes exceeds a certain limit, governments can implement stricter regulations and policies, which might have an adverse effect on the shadow economy size.

Meanwhile, the mean parameters of GDP growth in models 1 and 2 are −1.2211 and −0.9801, respectively. The negative impact probability of this variable on the extent of the informal economy is above 98% in both cases. The results indicate that GDP growth negatively affects the informal economy magnitude. Hence, hypothesis H2 is validated for the BRICS economies.

Likewise, the rule of law (RL) and corruption control (CC) decrease the informal economy size in the BRICS as their mean parameters are negative in all analysis. Therefore, for these countries, the hypotheses H3 and H8 are valid.

Another factor that can impede the development of the underground economy is government effectiveness (GE). Particularly, the mean parameters of the variable in

Table 3 Results of Bayesian linear regression

Independent variables	Model (1)		Model (2)	
	Mean coefficient	Probability of mean coefficient	Mean coefficient	Probability of mean coefficient
CI	0.2633 [0.0872; 0.4527]	0.9982*	0.9952 [0.4292; 1.5552]	0.9994*
CIsquare	–	–	−0.0090 [−0.0140; −0.0041]	0.9996**
GDPg	−1.2211 [−2.2938; −0.0739]	0.9811**	−0.9801 [−1.5951; −0.3573]	0.9984**
TO	0.0704 [−0.1839; 0.3163]	0.7181*	0.1464 [0.0179; 0.2745]	0.9871*
FDI	1.2173 [−0.4417; 2.8375]	0.9275*	0.2363 [−0.8990; 1.3676]	0.6616*
GovE	0.1039 [−0.1306; 0.3370]	0.8138*	0.3077 [−0.2670; 0.8950]	0.8556*
GovEsquare	–	–	−0.0054 [−0.0107; −0.0003]	0.9803**
RL	−1.0246 [-3.0110; 0.9042]	0.8497**	−0.2977 [−2.1283; 1.5389]	0.6252**
CC	−1.2497 [−3.2695; 0.7328]	0.8904**	−0.4528 [−2.3497; 1.4317]	0.6818**
GE	−0.8318 [−2.8046; 1.1145]	0.7986**	−0.5898 [−2.4918; 1.3174]	0.7274**
_cons	0.3877 [−1.5848; 2.3484]	0.6506*	0.0030 [−1.9399; 1.9477]	0.5016*
var	40.4287 [20.8605; 74.6221]	–	7.7315 [4.2749; 13.7769]	–
Turning point	–		–	
CI*	–		55.13	
GovE*	–		28.41	

Notes Dependent variable: Shadow economy; 95% credible intervals in parentheses; * The probability of the mean coefficient is positive; ** The probability of the mean coefficient is negative
Source Authors' estimations

models 1 and 2 are −0.8318 and −0.5898, respectively. We find strong disadvantage effects of this factor on the informal economy in the BRICS. Thus, we suggest that for these economies, hypothesis H4 is also validated.

There is very strong evidence that an inverted U-shaped relationship links government expenditure (GovE) and the informal economy regarding the government expenditure variable. Moreover, the turning point is 28.41% of GDP. The results discovered that government expenditure might promote the level of the informal economy as long as it does not exceed the turning point. This finding is different from our original hypothesis H7.

Another factor that is also contrary to our initial prediction is FDI. The results reveal that FDI has a positive effect on informality in the BRICS economies. Thus, hypothesis H6 is rejected. As described in the literature review, the influence of government spending and FDI on the informal sector is still not unequivocal. We, therefore, suggest that further research is needed to gain a deeper understanding of the nature of these relationships.

Finally, the mean coefficients of trade openness (TO) in both cases are positive. The posterior probability of this variable in models 1 and 2 are 71.81% and 98.71%, respectively. Otherwise stated, a rise in trade openness can decrease the dimension of informality in the BRICS economies. Hence, hypothesis H5 is valid for the economies examined.

5 MCMC Diagnostics

As mentioned in Sect. 3.2, in Bayesian estimation, the posterior distributions of the parameters are generated through the MCMC method. Kruschke (2015) contends that stability and convergence should be considered when using the MCMC sampling algorithm to generate posterior distributions. The stability of the Bayesian linear regression is decided by the effective sample size (ESS). To get the best ESS, the researchers believe that it can be based on the effective sampling index with a threshold value of 0.01 (Nguyen and Duong 2021; Thach 2021).

Meanwhile, the famous test when performing multi chains is the convergent diagnosis Rc of Gelman-Rubin. Rc compares the variance between the chains against the variance within the chains—the more similar the variance of each chain, the greater the degree of convergence. The chains fully converge when Rc is less than 1.1 (Gelman and Rubin 1992; Kruschke 2015; Oanh et al. 2022).

Table 4 presents the MCMC diagnostic results for models 1 and 2. The results show that the sampling efficiency indexes of the parameters in both models are above 0.01, so the Gibbs sampling algorithm is stable. Besides, the Rc values of all parameters in both models are less than 1.1, so the MCMCs converge. Therefore, the diagnostic results indicate that the MCMC algorithm is stable and convergent (Table 4).

Table 4 Results of MCMC diagnostics

Independent variables	Model (1)			Model (2)		
	ESS	Efficiency	Rc	ESS	Efficiency	Rc
CI	27,598	0.9199	1.00003	30,000	1.0000	1.00001
CIsquare	–	–	–	30,000	1.0000	1.00000
GDPg	29,160	0.9720	1.00002	29,411	0.9804	1.00002
TO	29,304	0.9768	1.00016	30,000	1.0000	1.00001
FDI	27,346	0.9115	1.00004	30,000	1.0000	0.99999
GovE	30,000	1.0000	1.00012	30,000	1.0000	1.00011
GovEsquare	–	–	–	30,000	1.0000	1.00013
RL	25,711	0.8570	1.00007	30,000	1.0000	0.99999
CC	22,547	0.7516	1.00017	30,000	1.0000	1.00000
GE	26,729	0.8910	1.00000	29,166	0.9722	1.00002
_cons	29,766	0.9922	1.00014	30,000	1.0000	1.00023
var	14,842	0.4947	1.00000	18,748	0.6249	1.00029

Notes Dependent variable: Shadow economy
Source Authors' estimations

6 Concluding Remarks

Crime and the informal economy occur in many countries worldwide and are increasingly attracting the attention of researchers. As suggested by Monticelli (2014, p. 1), criminology and economics "are irretrievably intertwined". However, there is still little research on the relationship between these two indicators. Therefore, the present paper complements the literature by exploring the association between crime and the shadow economy size in the BRICS countries for the period 2001–2017 using Bayesian regression approach.

The findings indicate that crime is positively associated with the dimension of the informal economy as long as the crime value does not exceed a turning point. Once this value is higher than the turning point, the effect of crime on the informal sector will change. We suppose that this can be because governments may implement more stringent measures that impact the shadow economy when the extent of crime exceeds a certain level.

Moreover, we find that factors including GDP growth, the rule of law, corruption control, and government effectiveness have a detrimental effect on the informal economy magnitude in the BRICS. This result of the paper suggests that governments may need to promote economic growth, improve governance and institutional quality, and reduce corruption to curb the shadow economy. In contrast, trade openness and FDI are positively related to the dimension of the informal economy. Meanwhile, government expenditure has a positive impact on this informal sector when the spending is below a certain level.

Finally, both crime and the shadow economy are complex phenomena. The relationship is therefore probably complicated and may involve numerous motives. This study uses the crime index derived from the Numbeo organization. The index ranges from 0 (i.e., very low crime rate) to 100 (i.e., very high crime rate). Nevertheless, further studies can compare the impact of certain criminal groups, such as the white-collar criminals, the organized crime, and the armed robbery, on the size of the underground economy between continentals and countries. Moreover, we believe that more studies will be needed to further explore other determinants and the nature of the shadow economy.

References

Achim, M.V., Borlea, S.N., Găban, L.V., Cuceu, I.C.: Rethinking the shadow economy in terms of happiness. Evidence for the European union member States. Technol. Econ. Dev. Econ. **24**(1), 199–228 (2018). https://doi.org/10.3846/20294913.2016.1209250

Akerlof, G.A., Dickens, W.T.: The economic consequences of cognitive dissonance. Am. Econ. Rev. **72**(3), 307–319 (1982)

Arrow, K.J.: Social responsibility and economic efficiency. Public Policy **21**(3), 303–317 (1973)

Bayes, T.: LII. An essay towards solving a problem in the doctrine of chances. By the late Rev. Mr. Bayes, FRS communicated by Mr. Price, in a letter to John Canton, AMFR S. Philos. Trans. Royal Soc. Lond. **53**, 370–418 (1763)

Becker, G.S.: Crime and punishment: an economic approach. J. Polit. Econ. **76**(2), 169–217 (1968)

Becker, G.S., Stigler, G.J.: Law Enforcement, malfeasance, and compensation of enforcers. J. Legal Stud. **3**(1), 1–18 (1974). Retrieved from http://www.jstor.org/stable/724119

Block, J.H., Jaskiewicz, P., Miller, D.: Ownership versus management effects on performance in family and founder companies: a Bayesian reconciliation. J. Fam. Bus. Strat. **2**(4), 232–245 (2011). https://doi.org/10.1016/j.jfbs.2011.10.001

Buehn, A., Schneider, F.: Corruption and the shadow economy: like oil and vinegar, like water and fire? Int. Tax Public Financ. **19**(1), 172–194 (2012). https://doi.org/10.1007/s10797-011-9175-y

Dell'Anno, R.: The shadow economy in portugal: an analysis with the mimic approach. J. Appl. Econ. **10**(2), 253–277 (2007). https://doi.org/10.1080/15140326.2007.12040490

Dreher, A., Schneider, F.: Corruption and the shadow economy: an empirical analysis. Public Choice **144**(1), 215–238 (2010). https://doi.org/10.1007/s11127-009-9513-0

Duho, K.C.T., Amankwa, M.O., Musah-Surugu, J.I.: Determinants and convergence of government effectiveness in Africa and Asia. Public Administr. Policy **23**(2), 199–215 (2020). https://doi.org/10.1108/PAP-12-2019-0039

Esaku, S.: Does corruption contribute to the rise of the shadow economy? empirical evidence from Uganda. Cogent Econ. Finan. **9**(1), 1932246 (2021). https://doi.org/10.1080/23322039.2021.1932246

Fleisher, B.M.: The effect of unemployment on juvenile delinquency. J. Polit. Econ. **71**(6), 543–555 (1963)

Fugazza, M., Fiess, N.M.: Trade Liberalization and Informality: new Stylized Facts. UN Policy Issues Int. Trade Commodities Study Series **43**, 1–47 (2010)

Gelman, A., Rubin, D.B.: Inference from iterative simulation using multiple sequences. Stat. Sci. **7**(4), 457–472, 416 (1992). doi:https://doi.org/10.1214/ss/1177011136

Giroud, A., Ivarsson, I.: World Investment Report 2020: International production beyond the pandemic. J. Int. Bus. Policy **3**(4), 465–468 (2020). https://doi.org/10.1057/s42214-020-00078-2

Goel, R.K., Saunoris, J.W.: Military versus non-military government spending and the shadow economy. Econ. Syst. **38**(3), 350–359 (2014). https://doi.org/10.1016/j.ecosys.2013.12.004

Gottschalk, P., Gunnesdal, L.: White-Collar Crime in the Shadow Economy: Lack of Detection, Investigation and Conviction Compared to Social Security Fraud: Springer Nature (2018)

Habibullah, M., Eng, Y.K.: Crime and the underground economy in Malaysia: are they related? J. Global Bus. Manag. **2**(1), 138–154 (2011)

Jamalmanesh, A., Meidani, A.A.N., Mashhadi, M.K.: Government effectiveness, rule of law and informal economy in Asian developing countries. Int. J. Econ. Manag. Soc Sci. **3**(10), 551–555 (2014)

Johnson, S., Kaufmann, D., Shleifer, A., Goldman, M.I., Weitzman, M.L.: The unofficial economy in transition. Brook. Pap. Econ. Act. **1997**(2), 159–239 (1997). https://doi.org/10.2307/2534688

Katsios, S.: Tourism in 'Yellow Times': The De-formalisation of the Greek Economy and Its Impact on Tourism. In: Papathanassis, A., Katsios, S., Dinu, N.R. (eds.) Yellow Tourism: Crime and Corruption in the Holiday Sector, pp. 209–225. Springer International Publishing, Cham (2019)

Kaufmann, D., Kraay, A., Massimo, M.: Governance Matters III: Governance Indicators for 1996, 1998, 2000, and 2002. World Bank Econ. Rev. **18**(2), 253–287 (2004)

Kruschke, J.K.: Doing Bayesian Data Analysis, 2nd edn. Academic Press, Boston (2015)

Lee, D.: How does social capital reduce the size of the shadow economy? Glob. Econ. Rev. **42**(3), 251–268 (2013). https://doi.org/10.1080/1226508X.2013.833846

Li, L.: BRICS: a limited role in transforming the world. Strateg. Anal. **43**(6), 499–508 (2019). https://doi.org/10.1080/09700161.2019.1677017

Luong, T.T.H., Nguyen, T.M., Nguyen, T.A.N.: Rule of law, economic growth and shadow economy in transition countries. J. Asian Finan. Econ. Bus. **7**(4), 145–154 (2020). https://doi.org/10.13106/jafeb.2020.vol7.no4.145

Malaczewska, P.: Useful government expenditure influence on the shadow economy. Quant. Methods Econ. **14**(2), 61–69 (2013)

Medina, L., Schneider, F.: Shedding Light on the Shadow Economy: A Global Database and the Interaction with the Official One. Center for Economic Studies and ifo Institute (CESifo). Munich, Germany (2019)

Monticelli, D.M.: Crime and the Economy. In: Albanese, J.S. (ed.) The Encyclopedia of Criminology and Criminal Justice, pp. 1–5. Blackwell Publishing Ltd., New Jersey (2014)

Naylor, R.T.: From underworld to underground. Crime Law Soc. Chang. **24**(2), 79–150 (1995). https://doi.org/10.1007/BF01298379

Nguyen, D.V., Duong, M.T.H.: Shadow economy, corruption and economic growth: an analysis of BRICS countries. J. Asian Finan. Econ. Bus. **8**(4), 665–672 (2021). https://doi.org/10.13106/jafeb.2021.vol8.no4.0665

Nguyen, H.T., Sriboonchitta, S., Thach, N.N.: On Quantum Probability Calculus for Modeling Economic Decisions. In: Kreinovich, V., Sriboonchitta, S. (eds.) Structural Changes and their Econometric Modeling, pp. 18–34. Springer International Publishing, Cham (2019a)

Nguyen, H.T., Thach, N.N.: A Closer Look at the Modeling of Economics Data. In: Kreinovich, V., Thach, N.N., Trung, N.D., Van Thanh, D. (eds.) Beyond Traditional Probabilistic Methods in Economics, pp. 100–112. Springer International Publishing, Cham (2019)

Nguyen, H.T., Trung, N.D., Thach, N.N.: Beyond Traditional Probabilistic Methods in Econometrics. In: Kreinovich, V., Thach, N.N., Trung, N.D., Van Thanh, D. (eds.) Beyond Traditional Probabilistic Methods in Economics, pp. 3–21. Springer International Publishing, Cham (2019b)

Nikopour, H., Shah Habibullah, M., Schneider, F., Law, S.H.: Foreign Direct Investment and Shadow Economy: A Causality Analysis Using Panel Data. University Library of Munich, Germany (2009)

Oanh, T.T.K., Diep, N.V., Truyen, P.T., Chau, N.X.B.: The impact of public expenditure on economic growth of provinces and cities in the southern key economic zone of Vietnam: bayesian approach. In: Ngoc Thach, N., Ha, D.T., Trung, N.D., Kreinovich, V. (Eds.), Prediction and Causality in Econometrics and Related Topics, pp. 328–344. Springer International Publishing, Cham (2022)

Petersen, H.G., Thießen, U., Wohlleben, P.: Shadow economy, tax evasion, and transfer fraud—definition, measurement, and data problems. Int. Econ. J. **24**(4), 421–441 (2010). https://doi.org/10.1080/10168737.2010.525973

Rocque, M., Saunoris, J.W., Marshall, E.C.: Revisiting the relationship between the economy and crime: the role of the shadow economy. Justice q. **36**(4), 620–655 (2019). https://doi.org/10.1080/07418825.2018.1424230

Schneider, F.: Illegal activities and the generation of value added: size, causes and measurement of shadow economies. Bull. Narc. **52**(1/2), 99–134 (2000)

Schneider, F.: The influence of public institutions on the shadow economy: An empirical investigation for OECD countries. Rev. Law Econ. **6**(3), 441–468 (2010). https://doi.org/10.2202/1555-5879.1542

Schneider, F., Enste, D.H.: The Shadow Economy: An International Survey, 2nd edn. Cambridge University Press, Cambridge (2013)

Thach, N.N.: How values influence economic progress? evidence from south and Southeast Asian Countries. In: Ngoc Thach, N., Kreinovich, V., Trung, N.D. (Eds.), Data Science for Financial Econometrics, Vol. 898, pp. 207–221. Springer International Publishing, Cham (2021)

Torgler, B., Schneider, F.: The impact of tax morale and institutional quality on the shadow economy. J. Econ. Psychol. **30**(2), 228–245 (2009). https://doi.org/10.1016/j.joep.2008.08.004

Williams, C.C., Horodnic, I.A.: Explaining and tackling the shadow economy in Estonia, Latvia and Lithuania: a tax morale approach. Baltic J. Econ. **15**(2), 81–98 (2015). https://doi.org/10.1080/1406099X.2015.1114714

Witte, A.D.: Urban crime: issues and policies. Hous. Policy Debate **7**(4), 731–748 (1996). https://doi.org/10.1080/10511482.1996.9521241

Impact of Financial Leverage on Investment Decision: Case of Enterprises Listed on Ho Chi Minh Stock Exchange (Hose) in Vietnam

Nguyen Ngoc Thach and Nguyen Thi Nhu Quynh

Abstract Investment decision is the most important among the financial decisions of an enterprise. The purpose of this paper is to analyze the impact of financial leverage on the investment decision of enterprises listed on Ho Chi Minh Stock Exchange (HOSE) in Vietnam for the period 2011–2018. Different from a large number of previous studies, this paper employs Bayesian mixed-effects regression and finds that financial leverage has a positive relationship with investment activities. Besides, the results show that cash flow, sales and GDP growth also exert positive effects on investment decisions whereas inflation rate is likely to decrease the investment of enterprises.

Keywords Bayesian mixed-effects · Financial leverage · Investment decision · Non-financial enterprises · HOSE

1 Introduction

According to Damodaran (2010), investment decision is the most important among the three decisions of enterprises including investment, financing, and asset management to increase firm value. The author argued that enterprises whether large or small, listed or unlisted when they had the right investment decision, the firm value would increase. On the other hand, a wrong investment decision may cause a decrease in the value of corporate. In addition, corporate investment activity is one of the main factors to promote economic growth. It enhances capital accumulation, employment, and reduction in the poverty rate, which improves the quality of life for citizens in a country (Onwe and Olarenwaju 2014).

N. Ngoc Thach · N. T. N. Quynh (✉)
Banking University Ho Chi Minh City, 36, Ton That Dam Street, District 1, Ho Chi Minh City 700000, Vietnam
e-mail: quynhntn@buh.edu.vn

N. Ngoc Thach
e-mail: thachnn@buh.edu.vn

N. Ngoc Thach et al. (eds.), *Financial Econometrics: Bayesian Analysis, Quantum Uncertainty, and Related Topics*, Studies in Systems, Decision and Control 427,
https://doi.org/10.1007/978-3-030-98689-6_18

Like all other economies, Vietnam has been much affected by the current Covid-19 pandemic. The epidemics and natural disasters in 2020 have severely influenced Vietnam's economy and society. It has made production activities and supply chains disrupted. In particular, businesses are heavily affected with many enterprises having to suspend operations, even going bankrupt, dissolving, or reducing production. Hence, enterprises need to take measures to maintain business operation, recover and expand corporate investment activities.

Among many factors determining the investment decisions of enterprises, financial leverage has drawn a great deal of interest among scholars. According to Aivazian et al. (2005), a central issue in corporate finance is the relationship between financial leverage and investment decision. There are some previous investigations conducted on this topic such as Vo (2019), Aivazian et al. (2005), Umutlu (2010). However, these research results are inconclusive. Aivazian et al. (2005) and Vo (2019) demonstrate that financial leverage negatively affects firm investment decisions, whereas Umutlu (2010) shows that the influence of financial leverage on investment largely depended on estimation approaches. Furthermore, most previous studies have used frequentist estimation methods. In applying these methods, the accuracy of estimation results is dependent on only data without incorporating prior information (Thach 2019). Besides, these studies are dedicated to advanced countries. Some works analyze the linkage between financial leverage and investment decision of Vietnamese enterprises (for example, Vo (2019), Nguyen et al. (2019a, b)), but they employed outdated data.

This paper aims to investigate the impact of financial leverage on investment decisions in the enterprises listed in the Ho Chi Minh Stock Exchange market (HOSE) in Vietnam during 2011–2018. The research has some important contributions to the existing literature. Firstly, it conducted an in-depth analysis of the influence of financial leverage on investment decisions in the listed firms of an emerging economy like Vietnam with a data set updated up to 2018. Secondly, distinguishing from the earlier works, our analysis is carried out within the Bayesian framework, which provides a robust empirical foundation for policymaking. The Bayesian approach has several advantages over more traditional frequentist methods that will be presented in the methodology section. The third is that we intend to classify the key factors impacting firms' investment into two groups, including firm-specific factors and macroeconomic conditions.

The remainder of this paper is organized as follows. Section 2 presents the literature review; Sect. 3 describes data, model, and methodology; Sect. 4 analyzes empirical results, and finally, we provide main conclusions and suggest some policy recommendations in Sect. 5.

2 Literature Review

According to Ross et al. (2014), financial leverage indicates the extent to which debt is used in a firm's capital structure. The more debt financing a firm, the more financial leverage it uses. Up to now, there are a large number of analyses performed on the link between financial leverage and a firm's investment. In 1958, proposing the M&M theory, Modigliani and Miller (1958) indicated that leverage is unrelated to investment choice and the value of a firm. However, Myers (1977) implied that increasing debt levels reduces the motivation of the management and shareholder coalition in the regulation of enterprises to invest in positive NPV (net present value) investment opportunities because the outcome of the investment must be shared more with the bondholders. For this reason, the management can miss the profitable projects (underinvestment) which reduces the firm value when the debt ratio is higher. The theory about the relationship between leverage and investment commences the conflict of interests between managers and shareholders (the alternative theory) and documents that managers tend to expand the firm size at the cost of reducing the wealth of shareholders (overinvestment). This implies that when the firm takes on poor projects although their managers can make more work for themselves and raise their power in the enterprise it decreases the firm's value. Considering the relation between leverage and growth, Lang et al. (1996) use a large sample of US industrial firms from 1970 to 1989. The research reveals a strong negative between leverage and investment in subsequent years. At the same time, the paper distinguishes between the influence of leverage on growth in a firm's core business and its non-core business. The results show that leverage is not the cause of reducing growth for firms that have good investment opportunities, but its negative impact on growth for firms whose growth opportunities are either not recognized by the capital markets or are not sufficiently valuable to solve the consequences of excessive debt. Furthermore, Aivazian et al. (2005) use the information on 1035 major Canadian industrial companies over the period from 1982 to 1999 to investigate the impact of financial leverage on investment decisions. The result shows that leverage is strongly negatively related to firms' investment. This negative impact is stronger for firms with lower growth opportunities than ones with higher growth opportunities. Lin and Wong (2008) access 1203 companies listed on the Shenzhen Stock Exchange or Shanghai Stock Exchange in China in the period from 1991 to 2004 to examine the link between leverage and investment activities. The authors discover that leverage has a negative impact on investment. Besides, leverage has a weaker negative impact on investment in firms that have low growth opportunities and poor operating performance. This relationship is weaker in a firm with a higher level of state shareholding. Umutlu (2010) uses data on Turkish non-financial firms listed on the Istanbul Stock Exchange showing that leverage has a negative impact on investment for firms with low Tobin's Q. However, under the adjustment of the model, the link between financial leverage and firm investment disappears. Franklin and Muthusamy (2011) use a data sample of Indian pharmaceutical companies for 1998–2009. The results indicate that leverage has a significant positive influence on investment. However, a negative relationship between leverage and investment exists for medium firms, whereas a positive relationship is discovered for large firms. Exploring the influence of corporate debt on investment in 920 thousand firms of five peripheral euro area countries (including listed and unlisted in Italy, Spain, Portugal, Greece, and Slovenia) during 2005–2014, Gebauer et al. (2018) reveal a non-linear relationship between firms' leverage and investment. The authors identify a threshold beyond which leverage has a significant and negative impact on investment. Using a sample of 280 non-financial firms listed in the Pakistan Stock Exchange during 2000–2018, Ahmad et al. (2021) indicated that leverage is the main factor affecting investment decisions and asymmetric information increases its adverse effect on the investment of corporates.

In Vietnam, using a data sample of 2500 small and medium enterprises, Nguyen et al. (2019a, b) find a positive relationship between financial leverage and investment decisions. This paper shows that the firms with higher financial leverage are more likely to take external financing sources than internal ones. Contrary to Nguyen et al. (2019a, b), Vo (2019) reveals a negative relationship between financial leverage and firm investment analyzing a panel of enterprises listed in HOSE during 2006–2015.

In sum, the topic of the link between financial leverage and corporate investment has attracted lots of attention from many scholars. However, the empirical results are inconclusive. In Vietnam, this relation is controversial as well. That is why the research is conducted to clarify this relation by supplementing new empirical evidence on the impact of financial leverage on firm investment decisions in an emerging country—Vietnam. From the reliable results obtained through a Bayesian approach, the authors will propose some useful recommendations to enhance investment activity of the firms listed on HOSE.

3 Method, Model and Data

3.1 Methodology

The current research applies the multilevel (mixed-effects) perspective in a Bayesian framework to estimate the effect of financial leverage on the investment decision of enterprises listed on HOSE.

Firstly, social sciences research has increasingly involved multilevel or hierarchically nested data sets. That is because units of analysis or observations at level 1 are nested within units of analysis at level 2, and so forth. When hierarchically nested data are encountered, we need to employ econometric techniques that account for such nesting to analyze them. As recommended by Nezlek (2001), the outcomes of modeling hierarchically nested data set not taking into consideration the multilevel nature of the data might be imprecise. The multilevel framework is suitable when data are gathered at multiple levels simultaneously. In this regard, "levels" relate to how the data are organized, but more important statistically, to whether units of analysis are dependent. This is the one of major advantages of multilevel (or mixed-effects) analysis over the simple-level OLS regression. In our current study on the RCEP, countries in this bloc share whatever features the bloc has, and the financial development observations the countries provide have the individual features of countries. In this case, the financial development observations, i.e., individual-level or level 1 observations are not independent. The absence of independence implies that a traditional OLS regression where individual-level observations should be referred to as independent cannot be applied since a fundamental assumption of observation independence is violated. The details of multilevel modeling are presented in Raudenbush and Bryk (2002), Snijders and Boskers (1999), Kreft and de Leeuw (1998), and Nezlez (2001).

Secondly, in recent decades, the Bayesian probabilistic approach has been popularized in the context of a crisis deepening in frequentist statistics (Anh et al. 2018; Nguyen and Thach 2018; Nguyen et al. 2019a, b; Kreinovich et al. 2019; Tuan et al. 2019; Thach 2020; Thach et al. 2020, 2021; Thach and Ngoc 2021). The Bayesian framework is argued to have strong advantages over standard inference. First, unlike point estimates in frequentist inference, Bayesian outcomes are the entire posterior distribution of a particular parameter and the Bayesian approach thus allows for probabilistic statements such as a variable is likely to impact on another one or a 90% probability of the true value of a parameter falling into an interval. Second, frequentist methods drop non-significant though potentially impacting variables out of the analysis, while Bayesian ones take into account all the variables. Third, thanks to combining prior information with available data, Bayesian inferential results are more accurate as a limit is not set to data sample size.

In order to estimate the effect of financial leverage on investment decisions in the listed enterprises of interest, this work employs a Bayesian mixed-effects regression, in which both models with and without random effects (intercepts) are estimated. Random intercepts reflect variations across countries in initial investment level. We incorporate GDP and CPI as two control variables in the research model. A Bayes factor test and a model test will be used to choose the more appropriate model. Concerning prior distributions, Lemoine (2019) strongly proposes informative priors, and Block et al. (2011, 2012) recommend standard Gaussian distributions for model parameters.

3.2 Model

To examine the impact of financial leverage on investment decisions in the enterprises listed on HOSE, according to Ahmad et al. (2021), Vo (2019), Aivazian et al. (2005), Lang et al. (1996), the authors estimate a reduced investment equation as follows:

$$\begin{aligned} INV_{i,t} =& \alpha_0 + \alpha_1 Leverage_{i,t-1} + \alpha_2\left(\frac{CF_{i,t}}{K_{i,t-1}}\right) + \alpha_3\left(Sale_{i,t-1}/K_{i,t-1}\right) \\ &+ \alpha_4 GDP_t + \alpha_5 CPI + \mu_i + \varepsilon_{i,t} \end{aligned} \tag{1}$$

where i and t refer to enterprise and year, respectively; α_0 is the constant, μ_i and$\varepsilon_{i,t}$ are firm and time fixed effect.

In Eq. (1), $INV_{i,t}$ is the dependent variable, as a proxy for the investment policy of firm i in year t. This variable is calculated by using the formula: $INV_{i,t} = \frac{I_{i,t}}{K_{i,t-1}}$ where $I_{i,t}$ is the net investment of firm i in year t, which is calculated by a book value of tangible fixed assets in year t minus book value of its tangible in year t—1 plus depreciation at the t (Ahmad et al. 2021; Vo 2019); $K_{i,t-1}$ is lagged net fixed assets; $Leverage_{i,t-1}$ is the lagged leverage, which is calculated by the ratio of book value of total liabilities to book value of total assets of firm i at time t—1 (Lang et al. 1996), (Umutlu 2010); $CF_{i,t}$ is cash flow of firm i at time t; $Sale_{i,t-1}$ is lagged net sales of firm i, GDP_t and CPI_t are control variables reflecting economic growth and inflation rate at time t.

3.3 Data Description

In this paper, we use a data set of 148 non-financial enterprises listed at HOSE in the period 2011–2018 to analyze the impact of financial leverage on the investment decisions of the enterprises. To achieve the research goal, we use both firm-level and country-level data. Firm-level data are taken from the audited financial statements whereas country-level data from the database of the International Monetary Fund (IMF).

Table 1 provides the descriptive statistics of the key variables in the model, the research omitted observations with non-available data, so the total observations are 1,112. In which, the average value of firm investment (INV) is 1.524, in which the lowest and the highest value is −1.901 and 47.151, respectively. This indicates a diversity of the investment activities of the listed firms and the reason is that these enterprises operate in various industries. The mean value of financial leverage is 0.474 with a standard deviation of 0.210, which shows that the level of debt varies across firms and these firms are facing financing constraints in Vietnam. Cash flow (CF/Fixed assets) and revenue (Sale/Fixed assets) have an average value of 2.276 and 10.134, respectively. In the period 2011–2018, the average GDP and CPI in Vietnam are 0.062 and 0.060, respectively.

Table 1 Summary statistics of the main variables

Variable	Mean	Std. Dev	Min	Max
INV (Net investment$_{i,t}$/Fixed assets $_{i,t—1}$)	1.524	3.628	−1.901	47.151
Leverage$_{i,t—1}$	0.474	0.210	0.002	0.931
CF$_{i,t}$/ Fixed assets $_{i,t—1}$	2.276	5.070	−4.442	55.763
Sale$_{i,t-1}$ / Fixed assets $_{i,t—1}$	10.134	18.172	0.000	183.913
GDP	0.062	0.006	0.052	0.071
CPI	0.060	0.052	0.006	0.187

Source The authors' calculations

4 Bayesian Simulation Results

4.1 MCMC Convergence Diagnostics

Comparison results of the Bayes factor test and model test show that the model without random effects better performs, so it is selected for further analysis. Let us inspect its MCMC convergence before proceeding to inference. The initial indicators such as efficiency and acceptance rate greatly influence chain convergence. Efficiency indicates the mixing properties of MCMC sequences. High-efficiency rate shows that MCMC sequences mix well, whereas low efficiency implies bad mixing in the simulated MCMC sample. A rate of acceptance is the proportion of the accepted proposals of model parameters to total proposals. An efficient MCMC algorithm has an acceptance rate between 15 and 50% (Roberts and Rosenthal 2001) and thus sufficiently large effective sample size (ESS) for all model parameters (all efficiencies should be more than 0.01). In our case, the acceptance rate is 33% and all the efficiencies are higher than 0.13, which are certainly acceptable for an MCMC sampler.

To more fully explore the convergence issues of MCMC, we need to practice trace plots and autocorrelation plots (Kass 1997). Besides these visual tests, ESS as a numeric method of testing can be adopted. Figure 1a, b demonstrate that trace plots for all the model parameters traverse quickly through a posterior domain; all the autocorrelation plots have no lags or die off after a few positive lags.

Table 2 denotes that all the parameters of the model have an efficiency of more than 0.13, while the warning level is 0.1. Furthermore, all the correlation times are relatively small.

Thus, all the above allows for a conclusion that MCMC sequences have converged to the desired distribution and we can proceed to inference.

(a) (b)

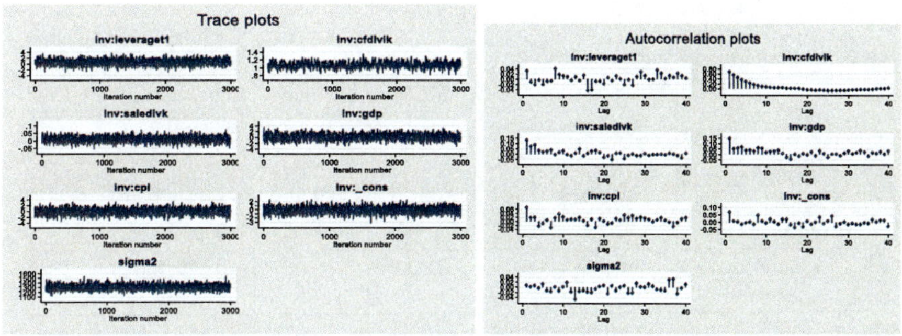

Fig. 1 Graphical tests for MCMC convergence. Source The authors' calculations

Table 2 Effective sample size

	ESS	Corr. time	Efficiency
INV			
$Leverage_{i,t-1}$	3000.000	1.000	1.000
$CF_{i,t}$/ Fixed assets $_{i,t-1}$	401.880	7.460	0.134
$Sale_{i,t-1}$ / Fixed assets $_{i,t-1}$	1669.150	1.800	0.556
GDP	1288.760	2.330	0.430
CPI	2617.360	1.150	0.873
_cons	2631.360	1.140	0.877
sigma2	3000.000	1.000	1.000

Source *The authors' calculations*

4.2 Interpretation of Empirical Results

The model summary in Table 3 reports that for all the model parameters, Monte Carlo chain standard error (MCSE) estimates are close to one decimal, which is reasonable for an MCMC algorithm. In general, the lower MCSE, the greater preciseness mean estimates obtain. Contrary to frequentist inference, in Bayesian inference, 95% credible intervals indicate which range the true value of a certain parameter belongs to, e.g., the mean value of the variable $Leverage_{t-1}$ lies in an interval between -1.44 and 2.27 with a 95% probability, and so on.

In view of probability, we can state that variables financial leverage of lag year ($Leverage_{i,t-1}$), and sales ($Sale_{i,t-1}$ /Fixed $assets_{i,t-1}$) have positive effects on the investment decisions (INV) of the enterprises, in particular, variable cash flow ($CF_{i,t}$/Fixed $assets_{i,t-1}$) strongly positively contributes to investment decision. The effect of variables GDP growth rate (GDP) is positive, whereas, in contrast, inflation rate (CPI) exerts a negative impact on the investment decisions of the enterprises. The mean value of the intercept takes the negative sign.

Table 3 Posterior model summary

	Mean	Std. Dev	MCSE	Median	Equal-tailed [95% Cred. Interval]
INV					
$Leverage_{i,t-1}$	0.390	0.947	0.017	0.389	−1.441; 2.270
$CF_{i,t}$/ Fixed assets $_{i,t-1}$	1.058	0.078	0.004	1.058	0.906; 1.210
$Sale_{i,t-1}$ / Fixed assets $_{i,t-1}$	0.008	0.019	0.000	0.007	−0.029; 0.043
GDP	0.020	0.994	0.028	0.021	−1.939; 1.995
CPI	-0.042	1.008	0.020	-0.016	−2.075; 1.955
_cons	-0.161	0.766	0.015	-0.161	−1.659; 1.371
sigma2	1311.433	55.293	1.010	1310.751	1206.124; 1424.788

Source *The authors' calculations*

The empirical results obtained the following economic meanings:

Firstly, the results demonstrate that financial leverage positively affects the firms' investment decisions. That implies that when firms increase liabilities, their investment activities are improved. Although this result contradicts the financial theory of Myers (1977) as well as some previous studies (for example, Vo (2019), Aivazian et al. (2005)). However, this finding is consistent with Trinh et al. (2017) and Nguyen et al. (2019a, b) and practice in Vietnam. In Vietnam where most small and medium enterprises are operating, according to the Vietnam Association of Small and Medium Enterprises, small and medium firms account for 97% of total enterprises in Vietnam.[1] Since these enterprises do not have capital resources enough to make an optimal investment, so they tend to use external capital to expand their investment activities (Franklin and Muthusamy 2011). Hence, in order to enhance investment activities, they use debt or non-interest financing sources to finance projects.

Secondly, besides financial leverage, from the results in Table 3, cash flow has a strong positive on investment decisions, implying that firms with higher cash flow invest more. This finding is consistent with Aivazian et al. (2005), Franklin and Muthusamy (2011), Vo (2019). As usual, enterprises prefer internal to external capital because of the low cost of the former as well as does not dilute control of the businesses. For this reason, when more internal capital encourages businesses to expand their investment. In addition, the sales variable is also positively related to investment decisions of the enterprises listed on HOSE. This is consistent with the Vietnamese circumstances, where the firms that achieve large sales are able to increase earning, which will create incentives for them to seek more investment projects.

Thirdly, regarding the macroeconomic conditions, they also affect the firms' investment. GDP growth exerts a positive effect whereas the inflation rate has a negative one on firms' investment decisions. These findings are consistent with Vietnam. According to Anwer and Sampath (1999), in case of GDP increases, a government tends to spend more on building infrastructure. This leads to an increase in the marginal productivity of capital and labor in the private sector, which in turn encourages firms to invest more. Besides, when inflation rate rises, the real income of the economy reduces. This may be the reason why businesses shrink investments.

In sum, among the factors influencing the investment decisions of the enterprises listed in Vietnam, debt motivates the enterprises to increase investment and so is used as a cushion for financing the projects. Like leverage, other factors, cash flow, sales, and economic conditions, similarly affect the firms' investment decisions.

---

[1] Access from http://www.mpi.gov.vn/en/Pages/tinbai.aspx?idTin=49802&idcm=133, on 4:40, 5th, July, 2021.

5 Conclusion

This paper analyses the influence of financial leverage on firms' investment decisions. Based on a data sample consisting of 147 non-financial enterprises listed on HOSE during 2011–2018, by using the Bayesian mixed-effects regression, the results point out that liabilities have a positive effect on the investment decisions of the studied firms. This result contradicts the theory of Myers (1977) as well as some previous studies (for example, Vo (2019), Aivazian et al. (2005)), but it is consistent with Nguyen et al. (2019a, b) and the private sector in Vietnam, where small and medium enterprises make up the economy. Besides, cash flows, sales, and macroeconomic conditions also affect the firms' investment decisions.

From the above results, the authors suggest some important recommendations as follows: (1) in order to increase access to debts, the corporate financial management needs to develop a reasonable business plan to enhance the reliability of the owners; (ii) the research results indicate that cash flow is strongly and positively related to the firms' investment decisions and this requires financial managers to take measures to manage cash flow well for business; and (iii) regarding the Government and government organizations, they should encourage banks to finance the businesses, on the one hand, to increase investment, on the other hand, banks can control the usage of capital for the right purpose to increase the efficiency of investment projects. The Government needs to maintain sustained GDP growth and curb the inflation rate to stimulate the production activities of enterprises.

References

Ahmad, M.M., Hunjra, A.I., Taskin, D.: Do asymmetric information and leverage affect investment decisions?. Quart. Rev. Econ. Finan. (2021)

Aivazian, V.A., Ge, Y., Qiu, J.: The impact of leverage on firm investment: Canadian evidence. J. Corporate Finan. **11**(1–2), 277–291 (2005)

Anwer, M.S., Sampath, R.: Investment and Economic growth. Paper presented at the Presented at Western Agricultural Economics Association Annual Meeting (1999)

Block, J.H., Jaskiewicz, P., Miller, D.: Ownership versus management effects on performance in family and founder companies: a Bayesian reconciliation. J. Fam. Bus. Strat. **2**, 232–245 (2011)

Block, J.H., Hoogerheide, L., Thurik, R.: Are education and entrepreneurial income endogenous? a bayesian analysis. Entrep. Res. J (2012) Manuscript 1051. https://doi.org/10.1515/2157-5665.1051

Damodaran, A.: Applied Corporate Finance. Wiley (2010)

Firth, M., Lin, C., Wong, S.M.L.: Leverage and investment under a state-owned bank lending environment: evidence from China. J. Corp. Finan. **14**(5), 642–653 (2008)

Franklin, J.S., Muthusamy, K.: Impact of leverage on firms investment decision. Int. J. Scient. Eng. Res. **2**(4) (2011)

Gebauer, S., Setzer, R., Westphal, A.: Corporate debt and investment: a firm-level analysis for stressed euro area countries. J. Int. Money Financ. **86**, 112–130 (2018)

Kass, R.E., Carlin, B.P., Gelman, A., Neal, P.M.: Markov Chain Monte Carlo in Practice: A Roundtable Discussion (1997). Retrieved from http://www.stat.columbia.edu/~gelman/research/published/kass5.pdf. Accessed April 18, 2020

Kreft, I.G.G., Leeuw, J.D.: Introducing Multilevel Modeling. Sage Publications, Newbury Park, CA (1998)

Kreinovich, V., Nguyen, T.N., Nguyen, T.D., Dang, T.V. (eds.): Beyond Traditional Probabilistic Methods in Economics. Springer, Cham (2019). https://doi.org/10.1007/978-3-030-04200-4

Lang, L., Ofek, E., Stulz, R.: Leverage, investment, and firm growth. J. Financ. Econ. **40**(1), 3–29 (1996)

Lemoine, P.N.: Moving beyond noninformative priors: Why and how to choose weakly informative priors in Bayesian analyses. Oikos **128**, 912–928 (2019)

Ly, A.H., Le, D.S., Kreinovich, V., Nguyen Ngoc, T.N. (eds.) Econometrics for Financial Applications. Springer, Cham (2018)

Modigliani, F., Miller, M.H.: The cost of capital, corporation finance and the theory of investment. Am. Econ. Rev. **48**(3), 261–297 (1958)

Myers, S.C.: Determinants of corporate borrowing. J. Finan. Econ. **5**(2), 147–175 (1977)

Nezlek, J.B.: Multilevel random coefficient analyses of event and interval contingent data in social and personality psychology research. Pers. Soc. Psychol. Bull. **27**, 771–785 (2001)

Nguyen, H.T., Nguyen, T.N.: A panorama of applied mathematical problems in economics. Thai J. Math. Spe. Issue Ann. Meeting Math., 1–20 (2018)

Nguyen, H.T., Nguyen, T.D., Nguyen, T.N.: Beyond traditional probabilistic methods in econometrics. In: Kreinovich, V., Nguyen, T., Nguyen, T., Dang, T.V. (eds.) Beyond Traditional Probabilistic Methods in Economics. ECONVN 2019a. Studies in Computational Intelligence, Vol. 809. Springer, Cham (2019)

Nguyen, T.N.: A Bayesian approach in the prediction of the Us's gross domestic products. Asian J. Econ. Bank. **163**, 5 (2019)

Nguyen, N.B.T., Ichihashi, M., Kakinaka, M.: The link between financial leverage and investment decisions in Vietnam's small and medium-sized enterprises. Asia-Pacific J. Account. Econ. (2019b). https://doi.org/10.1080/16081625.2019.1673196

Nguyen, T.N.: How to explain when the ES is lower than one? a bayesian nonlinear mixed-effects approach. J. Risk Finan. Manag. (2020). Retrieved February 20 from https://www.mdpi.com/1911-8074/13/2/21

Nguyen, T.N., Bui, N.H.: Impact of economic freedom on corruption revisited in ASEAN countries: a bayesian hierarchical mixed-effects analysis. Economies **9**(1), 3 (2021). https://doi.org/10.3390/economies9010003

Nguyen, T.N., Kreinovich, V., Nguyen, T.D. (eds.): Data science for financial econometrics. Proceedings of the ECONVN 2020, Studies in Computational Intelligence, Ho Chi Minh City, Vietnam, January 14–16. Springer, Cham (2020)

Nguyen, T.N., Doan, H.T., Nguyen, T.D., Kreinovich Vladik, V. (eds.): Prediction and causality, and related topics. Proceedings of the ECONVN 2021, Studies in Computational Intelligence, Ho Chi Minh City, Vietnam, January 11–13. Cham: Springer (to appear) (2021)

Onwe, O.J., Olarenwaju, R.R.: Impact of inflation on corporate investment in the sub-Saharan African Countries: An empirical analysis of the west-African monetary zone. Int. J. Bus. Soc. Sci. **5**(1), 189–199 (2014)

Raudenbush, S.W., Bryk, A.S.: Hierarchical Linear Models, 2nd edn. Sage Publications, Newbury Park, CA (2002)

Roberts, G.O., Rosenthal, J.S.: Optimal scaling for various Metropolis-Hastings algorithms. Stat. Sci. **16**, 351–367 (2001)

Ross, S.A., Westerfield, R., Jordan, B.D.: Fundamentals of Corporate Finance. Irwin, New York, NY, USA (2014)

Snijders, T., Bosker, R.: Multilevel Analysis. Sage Publications, London (1999)

Trinh, H.T., Kakinaka, M., Kim, D., Jung, T.Y.: Capital structure and investment financing of small and medium-sized enterprises in Vietnam. Glob. Econ. Rev. **46**(3), 325–349 (2017)

Tran, T.A., Kreinovich, V., Nguyen, T.N.: Decision making under interval uncertainty: beyond Hurwicz pessimism-optimism criterion. In: Kreinovich V., Nguyen, T., Nguyen, T., Dang, T.V. (eds.) Beyond Traditional Probabilistic Methods in Economics. ECONVN 2019. Studies in

Computational Intelligence, 2019. Vol. 809. Springer, Cham (2019). https://doi.org/10.1007/978-3-030-04200-4_14

Umutlu, M.: Firm leverage and investment decisions in an emerging market. Qual. Quant. **44**(5), 1005–1013 (2010)

Vo, V.X.: Leverage and corporate investment–Evidence from Vietnam. Financ. Res. Lett. **28**, 1–5 (2019)

What Affects the Capital Adequacy Ratio? A Clear Look at Vietnamese Commercial Banks

Pham Hai Nam, Nguyen Ngoc Tan, Nguyen Ngoc Thach, Huynh Thi Tuyet Ngan, and Nguyen Minh Nhat

Abstract This article studies the impact of factors on the capital adequacy ratio (CAR) of commercial banks in Vietnam. The authors use the Bayesian regression method via Gibbs sampling. The data used in the study is secondary sources from the financial statements of 30 Vietnamese commercial banks and the General Statistics Office of Vietnam for the period 2012–2018. The research results show that bank loan, loan loss provision, liquid assets, profitability are the factors that negatively affect CAR. On the other hand, factors that have a positive impact on bank CAR are bank size, inflation, and GDP growth.

Keywords Bayesian · Capital adequacy ratio · Commercial bank

1 Introduction

Commercial banks play a crucial role in the development and stability of businesses in particular and the economy in general (Anggono 2014; Iacobelli 2016). However, banks are unique businesses, susceptible to the economic cycle and fluctuations, and greatly influenced by the central bank's monetary policy. Therefore, the safety of the banking system is closely related to the safety of the national financial system. Therefore, banks always receive the attention and regular supervision of the central bank

P. H. Nam (✉) · N. N. Thach · N. M. Nhat
Banking University HCMC, Ho Chi Minh City, Vietnam
e-mail: namph@buh.edu.vn

N. N. Thach
e-mail: thachnn@buh.edu.vn

N. M. Nhat
e-mail: nhatnm@buh.edu.vn

N. N. Tan
Office of People's, Committee of Ho Chi Minh City, Ho Chi Minh City, Vietnam

H. T. T. Ngan
Vietnam Maritime Commercial Joint Stock Bank (MSB), Ho Chi Minh City, Vietnam

N. Ngoc Thach et al. (eds.), *Financial Econometrics: Bayesian Analysis, Quantum Uncertainty, and Related Topics*, Studies in Systems, Decision and Control 427,
https://doi.org/10.1007/978-3-030-98689-6_19

in all aspects of business activities. To ensure the safety for commercial banks, one of the critical criteria that banks need to meet is the capital adequacy ratio (CAR), which is also an indicator to evaluate the risk for each bank, helping investors, customers, and policymaker assess the health of each bank (Bhattarai 2020; Vo et al. 2014). In addition, Vietnam's economy is increasingly integrating deeply into the world economy, such as joining the European-Vietnam Free Trade Agreement (EVFTA) and the Comprehensive and Progressive Agreement for Trans-Pacific Partnership (CPTPP), which requires banks to change their management methods, strengthen internal resources, and apply technology to be able to compete better on the international playing field. Significantly, the period 2012–2018 is the period when banks restructure in a safe and healthy direction for the entire banking system and meet the CAR according to international standards, which is the goal that banks must achieve.

In Vietnam, the CAR according to international standards of Basel was first introduced in Decision No. 297/1999/QD-NHNN regulating CAR in the operation of credit institutions, corrected and supplemented according to Decision No. 457/2005/QD-NHNN and Circular 13/2010/TT-NHNN on prudential ratios in operations of credit institutions, to update and respond to the changes of the world economy in general and the banking sector in particular. The SBV continues to issue regulations related to CAR of Vietnamese commercial banks such as Circular No. 36/2014/TT-NHNN, Circular No. 06/2016/TT-NHNN amending some articles of Circular 36/2014/ TT-NHNN, Circular No. 41/2016/TT-NHNN, Circular No. 13/2018/TT-NHNN, Circular No. 22/2019/TT-NHNN. In particular, the new point of Circular No. 41/2016/TT-NHNN is that the CAR takes into account credit risk, market risk, operational risk, and the CAR must reach a minimum value of 8%, consistent with recommendations of the Basel Committee. As of December 31, 2019, 18 Vietnamese commercial banks have been approved to apply Circular No. 41/2016/TT-NHNN—Circular regulating capital adequacy ratio for banks and foreign bank branches. In which, Circular No. 41/2016/TT-NHNN, Circular No. 13/2018/TT-NHNN are legal documents built based on risk management under the Basel II Treaty. However, in Vietnam, the regulation of the minimum capital adequacy ratio is not uniform. According to Circular No. 41/2016/TT-NHNN, the CAR must be at least 8% for banks applying the minimum capital adequacy ratio. Meanwhile, for banks applying CAR calculation according to Circular No. 22/2019/TT-NHNN, the CAR must be at least 9%. In fact, CARs of banks vary widely, can be very high (over 15%) or only a few percentage points higher than the regulated rate (Le 2020). Maintaining too high a CAR is also not good because the bank has to maintain more capital or the asset portfolio is too safe, leading to low profitability. In contrast, CAR is too low, only slightly higher than the regulated rate, which means that the bank has to face more risks, its ability to withstand shocks is reduced. Therefore, maintaining CAR at a moderate level through controlling CAR's factors can help banks both achieve good profitability and achieve safety and soundness in operations. Therefore, studying and understanding the factors affecting the CAR of Vietnamese commercial banks can help bank managers identify the factors affecting the CAR, thereby offering solutions to achieve a reasonable CAR, enhancing the bank's stability and business efficiency. In addition, the novelty of this study compared with previous studies on the approach.

Previous studies applied traditional methods such as Ordinary Least Squares (OLS), Fixed Effects Model (FEM), Random Effects Model (REM), Feasible Generalized Least Squares (FGLS) and Generalized Method of Moments (GMM), which caused much controversy and criticism (Anh et al. 2018; Hung et al. 2020). In this study, the authors apply the Bayesian method, which is a modern approach related to mathematics, statistical probability or econometrics applied in the field of economics or finance, such as Wang et al. (2019), Hung (2020) and Thach et al. (2019).

2 Literature Review

Banks need to ensure a minimum capital adequacy ratio to ensure a healthy and stable operation of the banking system, to help banks avoid chasing excessive profits and ignoring regulations on risk control. Unlike conventional enterprises, the capital structure of banks is very different because it is financed mainly by debt (Berlin 2011). That means banks use their financial leverage to create financial leverage for businesses in the economy (Hoque and Pour 2018). However, for a bank's capital structure, the standard view of required capital is that banks must also keep their capital buffers above the regulatory minimum to avoid the costs of issuing new equity in a short time. Therefore, the bank's capital structure decisions are closely tied to country-specific regulations on minimum capital ratios. Berger and Herring (1995) argue that two factors influence the capital structure of a bank. The first factor is the capital requirements of banks according to the market so that banks can take advantage when there are profitable opportunities or hedge against possible risks. The second factor is regulatory safeguard requirements (deposit insurance, access to a discount scheme, etc.), potentially reducing bank capital. Berger and Herring (1995) also emphasize the importance of legal capital requirements, as do Osterberg and Thompson (1990) in analyzing optimal leverage ratios that consider the trade-off between the tax advantages of debt and bankruptcy costs. However, regulatory pressure on banks that maintain capital is asymmetric, with regulators sounding the alarm only when capital adequacy ratios are too low, but often with little or no query regarding capital adequacy ratio is too high (Pham and Nguyen 2017). For banks themselves, maintaining a CAR that is too low or too high is not good; that is, too risky or too safe, which has a negative impact on the bank's performance. Therefore, banks need to identify the influencing factors and maintain the CAR at a reasonable level. Empirical studies to determine the factors affecting CAR are interesting topic, specifically as follows.

Aktas et al. (2015) studied the factors affecting the CAR of commercial banks in 10 Southeast European countries. The data set was collected from 71 commercial banks in the period 2007–2012. By the FGLS regression method, the research results show bank size, profitability, financial leverage, liquid assets, and bank risk are factors belonging to bank characteristics that affect CAR. Besides, environmental factors affecting CAR include economic growth, volatility index of the European stock market, deposit insurance ratio, governance index.

Aspal and Nazneen (2014) investigate the factors affecting the CAR of private banks in India. Secondary data was collected by the authors from 20 banks from 2007 to 2012. By the OLS regression method, the research results show that the factors that have a negative impact on CAR include bank loan, asset quality, management efficiency. The opposite influencing factors are liquidity assets, asset sensitivity. In addition, the authors also found that the banks in the sample maintain a higher CAR than required by the Reserve Bank of India.

Charitou's study (2019) uses a dataset including 2135 observations of banks in the US from 2012 to 2017 to study the factors affecting the CAR of these banks. The study shows that the factors that have a positive impact on CAR are ROA, the ratio of income to net debt, and loan loss provision. Conversely, the opposite impact factors are operating costs, bank assets.

Bhattarai's study (2020) was conducted to determine the factors affecting the CAR of Nepalese commercial banks from 2013 to 2018. The study used data sets of 11 commercial banks and Pooled OLS, FEM, REM regression methods. The research focuses on micro factors, including credit risk, asset quality, management quality, ROA, liquid assets, bank size. The macro factors are GDP growth and inflation. The author concludes that liquid assets have a positive effect on CAR. Meanwhile, bank size and inflation have opposite effects.

The study by El-Ansary et al. (2019) included 38 Islamic banks and 75 conventional banks in 10 countries in the Middle East and North Africa during 2009–2013. Independent measures in the study include profitability, liquidity risk, credit risk, bank size, customer deposits, operational efficiency, portfolio risk, and two macroeconomic variables (GDP growth rate and average world governance indicators for each country). Using the GMM regression method, the research results show that the CARs of both Islamic banks and conventional banks are affected by bank size, performance, and GDP growth. In addition, for Islamic banks, CAR is affected by customer deposits. Meanwhile, the CAR of conventional banks is affected by profitability, credit risk, portfolio risk.

Kartal (2019) studies the factors affecting the CAR of Turkish banks from 2006 to 2019. Using the Multivariate Adaptive Regression Splines (MARS) method, the research results show that the CARs of banks in Turkey are affected by the ratio of credit to total assets, risky assets, legal capital, non-performing loans, non-performing loans to total loans, credit to total deposit ratio.

In Vietnam, to find out the factors affecting CAR, Pham and Nguyen (2019) used the FEM regression method for 29 commercial banks in 2013–2017. The estimated results show that the bank size and the equity-to-asset ratio are positively related to CAR. In contrast, the return on total assets and the loan loss provisions ratio are negatively related to the CAR. Howerver, this study has not found any evidence of the impact of net interest income, loan-to-total assets ratio and non-performing loans ratio on the dependent variable.

Vo et al. (2014) studied the factors affecting CAR with data from the annual reports of 28 banks selected as a sample in the period from 2007 to 2012. By the FGLS method, the research results show that the liquid assets and the ratio of loan loss provisions have a positive impact on CAR. Meanwhile, bank size, capital mobilization ratio, ROE has a negative impact on CAR. However, this study has not found quantitative evidence from the impact of leverage ratio (LEV) and loan ratio (LOA) on the capital adequacy ratio.

Also studying the CAR of Vietnamese commercial banks, Pham and Nguyen (2017) collect data from 29 commercial banks in 2011–2015. The authors use independent variables including bank size, leverage ratio, loan loss provisions ratio, NIM, bank loan, liquid assets; the research shows that NIM, liquid assets have a positive impact on CAR; loan loss provisions ratio, bank loans have a negative impact on CAR.

Research hypothesis

Bank size

Bank size is one of the important factors affecting CAR. Bank size can be expressed through the number of branches and transaction offices or the total assets on the balance sheet. Large banks can easily access external capital at a low cost through the branch network, leading to reduced CAR (Ansary et al. 2015). In addition, the larger the size of the bank, the lower the CAR because the large size is a guarantee for the safety and reputation of the bank (Gropp and Heider 2007). Therefore, the authors hypothesize.

Hypothesis H1: Bank size has a negative effect on CAR.

Bank loan

Banks use the money raised to lend and invest, generating profits for the bank. When bank loan increases, it means that the bank can achieve higher profits, but at the same time, the risk also increases, reducing the CAR. Previous studies show different results on the direction of the impact of bank loan on CAR. Some studies show that bank loan has a positive effect on CAR as studied by Aspan and Nazneen (2014), Bhattarai (2020) or has the opposite effect as research by Pham and Nguyen (2017), Polat and Al-Khalaf (2014). Therefore, the authors hypothesize.

Hypothesis H2: Bank loan has a negative impact on CAR.

Loan loss provision

As banks make it easier to lend, non-performing loans can increase. At that time, banks need to make more provision for credit risks, depending on the group of overdue loans. In addition, when loan loss provision increases, it may also be due to the fact that banks invest in a higher risk asset portfolio, leading to a decrease in CAR (Bhattarai 2020). Therefore, the authors hypothesize:

Hypothesis H3: Loan loss provision has a negative impact on CAR.

Liquid assets

Holding liquid assets allows the bank to quickly respond to customers' withdrawal and other short-term obligations (Bhattarai 2020). These banks are also better able to respond to shocks as well as operate more safely. Studies by Vo et al. (2014), Angbazo (1997), Büyükşalvarcı and Abdioğlu (2011) show that liquid assets and CAR have a positive relationship. Therefore, the authors hypothesize.

Hypothesis H4: Liquid assets have a positive effect on CAR.

Profitability

Banks have better profitability due to business performance. Besides, when there are higher profits, banks can retain profits to increase capital and the bank's CAR increases accordingly (Gropp and Heider 2007). Furthermore, studies by Aktas et al. (2015), Charitou (2019), El-Ansary et al. (2019) show that the profitability of commercial banks has a positive impact on CAR. Therefore, the authors hypothesize.

Hypothesis H5: The profitability of commercial banks has a positive impact on CAR.

Inflation

High inflation can erode bank capital, leading to a negative relationship between inflation and CAR (Williams 2011). However, the research results of Ogere et al. (2013) show a positive relationship between inflation and CAR. In this study, the authors expect inflation to have a negative impact on the CAR of commercial banks.

Hypothesis H6: Inflation has a negative effect on CAR.

GDP growth

When the economic growth is high, banks can lend more and customers can pay better, leading to a decrease in non-performing loans and loan loss provisions. In addition, when the economy grows well, banks do not need to increase their capital because of the high safety in banking operations. However, when the economic growth is low, the level of risk increases, banks need to increase capital reserves to hedge against future losses (Bhattarai 2020). Therefore, the authors hypothesize.

Hypothesis H7: GDP growth has a negative effect on CAR.

3 Data and Methodology

Research model

Based on the research models of Bhattarai (2020), Aktas et al. (2015), Vo et al. (2014), the authors propose a multivariate regression model as follows:

$$\text{CAR} = \alpha_0 + \alpha_1\text{SIZE} + \alpha_2\text{LOAN} + \alpha_3\text{LLP} + \alpha_4\text{LIQUI} + \alpha_5\text{ROA}$$

$$+ \alpha_6\text{INFLAT} + \alpha_7\text{GGDP} + \varepsilon$$

where:

$\propto_0$: Constant terms;

$\propto_i$ (i = 1,7): Coefficients of variables;

ε: Error terms.

The descriptions of the variables are provided in Table 1.

Data collection

The study was conducted on a sample of 30 Vietnamese commercial banks in the period 2012–2018. This is the period when Vietnamese commercial banks implemented financial restructuring, rearranged the banking system, and renewed the banking administration system in a modern direction, in line with international practices and standards (Government 2012). In addition, the total assets of 30 banks in the sample accounted for 86% of the total assets of Vietnam's commercial banking system, ensuring the representativeness of commercial banks. Micro data is collected from the financial statements of banks; macro data is collected from the General Statistics Office.

Methodology

In this article, the authors apply the Bayesian regression method via Gibbs sampling algorithm, which is a new approach compared to previous studies on the factors affecting CAR of commercial banks. The Bayesian method has many advantages over the traditional approach, such as a simpler and more intuitive interpretation of the results, not limited by sample size, or applied when traditional methods fail. In general, this study is similar to several works using modern research methods related to mathematics, statistical probability or econometrics applied in the field of economics or finance, such as Wang et al. (2019), Hung (2020), Thach et al. (2019), Galindo et al. (2020), and Khrennikova,(2019).

The Gibbs sampling algorithm is a special case of the Metropolis–Hastings algorithm. Metropolish et al. (1953) were the first to build the algorithm, and then, Hastings (1970) developed a more efficient algorithm. Because previous studies on the factors affecting the CAR of commercial banks all applied the traditional method, so we do not have information about the a priori distribution of the variables in the model. Furthermore, the sample size in this study is large (over 300 observations), so a priori information did not have much influence on the results. Therefore, the authors propose the normal distribution for the regression coefficients of the observed variables and the Igamma distribution for the variances in the model, specifically as follows:

Table 1 Variable definitions

Variables		Formula	Notation	Previous studies
Dependent	Capital adequacy ratio	(Tier 1 capital* + Tier 2 capital**)/Risk-weighted assets	CAR	Bhattarai (2020), Charitou (2019), El-Ansary et al. (2019), Kartal (2019), Aktas et al. (2015), Aspal and Nazneen (2014), Pham and Nguyen (2019), Pham and Nguyen (2017), Vo et al. (2014)
Independent	Bank size	The logarithm of total assets	SIZE	Bhattarai (2020), Charitou (2019), El-Ansary et al. (2019), Aktas et al. (2015), Pham and Nguyen (2019), Pham and Nguyen (2017), Vo et al. (2014)
	Bank loan	Total loans/Total assets	LOAN	Bhattarai (2020), Aspal and Nazneen (2014), Pham and Nguyen (2017), Vo et al. (2014)
	Loan loss provision	Loan loss provision/Total loan	LLP	Bhattarai (2020), Charitou (2019), Pham and Nguyen (2019), Pham and Nguyen (2017), Vo et al. (2014)
	Liquid assets	Liquid assets/Total assets	LIQUI	Aktas et al. (2015), Aspal and Nazneen (2014), Pham and Nguyen (2017), Vo et al. (2014)
	Return on assets	Net income/Total assets	ROA	Pham and Nguyen (2019)
	Inflation		INFLAT	Bhattarai (2020), Aktas et al. (2015)

(continued)

Table 1 (continued)

Variables		Formula	Notation	Previous studies
	GDP growth		GGDP	Bhattarai (2020), El-Ansary et al. (2019), Aktas et al. (2015)

*Tier 1 capital is a type of equity capital of a bank. Basically, Tier 1 capital includes the charter capital and reserve funds of that bank as well as the retained earnings
**Tier 2 capital reflects a bank's financial capacity in relation to less reliable forms of financial resources than Tier 1 capital. It consists of the bank's supplementary capital including undisclosed reserves, revaluation reserves, and subordinate debt

$$\alpha - N(0; 100)$$

$$\sigma^2 \sim \text{Ingamma}(0.01; 0.01)$$

4 Results and Discussion

Table 2 presents the results of multivariate regression with panel data by Bayesian method via Random-walk Metropolis–Hastings algorithm and Gibbs sampling.

To ensure that Bayesian inference based on MCMC simulation is reasonable, the authors perform convergence diagnosis of MCMC chains. It can conclude that the model is robust if the MCMC chains converge (Nguyen 2020). Testing the convergence of MCMC chains is done through trace plots, histograms, and autocorrelation charts.

The test results from Fig. 1 show that the histograms run fast through the distribution and do not create trends, the autocorrelation drop quickly (die off after 40

Table 2 Summary of regression results for the dependent variable CAR

	Mean coefficient	Std	MCSE
SIZE	−0,0289	0,0031	0,0000
LOAN	0,0302	0,0429	0,0004
LLP	1,12,833	0,7165	0,0071
LIQUI	0,0397	0,0458	0,0004
ROA	1,8322	0,6685	0,0069
INFLAT	−0,0815	0,2201	0,0022
GGDP	−1,4158	0,8739	0,0087
_cons	1,1129	0,1130	0,0011
var	0,0015	0,0001	0,0000

Source *Authors' calculation*

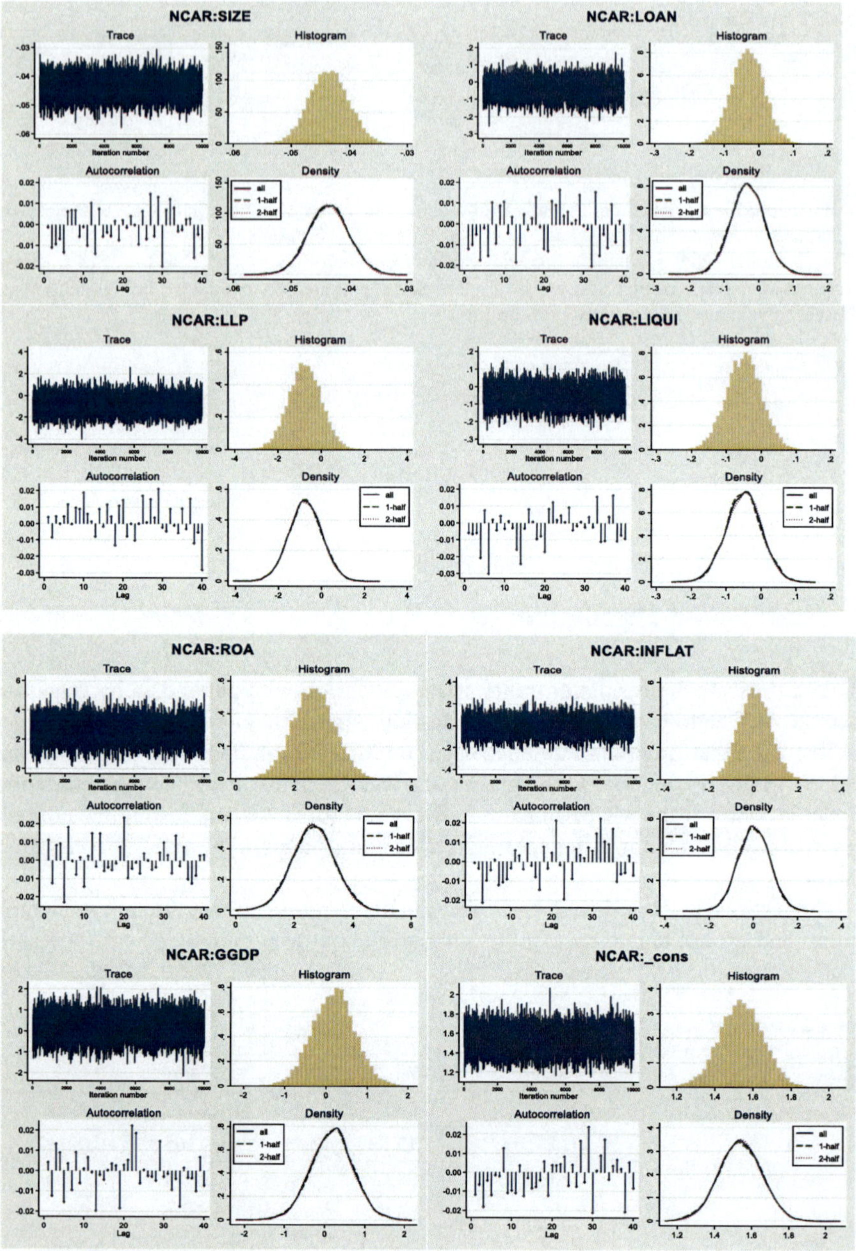

Fig. 1 Graphical diagnostics for MCMC convergence

lags), showing low autocorrelation. The histograms resemble the shape of probability distributions. From this, we can conclude that Bayesian inference is robust and the results can be used for analysis.

The results from Table 2 show that the variables that have a positive impact on CAR are LOAN, LLP, LIQUI, ROA. Variables with opposite effect are SIZE, INFLAT, GGDP.

Bank size (SIZE) has a negative effect on CAR, consistent with the research hypothesis and studies of Bhattarai (2020), Charitou (2019), El-Ansary et al. (2019), Aktas et al. (2015), Vo et al. (2014), but contrary to the results of Pham and Nguyen (2019). Large banks can rely on their reputation to mobilize external capital at a low cost. Moreover, the large scale is the guarantee of the bank's safety. Therefore, large-sized banks often take on higher risks by increasing lending with easy terms, leading to an increase in non-performing loans. These banks also believe that the government will intervene in the event of a bank failure due to the view that it is "too big to fail".

Bank loan (LOAN) has a positive effect on CAR, consistent with the studies of Bhattarai (2020), Aspal and Nazneen (2014), but contrary to the results of Pham and Nguyen (2017), Polat and Al- Khalaf (2014) and research hypothesis. The more banks lend, the more profitable they are to use retained earnings to raise capital. In addition, the peculiarity of Vietnamese banks is that profits are heavily dependent on lending. Therefore, an increase in lending will have the effect of increasing the bank's CAR.

Loan loss provision (LLP) has a positive effect on CAR, contrary to the research hypothesis and studies of Pham and Nguyen (2019), Pham and Nguyen (2017). However, the research results are similar to those of Charitou (2019), Vo et al. (2014). When banks have to make higher-risk provisions, it means that banks lend more and achieve higher profits, leading to an increase in CAR.

Liquid assets (LIQUI) have a positive effect on CAR, consistent with the research hypothesis and the studies of Aktas et al. (2015), Aspal and Nazneen (2014), Vo et al. (2014)) but contrary to the research results of Pham and Nguyen (2017). Holding highly liquid assets helps banks quickly meet customers' withdrawal needs and other debt obligations, increasing the bank's reputation. Moreover, customers often tend to deposit money in banks with high reputation, strong brand. Therefore, when a bank regularly receives deposits from customers, not only the liquidity of that bank is better, but the CAR of that bank also increases.

Profitability (ROA) has a positive effect on CAR, consistent with the research hypothesis and studies of Charitou (2019), El-Ansary et al. (2019), Aktas et al. (2015). Banks with high profitability can use retained earnings to raise capital or issue securities to investors due to good financial results, resulting in an increased CAR.

Inflation (INFLAT) has a negative effect on CAR, similar to the research hypothesis and studies of Bhattarai (2020), Aktas et al. (2015). When inflation is high, customers often need to withdraw money from banks to invest in other safer and more profitable channels. Besides, when inflation is high, customers' demand for loans will decrease due to rising interest rates. From there, bank profits will decrease, and CAR will decrease accordingly.

Economic growth (GGDP) has a negative effect on CAR, consistent with the hypothesis and studies by El-Ansary et al. (2019), Aktas et al. (2015). When economic growth is high, banks do not need to increase capital buffer due to high safety in banking operations. As a result, the bank's CAR decreased.

5 Conclusion and Policy Implications

This study aims to find out the factors affecting the CAR of Vietnamese commercial banks in the period 2007–2018. By using Bayesian method via Gibbs sampling algorithm, the research results show the factors that positively affect the CAR of Vietnamese commercial banks include bank loan, loan loss provision, liquid assets, and profitability. Negative factors include bank size, inflation, GDP growth. The research results provide evidence for bank managers to have a more reasonable basis for adjusting CAR for each bank, specifically as follows.

First, it is necessary for banks with low CAR to carefully consider increasing the size because it will lower the CAR further. The increase in scale should be considered carefully, corresponding to the capacity of the bank. In addition, it is necessary to improve profitability so that retained earnings can be used and attract investors to issue securities to increase equity, improve financial capacity and capital adequacy ratio.

Second, banks with too high CAR should be more cautious when lending, ensure loan quality, and expand the bank's scale to better approach customers and dominate the market. However, the expansion should be closely tied to the bank's business performance. It is unnecessary to hold too many liquid assets but to allocate them to more profitable investments.

References

Anh, L.H., Le, S.D., Kreinovich, V., Thach, N.N. (eds.): Econometrics for Financial Applications. Springer, Cham (2018). https://doi.org/10.1007/978-3-319-73150-6(2018)

Büyükşalvarcı, A., Abdioğlu, H.: Determinants of capital adequacy ratio in Turkish Banks: a panel data analysis. Afr. J. Bus. Manage. **5**(27), 11199–11209 (2011)

Aktas, R., Acikalin, S., Bakin, B., Celik, K.: The determinants of banks' capital adequacy ratio: some evidence from south Eastern European countries. J. Econ. Behav. Stud. **7**(1), 79–88 (2015)

Angbazo, L.: Commercial bank net interest margins, default risk, interest rate risk, and off—balance sheet banking. J. Bank. Finance **21**(1), 55–87 (1997)

Anggono, N.A.H.: Determinants of capital adequacy ratio in 19 commercial banks. J. Bus. Manag. **3**(7), 752–764 (2014)

Aspal, P.K., Nazneen, A.: An empirical analysis of capital adequacy in the Indian private sector banks. American J. Res. Com. **2**(11), 28–42 (2014)

Bhattarai, B.: Determinants of capital adequacy ratio of commercial banks in Nepal. Asian J. Finan. Accoun. **6**(1), 194–213 (2020)
Berger, A.N., Herring, R.J.: The role of capital in financial institutions. J. Bank. Finance **19**(3), 393–430 (1995)
Berlin, M.: Can we Explain Banks' Capital Structure? Business Review, Q2 (2011), 1–11. Federal Reserve Bank of Philadelphia (2011)
Charitou, M.: Determinants of the capital adequacy of U.S financial institutions. Int. Finan. Bank. **6**(1), 31–39 (2019)
El-Ansary, O., El-Masry, A.A., Yousry, Z.: Determinants of capital adequacy ratio (CAR) in MENA region: Islamic vs. conventional banks. Int. J. Account. Finan. Report. **9**(2), 287–313 (2019)
Galindo, O., Svitek, M., Kreinovich, V.: Quantum (and More General) models of research collaboration. Asian J. Econ. Bank. **4**(1) (2020)
Gropp, R., Heider, F.: What can corporate finance say about banks' capital structures? Working Paper. SSRN (2007)
Government: Project on restructuring the system of credit institutions for the period 2011–2015, Hanoi (2012)
Hung, N.T.: On the calculus of subjective probability in behavioral economics. Asian J. Econ. Bank. **4**(1) (2020)
Hastings, W.K.: Monte Carlo sampling methods using Markov chains and their applications. Biometrika **57**, 97–109 (1970)
Hoque, H., Pour, E.K.: Bank-level and country-level determinants of bank capital structure and funding sources. Int. J. Financ. Econ. **23**(4), 504–532 (2018)
Iacobelli, A.: Determinants of profitability: empirical evidence from the largest global banks. Finan. Anal., 11 (2016)
Kartal, M.T.: Defining influential factors of capital adequacy ratio: an examination upon turkish banking sector (2006/Q1-2019/Q1). Emerg. Markets J. **9**(2), 17–26 (2019)
Khrennikova, P.: Quantum probability based decision making in finance: from individual preferences to market outcomes. Asian J. Econ. Bank. **3**(1) (2019)
Le, H.T.: Factors affecting capital adequacy ratio of Vietnamese commercial banks. J. Indus. Trade, No. 29+30 (2020)
Metropolis, N., Rosenbluth, A.W., Rosenbluth, M.N., Teller, A.H., Teller, E.: Equation of state calculations by fast computing machines. J. Chem. Phys. **21**, 1087–1092 (1953)
Nguyen, N.T.: How to explain when the ES is lower than one? A Bayesian nonlinear mixed-effects approach. J. Risk Finan. Manag. **13**(2), 1–17 (2020). https://doi.org/10.3390/jrfm13020021
Pham, P.T., Nguyen, T.K.N.: Factors affecting the minimum capital adequacy ratio of commercial banks in Vietnam. Scient. J. Can Tho Univ. **55**, 78–84 (2019)
Pham, T.X.T., Nguyen, N.A.: The determinants of capital adequacy ratio: the case of the Vietnamese banking system in the period 2011–2015. J. Sci. Econ. Bus. **33**(2), 49–58 (2017)
Ogere, G.A., Peter, Z., Inyang, E.E.: Capital adequacy ratio and banking risks in the nigeria money deposit banks. Res. J. Finan. Accoun. **4**(17), 17–25 (2013)
Osterberg, P.W., Thompson, B.J.: Optimal Financial Structure and Bank Capital Requirements: An Empirical Investigation, Federal Reserve Bank of Cleveland Working Paper, no 9007 (1990)
Polat, A., Al-khalaf, H.: What determines capital adequacy in the banking system of kingdom of Saudi Arabia? a panel data analysis on Tadawul banks. J. Appl. Finan. Bank. **4**(5), 27–43 (2014)
Thach, N.N., Anh, L.H., An, P.T.H.: The effects of public expenditure on economic growth in Asia countries: a Bayesian model averaging approach. Asian J. Econ. Bank. **3**(1), 126–149 (2019)
Vo, H.D., Nguyen, M.V., Do, T.T.: Determinants of capital adequacy ratio: empirical evidence from the Vietnamese commercial banking system. Sci. J. Ho Chi Minh City Open Univ. **9**(2), 87–100 (2014)
Wang, C., Wang, T., Trafimow, D., Chen, J.: Extending a Priori Procedure to Two Independent Samples under Skew Normal Settings. Asian J. Econ. Bank. **3**(2) (2019)
Williams, H.T.: Determinants of capital adequacy in the banking sub-sector of the Nigeria economy: efficacy of CAMELS. Int. J. Acad. Res Bus. Soc. Sci. **1**(3), 233–248 (2011)

BIC Algorithm for Word of Mouth in Fast Food: Case Study of Ho Chi Minh City, Vietnam

Nguyen Thi Ngan, Bui Huy Khoi, and Ngo Van Tuan

Abstract The paper uses BIC Algorithm for Word of Mouth in Fast Food: Case Study of Ho Chi Minh City (HCMC), Vietnam. It is used to discover the factors affecting the word of mouth (WOM) of customers in Vietnam. The official study was conducted on 183 customers living in HCMC to accurately identify the factors that affect word of mouth in fast food. This paper found six factors affecting the word of mouth of customers in fast food including brand image, product innovation, selling price, brand preference, reputation, and credibility. The paper explore price has the highest influence and the lowest is reputation and all factors positively affect the word of mouth of customers. Previous studies revealed that using linear regression. This study uses the optimal choice by the BIC algorithm.

1 Introduction

Nowadays, the foodservice industry in Vietnam has expanded greatly compared to 10 years ago. Vietnam has about 540,000 stores, of which about 430,000 are street restaurants, 80,000 are eat-in restaurants, but there are only about 7,000 fast food restaurants. This comparison shows how modest the fast food industry is Dat Le (2019). Fast lifestyle along with the changing consumption trends of the young generation is becoming the driving force behind the booming fast food market. Many big brands in the fast food sector have been present in the COVID-19 pandemic that has spurred the growth of three food and beverage industries: fast food, takeaway, and car food so They are seen as pioneers in the word of mouth development in the context

N. T. Ngan · B. H. Khoi (✉)
Industrial University of Ho Chi Minh City, Ho Chi Minh City, Vietnam
e-mail: buihuykhoi@iuh.edu.vn

N. T. Ngan
e-mail: nguyenthingan@iuh.edu.vn

N. Van Tuan
Banking University of Ho Chi Minh City, Ho Chi Minh City, Vietnam
e-mail: tuannv@buh.edu.vn

N. Ngoc Thach et al. (eds.), *Financial Econometrics: Bayesian Analysis, Quantum Uncertainty, and Related Topics*, Studies in Systems, Decision and Control 427,
https://doi.org/10.1007/978-3-030-98689-6_20

of Vietnam presently. The trend in 2021 remains the same. The COVID-19 pandemic is like a "bump" for the takeaway food industry. The take-out service has expanded its market share by 7%, garnering 43% of all buyers in 2019, compared to 36% in 2019. According to the AFP news agency, NPD Group expert Maria Bertoch stated that the take-out sector accounted for 15% of food and beverage services in 2019 but will climb to 30% in 2020, indicating that "many establishments have restricted their losses" Minh (2021). So many prominent fast food brands took advantage of the pandemic's word of mouth.

The above figures partly reflect the "rapid" growth of the fast food sector. This said the increasing demand of customers means that the challenge for the Fast Food sector is increasing. So what factors do these businesses have to meet the increasing needs of customers? How to keep customers satisfied and spread the word of positive things to the business?

Existing studies in the world have confirmed that word of mouth plays a very important role in shaping consumer attitudes and behaviors, even this form is many times more effective than traditional methods as other forms of marketing Sandes and Urdan (2013). According to Hanaysha (2016), customers can trust a higher level of the brand. These findings also confirm the significant positive relationship between word of mouth and brand image. According to Taghizadeh et al. (2013), product innovation has a positive effect on word of mouth communication. In general, there are many studies on word of mouth in the field of fast food but this chapter uses the BIC Algorithm for WOM in Fast Food. It is different from the previous study.

2 Literature Review

2.1 Brand Image (BI)

It is the customer's overall perception of the value of a brand, which motivates them to buy, maintain or increase ownership of something to satisfy a need, want, or purpose, or a certain desire. Brand equity has been formulated and measured in previous studies using a range of dimensions Kaushik and Soch (2021). In general, the most frequently used dimensions including brand awareness, perceived quality, brand positioning, and brand loyalty were suggested by Aaker (1991). Brand equity is also measured in the literature using brand distinctiveness, brand awareness, and brand loyalty Yasin et al. (2007). Similarly, Taleghani and Almasi (2011) used brand awareness, perceived quality, brand association, and brand image to formulate ideas and measure brand equity. Another contribution to brand equity was provided by Hanaysha and Hilman (2015), who integrated brand loyalty, brand awareness, brand leadership, and brand image to measure brand equity.

2.2 *Product Innovation (PI)*

It is the adoption of new internally created or purchased equipment, systems, policies, programs, processes, products, or services to the adopting organization Baba (2012). Service or product variation will affect customer intentions towards the manufacturer or service provider over time Bustinza et al. (2018). Ishaq (2011) has shown that product innovation has a positive effect on behavioral responses such as word of mouth and intention to switch.

2.3 *Selling Price (SP)*

It is an attribute that must be given up or sacrificed to obtain some type of product or service Erkoyuncu (2013). Customers are often price conscious in their purchasing behavior [?]. Price is also an important factor in choice situations because consumer choice often depends heavily on the price of alternatives. [?] determined that the role of price, as an attribute of performance, can have a direct influence on customer satisfaction and behavioral intentions.

2.4 *Brand Preference (BP)*

Long et al. (2012) show the consumer's decision to choose or purchase a certain brand in the presence of other competing brands and it can be formed based on experience in the industry past or suggested by others.

2.5 *Reputation (RE)*

According to Hatch and Schultz (1997), organizational reputation is the keen perception of an organization by stakeholders. It is formed through a long period of understanding and evaluation of the success of that organization Melanthiou (2014).

2.6 *Credibility (CR)*

Online word of mouth information can be generated by most network users; therefore, the quality and reliability of information are increasingly concerned Brian Lee and Li (2018). According to Wathen and Burkell (2002), credibility is a prerequisite in the process of persuading an individual. Therefore, this study places an expectation that credibility is positively related to information usefulness and leads to the information source application.

2.7 Word of Mouth (WOM)

Arndt (1967), a pioneer in word of mouth (WOM) research, gave the following definition of WOM: "Direct verbal communication between a receiver and a communicator concerning a brand, product or service, and the recipient perceives that the sender's messages are non-commercial". Anderson (1998) stated the definition of Word of Mouth as follows: "is an informal form of communication between two parties related to the evaluation of certain products or services". According to Kirby and Marsden (2006) word of mouth is "word of mouth, person-to-person communication, between recipients and transmitters related to a brand, a product, service or information on the market." According to Silverman (2011), word of mouth is the communication of products and services between people, who are independent of the business providing the product or service, etc. Word of mouth originates from a third party and is spontaneous transmission in a manner independent of manufacturer or seller Zhou and Duan (2015). Word of mouth (WOM) is a form of communication between people C2C (Consumer To Consumer) directly or indirectly via phone, email, forums, blogs, social networks, etc. through WOM, the information is "spread" quickly in the community Kasa et al. (2018). Thus, businesses by using and managing active word of mouth can attract many new customers of significant significance over the lifetime of their operations. Word of mouth communication is seen as the key to profitability ZorBari-Nwitambu (2017).

3 Methodology

3.1 Sample Approach

Table 1 shows that there are 99 female respondents (54.1%) and 84 male respondents (46.24%). The age of respondents to the questionnaire is under 25 years old with 46 respondents (25.1%). From 25 to 34 years old, there were 70 respondents (38.3%), from 35 to 44 years old had 57 respondents (31.1%), and over 45 years old had 10 respondents (5.5%).

Based on Table 1, there are mainly 82 respondents (44.8%) with unskilled, diploma 84 respondents (45.9%) degree 17 (9.3%) university level of education. According to Table 1, there are 59 respondents (32.2%) with monthly income from 5.000.000 VND—10,000,000 VND, 57 respondents (31.1%) with income less than 5,000,000 VND, 57 respondents (31.1%) with income over 11,000,000 VND - 15,000,000 VND and over 15,000,000 VND are 10 respondents (5.5%).

Table 1 Statistics of sample

Characteristics		Amount	Percent (%)
Sex and Age	Female	99	54.1
	Male	84	45.9
	Below 25	46	25.1
	25–34	70	38.3
	35–44	57	31.1
	Above 45	10	5.5
Education	Unskilled	82	44.8
	Diploma	84	45.9
	Degree	17	9.3
Income/Month	Below 5 million VND	57	31.1
	5–10 million VND	59	32.2
	11–15 million VND	57	31.1
	Over 15 million VND	10	5.5

3.2 Reliability Test

Cronbach's Alpha coefficient of 0.6 or more is acceptable (Nunnally 1978; Peterson 1994; Slater 1995) in case the concept being studied is new or new to the subject with respondents in the research context. However, according to Nunnally (1994), Cronbach's Alpha (α) does not indicate which variables should be discarded and which should be kept. Therefore, besides Cronbach's Alpha coefficient, one also uses Corrected item-total Correlation (CITC) and those variables with Corrected item-total Correlation greater than 0.3 will be kept.

4 Results

4.1 Reliability

Factors and items are in Table 2, Cronbach's Alpha coefficient of greater than 0.6 and Corrected Item—Total Correlation is higher than 0.3 reliable enough to carry out further analysis and mean of item from 3.25 to 4.62 is good.

Table 2 Factor and item

Factor	α	CITC	Item	Code	Mean
BI	0.871	0.730	The brand of the fast food business has a trendy and trendy image.	BI1	3.76
		0.762	The brand of the fast food business has attractive features.	BI2	3.80
		0.767	The brand of the fast food business has unique features.	BI3	3.79
PI	0.852	0.744	A fast food business that offers flexible products that meet my needs	PI1	4.62
		0.759	The new products the fast food business offers meet my needs	PI2	4.28
		0.718	Constantly renewing models	PI3	3.96
SP	0.881	0.794	The selling price of a fast food business is the same as that of other businesses	SP1	3.33
		0.774	Fast food businesses have more promotions and incentives than other businesses	SPI2	3.25
		0.744	Customers are willing to pay more for the business if the fast food product is of better quality.	SP3	3.43
BP	0.863	0.716	I prefer going to this business over others.	BP1	3.61
		0.812	I usually buy from this business.	BP2	3.69
		0.696	This business is my choice over others.	PB2	3.48
RE	0.839	0.705	Fast food business with outstanding popularity	RE1	4.02
		0.706	Fast food businesses are rated better than other businesses	RE2	3.99
		0.697	Fast food businesses are preferred by more people than other businesses	RE3	4.07
CR	0.885	0.863	Information is shared through word of mouth persuasive, can refer	CR1	3.87
		0.707	Information shared by word of mouth is verified	CR2	3.93
		0.725	Information shared through word of mouth can be trusted	CR3	3.93
		0.856	Information shared through word of mouth is correct	CR4	3.86
WOM	0.843	0.684	I have recommended the fast food business to so many people.	WOM1	3.94
		0.607	I talk about fast food business with my colleagues	WOM2	3.90
		0.608	I try to spread good words about the fast food business	WOM3	3.92
		0.841	I provide the fast food business with a lot of aggressive word-of-mouth advertising.	WOM4	3.87

4.2 *BIC Algorithm*

BIC (Bayesian Information Criteria) was utilized to choose the best model by R software. BIC has been used in the theoretical context for model selection. As a regression model, BIC can be applied, estimating one or more dependent variables from one or more independent variables Raftery et al. (1997). An essential and useful measurement for deciding a complete and straightforward model is the BIC. Based on the BIC information standard, a model with a lower BIC is selected. The best model will stop when the minimum BIC value Kaplan (2021). R report shows every step of searching for the optimal model. BIC selects the best 1 model as in Table 3. There is only one model in this study so it is the optimal model.

There are six independent and one dependent variable. Brand image (BI), Product Innovation (PI), Selling Price (SP), Brand Preference (BP), Reputation (RE), Credibility (CR) influence Word of Mouth (WOM) with a Probability is 100%. Other measurements are suitable with research data.

1. a. ***Model Evaluation***

According to the results from Table 4, BIC shows model 1 is the optimal selection because BIC (−192.62409) is minimum. Brand image (BI), Product Innovation (PI), Selling Price (SP), Brand Preference (BP), Reputation (RE), Credibility (CR) impact Word of Mouth (WOM) is 70.6% in Table 4. There are other factors that the subject has not studied: economic situation, the influence of culture, society, the influence of technology, and other marketing agents. BIC finds model 1 is the optimal choice and six variables have a probability of 100%. The above analysis shows the regression equation below is statistically significant.

WOM = 0.48117 + 0.14271BI + 0.15026PI + 0.13975SP + 0.09484BP + 0.10046RE + 0.26385CR

Table 3 BIC model selection

WOM	Probability (%)	SD	model 1
Intercept	100	0.18356	0.48117
BI	100	0.02180	0.14271
PI	100	0.02144	0.15026
SP	100	0.01846	0.13975
BP	100	0.02020	0.09484
RE	100	0.02700	0.10046
CR	100	0.04387	0.26385

Table 4 Model Test

Model	nVar	R^2	BIC	post prob
Model 1	6	0.706	−192.62409	1

4.3 Discussion

This study uses the optimal choice by the BIC algorithm in developed countries in a field and consumer market specifically. The study with 183 respondents was collected to study the factors affecting word of mouth of customers in the fast food in Ho Chi Minh City, Vietnam. These findings may not be research-intensive in this area due to factors that can influence customer word of mouth in the fast food sector in Ho Chi Minh City.

Although the R^2 of this study is not too large (70.6%), the regression test results also show that the hypotheses of the research model are not violated. R^2 is not too large which can be explained by the effects of the factors considered in the scope of this study, the remaining variability of the consumer's purchase intention may still be influenced by other factors that the subject has not studied: economic situation, the influence of culture, society, the influence of technology, and other marketing agents. Therefore, the study of the topic can be a reference for further studies with related content. In terms of research results, the topic is almost new in the fast food market and can be used for reference and comparison with other similar studies.

5 Conclusions

In this study, the factors affecting the word of mouth of customers in the fast food sector are built according to the proposed model of Taghizadeh et al. (2013), Hanaysha (2016), and Rodrigues and Brandão (2021). The official model includes 6 factors affecting the word of mouth of customers: Brand Image, Product innovation, Price, Brand Preference, Reputation, Credibility, Word of mouth. This model tests the influence of factors on word of mouth with a word-of-mouth scale consisting of 6 main factors measuring positive word of mouth. After the research and data processing of the BIC algorithm, 6 main factors that affect the word of mouth of customers include Brand image; Product innovation; Price; Like; Reputation; Reliability of information. The results of the linear regression analysis show that word of mouth has a positive relationship with these factors. Although the study was conducted differently in terms of business field and research scope, the obtained research results are somewhat consistent with previous research results. There are very few research projects on the factors affecting the word of mouth of customers, especially in Vietnam and in the fast food sector. In addition, fast food contributes a very important and meaningful part to the demand for eating. Therefore, if fast food makes customers happy, they will spread positive word of mouth about the fast food business. But on the contrary, the fast food business will lose a lot of advantages and prestige in the hearts of customers, which is difficult to build or recreate. In addition, the large number of customers of fast food is word of mouth from a variety of word of mouth. The closer they have a relationship, the more decisive their words are to the person being passed on. Therefore, with the results of this study, the great role of word of mouth is once again confirmed. Increasing word of mouth between customers will bring benefits to businesses. It helps fast food businesses in the Vietnamese market operate more smoothly and compete more fiercely. The perception of customers to the fast food business will be higher, the impressive features of the fast food business will also be left in the hearts of customers, especially word-of-mouth about the food. Fast food also increased significantly and of course when the main factors perceived by customers also increased. The research results also prove that using word of mouth in branding is the right thing to do.

Limitations

The sampling method is convenient random sampling, so the accuracy is not high. In addition, the sample size is small (183), although it satisfies the conditions for exploratory factor analysis and regression analysis, the generalizability is not high. Referring to the literature is difficult because there are not many studies on this topic in Vietnam, if any, it is just research on word of mouth activities in many different fields such as fast food businesses, banks, etc. There is no research in the field of fast food. This study uses simple linear regression analysis, which is easy to perform, so the reliability is not as high as today's modern analyzes such as SEM linear structure analysis with higher complexity. Therefore, further studies should conduct SEM analysis to increase the reliability of the study. As a result of the study, it was found that age demographics influence word of mouth in the fast food sector. In addition, age is inversely proportional to the mean, or in other words, the younger the age, the larger the mean. However, due to the limited time of the study, the number of respondents is not high, so it is not possible to make a profound and meaningful statement. This is also a limitation of the study. Therefore, the researcher wishes that the following studies can be based on this to broaden the scope of the study or carry out a more thorough study to get a deeper meaning for practice. It can be said that this is a scientific research paper on the field of fast food in Ho Chi Minh City. Also, an inherited research paper was developed from the previous research paper. References from researchers in this area also have limitations in the research process. In addition, inherited research papers of researchers from many different fields, depending on the time of research and the research market, can be applied as a research model. Further studies need to learn and study the theory carefully before making the model because sometimes disagreements lead to unsatisfactory results.

Authors Contribution

Three authors contribute to the paper. Bui Huy Khoi contributed to the study of data and the gathering of research-related references. Nguyen Thi Ngan contributed to the compilation of data and the manuscript was revised. Ngo Van Tuan conducted a data survey.

Acknowledgements This research is funded by the Industrial University of Ho Chi Minh City, Vietnam.

References

Aaker, D.A.: Managing brand equity. New York, Maxweel Macmillan-Canada. Inc. (1991)

Anderson, E.W.: Customer satisfaction and word of mouth. J. Serv. Res. **1**(1), 5–17 (1998)

Arndt, J.: Role of product-related conversations in the diffusion of a new product. J. Market. Res. **4**(3), 291–295 (1967)

Baba, Y.: Adopting a specific innovation type versus composition of different innovation types: case study of a Ghanaian bank. Int. J. Bank Market. (2012)

Brian Lee, H.C., Li, X.: Impact of online word of mouth on channel disintermediation for information goods. J. Manag. Inform. Syst. **35**(3), 964–993 (2018)

Bustinza, O.F., Vendrell-Herrero, F., Gomes, E., Lafuente, E., Opazo-Basáez, M., Rabetino, R., Vaillant, Y.: Product-service innovation and performance: unveiling the complexities. Int. J. Bus. Environ. **10**(2), 95–111 (2018)

Dat, L.: Many struggle fast-food chains in Vietnam: Due to facing a tough competitor is called lightning food (Vietnamese)? (2019). from https://doanhnhan.vn/nhieu-chuoi-thuc-an-nhanh-chat-vat-o-viet-nam-do-gap-phai-doi-thu-kho-xoi-la-loai-thuc-an-tia-chop-nay-23109.html

Erkoyuncu, J., Roy, R., Harrison, A.: Attribute based product-service system design: aerospace case study The Philosopher's Stone for Sustainability, pp. 55–60. Springer (2013)

Hanaysha, J.: Examining the link between word of mouth and brand equity: a study on international fast food restaurants in Malaysia. J. Asian Bus. Strat. **6**(3), 41 (2016)

Hanaysha, J., Hilman, H.: Product innovation as a key success factor to build sustainable brand equity. Manag. Sci. Lett. **5**(6), 567–576 (2015)

Hatch, M.J., Schultz, M.: Relations between organizational culture, identity and image. Europ. J. Market. (1997)

Ishaq, M.I.: An empirical investigation of customer satisfaction and behavioral responses in the Pakistani banking sector. Manag. Market. **6**(3), 457 (2011)

Kaplan, D.: On the quantification of model uncertainty: a bayesian perspective. Psychometrika **86**(1), 215–238 (2021)

Kasa, M., Negging, P., bin Yatim, A.: The mediating role of WOM (word of mouth) between antecedents and purchase intention among hotel guests in Sarawak, Malaysia. J. Soc. Sci. Res. **693–697**, 692 (2018)

Kaushik, P., Soch, H.: Interaction between brand trust and customer brand engagement as a determinant of brand equity. Int. J. Technol. Trans. Commercialisation **18**(1), 94–108 (2021)

Kirby, J., Marsden, P.: The viral. Buzz and Word of mouth revolution. Butterworth-Heinemann, Burlington, MA (2006)

Long, C., Gable, P., Boerstler, C., Albee, C.: Brands can be like friends: goals and interpersonal motives influence attitudes toward preferred brands, pp. 279–297. Theory and Practice, Consumer-Brand Relationships (2012)

Long, H.D.: Catering business 2021: Fast food and takeaway take the throne (Vietnamese) (2021). from https://tuoitre.vn/kinh-doanh-an-uong-2021-thuc-an-nhanh-va-thuc-an-mang-di-len-ngoi-20210210182252187.htm

Melanthiou, Y.: Success through innovation, reputation, and location. J. Promot. Manag. **20**(4), 411–412 (2014)

Minh, A.: Fast food is on the throne (Vietnamese) (2021). from http://hiephoibanle.com.vn/linh-vuc-thuc-nhanh-dang-len-ngoi/

Nunnally, J. C.: Psychometric Theory: 2d Ed: McGraw-Hill (1978)

Nunnally, J. C. (1994). Psychometric theory 3E: Tata McGraw-hill education

Peterson, R.A.: A meta-analysis of Cronbach's coefficient alpha. J. Cons. Res. **21**(2), 381–391 (1994)

Raftery, A.E.: Bayesian model selection in social research. Sociol. Methodo., 111–163 (1995)

Raftery, A.E., Madigan, D., Hoeting, J.A.: Bayesian model averaging for linear regression models. J. Amer. Stat. Assoc. **92**(437), 179–191 (1997)

Rodrigues, C., Brandão, A.: Measuring the effects of retail brand experiences and brand love on word of mouth: a cross-country study of IKEA brand. Int. Rev. Retail Distrib. Cons. Res. **31**(1), 78–105 (2021)

Sandes, F.S., Urdan, A.T.: Electronic word-of-mouth impacts on consumer behavior: exploratory and experimental studies. J. Int. Cons. Market. **25**(3), 181–197 (2013)

Septiani, D.I., Chaerudin, R.: The effect of customers' price perception, perceived quality and brand image toward purchasing intention in bandung local shoe brand. KnE Soc. Scie., 1242-1254–1242-1254 (2020)

Silverman, G.: Secrets of word-of-mouth marketing: how to trigger exponential sales through run-away word of mouth. Amacom Books (2011)

Slater, S.F.: Issues in conducting marketing strategy research. J. Strat. Market. **3**(4), 257–270 (1995)

Taghizadeh, H., Taghipourian, M.J., Khazaei, A.: The effect of customer satisfaction on word of mouth communication. Res. J. Appl. Sci. Eng. Technol. **5**(8), 2569–2575 (2013)

Taleghani, M., Almasi, M.: Evaluate the factors affecting brand equity from the perspective of customers using Aaker's model. Kuwait Chapter Arabian J. Bus. Manag. Rev. **33**(832), 1–13 (2011)

Varki, S., Colgate, M. (Varki): The role of price perceptions in an integrated model of behavioral intentions. J. Serv. Res. **3**(3), 232–240

Wathen, C.N., Burkell, J.: Believe it or not: factors influencing credibility on the Web. J. Amer. Soc. Inform. Sci. Technol. **53**(2), 134–144 (2002)

Yasin, N. M., Noor, M.N., Mohamad, O.: Does image of country-of-origin matter to brand equity? J. Product Brand Manag. (2007)

Zhou, W., Duan, W.: An empirical study of how third-party websites influence the feedback mechanism between online word-of-mouth and retail sales. Decis. Support Syst. **76**, 14–23 (2015)

ZorBari-Nwitambu, B.: Positive word of mouth and profitability: the experience of banks in Port Harcourt-Nigeria. Int. J. Manager. Stud. Res. **5**(5), 42–48 (2017)

The Presence of Market Discipline: Evidence from Commercial Banking Sector

Le Ngoc Quynh Anh and Pham Thi Thanh Xuan

Abstract The study selected the model with the lowest error outcomes utilizing modest observational data from 272 banks in the Asia–Pacific area from 2015 to 2019 and a large number of characteristics using the Lasso regression approach. The study reached the following outcomes: (i) The study reveals the presence of market discipline and its sensitivity to the bank's risks, particularly when enforcing Basel III capital standards. (ii) Our research reveals signs of a deterioration in market discipline under regulatory scrutiny, notably in emerging-market banks. (iii) Finally, the findings of the study, in particular, indicate the reasons why banks fail to report bank risks in compliance with the third pillar of the Basel III framework. When the Basel III framework is followed, depositors can expect higher interest rates or more market discipline towards risky banks. Our research has implications for bank supervisors, policymakers and bank managers.

Keywords Market discipline · Bank risk · Bank regulations · Lasso regression

1 Introduction

Economists and bank executives are becoming increasingly concerned about market forces and support increased participation of private agents (Flannery 2001). Market discipline in the banking industry can be defined as a situation in which private agents, such as depositors, creditors, and stockholders, incur rising costs as a result of banks' large-scale risk management efforts. Depositors can penalize higher-risk institutions by demanding higher interest rates or withdrawing their funds, according to research by Soledad et al. (2001). Because depositors are extremely risk averse, banks will be penalized by paying higher interest rates on deposits or attracting fewer

L. N. Q. Anh (✉)
University of Economics – Hue University, Hue City, Vietnam
e-mail: lengocquynhanh@hueuni.edu.vn

P. T. T. Xuan
University of Economics and Law, Vietnam National University, Ho Chi Minh City, Vietnam
e-mail: xuanptt@uel.edu.vn

N. Ngoc Thach et al. (eds.), *Financial Econometrics: Bayesian Analysis, Quantum Uncertainty, and Related Topics*, Studies in Systems, Decision and Control 427,
https://doi.org/10.1007/978-3-030-98689-6_21

deposits if they take on excessive risk. Depositors are acutely aware of the bank's hazards, potentially limiting the bank's excessive risk-taking. Market discipline is the third of the three pillars of the regulatory system, according to the Basel Committee on Banking Supervision. Market discipline is defined as a market-based push for banking-related risk transparency and disclosure, as well as regulatory procedures to promote market safety and soundness. As a result, several studies in both developed and developing nations have underlined the need of market discipline in limiting banking risk (Hamalainen et al. 2005; Nier and Baumann 2006).

The economic and business environment of emerging countries is quite different from that of developed countries (Hunjra et al. 2020). This leads to differences in sensitivity to the market environment. In addition, Asia–Pacific banks are facing strong competition, which has a negative effect on their risk taking. The principle "too—big—to—fail"[1] also applies to Asia–Pacific banks, mainly as they face poor regulation and supervision (Ahmad and Albaity 2019). For these reasons, the research has selected banks with large capital and adequate risk disclosure data in the Asia–Pacific region as a sample for research and comparison between the two groups. Based on our model, we examine the existence of depositors' discipline for disclosing bank risks and the impact of the Basel III framework on depositors' sensitivity. This shows the behavior of banks in delaying disclosure of risks when completing the safety framework as proposed by Basel.

This study contributes to the literature in a number of ways. *Firstly,* the market discipline literature is dominated by studies in developed economies in the US and Europe (Demirgüç-Kunt and Huizinga 2004; Distinguin et al. 2012; Hasan et al. 2013). Evidence of market discipline in the Asia–Pacific region is limited (Afzal et al. 2021). South Asia is an exception. Besides, there are also some studies in this area, but only conducted within one country, such as (Hadad et al. 2011) in Indonesia or (Le 2020) in Vietnam. In the Asia–Pacific region, 272 banks with large capital in Tier 1 will conduct our research. The study will add an overview of market discipline for this region, while comparing emerging and developed economies in terms of market discipline sensitivity in the study area. *Secondly,* compared with previous studies, our study focuses on four types of risks that the Basel regulatory framework requires to improve information disclosure, namely credit risk, market risk, operational risk and counterparty credit risk. From there, to explain the behavior of banks in delaying disclosure of risks when implementing the Basel framework. *Thirdly,* this study is the first attempt to assess market discipline under the new capital and liquidity regulations using the Lasso regression model as suggested by (Tibshirani 1996, 2011; Hastie et al. 2009, 2015). *Finally,* we find evidence of evidence for the presence of market discipline under the economic freedom index of countries. This is an important result that complements the theoretical and experimental evidence of sensitivity between the two factors compared with previous studies.

[1] The "too big to fail" (TBTF) theory asserts that certain corporations, particularly financial institutions, are so large and so interconnected that their failure would be disastrous to the greater economic system, and that they therefore must be supported by governments when they face potential failure.

The remainder of this paper is structured as follows: Sect. 2 provides a literature review. Section 3 discusses the methods and data used. Section 4 discusses the findings while Sect. 5 concludes.

2 Literature Review

Market discipline is a self-regulating process in which participants assess a bank's risk tolerance and take appropriate steps to prevent them from engaging in excessive risk-taking activities (Balasubramnian and Cyree 2014). Since banks operate efficiently in a "market economy", risk regulation can be determined by market participants (BIS 2008). When investors face the prospect of loss, they will be more cautious about their investments. Many empirical studies have addressed the existence and effectiveness of market discipline. Many private agents can create market discipline as depositors, creditors, and equity holders. Some of the articles focus on the discipline of depositors. Soledad et al. (2001), for example, show that depositors withdraw their deposits or demand higher deposit rates from bad banks. Their study also revealed the presence of market discipline in Argentine, Chilean and Mexican banks, including for small, insured depositors. Ghosh and Das (2005), Hasan et al. (2013) and Karas et al. (2019) used and validated the idea that depositors are central to market discipline. The study will focus on the behavior of depositors when considering market discipline.

Research on market discipline from the perspective of depositors' behavior is divided into three groups. *Firstly,* studies are interested in the number of uninsured deposits (Karas et al. 2019; Khorassani 2000). *Secondly,* studies interested in the interest rates payable to depositors create market discipline for banks (Hadad et al. 2011; Le 2020 and Afzal et al. 2021). *Finally,* the group of studies used both the consideration of factors affecting the amount of uninsured deposits and the interest paid to depositors (Jordan 2000; Soledad et al. 2001; Ghosh and Das 2003; Ioannidou and Dreu 2011 and Demirgüç-Kunt and Huizinga 2004). *Overall,* these studies have demonstrated that in the presence of market discipline, uninsured deposits penalize riskier banks by withdrawing their funds and/or demanding higher deposit interest rates. In addition, these studies also showed the effect of increasing capital and meeting capital regulations that led to a weakening of market discipline.

However, these studies have the following gaps: (i) The bank risk measures used are mainly risks calculated from financial reporting metrics and not disclosed directly by banks, such as loan loss reserves over total assets, non-performing loans over total assets, the log of loan loss reserves over capital, the log of non-performing loans over capital, and the bank's Z-score. Within the scope of the reference, there is no research examining the impact of the group of risks that Basel III requires to be disclosed in the response process, including credit risk, market risk, operational risk and counterparty credit risk. Therefore, the study will focus on these risks, thereby explaining the risk disclosure behavior of banks in the process of implementing the new regulatory framework. (ii) These studies have shown that increasing capital and implementing capital regulation under the Basel framework weakens market discipline, but they have not shown an increase in depositors' response to riskier banks. In particular, previous studies did not address the effect of meeting liquidity regulations under the Basel III. In addition, our research also shows that high economic freedom weakens market discipline. It is explained that countries with a higher index of economic freedom will create a better economic environment and increase economic growth, reduce financial crises, financial crises, etc. banking crisis (Ahmed and Ahmad 2020), thereby weakening market discipline. This is a macro variable that has not been mentioned in previous studies on market discipline.

3 Methodology

The study will use a lasso regression model, which is a selection procedure that performs both transformation and regularization selection to improve prediction accuracy and interpretability. It combines the least square method with a constraint on the sum of the absolute values of the coefficients (Tibshirani 1996, 2011; Hastie et al. 2009, 2015). Lasso can be useful in estimating regression coefficients and performing variable selection. This method is useful when demonstrating the presence of market discipline through a selection of factors. By using Lambda penalties to select a suitable model with minimal risk, the prediction accuracy will be increased compared to previous traditional models. In addition, it is suitable for factor analysis with many endogenous factors and big data.

Applications of the models take the following form:

$$\begin{aligned}\hat{\beta}^{lasso} = \arg\min_{\beta} \sum_{i=1}^{N} (MD_i - \beta_0 - \beta_1 CR_i - \beta_2 MR_i - \beta_3 OR_i - \beta_4 CCR_i - \beta_5 RDI_i \\ - \beta_6 CAR_i - \beta_7 LCR_i - \beta_8 LNTA_i - \beta_9 GDP_i - \beta_{10} INF_i - \beta_{11} ECO\_FREE_i \\ - \beta_{12} CAR(\geq 10.5)_i - \beta_{13} LCR(\geq 100_i))^2 + \lambda \sum_{J=1}^{13} |\beta_j| \end{aligned} \quad (1)$$

where β_0 an intercept; β_j the corresponding coefficient and $\hat{\beta}^{lasso}$ represents the vector of fitted regression coefficients on factors. λ is a positive weighting parameter on the L1 penalty, which encourages sparsity in the resulting set of fitted regression coefficients.

MD_i the i-*th* observation of the maket discipline variable. To investigate market discipline, numerous studies have provided several indicators of market discipline from the perspective of depositors, such as based on the uninsured debt (Nier and Baumann 2006), subordinated debt (Gropp et al. 2005 and Sironi 2005), interest rate (Demirgüç-Kunt and Huizing 2004 and Hadad et al. 2011). For the first measure, interbank deposits are not sufficiently representative of depositors, on the other hand, the interbank deposit market is not very transparent for emerging countries. As for the second measure in our sample data, only about 50% of banks issue secondary debt and mainly large banks. Therefore, the study will use the third measure, *which is calculated by dividing the total interest expense by the total deposit.*

Bank risks including Credit risk (CR), Market risk (MR), operational risk (OR) and counterparty credit risk (CCR)[2] with the i-*th* observation. According to the provisions of the 3 pillars of the Basel framework, 4 types of risks need to be fully informed. Therefore, the study will mainly focus on these types of risks to discipline from depositors and *expects a positive sign among risks to market discipline* (Hadad et al. 2011; Le 2020 and Afzal et al. 2021).

Bank-*specific variables: RDI* is a revenue diversification index.[3] In theory, the combined cash flows from non-correlated revenue sources should be more stable than the individual sources (Stiroh and Rumble 2006). In contrast, Baele et al. (2007) and Williams and Rajaguru (2013) provide evidence that diversification decreases a bank's idiosyncratic risk. Therefore, depositors may have different responses to the diversification of the Bank's revenue. *CAR* (Capital Adequacy Ratio—Tier 1) is calculated by Tier 1 Capital, net of deductions, divided by Total Risk Weighted Assets, expressed as a %. *LCR* (The Liquidity Coverage Ratio) is calculated by The value of High Quality Liquid Assets divided by the total net cash outflows over the next 30 calendar days, expressed as a %. Research of Abbas et al. (2019) suggests that high-capitalized banks tend to have better access to financing sources with lower cost and risk and better access to higher quality asset markets than low capitalized banks. As a result, depositors will reduce their deposit interest rate requirements and increase deposits in capitalized and liquid banks as a result of (Ghosh and Das 2003 and Afzal et al. 2021). In contrast, banks will be holding more capital aside, they need to be more selective when they are allocating their resources to project financing. This can ultimately lead project financing to be more difficult to obtain from banks (Ozkan and Iqbal 2015). Thus, well-capitalized banks appear to have lower profitabiity (Goddard et al. 2013). In addition to those arguments, Baker and Wurgler (2015) demonstrated that higher capital requirements are also related to higher cost of bank equity. Therefore, depositors will demand a higher interest rate for banks with higher capital (Hadad et al. 2011). From the conflicting results of previous studies, we will examine the effect of these two variables on depositors' discipline. *LNTA,* the natural logarithm of total assets, is used to control for the effect

---

[2] Credit risk (CR) is "the possibility that the bank borrower or counterparty will not meet its obligations under the agreed terms". Counterparty credit risk (CCR) that one party to a financial transaction will default. Market risk (MR) is "the risk of loss in on-balance sheet and off-balance sheet positions arising from fluctuations in market prices". Operational risk (OR) is "the risk of loss due to inadequate or failed internal processes, people and systems, or from external events". BIS (2008).

[3] Following the work of Stiroh and Rumble (2006), we calculate a revenue diversification index (RDI). The RDI is estimated as: $RDI = 1-(SHNET^2 + SHNON^2)$; where SHNET is the share of net operating revenue from net interest sources, SHNET = NET/(NET + NON) and NET is net interest income and NON is non - interest income; SHNON is the share of net operating revenue from non-interest sources, where SHNON = NON/(NET + NON).

of bank size. Mixed results were found in some studies as a negative sign (Flannery and Sorescu 1996) or a positive sign (Hadad et al. 2011). *Therefore, we do not have any a priori indication on this variable.*

General macroeconomic conditions including gross domestic product growth rate *(GDP),* inflation rate *(INF)* and economic freedom index *(ECO_FREE)* were also used as control variables. Since studies conducted for Asia–Pacific countries differ in the characteristics and variability of the indicators, the specific expected signs for these variables are not clear.

Furthermore, other control variables are used: *CAR(≥10.5)* and *LCR (≥100)* are dummy variables—specifying the minimum capital adequacy requirement 10.5% and liquidity coverage ratio account greater than 100%. These are the two regulations on capital and liquidity according to the Basel III framework. The study does not consider two other important regulations, NFSR and LEV, because these are two new regulations, so the data provided by banks is lacking. The study expects a negative sign for these two variables to prove that the theory of regulatory discipline can replace market discipline. This result is also corroborated by previous studies such as Hadad et al. (2011), Le (2020).

4 Empirical Results and Diagnostic Test

4.1 Data

Our sample is a balanced panel that includes the financial data of the top 272 commercial banks in four developed countries (Australia, Hong Kong, Japan and New Zealand) and ten emerging countries (China, India, Indonesia, Malaysia, Philippines, Singapore, South Korea, Taiwan, Thailand and Vietnam) in the Asia–Pacific region. A developed market is a country that is most developed in terms of its economy and capital markets. The country must have a high income, but this also includes openness to foreign ownership, ease of capital movement, and the efficiency of the market institutions. An emerging market is a market that has some characteristics of a developed market but does not fully meet its standards (MSCI—Market Classification Framework[4]). We limited our sample to large banks due to the numerous differences associated with bank size and the ability to implement Basel III regulations. Furthermore, sufficient data is only available for large capital tier 1 banks. All financial items related to bank characteristics are primarily drawn from the link: https://www.Banker

[4] MSCI—Morgan Stanley Capital International, is an American finance company headquartered in New York City and serves as a global provider of equity, fixed income, hedge fund stock market indexes, and multi-asset portfolio analysis tools. The MSCI Market Classification Framework was developed by MSCI, which aims to reflect the views and practices of the international investment community by striking a balance between a country's economic development and the accessibility of its market while preserving index stability.

database.com and the statements of various banks in annual financial reports (2015–2019). In addition, data on disclosed bank risks is applied according to calculation methods such as: credit risk, counterparty credit risk, market risk—Standardized approach (SA), operational risk—Basic indicator approach (BIA).

Table 1 presents the descriptive statistics for the raw variables and Table 2 reports the Pearson correlation coefficient during the study period. To ensure that these correlations will not lead to multicollinearity, we proceed with the variance inflation test (VIF). The VIFs of the variables are below 10 and the mean VIF of the model regression is below 2, indicating that multicollinearity is not a serious problem.[5]

4.2 Diagnostic Test

This study divides the dataset into two parts: using 80% for training and 20% for testing. The training split is used to fit the model (i.e. estimate the parameters of the model for a fixed set of tuning parameters). The test split is used to estimate the performance of the final chosen model.

Figure 1 shows the statistical pattern of models 1, 2, 3. In Table 3, there is a clear trend of increasing error levels when increasing the lambda penalty. Therefore, the statistical model eliminates the case of the "underfitting" model. In addition, statistical models can only choose one model with the lowest error level. This eliminates the case where the statistical model is said to be "overfitting"—there are many options for building models based on the dataset (Czum 2020).[6] Thus, this result is consistent and reliable. In order to get a good fit, we will stop at a point just before where the error starts increasing. At this point, the model is said to have good skills on both training and unseen testing datasets. Table 3 displays the models that were chosen.

[5] Gujarati (2003) indicates the VIF cut-off is 10. If the calculated VIF is more than 10, it can be an indication of multicollinearity.

[6]

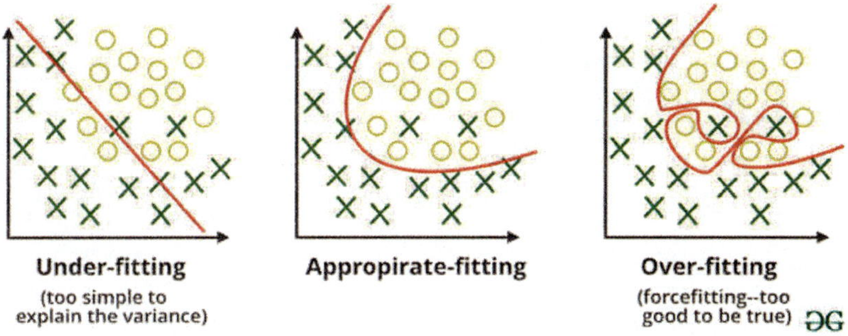

Table 1 Descriptive statistics

Variables	Obs	Mean	Std. Dev	Min	Max
Panel A: Full sample (N = 1360 bank years)					
MD	1360	1.830	2.637	1.085	55.608
CR	1360	30.453	29.256	0.672	104.812
MR	1360	1.221	2.313	0.002	22.434
OR	1360	2.479	2.746	0.105	18.965
CCR	1360	0.126	0.571	7.605E-05	8.291
CAR	1360	12.896	4.267	6.1	75.4
LCR	1360	158.757	75.342	40	818.7
LNTA	1360	1.176	1.176	9.007	15.276
INF	1360	1.843	1.843	−0.922	8.639
GDP	1360	5.891	5.660	−21.595	30
ECO_FREE	1360	65.486	10.942	51.7	90.2
RDI	1360	0.338	0.136	−0.403	0.500
CAR(≥10.5)	1360	0.569	0.495	0	1
LCR(≥100)	1360	0.368	0.482	0	1
Panel B: Emerging and developed countries					
Emerging (N = 998 bank years)					
MD	998	2.286	2.818	1.085	55.608
CR	998	36.029	29.457	0.672	104.812
MR	998	1.464	2.586	0.002	22.434
OR	998	2.932	2.883	0.441	18.965
CCR	998	0.097	0.565	7.605E-05	8.291
CAR	998	12.581	4.563	6.1	75.4
LCR	998	159.154	77.257	40	818.7
LNTA	998	11.046	1.174	9.007	15.276
INF	998	2.544	1.848	−0.922	6.499
GDP	998	7.633	5.693	0.522	30
ECO_FREE	998	61.105	8.857	51.7	89.4
RDI	998	0.330	0.145	−0.403	0.500
CAR(≥10.5)	998	0.577	0.494	1	0
LCR(≥100)	998	0.425	0.495	1	0
Developed (N = 362 bank years)					
MD	362	1.926	2.109	0	15.116
CR	362	36.79	17.025	6.52	72.465
MR	362	1.141	1.085	0.086	5.055
OR	362	3.474	1.689	0.105	6.124

(continued)

Table 1 (continued)

Variables	Obs	Mean	Std. Dev	Min	Max
CCR	362	0.706	0.927	0.002	3.493
CAR	362	14.38	2.894	10.4	23
LCR	362	153.859	66.995	96.9	607.9
LNTA	362	12.287	1.338	10.443	14.946
INF	362	1.963	1.513	−0.55	3.735
GDP	362	1.864	1.415	−1.249	3.791
ECO_FREE	362	82.833	6.473	69.6	90.2
RDI	362	0.371	0.11	0.085	0.499
CAR(≥10.5)	362	0.987	0.116	1	0
LCR(≥100)	362	0.987	0.115	1	0

Note This table presents the correlations between the variables included in this study. The dependent variable is Market discipline (MD). The bank risk variables include: Credit risk (CR), Market risk (MR), Operational risk (OR), Counterparty credit risk (CCR). The bank—specific variables include: Capital Adequacy Ratio (CAR), Liquidity Coverage Ratio (LCR), the natural logarithm of total assets (LNTA), Revenue diversification index (RDI). Control variables for general macroeconomic conditions include: Gross domestic product growth rate (GDP), Inflation rate (IMF), Economic freedom index (ECO_FREE.) The bank regulation variables are CAR(≥10.5), LCR(≥100)

Source Authors' own estimation

4.3 *Empirical Result*

The study examines the presence of market discipline in the Asia–Pacific region using Lasso regression. We use 3 models to test our hypotheses. After meeting the requirements for the diagnostic test and performing the selection of Lamdba fines, the study selected 3 models with the lowest error level and the regression results are reported in Table 3. In addition, Fig. 2 shows the impact of disclosed banking risks on market discipline by country group.

Presence *of market discipline, its sensitivity to risks through capital and liquidity regulations under the Basel III Framework.*

The positive and significant signals of MR and OR suggest the presence of market discipline under the influence of disclosed banking risks (Fig. 2). According to Cheng et al. (2018), operational risk increases cost inefficiencies while decreasing economic efficiency. According to Ekinci (2016), the bigger the market risk, the larger the volatility of the bank's stock returns. Depositors want a greater rate of interest than banks with larger market and operational risk, which explains why. CCR, on the other hand, has a large negative impact on depositor interest rates. When the counterparty's credit risk is significant, depositors will accept a low deposit rate. Because, according to Groups (2013), this risk indicates that banks are adjusting credit prices. Even when we employ two dummy variables to indicate capital and liquidity rules in accordance

Table 2 Correlation matrix

	1	2	3	4	5	6	7	8	9	10	11	12	13	14
1	1.000													
2	0.298*	1.000												
3	0.254*	0.472*	1.000											
4	0.317*	0.733*	0.468*	1.000										
5	−0.031	0.136*	0.173*	0.097*	1.000									
6	0.011	0.012	0.149*	0.201*	0.063*	1.000								
7	0.071	−0.107*	−0.051	−0.033	−0.093*	0.213*	1.000							
8	−0.054*	0.057	−0.055*	0.026	0.274*	−0.136*	−0.206*	1.000						
9	0.201	0.346*	0.232*	0.248*	0.037	−0.094*	−0.040	−0.077*	1.000					
10	−0.008	0.277*	0.147*	0.096*	0.039	−0.244*	−0.103*	−0.090*	0.629*	1.000				
11	−0.222*	0.019	0.019	−0.040	0.189*	0.222*	−0.106*	0.126*	0.056*	−0.104*	1.000			
12	−0.033	0.129*	0.215*	0.142*	0.079*	0.082*	−0.054	0.244*	0.125*	0.110*	0.199*	1.000		
13	0.127*	0.339*	0.270*	0.393*	0.159*	0.460*	0.095*	0.097*	0.101*	−0.032*	0.292*	0.146*	1.000	
14	0.207*	0.386*	0.180*	0.377*	0.202*	−0.027	0.238*	0.309*	0.216*	0.118*	0.036	0.078*	0.258*	1.000

(continued)

Table 2 (continued)

	1	2	3	4	5	6	7	8	9	10	11	12	13	14
VIF		2.27	1.33	2.81	1.22	1.57	1.23	1.43	1.87	2.18	1.39	1.25	1.5	1.14
1/VIF		0.44	0.752	0.355	0.818	0.638	0.813	0.7	0.535	0.459	0.72	0.8	0.665	0.878
Mean VIF													**1.63**	

*Note: *significant at 5% level*
This table presents the correlations between the variables included in this study. The dependent variable is Market discipline (MD). The bank risk variables include: Credit risk (CR), Market risk (MR), Operational risk (OR), Counterparty credit risk (CCR). The bank—specific variables include: Capital Adequacy Ratio (CAR), Liquidity Coverage Ratio (LCR), the natural logarithm of total assets (LNTA), Revenue diversification index (RDI). Control variables for general macroeconomic conditions include: Gross domestic product growth rate (GDP), Inflation rate (IMF), Economic freedom index (ECO_FREE.) The bank regulation variables are CAR(≥10.5), LCR(≥100)
1: MD, 2:CR, 3:MR, 4:OR, 5:CCR, 6:CAR, 7: LCR, 8: LNTA, 9:INF, 10:GDP, 11:ECO_FREE, 12: RDI, 13: CAR(≥10.5), 14: LCR(≥100)

Source Authors' own estimation

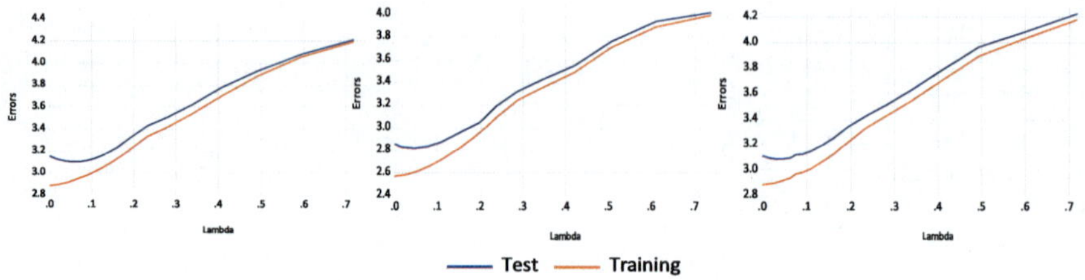

Fig. 1 Training/ Test error evolution

with Basel III, as illustrated in models 2 and 3, the outcome remains constant. The findings also revealed that, according to (Le 2020), credit risk had essentially no effect on depositor discipline.

When capital needs are taken into account, Model 1 shows how the impact of banking risks on market discipline increases. The study also discovered that the CAR (≥10.5) and MD variables had the opposite effect, but that the LCR (≥100) variable had no effect on MD, utilizing these two models. Depositors are more confident in a bank's safety when it fulfills the Basel 3 framework's minimum capital requirement of more than 10.5 percent. As a result, they will accept lower interest rates, weakening market discipline. This finding supports the concept that regulatory discipline can take the place of market discipline, as well as earlier research (Hadad et al. 2011; Le 2020; Afzal et al. 2021).

When it comes to capital requirements, the above statistics are similar for rising economies (model 5). Depositors are more interested in and respond strongly to the risks revealed by banks under Basel III when banks build capital to execute capital requirements. When enforcing liquidity limits, these institutions, on the other hand, clearly demonstrate market discipline. Model 6 indicates that if banks fail to achieve the Basel 3 liquidity criteria, depositors will demand higher deposit rates. This finding is in line with Abbas et al. (2019).'s study, which advocates for banks to comply with Basel III requirements.

Market discipline—the difference between emerging and developed economies.

When banking regulations are not taken into account, Models 4 and 7 illustrate the impact of factors on market discipline in emerging and developed countries, respectively. When these two models are compared, the regression results for credit risk are the opposite (Fig. 2). Depositors, in particular, expect a higher interest rate when credit risk for emerging countries is higher, according to Hadad et al. (2011). Depositors in industrialized countries, on the other hand, display credit risk-talking behavior, which is consistent with Ioannidou and Dreu's findings (2011). Because credit risk is so tightly linked to an investment's prospective return, the yields on bonds have a high correlation with their perceived credit risk, according to CFA Study Preparation research (2021).

When banks in emerging economies have smaller capital, a lower liquidity coverage ratio, and a larger revenue diversification rate, some disparities between the two sets of countries, such as market discipline, will be more obvious. The data,

Table 3 The result of lasso regression

Dependent variable	MD						
Independent variables	Model 1	Model 2	Model 3	Model 4	Model 5	Model 6	Model 7
Lambda at minimum error	0.051167	0.03548	0.06235	0.03645	0.03645	0.03645	0.09566
CR	0.000000	0.000000	0.000000	0.003859	0.003909	0.003741	−0.019599
MR	0.043550	0.049642	0.040031	0.019368	0.023217	0.019739	0.190726
OR	0.028559	0.032555	0.028571	0.000000	0.000000	0.000000	0.000541
CCR	−0.037183	−0.044407	−0.028691	0.000000	0.000000	0.000000	0.000000
CAR	0.000000	−0.001762	0.000000	−0.005590	−0.000966	−0.005334	0.007126
LCR	0.000000	−4.47E-05	0.000000	−0.000638	−0.000525	−0.000545	7.38E-06
LNTA	−0.250899	−0.262748	−0.247064	−0.370436	−0.364735	−0.370563	0.251718
INF	0.257402	0.279026	0.241537	0.362183	0.359766	0.362205	−0.071425
GDP	−0.131378	−0.141118	−0.126534	−0.170766	−0.171516	−0.171218	−0.028715
ECO_FREE	−0.057198	−0.057187	−0.056215	−0.054699	−0.053095	−0.054386	−0.018435
RDI	0.000868	0.200936	0.000000	1.360157	1.374628	1.355407	−11.72658
CAR(≥10.5)		−0.119790			−0.141985		
LCR(≥100%)			0.000000			−0.114291	
N	1360	1360	1360	998	998	998	362

Notes This table presents the correlations between the variables included in this study. The dependent

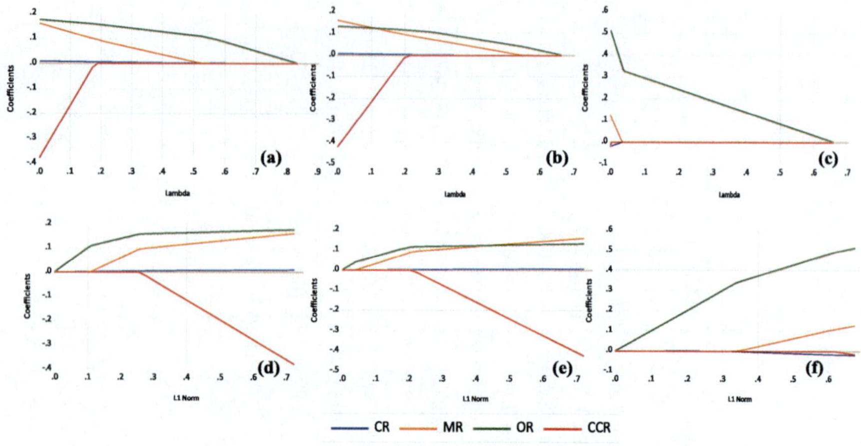

Fig. 2 Coefficient estimates for Lasso regression versus Lambda (fig a, b, c), L1 Norm (fig d, e, f)[7]

on the other hand, corroborate the "too-big-to-fail" theory in industrialized countries. Depositors will be more concerned and demand a higher interest rate from banks with higher capital, liquidity coverage ratios, bank size, and less diversified revenue, resulting in more market discipline.

Finally, the analysis reveals that countries with a low economic freedom rating have better market discipline. This is true regardless of whether or not laws within the Basel framework are considered, as well as for groups of developed and emerging countries.

5 Conclusion

We uncover evidence of market discipline using a balanced panel of 272 commercial banks since higher deposit rates are linked to bank risks such as credit risk, risk market risk, operational risk, and counterparty credit risk (these are the types of risks that Basel III requires banks to disclose). In addition, deposit rates and capital and liquidity regulations have an inverse connection, according to our findings. This is in line with high CAR and LCR signals, which lower the likelihood of a bank collapsing, resulting in a weakening of market discipline. Our research uncovers indications of a weakening of market discipline under regulatory oversight, particularly in banks from emerging nations.

The following are some of the significant results of our findings: (i) Assist regulators in developing strategies to manage banks' risk and achieve Basel III liquidity and capital requirements. Market discipline has become less effective, and banking

[7] Fig a, Fig d (Full sample); Fig b, Fig e (emerging countries) and Fig c, Fig f (developed countries).

authorities must compensate by expanding their supervisory powers. (ii) In particular, the study's findings reveal the reasons why banks fail to disclose bank risks in accordance with the Basel III framework's third pillar. Depositors expect higher interest rates or increased market discipline towards riskier banks when the Basel III framework is followed.

References

Abbas, F., Iqbal, S., Aziz, B.: The impact of bank capital, bank liquidity and credit risk on profitability in postcrisis period: a comparative study of US and Asia. Cogent Econ. Financ. **7**(1), 1–18 (2019). https://doi.org/10.1080/23322039.2019.1605683

Afzal, A., Mirza, N., Arshad, F.: Market discipline in South Asia: evidence from commercial banking sector. Int. J. Financ. Econ. **26**(2), 2251–2262 (2021). https://doi.org/10.1002/ijfe.1904

Ahmad, R., Albaity, M.: The determinants of bank capital for East Asian countries. Glob. Bus. Rev. **20**(6), 1311–1323 (2019). https://doi.org/10.1177/0972150919848915

Ahmed, S., Ahmad, H.K.: Impact of economic and political freedom on economic growth in Asian economies. Europ. Online J. Nat. Soc. Sci. **9**(1), 219–231 (2020)

Baele, L., De Jonghe, O., Vander Vennet, R.: Does the stock market value bank diversification? J. Bank. Financ. **31**(7), 1999–2023 (2007). https://doi.org/10.1016/j.jbankfin.2006.08.003

Baker, M., Wurgler, J.: Do strict capital requirements raise the cost of capital? bank regulation, capital structure, and the low risk anomaly. Amer. Econ. Rev. (2015). https://doi.org/10.1257/aer.p20151092

Balasubramnian, B., Cyree, K.B.: Has market discipline on banks improved after the Dodd-Frank Act? J. Bank. Finance **41**(1), 155–166 (2014). https://doi.org/10.1016/j.jbankfin.2014.01.021

Cheng, C.P., Phung, M.T., Hsiao, C.L., Shen, D.B., Chen, B.S.: Impact of operational risk toward the efficiency of banking-evidence from Taiwan's banking industry. Asian Econ. Financ. Rev. **8**(6), 815–831 (2018). https://doi.org/10.18488/journal.aefr.2018.86.815.831

CFA Level I Exam: CFA Study Preparation. (n.d.). Retrieved May 26, 2021, from https://analystnotes.com/cfa-study-notes-credit-risk-vs-return-yields-and-spreads.html

Counterparty Risk and CVA Survey Current market practice around counterparty risk regulation, CVA management and funding (2013)

Czum, J.M.: Dive into deep learning. J. Am. Coll. Radiol. **17**(5), 637–638 (2020). https://doi.org/10.1016/j.jacr.2020.02.005

Demirgüç-Kunt, A., Huizinga, H.: Market discipline and deposit insurance. Statist. Probab. Lett. **66**(4), 375–399 (2004). https://doi.org/10.1016/j.jmoneco.2003.04.001

Distinguin, I., Kouassi, T., Tarazi, A.: Interbank deposits and market discipline: evidence from central and eastern Europe. SSRN Electron. J., 1–38 (2012).https://doi.org/10.2139/ssrn.2119956

Ekinci, A.: The Effect of Credit and Market Risk on Bank Performance: Evidence from Turkey. Undefined (2016)

Flannery, M.J.: The faces of "market discipline." J. Financ. Serv. Res. **20**(2–3), 107–119 (2001). https://doi.org/10.1023/A:1012455806431

Flannery, M.J., Sorescu, S.M.: Evidence of bank market discipline in subordinated debenture yields: 1983–1991. J. Financ. **51**(4), 1347–1377 (1996). https://doi.org/10.1111/j.1540-6261.1996.tb04072.x

Ghosh, S., Das, A.: Market discipline in the Indian banking sector: An empirical exploration. NSE Research Initiative, NSE: Mumbai, May, 1–19 (2003). http://128.118.178.162/econ-wp/fin/papers/0410/0410020.pdf

Goddard, J., Liu, H., Molyneux, P., Wilson, J.O.S.: Do bank profits converge? Eur. Financ. Manag. (2013). https://doi.org/10.1111/j.1468-036X.2010.00578.x

Gropp, R., Vesala, J.M., Vulpes, G.: Equity and bond market signals as leading indicators of bank fragility. SSRN Electron. J. **38**(2), 399–428 (2005). https://doi.org/10.2139/ssrn.318359

Hadad, M.D., Agusman, A., Monroe, G.S., Gasbarro, D., Zumwalt, J.K.: Market discipline, financial crisis and regulatory changes: Evidence from Indonesian banks. J. Bank. Finance **35**(6), 1552–1562 (2011). https://doi.org/10.1016/j.jbankfin.2010.11.003

Hamalainen, P., Hall, M., Howcroft, B.: A framework for market discipline in bank regulatory design. J. Bus. Financ. Acc. **32**(1–2), 183–209 (2005). https://doi.org/10.1111/j.0306-686X.2005.00592.x

Hasan, I., Jackowicz, K., Kowalewski, O., Kozłowski, Ł: Market discipline during crisis: evidence from bank depositors in transition countries. J. Bank. Finance **37**(12), 5436–5451 (2013). https://doi.org/10.1016/j.jbankfin.2013.06.007

Hastie, T., Tibshirani, R., Friedman, J.: The Elements of Statistical Learning The Elements of Statistical LearningData Mining, Inference, and Prediction, Second Edition. In Springer series in statistics (2009). https://doi.org/10.1007/978-0-387-84858-7

Hastie, T., Tibshirani, R., Wainwright, M.: Statistical learning with sparsity: The lasso and generalizations. In Statistical Learning with Sparsity: The Lasso and Generalizations (2015). https://doi.org/10.1201/b18401

Hunjra, A.I., Zureigat, Q., Mehmood, R.: Impact of capital regulation and market discipline on capital ratio selection: a cross country study. Int. J. Financ. Stud. **8**(2), 1–13 (2020). https://doi.org/10.3390/ijfs8020021

Ioannidou, V., Dreu, J. de: The impact of explicit deposit insurance on market discipline. SSRN Electron. J. (2011).https://doi.org/10.2139/ssrn.888681

Jordan, J.S. (n.d.). John S. Jordan (2000)

Karas, A., Pyle, W., Schoors, K.: Deposit Insurance, Market Discipline and Bank Risk. U.S.E. Working Paper Series, 19(02) (2019)

Khorassani, J.: Khorassani J 2000—an empirical study of depositor sensitivity to bank risk, JEF.pdf. **24**(I), 15–27 (2000)

Le, T.D.: Market discipline and the regulatory change: evidence from Vietnam. Cogent Econ. Financ. 8(1) (2020). https://doi.org/10.1080/23322039.2020.1757801

Market Discipline, Capital Adequacy and Bank Behaviour on JSTOR (2005). Ghosh and Das. https://www.jstor.org/stable/4416369?seq=1

Nier, E., Baumann, U.: Market discipline, disclosure and moral hazard in banking. J. Financ. Intermed. **15**(3), 332–361 (2006). https://doi.org/10.1016/j.jfi.2006.03.001

Ozkan, C., Iqbal, Z.: Implications of Basel III for Islamic Banking- Opportunities and Challenges. Policy Research Working Paper (2015)

Sayyed, Q.: Basic Econometrics Fourth Edition. Retrieved May 23, 2021, from https://www.academia.edu/40263427/BASIC_ECONOMETRICS_FOURTH_EDITION

Sironi, A.: Testing for market discipline in the european banking industry: evidence from subordinated debt issues. SSRN Electron. J. **35**(3), 443–472 (2005). https://doi.org/10.2139/ssrn.249284

Soledad, M., Peria, M., Schmukler, S.L.: Do depositors punish banks for bad behavior? Market discipline, deposit insurance, and banking crises. J. Financ. **56**(3), 1029–1051 (2001). https://doi.org/10.1111/0022-1082.00354

Stiroh, K.J., Rumble, A.: The dark side of diversification: the case of US financial holding companies. J. Bank. Financ. **30**(8), 2131–2161 (2006). https://doi.org/10.1016/j.jbankfin.2005.04.030

Tibshirani, R.: Regression shrinkage and selection via the lasso. J. Roy. Stat. Soc.: Ser. B (methodol.) **58**(1), 267–288 (1996). https://doi.org/10.1111/j.2517-6161.1996.tb02080.x

Tibshirani, R.: Regression shrinkage and selection via the lasso: a retrospective. J. r. Stat. Soc. Ser. B Stat Methodol. (2011). https://doi.org/10.1111/j.1467-9868.2011.00771.x

Williams, B., Rajaguru, G.: The chicken or the egg? The trade-off between bank fee income and net interest margins. Aust. J. Manag. **38**(1), 99–123 (2013). https://doi.org/10.1177/0312896212440268

Bayesian Model Averaging Method for Intention Using Online Food Delivery Apps

Dam Tri Cuong

Abstract The quick development of e-commerce has produced new types of business, for example, online to offline, and has changed the traditional doing of business. Clients place orders for products/services online and afterward get the goods/services at an offline place. Furthermore, the capacity to share information rapidly has prompted the quick development of mobile trade associating providers and clients via cell phone applications. In the literature, many studies have mainly suggested the partial least squares (PLS) approach for factors that affect the usage intention of online food delivery apps. This approach commonly chooses only the best model among feasible selection models based on some model selection criteria. However, this method commonly neglects the uncertainty connected with the choice of model. So, the result of the estimated model could be biased and guide to incorrect inference in investigating for intention using online food delivery apps. Thus, it is important to examine the uncertainty between candidate models, especially when these models are seen as feasible models notwithstanding differences in estimation. The Bayesian model averaging (BMA) approach is one of the systematic techniques for resolving model uncertainty, which allows the estimation of the power of results to alternative models by measuring posterior distributions over coefficients and models. Therefore, this paper applies the Bayesian model averaging (BMA) method to choose the optimal models for factors that influence intention using online food delivery apps. The results showed that the best four models were selected for factors affecting intention using online food delivery apps.

Keywords Bayesian analyses · Bayesian model averaging · Usage intention · Online food delivery apps

D. T. Cuong (✉)
Industrial University of Ho Chi Minh City, Ho Chi Minh City, Vietnam
e-mail: damtricuong@iuh.edu.vn

N. Ngoc Thach et al. (eds.), *Financial Econometrics: Bayesian Analysis, Quantum Uncertainty, and Related Topics*, Studies in Systems, Decision and Control 427,
https://doi.org/10.1007/978-3-030-98689-6_22

1 Introduction

With the rapid progression of information communication technology and cell phones, intelligent technologies, and mobile application software get to become a wide and essential part of daily life (Alalwan 2020). Likewise, the quick development of e-commerce has produced new types of business, for example, online to offline, and has changed the traditional doing of business. Clients place orders for products/services online and afterward get the goods/services at an offline place. The capacity to share information rapidly has prompted the quick development of mobile trade associating providers and clients via cell phone applications (Lee et al. 2019). Furthermore, technology is used not only for interpersonal relations but also for trading goods/services. Apps depict trading opportunities and have been investigated in research to examine customer attitudes towards online services, to find how companies can develop this distribution system and connection with their clients (Christino et al. 2021). Besides, the mobile transaction is one of the commonly utilized communications globally. Five billion people have mobile devices and a sum of six billion people have used these devices throughout the world (Kiat et al. 2017). In Vietnam, with citizens of around 100 million people, of which 60% are the young, Vietnam is viewed as a potential market in online payments. Furthermore, during the Covid-19 pandemic period, the habit of cashless payment has increasingly grown. Accompanying with online payment services, online food ordering apps are also an area benefiting from the purchasing ways changing. The report of the IMARC group demonstrated that the online food delivery market in Vietnam increased 38% per year from 2014 to 2019 and to be estimated to develop double digits in the period 2020–2025 (Bao 2020).

In the literature, many studies have mainly suggested the partial least squares (PLS) approach for factors that impact the usage intention of online food delivery apps (e.g. see Christino et al. (2021), Alalwan (2020), San and Dastane (2021), Troise et al. (2020), Lee et al. (2017), Yeo et al. (2017)). This approach commonly chooses only the best model among feasible selection models based on some model selection criteria. However, this method commonly neglects the uncertainty connected with the choice of model. So, the result of the estimated model could be biased and guide to incorrect inference in investigating for intention using online food delivery apps. Therefore, it is important to examine the uncertainty between candidate models, especially when these models are seen as feasible models notwithstanding differences in estimation. The Bayesian model averaging (BMA) approach is one of the systematic techniques for resolving model uncertainty, which allows the estimation of the power of results to alternative models by measuring posterior distributions over coefficients and models (Montgomery and Nyhan 2010). Furthermore, it can perform more accurate and reliable predictions than other techniques (Li and Shi 2010).

Therefore, it is charming to apply the BMA method for the predictive models of factors affecting intention using online food delivery apps. Yet, up to now, not yet such studies can be found in the literature for intention using online food delivery apps by using BMA. Hence, this paper applies the Bayesian model averaging (BMA) method to choose the optimal models for factors affecting intention using online food delivery apps.

2 Literature Review

2.1 Unified Theory of Acceptance and Use of Technology (UTAUT)

Venkatesh et al. (2003) revealed the unified model, called the unified theory of acceptance and use of technology (UTAUT), with four key factors (i.e. performance expectancy, effort expectancy, social influence, and facilitating conditions) of behavioral intention and usage behavior. Besides, Venkatesh et al. (2003) applied four moderators (i.e. age, gender, experience, and voluntariness of use) of these essential relations. The theory was formed by the review and combination of eight common theories and models (Dwivedi et al. 2017). These eight common theories and models are the theory of reasoned action (TRA), technology acceptance model (TAM), motivational model (MM), theory of planned behavior (TPB), combined TAM and TPB (C-TAM-TPB), model of PC utilization (MPCU), innovation diffusion theory (IDT), and social cognitive theory (SCT) (Venkatesh et al. 2003).

The UTAUT is the theoretical framework for this research. Compared with other widely utilized theories, such as the theory of planned behavior (TPB) and the technology acceptance model (TAM), the UTAUT is a newer and more integrative theory, mixing factors from contemporary theories as predictors of technology acceptance and usage (Okumus et al. 2018).

2.2 Performance Expectancy

Performance expectancy relates to the ability of the new system and application to support clients with accomplishing what they need and want in a more advantageous manner (Alalwan 2020; Venkatesh et al. 2003). Clients are more expected to have a positive response and intention towards utilizing a new system if they notice that the system will save clients more time and exertion than traditional ones do (Alalwan 2020; Dwivedi et al. 2017). Prior researches disclose that performance expectancy affects positively intention using online food delivery apps (Christino et al. 2021; Okumus et al. 2018).

2.3 Effort Expectancy

Effort expectancy is the degree to which a person thinks that utilizing a system is without effort. It concerns perceived ease of use in TAM Davis et al. (1989), which indicates that a system perceived to be easier to use is more probable to cause behavioral intention (Chiu and Wang 2008). Some studies showed effort expectancy as one of the predictors of intention using online food delivery apps (Christino et al. 2021; Okumus et al. 2018).

2.4 Social Influence

Social influence is the extent to which a person notices that relevant people think a person should utilize a system (Venkatesh et al. 2003). This study illustrates social influence as the extent to which a cell phone person believes that influential people think a person should utilize the smartphone app. This construct is similar to the subjective norm in TPB (Ajzen (1991)), which declares that the more positive the social influence of behavior, the more influential an individual's intention to do it (i.e. use food delivery apps) (Okumus et al. 2018). Former studies demonstrate the social influence of behavior is an antecedent of intention using online food delivery apps (Christino et al. 2021; Okumus et al. 2018).

2.5 Facilitating Conditions

Facilitating conditions are expressed as the extent to which a person thinks that a firm with existing technical infrastructure to help the usage of the system (San Martín and Herrero 2012). The previous study signifies facilitating conditions factor has positively affected intention using online food delivery apps (Christino et al. 2021). Another study shows that facilitating conditions factor does not influence intention using online food delivery apps (Okumus et al. 2018).

2.6 Usage Intention

Usage intention of technology is acknowledged as behavioral intention to use, it is considered to be the motivational factor that individual plans to perform in certain behavior (Ajzen 1991). Previous research implied that intention to use a system was the most meaningful predictor of its selection and actual utilization (Venkatesh et al. 2003, 2012). This assumes that individuals' intentions seize the motivational factors that affect their behavior and show individuals' readiness to develop the action

(Okumus et al. 2018). Usage intention is the result of performance expectancy, effort expectancy, social influence, and facilitating conditions (Venkatesh et al. 2003).

3 Methodology

3.1 Analytical Technique

In this analysis, the BMA approach with RStudio software is implemented to measure the optimal models. The BMA not only selects the best model among feasible choice models but also can select many models to explain the dependable variable. In this study, the independent variables include four variables (such as performance expectancy, effort expectancy, social influence, and facilitating conditions), and the dependent variable is the usage intention of online food delivery apps.

BMA method uses posterior probability as the weight to average all plausible models considered (Zou et al. 2021). Thus, let $M = \{M_1, \ldots, M_K\}$ denotes the set of all models and let y signify the quantity of interest, as the future observed values, and then the posterior distribution of y, given the observed data D is

$$P(y|D) = \sum_{k=1}^{K} P(y|M_k,D)\, P(M_k|D) \tag{1}$$

where:

$P(y|M_k,D)$ is the mean of the posterior distribution of y based on the candidate model M_K, which is the output of the BMA approach.

$P(M_k|D)$ is the probability of the true prediction model M_K, which is also referred to as the posterior model probability.

The posterior probability of the model M_K is given by

$$P(M_k|D) = \frac{P(D|M_k)\, P(M_k)}{\sum_{l=1}^{K} P(D|M_l)\, P(M_l)}, \tag{2}$$

where

$$P(D|M_k) = \int P(D|\theta_k, M_k)\, P(\theta_k|M_k)\, d\theta_k \tag{3}$$

is the marginal likelihood of the model M_K, θ_k is the vector of parameters of the model M_K, $P(\theta_k|M_k)$ is the prior density of θ_k under model M_K, $P(D|\theta_k, M_k)$ is the likelihood, and $P(M_k)$ is the prior probability that M_K is the true model (Raftery et al. 1997).

In practice, the estimates of the above quantities utilize the Bayes factor (Montgomery and Nyhan 2010). Using the Bayes rule, the posterior odds of some model M_k to M_j can be estimated as follows, which is the Bayes factor (Montgomery and Nyhan 2010; Raftery 1995).

$$\frac{P(M_k|D)}{P(M_j|D)} = \frac{P(D|M_k)}{P(D|M_j)} x \frac{P(M_j)}{P(M_k)} \tag{4}$$

The posterior mean and variance of y are shown as follows:

$$E(y|D) = \sum_{k=1}^{K} E(y|D,M_k)\,P(M_k|D) \tag{5}$$

$$Var(y|D) = \sum_{k=1}^{K} Var(y,|D,M_k) + E(y|D,M_k)^2)P(M_k|D) - E(y|D)^2 \tag{6}$$

3.2 *Data and Sample*

In this study, analysis data is collected from clients in Ho Chi Minh City, Vietnam. The Likert scale of five-point was used to estimate the determinants (from 1 = completely object to 5 = fully agree). The convenient method through the online survey is conducted for getting the population sample.

4 Results and Discussion

4.1 *Descriptive Statistics*

After rejecting the questionnaire didn't have adequate information, 302 questionnaires are adopted for the analysis. Table 1 illustrates the demographic characteristics of the sample.

4.2 *Bayesian Model Averaging (BMA)*

The optimal models result for the BMA technique is exhibited in Table 2.

Table 1 Demographic characteristics of the sample

Characteristics	Classifications	Frequency	Percent
Gender	Male	120	39.7
	Female	182	60.3
	Total	302	100.0
Age	<18	9	3.0
	18–25	222	73.5
	26–40	54	17.9
	>40	17	5.6
	Total	302	100.0
Occupation	Student	184	60.9
	Civil servant	76	25.2
	Housewife	22	7.3
	Other	20	6.6
	Total	302	100.0

Table 2 The best 4 models were selected by using the BMA

	p! = 0	EV	SD	Model 1	Model 2	Model 3	Model 4
Intercept	100.0	1.05578	0.17348	1.05051	0.94571	1.21546	1.05707
PERF	100.0	0.24822	0.05198	0.25249	0.22652	0.26861	0.23305
EFFO	100.0	0.33747	0.05937	0.32677	0.31391	0.39167	0.36556
SOCI	75.9	0.09986	0.06930	0.13504	0.11925		
FACI	27.4	0.02551	0.04865		0.08396		0.10717
nVar				3	4	2	3
r2				0.511	0.517	0.497	0.506
BIC				−199.220	−196.697	−196.253	−195.796
Post prob				0.591	0.168	0.134	0.107

As displayed in Table 2, Table 2 lists the posterior effect probabilities (p! = 0), expected values (EV) or posterior means, standard deviations (SD), and the best 4 models were selected by using the BMA. The posterior effect probabilities (p! = 0) indicate the regression coefficient probabilities different zero (i.e. have affecting relate to the dependent variable). For example, the posterior affect probability of performance expectancy is 100, which implies the performance expectancy factor happens 100% in all models. Expected values (EV) or posterior mean of the regression coefficient are measured for all models. Standard deviations (SD) of the regression coefficient are measured for all models. Model 1 in Table 2 refers to the best model in 4 models, model 2 signifies the second-best model after model 1, etc. The best model is assessed by three indexes (r2, BIC, post prob), in which post prob is the most critical index (Tuan 2020).

where:

PERF: Performance expectancy, EFFO: Effort expectancy, SOCI: Social influence, FACI: Facilitating conditions.

nVar: Number of variables, r2: Determination coefficient, BIC: Bayesian information criterion, post prob: Posterior probability.

As shown in Table 2, the Bayesian model averaging results of independent variables for usage intention of online food delivery apps (dependent variable). The posterior affect probabilities of two predictor variables (performance expectancy and effort expectancy) are 100%. This result explains that these two predictor variables have in all the chosen models. Consequently, two variables are the essential factors impacting the usage intention of online food delivery apps.

As presented in Table 2, model 1 recommends three variables (nVar = 3) including performance expectancy, effort expectancy, and social influence. These three predictor variables explain 51.1% of the variance in the usage intentions of online food delivery apps. Model 1 has the lowest BIC index (BIC = −199.220) when compared with other models. Furthermore, the posterior probability that the model appears in the analyses is 59.1%, also the highest posterior probability when compared with other models. Model 2 consists of four variables (i.e. performance expectancy, effort expectancy, social influence, and facilitating conditions). These four predictor variables explain 51.7% of the variance in the usage intentions of online food delivery apps. Model 2 has a low BIC index, but the posterior probability is low (16.8%). Model 3 consists of two variables (i.e. performance expectancy and effort expectancy). These two predictor variables explain 49.7% of the variance in the usage intentions of online food delivery apps. Model 3 also has a low BIC index, but the posterior probability is small (13.4%). Model 4 consists of three variables (i.e. performance expectancy, effort expectancy, and facilitating conditions). These three predictor variables explain 50.6% of the variance in the usage intentions of online food delivery apps. Model 4 also has a low BIC index, but the posterior probability is small (only 10.7%).

These results are also described in Fig. 1 as follows.

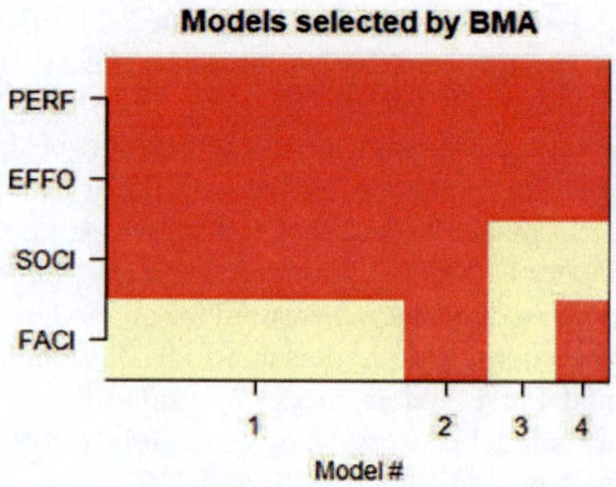

Fig. 1 The models were selected by using the BMA

As depicted in Fig. 1. Figure 1 has the vertical axis lists predictor variables (PERF: Performance expectancy, EFFO: Effort expectancy, SOCI: Social influence, FACI: Facilitating conditions), the horizontal axis as the model (only the best 4 models). The predictor variables have a red color that indicates the positive regression coefficients. Besides, the two variables (performance expectancy and effort expectancy) appear 100% in all models. The social influence factor appears 75.9%, and the facilitating conditions variable appears 27.4%.

5 Conclusions

This analysis has adopted Bayesian model averaging (BMA) to choose the optimal models for factors that influence intention using online food delivery apps. The findings identified the two key factors (performance expectancy and effort expectancy) that affect intention using online food delivery apps. Furthermore, the results also have revealed that the optimal 4 models were selected by using the BMA. The findings are considered to be suited with the practical. As in the real-life, this is logical because we don't have only 1 choice, we can have equivalent options. The Bayesian approach can give us opportunities in thought and assess the models' extent of uncertainty.

References

Ajzen, I.: The theory of planned behavior. Organ. Behav. Hum. Decis. Process. **50**, 179–211 (1991). https://doi.org/10.1016/0749-5978(91)90020-T

Alalwan, A.A.: Mobile food ordering apps: an empirical study of the factors affecting customer e-satisfaction and continued intention to reuse. Int. J. Inf. Manage. **50**, 28–44 (2020). https://doi.org/10.1016/j.ijinfomgt.2019.04.008

Bao, B: The battle of online food delivery applications in Vietnam 2020 (Vietnamese), https://doanhnghiephoinhap.vn/cuoc-chien-cua-cac-ung-dung-giao-do-an-truc-tuyen-tai-viet-nam-nam-2020.html

Chiu, C.M., Wang, E.T.G.: Understanding Web-based learning continuance intention: the role of subjective task value. Inf. Manag. **45**, 194–201 (2008). https://doi.org/10.1016/j.im.2008.02.003

Christino, J., Cardozo, É., Petrin, R., Pinto, L.: Factors influencing the intent and usage behavior of restaurant delivery apps. Rev. Bus. Manag. **23**, 21–42 (2021). https://doi.org/10.7819/rbgn.v23i1.4095

Davis, F.: Perceived usefulness, perceived ease of use, and user acceptance of information technology. MIS q. **13**, 319–340 (1989). https://doi.org/10.5962/bhl.title.33621

Dwivedi, Y.K., Rana, N.P., Janssen, M., Lal, B., Williams, M.D., Clement, M.: An empirical validation of a unified model of electronic government adoption (UMEGA). Gov. Inf. q. **34**, 211–230 (2017). https://doi.org/10.1016/j.giq.2017.03.001

Kiat, Y.C., Samadi, B., Hakimian, H.: Consumer behaviour towards acceptance of mobile marketing. Int. J. Bus. Soc. Sci. **8**, 92–105 (2017)

Lee, E.Y., Lee, S.B., Jeon, Y.J.J.: Factors influencing the behavioral intention to use food delivery apps. Soc. Behav. Pers. **45**, 1461–1474 (2017). https://doi.org/10.2224/sbp.6185

Lee, S.W., Sung, H.J., Jeon, H.M.: Determinants of continuous intention on food delivery apps: extending UTAUT2 with information quality. Sustainability (switzerland) **11**, 1–15 (2019). https://doi.org/10.3390/su11113141

Li, G., Shi, J.: Application of Bayesian model averaging in modeling long-term wind speed distributions. Renew. Energy **35**, 1192–1202 (2010). https://doi.org/10.1016/j.renene.2009.09.003

Montgomery, J.M., Nyhan, B.: Bayesian model averaging: theoretical developments and practical applications. Polit. Anal. **18**, 245–270 (2010). https://doi.org/10.1093/pan/mpq001

Okumus, B., Ali, F., Bilgihan, A., Ozturk, A.B.: Psychological factors influencing customers' acceptance of smartphone diet apps when ordering food at restaurants. Int. J. Hosp. Manag. **72**, 67–77 (2018). https://doi.org/10.1016/j.ijhm.2018.01.001

Raftery, A.E.: Bayesian model selection in social research. Sociol. Methodol. **25**, 111–163 (1995). https://doi.org/10.2307/271063

Raftery, A.E., Madigan, D., Hoeting, J.A.: Bayesian model averaging for linear regression models. J. Am. Stat. Assoc. **92**, 179–191 (1997). https://doi.org/10.1080/01621459.1997.10473615

San, S.S., Dastane, O.: Key factors affecting intention to order online food delivery (OFD)*. J. Indust. Distrib. Bus. **12**, 19–27 (2021)

San Martín, H., Herrero, Á.: Influence of the user's psychological factors on the online purchase intention in rural tourism: Integrating innovativeness to the UTAUT framework. Tour. Manage. **33**, 341–350 (2012). https://doi.org/10.1016/j.tourman.2011.04.003

Troise, C., O'Driscoll, A., Tani, M., Prisco, A.: Online food delivery services and behavioural intention–a test of an integrated TAM and TPB framework. Br. Food J. **123**, 664–683 (2020). https://doi.org/10.1108/BFJ-05-2020-0418

Tuan, N.V.: Regression Modeling and Scientific Discovery (Vietnamese). Ho Chi Minh City General Publishing House, Ho Chi Minh (2020)

Venkatesh, V., Morris, M., Davis, G., Davis, F.: User acceptance of information technology: toward a unified view. MIS q. **27**, 425–478 (2003). https://doi.org/10.1016/j.inoche.2016.03.015

Venkatesh, V., Thong, J.Y., Xu, X.: Consumer acceptance and use of information technology: extending the unified theory of acceptance and use of technology. MIS q. **36**, 157–178 (2012). https://doi.org/10.1109/MWSYM.2015.7167037

Yeo, V.C.S., Goh, S.K., Rezaei, S.: Consumer experiences, attitude and behavioral intention toward online food delivery (OFD) services. J. Retail. Consum. Serv. **35**, 150–162 (2017). https://doi.org/10.1016/j.jretconser.2016.12.013

Zou, Y., Lin, B., Yang, X., Wu, L., Muneeb Abid, M., Tang, J.: Application of the bayesian model averaging in analyzing freeway traffic incident clearance time for emergency management. J. Adv. Transp. (2021). https://doi.org/10.1155/2021/6671983

Factors Affecting the Downsizing of Small and Medium Enterprises in Vietnam

Nguyen Ngoc Thach and Nguyen Thi Ngoc Diep

Abstract This paper investigates the impact of selected factors on the downsizing of 476 small and medium enterprises (SMEs) in Vietnam in the period 2017–2019. By applying a variable selection technique, the Least Absolute Shrinkage and Selection Operator (LASSO), the findings suggest that return on assets (ROA), firm size (SIZE), ratio between debt and total assets (LEV), annual GDP growth rate (GDP), and political stability and absence of violence (PVE) exert a significant impact on the downsizing of SMEs. Based on the empirical findings, this study provides some relevant implications for SMEs managers. Also, policymakers need to improve macroeconomic policy orientations in order to promote sustained economic growth and maintain political stability and absence of violence, keeping SMEs business in good state.

Keywords Downsizing · SMEs · LASSO · VietNam

1 Introduction

Small and medium-sized enterprises (SMEs) are a part of the Vietnamese business system. According to MOPI (2020), there are 758,610 enterprises in the whole country, among which SMEs account for over 95% of operations. They contribute 45% of GDP, and 31% of annual government budget revenue. Although the number of SMEs is great, small and micro enterprises accounts for a very large proportion, while medium-sized ones only for 1.6% of the total number of SMEs (MOPI 2020).

N. N. Thach
Banking University of Ho Chi Minh City, 36 Ton That Dam Street, District 1, Ho Chi Minh City, Vietnam
e-mail: thachnn@buh.edu.vn

N. T. N. Diep (✉)
University of Economics and Law, Ho Chi Minh City, Vietnam
e-mail: diepntn@uel.edu.vn

Vietnam National University, Ho Chi Minh City, Vietnam

N. Ngoc Thach et al. (eds.), *Financial Econometrics: Bayesian Analysis, Quantum Uncertainty, and Related Topics*, Studies in Systems, Decision and Control 427,
https://doi.org/10.1007/978-3-030-98689-6_23

Most SMEs currently have only a charter capital of about 10 billion VND with a workforce of 10–20 employees (MOPI 2020). One of the main strengths of SMEs is quick and effective capital recovery thanks to not only their compact organizational structure but also their ability to penetrate into market niches. During 2017–2019, the Vietnam Government was focusing on many measures to support SMEs. However, the large amount of SMEs are still engaged in the low-technology service and trade sectors, with only 20% of them operating in high-tecnology manufacturing; labor productivity is low (MOPI 2020). The Covid-19 pandemic has put SMEs in an extremely difficult situation, with many of them at risk of bankruptcy. The COVID-19 has negatively impacted on sales of SMEs, considerably reducing their revenue. According to (MOPI 2020) small and micro enterprises obtained negative ROA and ROE in 2016–2018.

Many researchers has been discussing the various determinants of the downsizing of enterprises, for example, Alakent và Lee (2010); Mellahi and Wilkinson (2010); Vicente-Lorente and Zúñiga-Vicente (2018), Nguyen (2021). Datta and Basuil (2015) analysed the consequences of downsizing. Furthermore, some studies suggest that downsizing negatively affects firm performance (Gandolfi 2015; Schenkel and Teigland 2017). SMEs are particularly sensitive to downsizing, with both positive and negative effects (Sune and Lopez 2017; Vicente-Lorente and Zúñiga-Vicente 2018),. There are many indicators to measure the downsizing of SMEs, but they, in general, are based on two key factors: (i) the number of employees or (ii) the amount of capital of an enterprise available at a given time to measure whether the enterprise is downsizing or not. Nevertheless, to the best of our knowledge, there are few studies on firm downsizing in Vietnam. Furthermore, they employed outdated frequentist methods. Thus, in this paper, we analyze the factors affecting the downsizing of SMEs in Vietnam.

This paper is structured into five sections. The first section gives the introduction. The second section provides the literature review. Section 3 specifies the methodology, model, and data sampling. The forth section presents the results and discussions. Finally, Sect. 5 concludes the study.

2 Literature Review

2.1 *Role of Capital for Enterprises*

Capital is one of two indispensable factors of firm production, which form the Cobb–Douglas (1928) production function. In 1928, Charles Cobb and Paul Douglas published a paper where they considered that production is determined by labor and capital. Later, their modified functional form is the following:

$$Q(L, K) = AL^{\beta}K^{\alpha} \tag{1}$$

where Q is total production (the real value of all goods produced in a year), L is labor input (the total number of person-hours worked in a year), K is capital input (the real value of all machinery, equipment, and buildings), A is total factor productivity, β and α are the output elasticities of labor and capital, respectively.

The Cobb–Douglas production function shows the relationship between the quantities of the production factors (capital and labor) used and the quantity of output achieved. This also shows the importance of capital for SMES. Cascio et al. (1997) found that in the enterprises that cut down their labor size without combining with restructuring and optimizing capital, their production results are not effective as compared to other enterprises.

2.2 *Life Cycle Theory*

Churchill and Lewis (1983) mentioned that development process goes through many stages in a lifecycle. Cooper (1979) suggested that there are three stages, while Maruhn and Greiner (1972) pointed at five stages. Cranston and Flamholtz (1986) assumed seven stages. However, the study of Greiner (1972) ont the five stages of enterprise development is of significant interest and worthy of mention. Those are creativity, direction, empowerment, coordination, and cooperation. Thus, SMEs need to determine at what stage they are to know where they are, as well as recognize and prepare for inevitable changes. For example, if an enterprise which has reached the recession stage, an increase in factors affecting the development of the business will be ineffective or will not meet the set expectations.

The key factors determining what stage of the life cycle of a business is considered as follows:

Firm size: It is measured by sales, total assets or number of employees (Timmoms 1994; Serrasqueiro and Nunes 2008).

Firm age (number of years in operation): According to Timmons (1994), Covin et al. (2006), Soininen et al (2012), the age of a firm indicates which stage of the life cycle the firm is in.

Growth: the growth rate of enterprises in general is different at different growth stages (Chandler and Hanks 1993).

Organizational structure: according to Chandler (1962), enterprises adjust organizational structure to solve difficulties arising in production and business activities. There are several forms of organizational structure such as centralization, formalization, vertical differentiation and number of levels. Organizational structure and related management issues are interrelated, which makes organizational structure form relevant to certain stages in the life cycle of enterprises (Chandler and Hanks 1993).

The main difficulty faced by firms: in theory, firms at the same stage tend to deal with the same set of problems (Serrasqueiro and Nunes 2008).

3 Methodology

3.1 Model and Data

Based on Serrasqueiro and Nunes (2008), Kaufmann, Kraay and Mastruzzi (2006), the research model in the present study is specified as follows:

$$\begin{aligned} DC_{,t} = & \beta_0 + \beta_1 Age_{i,t} + \beta_2 Size_{i,t} + \beta_3 Foreign_{i,t} + \beta_4 Export_{i,t} + \beta_5 ROA_{i,t} \\ & + \beta_6 ROS_{i,t-1} + \beta_7 LEV_{i,t} + \beta_8 INF_{i,t} + \beta_9 GDP_t + \beta_{10} VAE_t + \beta_{11} PVE_t \\ & + \beta_{12} RLE_t + \beta_{13} RQE_t + \beta_{14} RLE_t + \beta_{15} CCE_t + \varepsilon_{i,t} \end{aligned} \quad (2)$$

where dependent variable Downsizing capital (DC) is calculated by the ratio of (Total $capital_t$—Total $capital_{t-1}$)/Total $capital_{t-1}$; Independent variables are the following:

Characteristics of firm: Similar to Covin et al. (2006), Soininen et al. (2012), we use firm-specific control variables, such as firm age and size, to control for their possible effect on firm's downsizing capital;

Foreign: Dummy variable that assumes the value of 1 if shareholder control belongs to foreigners and the zero value otherwise;

Intrinsic factors: Export is calculated by dividing total export value by total revenue from sales and service provision; ROA is return on assets calculated by dividing a firm's net income by total average assets; ROS is return on sales calculated by dividing operating profit by net sales; LEV is ratio between debt and total assets;

Macro factors: Similar to Kaufmann, Kraay và Mastruzzi (2006), we use The Worldwide Governance Indicators (WGI) report on six broad dimensions of governance for over 215 countries and territories: (I) Voice and accountability (VAE); (II) Political stability and absence of violence (PVE); (III) Government effectiveness (GEE); (IV) Regulatory quality (RQE); (V) Rule of law (RLE); (VI) Control of corruption (CCE); GDP is GDP growth rate ((GDP_t—GDP_{t-1})/GDP_{t-1}); INF is Inflation rate measured by the consumer price index ((CPI_t—CPI_{t-1}) / CPI_{t-1}).

This paper uses secondary data from the General Statistics Office's annual enterprise survey, which was conducted in Vietnam for the period 2017–2019. From the data of the survey, we filter out the firms with the average number of employees participating in social insurance from 50 to 100 employees and total annual revenue from 100 to 300 billion VND. However, SMEs which participated in the General Statistics Office's annual enterprise survey for less than 3 years or 3 full years, but do not satisfy the criteria that the total capital has decreased continuously over 3 years from 2017 to 2019 will be excluded. As a result, after filtering out the list of SMEs linked to the research criteria, our sample is consisted of 476 SMEs (1.428 observations). Besides, data on macro factors (GDP, CPI, and WGI) are collected from the World Bank database.

3.2 Method

Identifying the determinants significantly influencing a research subject is not an easy task for researchers. When shortlisting variables for linear models, researchers often look at p-values to make a decision. This can be misleading. For example, they could omit important variables that are highly correlated but also have high p-values. On the other hand, irrelevant variables can be included in the model, generating unnecessary complexity in handling it. In addition, overfitting issue can emerge if the number of observations is smaller than the number of variables in the model.

In an increasing crise of the frequentist approach (see, for instance, Nguyen and Thach 2018; Anh et al. 2018; Nguyen and Thach 2019; Hung et al. 2019a, b; Sriboonchitta et al. 2019; Svitek et al. 2019; Kreinovich et al. 2019; Tuan et al. 2019; Thach et al. 2019; Thach 2020a, b), along with other non-traditional probabilistic methods, LASSO, originally proposed by Tibshirani (1996), is an extension of OLS regression which performs both variable selection and regularization through a shrinkage factor. It is capable of enhancing the accuracy and the interpretability compared to classical regression methods (Tibshirani 1996). In the same "spirit" of ridge regression, i.e., shrinkage estimation, LASSO is an estimation method for estimating parameters in linear regression models, but by shrinking the parameters with respect to another norm, namely L^1—norm, rather than mean the L^1, L^2 norm on the space $\mathbb{R}^n$. L^2 norm (euclidean distance from the origin) is $\|\beta\|_2 = \sqrt{\sum_{j=1}^{k} \beta_j^2}$ is also known as "L^2 regularization" (making parameters smaller, control parameters, using L^2 norm). Specifically, LASSO offers a solution to the minimization under constraint problem $\min_{\beta \in \mathbb{R}^k}\left[\|Y - X\beta\|_2^2\right]$ subject to$\|\beta\|_1 \leq t$, where the constraint $\|\beta\|_1 \leq t$ on the parameters is usually not part of the model (unless there is some prior knowledge on the parameter), but only a statistical device used to improve the MSE of the estimator. The geometry of LASSO explains why LASSO does the covariate selection while performing the estimation. First, the (OLS) objective function $Q : \mathbb{R}^k \to \mathbb{R}$ $Q(\beta) = E\left(\|Y - X\beta\|_2^2\right)$ is a quadratic form inβ. As such, each level set, i.e., $L_c = \{\beta : Q(\beta) = c\}, c \in \mathbb{R}^+$ is an ellipsoid (this boundary and its interior form a convex set in$\mathbb{R}^k$, let $\tilde{\beta}$ be the point where Q(β) is minimum (a global minimum since Q(β) is convex inβ), with L_c being the minimum value of the LASSO objective function, i.e., the OLS solution. As c gets larger and larger (than $Q(\tilde{\beta})$), the corresponding level sets (ellipsoids), indexed by c, get larger and larger.

Unlike the "sphere" in$\mathbb{R}^k$, using Euclidean distance / L^2—norm, i.e., $\left\{\beta \in \mathbb{R}^k : \|\beta\|_2^2 = t\right\}$ with $\|\beta\|_2 = \sqrt{\sum_{j=1}^{k} \beta_j^2}$, the L^1—"sphere" (boundary of the L^1 constraint) is not a sphere (geometrically) but a diamond with "corners". As such, it is possible that a level set L_c could hit a corner first, i.e., the LASSO solution (as an estimate ofβ) could have some components (estimates of components of the model parameterβ) equal to zero exactly. When the LASSO algorithm produces these zero estimates, say, for β_j, the corresponding covariates X_j should be left out, as there is no contribution (to Y) from their part. This is a covariate selection procedure, based on estimation. The other non zero estimates are used to explain the model as well as for prediction of Y on new X. LASSO regression is performed by trading off a small increase in bias for a large decrease in variance of the predictions, hence may improve the overall prediction accuracy.

This study applies LASSO method, because this approach performs well in the presence of multicollinearity problem, and it displays the ideal properties to minimize numerical instability that may occur due to overfitting problem (Thach et al. 2020a, b, 2021). To improve the research accuracy, LASSO will minimize the parameter estimates to 0 and in some cases, equate the parameters close to zero and thus allow some variables to be excluded from the model.

4 Research Results

Table 1 shows the minimum, maximum, average value and standard deviation of these variables.

The regression results in Table 2 show that the LASSO method (LASSO regression) removes the variables with coefficients equal to or close to zero compared to the RIDGE regression method. This method also identifies five variables impacting on DC including: return on assets (ROA), rirms size (SIZE), ratio between debt and total assets (LEV), annual GDP growth rate (GDP) and political stability and absence of violence (PVE).

Table 1 Descriptive statistics of variables

Variable	Obs	Mean	Std. Dev	Min	Max
DC	1428	−0.156144	0.177085	−0.9900301	−0.004171
Age	1428	16.10433	10.875321	3	27
Size	1428	3.569055	0.981464	0	5.783825
Foreign	1428	0.287241	0.571949	0	1
Export	1428	0.155393	1.312681	0	34.90216
ROA	1428	−0.000128	0.001567	−0.016215	0.012064
ROS	1428	−0.065392	0.494313	−9.532191	6.842195
LEV	1428	0.04507	0.008061	−0.0031392	0.094306
INF	1428	0.03286	0.003510	0.02790	0.03530
GDP	1428	0.06970	0.001150	0.02790	0.03540
VAE	1428	−1.38021	0.020107	−1.40712	−1.358790
PVE	1428	0.07206	0.077957	−0.02235	0.168574
GEE	1428	0.00247	0.056601	−0.07045	0.067519
RQE	1428	−0.50811	0.057477	−0.58768	−0.453940
RLE	1428	−0.21705	0.187611	−0.35982	0.048006
CCE	1428	−0.41957	0.016943	−0.43620	−0.396320

Source Authors' calculation using Stata 16

Table 2 Regression results using LASSO

Variable	LASSO regression	RIDGE regression
	Coef	Coef
AGE		7.3540470
SIZE	1.0807005	0.9914139
FOREIGN		−0.0324640
EXPORT		−0.7628748
ROA	−0.0253531	−0.0395028
ROS		1.6445934
LEV	21.7566901	119.4375329
INF		−2.1106202
GDP	0.4996937	0.8480187
VAE		−13.0706793
PVE	0.0099692	0.1798264
GEE		−0.7583766
RQE		136.1736934
RLE		−0.0362315
CCE		0.0143286

Source Authors' calculation using Stata 16

The results indicate no significant effects of firm age (AGE), FOREIGN, EXPORT, return on sales (ROS), debt and total assets (LEV), inflation (CPI). Through these result, we see that downsizing of small and medium enterprises in Vietnam regardless of whether the company is newly established or has been operating for a long time. In fact, SMEs in Vietnam are mostly small businesses, which use less domestic raw materials to serve domestic demand, production technology. has not yet innovated to keep up with export standards of developed countries in the world. Currently, SMEs account for 98% of businesses in Việt Nam but only 21% of them are linked to foreign supply chains. This rate is lower than in many countries in the Southeast Asian region such as Thailand at 30% and Malaysia at 46% (Nguyen and etc. 2020). Therefore, the effects on FOREIGN, EXPORT, inflation as well as debt and total assets do not significantly affect the downsizing of small and medium enterprises in Vietnam. On the other hand, the elements of the WGI, voice and accountability (VAE), government effectiveness (GEE), regulatory quality (RQE), rule of law (RLE) and control of corruption (CCE), are not significant determinants of downsizing capital of SMEs. Realistically, with the current state management regime in Vietnam, the issues of voice and accountability (VAE), government effectiveness (GEE), regulatory quality (RQE), the rule of law (RLE) and control of corruption (CCE) have not been evaluated accurately. Therefore, these factors have not shown their influence on the downsizing of SMEs in Vietnam.

5 Conclusion

By performing the LASSO algorithm, the results show that the downsizing of the Vietnamese SMEs is affected by return on assets (ROA), firms size (SIZE), ratio between debt and total assets (LEV), annual GDP growth rate (GDP) and political stability and absence of violence (PVE).

Normally, from the perspective of SMEs and state management, non-downsizing of firms is generally considered to be a good sign. However, the current COVID-19 pandemic has impacted national economies worldwide and SMEs have been experiencing severe financial problems. Therefore, it is more important for SMEs to maintain stability in production and business activities than expansion of production and business.

In this context, the rise of assets and leverage ratio of SEMs might increase the shrinking of enterprises. However, despite the disadvantages of doing business in the context of the Covid-19 outbreak, SMEs themselves, because their apparatus is not cumbersome, still have the advantage of agility in innovation and creativity, which can be used to exploit business opportunities through connecting domestic supply chains and accessing foreign markets, maintaining old markets as well as developing new products.

For policymakers, it is necessary to develop appropriate macro policy orientations to make good use of the business agility of SMEs to promote economic growth, to maintain political stability and absence of violence, to improve legal frameworks to facilitate healthy competition among SMEs, thereby achieving the goal of sustainable and stable development in the next period.

Acknowledgements This work was supported by the University of Economics and Law, Viet Nam National University Ho Chi Minh City.

References

Alakent, E., Lee, S.H.: Do institutionalized traditions matter during crisis? Employee downsizing in Korean manufacturing organizations. J. Manage. Stud. **47**(3), 509–532 (2010)

Cascio, W.F., Young, C.E., Morris, J.R.: Financial consequences of employment-change decisions in major US corporations. Acad. Manag. J. **40**(5), 1175–1189 (1997)

Chandler, G.N., Hanks, S.H.: Measuring the performance of emerging businesses: a validation study. J. Bus. Ventur. **8**(5), 391–408 (1993)

Chandler, A.D.: Strategy and Structure. MIT Press, Cambridge, Mass (1962)

Cobb, C.W., Douglas, P.H.: A theory of production. Amer. Econ. Rev. **18**(Supplement), 139–165 (1928)

Cooper, R.G.: The dimensions of industrial new product success and failure. J. Mark. **43**(3), 93–103 (1979)

Covin, J.G., Green, K.M., Slevin, D.P.: Strategic process effects on the entrepreneurial orientation–sales growth rate relationship. Entrep. Theory Pract. **30**(1), 57–81 (2006)

Cranston, H.S., Flamholtz, G.: The role of management development in making the transition from an entrepreneurship to a professionally managed organisation. J. Manag. Dev. (1986)

Datta, D.K., Basuil, D.A.: Does employee downsizing really work?. In: Human Resource Management Practices, pp. 197–221, Springer, Cham (2015)
Hung, N.T., Songsak, S., Thach, N.N.: On quantum probability calculus for modeling economic decisions. In: Kreinovich, V., Sriboonchitta, S. (eds.) Structural Changes and their Econometric Modeling, TES 2019a. SCI. vol. 808, pp. 18–34. Springer, Cham (2019a). https://doi.org/10.1007/978-3-030-04263-9_15
Hung, N.T., Trung, N.D., Thach, N.N.: Beyond traditional probabilistic methods in econometrics. In: In: Kreinovich, V., Thach, N.N., Trung, N.D., Van Thanh, D. (eds.) Beyond Traditional Probabilistic Methods in Economics, ECONVN 2019b. SCI. vol. 809. Springer, Cham (2019b). https://doi.org/10.1007/978-3-030-04200-4_13
Gandolfi, F., Hansson, M.: A global perspective on the non-financial consequences of downsizing. Rev. Int. Comparat. Manag. **16**(2), 185–204 (2015)
Greiner, L.: Evolution and revolution as organizations grow. Harv. Bus. Rev., 37–46 (1972)
Kaufmann, D., Kraay, A., Mastruzzi, M.: Governance matters V: aggregate and individual governance indicators for 1996–2005, Vol. 4012, World Bank Publications (2006)
Kreinovich, V., Thach, N.N., Trung, N.D., Van Thanh, D. (eds.): ECONVN 2019. SCI, 'vol. 809. Springer, Cham (2019). https://doi.org/10.1007/978-3-030-04200-4
Kreinovich, V., Ly, A.H., Kosheleva, O., Sriboonchitta, S.: Efficient parameter-estimating algorithms for symmetry-motivated models: econometrics and beyond. International Econometric Conference of Vietnam, pp. 134–145. Springer, Cham (2018)
Lewis, V.L., Churchill, N.C.: The five stages of small business growth. University of Illinois at Urbana-Champaign's Academy for Entrepreneurial Leadership Historical Research Reference in Entrepreneurship (1983)
Maruhn, J., Greiner, W.: The asymmetrie two center shell model. Z. Phys. **251**(5), 431–457 (1972)
Mellahi, K., Wilkinson, A.: Managing and coping with organizational failure: introduction to the special issue (2010)
MOPI. The White Book on Vietnamese Businesses 2020. General Statistics Office of the Minis-try of Planning and Investment. Hanoi, Vietnam (2020)
Nguyen, H.T., Thach, N.N.: A closer look at the modeling of economics data. In: Kreinovich, V., Thach, N.N., Trung, N.D., Van Thanh, D. (eds.) ECONVN 2019. SCI, vol. 809, pp. 100–112. Springer, Cham (2019). https://doi.org/10.1007/978-3-030-04200-4_7
Nguyen, T.H.Y.: Financial performance and organizational downsizing: evidence from Smes in Vietnam. In Data Science for Financial Econometrics, pp. 407–416, Springer, Cham (2021)
Schenkel, A., Teigland, R.: Why doesn't downsizing deliver? A multi-level model integrating downsizing, social capital, dynamic capabilities, and firm performance. Int. J. Human Resour. Manag. **28**(7), 1065–1107 (2017)
Serrasqueiro, Z.S., Nunes, P.M.: Performance and size: empirical evidence from Portuguese SMEs. Small Bus. Econ. **31**(2), 195–217 (2008)
Soininen, J., Martikainen, M., Puumalainen, K., Kyläheiko, K.: Entrepreneurial orientation: growth and profitability of Finnish small-and medium-sized enterprises. Int. J. Prod. Econ. **140**(2), 614–621 (2012)
Sune, A., Lopez, L.: Forgetting-relearning cycles in organizational downsizing strategies. In European Conference on Knowledge Management, pp. 947–955. Academic Conferences International Limited (2017)
Sriboonchitta, S., Nguyen, H.T., Kosheleva, O., Kreinovich, V., Nguyen, T.N.: Quantum approach explains the need for expert knowledge: on the example of econometrics. In: Kreinovich, V., Sriboonchitta, S. (eds.) TES 2019. SCI, vol. 808, pp. 191–199. Springer, Cham (2019). https://doi.org/10.1007/978-3-030-04263-9_15
Svítek, M., Kosheleva, O., Kreinovich, V., Nguyen, T.N.: Why quantum (wave probability) models are a good description of many non-quantum complex systems, and how to go beyond quantum models. In: Kreinovich, V., Thach, N.N., Trung, N.D., Van Thanh, D. (eds.) ECONVN 2019. SCI, vol. 809, pp. 168–175. Springer, Cham (2019). https://doi.org/10.1007/978-3-030-04200-4_13

Thach, N.N., Anh, L.H., An, P.T.H.: The effects of public expenditure on economic growth in Asia countries: a Bayesian model averaging approach. Asian J. Econ. Bank. **3**, 126–149 (2019)

Thach, N.N.: How to explain when the ES is Lower than one? A Bayesian nonlinear mixed-effects approach. J. Risk Financ. Manag. **13**, 21 (2020a)

Thach, N.N. The variable elasticity of substitution function and endogenous growth: an empirical evidence from Vietnam. Int. J. Econ. Bus. Admin. VIII, pp. 263–277 (2020b)

Thach N.N., Anh L.H., Hoang N.K.: Applying lasso linear regression model in forecasting Ho Chi Minh City's public investment. In: Studies in Computational Intelligence, Data Science for Financial Econometrics, vol.898. Springer, Cham, pp. 245–253 (2020a). https://doi.org/10.1007/978-3-030-48853-6_17

Thach, N.N., Nguyen Diep, T.N., Doan Hung, V.: Non-interest income and competition: the case of Vietnamese commercial banks. In: Studies in Computational Intelligence, Data Science for Financial Econometrics, vol. 898, Springer, Cham, pp. 281–290 (2020b). https://doi.org/10.1007/978-3-030-48853-6_20

Thach, N.N., Nguyen Diep, T.N., Doan Hung, V.: The determinants of non-interest income of banks with dominant state capital in Vietnam. In: Studies in Computational Intelligence, Prediction and Casuality and Related Topics, vol. 983, Springer, Cham, pp. 229–240 (2021). https://doi.org/10.1007/978-3-030-77094-5_20

Tuan, T.A., Kreinovich, V., Nguyen, T.N.: Decision making under interval uncertainty: beyond Hurwicz Pessimism-optimism criterion. In: Kreinovich, V., Thach, N.N., Trung, N.D., Van Thanh, D. (eds.) ECONVN 2019. SCI, vol. 809, pp. 176–184. Springer, Cham (2019). https://doi.org/10.1007/978-3-030-04200-4_14

Tibshirani, R.: Regression shrinkage and selection via the lasso. J. Royal Stat. Soc. Ser. B (Stat. Methodol.) **58**, 267–288 (1996)

Timmons, J.A.: New Venture Creation. Irwin, Chicago (1994)

Vicente-Lorente, J.D., Zúñiga-Vicente, J.Á.: The U-shaped effect of R&D intensity on employee downsizing: evidence from Spanish manufacturing firms (1994–2010). Int. J. Human Resour. Manag. **29**(15), 2330–2351 (2018)

Impact of Microfinance Institutions' Lending Interest Rate on Their Financial Performance in Vietnam: A Bayesian Approach

Thuy T. Dang, Hau Trung Nguyen, and Ngoc Diem Tran

Abstract The research focuses on the impact of lending interest rate on financial performance in Vietnamese microfinance institutions (MFIs) in the period of 2014–2019. The data of 28 MFIs is collected through Microfinance Information Exchange (Mix market), State Bank of Vietnam and Vietnam Microfinance Working Group (VMFWG). Based on Bayesian approach in our research is more reliable and effective statistical inference than p-value hypothesis testing—traditional method. The research's results show that lending interest rate has positive impact on financial performance in Vietnamese MFIs.

Keywords Microfinance institution · Financial performance · Lending interest rate · Vietnam · Bayesian approach

1 Introduction

Asian Development Bank (2000) defined microfinance as "the provision of a broad range of financial service, such as deposits, loans, payment services, money transfer and insurance to poor and low-income households and their microenterprises". Thus, low-income individuals are capable of lifting themselves out of poverty if they are given access to financial services. Access to sustainable financial service could enable the poor to start-up a microenterprise and/or do business to increase their income.

T. T. Dang · N. D. Tran
Vietnam Institute for Indian and Southwest Asian Studies, Vietnam Academy of Social Sciences, 176 Thai Ha street, Dong Da Dist, Hanoi, Vietnam
e-mail: thuy0183@gmail.com

N. D. Tran
e-mail: ngocdiemtran1993@gmail.com

H. T. Nguyen (✉)
Banking Strategy Institute, State Bank of Vietnam, 504 Xa Dan street, Dong Da Dist, Hanoi, Vietnam
e-mail: hau.nguyentrung@sbv.gov.vn

N. Ngoc Thach et al. (eds.), *Financial Econometrics: Bayesian Analysis, Quantum Uncertainty, and Related Topics*, Studies in Systems, Decision and Control 427,
https://doi.org/10.1007/978-3-030-98689-6_24

Microfinance institutions (MFIs) engage in small financial transactions to serve low income households, micro enterprises, small scale farmers and others who are unbanked. International Fund for Agricultural Development—IFAD (2008) proved that MFIs (credit unions or non-governmental organizations) enable the poor to get access to small loans, receive deposits from abroad relatives and protect their savings. Experiences from Bangladesh, Benin and Dominica show that the poor tends to pay higher repayment rates than other borrowers.

The lending interest rates should enable MFIs to provide sustainable financial services to large numbers of poor clients while being independent of any form of subsidy (CGAP, 1997). Small Industries Development Bank of India (2011) reported that the high interest rates charged by MFIs aim at recovering the cost of the loans. Mwangi (2014) and Miriti (2014) showed a strong positive relationship between lending interest rates and MFIs' financial performance.

In late 1980s, microfinance in Vietnam has mainly been conducted in development projects of international organizations (UNDP, FAO, WB, ADB, etc.), international non-governmental organizations (INGOs), savings-credit projects or bilateral projects (Swedish SIDA…) targeting their selected groups. Partners of these projects were socio-political organizations, professional associations… in which Vietnam Women's Union was the largest partner. Up to now, microfinance institutions operate in two forms: licensed microfinance institutions (TYM, M7-MFI, Thanh Hoa-MFI, and CEP) and around 30 unlicensed microfinance institutions (semi-formal microfinance institutions or microfinance programs/projects) in Vietnam (Tam and Mai 2021). According to VMFWG (2013), interest rate from MFIs in Vietnam was higher than that of commercial banks because: (i) MFIs engaged in small loans with high administration cost per loan compared to commercial bank, (ii) MFIs provided door-to-door services, thus their operational costs should be higher, and (iii) MFIs customers in remote areas had no collateral for loans, so a premium should be added in interest rate to compensate for credit risks to MFIs (which have a positive impact on their financial performances).

This research found that loans to customers at high interest rates from MFIs in Vietnam may lead to their low outreach. Additionally, loans to customers with high interest rates will support MFIs to earn and boost profits, financial self-sufficiency (FSS) and operational self- sufficiency (OSS).

With Bayesian approach, this research evaluates the impact of MFIs' lending interest rate on their financial performance in Vietnam. We organize the remainder of this research as follows: The second section presents a brief literature review. The third section describes the data and model specification. The fourth section gives some discussions. The last section confers conclusions.

2 Literature Review

2.1 Research on Interest Rate of MFIs

Bogan (2012), Khachatryan and Hartarska (2017) and Dang (2020) investigated specific character of MFIs and concluded that since most MFIs' loans are uncollateral loans, their lending rates tend to be higher than other financial institutions and group lending activities will increase borrowers' debt repayment obligations.

In fact, lending interest rates of MFIs are normally higher 20–40% than other financial institutions (Morduch et al. 2012; Dehejia 2009). Microfinance outreach to the poor because of its convenience, simplicity, speed and addition non-financial services. Ramírez Rocha (2018) analyzed the differences in lending interest rates of MFIs in Latin America, Africa, Eastern Europe and Asia. He also investigated in operating expenses, average loan per borrower, real growth GDP, and government effectiveness which are key factors explaining differences in lending interest rates and found that operating expenses are essential drivers of lending interest rates and MFIs will benefit from an increase in lending interest rate margin.

Many researchers argued that lending interest rates of MFIs has a limited impact on credit demand and high subsidies, low interest rates will increase the sustainability of MFIs (Kar and Bali Swain 2014). Morduch (2000) pointed out that it is difficult for MFIs to balance two basic goals: (i) poverty alleviation such as supporting the poor, disadvantaged groups and small enterprises to access to basic financial products and services, and (ii) increase organizational sustainability through increased profitability. A lot of researches have focused on whether it is possible to coexist with two goals in MFIs (Kar 2013; Bos and Millone 2015; Cotler and Oreggia 2008; Vanroose and D'Espallier 2012) while other researches are only interested in determinants of lending interest rates (Roberts 2013; Dorfleitner et al. 2013; Basharat et al. 2015; Guo and Jo 2017).

Other researchers suggested that microfinance lending interest rates should be lower than lending interest rates of other credit institutions because the borrowers are the poor. MFIs aim at social issues, specifically poverty reduction by providing small, uncollateral loans to target customers in rural areas. With small loans from MFIs, many people succeeded in doing business and moved out of chronic poverty. However, Sun and Im (2015) affirmed that high lending interest rates from MFIs can increase the financial burden for borrowers. Ruan Rodrigo (2017) argued that slight fluctuations in interest rates also negatively affect the borowing decision of the poor and the disadvantaged in society. Thus, harmonizing benefit between customers and MFIs is still being concerned.

Cotler and Almazan (2013) stated that funding costs, loan size and efficiency are measured by operating costs. And operation costs are important factors in calculating lending interest rates. Dorfleitner et al. (2013) also argued that operating costs and management efficiency are important variables in determining lending interest rates. Cull et al. (2014) and Vanroose and D'Espallier (2012) pointed out the impact of bank penetration in the development of the financial sector and found a negative relationship between bank penetration and lending interest rate of MFI. Supporting this issue, Trujillo et al. (2014) pointed out that if a country has an effective regulatory policy framework and a developed financial supervision system, lending interest rates of MFIs tend to be lower. Xu et al. (2016) found that the smaller the loans of MFIs, the higher the lending interest rates.

In Vietnam, beside formal MFIs, the group of semi-formal MFIs (financial institutions, banks providing microfinance products and services) offer small loans to poor customers with lower interest rates than formal MFIs. Notably, lending interest rates of all MFIs are higher than those of comercial banks (Phuong and Van 2017).

2.2 Research on Financial Performance of MFIs

According to Kent and Dacin (2013), poorest customer segment is a difficulty niche market because customers have a lot of difficulties in accessing loans from formal financial institutions. Thus, such segment is the potential market segment for MFIs. Accessing this segment, MFIs face with relatively high management and operating costs due to a large number of small loans (Gutierrez-Nieto et al. 2007). Thus, MFIs' performance should not only be evaluated by financial efficiency or effectiveness like other financial institutions. Serrano-Cinca and Gutiérrez-Nieto (2014) and Nguyen and Le (2013) specified that MFIs should balance financial and social goals and step by step reach self-sustainability and suitable profit growth, while enabling poor people to access products and services. Piot-Lepetit and Nzongang (2014) showed that MFIs may face a dual situation: (i) financial results are not enough to warrant investments and (ii) the social role is not sustainable without returns on the costs involved. In this case, MFIs have to make use of their social funding resources.

Roberts (2013) affirmed that MFIs should maximize their profit to reach long-term sustainable objective. It is also important for-profit MFIs that investors perceive that MFIs are making profits whether with dividends from the profit derived from the loan operations carried out or with capital gains derived from an increased valuation of ownership rights (Tchakoute-Tchuigoua 2010).

Various factors that can affect an MFI's performance, namely lending interest rates, leverage, business size, economic growth, and risk.

MFIs produce a bigger share of their revenue from interest income on loans. This indicates that MFIs with high-quality loan portfolios are more likely to grow. Many investors choose to put their money into MFIs that are growing because they know they will get a good return (Fernando 2006). Lending interest rates, according to Saunder (1995), have an impact on general economic activity, including the flow of products, services, and financial assets within the economy and around the world. He explains that interest rates are proportional to the current worth of money and the future value of money.

Leverage has an impact on the performance of MFIs' businesses. Hirtle and Stiroh (2007) claims that MFIs with greater profit rates will remain low leveraged due to their ability to fund themselves. On the other side, a high level of leverage raises the chance of a company going bankrupt. Total assets are thought to have a beneficial impact on a company's financial performance, with more assets implying less risk (Barton and Gordon 1987).

The size of an MFI can have a favorable impact on financial performance since larger companies can take advantage of this advantage to gain financial advantages in business relationships. Large MFIs have easier access to the most crucial production variables, such as people resources. In addition, large MFIs frequently receive lower-cost support (Akhigbe and McNulty 2005). Large MFIs can also obtain cheaper sources of finance, allowing them to remain competitive. However, such funding come with criteria that huge MFIs can readily meet.

Other major aspects influencing a MFI's financial performance are risk and growth. The level of risk exposure can generate changes in the company's market value because market value is conditioned by its results. Economic growth is another factor that aids in achieving a higher financial market position, as market value takes into account predicted future profits (Bekan 2011). The level of risk exposure can affect changes in the MFI's market value and hence how it is regarded in the market, because market value is conditioned by the MFI's results (Gietzen 2017).

In microfinance, popular metrics to evaluate financial performance of MIFs are return on assets (ROA), return on net equity (ROE), operational self-sufficiency (OSS) and financial self-sufficiency (FSS) (Strøm et al. 2014), profitability, credit balance of MFIs (Louis et al. 2013), financial expenses to revenue ratio, and MFIs' asset portfolio (Sanfeliu et al. 2013). However, in this research, the authors use 7 metrics (profitability, economic efficiency, operational efficiency, cost of operation, asset quality and size) to analyze financial performance of Vietnam's MFIs.

2.3 Relation Between Lending Interest Rates and Financial Performance of MFIs

According to Fernando (2006), interest charged on loans is MFIs' main source of income. Thus, it should be high enough to cover operational costs. Since microlending remains a high-cost operation, interest rates remain high. For this reason, we should not comapare to rates charged by commercial banks. With economies of scale, larger loans refer to lower administrative costs per transaction and result in lower interest rates.

Were and Wambua (2013) investigated the determinants of interest rate spreads in Kenya's banking sector based on panel dataanalysis. His empirical results show that bank-specific factors play a significant role inthe determination of interest rate spread. These include bank size based on bank assets, credit risk as measured by non-performing loans to total loans ratio, liquidity risk, return on average assets and operating costs. A rise in lending interest rates is good for MFIs due to higher returns on new investments, increased profit margins on loans. As a result, an increase in interest rates which lead to good financial performance of financial institution

Table 1 Describing variable

Variables	Abbreviation	Description
Dependent variable		
Financial performance	ROA	ROA was measured by taking a ratio of net profit after taxes/total assets
	ROE	ROE was measured taking a ratio of net income after tax/average shareholder's equity
Independent variable		
Lending interest rate	LIR	Lending interest rate as measured by ratio of interest income to asset generating the income
Operation expense	OEA	Operation expense/ assets (%)
Management efficiency	ME	Management efficiency as measured by Non interest expense to total assets
Assest	ASS	Ln (Assest)
Offices	OFI	Ln (Offices)

Source The authors

indicates signal of good returns in the form of dividends. Interest on loans is behind a bank's dismal profitability (Njihia 2005).

3 Data and Model Specification

3.1 Research Data

The research uses panel data covering 28 Vietnamese MFIs in the period of 2014–2019. Data were collected from MixMarket through Annual Report, State Bank of Vietnam and Vietnam Microfinance Working Group (VMFWG).

3.2 Used Variables in the Study

Variables used in the model are shown in Table 1.

3.3 Methedology

This section examines the impact of lending interest rate on financial performance in Vietnam's MFIs. A multivariate regression model is:

Model 1: $\boldsymbol{ROA} = \alpha_1 + \alpha_2\text{LIR} + \alpha_3\text{OEA} + \alpha_4\text{ME} + \alpha_5\text{ASS} + \alpha_6\text{OFI}$

Model 2: $\boldsymbol{ROE} = \beta_1 + \beta_2\text{LIR} + \beta_3\text{OEA} + \beta_4\text{ME} + \beta_5\text{ASS} + \beta_6\text{OFI}$

Since previous studies used frequency approach, a prior information was not available. However, with a sample of 28 MFIs, study period of 6 years, number of observations is relatively large and prior information will have a small impact on posterior probability distribution. In this case, Block et al. (2011) proposed a standard Gaussian distribution with different a prior information, that is, simulations with different a prior information will be selected. After conducting regression of these simulations, Bayesian factor analysis and Bayesian test will be conducted to select the most suitable a prior information simulation.

Next step, the authors carry out Bayesian regression for the above simulations, and conducted Bayesian factor analysis and posterior Bayesian test to select the simulation with most appropriate prior information.

Simulations in Table 2 show decreasing prior information levels with simulation 1 having the strongest prior information and simulation 5 having the lowest prior information.

The next step, the authors carry out regression of above simulations, then conduct Bayesian factor analysis with posterior Bayesian test to select simulation with the most appropriate a prior information.

The order of priorty for Bayesian factor analysis is simulation with average Log BF, largest Log ML and minimum DIC average. Results of Bayesian factor test in Table 3 show that with model 1, simulation 3 has a great advantage when Log (ML) and Log (BF) of simulation are the largest, although the DIC of simulation 3 is not as good as simulation 4 and 5; however, the posterior probability P(M|y) of simulation 3 is the largest, which implies prior distribution of simulation 3 is the best fit. We also get the same results with model 2 when simulation 8 has Log (ML), Log BF and P(M|y) outperforming other simulations.

Besides, to ensure that Bayesian inference is plausible, the authors continue to conduct convergence analysis of MCMC series through convergence diagnostics by graph.

Results in Fig. 1 show that graphs of parameters in the model are quite reasonable, the trace plots fluctuate around mean value, and histogram shows low autocorrelation; shape of charts is homogenous and normal distribution. In addition, graphs show a good mix, autocorrelation coefficient t fluctuates around the level of less than 0.02, it proves to be consistent with distribution simulation density and reflects the lags within effective term. Therefore, MCMC series meets convergence condition.

Convergent diagnostic analysis by chart and Grubin shows characteristics of MCMC series, not the fitness of model with observed data. To perform Bayesian model fit test, Gelman et al (2014) proposed to use the smallest observation to measure the difference between observed and simulated data for estimated model.

Table 4 shows mean and error of ROA and minimum ROE values (min) in the simulation data (T) respectively as 7.446 and 1.616 for ROA; −6.767 and 1.958 for ROE. In observed sample (T_obs), minimum ROA and ROE are −6.7 and −5.92 respectively. Last column P(T> =T_obs) shows the probability of predicting mean of minimum ROA and ROE values in simulations greater than or equal to minimum ROA and ROE values of observed sample. The P(T> =T_obs) value, ideally, should

Table 2 Summary of simulations

Model 1	
Likelihood function	ROA ~ N (μ, σ)
Prior distribution	
Simulation 1	α_i ~ N (0; 1) σ^2 ~ Invgamma (0, 01; 0.01)
Simulation 2	α_i ~ N (0; 10) σ^2 ~ Invgamma (0, 01; 0, 01)
Simulation 3	α_i ~ N (0; 100) σ^2 ~ Invgamma (0, 01; 0, 01)
Simulation 4	α_i ~ N (0; 1000) σ^2 ~ Invgamma (0, 01; 0, 01)
Simulation 5	α_i ~ N (0; 10000) σ^2 ~ Invgamma (0, 01; 0, 01)
Model 2	
Likelihood function	ROE ~ N (μ, σ)
Prior distribution	
Simulation 1	β_i ~ N (0; 1) σ^2 ~ Invgamma (0, 01; 0.01)
Simulation 2	β_i ~ N (0; 10) σ^2 ~ Invgamma (0, 01; 0, 01)
Simulation 3	β_i ~ N (0; 100) σ^2 ~ Invgamma (0, 01; 0, 01)
Simulation 4	β_i ~ N (0; 1000) σ^2 ~ Invgamma (0, 01; 0, 01)
Simulation 5	β_i ~ N (0; 10000) σ^2 ~ Invgamma (0, 01; 0, 01)
i = 1, 2, 3, 4, 5, 6, 7	

Source Authors' own calucation

be close to 0.5. However, if these values are between 0.05 and 0.95, simulations of Bayesian regression model are considered appropriate (Gelman et al. 2014). In this research, posterior predictive p-value was 0.35 for ROA and 0.36 for ROE satisfying the above condition. Therefore, it can be said that simulation data in this research is suitable for forecasting.

Table 3 Results of a Bayesian factor test and Bayes model test

Model 1					
	Chains	Avg DIC	Avg log (ML)	Avg log BF	P(Mly)
Simulation 1	3	820.5677	−428.5663	1.000	0.0005
Simulation 2	3	790.3200	−422.3443	6.222	0.2393
Simulation 3	3	784.4202	−421.1918	7.3744	0.7576
Simulation 4	3	784.2224	−426.8647	1.7016	0.0026
Simulation 5	3	784.2655	−433.6741	−5.1078	0
Model 2					
	Chains	Avg DIC	Avg log (ML)	Avg Log BF	P(Mly)
Simulation 6	3	872.9887	−453.8668		1
Simulation 7	3	836.9475	−446.6297	7.2371	0.0107
Simulation 8	3	826.48	−442.1124	11.7544	0.9829
Simulation 9	3	826.2602	−447.1577	6.7091	0.0063
Simulation 10	3	826.2532	−453.9104	−0.0437	0

Source Authors' own calucation

4 Discussion

Tables 5 and 6 show that model's acceptance rate reaches 1, model's minimum efficiency is 0.93, far exceeding the minimum required to be 0.01; The maximum Rc value of coefficient is 1, lower than permissible level of 1.2. Besides standard error, regression results table also provides the Monte-Carlo standard error (MCSE) which indicates the stability of MCMC series (Flegal et al 2008), and the closer the MCSE is to zero, the stronger the MCMC series. These authors also argured that MCSE values are: (i) acceptable if less than 6.5% of standard deviation, and (ii) optimal if less than 5% standard deviation is. Thus, these values in Tables 5 and 6 meet convergence requirements.

In the above section, the authors performed necessary tests to ensure the robustness of Bayesian model. However, Bayesian approach is a relatively new method, so in this study, the authors will compare estimation results of Bayesian approach to those of another approach. For the purpose of comparing the regression coefficients of two approaches, the authors compare regression coefficients of Bayesian model to those of Ordinary Least Square (OLS) method (Table 7).

The results of Table 7 show that the sign of regression coefficients from Bayesian method and OLS method are almost the same. The sign of regression coefficients LIR and OFI values have a positive effect on the ROE and ROA variables, while remaining values have a negative effect on ROA and ROE. Thus, estimation results of regression coefficients from Bayesian approach and OLS approach are relatively similar. However, in Bayesian approach, we can estimate the probability that independent variables could have impact on operational efficiency of microfinance institutions (Table 8).

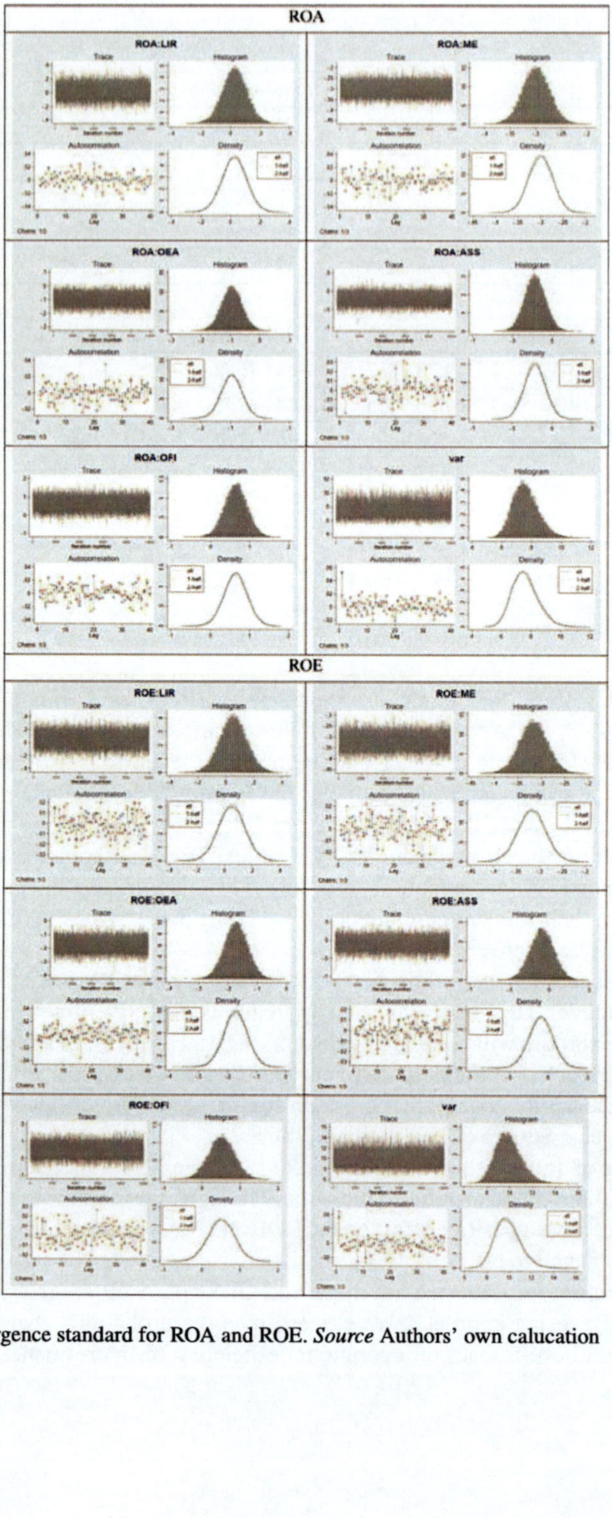

Fig. 1 Convergence standard for ROA and ROE. *Source* Authors' own calucation

Table 4 Posterior predictive sumary

Minsl	Mean	Std. Dev.	E(T_obs)	P(T > = T_obs)
ROA	−7.446	1.616	−6.7	0.3473
ROE	−6.767	1.958	−5.92	0.3598

Source Authors' own calucation

Table 5 Bayesian regression results of ROA

ROA	Mean	Std. Dev.	MCSE	Median	Equal-tailed	
					[95% Cred. Interval]	
LIR	0.0260	0.0841	0.0005	0.0264	−0.1398	0.1922
ME	−0.2999	0.0263	0.0002	−0.2999	−0.3513	−0.2486
OEA	−0.0961	0.0352	0.0002	−0.0961	−0.1655	−0.0270
ASS	−0.2172	0.1312	0.0008	−0.2176	−0.4730	0.0405
OFI	0.6416	0.2958	0.0017	0.6423	0.0598	1.2240
_cons	15.9005	2.1924	0.0127	15.8995	11.6039	20.2129
Var	7.5928	0.8678	0.0052	7.5244	6.0832	9.4689
Avg acceptance rate	1.000					
Avg efficiency: min	0.921					
Max Gelman-Rubin Rc	1.000					

Source Authors' own calucation

Table 6 Bayesian regression results of ROE

ROE	Mean	Std. Dev.	MCSE	Median	Equal-tailed	
					[95% Cred. Interval]	
LIR	0.0523	0.0959	0.0006	0.0515	−0.1347	0.2432
ME	−0.3172	0.0298	0.0002	−0.3172	−0.3764	−0.2587
OEA	−0.1734	0.0400	0.0002	−0.1735	−0.2517	−0.0950
ASS	−0.0919	0.1506	0.0009	−0.0917	−0.3885	0.2023
OFI	0.5072	0.3400	0.0020	0.5060	−0.1556	1.1793
_cons	19.7116	2.4949	0.0145	19.7203	14.7847	24.5657
Var	9.8844	1.1476	0.0069	9.7914	7.8894	12.3783
Avg acceptance rate	1.000					
Avg efficiency: min	0.921					
Max Gelman-Rubin Rc	1.000					

Source: Authors' own calucation

Table 7 Ordinary Least Squares (OLS) regression results

ROA	Coef	Std. Err.	t	P > |t|	[95% Conf. Interval]	
LIR	0.0307	0.1137	0.2700	0.7880	−0.1939	0.2552
ME	−0.3019	0.0264	−11.4500	0.0000	−0.3540	−0.2498
OEA	−0.0969	0.0348	−2.7800	0.0060	−0.1657	−0.0282
ASS	−0.2234	0.1309	−1.7100	0.0900	−0.4820	0.0351
OFI	0.6415	0.2940	2.1800	0.0310	0.0608	1.2222
_cons	15.9156	2.8865	5.5100	0.0000	10.2134	21.6178
ROE	Coef	Std. Err.	t	P > |t|	[95% Conf. Interval]	
LIR	0.0411	0.1296	0.3200	0.7510	−0.2149	0.2972
ME	−0.3212	0.0301	−10.6800	0.0000	−0.3806	−0.2618
OEA	−0.1750	0.0397	−4.4100	0.0000	−0.2534	−0.0965
ASS	−0.1067	0.1492	−0.7100	0.4760	−0.4015	0.1881
OFI	0.5077	0.3351	1.5100	0.1320	−0.1544	1.1697
_cons	20.1904	3.2907	6.1400	0.0000	13.6896	26.6912

Source Author's own caculation

Table 8 Posterior probability

ROA			
	Mean	Std. Dev.	MCSE
{ROA:LIR} > 0	0.6210	0.4851	0.0028
{ROA:ME} < 0	1.0000	0.0000	0.0000
{ROA:OEA} < 0	0.9971	0.0541	0.0003
{ROA:ASS} < 0	0.9484	0.2213	0.0013
{ROA:OFI} > 0	0.9847	0.1228	0.0007
ROE			
	Mean	Std. Dev.	MCSE
{ROE:LIR} > 0	0.7101	0.4537	0.0026
{ROA:ME} < 0	1.0000	0.0000	0.0000
{ROE:OEA} < 0	0.9999	0.0100	0.0000
{ROE:ASS} < 0	0.7346	0.4416	0.0026
{ROA:OFI} > 0	0.9340	0.2483	0.0014

Source Authors' own calucation

Lending interest rates have same direction movement with profitability of MFIs; however, this relation is not significant since probability of lending interest rates' positive effect on ROA and ROE are 62% and 71% respectively. Meanwhile, impact of operating costs on profitability for MFIs in Vietnam is quite clear, specifically ME and OEA has a respectively negative impact on ROA and ROE of nearly 100 and 99%. Therefore, to improve profitability, MFIs should aim to reduce costs instead of increasing interest rates, which support borrowers to repay loans and decrease bad debt. Like other organization, MFIs needs to maintain a reasonable interest rate for following purposes: (i) To satisfy borrowing needs of the poor and protect them from informal loans; (ii) To maintain and ensure sustainability of MFIs to cover all expenses. Similarly, in Vietnam, MFIs were officially regulated under the Law on Credit Institutions 2010 as a type of credit institutions that mainly performs a number of banking activities to satisfy the needs of low-income individuals, households and micro enterprises.

MFIs in Vietnam tried to access their customers in many ways, especially customers in isolated and remote areas. In fact, to approach and serve customers in the best way, MFIs establish their transaction system in 3 ways: branches located in central areas; transaction offices under the management of the branch; clusters which are groups of 10 to 80 members, normally have weekly activities at village cultural house in member customers' living area. It is the access to many customers through this method that helps MFIs in Vietnam gradually increase their profitability.

In addition, assets (with negative impact on financial performance) indicates that large MFIs are using assets inefficiently and need to be improved. The starting point of MFIs in Vietnam is usually from socio-political organizations such as Tinh Thuong One-member Limited Liability Microfinance Institution (TYM MFI) under the Vietnam Women's Union, Capital aid for employment of the Poor Microfinance Institution (CEP) under the Ho Chi Minh City Labor Confederation. Thanh Hoa Microfinance Institution started as a savings loan program of Thanh Hoa Provincial Women's Union… Therefore, when the organizations grow up, their management and organization skills do not meet the requirements.

5 Conclusion

In May 2020, State Bank of Vietnam adjusted the maximum interest rate for deposits with terms from 1 to less than 6 months at MFIs from 5.25%/year to 4.75%/year to support both poor customers and micro-enterprises which had been hurt by the Covid-19 pandemic. This adjustment was proved to have negative impacts on MFIs to maintain their operations as well as improve their financial performance in short-term. However, in long-term, MFIs should reach the balance between their financial performance and social benefit via suitable lending interest rate to their target customers and micro-enterprises.

This research evaluates impact of lending interest rates on financial performance of MFIs in Vietnam. Research results show that lending interest rate and offices have positive impact on financial performance of MFIs, while other variables (operation expense, management efficiency and asset) have negative impact on financial performance of MFIs. However, MFIs are now facing difficulties because of the COVID-19 pandemic and it is important to control operational costs effectively, apply new technology in monitoring provision of products and services so as to improve their financial efficiency. In short term, MFIs should cooperate with fintech company develop customer-oriented solutions such as banking/mobile payment products.

This research also indicates that lending interest rates have a positive impact on financial performance of microfinance institutions. When interest rates increase, MFIs' profits will increase corespondingly, and MFIs tend use a part of their profits to fulfill social goals, provide non-financial support activities to target customers. This activity will bring synergistic benefits to individual customers and MFIs in Vietnam.

References

Akhigbe, A., McNulty, J.: Profit efficiency sources and differences among small and large U.S. Commercial Banks. J Econ Finance 29 (2005)

Asian Development Bank (ADB): Finance for the Poor: Microfinance Development Strategy. Manila (2000)

Barton, S.L., Gordon, P.J.: Corporate strategy: useful perspective for the study of capital structure? Acad. Manag. Rev. 12(1) (1987)

Basharat, B., Hudon, M., Nawaz, A.: Does efficiency lead to lower prices? a new perspective from microfinance interest rates. Strategic Change **24**(3), 49–66 (2015)

Bekan, S.: Measurement of financial performance. Pakistan Institute of Development Economics, Islamabad (2011)

Block, J.H., Jaskiewicz, P., Miller, D.: Ownership versus management effects on performance in family and founder companies: A Bayesian reconciliation. J. Family Bus. Strategy **2**, 232–45 (2011)

Bogan, V.L.: Capital structure and sustainability: An empirical study of microfinance institutions. Rev. Econ. Stat. (2012)

Bos, J.W.B., Millone, M.: Practice what you preach: Microfinance business models and operational efficiency. World Development (2015)

CGAP.: Les taux d'intérêt applicables aux microcredits. The World Bank, Washington DC, USA (1997)

Cotler, P., Rodríguez Oreggia, E.: Rentabilidad y tamano de préstamo de las microfinanzas en Mexico. Economía Mexicana (2008)

Cotler, P., Almazan, D.: The lending interest rates in the microfinance sector: searching for its determinants. J. CENTRUM Cathedra. Bus. Econ. Res. J. (2013)

Cull, R., Demirguc-Kunt, A., Morduch, J.: Banks and microbanks. J Financ. Serv. Res (2014)

Dang, T.T., Vu, Q.H., Hau, N.T.: Impact of outreach on operational self-sufficiency and profit of microfinance institutions in Vietnam. Data Sci. Financ. Econom. Stud. Comput. Intell. (2020)

Dehejia, R.H.: Jonathan morduch and heather anne montgomery: do interest rates matter? Credit Demand in the Dhaka Slums. Financial Access Initiative (2009)

Dorfleitner, G., Leidl, M., Priberny, C., von Mosch, J.: What determines microcredit interest rates? Appl. Financ. Econ. (2013)

Fernando, N.A.: Understanding and dealing with high interest rates on microcredit. Asian Development Bank (2006)

Flegal, J.M., Haran, M., Jones, G.L.: Markov chain monte carlo: can we trust the third significant figure? Stat. Sci. **23** (2008)

Gelman, A., Carlin, J.B., Stern, H.S., Dunson, D.B., Vehtari, A., Rubin, D.B.: Bayesian data analysis, 3rd edn. Chapman & Hall/CRC, Boca Raton, FL (2014)

Gietzen, T.: The exposure of microfinance institutions to financial risk. Rev. Dev. Financ. (2017)

Guo, L., Jo, H.: Microfinance interest rate puzzle: price rationing or panic pricing? Asia-Pacific J. Financ. Stud. (2017)
Gutierrez-Nieto, B., Serrano-Cinca, C., Molinero, C.M.: Microfinance institutions and efficiency. Omega **35**(2) (2007)
Hirtle, B., Stiroh, J.: The return to retail and the performance of US Banks. J. Banking Financ. 31(4) (2007)
International Fund for Agricultural Development (IFAD).: Microfinance: a lifeline for poor rural people (2008)
Kar, A.K.: Mission drift in microfinance: Are the concerns really worrying? Recent cross-country results. Int. Rev. Appl. Econ. (2013)
Kar, A.K., Bali Swain, R.: Interest rates and financial performance of microfinance institutions: recent global evidence. Europ. J. Dev. Res. (2014)
Kent, D., Dacin, M.T.: Bankers at the gate: microfinance and the high cost of borrowed logics. J. Bus. Ventur. **28**(6) (2013)
Khachatryan, K., Hartarska, V., Grigoryan, A: Performance and capital structure of microfinance institutions in Eastern Europe and Central Asia. East. Eur. Econ. (2017)
Louis, P., Seret, A., Baesens, B.: Financial efficiency and social impact of microfinance institutions using self-organizing maps. World Dev. **46** (2013)
Miriti, J.M.: Factors influencing financial performance of savings and credit cooperative societies, Kenya. Unpublished Master of Arts of University of Nairobi, A case of Capital SACCO, Meru County (2014)
Morduch, J.: The microfinance schism. World Dev. **28**(4) (2000)
Morduch, J., Dehejia, R., Montgomery, H.: Do interest rates matter? Credit demand in the Dhaka slums. J. Dev. Econ. (2012)
Mwangi, N.S.: The effect of lending interest rates on financial performance of deposit taking micro finance institutions in Kenya. Unpublished Master of Science in Finance thesis of the University of Nairobi (2014)
Nguyen, K.A., Le, T.T: Sustainability of microfinance institutions in Vietnam. Transport Publishing House. Hanoi (2013)
Njihia, J.K.: The Determinants of banks profitability. Unpublished MBA project, University of Nairobi, The case of Kenyan quoted Banks (2005)
Piot-Lepetit, I., Nzongang, J.: Financial sustainability and poverty outreach within a network of village banks in Cameroon: a multi-DEA approach. Eur. J. Operat. Res. **234**(1) (2014)
Phuong, D.L., Van, D.T.: Implementation and solutions to develop microfinance activities in Vietnam. Journal of Forestry Science and technology. (2017)
Ramírez Rocha, A., Bernal Ponce, L.A., Cervantes Zepeda M.: Differences in the interest rates of microfinance institutions in some markets economies: An HLM approach. Estud. Econ. (2018)
Roberts, P.W.: The profit orientation of microfinance institutions and effective interest rates. World Dev. (2013)
Rodrigo, R.: The relationship between the performance and legal form of microfinance institutions. Revista Contabilidade & Finanças (2017)
Saunder, A.: The Determinants of bank interest. J. Int. Money Finance 19 (1995)
Sanfeliu, C.B., Royo, R.C., Clemente, I.M.: Measuring performance of social and non-profit microfinance institutions (MFIs): an application of multicriteria methodology. Math. Comput. Model. **57**(7) (2013)
Serrano-Cinca, C., Gutiérrez-Nieto, B.: Microfinance, the long tail and mission drift. Int. Bus. Rev. 23(1) (2014)
Small Industries Development Bank of India.: Study on Interest Rates and Costs of Microfinance Institutions. Access Development Services (2011)
Sun, S.L., Im, J.: Cutting microfinance interest rates: an opportunity co-creation perspective. Entrep. Theory Pract. (2015)
Strøm, R.Ø., D'Espallier, B., Mersland, R.: Female leadership, performance, and governance in microfinance institutions. J. Bank. Financ **42**(C) (2014)

Tchakoute-Tchuigoua, H.: Is there a difference in performance by the legal status of microfinance institutions? q. Rev. Econ. Financ. **50**(4) (2010)
Thanh Tam, L., Thi Thu Mai, N.: Vietnamese microfinance institutions' activities in the context of financial inclusion development: 10 years in retrospect. Financ. Monet. Mark. Rev. (2021)
Trujillo, V., Rodriguez-Lopez, F., Muriel-Patino, V.: Microfinance regulation and market development in Latin America. B.E. J. Econ. Anal. Policy. (2014)
Vanroose, A., D'Espallier, B.: Do microfinance institutions accomplish their mission? Evidence from the relationship between traditional financial sector development and microfinance institutions' outreach and performance. Appl. Econ. (2012)
VMFWG.: Microfinance-Policy framework for microfinance activities in Vietnam . No 19 (2013)
Were, R., Wambua, S.: Determinant of Interest Rate Spread. Kenya Institute for Public Policy Research and Analysis, Nairobi, Kenya (2013)
Xu, S., Copestake, J., Peng, X.: Microfinance institution's mission drift in macroeconomic context. J. Int. Dev. (2016)

Factors Influencing the Financial Development—A Metadata Analysis

Van Dung Ha, Thi Hoang Yen Nguyen, and van Chien Nguyen

Abstract Financial markets have greatly played an important role in allocating capital, connecting savings and investment, and promoting economic development in each country. The study uses Bayesian estimation method on data in 26 Asian countries from 2009 to 2019, the research results confirm that: public debt, economic growth, CO_2 emissions and trade openness have a significant and positive impact on financial development; meanwhile, foreign direct investment shows a negative impact on financial development.

Keywords Stock · Financial development · Bayes · Debt

1 Introduction

Financial markets play an important role in attracting and mobilizing major domestic and foreign sources, as well as encouraging and investing. The more developed the financial market, the more it will help to promote and improve the efficiency of financial use. In other words, the document field itself has the tools shown in the primary resource work to ensure the promotion of related element links. Basically, the financial market has three main functions: one is as a channel for capital from savers to traders, which is understood to help transfer capital from people who do not have profitable investment opportunities to those with profitable investment opportunities. Secondly, promoting the accumulation and concentration of capital to meet the needs

V. D. Ha (✉)
Banking University Hochiminh, Ho Chi Minh City, Vietnam
e-mail: dunghv@buh.edu.vn

T. H. Y. Nguyen · van C. Nguyen
Thu Dau Mot University, Thu Dau Mot City, Vietnam
e-mail: hoangyennt@tdmu.edu.vn

van C. Nguyen
e-mail: chiennv@tdmu.edu.vn

N. Ngoc Thach et al. (eds.), *Financial Econometrics: Bayesian Analysis, Quantum Uncertainty, and Related Topics*, Studies in Systems, Decision and Control 427,
https://doi.org/10.1007/978-3-030-98689-6_25

of building technical and material foundations as well as production and business. Third, it makes the use of capital more effective for those who have money to invest and those who borrow money to invest. Accordingly, the lender will earn interest through the lending rate, and the borrower will have to calculate the use of the loan capital to repay both capital and interest to the lender and generate income and accumulation for himself. Indeed, these arguments are summarized by many previous studies, such as Ismihan and Ozkan (2012), Man et al. (2012), Fernandez-Cuesta et al. (2019), Zhu et al. (2020), Funashima et al. (2020), Song et al. (2021). At the same time, these studies have also discovered many factors affecting financial development.

The development of the financial market is seen through positive signals from the stock market (Song et al. 2021; Funashima et al. 2020). This evidence from previous studies is consistent. Many financial theories accept that the stock market is an integral part of the financial market. It is classified according to the method of financing, where the trading of medium and long-term financial instruments takes place. In addition, from a practical perspective, Asian countries are located in the region of rapid economic growth, a number of countries in Asia are becoming one of the pillars of the regional economy such as China, Japan, Korea, India, UAE, Singapore; while some other countries are also maintaining high growth rates and are becoming emerging economies such as Vietnam, Thailand, Malaysia, Pakistan, Bangladesh, Saudi Arabia. Currently, Asia is home to 4.6 billion people, equivalent to nearly 60% of the world's population. According to the assessment of the World Economic Forum (2019), GDP in Asia is likely to overtake the GDP of the rest of the world and is expected to contribute approximately 60% of economic growth in the world. In other words, there exists a relationship between economic growth and financial development both from the evidence of actual statistics and the results of previous research (Al-mulali and Sab 2012; Funashima et al. 2020; Song et al. 2021).

Currently, factors affecting financial development have been investigated in the world, such as the study of Fernandez-Cuesta et al. (2019) in Europe, the study of Fernandez-Cuesta et al. (2019) in 16 EU countries, Man et al. (2012) in Romania, Funashima et al. (2020) in Japan. Although other studies confirm that FDI, GDP, human capital, public debt, and trade, environmental quality can affect financial development, there is no specific research that has been conducted in Asia recently. Therefore, this existing gap should be solved by investigating the relationship regarding financial development in Asia.

The remainder of this study will be shown as follows: Sect. 2 depicts the literature review. Section 3 discusses data collection, and methodology, especially briefly analyses the Bayes approach. Sects. 4, and 5 discuss the results and conclusions.

2 Literature Review

In the past, the public debt crisis was also known in the early 1980s of the twentieth century. Until the end of 2008, early 2009, the public debt crisis in Greece was spreading to a number of European countries. At that time, public debt and public debt management became hot issues that were of special interest to the leaders of countries around the world. Public debt comes from government spending needs; when government spending is greater than the amount of taxes, fees, and charges collected, the State must borrow (domestic or foreign) to cover the budget deficit. In general, covering the budget deficit with domestic or foreign borrowing can be detrimental to the macroeconomic environment, which potentially affects financial development. Furthermore, the lower the financial depth, the greater the adverse impact of public debt on financial development and macroeconomic outcomes (Ismihan and Ozkan 2012).

In developing countries, where the private sector is small and incompetent to promote economic development, the state economy plays a particularly important role. To achieve rapid growth, governments of developing countries often use expansionary fiscal policies, increasing government spending. Furthermore, tax cuts will boost aggregate demand, production, and economic growth. However, implementing an expansionary fiscal policy means an increase in the budget deficit, and the government must borrow to cover the deficit. The long-term use of expansionary fiscal policy will increase the debt burden. In the event that the growth rate of budget revenue cannot keep up with the growth rate of debt repayment obligations, the government is forced to use new borrowing measures to repay old debts. This situation will increase the risk of insolvency for the government, if the total liabilities exceed the collection capacity of the state budget. In the case study of Man et al. (2012) in Romania, in that situation, Romania always balances sustainable public debt growth. That is to balance the benefits of borrowing as well as the risks of debt that the government may face. This article also suggested that Romania needs to develop a strategy to maintain an appropriate public debt ratio and manage public debt to adapt to the financial crisis.

In addition to the specific factors discovered by previous studies, such as public debt, more recently, in the context of technology 4.0 gradually spreading in many countries, the study of Zhu et al. (2020) argues that financial development can adversely affect technological innovation and support economic growth. Among them, countries with higher financial development have a lower positive impact on technological innovation and even none. In other words, there exists a nonlinear relationship between financial development, innovation, and growth in which innovation has no effect on growth when financial development crosses the threshold of 60% of GDP (Zhu et al. 2020). In this case, financial development can be measured by stock market development, as suggested by Funashima et al. (2020). There is a relationship between economic performance and stock performance in the case of Japan. Specifically, the effect of the announcement of the initial GDP on the stock price is modest. Additionally, the stock market is likely to respond positively to the first revision of the GDP indication but negatively to the second revision. It is evident that GDP has greatly impacted on financial development.

In addition, broader empirical evidence to assess the impact of growth on financial development, Song et al. (2021) conducted a study in 142 countries between 2002 and 2016. According to the author, there exists a long-term relationship between economic growth and economic development on a global scale and within developing countries. However, no long-term relationship has been found in developed countries. Specifically, there is a positive relationship between economic growth and financial development, while corruption has a negative effect. This supports the view that boosting economic growth in developing countries will help expand financial development. Indeed, developing countries have underdeveloped financial markets. The expansion of financial development will create more opportunities for the development of enterprises and production and business activities, creating competitive advantages for products and national competitiveness. Moreover, in the current context of economic development, countries are increasing energy use. The process of energy consumption is also an important factor in promoting growth and especially promoting financial development. It corresponds to an economy where there is a high level of pollution (Al-mulali and Sab 2012). Therefore, countries should work towards energy efficiency. Priority should be given to the use of energy-saving projects to both achieve a reduction in environmental pollution while maintaining economic growth and financial development.

Finally, a theoretical overview of the factors affecting financial development has also identified a recently recognized line of research. Reducing CO_2 emissions has also become an important tool in managing financial debt in businesses with sustainable development strategies. In another study, Fernandez-Cuesta et al. (2019) using data from 16 stock markets in the EU indicated that carbon risk is considered to be one of the main drivers of the selection of financial debt in the business. Therefore, there is a positive effect between carbon dioxide emissions and financial debt, meaning that reducing carbon dioxide emissions should be a preferred factor that positively impacts sustainable development and firm development.

3 Data and Methodology

3.1 Data

The study uses unbalanced panel data in Asian countries in the period from 2009 to 2019. The data was gathered by the World Bank's World Development Indicators and the International Monetary Fund (IMF). Furthermore, the data were also collected by the Departments of Statistics in the relevant countries.

The countries in the study were selected from emerging Asian countries, including Qatar, Saudi Arabia, China, Pakistan, Philippines, Vietnam, Brunei, Sri Lanka, Malaysia, Myanmar, Indonesia, Iran, Israel, Japan, Kazakhstan, Korea, Kuwait, Laos, Bhutan, Turkey, Bahrain, Bangladesh, Thailand, India, Cambodia, and Afghanistan.

3.2 Methodology

This study employs Bayesian estimations, which is considered an advantage over the frequentist approach method to get the research objectives (Briggs and Nguyen 2019); Kreinovich et al. 2019).

Based on previous studies by Ismihan and Ozkan (2012), Man et al. (2012), Fernandez-Cuesta et al. (2019), Zhu et al. (2020), Funashima et al. (2020), Song et al. (2021), this study uses the Bayesian regression method to assess the change in financial development in Asian countries. The proposed research model has the following form:

$FDit = \beta_0 + \beta_1 DEBT\mathrm{it} + \beta_2\mathrm{GDPit} + \beta_3\mathrm{FDIit} + \beta_4\mathrm{EPit} + \beta_5\mathrm{TOit} + \mu_i + \mu'_{it}$
where,

- FDit, indicating the financial development at the country i at the time t, is counted by the broad money supply (M2/GDP);
- DEBTit, is measured the public debt of country i at year t, by the ratio of public debt to GDP;
- GDPit, is measured the size of the economy of country i at year t, measured log of real GDP;
- FDIit, indicating for the foreign direct investment of country i at year t, measured by the ratio of foreign investment to GDP,
- EPit, indicating for per capita carbon dioxide emissions in the country i at the time t;
- TOit, reflecting the trade openness of country I at year t, measured by % of total trade to GDP;
- β_0 is the intercept;
- β_{1i}, β_{2i}, β_{3i}, β_{4i}, β_{5i} is the estimated coefficient for the variables: public debt, GDP, foreign direct investment, CO_2 emissions, trade openness;
- μ'_{it} is error terms (Table 1).

This paper will select the most appropriate prior by taking 5 candidate simulations. Through these simulations, the prior information can be used to remove the biases of large sample size. More or less informative normal prior will be used and processed as follow (Table 2).

Log BF, Log(ML), and DIC criteria will be used to select appropriate simulations. After that, some post-simulation tests such as autocorrelation, normal distribution, and the Max Gelman-Rubin Rc test will be carried out to verify the validity of Bayesian inference.

Table 1 Variables used in the study

Variable	Abbreviation	Expected sign
Dependent variable		
Financial development	FD	
Independent variables		
Public debt	DEBT	+
Economic performance	GDP	+
Foreign direct investment	FDI	+
Environmental pollution	EP	−
Trade openness	TO	±

Table 2 Likelihood model

	$FD \sim N(\mu, \delta)$
Prior distributions	
Simulation 1	$\alpha_i \sim N(0, 1)$ $\delta^2 \sim Invgamma(0.01, 0.01)$
Simulation 2	$\alpha_i \sim N(0, 10)$ $\delta^2 \sim Invgamma(0.01, 0.01)$
Simulation 3	$\alpha_i \sim N(0, 100)$ $\delta^2 \sim Invgamma(0.01, 0.01)$
Simulation 4	$\alpha_i \sim N(0, 1000)$ $\delta^2 \sim Invgamma(0.01, 0.01)$
Simulation 5	$\alpha_i \sim N(0, 10000)$ $\delta^2 \sim Invgamma(0.01, 0.01)$
i = 1, 2, 3, 4, 5	

4 Results and Discussions

4.1 Descriptive Statistics

See Table 3.

4.2 Matrix of Correlation

According to the study of Gujarati (2004), correlation matrix analysis aims to evaluate the degree of correlation between the independent variables in the model in order to avoid possible multicollinearity in the estimation. Gujarati (2004) said that multicollinearity will occur when the correlation coefficient between the independent variables is greater than or equal to 0.85 (Table 4).

Table 3 Descriptive statistics

Variable	Mean	Std. Dev.	Max	Min	Skewness	Kurtosis
FD	81.541	49.159	19.412	252.647	1.583	5.190
DEBT	46.239	43.175	1.109	249.113	3.010	14.017
GDP	3.777	0.595	2.641	4.929	0.107	1.851
FDI	2.524	2.986	−4.336	14.145	1.689	6.545
SES	82.844	21.180	32.856	120.651	−0.576	2.473
EP	7.415	8.334	0.151	34.544	1.426	4.405

Source Authors' calculation

Table 4 Correlation matrix

Variable	FD	DEBT	GDP	FDI	EP	TO
FD	1					
DEBT	0.137	1.000				
GDP	0.365	−0.265	1.000			
FDI	0.029	−0.034	0.282	1.000		
EP	0.197	−0.447	0.838	0.267	1.000	
TO	0.582	0.096	0.580	0.396	0.523	1.000

Source Authors' calculation

4.3 Empirical Results and Discussions

According to the Log BF, Log (ML), and DIC criteria in the Bayesian factor test and the Bayes model test, the first simulation will be selected to investigate the determinants of financial development (Table 5).

The simulation results indicate that the Max Gelman-Rubin Rc value is lower than 1.1, which implies the convergence of MCMC. The other tests of autocorrelation, normal distribution, and stationary conditions can be visually drawn from Fig. 1. Autocorrelation graphs indicate that most values fluctuate around [−0.2; 0.2] and they seem to be randomly distributed. So, it can be concluded that autocorrelation

Table 5 Results of the Bayesian factor test and the Bayes model test

	Chains	Avg DIC	Avg log (ML)	Avg log BF	P(Mly)
Simulation 1	3	1848.4336	−935.7037	1	0.5646
Simulation 2	3	1842.2651	−935.9972	−0.2935	0.4209
Simulation 3	3	1840.5372	−939.3827	−3.6790	0.0143
Simulation 4	3	1836.2410	−943.4321	−7.7284	0.0002
Simulation 5	3	1835.2003	−948.3979	−12.6942	0.0000

Source Authors' calculation

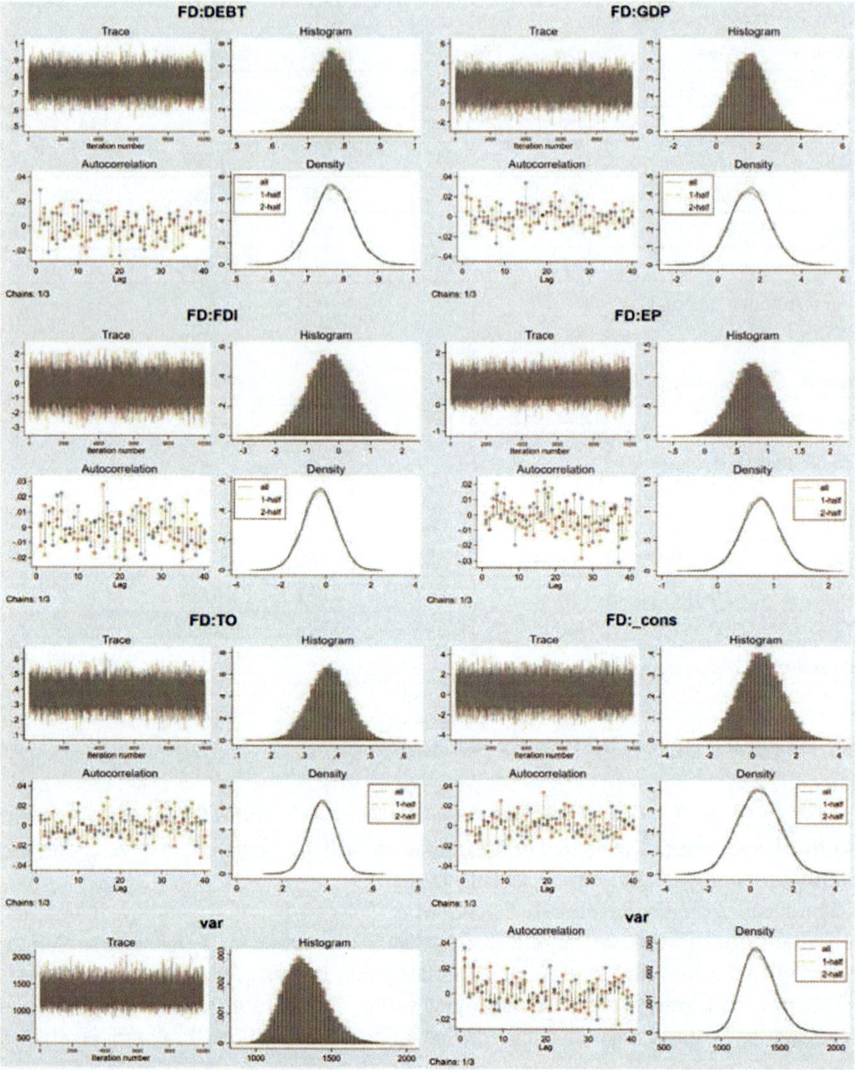

Fig. 1 Tests for MCMC convergence. *Source* Authors' calculation

does not exist. The trace plots indicate all well-mixing parameters since the plots have nearly constant means and variances. The histograms show that the distributions of parameters are normalized. The Kernel density plots show three density curves close to each other, which indicates the MCMC chain has converged and is well-mixed. So, it can be said that the parameters of the simulation are convergent.

Table 6 shows that the variables DEBT, GDP, EP, and TO have a positive impact on financial development (FD), while only the variable FDI has a negative effect on FD.

Table 6 Bayesian simulation results

Variable	Mean	Std. Dev	MCSE	Median	Equal-tailed	
					[95% Cred. Interval]	
FD						
DEBT	0.7681433	0.0556998	0.000323	0.7679611	0.6584224	0.877429
GDP	1.593398	0.922929	0.005413	1.596261	−0.2209185	3.393639
FDI	−0.0601383	0.7389234	0.004283	−0.0591669	−1.506194	1.397121
EP	0.8042013	0.3297218	0.001904	0.8037293	0.1594054	1.451861
TO	0.3610658	0.0599749	0.000349	0.361077	0.2426669	0.4789554
_cons	0.7072854	0.9906083	0.005719	0.7046068	−1.240607	2.656446
var	1348.954	147.6371	0.862398	1337.963	1090.504	1670.611
Avg acceptance rate		1				
Avg efficiency: min		0.9689				
Max Gelman-Rubin Rc		1				

Source Authors' calculation

Simultaneously, Table 6 also shows that the Monte-Carlo Standard Error (MSCE) has a very small value, showing the high accuracy of the estimated parameter in the model. Therefore, the model results are reliable.

Research results show that public debt has a positive impact on financial development in 26 Asian countries. It indicates that countries that increase public debt to serve economic development will have a positive impact on financial development. When the country borrows, the loan sources can come from the domestic or foreign sector. Particularly, foreign capital has greatly promoted the development of the domestic capital market, investment, and economic development. In addition, the capital flows are circulated in the financial system and help to develop the financial markets. This finding is also supported by Man et al. (2012) in Romania, who believe that public debt can only be effective if this capital source is used effectively in the financial market. However, Ismihan and Ozkan (2012) argued that countries with low financial depth have negative effects on their financial development and lead to macroeconomic instability in light of increasing public debt.

Economic growth is an important factor in increasing people's incomes, living standards, and financial markets. Research results show that a country with higher economic growth will significantly enhance its financial development. In addition, foreign direct investment has a negative effect on financial development. It is easy to explain that a country with the greater foreign direct investment will negatively reduce financial development, and conversely, a country with shrinking foreign investment will be similar to a more developed financial market (Nguyen 2020). The results of this study can be explained by the fact that a country always needs capital for economic development. If foreign capital decreases, the country must seek capital from the financial market. In contrast, if a country easily attracts domestic and foreign capital, the role of financial markets may not become so important.

CO_2 emissions have a positive impact on financial development. It indicates that a higher level of environmental pollution is positively linked to a higher level of financial development in Asia. This finding can be explained by the fact that Asian countries are likely to accept environmental pollution to promote economic growth and financial development. According to the Kuznets curve theory, developing countries often accept high environmental pollution in the early stages of economic development. When development reaches a certain level, environmental pollution will decrease. Asia has a huge number of developing or emerging economies. Thus, increasing CO_2 emissions will help countries get investment capital and promote financial development. However, this research result is not supported by Al-mulali and Sab (2012) that the process of reducing environmental pollution will have an impact on promoting economic growth and financial development in Sub-Saharan African countries. Although these countries have very low levels of development, CO_2 emissions have also been increasing gradually in recent years. Therefore, investing in eco-friendly, energy-saving projects will greatly promote financial development in Sub-Saharan African countries (Wang et al. 2020, 2021).

The study also found that trade openness has a positive impact on financial development. Specifically, a country with greater trade openness is consistent with greater financial development. The results of this study are similar to those of Khan et al. (2020). China's economy has made many achievements in the period of 1987–2017 through the economic reforms implemented since 1977. China has also robustly invested in human capital by increasing the quality of training in the education system and human health. Therefore, China has significantly improved labor productivity in order to promote financial development as well as financial markets.

5 Conclusions

Research on factors affecting financial development has been carried out in several studies in recent years, but it has not been analyzed in Asia. Using panel data of 26 emerging Asian countries, including Qatar, Saudi Arabia, China, Pakistan, Philippines, Vietnam, Brunei, Sri Lanka, Malaysia, Myanmar, Indonesia, Iran, Israel, Japan, Kazakhstan, Korea, Kuwait, Laos, Bhutan, Turkey, Bahrain, Bangladesh, Thailand, India, Cambodia, and Bayesian estimation method is concerned, the research results confirm that public debt, economic growth, CO_2 emissions, and trade openness have positive effects on financial development in 26 Asian countries. Meanwhile, foreign direct investment has a negative impact on financial development.

The study also has some recommendations for Asian countries. First, countries should continue to create a favorable legal environment and improve the quality of using public debt to promote financial markets. In the digital economy, all countries are pursuing a green growth model through improving their strategies to attract FDI, especially clean FDI and the use of renewable energy to reduce toxic substances into the environment and sea-level rise, using more renewable energy and maintaining a sustainable environment. Third, countries need to continue to improve the quality of human resources and adapt to the industrial revolution 4.0 for economic and financial development. In the context of the Fourth Industrial Revolution, digital technology necessitates the availability of human resources capable of grasping science and technology in order to serve socio-economic development. Finally, countries need to take advantage of the benefits of economic integration and free trade agreements to promote financial development.

References

Al-mulali, U., Sab, C.N.B.C.: The impact of energy consumption and CO_2 emission on the economic growth and financial development in the Sub Saharan African countries. Energy **39**(1), 180–186 (2012)

Briggs, W.M., Nguyen, T.H.: Clarifying ASA's view on P-values in hypothesis testing. Asian J. Econ. Bank. **3**(2), 1–16 (2019)

Funashima, Y., Lizuka, N., Ohtsuka, Y.: GDP announcements and stock prices. J. Econ. Bus. **108**(2), 105872 (2020)

Fernandez-Cuesta, C., Castro, P., Tascon, M.T., Castano, F.J.: The effect of environmental performance on financial debt. European evidence. J. Cleaner Produc. **207**, 379–390 (2019)

Gujarati, D.: Basic Econometrics, 4th edn. McGraw-Hill Companies, New York

Ismihan, M., Ozkan, F.G.: Public debt and financial development: a theoretical exploration. Econ. Lett. **115**(3), 348–351 (2012)

Khan, Z., Hussain, M., Shahbaz, M., Yang, S., Jiao, Z.: Natural resource abundance, technological innovation, and human capital nexus with financial development: a case study of China. Resour. Policy **65**, 101585 (2020)

Kreinovich, V., Nguyen, N.T., Nguyen, D.T., Dang, V.T.: eds. Beyond Traditional Probabilistic Methods in Economics. Springer, Cham (2019)

Man, M., Macris, M., Boca, I.S.: Dynamics of public debt service in the context of romania's current financial crisis. Procedia. Soc. Behav. Sci. **62**, 462–467 (2012)

Nguyen, V.C.: Trade liberalization, economic reforms and foreign direct investment–a critical analysis of the political transformation in vietnam. international journal of advanced science and technology. Int. J. Adv. Sci. Technol. 29(3), 6837–6850 (2020)

Song, C.Q., Chang, C.P., Gong, Q.: Economic growth, corruption, and financial development: global evidence. Econ. Model. **94**, 822–830 (2021)

Wang, Z., Bui, Q., Zhang, B., Le, T., Pham, H.: Biomass energy production and its impacts on the ecological footprint : an investigation of the G7 countries. Sci. Total Environ. **743**, 140741(2020) https://doi.org/10.1016/j.scitotenv.2020.140741

Wang, Z., Bui, Q., Zhang, B., Nawarathna, C.L.K., Mombeuil, C.: The nexus between renewable energy consumption and human development in BRICS countries: the moderating role of public debt. Renew. Energy **165**, 381–390 (2021). https://doi.org/10.1016/j.renene.2020.10.144

World Economic Forum.: In 2020 Asia will have the world's largest GDP. Here's what that means. https://www.weforum.org/agenda/2019/12/asia-economic-growth/ (2019)

Zhu, X., Asimakopoulos, S., Kim, J.: Financial development and innovation-led growth: Is too much finance better?. J. Int. Money Financ 100, 102083 (2020)

The Role of Liability in Managing Financial Performance: The Investigation in the Vietnamese SMEs Context

Truong Thanh Nhan Dang, Van Dung Ha, and Van Tung Nguyen

Abstract The effectiveness in liability management is truly important for the liquidity and survival of firms such as small and medium enterprise (SMEs). However, there are still inconsistencies of results from previous empirical studies on the effect of liabilities on financial performance of nonfinancial firms. Based on this gap, this study aims to investigate the influence of total liabilities and different controlled variables including total assets (indicating firm size), gender of firm manager, firm age and exporting experience on the financial performance of SMEs. This research focused on SMEs within the Vietnamese context and used Bayesian estimations. The research results demonstrate that liability has positive impacts on firm financial performance. This study also found a negative effect of total assets (firm size), manager gender as male and firm age on financial performance, and a positive impact of export experience on financial performance. It is suggested that practitioners in SMEs should consider employing liability as a financing method and also consider participating in export activities.

Keywords Liability · SMEs · Total assets · Firm age · Financial performance · Vietnam

1 Introduction

In most countries, SMEs play a very important role in driving economic development and have also employed a growing proportion of the workforce. The number of privately owned small and medium-sized companies all over the world has been

T. T. N. Dang · V. D. Ha (✉) · V. T. Nguyen
Banking University Hochiminh City, Ho Chi Minh City, Vietnam
e-mail: dunghv@buh.edu.vn

T. T. N. Dang
e-mail: nhandtt@buh.edu.vn

V. T. Nguyen
e-mail: tungnv@buh.edu.vn

N. Ngoc Thach et al. (eds.), *Financial Econometrics: Bayesian Analysis, Quantum Uncertainty, and Related Topics*, Studies in Systems, Decision and Control 427,
https://doi.org/10.1007/978-3-030-98689-6_26

increasing rapidly; nonetheless, this firm type has faced several issues that deter the growth (Muriithi 2014). One of the main obstacles for most SMEs is the difficulty of financing, according to Da Silva et al. (2017), all small companies have to deal with liquidity constraints and limitation of resources.

As emphasized by Ogijiuba et al. (2004) and Muriithi (2014), the use of liabilities is a considerable factor for entrepreneurs to turn their idea into a real business model. Without the effective control of liabilities, entrepreneurs would struggle remarkably to make their dreams come true. In attempt to explore the nature and effect of SME financing, there have been various studies examining the influence of liability on financial performance of SMEs (Murigi 2014). However, there is still limited research in developing countries, if any, which examines the impact of financial obligations, especially with the focus on liability, and different other controlled variables (Murigi 2014; Muriithi 2014) on the financial performance of SMEs. There are also inconsistencies of results from previous empirical studies on the effect of liabilities on financial performance of nonfinancial firms (Menike and Prabath 2014; Muchiri et al. 2016). Based on this gap, this study aims to investigate the influence of total liabilities as a financing source and different controlled variables including total assets (indicating firm size), gender of firm manager and years of operation since the establishment and exporting experience on the financial performance of SMEs. This research focused on SMEs within the Vietnamese context.

The research findings are expected to be of significance to the various stakeholders.

including financial institutions, SMEs' managers and owners, the government, welfare organization, general public, researchers and academicians.

2 Financing Decisions

Financing decisions such as using debt, equity or even liability (a broader term of financial obligations) are helpful for the generation of business ideas and innovation. Financing methods also have a tight connection with business strategy; financial strategy also enables the creation of added value and raises the level of competitiveness (Mikic et al. 2016). Healthy financial conditions are essential for the realization of new good ideas and for the growth and development of enterprises. SMEs have to overcome many barriers in their business life cycle, among which financing is one of most common challenges (Mikic et al. 2016).

In order to meet financial requirements, SMEs can select among a variety of financing alternatives such as using personal sources of finance, bank loans, micro-finance funds, venture capital funds, leasing and negotiating with suppliers in terms of repayment of financial obligations (Muriithi 2014). It is challenging for business leaders of SMEs to manage finance, especially during the process of launching a new entrepreneurial project. Notably, continual changes in the business environment also makes it more difficult for SMEs' financial management. When considering financing decisions, SMEs' leaders may consider the following factors (Stokes and Wilson 2010): legal forms of conducting, business phase of enterprise life cycle and the nature of funds.

3 Liability Definitions

A liability is defined as a present obligation, the settlement of which is expected to result in an outflow from the entity embodying economic benefits (ISA 32, 2004). Liability is the obligation to deliver cash or another financial asset to another entity, or to exchange financial assets or financial liabilities with another entity under conditions that are potentially unfavourable to the entity, or a contract that will or may be settled in the entity's own equity instruments (ISA 32, 2004).

Liability is also a non-derivative for which the entity is or may be obliged to deliver a variable number of the entity's own equity instruments or a derivative that will or may be settled other than by the exchange of a fixed amount of cash or another financial asset for a fixed number of the entity's own equity instruments (Pro-active Accounting Activities in Europe 2008).

4 The Overview of Previous Research

The financial performance of SMEs can be measured through profitability. There are various profitability measures that are used to measure the organizational performance especially in SME studies such as Return on Assets (ROA) and the Return on Equity (ROE) (Mamaro and Legotlo 2020). The financial structure is typically the combination of both debt and equity financing (Mamaro and Legotlo 2020). Kajananthan (2012) mentioned that successful selection of capital financing may affect financial performance of organisations positively or negatively.

Some previous studies have examined the relationship between liability and financial performance of SMEs. For instances, the study of Kajirwa (2015) showed that liability does not influence the financial performance of commercial banks; Nwude et al. (2016) found that liability has negative and significant influence on the financial performance of Nigerian quoted firms. Many empirical studies such as Margaritis and Psillaki (2010) and Nimalathasan and Velnampy (2010) found that there is empirical relationship between liability and the financial performance. However, some academic researchers such as Iorpev and Kwanum (2012), Nwude et al. (2016) found the negative significant relationship between liability and financial performance.

Sadeghian et al. (2012) examined the relationship between capital structure and firm performance of Tehran Stock Exchange Companies for a period 2006–2011. The study found a positive significant relationship between return on asset, return on equity with total asset ratio and short-term debt but insignificant relationship with long term debt and gross profit margin.

Iorpev and Kwanum (2012) studied the impact of capital structure on the performance of manufacturing companies in Nigeria. The research found that total debt to equity was positively related with ROA and negatively related with profit margin. The findings reveal statistically that capital structure was not a major factor influencing firm performance.

5 The Role of Agency Theory in Explaining Financing Decisions

Jensen and Meckling (1976) introduced the Agency theory which states that, the Agency costs which arise from the conflicts between the managers and the owners of the firm can be reduced due to a certain proportion of liability in the financial structure of the firm (Leland 1998). The reduction of agency conflicts would help to reduce agency costs which can improve financial performance. The employment of liability as examined by Jensen and Meckling (1976) can also help to control and monitor managers in the firm to ensure that the business objectives can be fulfilled.

Buferna et al. (2005) supported this theory by demonstrating that the inclusion of liability in the financial structure would enhance motivation for managers to stimulate firm growth, improve cash flows which ensure repayments of financial obligations. This helps to enhance the firm's profitability (Dawar 2014).

This theory argues that any form of liability that a firm uses helps to reduce agency conflicts between managers and shareholders of the organization and hence boosts financial growth (Rashid 2015). The Agency theory plays a critical role in explaining financing decisions such as employing liabilities (Shikumo 2020).

6 The Controlled Variables' Impact on Financial Performance

As initially introduced, different controlled variables studied in this research include total assets (indicating firm size), gender of firm manager, years of operation since the establishment and whether the firm has export activity on the financial performance of SMEs.

Regarding total assets (indicating firm size), the significant connection between firm size and financial performance was also supported by Majumdar, Ammar et al. (2003), Papadognas and Vijayakumar and Tamizhselvan (2010).

Jónsson (2007) identified three categories of firm theories which argue that firm size is an impactful factor in financial performance related studies. They are Principal-agent theory, Strategic theories and Institutional theory.

Principal-agent theory recommends that the separation of corporate ownership and control potentially can lead to self-interested actions by managers who might expand their firm more to increase their own benefits, such as more prestige, better pay, and stock option. Strategic theories, meanwhile, proposes that firms can apply cost leadership, product differentiation, or focus-based domination for strategic competitiveness, with which firm size is associated. Institutional theory suggests that organizations (firms) seek to behave in ways which would not make them noticed as different and thereby singled out for criticism. Therefore, organizations will gradually become more similar in actions. As the perception of business growth and "larger is better than smaller" is embedded in the institutional environment of organizations, the pressure will force firms to comply with the institutional environment.

Until now, there is still limited research focusing on manager gender's impact on financial performance. Salloum et al. (2016) investigated the relationship between the presence of women in management and firms' financial performance, and found that the presence of women in management in Lebanon was not positively correlated with firms' financial performance. Satria et al. (2020) studied the effect of gender diversity on company financial performance and found that the proportion of women councils did not affect financial performance as measured by return on asset and return on equity.

Regarding years of operation since the establishment (firm age), there have been a few studies exploring the impact of this factor on financial performance. Gurbuz et al. (2010) could not demonstrate a significant relationship between firm age and return on assets. Meanwhile, Basti et al. (2011) found a positive relationship between age and profitability measures including return on assets, return on equity and basic earning power. On the contrary, Dogan (2013) found a negative relation between firm age and return on assets running a multiple regression on data from 200 listed companies between the years 2008–2011. Both theoretical studies and empirical evidence on firm age-performance relationship generated conflicting results which are highly dependent on the countries and periods under consideration as well as on the estimation methodologies employed (Akben-Selcuk 2016).

With regards to export activity, some previous studies such as Esteve-PeRez et al. (2013) examined the impact of export experience on SMEs' financial performance. The study found that export survival analysis was useful to obtain understand firms' export performance in order to come up with adequate export-promotion policies and interventions. Dhliwayo (2016) showed that export experience had a statistically significant impact on sales and profitability, but not on savings, and performance in sales and profitability increased with export experience.

7 Methodology

The data used to analysis is extracted from the survey by the Central Institute for Economic Management (CIEM) of the Ministry of Planning and Investment (Vietnam) in collaboration with UNU-WIDER University and the Institute of Labor Science and Society (ILSSA). There are totally 2,647 SMEs in mostly updated 2015 survey and only 994 SMEs sufficiently used in the model.

To reach the research objectives, this paper uses Bayesian estimations since Bayesian method is recently expanding its importance due to its advantages over the frequentist approach (Briggs and Hung 2019; Nguyen et al. 2019; Sriboonchitta et al. 2019; Svitek et al. 2019; Kreinovich et al. 2019).

Basing on the Bayesian framework, the paper focuses on the impact of liabilites on financial performance of SME in Vietnam. While the financial performance is measured by ROA and ROE, firm liability is measured as total liabilities at the end of previous year (as in the survey quesionnaire). As mentioned above, the majority of previous studies indicate the positive impact of liabilities on firm financial performance (Margaritis and Psillaki 2010; Nimalathasan and Valeriu 2010; and Kajirwa 2015), this study expects that higher liability will lead to higher ROA and ROE, and liability is measured by natural logarithm of firm total liabilities (LnLIB).

The group of controlled variables includes total assets, exports, number of established years, and gender of firm manager. Natural logarithm of total assets (LnAS) is a proxy of firm size. Exports (Ex) in this study is a binary variable, which recieves 1 if the firm has exported products and otherwise 0. The number of years established is used to estimate impact of firm age to firm financial performance and this variable is in natural logarithm. The gender of firm manager is another binary variable, which recieve 1 if the manager is male, otherwise 0. The research model has the following form:

$$ROA_i = \beta_1 + \beta_2 LnLIAB_i + \beta_3 LnAS + \beta_4 Ex_i + \beta_5 Year\_est_i + \beta_6 Gender_i + e_i$$

$$ROE_i = \alpha_1 + \alpha_2 LnLIAB_i + \alpha_3 LnAS + \alpha_4 Ex_i + \alpha_5 Year\_est_i + \alpha_6 Gender_i + v_i$$

where ROA and ROA are used to measure firm financial performance; e_i and v_i are error terms.

In order to choose the most appropriate priors for large sample size, the paper analyzes the sensitivity via simulations. Five candidate simulations with normal prior from more to less informative will be processed as following (Table 1).

After that, the Bayes factor test and the Bayes model test will be carried out to choose the most appropriate simulation regarding to Log BF, Log(ML), and DIC criteria. For post-estimation test for the validity of Bayesian inference, the paper will use convergence diagnostics via such tests as autocorrelation, normal distribution, stationary, and Max Gelman-Rubin Rc test.

Last but not least, to check the robustness of parameter spaces of posterior simulations, the reseach specify prior means from −0.5 to 0.5.

8 Reseach Results

The results of the Bayesian factor test and the Bayes model test for ROA.

Table 1 Likelihood model

	ROA, ROE $\sim$ N(μ, δ)
Prior distributions	
Simulation 1	$\alpha_i \sim$ N(0, 1) $\delta^2 \sim$ Invgamma(0.01, 0.01)
Simulation 2	$\alpha_i \sim$ N(0, 10) $\delta^2 \sim$ Invgamma(0.01, 0.01)
Simulation 3	$\alpha_i \sim$ N(0, 100) $\delta^2 \sim$ Invgamma(0.01, 0.01)
Simulation 4	$\alpha_i \sim$ N(0, 1000) $\delta^2 \sim$ Invgamma(0.01, 0.01)
Simulation 5	$\alpha_i \sim$ N(0, 10000) $\delta^2 \sim$ Invgamma(0.01, 0.01)
i = 1, 2, 3, 4, 5	

To choose the most appropriate simulation, the higher Log BF, higher Log(ML) and lower DIC are selected. The results from Table 2 reveal that for ROA the first simulation would be fitted while the second simulation is appropriate for ROE.

The Bayesian estimation results based on the above selection procedure and presented in Table 3.

The Max Gelman-Rubin Rc values are stringently lower than 1.1 in both cases of ROA and ROE indicating that the convergence of MCMC, which is an importantly approproriate indicator for Bayesian analysis.

Post estimation tests of autocorrelation, normal distribution, stationary will be done via diagnostic graphs (Figs. 1 and 2).

In both cases of ROA and ROE, the plots indicate low autocorrelation while the trace plots depict good mixing. The normal distribution can be drawn from the kernel density plots and histograms. Therefore, the MCMC convergence is hold (Table 4).

The robustness check indicates that the posterior estimators are insignificant different in posterior means, MCSEs, and credible intervals when the normal priors for all parameters are adjusted from −0,5 to 0.5 with 0.1 spacing. So, it can be said that the results are robust.

9 Discussion

The Bayesian inferences indicate that the liability has positive impacts on firm financial performance. This results are strongly supported by previous studies such as Vintila and Nenu (2015), Margaritis and Psillaki (2010) and Nimalathasan and Velnampy (2010). The higher level of liability will lead to higher ROA and ROE. It proves that the use of liability can not only increase firm capital sources but also increase firm financial performance. This finding can also be enlightened by

Table 2 Results of a Bayesian factor test and a Bayes model test (ROA)

	Chains	ROA				ROE			
		Avg DIC	Avg log (ML)	Avg Log BF	P(M|y)	Avg DIC	Avg log (ML)	Avg Log BF	P(M|y)
Simulation 1	3	2347.64	−1.20e + 03	1	0.978	4306.11	−2.18e + 03	1	0.021
Simulation 2	3	2347.32	−1.20e + 03	−3.812	0.021	4297.71	−2.17e + 03	3.818	0.973
Simulation 3	3	2347.28	−1.20e + 03	−10.412	0	4297.64	−2.18e + 03	−1.325	0.005
Simulation 4	3	2347.29	−1.20e + 03	−17.319	0	4297.59	−2.19e + 03	−8.105	0
Simulation 5	3	2347.29	−1.20e + 03	−24.197	0	4297.63	−2.19e + 03	−14.940	0

Source Authors' calculation

Table 3 Bayesian simulation results

	ROA						ROE					
	Mean	Std. Dev	MCSE	Median	Equal-tailed [95% Cred. Interval]		Mean	Std. Dev	MCSE	Median	Equal-tailed [95% Cred. Interval]	
lnLIAB	0.06462	0.01810	0.00010	0.06459	0.02929	0.10013	0.20246	0.04897	0.00028	0.20208	0.10704	0.29929
lnAS	−0.15778	0.02178	0.00012	−0.15776	−0.20060	−0.11480	−0.43776	0.05863	0.00034	−0.43748	−0.55261	−0.32321
Ex	0.08926	0.08487	0.00049	0.08968	-−0.07652	0.25456	0.22146	0.22844	0.00131	0.22136	−0.22602	0.67178
Year_est	−0.00372	0.00284	0.00001	−0.00370	−0.00932	0.00184	−0.00598	0.00759	0.00131	−0.00601	−0.02083	0.00899
Gender	−0.11659	0.05204	0.0003	−0.11672	−0.21948	−0.01472	−0.02740	0.13883	0.00080	−0.02775	−0.29752	0.24447
_cons	2.49706	0.27751	0.00160	2.49657	1.94762	3.04058	5.90534	0.74519	0.00430	5.9056	4.43792	7.3481
Var	0.61788	0.02782	0.00016	0.61706	0.56583	0.67498	4.39201	0.19863	0.00114	4.38698	4.02342	4.80126
Number of obs			994						994			
Avg acceptance rate			1						1			
Avg efficiency: min			0.9896						0.9908			
Max Gelman-Rubin Rc			1						1			

Source Authors' calculation

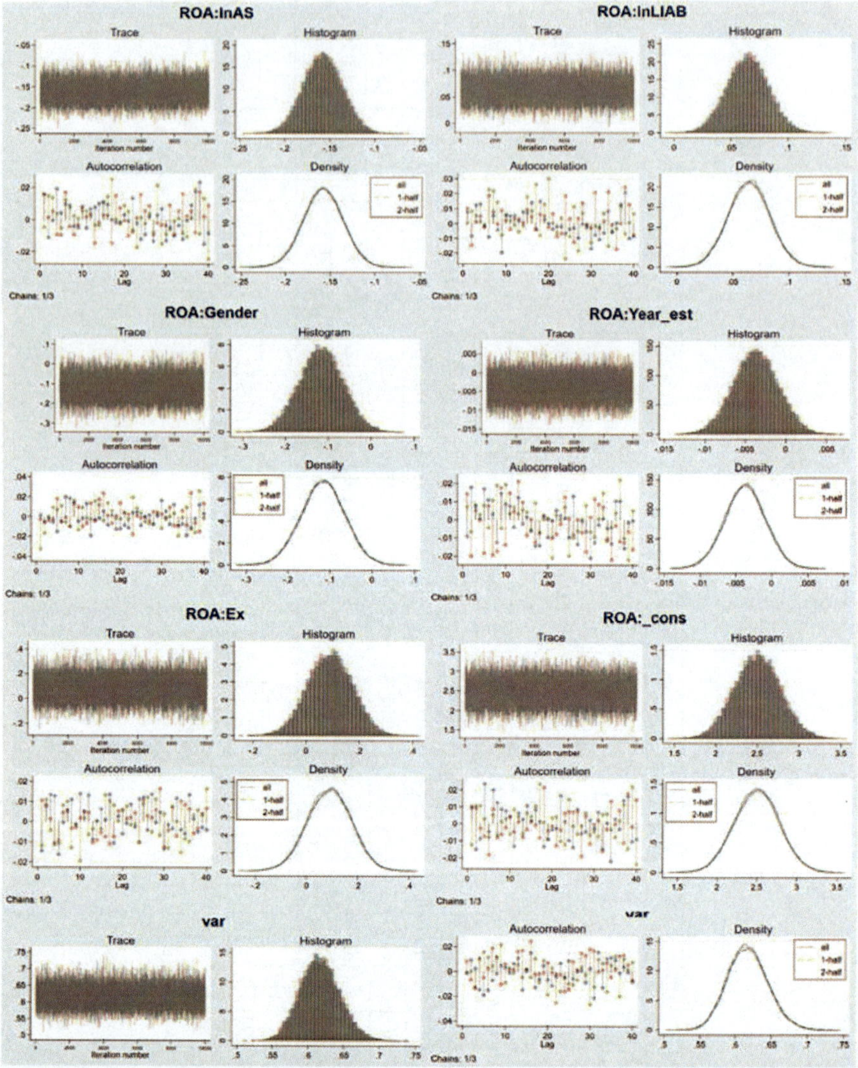

Fig. 1 Visual tests for MCMC convergence of ROA

the Agency theory (Jensen and Meckling 1976), which explains the mechanism in which a certain proportion of liability in the financial structure of the firm helps to improve financial performance through the reduction of the conflicts between the managers and the firm owners. The role of this theory in reasoning the positive impact of liability on financial performance was also supported by Buferna, Bangassa and Hodgkinson (2005), Rashid (2015) and Shikumo et al. (2020).

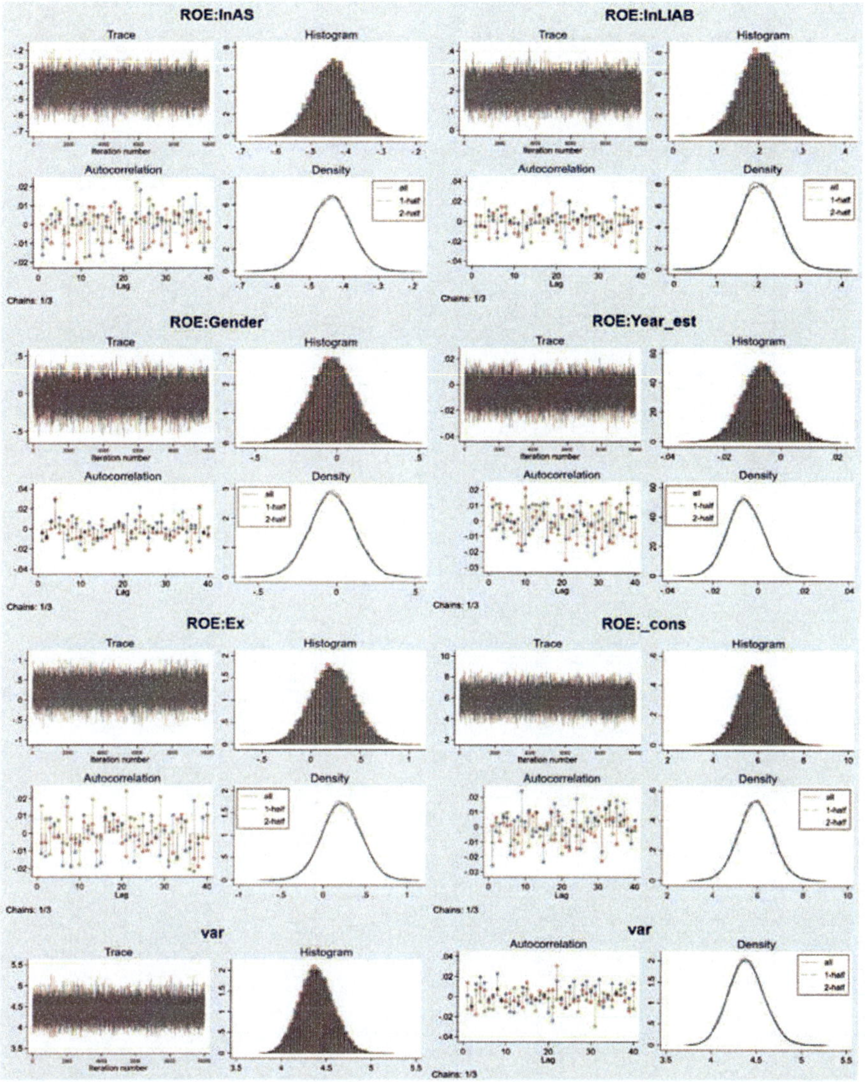

Fig. 2 Visual tests for MCMC convergence of ROE

The finding, nevertheless, is different from some results of previous studies which found an insignificant relationship or a negative relationship between liability and financial performance such as Iorpev and Kwanum (2012) and Nwude et al. (2016). The conflicts in such findings can be reasoned by the distinctive features of contextual elements in the studied settings or the employed methodology.

Regarding the controlled variables studied in this paper, the findings of their impact on firm financial performance can be compared with some previous studies.

Table 4 Sensitivity analysis with respect to prior choice

	ROA					ROE				
	LnLIAB	LnAS	Ex	Year_est	Gender	LnLIAB	LnAS	Ex	Year_est	Gender
Mean										
$\alpha_i \sim$ N(−0.5, 1)	0.0635	−0.1552	0.0848	−0.0036	−0.1150	0.2019	−0.4362	0.2148	−0.0058	−0.0265
$\alpha_i \sim$ N(−0.4, 1)	0.0639	−0.1557	0.0857	−0.0036	−0.1154	0.2019	−0.4363	0.2156	−0.0059	−0.0266
$\alpha_i \sim$ N(−0.3, 1)	0.0641	−0.1565	0.0869	−0.0036	−0.1160	0.2021	−0.4366	0.2179	−0.0058	−0.0278
$\alpha_i \sim$ N(−0.2, 1)	0.0644	−0.1570	0.0879	−0.0036	−0.1162	0.2024	−0.4364	0.2162	−0.0058	−0.0263
$\alpha_i \sim$ N(−0.1, 1)	0.0644	−0.1574	0.0888	−0.0036	−0.1163	0.2025	−0.4376	0.2174	−0.0058	−0.0282
$\alpha_i \sim$ N(0, 1)	0.0649	−0.1582	0.0904	−0.0036	−0.1162	0.2025	−0.4377	0.2202	−0.0059	−0.0255
$\alpha_i \sim$ N(0.1, 1)	0.0651	−0.1587	0.0910	−0.0037	−0.1163	0.2032	−0.4383	0.2203	−0.0058	−0.0272
$\alpha_i \sim$ N(0.2, 1)	0.0651	−0.1590	0.0916	−0.0037	−0.1169	0.2027	−0.4381	0.2194	−0.0059	−0.0277
$\alpha_i \sim$ N(0.3, 1)	0.0653	−0.1596	0.0930	−0.0037	−0.1173	0.2035	−0.4391	0.2194	−0.0059	−0.0281
$\alpha_i \sim$ N(0.4, 1)	0.0656	−0.1601	0.0942	−0.0037	−0.1173	0.2030	−0.4390	0.2213	−0.0058	−0.0285
$\alpha_i \sim$ N(0.5, 1)	0.0658	−0.1607	0.0951	−0.0037	−0.1177	0.2033	−0.4396	0.2236	−0.0058	−0.0291
Std. Dev										
$\alpha_i \sim$ N(−0.5, 1)	0.0181	0.0217	0.0851	0.0028	0.0518	0.0485	0.0590	0.2268	0.0075	0.1376
$\alpha_i \sim$ N(−0.4, 1)	0.0182	0.0218	0.0853	0.0028	0.0519	0.0485	0.0588	0.2252	0.0075	0.1386
$\alpha_i \sim$ N(−0.3, 1)	0.0182	0.0218	0.0850	0.0028	0.0518	0.0487	0.0584	0.2282	0075	0.1397
$\alpha_i \sim$ N(−0.2, 1)	0.0183	0.0219	0.0844	0.0028	0.0520	0.0489	0.0589	0.2270	0.0075	0.1388
$\alpha_i \sim$ N(−0.1, 1)	0.0183	0.0218	0.0852	0.0028	0.0519	0.0487	0.0590	0.2286	0.0075	0.1390
$\alpha_i \sim$ N(0, 1)	0.0183	0.0217	0.0849	0.0028	0.0522	0.0483	0.0582	0.2271	0.0075	0.1389
$\alpha_i \sim$ N(0.1, 1)	0.0182	0.0218	0.0848	0.0028	0.0518	0.0487	0.0586	0.2278	0.0075	0.1386

(continued)

Table 4 (continued)

	ROA					ROE				
	LnLIAB	LnAS	Ex	Year_est	Gender	LnLIAB	LnAS	Ex	Year_est	Gender
$\alpha_i \sim$ N(0.2, 1)	0.0182	0.0218	0.0852	0.0028	0.0520	0.0485	0.0587	0.2263	0.0074	0.1387
$\alpha_i \sim$ N(0.3, 1)	0.0182	0.0217	0.0853	0.0028	0.0520	0.0484	0.0583	0.2270	0.0075	0.1389
$\alpha_i \sim$ N(0.4, 1)	0.0181	0.0216	0.0846	0.0028	0.0521	0.0489	0.0590	0.2274	0.0075	0.1399
$\alpha_i \sim$ N(0.5, 1)	0.0181	0.0218	0.0851	0.0028	0.0518	0.0489	0.0588	0.2272	0.0074	0.1399
MCSE										
$\alpha_i \sim$ N(−0.5, 1)	0.0001	0.0001	0.0005	0.00002	0.0003	0.0003	0.0003	0.0013	0.00004	0.0008
$\alpha_i \sim$ N(−0.4, 1)	0.0001	0.0001	0.0005	0.00002	0.0003	0.0003	0.0003	0.0013	0.00004	0.0008
$\alpha_i \sim$ N(−0.3, 1)	0.0001	0.0001	0.0005	0.00002	0.0003	0.0003	0.0003	0.0013	0.00004	0.0008
$\alpha_i \sim$ N(−0.2, 1)	0.0001	0.0001	0.0005	0.00002	0.0003	0.0003	0.0003	0.0013	0.00004	0.0008
$\alpha_i \sim$ N(−0.1, 1)	0.0001	0.0001	0.0005	0.00002	0.0003	0.0003	0.0003	0.0013	0.00004	0.0008
$\alpha_i \sim$ N(0, 1)	0.0001	0.0001	0.0005	0.00002	0.0003	0.0003	0.0003	0.0013	0.00004	0.0008
$\alpha_i \sim$ N(0.1, 1)	0.0001	0.0001	0.0005	0.00002	0.0003	0.0003	0.0003	0.0013	0.00004	0.0008
$\alpha_i \sim$ N(0.2, 1)	0.0001	0.0001	0.0005	0.00002	0.0003	0.0003	0.0003	0.0013	0.00004	0.0008
$\alpha_i \sim$ N(0.3, 1)	0.0001	0.0001	0.0005	0.00002	0.0003	0.0003	0.0003	0.0013	0.00004	0.0008
$\alpha_i \sim$ N(0.4, 1)	0.0001	0.0001	0.0005	0.00002	0.0003	0.0003	0.0003	0.0013	0.00004	0.0008
$\alpha_i \sim$ N(0.5, 1)	0.0001	0.0001	0.0005	0.00002	0.0003	0.0003	0.0003	0.0013	0.00004	0.0008
[95% Cred. Interval]										
$\alpha_i \sim$ N(−0.5, 1)	0.0283 0.0992	−0.1979 −0.1126	−0.0821 0.2507	−0.0092 0.0018	−0.2163 −0.0128	0.1074 0.2969	−0.5522 −0.3210	−0.2280 0.6539	−0.0207 0.0089	−0.2950 0.2447

(continued)

Table 4 (continued)

	ROA					ROE				
	LnLIAB	LnAS	Ex	Year_est	Gender	LnLIAB	LnAS	Ex	Year_est	Gender
$\alpha_i \sim$ N(−0.4, 1)	0.0282 0.0992	−0.1982 −0.1125	−0.0804 0.2513	−0.0092 0.0018	−0.2163 −0.0134	0.1068 0.2975	−0.5503 −0.3209	−0.2233 0.6544	−0.0208 0.0087	−0.2992 0.2453
$\alpha_i \sim$ N(−0.3, 1)	0.0284 0.1001	−0.1989 −0.1136	−0.0802 0.2532	−0.0092 0.0018	−0.2179 −0.0147	0.1066 0.2973	−0.5520 −0.3223	−0.2350 0.6681	−0.0205 0.0089	−0.3015 0.2469
$\alpha_i \sim$ N(−0.2, 1)	0.0282 0.1003	−0.1995 −0.1137	−0.0788 0.2530	−0.0092 0.0018	−0.2181 −0.0143	0.1065 0.2988	−0.5513 −0.3204	−0.2314 0.6576	−0.0205 0.0089	−0.2974 0.2454
$\alpha_i \sim$ N(−0.1, 1)	0.0286 0.1000	−0.2004 −0.1148	−0.0775 0.2550	−0.0091 0.0018	−0.2177 −0.0145	0.1068 0.2974	−0.5533 −0.3212	−0.2277 0.6638	−0.0208 0.0088	−0.3051 0.2436
$\alpha_i \sim$ N(0, 1)	0.0291 0.1012	−0.2011 −0.1157	−0.0746 0.2581	−0.0091 0.0018	−0.2196 −0.0142	0.1081 0.2970	−0.5531 −0.3230	−0.2259 0.6639	−0.0207 0.0087	−0.2979 0.2459
$\alpha_i \sim$ N(0.1, 1)	0.0291 0.1012	−0.2017 −0.1155	−0.0752 0.2561	−0.0092 0.0018	−0.2173 −0.0143	0.1071 0.2981	−0.5529 −0.3236	−0.2271 0.6661	−0.0207 0.0090	−0.2982 0.2484
$\alpha_i \sim$ N(0.2, 1)	0.0292 0.1007	−0.2019 −0.1167	−0.0752 0.2581	−0.0092 0.0019	−0.2192 −0.0155	0.1089 0.2984	−0.5535 −0.3238	−0.2278 0.6593	−0.0205 0.0089	−0.2992 0.2437
$\alpha_i \sim$ N(0.3, 1)	0.0289 0.1011	−0.2019 −0.1165	−0.0750 0.2619	−0.0092 0.0018	−0.2190 −0.0148	0.1096 0.2977	−0.5531 −0.3234	−0.2222 0.6660	−0.0206 0.0087	−0.3031 0.2431
$\alpha_i \sim$ N(0.4, 1)	0.0304 0.1016	−0.2027 −0.1178	−0.0712 0.2595	−0.0093 0.0018	−0.2192 −0.0153	0.1066 0.2986	−0.5547 −0.3232	−0.2249 0.6720	−0.0207 0.0090	−0.3065 0.2430
$\alpha_i \sim$ N(0.5, 1)	0.0303 0.1013	−0.2033 −0.1179	−0.0726 0.2617	−0.0093 0.0017	−0.2188 −0.0161	0.1078 0.3003	−0.5550 −0.3242	−0.2173 0.6708	−0.0205 0.0087	−0.3066 0.2451

Source Authors' calculation

This study found a negative effect of total assets (firm size) on financial performance, which is different from previous studies' results such as Papadognas and Vijayakumar and Tamizhselvan (2010), but in line with the study of Ammar et al. (2003). This type of relationship can also be explained by firm theories (Jónsson 2007) such as Institutional theory. The pressure from the institutional environment of organizations force them to grow larger while their management capacity is not yet ready, which may make the firms encounter various issues which result in reduced financial performance.

Besides, this research found a negative impact of manager gender as male on SMEs' financial performance, which means the more number of male managers a firm has, the lower financial performance would be generated. This may be reasoned that more gender diversity would be more helpful to increase financial performance of firms (Satria et al. 2020). It is recommended that future studies, both theoretically and empirically, should explore this relationship as a research gap.

Besides, years of operation since the establishment (firm age) had a negative influence on financial performance, which is similar to the finding of Dogan (2013), but different from the findings of Gurbuz et al. (2010) and Basti et al. (2011). As argued by Akben-Selcuk(2016), the conflicting results regarding firm age-performance relationship are related to the studied contexts such as countries, time periods as well as the used research methodologies. The negative impact of firm age on financial performance in this study may be caused by different factors such as market competition, mark trends and firm size.

Next, export experience was found to have a positive impact on financial performance, which is similar to the finding of Dhliwayo (2016). This can be explained that export experience would enable SMEs to increase foreign sales, diversify revenue flows and consequently achieve better financial performance.

It is highly recommended that future research should justify these findings in different contexts such as other countries or specific industries by using other kinds of methodology. A combination of qualitative and quantitative research methods would be helpful to investigate the research topic in more depth.

10 Conclusion

To recapitulate, the research results demonstrate that the liability has positive impacts on firm financial performance. This study also found a negative effect of total assets (firm size), manager gender as male and firm age on financial performance, and a positive impact of export experience on financial performance. This research has both of theoretical and practical contributions. Theoretically, this study can contribute to the literature development regarding the examination of the role of liability in improving financial performance of firms. Practically, this paper can help business leaders of not only SMEs but other type of firms to pay more attention to the significance of using liability as well as other influential factors such as firm size, firm age and export activity. The research findings would be meaningful to the various stakeholders such as financial institutions, SMEs' managers and owners, the government, welfare organization, general public, researchers and academicians.

In terms of managerial implications, it is suggested that practitioners in SMEs should consider employing liability as a financing method and also consider participating in export activities. Financial analysis is certainly needed to help business leaders evaluate the overall financial situations in order to make timely and appropriate decisions. SMEs can cooperate with external financial advisory firms to obtain useful counselling services, which would assist them in their strategic planning. Last but not least, more gender diversity might be helpful to create friendly and cooperative working environments, which would subsequently foster financial performance.

This research still involves some limitations. Firstly, the study did not focus on specific types of liabilities such as short term or long term liabilities. Secondly, this research concentrated on SMEs in general without studying a specific area or industry. Thirdly, some other factors which might be related to financial performance such as organisational culture, organisational structure and number of workforce have not yet been considered. It is recommended that future research should take into accounts these limitations as potential research gaps and investigate the issue with other alternative of methodologies.

References

Akben-Selcuk, E.: Does Firm Age Affect Profitability? Evidence from Turkey. International Journal of Economic Sciences **3**, 1–9 (2016)

Ammar, A., et al.: Indicator Variables Model of Firm's Size-Profitability Relationship of Electrical Contractors Using Financial and Economic Data. J. Constr. Eng. Manag. **129**, 192–197 (2003)

Basti, E., Bayyurt, N., Akın, A.: A Comparative Performance Analysis of Foreign and Domestic Manufacturing Companies in Turkey. European Journal of Economic and Political Science **4**(2), 125–137 (2011)

Briggs, W.M., Nguyen, T.H.: Clarifying ASA's view on P-values in hypothesis testing. Asian Journal of Economics and Banking **3**(2), 1–16 (2019)

Buferna, F. M., Bangassa, K., & Hodgkinson, L. (2005). Determinants of Capital Structure: Evidence from Libya. University of Liverpool, 8.

Da Silva, T.P., Leite, M., Guse, J.C., Gollo, V.: Financial and Economic Performance of Major Brazilian Credit Cooperatives. Contaduria Administracion **62**, 1442–1459 (2017)

Dawar, V.: Agency Theory, Capital Structure and Firm Performance: Some Indian Evidence. Manag. Financ. **40**(12), 1190–1206 (2014)

Dhliwayo, S. (2016). Export Experience and Financial Performance of Small and Medium Enterprises. Environmental Economics (open-access). 7(3).

Dogan, M.: Does Firm Size Affect the Firm Profitability? Evidence from Turkey. Research Journal of Finance and Accounting **4**(4), 53–59 (2013)

Esteve-Perez, S., Requena-Silvente, R., Pallardo-Lopez, V.J.: The Firm Duration of Firm- destination Export Relationships: Evidence from Spain, 1997–2006. Econ. Inq. **51**(1), 159–180 (2013)

Gurbuz, A.O., Aybars, A., Kutlu, O.: Corporate Governance and Financial Performance with a Perspective on Institutional Ownership: Empirical Evidence from Turkey. J. Appl. Manag. Account. Res. **8**(2), 21–37 (2010)

Iorpev, L., Kwanum, I.M.: Capital Structure and Firm Performance: Evidence from Manufacturing Companies in Nigeria. International Journal of Business and Management Tomorrow **2**(5), 1–17 (2012)

ISA 32 (2004). IAS 32: Financial Instruments: Disclosure and Presentation (1st ed). International Accounting Standards.

Jensen, M.C., Meckling, W.H.: Theory of the Firm: Managerial Behaviour, Agency Costs and Ownership Structure. J. Financ. Econ. **3**, 305–360 (1976)

Jónsson, B.: Does the Size Matter? The Relationship between Size and Profitability of Icelandic Firms. Bifröst Journal of Social Science **1**, 43–55 (2007)

Kajananthan, R.: Effect of Corporate Governance on Capital Structure: Case of the Srilankan Listed Manufacturing Companies. Res. World **3**(4), 63–71 (2012)

Kajirwa, H.I.: Effects of Debt on Firm Performance: A Survey of Commercial Banks Listed on Nairobi Securities Exchange. Global Journal of Advanced Research **2**(6), 1025–1029 (2015)

Kreinovich, V., Nguyen, N.T., Nguyen, D.T., Dang, V.T. (eds.): Beyond Traditional Probabilistic Methods in Economics. Springer, Cham (2019)

Leland, H.E.: Agency Costs, Risk Management, and Capital Structure. J. Financ. **53**(4), 1213–1243 (1998)

Mamaro, L. P., & Legotlo, T. (2020). The Impact of Dept Financing on Financial Performance: Evidence from Retail Firms Listed on JSE. Journal of Accounting and Management, 10(3).

Margaritis, D., Psillaki, M.: Capital Structure, Equity Ownership and Firm Performance. J. Bank. Finance **34**(3), 621–632 (2010)

Menike, M.G.P., Prabath, U.S.: The Impact of Accounting Variables on Stock Price: Evidence from the Colombo Stock Exchange. Sri Lanka: International Journal of Business and Management **9**(5), 210–221 (2014)

Mikic, M., Novoselec, T., Primorac, D.: Influence of Financing Source on the Small Business Performance. International Journal of Economic Perspectives **10**(2), 62–72 (2016)

Muchiri, M.J., Muturi, W.M., Ngumi, P.M.: Relationship between Financial Structure and Financial Performance of Firms Listed at East Africa Securities Exchanges. Journal of Emerging Issues in Economics, Finance and Banking **7**(5), 1734–1755 (2016)

Murigi, M. (2014). The Effect of Financial Access on the Financial Performance of Small and Micro Enterprises in Mukuru Slums [Master's thesis, University of Nairobi]. http://erepository.uonbi.ac.ke/bitstream/handle/11295/76221/Murigi_The%20effect%20of%20financial%20access%20on%20the%20financial%20performance%20of%20small%20and%20micro%20enterprises.pdf?sequence=3

Muriithi, S.N. (2014). The Effect of Financing Sources on the Financial Performance of Top 100 Mid-sized Companies in Kenya [Master's thesis, University of Nairobi]. http://erepository.uonbi.ac.ke/bitstream/handle/11295/78386/Muriithi_The%20Effect%20Of%20Financing%20Sources%20On%20The%20Financial%20Performance%20Of%20Top%20100%20Mid-Sized%20Companies%20In%20Kenya.pdf?sequence=3&isAllowed=y

Nguyen, H.T., Nguyen, D.T., & Nguyen, N.T.: Beyond Traditional Probabilistic Methods in Econometrics. In Beyond Traditional Probabilistic Methods in Economics, ECONVN 2019, Studies in Computational Intelligence. Edited by Vladik Kreinovich, Nguyen Ngoc Thach, Nguyen Duc Trung and Dang Van Thanh. Cham: Springer, **809** (2019)

Nimalathasan, B.: Working Capital Management and Its Impact on Profitability: A Study of Selected Listed Manufacturing Companies in Sri Lanka. Manager **12**, 76–82 (2010)

Nwude, E.C., Okpara, I.I., Agbadua, B.O., & Udeh, S.N.: The impact of debt structure on firm performance: Empirical evidence from Nigerian quoted firms. Asian Economic and Financial Review **6**(11), 647–660 (2016)

Ogijiuba, K., Oluche, F., & Adenuga, A. (2004). Credit Availability to Small and Medium Scale Enterprises in Nigeria: Importance of New Capital Base for BanksBackground Issues. Working Paper Department of International Economics Relations, Central Bank of Nigeria.

Pro-active Accounting Activities in Europe (2008). Distinguishing between Liabilities and Equity. Pro-active Accounting Activities in Europe (PAAinE). European Financial Reporting Advisory Group.
Rashid, A.: Revisiting Agency Theory: Evidence of Board Independence and Agency Cost from Bangladesh. J. Bus. Ethics **130**(1), 181–198 (2015)
Sadeghian, N.S., Latifi, M.M., Soroush, S., Aghabagher, Z.T.: Debt Policy and Corporate Performance: Empirical Evidence from Tehran Stock Exchange Companies. Int. J. Econ. Financ. **4**(11), 217–224 (2012)
Salloum, C., Mercier- Suissa, C. & Khalil, S. (2016). The Rise of Women and Their Impact on Firms' Performance. International Journal of Entrepreneurship and Small Business, 27(2/3).
Satria, Y., Mahadwartha, P.A. & Ernawat, E. (2020). The Effect of Gender Diversity on Company Financial Performance. 17th International Symposium on Management, 115.
Shikumo, D.H., Oluoch, O., Wepukhulu, M.J.: Effect of Short-Term Debt on Financial Growth of Non-Financial Firms Listed at Nairobi Securities Exchange. Research Journal of Finance and Accounting **11**(20), 1697–2847 (2020)
Sriboonchitta, Songsak, Hung T. Nguyen, Olga Kosheleva, Vladik Kreinovich, and Thach Ngoc Nguyen.: Quantum Approach Explains the Need for Expert Knowledge: On the Example of Econometrics. In Structural Changes and Their Econometric Modeling, TES 2019, Studies in Computational Intelligence. Edited by Vladik Kreinovich, Songsak Sriboonchitta. Cham: Springer **808** (2019)
Stokes, D., & Wilson, N. (2010). Small Business Management & Entrepreneurship. Cengage Lmg Business Press.
Svítek, Miroslav, Olga Kosheleva, Vladik Kreinovich, and Thach Ngoc Nguyen.: Why Quantum (Wave Probability) Models Are a Good Description of Many Non-quantum Complex Systems, and How to Go Beyond Quantum Models. In Beyond Traditional Probabilistic Methods in Economics, ECONVN 2019, Studies in Computational Intelligence. Edited by Kreinovich V., N. Thach, N. Trung and Van Thanh D. Cham: Springer **809** (2019)
Velnampy, T., Nimalathasan, B.: Firm Size on Profitability: A Comparative Study of Bank of Ceylon and Commercial Bank of Ceylon Ltd in Srilanka. Global Journal of Management and Business Research **10**(2), 96–100 (2010)
Vijayakumar, A. & Tamizhselvan, P. (2010). Corporate Size and Profitability-An Empirical Analysis. College Sadhana Journal for Bloomers of Research, 3(1), 44–53.
Vintila, G., Nenu, A.: An Analysis of Determinants of Corporate Financial Performance: Evidence from the Bucharest Stock Exchange Listed Companies. Int. J. Econ. Financ. Issues **5**(3), 732–739 (2015)

A Bayesian Analysis of Tourism on Shadow Economy in ASEAN Countries

Duong Tien Ha My, Le Cat Vi, Nguyen Ngoc Thach, and Nguyen Van Diep

Abstract Tourism contributes to boost a country's economic growth, improve foreign currency supply, and engender employment and business opportunities in many ASEAN countries. Previous studies suggest that tourism can be associated with the shadow economy. Hence, the current paper examines the impact of tourism on the informal economy in the ASEAN countries from 1999 to 2017. To accomplish the research objective, we utilize the Bayesian linear regression technique in order to analyze collected data. Specifically, we apply a normal prior distribution, while the posterior distribution is generated using the Markov Chain Monte Carlo (MCMC) method via the Gibbs sampler algorithm. The main finding of the paper is that tourism represented by tourism expenditures, tourism receipts, and tourist arrivals positively influence the shadow economy size. Moreover, the posterior probability of tourism receipts and tourist arrivals is higher than that of tourism expenditures. Other factors that also promote the shadow economy are the government size and tax burden. Indeed, we find extreme evidence for the impact of these factors on informality in all models. Conversely, GDP per capita, trade openness, rule of law, and corruption control adversely affect the shadow economy in the ASEAN countries between 1999–2017.

Keywords ASEAN · Tourism · Shadow economy · Bayesian approach

D. T. H. My · N. Van Diep (✉)
Ho Chi Minh City Open University, 35-37 Ho Hao Hon, Co Giang Ward, District 1, Ho Chi Minh City, Vietnam
e-mail: diep.nv@ou.edu.vn

L. C. Vi
University of Economics and Law, Ho Chi Minh City, Vietnam

Vietnam National University, Ho Chi Minh City, Vietnam

N. N. Thach
Banking University of Ho Chi Minh City, Ho Chi Minh City, Vietnam

N. Ngoc Thach et al. (eds.), *Financial Econometrics: Bayesian Analysis, Quantum Uncertainty, and Related Topics*, Studies in Systems, Decision and Control 427,
https://doi.org/10.1007/978-3-030-98689-6_27

1 Introduction

Tourism is an important driving force for promoting economic growth, increasing wealth, providing foreign currency, and creating livelihood and business opportunities for many countries. The critical role of tourism is limited to businesses that directly contribute accommodation, transportation or hospitality, entertainment sector, etc., to tourists and closely connects to other tourism-supporting industries. In practice, the United Nations Environment Programme and World Tourism Organization (UNEP and UNWTO 2005) and United Nations Conference on Trade and Development (UNCTAD 2013) show that tourism is linked to numerous sectors, including but not limited to taxes and charges; land and infrastructure provision and management; security and emergency services; and transportation, hospitality, telecommunications, financial, health, retail, recreational, and other services.

The World Travel and Tourism Council (WTTC 2020) indicates that the direct and indirect impacts of tourism generated US$8.9 trillion in 2019, representing about 10.3% of global GDP. During the same period, this sector helped create 330 million jobs, accounting for 1/10 jobs worldwide. Before the pandemic, tourism establishes as one of the fastest-growing economic industries in the globe. Based on the World Tourism Organization (UNWTO 2021), real growth in international tourism revenue reached 54%, higher than the global GDP growth (44%) between 2009–2019. Furthermore, tourism is an essential source of foreign currency, decreasing trade deficits for various countries. Indeed, the total export earnings from international tourism were more than 1.7 trillion USD in 2019 (UNWTO 2021).

As shown above, tourism has complex and extensive relationships with other sectors, in which service provision considerably prevails. Therefore, previous studies suggest that tourism can be related to the shadow economy (Schneider 2013; Xu and Lv 2021). Notably, Schneider (2013) examines data of some European Union countries and reveals that hotels and restaurants account for the highest share of the shadow economy (nearly 20% of the total GDP produced in this sector). While tourism involves more than just hotel and restaurant activities, this figure indicates a potentially significant effect of the shadow economy. Interestingly, Xu and Lv (2021) claim that tourism is always negatively related to the informal economy. There is exists a threshold level where tourism can have a more effect on informality. The impact of tourism on the shadow economy is still controversial, which requires further studies on this relationship.

The present study, employing a Bayesian approach, aims to investigate the influence of tourism on the shadow economy size in Association of Southeast Asian Nations (ASEAN) between 1999–2017. ASEAN musters ten countries, namely Brunei Darussalam, Cambodia, Indonesia, Laos, Malaysia, Myanmar, Philippines, Singapore, Thailand, and Vietnam. Both tourism and the shadow economy are crucial economic phenomena of many ASEAN countries. On the one side, according to UNWTO (2021), Thailand is one of the world's top tourism earners in 2019. Philippines, Indonesia, Thailand, and Vietnam are among the countries with the highest number of jobs created by travel and tourism over 2014–2019. Moreover, the growth rates of travel and tourism in countries such as the Philippines (8.6%), Vietnam (7.7%), and Malaysia (6.6%) in 2019 are very high (WTTC 2020). On the other side, the shadow economy size in this region ranges from 11.46 to 48.89% between 1999–2017 (Medina and Schneider 2019). It can harm the region's economic growth (Nguyen and Luong 2020). Therefore, it is essential to uncover the relationship between tourism and informality, whereby countries can apply appropriate measures to foster tourism and better control informal activities.

The rest of the paper proceeds as follows. In the next section, we present the conceptual framework and previous studies. Section 3 explains research models, data sources and description, and econometric methodology. Section 4 highlights the main findings and discussion. Finally, Sect. 5 provides the conclusion.

2 Theoretical and Empirical Background

2.1 Conceptual Framework of the Shadow Economy

Tan, Habibullah, and Yiew (2016) reveal that every country exists a dual economy: an official economy and a shadow sector. Fleming, Roman, and Farrell (2000) argue that when attempting to find about the world economy size, we mostly depend on official statistics such as trade, profits and investment to measure basic economic indicators, but skip to measure the shadow economy size because of the complex methodology as well as the lack of statistical data. According to Bajada (1999) and Caridi and Passerini (2001), there is no approval about the definition of informality due to the various economic activities of the shadow economy. Therefore, the shadow economy's definitions and concepts are determined by the chosen estimation methodology and measurements (Schneider and Enste 2013).

Katsios (2019), Kesar and Čuić (2017) suggest that there are many terms such as unreported, informal, shadow, black, unrecorded, parallel, hidden, clandestine, underground, subterranean, grey or offshore economy when seeking about economic activities which fall outside the National Accounting Systems' scope. Eilat and Zinnes (2000) define that shadow activity is the trend of individuals and firms without the government interference or collaboration to effectuate their economic activities. These involve activities that avoid taxes, disregard the currency requirements, shunt regulatory requirements, disappear in reporting mechanism, and are essentially illegal.

2.2 Previous Empirical Studies and Research Hypotheses

In general, to understand the complexity of economic as well as social impacts, it is essential to compass the reasons causing the advent of the shadow economy. There are two main reasons for the shadow economy's driving forces: (1) the principal causes are the increasing social security contributions and burden of taxation and (2) the increasing of regulatory discretion, bureaucracy, corruption, and a weak rule of law leading to a decline in the institutional quality (Dreher et al. 2009). Based on Schneider, Buehn, and Montenegro (2010), there are four major driving forces comprising: (1) social security contribution and tax burdens, (2) enhanced regulations' authorities, (3) debasement of the public goods and services' quality, and (4) the official economy's state. Specifically, individuals have the motivation to join the shadow economy if: they have chances to save more money when joining in the shadow economy; the community agrees to work outside of the official economy; the probability of being caught for various form of shadow activity is low, in addition, goods or services' payments are transacted by cash (especially in tourism-related businesses). Most empirical studies reveal that the rising tax burden and social security contributions are the strongest prompts of the shadow economy's broadening.

In addition to the above factors, Ott (2002) also demonstrates some other factors: the lack of tax payment, discretionary rights of public servants' experience, absence of democratic, economic, regulative institutions, and government profligacy. Besides the internal social and economic factors affecting the shadow economy's size, these issues are increased with the eruption of international free trade as well as economic integrations around the globe, which leads to the appearances of international harmonization and the call for strategic worldwide law enforcement policies to combat them (Katsios 2006).

Despite the dimension of the global tourism economy, with regards to tourist expenditure, arrivals, exports, and employment, many countries depend heavily on tourism as a spring of benefits. Surprisingly, so far, there has been very little attention to understand deeper or analyze and measure the shadow economy in tourism across countries. Based on Kesar and Čuić (2017), the tourism-related shadow economy has concerned such rare scientific attention because of three reasons: firstly, tourism business structure is too fragmented and complex to expand the research scope; secondly, the existence of spatial distribution of tourism activities is highly hard to operationalize study; finally, the topic is a heightened sensitivity phenomenon of society, especially in underdeveloped areas.

Despite the plenty of empirical evidence which demonstrates that the shadow economy in tourism is popular and manifest through numerous economic activities as well as different types of tourist destinations, most of empirical studies are converged around two aspects: 1) identify the tourism economic activities influenced by the underground economy, 2) measure the unregistered workforce's size in tourism.

The very early research by Wahnschafft (1982) reveals the separation of formal, informal economies and their impacts on tourism in Thailand. In 2007, Skalpe (2007) identified more studies about the national shadow economy than industry sectors two decades later. Schneider (2002) demonstrates that a macro-economic framework can be transferred to the industrial sector. In that way, Skalpe (2007) points out that in Norway, the hostess must handle cash and be in charge of the cash registers to keep additional revenue from celebrating special deals away from official tax or maintain the illegal alcoholic drinks incomes.

In Denmark, Hjalager (2008) evaluates the economic significance and impacts of tax fiddles, illegal/undeclared employment, and illegal supplies report in the restaurant sector. Slocum et al. (2011) conducted a study on the shadow economy's aspects in tourism of Tanzanian, which is investigated to reduce poverty and leads to lots of net advantages to the poorest and the most vulnerable resident groups. The underground economy in tourism in small and medium enterprises (SMTE) of Bulgaria shows that the spreading up of the shadow economy nearly works against the adoption of e-business by small tourism enterprises (Vladimirov 2015). Badariah et al. (2015) and Pavicevic (2014) argue that the tourism development in Montenegro is impeded by the impact of uncontrolled activities in the informal economy. For instance, there is a rapid growth in the number of uncontrolled private accommodation facilities in the Montenegro coastal region; hence the tourism destination's quality is compromised, and the development of basic tourism industries is destroyed. The significant issues of the shadow economy generated by tourism-related economic activities are not announced seasonal accommodation capacities and related revenue, non-invoicing, no cash registers in food and beverage facilities, unregistered workers, or deficiency in tourist taxes collection.

Based on Zhang (2017), in both developed and developing economies, the tourism-related shadow economy is a growing phenomenon, emphasizing informal employment as its central pillar due to the lack of formal jobs for all job seekers. Among all these recent studies on the tourism-related shadow economy, the research gap in this field is the lack of empirical evidence, literature, and data. Therefore, it needs to be refilled these gaps with new insights and considerations. In the financial market, tourism (restaurants, tour operators, travel agencies, hospitality firms, and transport carriers) and tourists perform two basic types of transactions: transactions between tourism firms and tourists and tourism firms. Those tourisms transactions are specific types of export as part of tourism receipts.

Moreover, there is a higher illegal transaction than in the first type, which probably occurs in the communication line between SMTE. Hence, a part of the unlawful transaction is an impact from SMTE and tourist exchange line. Šergo et al. (2017) show that the connection between tourism receipts and the shadow economy is neither absolutely negative nor undoubtedly clear under the transaction's presence among big hospitality enterprises. However, on the total level of receipts, when we compound SMTEs transactions, it somehow declines the regulatory vision of the economy; therefore, concentration in tourism promotes the shadow economy. Additionally, the shadow economy may decrease the quality of tourism services and the debasement of the overall destination images (Badariah et al. 2015; Pavicevic 2014). Despite the lack of data regarding the consequence of the shadow economy on the development of the tourism industry, Din et al. (2016) indicate a positive relationship between tourism receipts and some governance effectiveness indicators as corruption control and regulatory quality. Therefore, it may be evaluated that better enforcement of regulations decreases the informal economy size. In the long run, there is a negative relationship between tourism receipts and the shadow economy size. Based on the above arguments, the following hypothesis is:

H1: Tourism receipts have an adverse impact on shadow economy size.

Most studies about tourism demand apply the demand theory to explain or interpret the individual differences in tourism participation behavior. Based on the demand theory, the individuals' income affects their travel decisions, which is a limited budget and depends on the spending capacity and utility (Crawford et al. 1991). Most empirical studies reveal that tourism expenditure positively affects travel decisions (Alegre et al. 2009; Eugenio-Martin and Campos-Soria 2011). In general, according to Alegre et al. (2009), the income elasticity is under the unit value and has greater value than the decision to travel abroad (Bernini and Cracolici, 2016). Tourism indeed is one of the essential industries in many economies by foreign currency earnings, creating tax revenues, and employment opportunities for the target country (Alam and Paramati 2016). Based on the above arguments, the following hypothesis is:

H2: Tourism expenditure has an adverse impact on shadow economy size.

Slocum et al. (2011) agree that the shadow economy is considered a way to lessen poverty and provide net gains to the most vulnerable individual—the poorest. As cited by Pavicevic (2014), the development of the informal economy is more probably to work against the e-business adoption of small tourism enterprises in Bulgaria. According to Pavicevic (2014) and Badariah et al. (2015), the spread up of tourism in Montenegrin is halt as a consequence of the unrestrained activities that include the uncontrolled development in the quantity of unregistered private accommodation facilities in coastal region which affects the tourism destinations' quality and demolish the evolution of primary tourism industries. Similarly, Milić (2014) claims that the most extensive problems of the underground economy in tourism activities are unregistered seasonal accommodation capacities or income, non-invoicing, absence of cash registers in restaurants, and failure in tourist taxes collection, unregistered workers, etc.

The linkage between tourist arrivals and the shadow economy size in Thailand is dissociated using wavelet coherence and causality by Sinha, Sengupta, and Mehta (2021). Two essential findings of Sinha et al. (2021) are: (a) in the short run, there is a significant relationship between tourist arrivals and the spread up of the shadow economy, and (b) in the long run, there is bidirectional causality between tourist arrivals and the spread up of the shadow economy across the frequency levels. The results show that tourist arrival causes the development of the underground economy in the short and medium run. Smith (2011) states that there might be a two-way linkage between the underground economy and tourism, as tourism enhances the growth of the shadow economy in the presence of poor legal enforcement and high corruption. The tourist arrivals may have a bidirectional causal association with the shadow economy. For example, a two-way coherence exists between tourist arrivals and organized crime. Organized crime is catalyzed by tourism, which creates opportunities to invest black money in tourism activities and facilities, such as prostitution and gambling (Mekinc et al. 2013). Obviously, the safety conditions of the destination countries are the prior concern of tourists, and therefore, the safer countries will attract a greater number of tourists (Altindag 2014). Furthermore, organized crime is well connected to the shadow economy because organized crime groups manage the central part of the shadow economy (Schneider and Enste 2000). It increases tax evasion activities leading to the loss of revenue for the government, which could have been invested in the growth and development of the tourism sector (Webb et al. 2013). Based on the above arguments, the following hypothesis is:

H3: Tourist arrivals have an adverse impact on shadow economy size.

In summary, previous studies indicate that the connection between tourism and the shadow economy is mixed and inconclusive. Some authors suppose that tourism is adversely related to the informal sector. Specifically, Xu and Lv (2021) claim that tourism always has a detrimental effect on the shadow economy size. The impact increases if the tourism development level is higher than a threshold value. By contrast, some authors argue that this linkage can be positive under certain circumstances. For instance, Jaliu and Răvar (2019) suggest that tourism in Romania contributes considerably to the informal economy.

Meanwhile, Lv (2020) uncovers a U-shaped connection between tourism and informality. The author states that tourism negatively affects informality when the tourism development level is low, while positively affects informality when this level is high. In such a context, we supplement the literature by exploring the influence of tourism on the underground economy using data from the ASEAN countries. The methodology and main empirical findings are present in the following sections.

3 Methodology

3.1 Model and Data

To examine the effect of the tourism on the shadow economy size of ASEAN countries, we estimate the following econometric model:

$$SSE_i = \alpha + \beta Tourism_i + \lambda^T CV_i + \varepsilon_i \quad (1)$$

where SSE is the size of the shadow economy, Tourism is the primary explanatory variable representing tourism development, CV is a set of control variables suggested by the empirical literature (i.e., GDP per capita, trade openness, tax burden, government size, and two institutional-quality variables, namely the rule of law and control of corruption indices), and ε represents error term.

The data cover the ten ASEAN member countries from 1999 to 2017. Using available data, we collect SSE data based on data generated by Medina and

Schneider (2019). They used the multiple indicator-multiple cause (MIMIC) estimation approaches to calculating the shadow economy size as a percentage of GDP for 157 economies over the period 1991–2017.

In this paper, we use the international tourism expenditures (TEXPE), international tourism receipts (TREC), and the total number of international tourist arrivals (TARR) to represent tourism development. Specifically, (Laimer et al. 2006, p. 8) argue that tourism expenditures are "expenditure of outbound visitors in other countries including their payments to foreign carriers for international transport." Tourism receipts are defined as "expenditure of international inbound visitors including their payments to national carriers for international transport. They should also include any other prepayments made for goods/services received in the destination country" (Laimer et al. 2006, p. 8). And finally, foreign inbound tourists (overnight visitors) are the number of tourists visiting a country other than the country where they have their usual place of residence. Data on tourism development comes from the World Development Indicators (WDI), which the World Bank produces. We use the natural log of international tourism expenditures, international tourism receipts, and the total number of international tourist arrivals.

In addition, we also have incorporated some control variables in Eq. (1) according to the empirical literature. We use the natural log of GDP per capita and extract the data from the WDI of the World Bank. Similarly, we obtain the trade openness and government size data from WDI. Meanwhile, the tax burden data are derived from the Heritage Foundation. We use the Global Governance Indicators (WGI)'s control of corruption index provided by the World Bank to measure corruption. Likewise, the index of the rule of law is obtained from WGI, as well. These two variables aim to measure the institutional quality of a country. Larger values of these indicators indicate that the government has better institutional quality. The variable descriptions are presented in Table 2.

Table 3 describes the average size of ASEAN member countries' shadow economy and tourism status throughout the study. We can see in this table that Singapore's average shadow economy size is the lowest (around 11% of GDP). In contrast, Cambodia and Thailand are the two economies with the largest average shadow economies in ASEAN (about 49% of GDP). Overall, the shadow economy size of ASEAN countries averaged 32.38% GDP. This result is concordant with the comments of Vorisek et al. (2021), who argue that shadow economic activity is common in developing countries and emerging markets, accounting for an average of more than 30% of GDP. In terms of international tourism expenditures, Myanmar is the lowest spender with approximately US$90 million, and Singapore is the top spender of the ten ASEAN countries with a spending of US$14,010 million. Thailand is the country with the highest international tourism receipts (US$25,490 million), while Brunei Darussala is the lowest with about US$165 million. Finally, the number of tourists to Malaysia is the highest in ASEAN with 18,214,120 arrivals, and Brunei Darussala has the lowest number of international tourists (210,188 arrivals).

Table 1 The types of shadow economic activities

<table>
<tr><th>Type of activity</th><th colspan="2">Monetary transaction</th><th colspan="2">Non-monetary transaction</th></tr>
<tr><td>Unlawful activities</td><td colspan="2">– Trade in stolen-goods
– Drug manufacturing and dealing
– Prostitution
– Smuggling
– Gambling
– Fraud
– etc</td><td colspan="2">– Bartering of drugs, stolen goods, smuggling, etc
– Producing or growing drugs for personal use
– Theft for own use</td></tr>
<tr><td>Lawful activities</td><td>Tax evasion</td><td>Tax avoidance</td><td>Tax evasion</td><td>Tax avoidance</td></tr>
<tr><td></td><td>– Unpublicized income from self-employment
– Salaries, wages, and assets from unpublicized work related to legal services and goods</td><td>– Employee discounts, fringe benefits</td><td>– Exchange of legal services and goods</td><td>– All do-it-yourself work and neighbor help</td></tr>
</table>

Source Schneider and Enste (2000)

3.2 *Econometric Methodology*

Previous studies used a frequency method to analyze the influence of tourism on the shadow economy size (Lv 2020; Sinha et al. 2021; Xu and Lv 2021). However, the interpretation of the results by the frequency school is no longer reliable in many cases (Amrhein et al. 2019; Briggs 2019; Fricker et al. 2019; Nguyen 2016; Nguyen and Thach 2019; Nguyen et al. 2019; Trafimow 2019; Wasserstein et al. 2019). Therefore, many researchers believe that it is necessary to change from dependent on frequency to other methods with higher reliability. Recently, some statistical scientists have advocated the Bayesian school as an alternative to tests of statistical significance (Gelman 2015, 2018; McShane et al. 2019; Ruiz-Ruano García and López Puga 2018; Wasserstein et al. 2019).

Therefore, in this paper, we will use a Bayesian approach to examine the influence of tourism on the shadow economy size of the ASEAN countries. The Bayesian analysis combines prior information through conditional probability, known as Bayesian rule:

$$Pr(\theta|Y) = \frac{Pr(\theta)Pr(Y|\theta)}{Pr(Y)} \quad (2)$$

where $\Pr(\theta|Y)$ is the posterior probability of the hypothesis adjusted based on observational data. Especially, posterior distributions describe the a priori probability of

Table 2 Summary of variables, descriptions and their sources

Variable	Legend	Description	Source
Dependent variable			
Shadow economy	SSE	The shadow economy size as a percentage of GDP	Medina and Schneider (2019)
Tourism variables			
Tourism expenditures	LogTEXPE	The natural log of international tourism expenditures (current US$)	WDI
Tourism receipts	LogTREC	The natural log of international tourism receipts (current US$)	WDI
Tourist arrivals	LogTARR	The natural log of international tourism number of arrivals	WDI
Other variables			
GDP per capita	LogGDP	The natural log of GDP per capita (constant 2010 US$)	WDI
Trade openness	OPEN	The sum of a country's exports and imports as a percentage of GDP	WDI
Tax burden	TAXB	Total tax burden as a percentage of GDP	Heritage Foundation
Government size	GSIZE	General government final consumption expenditure as a percentage of GDP	WDI
Control of Corruption	CCOR	The index ranges approximately from −2.5 (weak) to 2.5 (strong)	WGI
Rule of law	RLAW	The index ranges approximately from −2.5 (weak) to 2.5 (strong)	WGI

Source The authors

the hypothesis or parameter updated with new information. $\Pr(\theta)$ is the probability of the parameter or hypothesis θ based on a priori information. $\Pr(Y|\theta)$ is the likelihood of the hypothetical conditioned data. $\Pr(Y)$ is the normalization constant.

For simplicity, Bayes' rule is expressed as a proportion:

$$\Pr(\theta|Y) \propto \Pr(\theta)\Pr(Y|\theta) \tag{3}$$

Equation (3) says that the posterior distribution describes the weighted average of the information about the parameters in the observed data and the knowledge of the a priori parameters for the data observation (McShane et al. 2019; Thach 2021).

Table 3 Average size of the shadow economy and tourism development in the sample

Country	SSE (% of GDP)	TEXPE (US$ million)	TREC (US$ million)	TARR (number of tourists)
Brunei Darussala	28.33	483	165	210,188
Cambodia	48.76	240	1,421	2,063,826
Indonesia	23.18	8,031	9,155	7,379,615
Laos	29.13	377	542	2,252,615
Malaysia	30.17	7,348	14,040	18,214,120
Myanmar	46.51	90	651	1,398,320
Philippines	40.00	5,377	4,188	3,583,520
Singapore	11.46	14,010	10,930	11,099,860
Thailand	48.89	6,909	25,490	17,759,640
Vietnam	17.33	2,700	5,724	5,497,880
ASEAN	32.38	4,792	7,551	7,220,342

Source The authors

In the first step, we use a normal prior distribution because researchers consider it to be a good depiction of the distribution of influence (Block et al. 2011; Nguyen and Duong 2022; Oanh et al. 2022; Permai and Tanty 2018). We, therefore, use an a priori distribution $N(0, 1)$, which describes the previous default normal with a mean of 0 and a variance of 1 (Kosheleva et al. 2021). According to Block et al. (2011), using a normal prior distribution will not bias the results of Bayesian analysis of the hypotheses in the paper in a positive or negative direction. In the next step of the process, for the respective likelihood functions of the coefficients, we assume normal distributions with the parameters deduced from the econometric model in Eq. (1). Finally, posterior distributions of the parameters are generated through the MCMC and Gibbs sampler technique in the Bayesian estimator. Also, researchers need to check both the stability and convergence of MCMC when using the Gibbs sampling algorithm to generate posterior distributions (Kruschke 2015; Nguyen and Duong 2022; Oanh et al. 2022).

4 Results and Discussion

4.1 Correlation Analysis

We use the Bayesian Pearson correlation to analyze the correlation between the shadow economy size and the tourism development status of ASEAN countries. We estimate the Bayesian factor (BF), specifically BF_{10} to look for correlation (BF_{10} is the BF form for the likelihood of the data occurring under the null hypothesis divided by the likelihood of the data under the alternative hypothesis). According to Nuzzo

(2017), Ruiz-Ruano García and López Puga (2018), BF_{10} values greater than one signal more evidence in favor of the alternative, and a value less than one signals more evidence in favor of null.

The above-mentioned Bayesian analysis is performed with a "noninformative" prior, where all correlations between − 1 and + 1 are considered to be of equal a priori probability (Nuzzo 2017). Table 4 reports the Bayesian Pearson correlation matrix between ten variables. Table 4 shows that SSE negatively correlates with LogTEXPE, LogTREC, LogTARR, LogGDP, CCOR, and RLAW, while others are positive. Specifically, BF_{10} values are greater than 100, indicating extreme evidence of the correlation of LogTEXPE, LogTREC, LogTARR with SSE. Similarly, Table 4 also provides extreme evidence of a negative relationship between SSE and LogGDP, CCOR, and RLAW. The Bayesian correlation between SSE and OPEN/TAXB report BF_{10} values less than one, so there is no evidence of a positive correlation between these factors.

However, all correlation coefficients between LogTEXPE and LogTREC and LogTREC; LogTARR and LogTREC; LogGDP with CCOR and RLAW; OPEN and RLAW; CCOR and RLAW are above 0.80 with BF_{10} values all greater than 100, so there is extreme evidence of the correlation between these factors. This result shows that there is multicollinearity in the model (Kim, 2019). Based on this result, the linear regression using the frequency method did not fit the results because the assumptions of the linear regression based on the frequency school were not met. According to Permai and Tanty (2018), one of the parameter estimation methods that can be used is the Bayesian method. The Bayesian approach has several other attractive properties. According to Leamer (1973), the Bayesian method helps to solve multicollinearity problems. Therefore, this paper employs multiple linear regression model analysis with Bayesian parameter estimation.

4.2 Results of Bayesian Analysis

As described above, this paper discusses the parameter estimation for the dynamic system by a Bayesian analysis associated with MCMC. To perform Bayesian regression, we conducted with totals of MCMC 12,500 iterations. In which the first 20% of the iterations were discarded (2,500 iterations). Tables 4 and 5 show the post-estimation results of Bayesian linear regression. Hence, both stability and convergence are needed when using the MCMC sampler to produce the posterior distributions. The stability of the estimates is determined by the effective sample size (ESS). Table 5 indicates that the smallest ESS of all parameters is 24,855, so the closer the ESS values are to MCMC sample size and MCMC efficiencies for all model parameters is close to 1 (Kruschke 2015; Nguyen and Duong, 2022; Oanh et al. 2022). Therefore, the estimates are stable. In the current paper, we use a standard test that applies multiple series, namely Gelman-Rubin (Rc) convergence diagnostic. Rc compares the variance between chains relative to variance within chains. Completely

Table 4 Bayesian Pearson correlation matrix

	SSE	LogTEXPE	LogTREC	LogTARR	LogGDP	OPEN	TAXB	GSIZE	CCOR	RLAW
SSE	–	–	–	–	–	–	–	–	–	–
LogTEXPE	−0.430*** [4.301e + 7]	–	–	–	–	–	–	–	–	–
LogTREC	−0.267*** [157.713]	0.888*** [1.464e + 71]	–	–	–	–	–	–	–	–
LogTARR	−0.270*** [315.739]	0.803*** [2.287e + 46]	0.940*** [∞]	–	–	–	–	–	–	–
LogGDP	−0.507*** [2.998e + 13]	0.641*** [3.315e + 23]	0.422*** [1.880e + 8]	0.347*** [31,4467.47]	–	–	–	–	–	–
OPEN	−0.544*** [4.749e + 15]	0.472*** [1.850e + 10]	0.432*** [2.543e + 8]	0.429*** [9.550e + 8]	0.614*** [1.813e + 23]	–	–	–	–	–
TAXB	0.066 [0.136]	0.234* [15.339]	0.222* [10.413]	0.210 [8.618]	0.326*** [8,013.756]	0.276*** [228.9]	–	–	–	–
GSIZE	0.078 [0.161]	0.062 [0.129]	−0.153 [0.899]	−0.161 [1.382]	0.586*** [1.544e + 19]	−0.120 [0.413]	0.083 [0.172]	–	–	–
CCOR	−0.553*** [5.287e + 13]	0.626*** [1.209e + 19]	0.444*** [1.186e + 8]	0.335*** [13,740.822]	0.914*** [5.991e + 79]	0.778*** [1.739e + 39]	0.213 [4.872]	0.341*** [10,869.36]	–	–
RLAW	−0.595*** [4.285e + 16]	0.589*** [1.532e + 16]	0.409*** [3.634e + 6]	0.309*** [2,079.587]	0.890*** [1.200e + 69]	0.845*** [2.778e + 53]	0.257** [31.09]	0.262** [73.976]	0.950*** [∞]	–

Notes Bayes factors (BF) in brackets, * $BF_{10} > 10$, ** $BF_{10} > 30$, *** $BF_{10} > 100$
Source The authors

Table 5 MCMC diagnostics

Models	Model 1		Model 2		Model 3	
Independent variables	ESS	Rc	ESS	Rc	ESS	Rc
LogTEXPE	29,695 [0.9898]	1.00004	–	–	–	–
LogTREC	–	–	29,934 [0.9978]	0.99998	–	–
LogTARR	–	–	–	–	29,134 [0.9711]	1.00000
LogGDP	28,550 [0.9517]	0.99999	27,262 [0.9087]	1.00003	27,262 [0.9087]	1.00003
OPEN	27,017 [0.9006]	1.00004	28,720 [0.9573]	0.99999	28,720 [0.9573]	0.99999
TAXB	30,000 [1.0000]	0.99997	30,000 [1.000]	1.00002	30,000 [1.000]	1.00002
GSIZE	29,619 [0.9873]	1.00005	28,251 [0.9417]	1.00003	28,251 [0.9417]	1.00003
CCOR	30,000 [1.0000]	1.00006	30,000 [1.0000]	0.99996	30,000 [1.000]	0.99996
RLAW	28,919 [0.9640]	1.00004	28,822 [0.9607]	1.00005	28,822 [0.9607]	1.00005
_cons	29,525 [0.9842]	1.00004	29,779 [0.9926]	0.99997	29,779 [0.9926]	0.99997
variance	24,855 [0.8285]	1.00000	23,949 [0.7983]	0.99996	23,949 [0.7983]	0.99996

Notes Dependent variable is SSE, MCMC efficiencies in brackets, MCMC sample size are 30,000
Source The authors

converged chains have a value below 1.1 (Nguyen and Duong 2022; Oanh et al. 2022; Thach 2021). Table 5 reports that the Rc values are less than 1.1; thus, the converged MCMC indicates that chain values represent the posterior distribution. These results show that the posterior simulation satisfies the stipulations of Bayesian analysis.

From the Bayesian statistics point of view, we present a credible interval, which accommodates the main parameter with a certain probability, to analyze an effect. Specifically, Table 6 documents the mean coefficients and 95% credible intervals for the mean coefficients regarding the hypotheses.

The findings reveal that all three variables that represent tourism have beneficial impact on the shadow economy size. In other words, hypotheses H1, H2, and H3 are rejected for the ASEAN countries between 1999–2017. As mentioned in the literature review, the impact of tourism on the shadow economy is still mixed and inconclusive. Thus, our findings can supplement the literature on the relationship between these factors. Specifically, the mean parameter of tourism expenditures is 0.1608 with the posterior probability of 65.91%. Meanwhile, the mean coefficients of tourism receipts and tourist arrivals are 0.9044 and 0.7228, respectively. While

Table 6 Bayesian random-effects regression

Models	Model 1		Model 2		Model 3	
Independent variables	Mean	Probability of mean	Mean	Probability of mean	Mean	Probability of mean
LogTEXPE	0.1608 [−0.6075; 0.9210]	65.91%*	–	–	–	–
LogTREC	–	–	0.9044 [0.1526; 1.6390]	99.08%*	–	–
LogTARR	–	–	–	–	0.7228 [-0.1604; 1.6223]	94.47%*
LogGDP	−1.7219 [−3.4602; −0.0126]	97.59%**	−2.4744 [−4.1263; − 0.8165]	99.82%**	−2.2393 [−3.8233; −0.6460]	99.72%**
OPEN	−0.0448 [−0.0756; −0.0144]	99.82%**	− .0417 [−0.0717; − 0.0126]	99.74%**	−0.0409 [−0.0702; −0.0124]	99.74%**
TAXB	0.4969 [0.3070; 0.6862]	100%*	0.3612 [0.1673; 0.5568]	100%*	0.4343 [0.2653; 0.6016]	100%*
GSIZE	0.6606 [0.2002; 1.1163]	99.70%*	0.7040 [0.2676; 1.1359]	99.90%*	0.7486 [0.3117; 1.1730]	99.93%*
CCOR	−0.5828 [−2.3892; 1.2066]	73.66%**	−0.5056 [−2.3080; 1.2766]	70.66%**	−0.5279 [−2.3121; 1.2581]	72.03%**
RLAW	−1.6542 [−3.4716; 0.1455]	96.34%**	−1.4255 [−3.2231; 0.3674]	93.93%**	−1.5058 [−3.3120; 0.3183]	94.69%**
_cons	0.1408 [−1.8224; 2.1005]	55.63%*	0.0096 [−1.9410; 1.9758]	50.48%*	0.0778 [−1.8616; 2.0225]	52.90%*
variance	108.5641 [85.42; 137.95]	–	100.5104 [79.16; 127.33]	–	100.0068 [79.39; 125.45]	–

Notes Dependent variable is SSE, 95% credible intervals in brackets, * Probability of mean > 0, ** Probability of mean < 0
Source The authors

the positive effect probability of the mean parameter of tourism expenditures is only moderate, we find extreme evidence for the positive impact of tourism receipts and tourist arrivals on the informal economy.

Besides the above variables, we also analyze the effect of other factors on the shadow economy in the ASEAN countries. Notably, we uncover that GDP per capita, trade openness, and the rule of law are negatively linked to the informal economy, and this result is robust across three models. Furthermore, we report extreme negative effects of these factors on informality because their posterior probabilities in regressions are greater than 90%. This result posits that countries should promote GDP per capita and trade openness, as well as improve the rule of law to curb the informal sector size.

Likewise, control of corruption can hinder the expansion of the shadow economy in the ASEAN countries. The mean parameters of this factor in the three models are −0.5828, −0.5056, −0.5279, respectively. Moreover, we find strong evidence for the relationship between control of corruption and the shadow economy size. This result implies that governments can reduce the shadow economy size by controlling corruption.

By contrast, tax burden positively affects the informal economy size. The mean coefficient of this factor in model 1, model 2, and model 3 are 0.4969, 0.3612, and 0.4343, respectively. Since the posterior probabilities of this factor in all three models are 100%, we suggest extreme positive effects of the tax burden on the shadow economy. Similarly, we discover that the government size can promote informality because the mean coefficients of the variable are positive in all regressions. Also, we find extreme evidence for the positive impact of the government size on the shadow economy. These findings suggest that an increase in tax burden and the government size can lead to a larger size of the shadow economy in the ASEAN countries.

5 Conclusion

The shadow economy is a widespread economic phenomenon in many countries worldwide. Meanwhile, tourism is an important sector that can impact various industries. It also contributes to economic growth, improves wealth, provides foreign exchange, and creates employment and business opportunities for many countries. However, there are relatively few studies on the relationship between informality and tourism in the ASEAN countries. Therefore, the present paper investigates the effect of tourism on the shadow economy size in these countries between 1999–2017.

Our models examine three variables that represent tourism, namely tourism expenditures, tourism receipts and tourist arrivals. The results reveal that they are all positively linked to the shadow economy size in ASEAN. Nevertheless, whilst we discover extreme evidence for the effect of tourism receipts and tourist arrivals on the informal sector, the linkage between tourism expenditures and informality is only moderate. This result posits that tourism receipts and tourist arrivals play a more significant role in promoting the shadow economy than tourism spendings.

Other factors, namely tax burden, and government size, also positively affect the shadow economy size in the ASEAN countries. We suggest extreme positive impacts of these two factors on the shadow economy because their posterior probabilities in all models are greater than 99%.

In contrast, GDP per capita, trade openness, and the rule of law have detrimental effects on the informal economy size. Indeed, we find extreme evidence for such impacts on informality. Likewise, control of corruption can adversely affect the informal economy, and the negative effect probability of this factor is extreme. These findings suggest that governments should foster GDP per capita and trade openness, enhance the rule of law, and control corruption because this helps reduce the incentives for people to go underground.

Finally, tourism is a multidimensional concept that can involve various aspects. In this paper, we concentrate on tourism expenditures, tourism receipts, and tourist arrivals. Nevertheless, more studies can analyze the connection between other dimensions of tourism and the shadow economy size to elucidate this relationship further.

References

Alam, M.S., Paramati, S.R.: The impact of tourism on income inequality in developing economies: does Kuznets curve hypothesis exist? Ann. Tour. Res. **61**, 111–126 (2016). https://doi.org/10.1016/j.annals.2016.09.008

Alegre, J., Mateo, S., Pou, L.: Participation in tourism consumption and the intensity of participation: an analysis of their socio-demographic and economic determinants. Tour. Econ. **15**(3), 531–546 (2009). https://doi.org/10.5367/000000009789036521

Altindag, D.T.: Crime and international tourism. J. Lab. Res. **35**(1), 1–14 (2014). https://doi.org/10.1007/s12122-014-9174-8

Amrhein, V., Trafimow, D., Greenland, S.: Inferential statistics as descriptive statistics: there is no replication crisis if we don't expect replication. Am. Stat. **73**(sup1), 262–270 (2019)

Badariah, H., Habibullah, M.S., Baharom, A.: Does the shadow economy matter for tourism? Int. Evidence. Taylor's Bus. Rev. **5**(1), 1–9 (2015)

Bajada, C.: Estimates of the underground economy in Australia. Econ. Rec. **75**(4), 369–384 (1999). https://doi.org/10.1111/j.1475-4932.1999.tb02573.x

Bernini, C., Cracolici, M.F.: Is participation in the tourism market an opportunity for everyone? Some Evid. Italy Tour. Econ. **22**(1), 57–79 (2016). https://doi.org/10.5367/te.2014.0409

Block, J.H., Jaskiewicz, P., Miller, D.: Ownership versus management effects on performance in family and founder companies: a bayesian reconciliation. J. Fam. Bus. Strat. **2**(4), 232–245 (2011). https://doi.org/10.1016/j.jfbs.2011.10.001

Briggs, W.M.: Everything wrong with p-values under one roof. In: Kreinovich, V., Thach, N.N., Trung, N.D., Van Thanh, D. (eds.) Beyond Traditional Probabilistic Methods in Economics, pp. 22–44. Springer International Publishing, Cham (2019)

Caridi, P., Passerini, P.: The underground economy, the demand for currency approach and the analysis of discrepancies: some recent european experience. Rev. Income Wealth **47**(2), 239–250 (2001). https://doi.org/10.1111/1475-4991.00014

Crawford, D.W., Jackson, E.L., Godbey, G.: A hierarchical model of leisure constraints. Leis. Sci. **13**(4), 309–320 (1991). https://doi.org/10.1080/01490409109513147

Din, B.H., Habibullah, M.S., Baharom, A.H., Saari, M.D.: Are shadow economy and tourism related? International evidence. Procedia Econ. Financ. **35**, 173–178 (2016). https://doi.org/10.1016/S2212-5671(16)00022-8

Dreher, A., Kotsogiannis, C., McCorriston, S.: How do institutions affect corruption and the shadow economy? Int. Tax Public Financ. **16**(6), 773–796 (2009). https://doi.org/10.1007/s10797-008-9089-5

Eilat, Y., Zinnes, C.: The evolution of the shadow economy in transition countries: consequences for economic growth and donor assistance. Harv. Inst. Int. Dev. CAER II Discuss. Pap. **83**, 1–70 (2000)

Eugenio-Martin, J.L., Campos-Soria, J.A.: Income and the substitution pattern between domestic and international tourism demand. Appl. Econ. **43**(20), 2519–2531 (2011). https://doi.org/10.1080/00036840903299698

Fleming, M.H., Roman, J., Farrell, G.: The shadow economy. J. Int. Aff. **53**(2), 387–409 (2000)

Fricker, R.D., Burke, K., Han, X., Woodall, W.H.: Assessing the statistical analyses used in basic and applied social psychology after their p-value ban. Am. Stat. **73**(sup1), 374–384 (2019). https://doi.org/10.1080/00031305.2018.1537892

Gelman, A.: The connection between varying treatment effects and the crisis of unreplicable research: a bayesian perspective. J. Manag. **41**(2), 632–643 (2015). https://doi.org/10.1177/0149206314525208

Gelman, A.: Ethics in statistical practice and communication: five recommendations. Significance **15**(5), 40–43 (2018). https://doi.org/10.1111/j.1740-9713.2018.01193.x

Hjalager, A.M.: The illegal economy in the restaurant sector in Denmark. Tour. Hosp. Res. **8**(3), 239–251 (2008)

Jaliu, D.D., Răvar, A.S.: Informal tourism economy and EU funding: the case of Romania. In: Papathanassis, A., Katsios, S., Dinu, N.R. (eds.) Yellow Tourism: Crime and Corruption in the Holiday Sector, pp. 193–207. Springer International Publishing, Cham (2019)

Katsios, S.: The shadow economy and corruption in Greece. South-East. Eur. J. Econ. **1**, 61–80 (2006)

Katsios, S.: Tourism in 'Yellow Times': the de-formalisation of the Greek economy and Its impact on tourism. In: Papathanassis, A., Katsios, S., Dinu, N.R. (eds.) Yellow Tourism: Crime and Corruption in the Holiday Sector, pp. 209–225. Springer International Publishing, Cham (2019)

Kesar, O., Čuić, K.: Shadow economy in tourism: some conceptual considerations from Croatia. Zagreb. Int. Rev. Econ. Bus. **20**(2), 65–86 (2017). https://doi.org/10.1515/zireb-2017-0018

Kim, J.H.: Multicollinearity and misleading statistical results. Korean J Anesth. **72**(6), 558–569 (2019). https://doi.org/10.4097/kja.19087

Kosheleva, O., Kreinovich, V., Autchariyapanitkul, K.: Why beta priors: invariance-based explanation. In: Sriboonchitta, S., Kreinovich, V., Yamaka, W. (eds.) Behavioral Predictive Modeling in Economics, pp. 141–144. Springer International Publishing, Cham (2021)

Kruschke, J.K.: Doing Bayesian Data Analysis, 2nd edn. Academic Press, Boston (2015)

Laimer P, Weiß, J, Austria, S.: Data sources on tourism expenditure: The Austrian experiences taking into account the TBoP requirements. In: Paper presented at the International Workshop on Tourism Statistics. Madrid, Spain (2006)

Leamer, E.E.: Multicollinearity: a bayesian interpretation. Rev. Econ. Stat. **55**(3), 371–380 (1973). https://doi.org/10.2307/1927962

Lv, Z.: Does tourism affect the informal sector? Ann. Tour. Res. **80**, 102816 (2020). https://doi.org/10.1016/j.annals.2019.102816

McShane, B.B., Gal, D., Gelman, A., Robert, C., Tackett, J.L.: Abandon statistical significance. Am. Stat. **73**(sup1), 235–245 (2019). https://doi.org/10.1080/00031305.2018.1527253

Medina, L., Schneider, F.: Shedding light on the shadow economy: a global database and the interaction with the official one. Center for Economic Studies and ifo Institute (CESifo). Munich, Germany (2019)

Mekinc, J., Kociper, T., Dobovšek, B.: The impact of corruption and organized crime on the development of sustainable tourism. Varst.Slovje: J. Crim. Justice Secur. **15**(2), 218–239 (2013)

Milić, M.: Phenomenon of grey economy and its forms in the Montenegro economy. Poslovna Izvrsnost **8**(1), 65–80 (2014)

Nguyen, H.T.: Why P-values are banned? Thail. Stat. **14**(2), i–iv (2016)

Nguyen, H.T., Thach, N.N.: A closer look at the modeling of economics data. In: Kreinovich, V., Thach, N.N., Trung, N.D., Van Thanh, D. (eds.) Beyond Traditional Probabilistic Methods in Economics, pp. 100–112. Springer International Publishing, Cham (2019)

Nguyen, H.T., Trung, N.D., Thach, N.N.: Beyond traditional probabilistic methods in econometrics. In: Kreinovich, V., Thach, N.N., Trung, N.D., Van Thanh, D. (eds.) Beyond Traditional Probabilistic Methods in Economics, pp. 3–21. Springer International Publishing, Cham (2019)

Nguyen, T.A.N., Luong, T.T.H.: Corruption, shadow economy and economic growth: evidence from emerging and developing asian economies. Montenegrin J. Econ. **16**(4), 85–94 (2020). https://doi.org/10.14254/1800-5845/2020.16-4.7

Nguyen, V.D., Duong, T.H.M.: Corruption, shadow economy, FDI, and tax revenue in BRICS: a bayesian approach. Montenegrin J. Econ. **18**(2), 85–94 (2022). https://doi.org/10.14254/1800-5845/2022.18-2.8

Nuzzo, R.L.: An introduction to bayesian data analysis for correlations. PM&R **9**(12), 1278–1282 (2017). https://doi.org/10.1016/j.pmrj.2017.11.003

Oanh, T.T.K., Diep, N.V., Truyen, P.T., Chau, N.X.B.: The impact of public expenditure on economic growth of provinces and cities in the Southern Key economic zone of Vietnam: bayesian approach. In: Ngoc Thach, N., Ha, D.T., Trung, N.D., Kreinovich, V. (eds.) Prediction and Causality in Econometrics and Related Topics, pp. 328–344. Springer International Publishing, Cham (2022)

Ott, K.: The underground economy in Croatia 1990–2000. In: Paper Presented at the 58th Congress of the International Institute of Public Finance, Helsinki, Finland (2002)

Pavicevic, R.: Tourism Destroys the Shadow Economy (2014). http://www.cgekonomist.com/?broj=12&clanak=641&lang=en

Permai, S.D., Tanty, H.: Linear regression model using bayesian approach for energy performance of residential building. Procedia Comput. Sci. **135**, 671–677 (2018). https://doi.org/10.1016/j.procs.2018.08.219

Ruiz-Ruano García, A.M., López Puga, J.: Deciding on Null hypotheses using P-values or Bayesian alternatives: a simulation study. Psicothema **30**(1), 110–115 (2018). https://doi.org/10.7334/psicothema2017.308

Schneider, F.: The size and development of the shadow economies of 22 transition and 21 OECD countries. In: Institute for the Study of Labor (IZA). Bonn, Germany (2002)

Schneider, F.: The shadow economy in Europe, 2013. Johannes Kepler University Linz & A.T. Kearney, Inc., Chicago, Illinois (2013)

Schneider, F., Buehn, A., Montenegro, C.E.: New estimates for the shadow economies all over the world. Int. Econ. J. **24**(4), 443–461 (2010). https://doi.org/10.1080/10168737.2010.525974

Schneider, F., Enste, D.H.: Shadow economies: size, causes, and consequences. J. Econ. Lit. **38**(1), 77–114 (2000)

Schneider, F., Enste, D.H.: The Shadow Economy: An International Survey, 2nd edn. Cambridge University Press, Cambridge (2013)

Šergo, Z., Gržinić, J., Zanini-Gavranić, T.: Shadow economy and tourism receipts: evidence from Europe. Paper presented at the Interdisciplinary Management Research XIII, Opatija, Croatia (2017)

Sinha, A., Sengupta, T., Mehta, A.: Tourist arrivals and shadow economy: Wavelet-based evidence from Thailand. Tourism analysis, pp. 26(2–3) (2021). https://doi.org/10.3727/108354221X16186396395764

Skalpe, O.: The hidden economy in tourism (circumstantial evidence from the hotels and restaurants in Norway). In: Chang, P.R. (ed.) Tourism Management in the 21st Century, pp. 337–351. Nova Science Publishers, New York (2007)

Slocum, S.L., Backman, K.F., Robinson, K.L.: Tourism pathways to prosperity: perspectives on the informal economy in Tanzania. Tour. Anal. **16**(1), 43–55 (2011). https://doi.org/10.3727/108354211X12988225900045

Smith, A.: Obstacles to the growth of alternative tourism in Greece. Afr. J. Hosp. Tour. Leis. **1**(3), 1–8 (2011)

Tan, Y.L., Habibullah, M.S., Yiew, T.H.: The shadow economy in Malaysia: evidence from an ARDL model. Int. J. Econ. Manag. **10**(2), 261–277 (2016)

Thach, N.N.: How values influence economic progress? Evidence from South and Southeast Asian Countries. In: Ngoc Thach, N., Kreinovich, V., Trung, N.D. (eds.) Data Science for Financial Econometrics, pp. 207–221. Springer International Publishing, Cham (2021)

Trafimow, D.: A taxonomy of model assumptions on which P is based and implications for added benefit in the sciences. Int. J. Soc. Res. Methodol. **22**(6), 571–583 (2019). https://doi.org/10.1080/13645579.2019.1610592

UNCTAD. Sustainable tourism: Contribution to economic growth and sustainable development. United Nations Conference on Trade and Development, Geneva, Switzerland (2013)

UNEP, UNWTO. Making Tourism More Sustainable—A Guide for Policy Makers. United Nations Environment Programme, Division of Technology, Industry and Economics & World Tourism Organization, Paris, France and Madrid, Spain (2005)

UNWTO: International Tourism Highlights, 2020th edn. World Tourism Organization, Madrid (2021)

Vladimirov, Z.: Factors for the e-business adoption by small tourism firms and the role of shadow economic practices. Eur. J. Tour. Res. **10**, 5–34 (2015)

Vorisek, D., Kindberg-Hanlon, G., Koh, W.C., Okawa, Y., Taskin, T., Vashakmadze, E., Ye, S.L.: Informality in emerging market and developing economies: regional dimensions. In Ohnsorge, F., Yu, S. (eds.) The Long Shadow of Informality: Challenges and Policies. Washington, DC, World Bank (2021)

Wahnschafft, R.: Formal and informal tourism sectors: a case study in Pattaya Thailand. Ann. Tour. Res. **9**(3), 429–451 (1982). https://doi.org/10.1016/0160-7383(82)90022-6

Wasserstein, R.L., Schirm, A.L., Lazar, N.A.: Moving to a world beyond "$p < 0.05$". Am. Stat. **73**(sup1), 1–19 (2019). https://doi.org/10.1080/00031305.2019.1583913

Webb, J.W., Bruton, G.D., Tihanyi, L., Ireland, R.D.: Research on entrepreneurship in the informal economy: framing a research agenda. J. Bus. Ventur. **28**(5), 598–614 (2013). https://doi.org/10.1016/j.jbusvent.2012.05.003

WTTC. Travel & Tourism—Global economic impact trends 2020. London, United Kingdom: World Travel and Tourism Council (2020)

Xu, T., Lv, Z.: Does too much tourism development really increase the size of the informal economy? Curr. Issues Tour. 1–6 (2021). https://doi.org/10.1080/13683500.2021.1888898

Zhang, H.: Understanding the informal economy in hospitality and tourism: a conceptual review. California State Polytechnic University, Pomona, California (2017)

Fraud Identification of Financial Statements by Machine Learning Technology: Case of Listed Companies in Vietnam

Nguyen Anh Phong, Phan Huy Tam, and Ngo Phu Thanh

Abstract The study uses data of listed non-financial companies in 2018 and 2019, combining M-Score and Z-Score models, applying ANN and SVM machine learning techniques in forecasting evidence of fraud in financial statements. Research results show that using SVM technique and M-Score index has high accuracy in predicting.

1 Introduction

Financial statements are fundamental and important documents of a business, as it reflects the financial position and health of companies (Beaver 1966; Ravisankar et al. 2011). Financial statements provide important information for investors, creditors, and others to make decisions. They also contain information about business performance, financial status, and social responsibility of officially listed companies and unofficially listed companies (OTC). However, in recent years, financial fraud reporting cases have become increasingly serious (Wells 1997; Spathis et al 2002; Kirkos et al 2007; Yeh et al 2010; Humpherys et al. 2011; Kamarudin et al. 2012). Since the Asian Financial Crisis of 1997, there have been many cases of financial fraud reporting detected, however its effects are out of control, such as the Enron case in 2014. 2001, the WorldCom case in 2003 in the United States, and other cases such as ABIT Computer, Procomp, Infodisc and Summit Technology in 2004 in Taiwan. With these incidents, it is important to prevent negative effects to detect fraud before it occurs.

N. A. Phong (✉) · P. H. Tam · N. P. Thanh
University of Economics and Law, Ho Chi Minh City, Vietnam
e-mail: phongna@uel.edu.vn

P. H. Tam
e-mail: tamph@uel.edu.vn

N. P. Thanh
e-mail: thanhnp@uel.edu.vn

Vietnam National University, Ho Chi Minh City, Vietnam

N. Ngoc Thach et al. (eds.), *Financial Econometrics: Bayesian Analysis, Quantum Uncertainty, and Related Topics*, Studies in Systems, Decision and Control 427,
https://doi.org/10.1007/978-3-030-98689-6_28

Nowadays, under the development and application of algorithms and technology, fraudulent financial statements can be detected early by machine learning techniques, in which the typical classifcation problem (Kirkos et al. 2007) is considered a suitable problem in early detection and warning. The classification problem involves performing computations using the variable properties of some known categorical data, to classify we need classification rules. Then, the unclassified data is entered into the rules to obtain the final classification result. Regarding the problem of fraudulent financial reporting, many previous studies have suggested to use data mining method because of its superiority in forecasting after training a machine learning model with a large amount of data, as well as accuracy in both classification and prediction, which is much higher than that of conventional regression analysis. Effective methods applied in this problem include artificial neural network (ANN), decision tree (DT), Bayesian belief network (BBN), and support vector machine (SVM) methods.

In Vietnam, there were many cases where the inspection agency later discovered incorrect accounting practices, violations of information disclosure, falsifying financial results on audited financial statements. Recent examples such as Quang Nam Rubber Investment Joint Stock Company announced its separate and consolidated financial statements for the fourth quarter of 2019 with incorrect information about the target profit after corporate income tax compared to separate audited consolidated financial statements of year 2019. Or the case that Khang Minh Group got fined of 100 million VND for publishing false information. In the separate audited and consolidated financial statements for 2018 announced on March 27, 2019 and in the report on progress of using capital obtained from the public offering of securities on August 27, 2018, there is information that GKM used 79 billion VND of the total proceeds from the offering to increase capital from 45 billion VND to 135 billion VND in 2018 to invest in Khang Minh Brick Manufacturing Co., Ltd. (also known as Khang Minh Quartz Conslab) built an unburnt brick factory in accordance with the purpose of using the capital stated in the BOD Resolution No. 10/NQ-HĐQT-KM/2017. However, according to Khang Minh Group's report and documents, the 79 billion VND received from the offering was used to invest in Khang Minh Brick Manufacturing Co., Ltd. To build a factory producing paving stones with unburnt technology, quartz conslab…

To the best of author's knowledge, this is the first research applying machine learning algorithm namely ANN & SVM in the context of identify fraudulent financial statements in Viet Nam. Besides, Vietnamese corporates have some unique financial characteristics and accounting standards due to the regulations in Vietnam market, leading to differences when compared with businesses in other markets. As for ANN and SVM methods are computational methods based on machine learning algorithm, the authors expect to provide efficiency, accuracy, and timeliness in identifying fraudulent financial statements.

2 Empirical Studies

The US Committee of Sponsoring Organizations of the Treadway Commission (COSO) (Beasley et al. 1999) and SAS No. 99, 2002 have defined a concept of financial reporting fraud as intentional or reckless conduct, based on false information or omissions lead to significant error of the financial statements. The cost of preventing fraudulent financial statements in the United States is estimated to run into the billions of dollars each year (Humphreys et al. 2011). The US Association of Certifed Fraud Examiners (ACFE) classifies financial reporting fraud into six categories: (1) providing false financial information; (2) misuse or misappropriation of company property; (3) inappropriate support or lending; (4) improperly acquired property or income; (5) improper accounting of fraudulent costs or fees; and (6) improper financial management by an executive or board member.

The Taiwan Accounting Research and Development Organization released Auditing Standard 43 Joint Communique in 2006, where fraud is the purpose of management, one or more employees using deception and other methods to gain unjust or illegal gain. Therefore, it can be concluded that fraud has four elements including: (1) serious misrepresentation of the nature of transactions, (2) intentional violation of the rules, (3) victim acceptance receives an untrue statement; and (4) financial loss caused by the above three cases. Fraud related to audits of financial statements includes fraudulent financial statements and misappropriation of assets. Financial reporting fraud refers to dishonest financial statements intended to mislead users. The U.S. Securities Exchange Commission (SEC) states that financial statements must provide a comprehensive overview of a company's business and financial position and be audited.

Financial reporting fraud is the intentional and illegal conduct that results in misleading reports or misleading financial disclosures (Beasley 1996; Rezaee 2005; Ravisankar et al. 2011) and stakeholders adversely affected by these misleading financial statements (Elliot and Willingham 1980). Most of the previous studies used conventional multivariate statistical analysis, especially logistic regression analysis (Beasley 1996; Summers and Sweeney 1998; Bell and Carcello 2000; Spathis et al. 2002; Sharma 2004; Uzun et al. 2004; Chen et al. 2006; Humpherys et al. 2011). Conventional statistical methods require adherence to limitations of specific hypotheses, such as multicollinearity, heteroskedasticity, and normal distribution of data (Chiu et al. 2002). According to Chen (2005), the empirical financial variables often cannot comply with the relevant statistical conditions, such as the requirement for a normal distribution. The machine learning method, meanwhile, does not require traditional statistical hypotheses and is used by scholars as a classification tool. The results of experimental studies show that the machine learning method is positive. Most of the previous studies have used conventional statistical methods to make decisions regarding the continuity of activities while reviewing on the background of fixed and discriminant data. However, this method causes some limitations in terms of judgment and relatively high error rate. In recent years, several studies have applied data mining techniques and machine learning methods to detect fraudulent financial statements and thus minimize errors.

Studies applying DT technique to detect fraudulent financial statements include Hansen et al. (1992), Koh (2004), Kotsiantis et al. (2006), Kirkos et al. (2007) and Salehi and Fard (2013). Studies applying the BBN technique to detect fraudulent financial statements include: Kirkos et al. (2007) and Nguyen et al. (2008). Studies applying SVM technique to detect fraudulent financial statements include Zhou and Kapoor (2011), Shin et al. (2005), Chen et al. (2006), Yeh et al. (2010), Ravisankar et al. (2011) and Pai et al. (2011). Studies applying ANN techniques to detect fraudulent financial statements include Hansen et al. (1992), Coats and Fant (1993), Fanning and Cogger (1998), Koh (2004), Chen et al. (2006), Kirkos et al. (2007), Ravisankar et al. (2011) and Zhou and Kapoor (2011). Studies applying other techniques to detect fraudulent financial statements include Hajek and Henriques (2017), Yao et al. (2018), Dong et al. (2016), Tang et al. (2018), Craja et al. (2020), Sadgali et al. (2019), Mohammadi et al. (2020). The accuracy rate of using data mining techniques to detect fraudulent financial statements is varied. As stated above, most studies only use 1–2 data mining techniques, there is no comparison of models; and most use single-stage statistical processing to establish predictive models, such a review is often not prudent.

The research on reporting fraud and financial information disclosure in Vietnam at present only focus on using descriptive statistical methods through the evaluation of indicators, or only using traditional statistical models without descriptive statistics. Modern techniques have not been applied in assessment and prediction, such as research by Ha Thi Thuy Van (2016), Pham Minh Vuong and Nguyen Thi Ha Vy (2020).

3 Research Methodology and Data

ANN Algorithm

ANN is a system that simulates the computational power of a biological neural network (ANN concept diagram is shown in Fig. 1) because the biological vision

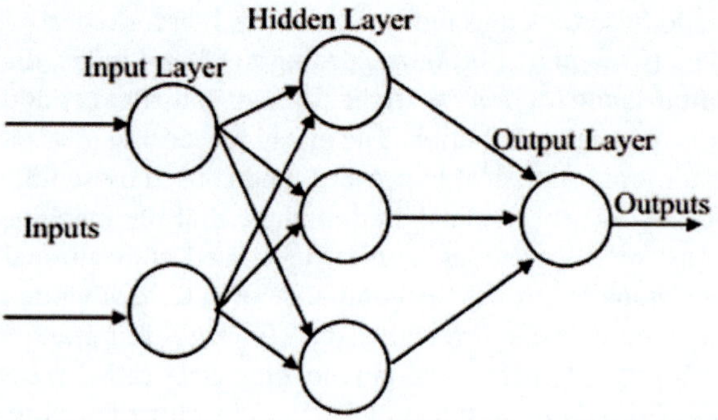

Fig. 1 ANN algorithm diagram

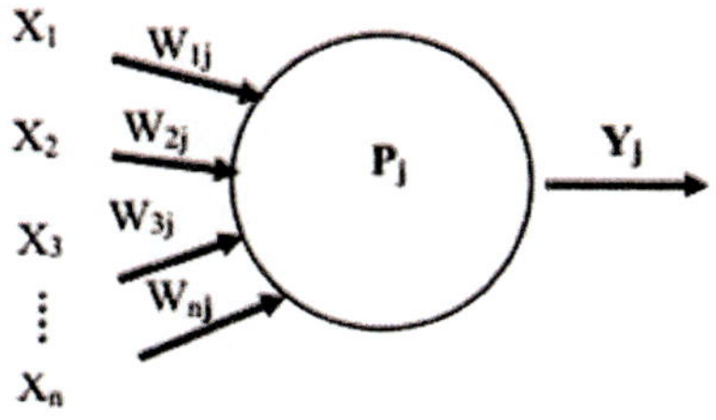

Fig. 2 Neuron in ANN network

and hearing capabilities are superior to computer systems at that time. Before that, it was expected to achieve powerful computing power by imitation. In the 1940s, scientists mimicked the simplest neuronal modes to establish the most primitive ANN. McCulloch and Pitts (1943) proposed a simplified neuronal mathematical model to simulate the computing power of the human brain. The neuron is the most basic unit of the ANN. Suppose there is a neuron j, get a corresponding input variable x, from n neurons in the upper layer, each input variable has a link weight w. The neuron combines all input variables by their weights into the combination function. The neuron activation function reuses the associative function to convert it into an output signal. The theory of the ANN neuron is shown in Fig. 2, where n represents the number of input variables, Xi is the ith input variable, Wij is the weight of the ith variable of the jth neuron, and Pj is association function of the jth neuron. If represented by the activation function f(x), then Yj = f(Pj) is the output value of the jth neuron.

To develop a neural network, it is necessary to have a set of parameters and a training model to train the necessary weights for the predictor variable from the beginning and those parameters are randomly generated. Therefore, the parameters used in each training model are different. Finally, the weights of the predictors generated by the neural network are also different, but the error will reach a minimum value, so the neural network is a deduction obtained through the trial-and-error method and its purpose is to minimize error in model prediction results. For the same data, the weights to be trained are not equal, and so the nature of the neural network is to emphasize the training model. Therefore, there will be no predictive formula for the results of the neural network model, only predictive results are obtained instead.

SVM Algorithm (Support Vector Machine)

SVM is a set of artistically intelligent learning methods proposed by Vapnik (1995). It is a machine learning method based on statistical learning theory and SRM (structural risk minimization). It mainly depends on using input training data to generate an optimal decomposition hyperplane that can distinguish two or more types (classes) of data through learning mechanism. It is a supervised learning, applicable for both prediction and classification method for data mining. The SVM concept diagram is shown in Fig. 3. SVM algorithm could be classified into linear and non-linear.

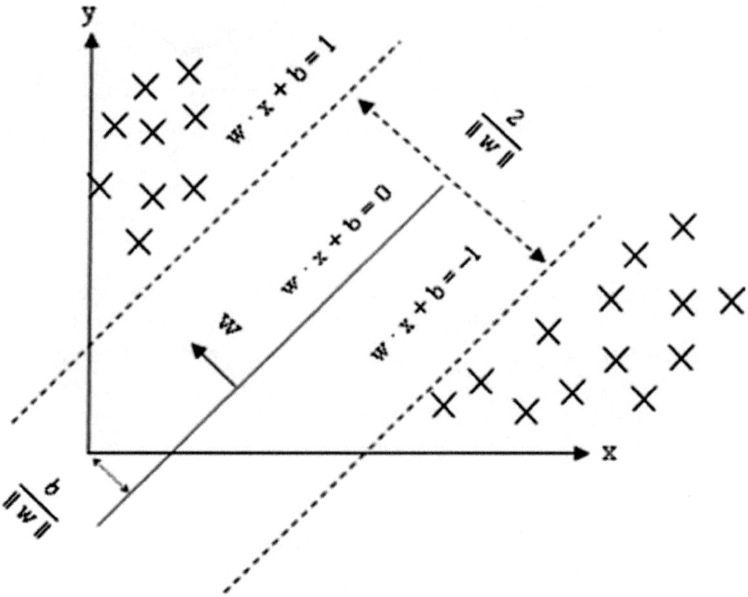

Fig. 3 SVM algorithm diagram

Research data

Table 1 lists the number of firms that the authors use in this study, the data includes non-financial companies (excluding firms in the financial, banking, securities, insurance sectors and all kinds of investment fund). The control sample consists of 2 years 2018 and 2019, the study processes the calculation results also based on the two years in analyzing, comparing, and controlling. The study includes firms listed on 3 exchanges, Hose, Hnx and Upcom, in which there are a total of 1424 and 1456 firms in the observed sample of 2018 and 2019, respectively.

Models and variables

This study uses 2 measures M-Score and Z-Score and then convert these scales into signs of fraud or non-fraud, ranking the degree of fraud based on conversion scales.

Table 1 Number of research firms

	2018	2019
HOSE	340	346
HNX	321	326
UPCOM	763	784
Total	1424	1456

Source Author's summary

The Beneish M-Score is a method which can be used to detect companies that are prone to fraud on their financial statements (Beneish 2012). From experience, companies with higher M Scores are more likely to fraud. Beneish M-Score is a probabilistic model, that is also a limitation in fraud detection since 100% accurate is not practically reachable. The formula of the Beneish M-Score according to Eq. (1) is as follows:

$$\begin{aligned} \boldsymbol{M} &= \mathbf{-4.840 + 0,920}\boldsymbol{DSRI} + \mathbf{0,528}\boldsymbol{GMI} + \mathbf{0,0404}\boldsymbol{AQI} + \mathbf{0,892}\boldsymbol{SGI} \\ &+ \mathbf{0,115}\boldsymbol{DEPI} - \mathbf{0,172}\boldsymbol{SGAI} + \mathbf{4,679}\boldsymbol{TATA} - \mathbf{0,327}\boldsymbol{LVGI} \end{aligned} \quad (1)$$

M-score higher than −2,22 shows signs of financial fraud for that companies, in which:

FRAUD1 = Dummy variable (1 for Manipulator and 0 for Non-Manipulator), this variable is converted from M-score, any observation with M-score higher than −2.22 will have a value of 1, other cases will have a value of 0.

DSRI = Sales index (Accounts Receivables$_t$/Sales$_t$*Number of Days) / (Accounts Receivables$_{t-1}$/Sales$_{t-1}$*Number of Days).

GMI = Gross Margin Index ([Sales$_{t-1}$–COGS$_{t-1}$)/Sales$_{t-1}$] / [Sales$_t$–COGS$_t$)/Sales$_t$]).

AQI = Asset Quality Index ([1 – ((Current Assets$_t$ + PP&E$_t$ + Total Long-term Investments$_t$)/Total Assets$_t$)]/[1 – ((Current Assets$_{t-1}$ + PP&E$_{t-1}$ + Total Long-term Investments$_{t-1}$)/Total Assets$_{t-1}$)]).

SGI = Sales Growth Index (Sales$_t$ /Sales$_{t-1}$).

DEPI = Depreciation Index (Depreciation$_{t-1}$/(PP&E$_{t-1}$ + Depreciation$_{t-1}$))/ (Depreciation$_t$/(PP&E$_t$ + Depreciation$_t$)).

SGAI = Sales and General Administration Expenses Index (SG&A Expense$_t$/Sales$_t$)/(SG&A Expense$_{t-1}$/Sales$_{t-1}$).

TATA = Total Accrual ((Income from Continuing Operations$_t$ – Cash Flow from Operations$_t$) / Total Assets$_t$).

LVGI = Leverage Index ([(Current Liabilities$_t$ + Total Long-term Debt$_t$)/Total Assets$_t$]/[(Current Liabilities$_{t-1}$ + Total Long-term Debt$_{t-1}$)/Total Assets$_{t-1}$]).

εi = residuals.

Altman Z-Score: The Z-score formula was devised in 1968 by Edward I. Altman to predict the bankruptcy of companies. The score outlines five financial ratios with all data derived from the company's financial statements. A score below 1.8 means the company is likely to go bankrupt, while companies with a score above 3.0 are unlikely to go bankrupt. In short, the lower the score, the higher the probability of bankruptcy. This formula can be used to predict the probability that a company will become insolvent within two years. The Z-score is used to predict corporate defaults and is an easy-to-calculate control for firms' financial distress in academic studies. The Z-score uses a variety of income statement and balance sheet values to measure the financial health of a company. Altman's Z-score is calculated through Eq. (2) as follows:

$$Z - Score = 1.2X1 + 1.4X2 + 3.3X3 + 0.6X4 + 0.999X5 \quad (2)$$

In initial empirical research, the Altman Z-Score was found to be 72% accurate in predicting bankruptcy two years prior to the review year, with a Type II error of 6. % (Altman 1968). In a series of follow-up tests covering three periods over the next 31 years until 1999, the model was found to be about 80%–90% accurate in predicting bankruptcy a year before the event occurs, with a Type II error of about 15%–20% (Altman 2000). Specifically, the measurement variables are as follows:

FRAUD2 = Dummy variable, takes 3 values as follows:

Z > = 2.99: Businesses with healthy finances (Safe).

1.81 < Z < 2.99: The business has no problems in the short term, but it is necessary to consider its financial condition carefully (Gray).

Z < = 1.81: The business has serious financial problems (Distress).

X1: Working Capital/Total Assets.

X2: Retained Earnings/Total Assets.

X3: EBIT/Total Assets.

X4: Market Value of Equity/Total Liabilities.

X5: Sales/Total Assets.

4 Results

Table 2 shows the mean values of the variables in the two calculation years (2018 and 2019). In the group of M-Score calculated indexes, there are two indicators that have an average value lower than the required value (standard value, benchmark), which is SGI and TATA. The sales growth index is not satisfactory in terms of the average value, like the accrual value, which is also lower than the standard value. In the group of Z-Score calculated indexes, there are no standard values (benchmark), but the value X2 lower than 0, this negative value indicates that the company pays high dividends on average, but the business result in the years is not good (loss), leading to this ratio lower than 0.

This study uses traditional 3-layer networks for ANN algorithm with ten thousand epochs, 6 neurons at middle layer, activation function Relu. Besides, SVM algorithm uses linear kernel. This paper uses Python programming language with built-in functions in library tensorflow.keras for ANN algorithm and library sklearn.svm.SVC for SVM algorithm. The performances of ANN and SVM algorithm are measured based on the accuracy in prediction for test set, after the models are buit on train set. The accuracy is the fraction of correct predictions. The accuracy is calculated based on the formular: Accuracy = Number of correct predictions / Total number of predictions.

Table 3 shows the prediction results of classifying firms which are under warning or subject to signs of financial statement fraud by machine learning method with high accuracy. For the ANN algorithm, if predicting using the M-Score, the accuracy is 97%, while if using the Z-Score group, the prediction accuracy is only 89%. For the SVM algorithm, in both groups of forecasting indexes, there are high accuracy

Table 2 Variables summary

Index	Benchmark	Average value in M-Score	
		2018	2019
DSRI	>1.165	2.18	1.77
GMI	>1.193	*0.64*	1.52
AQI	>1.254	2.15	1.72
SGI	>1.607	*1.13*	*1.11*
DEPI	>1.00	1.02	1.34
SGAI	>1.041	1.62	1.63
TATA	>0.018	*−0.01*	*−0.03*
LVGI	>1.00	1.03	1.05
M-Score	**>−2.20**	**−0.4373**	**−1.32**
		Average value in Z-Score	
X1		0.09	0.06
X2		−0.14	−0.16
X3		0.05	0.06
X4		4.23	4.59
X5		1.19	1.16
Z-Score		**3.82**	**3.93**

Source Author's summary

results, 97% and 95% for M-Score and Z-Score, respectively. Thus, in terms of the accuracy, the SVM method proved to be more effective in both index groups and the M-Score should be used for better accuracy than the Z-Score in forecasting.

5 Conclusion and Recommendations

With the above research results, using machine learning methods in forecasting and warning businesses with signs of fraud in financial statements is recommended. Using machine learning techniques combined with indicators and mathematical models in forecasting provides high reliability, high accuracy, and speed. To implement the forecasting, we just need to find the optimal model, take financial metrics from any business, and apply analytical machine learning to reach the results. In this study, the results stated that the SVM technique was more effective than ANN and using the M-Score was better than using the Z-Score.

Table 3 Confusion matrix of ANN and SVM algorithm for M-Score and Z-Score

ANN (M-Score)					ANN (Z-Score)				
	Confusion matrix					Confusion matrix			
	0	1				Distress	Gray	Safe	
0	624	13			Distress	329	12	0	
1	13	214			Gray	48	182	9	
	Precision	Recall	f1-score	Support	Safe	8	15	261	
0	0.98	0.98	0.98	637		Precision	Recall	f1-score	Support
1	0.94	0.94	0.94	227	Distress	0.85	0.96	0.91	341
Accuracy			**0.97**	**864**	Gray	0.87	0.76	0.81	239
Macro avg	0.96	0.96	0.96	864	Safe	0.97	0.92	0.94	284
Weighted avg	0.97	0.97	0.97	864	**Accuracy**			**0.89**	**864**
SVM (M-Score)					Macro avg	0.90	0.88	0.89	864
	Confusion matrix				Weighted avg	0.90	0.89	0.89	864
	0	1			*SVM (Z-Score)*				
0	633	4				Confusion matrix			
1	19	208				Distress	Gray	Safe	
	Precision	Recall	f1-score	Support	Distress	256	17	9	
0	0.97	0.99	0.98	637	Gray	3	199	9	
1	0.98	0.92	0.95	227	Safe	0	9	362	
Accuracy			**0.97**	**864**		precision	recall	f1-score	support
Macro avg	0.98	0.96	0.96	864	Distress	0.99	0.91	0.95	282
Weighted avg	0.97	0.97	0.97	864	Gray	0.88	0.94	0.91	211
					Safe	0.95	0.98	0.96	371
					Accuracy			**0.95**	**864**
					Macro avg	0.94	0.94	0.94	864
					Weighted avg	0.95	0.95	0.95	864

Source Author's summary

Funding Information This research is funded by Vietnam National University Ho Chi Minh City (VNU-HCM) under grant number DS2022-34-03.

References

Altman: Financial ratios, discriminant analysis and the prediction of corporate bankruptcy J. Financ. (September 1968). https://doi.org/10.1111/j.1540-6261.1968.tb00843.x

Altman: Predicting financial distress of companies: revisiting the Z-Score and Zeta (September 2000). https://doi.org/10.4337/9780857936097.00027

Beneish, M.D.: Fraud detection and expected return (2012). http://papers.ssrn.com/sol3/papers.cfm?abstract_id=1998387

Beasley, M.: An empirical analysis of the relation between the board of director composition and fnancial statement fraud. Account Rev. **71**(4), 443–466 (1996)

Beasley, M.S., Carcello, J.V., Hermanson, D.R.: Fraudulent fnancial reporting 1987–1997: an analysis of U.S. public companies. The Committee of Sponsoring Organizations of the Treadway Commission (COSO), New York (1999)

Beaver, W.H.: Financial ratios as predictors of failure. J. Account Res. **4**, 71–111 (1966)

Bell, T., Carcello, J.: A decision aid for assessing the likelihood of fraudulent financial reporting. Audit A J. Pract. Theory **9**(1), 169–178 (2000)

Chen, C.H.: Application of grey forecast theory and logit equation in financal crisis warning model from the preevent control viewpoint. Commer. Manag. q. **6**(4), 655–676 (2005)

Chen, G., Firth, M., Gao, D.N., Rui, O.M.: Ownership structure, corporate governance, and fraud: evidence from China. J. Corp. Financ. **12**(3), 424–448 (2006)

Chiu, C.C., Lee, T.S., Chou, Y.C., Lu, C.J.: Application of integrated identifcation analysis and ANN in data mining. J. Chin. Inst. Ind. Eng. **19**(2), 9–22 (2002)

Coats, P.K., Fant, L.F.: A neural network approach to forecasting financial distress. J. Bus. Forecast **10**, 9–12 (1993)

Craja, P., Kim, A., Lessmann, S.: Deep learning for detecting financial statement fraud. Decis. Support Syst. **139**, 113421 (2020).

Dong, W., Liao, S., Liang, L.: Financial statement fraud detection using text mining: a systemic functional linguistics theory perspective. In: Pacific Asia Conference on Information Systems (PACIS). Association For Information System (2016)

Elliot, R., Willingham, J.: Management Fraud: Detection and Deterrence. Petrocelli, New York (1980)

Fanning, K., Cogger, K.: Neural network detection of management fraud using published financial data. Int. J. Intell. Syst. Account Financ. Manag. **7**(1), 21–24 (1998)

Hà Thị Thuý Vân:Thủ thuật gian lận trong lập báo cáo tài chính các công ty niêm yết, tạp chí Tài chính tháng 4/2016 (2016)

Hansen, J.V., McDonald, J.B., Stice, J.D.: Artifcial intelligence and generalized qualitative-response models: an empirical test on two audit decision-making domains. Decis. Sci. **23**(3), 708–723 (1992)

Hajek, P., Henriques, R.: Mining corporate annual reports for intelligent detection of financial statement fraud–a comparative study of machine learning methods. Knowl. Based Syst. **128**, 139–152 (2017)

Humpherys, S.L., Moftt, K.C., Burns, M.B., Burgoon, J.K., Felix, W.F.: Identifcation of fraudulent financial statements using linguistic credibility analysis. Decis. Support Syst. **50**, 585–594 (2011)

Kamarudin, K.A., Ismail, W.A.W., Mustapha, W.A.H.W.: Aggressive fnancial reporting and corporate fraud. Procedia Soc. Behav. Sci. **65**, 638–643 (2012)

Kirkos, S., Spathis, C., Manolopoulos, Y.: Data mining techniques for the detection of fraudulent financial statements. Expert Syst. Appl. **32**(4), 995–1003 (2007)

Koh, H.C.: Going concern prediction using data mining techniques. Manag. Audit J. **19**, 462–476 (2004)

Kotsiantis, S., Koumanakos, E., Tzelepis, D., Tampakas, V.: Forecasting fraudulent financial statements using data miming. World Enformatika Soc. **12**, 283–288 (2006)

Mohammadi, M., Yazdani, S., Khanmohammadi, M.H., Maham, K.: Financial reporting fraud detection: An analysis of data mining algorithms. Int. J. Financ. Manag. Acc. **4**(16), 1–12 (2020)

McCulloch, W.S., Pitts, W.H.: A logical calculus of the ideas immanent in nervous activity. Bull. Math. Biophys. **5**, 115–133 (1943)
Nguyen, M.N., Shi, D., Quek, C.: A nature inspired Ying-Yang approach for intelligent decision support in bank solvency analysis. Expert Syst. Appl. **34**, 2576–2587 (2008)
Phạm Minh Vương và Nguyễn Thị Hà Vy: Dự báo gian lận báo cáo tài chính bằng các chỉ số tài chính cho các doanh nghiệp niêm yết tại Việt Nam, Tạp chí công thương 10/2020 (2020)
Pai, P.F., Hsu, M.F., Wang, M.C.: A support vector machine-based model for detecting top management fraud. Knowl. Based Syst. **24**, 314–321 (2011)
Quinlan, J.R.: C5.0: programs for machine learning. Morgan Kaufmann Publishers, Burlington (1986b)
Ravisankar, P., Ravi, V., Rao, G.R., Bose, I.: Detection of financial statement fraud and feature selection using data mining techniques. Decis. Support Syst. **50**, 491–500 (2011)
Rezaee, Z.: Causes, consequences, and deterrence of financial statement fraud. Crit. Perspect. Account. **16**(3), 277–298 (2005)
Salehi, M., Fard, F.Z.: Data mining approach to prediction of going concern using classifcation and regression tree (CART). Glob. J. Manag. Bus. Res. **13**(3), 24–30 (2013)
Sadgali, I., Sael, N., Benabbou, F.: Performance of machine learning techniques in the detection of financial frauds. Proc. Comput. Sci. **148**, 45–54 (2019)
Sharma, V.D.: Board of director characteristics, institutional ownership, and fraud: evidence from Australia. Audit A J. Pract. Theory **23**(2), 105–117 (2004)
Shin, K.S., Lee, T.S., Kim, H.J.: An application of support vector machines in bankruptcy prediction model. Expert. Syst. Appl. **28**, 127–135 (2005)
Spathis, C., Doumpos, M., Zopounidis, C.: Detecting false financial statements: a comparative study using multicriteria analysis and multivariate statistical techniques. Eur. Account Rev. **11**(3), 509–535 (2002)
Summers, S.L., Sweeney, J.T.: Fraudulently misstated financial statements and insider trading: an empirical analysis. Account Rev. **73**, 131–146 (1998)
Tang, X.B., Liu, G.C., Yang, J., Wei, W.: Knowledge-based financial statement fraud detection system: based on an ontology and a decision tree. KO Knowl. Org. **45**(3), 205–219 (2018)
Uzun, H., Szewczyk, S.H., Varma, R.: Board composition and corporate fraud. Financ. Anal. J. **60**(3), 33–43 (2004)
Vapnik, V.: The Nature of Statistical Learning Theory. Springer, Berlin (1995)
Wells, J.T.: Occupational Fraud and Abuse. Obsidian Book Publishing, Nottingham (1997)
Yao, J., Zhang, J., Wang, L.: A financial statement fraud detection model based on hybrid data mining methods. In: 2018 International Conference on Artificial Intelligence and Big Data (ICAIBD), pp. 57–61. IEEE (2018, May)
Yeh, C.C., Chi, D.J., Hsu, M.F.: A hybrid approach of DEA, rough set and support vector machines for business failure prediction. Expert Syst. Appl. **37**, 1535–1541 (2010)
Zhou, W., Kapoor, G.: Detecting evolutionary financial statement fraud. Decis. Support Syst. **50**, 570–575 (2011)

Financial Risks in the Construction Enterprises: A Comparison Between Frequency Regression and Bayesian Econometric

Thi Hong Le Hoang, Thuy Duong Phan, and Thi Huyen Do

Abstract The article analyzes the factors affecting the financial risks faced by construction companies listed on the stock markets in Vietnam. The data is collected from the financial statements of 133 construction companies listed on the stock markets in Vietnam from 2009 to 2019. In this study, the dependent variables of financial risks are measured by Debt structure, Solvency, Profitability, Operational ability, Capital structure that are independent variables in the study. Company size, company age, and growth rate are controlled variables. Research results from using Bayes method shows that companies have higher profitability, solvency, operational efficiency, self-financing ratio and fixed asset ratio, they will be less exposed to financial risks.

1 Introduction

Construction companies express their own typical features during their production and business activities as follows: production activities take place in a large scale, in many different areas; construction and installation works are usually of large volumes, great values, and require relatively long construction time; construction and installation works are highly specific and unique. As a result of that, construction companies are greatly affected by fluctuations in the economy such as fiscal policies, monetary policies set by the Government, interest rates, exchange rates, and the movement of cash flows; which makes the financial management of companies become complicated. In addition, a construction company may implement multiple

T. H. Le Hoang · T. D. Phan (✉) · T. H. Do
Faculty of Transport Economics, University of Transport Technology, Hanoi, Vietnam
e-mail: duongpt@utt.edu.vn

T. H. Le Hoang
e-mail: lehth@utt.edu.vn

T. H. Do
e-mail: huyendt@utt.edu.vn

N. Ngoc Thach et al. (eds.), *Financial Econometrics: Bayesian Analysis, Quantum Uncertainty, and Related Topics*, Studies in Systems, Decision and Control 427,
https://doi.org/10.1007/978-3-030-98689-6_29

construction projects of large investment in many locations with different complexities, so its capital demand is huge. To meet the capital needs for such projects, the construction company must secure sufficient capital that is mobilized from multiple different sources, which makes it difficult for the company to raise funds and repay debts. Although unexpected, there are always financial risks to any financial investment decision or business transaction of a construction company. Such risks can cause financial losses subject to the risk levels, but they may also push companies into insolvency or even bankruptcy. Therefore, construction companies are required to pay attention to financial risk management during their operations. Good financial risk management will help a company minimize the damage caused by a financial risk.

Financial risks faced by companies have been mentioned in many different studies.

Altman (1968) used the multifactorial discriminant analysis (MDA) to build a Z model for the purpose of measuring financial risks.

Alexander Bathory's research has provided a model to measure financial risks in enterprises in 1984. Studies applying Alexander Bathory's model can include: Bhunia and Mukhuti (2012) studying financial risks based on secondary data collected from financial reports of 513 enterprises listed at Bombay Stock Exchange, India for the period 2010–2011; Gang and Dan (2012) studied financial risks using financial statements data of 216 listed enterprises at Shenzhen Stock Exchange, China in 2010; Simantinee and Kumar (2015) measured and identified factors affecting financial risk of 50 enterprises including banks, analytical data collected from financial statements of 41 enterprises listed on NIGTY stock exchange from 2014 to 2015; Cao and Zen (2005) studied the factors affecting the financial risks of listed enterprises in China; Zhou and Zhao (2006) studied the financial risks of private enterprises in China.

Alexander Bathory's research introduced a model to measure financial risk in businesses in 1984. Studies applying Alexander Bathory's model can include: Bhunia and Mukhuti (2012) study financial risk based on Alexander Bathory's model. on secondary data collected from financial statements of 513 companies listed on Bombay Stock Exchange, India for the period 2010–2011; Gang and Dan (2012) studied financial risk using financial statement data of 216 companies listed on the Shenzhen Stock Exchange, China in 2010; Simantinee and Kumar (2015) measure and identify factors affecting financial risks of 50 enterprises including banks, analytical data collected from financial statements of 41 listed companies on the stock exchange. NIGTY contract from 2014 to 2015; Cao and Zen (2005) studied the factors affecting the financial risk of listed companies in China; Zhou and Zhao (2006) studied the financial risks of private enterprises in China.

In Vietnam, Hau (2013) identified financial risks through a group of indicators reflecting solvency with five indicators: General solvency, short-term solvency, quick ratio, solvency, and interest coverage. In addition, the author mentioned a group of indicators used to assess financial risks including: debt structure, operational performance, asset structure and capital structure, profitability, financial leverage, variance, and standard deviation. Lan (2013), also agreed with Hau (2017) when identifying financial risks through financial leverage and liquidity risks. In addition, the author mentioned other indicators such as exchange rate risk, interest rate risk, trade credit risk, risk in relation to raw material price volatility, and bankruptcy risk.

However, the current studies in Vietnam have only looked into financial risks occurring during a short period of time. Meanwhile, the financial risks of construction companies are required to be looked over for a sufficiently long period of time, at least ten years. This study extends the period of time to be researched and uses a different approach named Bayesian regression model to assess financial risks for listed construction companies in Vietnam during the period of 2009–2019.

2 Literature Review and Hypotheses

2.1 Literature Review

Since the 1960s, financial risk assessment models have been a major concern among researchers.

Z-score model: Altman (1968) was the first scientist to use multifactorial discriminant analysis (MDA) to build a Z model to measure the financial risks. The original Z model illustrated the fluctuations of the factors of asset investment structure, profitability, capital structure and operating performance to the financial risk of enterprises through 5 indicators: (i) working capital/total assets (X1 = WC/TA), (ii) retained earnings/total assets (X2 = RE/TA), (iii) earnings before interest and taxes/total assets assets (X3 = EBIT/TA), (iv) market value of equity/book value of liabilities (X4 = ME/TL), (v) sales/total assets (X5) = S/TA). Z = 1.2X1 + 1.4X2 + 3.3X3 + 0.64X4 + 0.999X. Research results of Altman (1968) show that the model has a reliability of 72–80%. Altman's Z model has been applied in many later studies. Napp's study (2011) also mentions other studies using the Z model in research on risk, recognizing signs of bankruptcy, and financial structure. Napp (2011), has applied a combination of financial ratios in TWINTECH—a listed company on the German stock market—from 2006 to 2010. Financial ratios have been calculated and approved to make a comparison between the bankrupt and the non-bankrupt group because the index results by themselves are not meaningful if not compared. Altman (1968), Ohlson (1980), Sandin and Porporato (1997), Ong et al. (2011), Ivicic and Cerovac (2009) concluded that there is a huge difference in the value of the debt ratio for two groups of firms (bankrupt and non-bankrupt) and the difference in the value of the debt ratio in time-series debt.

Alexander model: Researcher Alexander (1987) developed a model to measure the financial risks of enterprises with five financial indicators such as capital structure, profitability, and proportion of investments in financial assets, etc. and five financial indicators to identify financial risk. Bathory's model is as follows:

$$FRit = SZLit + SYit + GLit + YFit + YZit \quad (1)$$

In which:

FRit: the value measuring financial risk of index (dependent variable)

$$\text{SZLit} = (\text{Profit before tax} + \text{Depreciation} + \text{Deferred tax}) / \text{Current liabilities} \quad (2)$$

$$\text{SYit} = \text{Pre} - \text{tax profit/Operating capital} \quad (3)$$

$$\text{GLit} = \text{Shareholders}' \text{ interests /Current liabilities} \quad (4)$$

$$\text{YFit} = \text{Net tangible assets /Total liabilities} \quad (5)$$

$$\text{YZit} = \text{Working capital /Total assets} \quad (6)$$

According to Alexander's model, the higher the value of FRit is, the higher the performance of the enterprise is, and the lower its financial risks are.

Bhunia and Mukhuti (2012) also adopted Bathory's model to conduct their research with the analysis data based on the annual financial statements of the companies listed on the Bombay Stock Exchange, India. These two authors used descriptive statistics and correlative regression analysis to assess the impact of independent variables on Frit, the main dependent variable in Bathory's model. The study has examined five groups of factors affecting financial risks, including: Debt structure, solvency, profitability, operational performance, and capital structure.

Cao and Zen (2005) used financial leverage as a dependent variable to study the financial risks of large firms. According to their research results, financial risks are positively correlated with debt size and structure, negatively correlated with operational efficiency and profitability, and not correlated with interest rate and solvency.

The theory regarding this issue applied to Vietnam has been verified by Hau (2017) and Chi (2020). Hau (2017) analyzed the factors affecting financial risks of 34 real estate companies listed on the HSX for the period of 2013–2015. The study shows the financial risks of real estate companies listed on the Ho Chi Minh Stock Exchange have a significant correlation with liquidity ratio, current payment, quick ratio, general solvency, fixed capital; and has no correlation with debt structure, turnover ratio, profitability ratio of assets, inventory turnover, fixed asset turnover, total asset turnover, receivable turnover, core capital. Chi (2020) applied Alexander Bathory's model to measure financial risks from the data of 12 companies listed on HNX and HOSE. The research conducted by Chi (2020) shows that the variables including solvency, profitability, operational performance, financial structure are negatively correlated with financial risks and interest rate is positively correlated with financial risks.

Alexander Bathory's model is used as a dependent variable in multiple empirical studies conducted in India, China, Kenya, etc. through secondary databases of all types of businesses and industries collected at Stock Exchanges. In Vietnam, there are a big number of researches verify the database of enterprises listed on the stock markets, especially the database of state-owned enterprises. As a result of that, no explanation and recommendation are available. Therefore, within the scope of the research, the Alexander Bathory's model continues to be tested in Vietnam to identify the factors affecting financial risks, thereby contributing to improving the effectiveness of financial risk management in companies.

2.2 Research Hypotheses

Based on the available theory and experiments on financial risks, the group of authors have proposed 5 research hypotheses about the factors affecting financial risks of construction enterprises listed on the Vietnamese stock market as follows:

Debt structure: The debt structure of a construction enterprise shows the debt-to-equity ratio. If the debt-to-equity ratio of a company is too high, it will put pressure on the company to pay its due debts, thereby triggering financial risks. Bathory's research results demonstrate that debt structure is not correlated with financial risks. This result is also proved by the study of Gang and Liu (2012), Bhunia and Mukhuti (2012), and Hau (2017). Regarding the companies listed on the Vietnamese stock market, how are their financial risks and debt structures related? This question raises the first hypothesis:

H_1: The debt structures of the listed construction companies are positively correlated with their financial risks.

Solvency: The solvency reflects clearly the financial situation of a construction company. Solvency is the ability to pay due debts at any time (including both principal and interest). Good financial situation and high solvency will ensure the payment for the due debts and low financial risks. In Bathory's model, financial risks are negatively correlated with solvency. This result has been proved through the results of the researches done by: Gang and Dan (2012), Bhunia and Mukhuti (2012), and Hau (2017). The above arguments bring about the following hypothesis about the relationship between solvency and financial risks:

H2: The solvency of the listed companies in the construction industry is negatively correlated with financial risks.

Profitability: Profitability reflects the ability to create a profit from a unit of cost or input, which is the output reflecting business performance. When business activities of an industrial company are being well conducted, creating favorable conditions to increase profits, the profits of that company also increase. Therefore, the company has opportunity to increase accumulated profits, improve solvency, repay due debts, and reduce financial risks. Gang and Liu (2012), and Bhunia and Mukhuti (2012) have shown an inverse relationship between financial risks and profit. However, the research results of Hau (2017) do not confirm this hypothesis. The hypothesis put forward is:

H_3: The profitability of listed construction companies is negatively correlated to financial risks.

Operational Capability: Operational capability is an industrial enterprise's ability to achieve operational results after consuming the inputs during the business operation. The higher the growth of revenue or payment speed is, the further the development of business activities of the enterprise is and the lower financial risks are. The results of Bathory's study and subsequent studies failed to demonstrate a relationship of performance with financial risk (Gang and Dan 2012; Bhunia and Mukhuti 2012; Hau 2017). It is hypothesized that:

H_4: The operational capability of listed construction companies is negatively correlated with financial risks.

Capital structure: Capital structure shows the proportion of each type of capital in total capital. When the self-financing ratio is equal to the debt-to-equity ratio, it is likely that the company can repay due debts, the creditors can feel more secured, and the company's financial risks are lower. The higher the value of fixed assets is, the more secured the loans of the lenders are, the lower the financial risks of the business are and vice versa. The results of Bathory's study and subsequent studies show that there is a negative relationship between financial risks and capital structure (Gang and Liu 2012; Bhunia and Mukhuti 2012; Hau 2017). Below is the hypothesis:

H_5: The capital structure of the listed companies in the industry is negatively correlated with financial risks.

3 Research Model and Research Method

Based on the empirical studies in Sect. 2, we propose the following research mode:

$$\begin{aligned} \mathrm{FR_{it}} = {} & \beta_1 + \beta_2\mathrm{DS_{it}} + \beta_3\mathrm{QR_{it}} + \beta_4\mathrm{ROS_{it}} + \beta_5\mathrm{ROA_{it}} \\ & + \beta_6\mathrm{IT_{it}} + \beta_7\mathrm{FAT_{it}} + \beta_8\mathrm{TAT_{it}} + \beta_9\mathrm{RT_{it}} + \beta_{10}\mathrm{ES_{it}} \\ & + \beta_{11}\mathrm{FAS} + \beta_{12}\mathrm{AGE_{it}} + \beta_{13}\mathrm{SZ_{it}} + \beta_{14}\mathrm{GR_{it}} + \varepsilon_{\mathrm{it}} \end{aligned}$$

From the previous research, methods of measuring financial risk include the asset-liability ratio, probabilistic analysis, financial leverage coefficient, etc. Financial risk measured by asset -liability ratio method is vague, and it still needs to combine with return on assets. In this paper, Alexander Bathory model was used to measure the financial risk. The model can be expressed as below: $FR_{it} = SZL_{it} + SY_{it} + GL_{it} + YF_{it} + YZ_{it}$.

$\mathrm{FR_{it}}$ is the value measuring financial risk of index.

According to Alexander Bathory's view, the smaller the value of $\mathrm{FR_{it}}$ is, the weaker the enterprise strength is, and the higher financial risk of enterprise is (Table 1).Source: Compiled by the authors

Table 1 Description of variables in the research model

Variable name	Code	Related definition	Notes
Financial risk	FR_{it}	Bathory's model metrics	Financial risk metric values
Debt structure	DS	Liability structure ratio	Current liabilities/non-current liabilities
Solvency	QR	Quick Ratio	(Current assets-inventory)/current liabilities
Profitability	ROA	Return on Assets	ROA = Net income/Average total assets
	ROS	Return on Sales	ROS = Net income/Sales
Operation ability	IT	Inventory turnover	Cost of goods sold/inventories
	FAT	Fixed asset turnover	Sales/net fixed assets
	TAT	Total asset turnover	Sales/total assets
	RT	Accounts receivable turnover	Annual credit sales/accounts receivable
Capital structure	ES	Net assets ratio	Equity/total assets
	FAS	Fixed assets ratio	Fixed assets/total assets
Sale Growth	GR	Sale Growth	(Current period net sales—previous period net sales)/ previous period net sales
Firm size	SIZE	Firm size	Firm size measured by log of total assets
Firm age	AGE	Firm age	Year of research—year of establishment

The study used panel data collected from 133 construction enterprises listed on Vietnam Stock Market from 2009 to 2019, provided by FiinGroup JSC. Research data is extracted from the audited financial statements of these enterprises.

The baseline analysis was first performed to screen the sample, to eliminate observations that were too large, too small, or too different from the sample size. This basic analysis step helps to check the suitability of the sample before performing regression analysis OLS, FEM, REM, to ensure the reliability of quantitative research results. Specifically, the author group conduct statistical description analysis, correlation analysis to eliminate multi-collinear phenomena between independent variables. After selecting the appropriate method to run the model, the author examine the variance of variance, multicollinearity, autocorrelation, endogeneity of the model. In case the model has a defect, the author will use the GLS (Generalized least squares) method to overcome.

In frequency statistics, the parameters of the population are deemed as constants that are fixed but unknown. But for time series data, these parameters will vary. Therefore, it is not recommended to consider parameters as constants any more. In recent years, scientists have been convinced more and more about the disadvantages of the frequency statistical method (Frequentist) because this method leads to many wrong conclusions in scientific research (Tuan 2011). Scientific conclusions in frequency statistics are all based on data sets without regard to known information (Thach 2019). In Bayesian statistics, the parameters are assumed to be random variables and follow a distribution rule (van de Schoot and Depaoli 2014; Bolstad and Curran 2016). Conclusions based on Bayes method rely on a priori information combined with the collected data; therefore, they should be more accurate. In this study, the authors compare the results of frequency regression with those of Bayesian regression method to understand the impact of factors on the financial risks of the companies listed on the Vietnam Stock Exchange from 2009 to 2019.

4 Empirical Results

4.1 Frequency Regression Method

Table 2 presents models describing regression results and test results when choosing appropriate models corresponding to the frequency regression method.

The sampling was quite large with 1463 observations so the priori information did not significantly affect the accuracy of the model. Therefore, the authors adopted Normal priori distribution (1, 100) for the regression coefficient and Invagamma (2.5, 2.5) for variance of observational error.

Table 2 Frequency regression result

Varible	VIF	FR			
		POLS	FEM	REM	FGLS
DS	1.00	0.000004	−0.00001	−0.00001	−0.000003
QR	2.14	2.310***	2.115***	2.219***	1.061***
ROS	1.08	0.0975	0.0724	0.0904	0.164
ROA	1.64	6.534***	2.991***	4.079***	2.938***
IT	1.44	0.0343***	−0.0117***	−0.00243	−0.0100***
FAT	1.18	0.0019	0.0026**	0.00258***	0.0006
TAT	1.52	−1.123***	−0.224	−0.502***	−0.149***
RT	1.52	0.0902***	0.0305**	0.0514***	0.0525***
ES	2.42	6.551***	7.843***	7.859***	6.546***
FAS	1.39	3.821***	2.819***	3.446***	0.938***
AGE	1.02	−0.0076	−0.0067	−0.0102	−0.0023
SIZE	1.34	−0.131**	−0.102	−0.0485	−0.0186
GR	1.04	−0.0422	−0.0184	0.00184	−0.0360
Cons		3.007*	2.003	0.421	0.877
N		1247	1247	1247	1247
Significance		F (13, 1233) = 167.90	F(13,1102) = 62.65	Wald chi2(13) = 1191.23	Wald chi2(13) = 2311.96
White test		Chi2 (103) = 365.03 Prob > Chi2 = 0.0000			
Wooldridge test		F(1,126) = 27.550 Prob > Chi2 = 0.0000			
Hausman test			chi2(13) = 143.14 Prob > chi2 = 0.0000		
Wald test			chi2 (132) = 5.2e + 30 Prob > chi2 = 0.0000		

* p < 0,1 ** p < 0,05 *** p < 0,01

The diagnosis of convergence of MCMC is carried out in order to ensure that MCMC-based Bayesian Inference is reasonable. The testing of convergence of MCMC is performed via Trace plot, Histogram, Autocorrelation and Density pot. The test result shows that the histogram quickly passes the distribution; a massive drop in the autocorrelation chart displays a low autocorrelation. Shape of Histogram stimulating the shape of probability distributions is consistent. Therefore, it can be concluded that the Bayesian inference is stable.

Table 3 Bayesian regression result

	Mean	Std. Dev	MCSE	Median	Equal-tailed [95% Cred. Interval]	
DS	4.99e-06	0.000018	1.1e-06	4.98e-06	−0.00003	0.00004
QR	2.2389	0.0560	0.0166	2.2397	2.1536	2.3366
ROS	0.1047	0.0200	0.0025	0.1047	0.0671	0.1431
ROA	6.5064	0.0241	0.0065	6.5064	6.4604	6.5502
IT	0.0353	0.0035	0.0003	0.0352	0.0283	0.0425
FAT	0.0019	0.0011	0.0000	0.0019	−0.0002	0.0040
TAT	−1.1174	0.0840	0.0062	−1.1193	−1.2819	−0.9559
RT	0.0885	0.0091	0.0006	0.0884	0.0706	0.1066
ES	6.5837	0.0325	0.0086	6.5833	6.5222	6.6459
FAS	3.9029	0.0802	0.0204	3.9050	3.7523	4.0418
AGE	−0.0079	0.0089	0.0006	−0.0076	−0.0261	0.0091
SIZE	−0.1279	0.0063	0.0009	−0.1279	−0.1398	−0.1149
GR	−0.0951	0.0547	0.1281	−0.0933	−0.2006	0.0044
_cons	2.9619	0.0328	0.0094	2.9600	2.9054	3.0220
Sigma2	7.0591	0.2810	0.0059	7.0480	6.5149	7.6196

Regression results collected by adopting Bayes method are presented in the following table:

Regression results collected by adopting Bayes method are presented in the following table:

The simulation result shows that MC errors (MCSE) of posterior average value are decimal numbers which are less than 1.

Table 3 shows that according to the regression model based on Bayes method, debt structure, quick ratio, profitability, capital structure, operating performance except for asset turnover are positively correlated with FR. Meanwhile, asset turnover, company age and size are negatively correlated with FR. Sales growth rate does not affect FR.

Regarding debt structure, debt structure has a negative relationship with financial risks of construction enterprises, but the level of impact is not significant. This result is consistent with the research results of Gang and Dan (2012), Bhunia and Mukhuti (2012). When the ratio of short-term debt to long-term debt of a construction enterprise increases, the financial risks of the enterprise decrease and vice versa, but the level of impact is not significant.

Regarding solvency, solvency has a negative relationship with financial risks of construction enterprises. When the solvency of a construction company increase by 1%, its financial risks decrease by 2.24%. This result is consistent with the results of Gang and Dan (2012), Bhunia and Mukhuti (2012).

Regarding profitability, profitability has a negative relationship with financial risks of construction enterprises, the level of impact of ROA on financial risk is larger than ROS. When ROA increases by 1%, the financial risk of construction enterprises

Table 4 The comparison of two methods

Hypothesis	Classical regression models		Bayesian models	
	Direction	Results	Direction	Results
H_1	Pos	Rejected	Neg	Not Rejected
H_2	Neg	Not Rejected	Neg	Not Rejected
H_3	Neg	Not Rejected	Neg	Not Rejected
H_4	Neg/Pos	Rejected	Neg/Pos	Rejected
H_5	Neg	Not Rejected	Neg	Not Rejected

decreases by 6.5%. This result is similar to the research results of Gang and Dan (2012), Bhunia and Mukhuti (2012). Businesses with high profitability have better financial positions and are less exposed to financial risks.

Regarding operational performance, inventory turnover, receivable turnover, and fixed asset utilization efficiency are negatively correlated to financial risks, but total asset turnover is positively correlateds with financial risks. To minimize financial risks, construction enterprises need to improve the efficiency of inventory management and management of receivables to avoid capital appropriation.

Regarding capital structure, capital structure has a negative relationship with financial risks, the impact of self-financing coefficient is greater than the impact of the ratio of investment in fixed assets. When the self-financing coefficient increases by 1%, the financial risks of construction enterprises decrease by 6.58%. This means that businesses with high financial autonomy are less likely to face financial risks. When the ratio of investment in fixed assets increases by 1%, the financial risks of construction enterprises decrease by 3.9%. This result is similar to the research results of Gang and Dan (2012), Bhunia and Mukhuti (2012).

In addition, the firm age (AGE) and the firm size (SIZE) have negative impacts on Frit and positive impacts on financial risks.

4.2 The Comparison of the Results of the Two Statistical Methods

The results of three statistical methods including are classical regression models and Bayesian models stated in Table 4. Table 4 reports the comparison of two statistical methods for five hypotheses.

According to the first hypothesis, there is a negative relationship between debt structure and financial risk of construction companies. According to Classical regression models' results, the debt structure is positively correlated to financial risks and does not have statistical significance. According to the results of Bayesian model, the debt structure is negative correlated with financial risks. Therefore, the first hypothesis is accepted.

In the second and third hypotheses, the solvency and profitability are inversely proportional to financial risk of construction companies. The results of 2 models show that the solvency and profitability are inversely proportional with financial risks. Therefore, these two hypotheses are accepted.

In the fourth hypothesis, it is stated that operational performance is inversely proportional to financial risks of construction companies. According to the result of Classical regression models, inventory turnover and asset turnover are proportional to financial risks while receivable turnover is inversely proportional to financial risks. The effects of those variables on financial risks are not consistent. Therefore, this hypothesis is not accepted. As per Bayesian model, inventory turnover, receivable turnover and fixed asset turnover are inversely proportional to financial risks while asset turnover is proportional to financial risks. This explains why the fourth hypothesis is not accepted.

The fifth hypothesis shows that the capital structure is inversely proportional to financial risks of construction companies. The results of the 2 models indicate that capital structure is inversely proportional to financial risks. Therefore, this hypothesis is accepted.

As result, it is recommended that the empirical results of Bayesian models are similar to classic regression results. In addition, the results of Bayesian methods indicate their more credibility than classic models, are similar to the classical regression methods.

5 Conclusion and Recommendation

The research shows empirical evidence proving the influence of multiple factors on the sfinancial risks of construction companies. By using Bayes method, the finding shows that if companies have higher profitability, solvency, operational performance, self-financing coefficient, and fixed asset investment ratio, they will be less likely to face financial risks.

The finding provides empirical evidence for the listed companies as well as state management authorities so that they can come up with appropriate business strategies and policies which can help the companies reduce financial risks and achieve sustainable development. The details are as follows:

Construction companies need to develop appropriate business strategies to increase their sales. Science and technology are also required to be adopted in production activities to reduce costs, thereby improving their profits. Increased profitability and good financial ability may contribute to the reduction of financial risks for a company. Moreover, construction companies need to enhance their financial autonomy so they will not depend largely on external finance during the implementation of business and production projects. Vietnamese construction companies are characterized by a high level of debt utilization so they need to develop suitable policies to improve their financial autonomy and reduce the financial risks. Furthermore, the companies need to improve their operational performance and avoid dead capital in receivables and inventories.

Acknowledgements The authors are thankful to University of Transport Technology for funding this research. We would like to thank the referees for their helpful comments and suggestions.

References

Alexander, B.: The Analysis of Credit: Foundations and Development of Corporate Credit Assessment. McGraw-Hill Press (1987)

Altman, E.: Financial ratio, discriminant analysis and prediction of corporate bankruptcy. J. Financ. **23**, 589–609 (1968)

Bhunia, A., Mukhuti, S.: Financial risk measurement of small and medium-sized companies listed in Bombay Stock Exchange. Int. J. Adv. Manag. Econ.**1**(3), 27–34 (2013)

Bolstad, W.M., Curran, J.M.: Introduction to Bayesian Statistics, 3rd edn. Wiley, New Jersey (2016)

Cao, D., Zen, M.: An empirical analysis of factors influencing financial risk of listed companies in China. Techno Economics & Management Research, vol. 6, pp. 37–38 (2005)

Chi, N.T.M.: Financial risk analysis at telecommunications companies listed on Vietnam's stock market. Doctoral dissertation, National Economics University, Vietnam (2020)

Gang, F., Liu, D.: Empirical study on the financial risk factors for small and medium-sized enterprise: The evidence from 216 companies of small plates stock market in China, J. Contemp. Res. Business, **3**(9), 380–387 (2012)

Hau, V.T.: An analysis of factors influencing financial risk of real estimate firms listed on Ho Chi Minh stock market. J. Econ. Dev. **240**, 86–93 (2017)

Ivicic, L., Cerovac, S.:Credit risk assessment of corporate sector in Croatia. Financ. Theory Pract. **33**(4), 373–399 (2009)

Lan, T.T.P.: Construction-estate enterprises financial leverage risks. VNU J. Econ. Bus. **29**(3), 68–74 (2013)

Napp, A.K.: Financial risk management in SME. The use of financial analysis for identifying, analyzing and monitoring internal financial risk. Master Thesis, Aarhus School of Business, Aarhus University (2011)

Ohlson, J.A.: Financial ratios and the probabilistic prediction of bankruptcy. J. Account. Res. **18**(1), 109–131 (1980)

Ong, S., Choong Yap, V., Khong, R.W.L.: Corporate failure prediction: a study of public listed companies in Malaysia. Manag. Financ. **37**(6), 553–564 (2011)

Sandin, A.R., Porporato, M.: Corporate bankruptcy prediction models applied to emerging economies: evidence from Argentina in the years 1991–1998. Int. J. Commer. Manag. **17**(4), 295–311 (2007)

Simantinee, S., Kumar, T. V. V.: Factors Influencing Financial Risk-A Case Study of NSE NIFTY Companies. Int. J. Manage. Social Sci, **3**(8), 132–137 (2015)

Thach, N.N.: A Bayesian approach to US gross domestic product forecasting. Asian J. Econ. Bank. **163**, 5–18 (2019)

Tuan, N.V.: Introduction to the Bayesian method. J. Med. News **63**, 26–34 (2011)

Van de Schoot, R., Depaoli, S.: Bayesian analyses: Where to start and what to report. Eur. Health Psychol. **16**(2), 75–84 (2014)

Zhou, C., Zhao, D.: Empirical research on financial risks of China's private enterprises. Soft Sci. China **4**, 45–51 (2006)

Implications for Bank Functions in Terms of Regulatory Quality and Economic Freedom: The Bayesian Approach

Le Ngoc Quynh Anh, Pham Thi Thanh Xuan, and Le Thi Phuong Thanh

Abstract This paper examines the relationship between banking functions under two instruments of national governance, regulatory quality and economic freedom of 267 banks in the Asia–Pacific region. According to this research findings, stricter regulations will disrupt bank functions, particularly liquidity functions and capital requirements under the Basel regulatory framework. Countries with a higher index of economic freedom, on the other hand, are more likely to generate liquidity and a higher proportion of Tier 1 capital, assisting banks in the process of enforcing Basel regulations. Furthermore, the study sheds light on the negative relationship between the functions of the bank, namely liquidity, capital, and credit risk. Finally, Bayesian multivariate regression and Markov Chain Monte Carlo sampling (MCMC) have been chosen to replace traditional regression methods in models with high interaction between banking functions, between two country governance tools, and the number of parameters greatly exceeds the sample size.

Keywords Regulatory quality · Economic freedom · Bank liquidity creation · Bayesian analysis · MCMC

1 Introduction

Regulatory quality is viewed as the ability of the government to formulate and implement sound policies and regulations that permit and promote private sector development. Improved regulatory quality can promote economic growth by creating

L. N. Q. Anh (✉) · L. T. P. Thanh (✉)
University of Economics – Hue University, Hue City, Vietnam
e-mail: lengocquynhanh@hueuni.edu.vn

L. T. P. Thanh
e-mail: lethiphuongthanh@hueuni.edu.vn

P. T. T. Xuan (✉)
University of Economics and Law, Vietnam National University, Ho Chi Minh City, Vietnam
e-mail: xuanptt@uel.edu.vn

N. Ngoc Thach et al. (eds.), *Financial Econometrics: Bayesian Analysis, Quantum Uncertainty, and Related Topics*, Studies in Systems, Decision and Control 427,
https://doi.org/10.1007/978-3-030-98689-6_30

effective and efficient incentives for the private sector. Conversely, burdensome regulations have a negative impact on economic performance through economic waste and decreased productivity. Worldwide Governance Indicators (WGI).[1] Economic freedom is basically the fundamental right of every human being to control his or her own labour and property. In an economically free society, individuals are free to work, produce, consume, and invest in any way they please; governments allow labour, capital, and goods to move freely; and refrain from coercion or constraint of liberty beyond the extent necessary to protect and maintain liberty (Nagaraj and Zhang 2019). Economic freedom itself is an important value, and regulatory quality is a vital engine for generating the wealth that makes possible a wide range of important economic and social achievements. In recent years, it has been confirmed that implementation of the fundamental tenets of economic freedom leads to rapid increases in incomes; dramatic drops in poverty; sustainable gains in education, health, and the environment; and improved conditions for democracy and peaceful cooperation among neighbours (Miller et al. 2019).

Regarding this issue in the banking industry, previous studies have recognized the importance of government regulations to protect and enhance the growth of the banking system (Asaftei and Kumbhakar 2008; Besley and Burgess 2004; Johnson et al. 1997; Beltratti and Stulz 2012) noted that lax regulation could contribute to excessive risk-taking by banks. Kamarudin et al. (2016) found both statistically and economically significant positive effects of regulatory quality on the revenue efficiency of Islamic banks. A number of studies have shown evidence of a strong relationship between regulatory quality and bank efficiency levels (Acemoglu et al. 2001; Easterly and Levine 2003; Gee et al. 2016; Easterly and Levine 2003) pointed out that better institutional environments resulted in lower bank earnings. This positive impact of regulatory quality on the overall banking industry is a reflection of the significant effect of a better institutional environment on facilitating greater market and financial development, that results in the efficiency of the banking industry (Levine et al. 2000).

This study contributes to the literature in a number of ways. Firstly, studies have been conducted explicitly to investigate the impact of regulatory quality on the banking industry, with no systematic evidence that stronger regulation leads to better performance of banks. Second, the functional impact of banks on higher quality government regulation cannot be reasonably investigated without explicitly taking economic freedom into account. Therefore, the main objective of this study

[1] The Worldwide Goveernance Indicators (WGI) is a research dataset summarizing the views of a large number of enterprise, citizen, and expert survey respondents in industrial and developing countries. This data is gathered from a number of research institutes, think tanks, non-governmental organizations, international organizations, and private sector firms. The WGI does not reflect the official views of the Natural Resource Governance Institute, the Brookings Institution, the World Bank, its Executive Directors, or the countries they represent. The WGI is not used by the World Bank Group to allocate resources.

is to focus on providing evidence on the effects of banking functions not only on the quality of government regulation but also on economic freedom. Third, the study focuses on a controversial issue, which is the impact of these macro factors on bank capital, bank liquidity creation, and bank lending risk. Previous studies only focused on efficiency and bank stability. Finally, this study is the first attempt to assess bank functions under regulation quality and economic freedom using Bayesian regression.

The remainder of this paper is organised as follows: Sect. 2 presents the literature review. Section 3 discusses methodology and data, respectively. Section 4 reports empirical findings while Sect. 5 draws a conclusion.

2 Literature Review

Several studies have discovered a strong interdependence between capital, liquidity creation, and risk-taking in banks. These variables interact with one another and can cause one another. Umar et al. (2017), for example, conducted empirical research on a negative and significant relationship between bank capital and bank liquidity creation. Horváth et al. (2014) investigated the causal relationship between liquidity creation and capital in the banking industry and discovered that the two variables are negatively related. Fu et al. (2016) conclude, after several econometric estimations, that bank liquidity creation has a significant and negative effect on the level of bank capital across Asia–Pacific, and vice versa.

Berger et al. (2016) demonstrated that as a bank's ability to create liquidity declines, so do its risks, resulting in safer banks. The reason for this is banks that generate more liquidity face higher losses when forced to sell illiquid assets to meet customer liquidity demands. Banks with higher capital levels, on the other hand, may be more willing to make higher non-performance loans (Concetta Chiuri et al. 2002). With more capital, the bank can take on more risk (Bhattacharya and Thakor 1993; Repullo 2004). Previously, systematic testing of this complex relationship had been hampered by the lack of a compelling metric for measuring regulatory quality and economic freedom.

Literature on the impact of the quality of regulation and economic freedom has gained importance in recent years. As noted by Belasen and Hafer (2013) a number of studies have found that financial development and higher levels of economic freedom are associated with (or cause) economic growth. Many recent studies, including Hall and Sobel (2008); Ashby et al. (2013); Bennett and Vedder (2013) and Belasen & Hafer (2013) attest to the power of economic freedom to encourage entrepreneurship, to increase efficiency and production, and to improve equity in a variety of economic and cultural circumstances. Ahmed (2013) also further highlights that institutional factors (including economic freedom) are important in explaining growth and financial development in SSA. In depth, Hafer (2013) argues that countries with higher levels of initial economic freedom, on average, exhibit greater levels of financial intermediary development in subsequent years. On the other hand, stricter regulations lead to higher risk-taking incentives by banking institutions, leading to a reduction in charter value and stability of the banking system (Fernández and González 2005). Detailed examinations of freedom in setting interest rates—a specific and significant aspect of freedom in the banking industry—by Fernández and González (2005), Beltratti and Stulz (2011), Delis et al. (2011), Sufian and Habibullah (2010a, b) showed that greater freedom in setting interest rates is positively correlated with bank cost-efficiency level, suggesting that countries with more financial freedom and independence are more cost-efficient. More recent evidence reveals that banks operating in countries with higher freedom for banking activities are more profitable than their counterparts (Chortareas et al., 2013; Sarpong-Kumankoma et al., 2018; Sufian and Habibullah, 2010a, b).

Thus far, these results have broadened the knowledge of the importance of regulation quality and economic freedom. However, in spite of these findings, conspicuously absent from the banking literature is an examination of the links between economic freedom, regulator quality and bank functions. The overall effect of regulatory quality and economic freedom on bank liquidity creation, bank loan risk taking and bank capital remains theoretically unknown. Therefore, this study revisits the interrelatedness between bank liquidity creation, bank loan risk-taking, and bank capital in the presence of regulatory quality differences and economic freedom differences across countries.

3 Methodology

Bayesian analysis is a statistical procedure that uses probabilities to express unknown parameters in order to answer research questions. It is predicated on the fundamental assumption that not only the outcome of interest, but also all unknown parameters in a statistical model, are essentially random and subject to prior beliefs. In terms of the concept of derivation, there are two different statistical schools of thought in science, namely frequency statistics and Bayesian statistics. The first school is based on sample data, whereas the second school is based on known and observed data to derive current statistics, a process known as inference. The frequency system's parameters are fixed, whereas the Bayesian parameters are random and follow the law of distribution. Prior information serves as the theoretical foundation for Bayesian statistics, and conclusions drawn from prior information are combined with observational data. As a result, the Bayesian approach is becoming increasingly popular, particularly in the social sciences. With the rapid advancement of data science, big data, and computational software, Bayesian statistics have become a popular tool. To make inference on a parameter of a Bayesian model, it will be necessary to integrate the joint posterior probability with respect to all the other parameters. Except in very simple cases where the solution is analytical, this integration is carried out through computer simulation. The idea of studying the stochastic properties of a random variable through computer simulation is not recent (Metropolis and Ulam 1949). Contributions from Metropolis et al. (1953) and Hastings (1970) led to a general method nowadays referred to as the Metropolis–Hastings algorithm.

We used the Bayesian multivariate regression method in this study because: (i) the models have high interaction between banking functions and between two countries' governance tools. Furthermore, when the number of parameters exceeds the sample size, the Ordinary Least Squares regression (OLS) is unclear, and the model has low reliability. (ii) The Lasso regression model can overcome the aforementioned drawback by penalizing the model with the L1 norm of the regression coefficient vector. Most effects unrelated to the dependent variable are forced to zero by the L1 penalty. The selection of L1 is, however, still manual. Therefore, the study has approached the Bayesian method to solve these problems. We apply the Bayesian method through the Random-walk Metropolis–Hastings algorithm and select the Markov Chain Monte Carlo sample (MCMC)[2] We propose the following research model based on the following research hypotheses:

$$\begin{aligned} LC_{i,t} = &\alpha_0 + \alpha_1 CAP_{i,t} + \alpha_2 NPL_{i,t} + \alpha_3 REG_{i,t} + \alpha_4 ECO_{i,t} \\ &+ \alpha_5 LNTA_{i,t} + \alpha_6 GDP_{i,t} + \alpha_7 INF + \varepsilon_{i,t} \end{aligned} \tag{1}$$

$$\begin{aligned} CAP_{i,t} = &\beta_0 + \beta_1 LC_{i,t} + \beta_2 NPL_{i,t} + \beta_3 REG_{i,t} + \beta_4 ECO_{i,t} \\ &+ \beta_5 LNTA_{i,t} + \beta_6 GDP_{i,t} + \beta_7 INF_{i,t} + \delta_{i,t} \end{aligned} \tag{2}$$

$$\begin{aligned} NPL_{i,t} = &\gamma_0 + \gamma_1 CAP_{i,t} + \gamma_2 LC_{i,t} + \gamma_3 REG_{i,t} + \gamma_4 ECO_t \\ &+ \gamma_5 LNTA_{i,t} + \gamma_6 GDP_{i,t} + \gamma_7 INF_{i,t} + \theta_{i,t} \end{aligned} \tag{3}$$

where $LC_{i,t}$ stands for the measure of liquidity creation (Nonfat liquidity/Total assets (%)), $CAP_{i,t}$ stands for capital tier 1 (%) and $NPL_{i,t}$ stands for non-performing loans/Gross loans (%). $ECO_{i,t}$ and $REG_{i,t}$ symbolizes economic freedom and regulatory level, respectively. The subscripts i and t denote bank and time period, i ranges from 1 to 267 and t varies from 2016 to 2018. The α, $\beta and \gamma$ represent the slope of the all variables and ε, δ, θ represent the error terms.

LNTA: Ln of Total assets—the absolute measure of the bank's size. In this study, we expect a negative relationship between liquidity and bank size. This result supports the theory of too big to fail. The larger the banks, the lower the liquidity, which is

[2] The Markov Chain Monte Carlo method (MCMC) is a post-estimation method based on distribution simulation.

consistent with previous studies (Horváth et al.2014; Lei and Song 2013; Fungacova et al. 2012). In addition, there are many conflicting studies on the relationship between capital and bank size. For example, research by Bateni et al. (2014) showed that the larger the bank size, the lower the capital generated for private banks. In contrast, Jackson et al. (2002) showed that large banks tend to have excess capital in reserve to maintain their good ratings. Therefore, the study will revisit this relationship. In addition to macro variables and country characteristics such as REG and ECO variables, the study considers two macro variables, annual economic growth rate (GDP) (%) and inflation rate (INF) (%) in its research model (Jokipii and Monnin 2013; Laeven and Majnoni 2003; Bikker and Metzemakers 2005).

4 Empirical Results and Discussion

4.1 Descriptive Statistics

Our sample is a balance sheet that includes financial data for the top 267 commercial banks in five developed countries (Australia, Hong Kong, Japan, Macau, and New Zealand) and ten emerging countries (China, India, Indonesia, Macau, Malaysia, Philippines, Singapore, Korea, Taiwan, Thailand, and Vietnam) in Asia–Pacific. A developed market is a country that is the most developed in terms of its economy and capital markets. The country must have a high income, but this also includes openness to foreign ownership, the easy movement of capital, and the efficiency of market institutions. An emerging market is a market that has some characteristics of a developed market but does not fully meet its criteria (Maket MSCI Classification Framework). We focused our sample on banks with large tier 1 capital because there are many differences related to bank size. Furthermore, the full data is only available to large banks. All financial items related to banking characteristics are mainly drawn from TheBankerdatabase and the reports of different banks in their annual financial statements (2016–2018).

Liquidity creation (LC) is computed by following Berger and Bouwman (2009). Although "cat fat" is Berger and Bouwman's preferred measure, we employ only the "non-fat" liquidity creation measure, because our sample dataset lacks detailed information on off-balance sheet activities. However, it is worth noting that the pattern of non-fat and cat fat measures over time are similar, only their dollar value are different (Berger and Bouwman 2009). The final liquidity creation measure employed in the empirical models is the ratio of "non-fat" to the bank's total assets. *The economic freedom (ECO)* measure initially adopted here is an average of all ten economic freedom sub-indexes developed by The Heritage Foundation. This indicator takes values on a scale from 0 to 100, with higher values indicating an economic environment or set of policies that are most conducive to economic freedom. Finally, *the regulatory quality index (REG)*, which is constructed by the Worldwide Governance Indicators (WGIs) of the World Bank (Kaufmann et al. 2011), is used to account for the quality of institutional development in each country. The Regulatory quality index ranges from approximately −2.5 (weak) to 2.5 (strong) governance performance.

Table 1 represents descriptive statistics of all the variables used in this study. Table 2 provides information on correlation coefficients between pairs of variables in the sample with the "p" values in parentheses. The correlation coefficients are not above 0.8, so it is difficult to have serious multicollinearity (Pliskin and Kennedy 1987).

Table 1 Descriptive statistics

Variable	Obs	Mean	Std	Min	Max
Liquidity creation (LC)	687	0.020	0.096	−0.342	0.388
Capital tier 1 (CAP)	687	7.441	2.770	2.9	20.6
Non-performing loans (NPL)	687	2.340	3.560	0.1	27.9
Regulatory quality (REG)	687	0.598	0.868	−0.45	2.21
Economic freedom (ECO)	687	65.973	11.136	52	90
Bank size (LNTA)	687	11.229	1.158	9.173	15.213
Economic growth rate (GDP)	687	4.640	2.408	−0.7	9.9
Inflation rate (INF)	687	1.998	1.451	−0.5	4.1

Source Authors' own estimation

Table 2 Correlation matrix of variables

	LC	CAP	NPL	REG	ECO	LNTA	GDP	INF
LC	1							
CAP	−0.194[a]	1						
NPL	−0.155[a]	−0.083[a]	1					
REG	−0.063	−0.161[a]	−0.310[a]	1				
ECO	−0.059	−0.001	−0.318[a]	0.954[a]	1			
LNTA	−0.014	−0.331[a]	−0.104[a]	0.132	0.086[a]	1		
GDP	0.057	0.259[a]	0.339[a]	−0.881[a]	−0.777[a]	−0.133[a]	1	
INF	0.053	0.302[a]	0.291[a]	−0.487[a]	−0.342[a]	−0.082[a]	0.695[a]	1

Notes [a]Significant at 5% level
Variables Liquidity creation (LC), Capital tier 1 (CAP), Non—performing loans (NPL), Regulatory quality (REG), Economic freedom (ECO), Bank size (LNTA), Economic growth rate (GDP), Inflation rate (INF)
Source Authors' own estimation

4.2 Bayesian Simulation Results

In order to test the effectiveness and validity of Bayesian inference based on the MCMC model, the study uses the MCMC series convergence diagnostic, with Fig. 1 showing the convergent series between bank functions and Fig. 2 showing the convergence chain between banking functions and economic freedom (ECO) and regulatory quality (REG). Checking the convergence of the MCMC series is done through 4

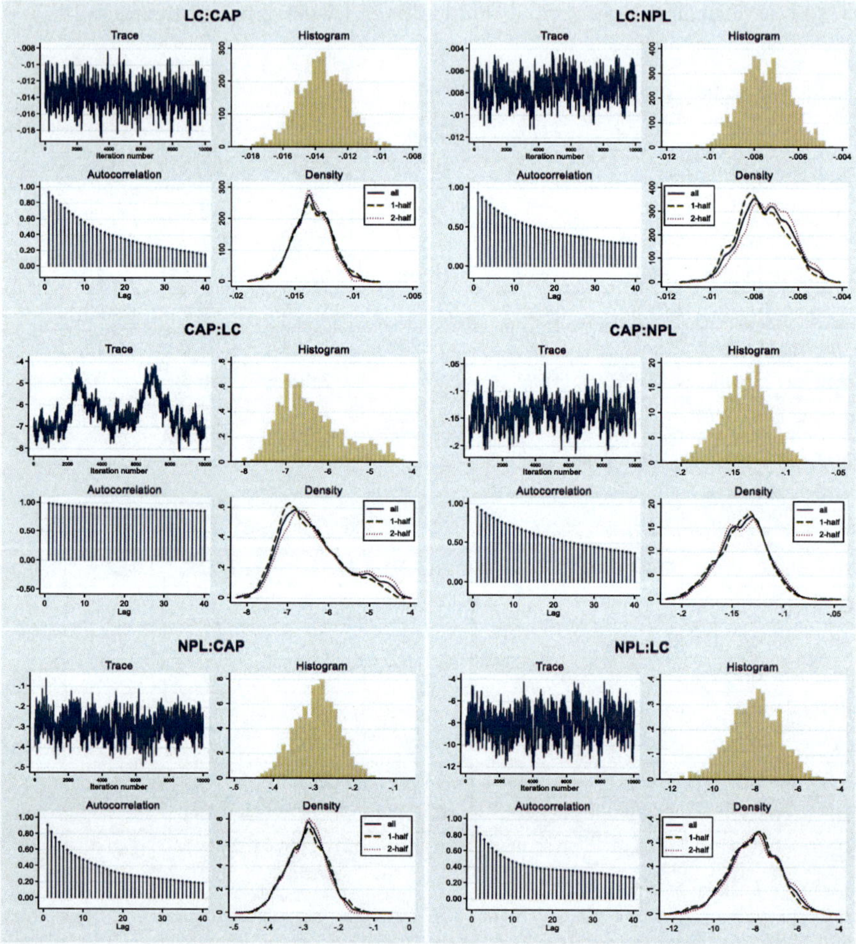

Fig. 1 Graphical diagnostics for MCMC convergence—Banking functions. *Source* Authors' own estimation

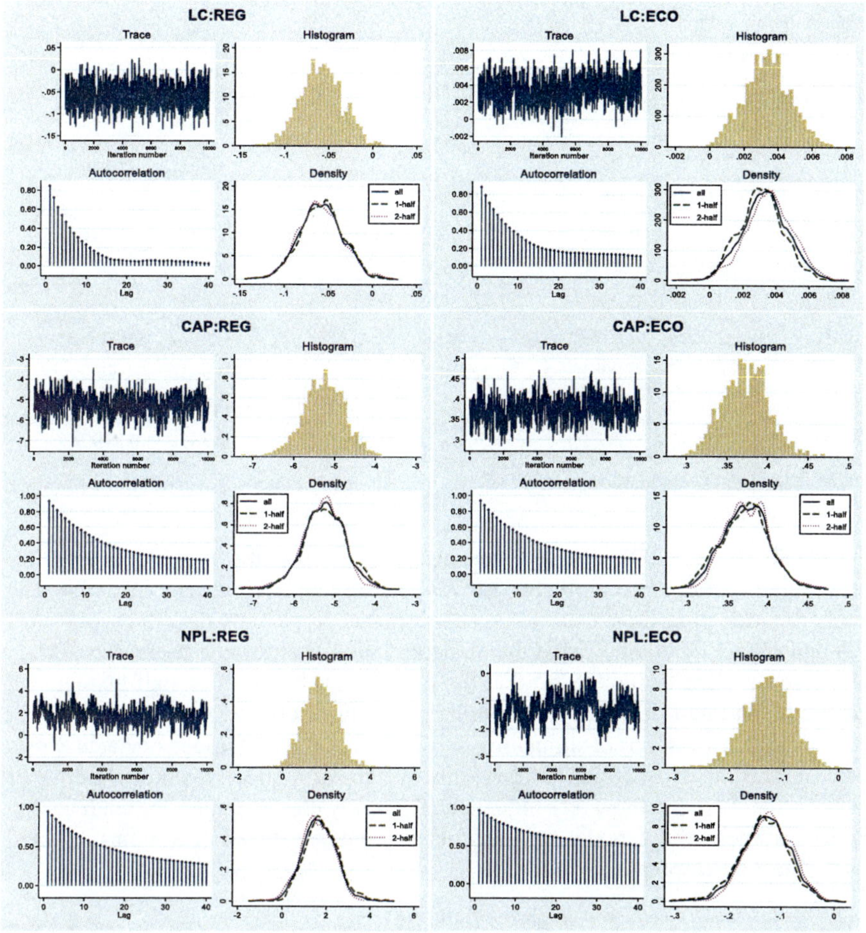

Fig. 2 Graphical diagnostics for MCMC convergence—Banking functions and regulation quality and economic freedom. *Source* Authors' own estimation

charts: through trace plots, a post-distribution chart (histogram), an autocorrelation chart, and kernel density (plot density). The test results from Figs. 1 and 2 show that the traces run quickly through the distribution. The correlation chart falls quickly, showing the low correlation. The shape of the histogram diagrams, simulating the shape of the precise distributions, is homogeneous. From this, it can be concluded that the Bayesian inference is efficient and reasonable. However, there are three relationships that need attention and, through the graphs, the estimation efficiency is not high. namely, the effect of liquidity generation on bank capital (CAP: LC) in Fig. 1, the regulatory quality effect on bad debt (NPL: REG) and the free economic influence

Table 3 The results

	LC		CAP		NPL	
	Mean (1)	Efficiency (2)	Mean (3)	Efficiency (4)	Mean (5)	Efficiency (6)
LC			−6.368	0.0015	−0.2915	0.0131*
CAP	−0.0136	0.0273*			−8.0511	0.0146*
NPL	−0.0075	0.0130*	−0.1394	0.0137*		
REG	−0.0583	0.0663*	−5.2648	0.0151*	1.6866	0.0098
ECO	0.0034	0.0120*	0.3731	0.0153*	−0.1289	0.0041
LNTA	−0.0114	0.0046	−0.6273	0.0352*	−0.4325	0.0426*
GDP	−0.0014	0.0127*	−0.1064	0.0347*	0.4012	0.0068
INF	0.0077	0.0082	0.2003	0.0286*	0.5616	0.0055
Constant	0.0706	0.0029	−6.428	0.0143*	13.9643	0.0031
Sigma_1_1	0.0087	0.2069*	4.5673	0.2097*	9.3453	0.2292*

Source Authors' own estimation

on bad debt (NPL: ECO) in Fig. 2. Sparseness and trends in the trace plot of a parameter suggest convergence problems. MCMC converged, but it did not mix well. The autocorrelation plots do not show good convergence, especially the effects of (CAP: LC) and (NPL: ECO) show that the autocorrelation is still quite high after 20 lag. Although the density plot found that the overall density, the first-half histograms, and the second-half densities were similar for all three effects, for the effect (CAP: LC), it showed a histogram simulating the shape of the distributions deviated. From there, the results of Bayesian inference for the above 3 effects are not effective or between them do not show any influence. The study checked again through testing the effectiveness of the model and the efficiency indexes shown in columns 2, 4, and 6 of Table 3 correspond to 3 models.

- *The relationship between the bank functions*

Research has shown the relationship between three functions of banks: "Higher bank capital is likely to reduce bad debt. The ability to create higher liquidity limits the capital and liquidity of the bank. Therefore, taking higher credit risk will hinder the liquidity creation and capital increase of the bank. "Specifically: *Firstly,* the results show that the bank's capital has a negative and significant effect on the bank's liquidity (Table 3, column 1), implying that the higher the bank's capital, the higher the liquidity, the larger the bank account. This result is consistent with the research results of Lei and Song (2013), Umar et al. (2017) and Horváth et al. (2014). These are studies that have provided evidence of a two-way effect between liquidity creation and bank capital. The study also showed a negative effect of bank capital on liquidity, but this result has a rather low efficiency level of 0.0015 (less than 0.01). *Secondly,* the results show a negative and two-way relationship between liquidity and bad debt of banks. This finding supports the results of Zhang et al. (2016). Bank liquidity creation is a more objective measure of risk taking than credit growth because liquidity provides us with an absolute amount of metabolic risk. According to the liquidity mechanism, the overall risk taken by banks shows a decreasing trend during this period. *Finally,* the study shows a negative and two-way relationship between bank capital and bad debt. The results are consistent with the results of Ben Jabra et al. (2017) as the study showed a negative relationship between credit risk and capital regulation. (ratio of tier 1 capital) of the bank. According to research by Ozili (2018), dormant loans can send unattractive signals to investors and can eventually cause banks to close. Therefore, to overcome this, banks often learn to manage bad debts by increasing the ratio of regulatory capital and loan growth.

- *The impact of regulation quality and economic freedom on the bank functions*

Research results show that regulatory quality negatively affects the liquidity and capital creation functions of banks but does not affect bad debts (Table 2, columns 1, 3 and 5). The results in column 6 show that the effectiveness of the estimate is below 0.01 or the regulatory quality does not affect bad debt. The explanations for these results are as follows: (i) the relationship between regulatory quality and liquidity creation is explained by the research Nagaraj and Zhang (2019) that better regulatory quality leads to a lower cost of capital. The presence of well-defined regulations makes it easier for foreign firms to conduct business in the host economy and promotes the host economy's integration with the rest of the world. Simple and well-defined regulations can act as a catalyst for financial integration, which in turn could lead to a lower cost of capital. A reduction in information asymmetry promotes credit extension (Avdjiev et al. 2021), leading to a reduced cost of capital. (ii) the relationship between regulatory quality and bank capital. This result is consistent with the study of Anginer et al. (2018) who found that bank capital is associated with a reduction in the systemic risk contribution of individual banks. This effect is more pronounced for banks located in countries with less efficient public and private monitoring of financial institutions and in countries with lower levels of information availability. Overall, their findings suggest that capital can act as a substitute for a weak institutional environment in reducing systemic risk.

The results of the economic freedom index have a positive impact on the liquidity and capital creation functions of banks. This result is supported by previous studies by Gropper et al. (2015) which found a link between bank performance, powerful politicians and levels of business freedom. economics They found that bank performance was positively related to a country's own economic freedom. Or, the findings from the study Sufian and Habibullah (2010a, b) show that overall economic freedom and entrepreneurial freedom have a positive impact on bank profitability. This is consistent with the view that raising the economic freedom index allows banks to engage in a variety of activities, helping them exploit economies of scale and scope, thereby generating higher income from non-traditional sources. Furthermore, greater economic freedom creates more jobs, avoiding corruption and moral hazard. They found that this had a positive effect on bank profitability, higher liquidity, and bank capital functions.

Some other findings of the study, such as: (i) LNTA was found to have a negative impact on bank functions. This result supports the theory of being too big to fail and is consistent with studies by Berger and Bouwman (2009), Fungacova et al. (2012), Lei and Song (2013), Roman Horváth et al. (2014), Chipeta and Deressa (2016) (ii) INF, GDP are used to control economic factors influencing banking sector stability. The study shows that these two variables positively and negatively affect the liquidity and capital functions of the bank, respectively, but do not affect the level of bad debt of the bank. This result is consistent with previous studies such as Jokipii and Monnin (2013), Laeven and Majnoni (2003), Bikker and Metzemakers (2005).

5 Conclusion

This study investigates the impacts of quality regulation and economic freedom on banking capital, banking liquidity creation, and banking credit risk in 267 large banks across Asia–Pacific developed and emerging countries over the 2016–2018 study period. Firstly, research has shown the relationship between three functions of banks: "Higher bank capital is likely to reduce bad debt. The ability to create higher liquidity limits the capital and liquidity of the bank. Therefore, taking higher credit risk will hinder the liquidity creation and capital increase of the bank. "Second, the study has provided empirical support for the current studies, namely that banks in countries with a higher free economy will have better liquidity creation functions and capital levels than in the rest of the countries. This is in line with the trend that the implementation of Basel regulations for banks will be relatively favorable in countries with high liberalization indexes or developed countries. However, the study also points out a point to note. That is, the quality of better and tighter regulation by countries will not only improve but also limit the functions of banks (liquidity creation and bank capital). This explains why the higher the quality of regulation, the more stringent the regulations, the more difficult it is to apply the regulations under the Basel framework. These results will help the functional units and bank managers need to consider and strengthen control of the economy or expand economic freedom.

Appendix: Bayesian Multivariate Regression

Bayesian multivariate regression
Random-walk Metropolis-Hastings sampling

MCMC iterations = 12,500
Burn-in = 2,500
MCMC sample size = 10,000
Number of obs = 687
Acceptance rate = .3064
Efficiency: min = .002898
avg = .03931
max = .2069

Log marginal-likelihood = 571.41484

	Mean	Std. Dev.	MCSE	Median	Equal-tailed [95% Cred.	Interval]
LC						
CAP	-.0135586	.0015751	.000095	-.0135709	-.0168148	-.0105247
NPL	-.0074945	.0011087	.000097	-.0075268	-.0095654	-.0052411
REG	-.0582629	.0243411	.000945	-.058494	-.1056345	-.0117825
ECO	.0033696	.0013898	.000127	.0033656	.0007286	.0061534
LNTA	-.0113675	.0028695	.000424	-.0113893	-.0169606	-.0060956
GDP	-.0014199	.0044475	.000395	-.0015432	-.0101084	.007767
INF	.007716	.0035514	.000393	.0078026	.0007662	.0147095
_cons	.0705848	.0667103	.012391	.0718357	-.0641281	.1934183
Sigma_1_1	.0087281	.0004753	.00001	.0087002	.0078435	.0097046

Efficiency summaries

MCMC sample size = 10,000
Efficiency: min = .002898
avg = .03931
max = .2069

	ESS	Corr. time	Efficiency
LC			
CAP	272.51	36.70	0.0273
NPL	130.46	76.65	0.0130
REG	663.33	15.08	0.0663
ECO	119.77	83.49	0.0120
LNTA	45.72	218.73	0.0046
GDP	126.64	78.96	0.0127
INF	81.82	122.22	0.0082
_cons	28.98	345.01	0.0029
Sigma_1_1	2068.88	4.83	0.2069

Bayesian multivariate regression
Random-walk Metropolis-Hastings sampling

MCMC iterations = 12,500
Burn-in = 2,500
MCMC sample size = 10,000
Number of obs = 687
Acceptance rate = .3562
Efficiency: min = .001477
avg = .04091
max = .2097

Log marginal-likelihood = -1550.1902

	Mean	Std. Dev.	MCSE	Median	Equal-tailed [95% Cred.	Interval]
CAP						
LC	-6.368042	.7899738	.205568	-6.517485	-7.54551	-4.558951
NPL	-.1394407	.0232785	.001989	-.1380379	-.1862481	-.0946569
REG	-5.264836	.5039689	.041038	-5.253453	-6.309022	-4.267407
ECO	.3730748	.0284875	.002304	.3726402	.3187105	.4305101
LNTA	-.6272576	.0695038	.003703	-.6294689	-.7643458	-.4905664
GDP	-.1063617	.1134473	.006087	-.1050995	-.3269096	.1153605
INF	.2003116	.0842473	.004979	.2006208	.0326367	.3617128
_cons	-6.427972	1.861427	.155477	-6.387466	-10.1194	-2.916348
Sigma_1_1	4.563703	.2502854	.005465	4.55848	4.09511	5.082069

Efficiency summaries

MCMC sample size = 10,000
Efficiency: min = .001477
avg = .04091
max = .2097

	ESS	Corr. time	Efficiency
CAP			
LC	14.77	677.15	0.0015
NPL	137.02	72.98	0.0137
REG	150.81	66.31	0.0151
ECO	152.85	65.42	0.0153
LNTA	352.29	28.39	0.0352
GDP	347.32	28.79	0.0347
INF	286.29	34.93	0.0286
_cons	143.34	69.77	0.0143
Sigma_1_1	2097.29	4.77	0.2097

Bayesian multivariate regression
Random-walk Metropolis-Hastings sampling

MCMC iterations = 12,500
Burn-in = 2,500
MCMC sample size = 10,000
Number of obs = 687
Acceptance rate = .3508
Efficiency: min = .003127
avg = .03654
max = .2292

Log marginal-likelihood = -1792.8413

	Mean	Std. Dev.	MCSE	Median	Equal-tailed [95% Cred.	Interval]
NPL						
CAP	-.2914795	.0550222	.004807	-.2888127	-.3997532	-.1837675
LC	-8.051125	1.215677	.100475	-8.008469	-10.49738	-5.753147
REG	1.686562	.7599344	.076703	1.664326	.2643425	3.221084
ECO	-.1288703	.0444778	.006926	-.1278881	-.2187045	-.0440372
LNTA	-.4325195	.1040185	.005041	-.4318277	-.6366284	-.2315655
GDP	.4012405	.1604625	.019488	.4037154	.0815935	.7243334
INF	.5616356	.1284964	.01738	.5640601	.3070631	.8144009
_cons	13.96433	2.542359	.454674	13.93486	9.139388	18.94638
Sigma_1_1	9.345299	.5191371	.010844	9.330098	8.348127	10.40896

Efficiency summaries

MCMC sample size = 10,000
Efficiency: min = .003127
avg = .03654
max = .2292

	ESS	Corr. time	Efficiency
NPL			
CAP	131.00	76.34	0.0131
LC	146.39	68.31	0.0146
REG	98.16	101.88	0.0098
ECO	41.24	242.46	0.0041
LNTA	425.80	23.49	0.0426
GDP	67.80	147.50	0.0068
INF	54.66	182.95	0.0055
_cons	31.27	319.84	0.0031
Sigma_1_1	2291.83	4.36	0.2292

References

Acemoglu, D., Johnson, S., Robinson, J.A.: The colonial origins of comparative development: an empirical investigation. Am. Econ. Rev. **91**(5), 1369–1401 (2001). https://doi.org/10.1257/aer.91.5.1369

Ahmed, A.D.: Effects of financial liberalization on financial market development and economic performance of the SSA region: an empirical assessment. Econ. Model. **30**(1), 261–273 (2013). https://doi.org/10.1016/j.econmod.2012.09.019

Anginer, D., Demirgüç-Kunt, A., Mare, D.S.: Bank capital, institutional environment and systemic stability. J. Financ. Stab. **37**, 97–106 (2018). https://doi.org/10.1016/j.jfs.2018.06.001

Asaftei, G., Kumbhakar, S.C.: Regulation and efficiency in transition: the case of Romanian banks. J. Regul. Econ. **33**(3), 253–282 (2008). https://doi.org/10.1007/s11149-007-9041-0

Ashby, N.J., Bueno, A., Martinez, D.: Economic freedom and economic development in the Mexican states. J. Reg. Anal. Policy **43**(1), 21–33 (2013). https://doi.org/10.22004/ag.econ.243945

Avdjiev, S., Binder, S., Sousa, R.: External debt composition and domestic credit cycles J. Int. Money Financ. **115** (2021). https://doi.org/10.1016/j.jimonfin.2021.102377

Bateni, L., Vakilifard, H., Asghari, F.: The influential factors on capital adequacy ratio in Iranian Banks. Int. J. Econ. Financ. **6**(11) (2014). https://doi.org/10.5539/ijef.v6n11p108

Belasen, A.R., Hafer, R.W.: Do changes in economic freedom affect well-being? J. Reg. Anal. Policy, **43**(1), 56–64 (2013). https://doi.org/10.22004/ag.econ.243948

Beltratti, A., Stulz, R.M.: Why did some banks perform better during the credit crisis? A cross-country study of the impact of governance and regulation. SSRN Electron. J. (2011). https://doi.org/10.2139/ssrn.1433502

Beltratti, A., Stulz, R.M.: The credit crisis around the globe: Why did some banks perform better? J. Financ. Econ. **105**(1), 1–17 (2012). https://doi.org/10.1016/j.jfineco.2011.12.005

Ben Jabra, W., Mighri, Z., Mansouri, F. (2017). Determinants of European bank risk during financial crisis. Cogent Econ. Financ. **5**(1). https://doi.org/10.1080/23322039.2017.1298420

Bennett, D.L., Vedder, R.K. (2013). A dynamic analysis of economic freedom and income inequality in the 50 U.S. States: Empirical evidence of a parabolic relationship. J. Reg. Anal. Policy **43**(1), 42–55. https://doi.org/10.2139/ssrn.2134650

Berger, A.N., Bouwman, C.H.S.: Bank liquidity creation. Rev. Financ. Stud. **22**(9), 3779–3837 (2009). https://doi.org/10.1093/rfs/hhn104

Besley, T., Burgess, R.: Can labor regulation hinder economic performance? Evidence from India. Quart. J. Econ. **119**(1), 91–134 (2004). https://doi.org/10.1162/003355304772839533

Bikker, J.A., Metzemakers, P.A.J.: Bank provisioning behaviour and procyclicality. J. Int. Financ. Mark. Inst. Money **15**(2), 141–157 (2005). https://doi.org/10.1016/J.INTFIN.2004.03.004

Chan, S.G., Mohd, M.Z.: Financial market regulation, country governance, and bank efficiency: evidence from East Asian countries. Contemp. Econ. **10**(1), 39–54 (2016). https://doi.org/10.5709/ce.1897-9254.197

Chipeta, C., Deressa, C.: Firm and country specific determinants of capital structure in Sub Saharan Africa. Int. J. Emerg. Mark. **11**(4), 649–673 (2016). https://doi.org/10.1108/IJoEM-04-2015-0082

Chortareas, G.E., Girardone, C., Ventouri, A.: Financial freedom and bank efficiency: evidence from the European Union. J. Bank. Financ. **37**(4), 1223–1231 (2013). https://doi.org/10.1016/j.jbankfin.2012.11.015

Delis, M.D., Molyneux, P., Pasiouras, F.: Regulations and productivity growth in banking. Business **44**(4), 1–29 (2011). https://mpra.ub.uni-muenchen.de/id/eprint/13891

Easterly, W., Levine, R.: Tropics, germs, and crops: How endowments influence economic development. J. Monet. Econ. **50**(1), 3–39 (2003). https://doi.org/10.1016/S0304-3932(02)00200-3

Fernández, A.I., González, F.: How accounting and auditing systems can counteract risk-shifting of safety-nets in banking: some international evidence. J. Financ. Stab. **1**(4), 466–500 (2005). https://doi.org/10.1016/j.jfs.2005.07.001

Fungacova, Z., Weill, L., Zhou, M.: Bank capital, liquidity creation and deposit insurance. SSRN Electron. J. 1–26 (2012). https://doi.org/10.2139/ssrn.2079656

Gropper, D.M., Jahera, J.S., Park, J.C.: Political power, economic freedom and Congress: effects on bank performance. J. Bank. Financ. **60**, 76–92 (2015). https://doi.org/10.1016/j.jbankfin.2015.08.005

Hall, J., Sobel, R.: Institutions, entrepreneurship, and regional differences in economic growth1. Am. J. Entrep. **1**(1), 69 (2008). https://www.ceeol.com/search/article-detail?id=152625

Hastings, W.K.: Monte carlo sampling methods using Markov chains and their applications. Biometrika **57**(1), 97–109 (1970). https://doi.org/10.1093/biomet/57.1.97

Horváth, R., Seidler, J., Weill, L.: Bank capital and liquidity creation: granger-causality evidence. J. Financ. Serv. Res. **45**(3), 341–361 (2014). https://doi.org/10.1007/s10693-013-0164-4

Hafer-Cato, R.W., undefined.: Economic freedom and financial development: international evidence. Cato J. **33**(1), 111–126 (2013). https://heinonline.org/hol-cgi-bin/get_pdf.cgi?handle=hein.journals/catoj33&section=8

Jackson, P., Perraudin, W., Saporta, V.: Regulatory and "economic" solvency standards for internationally active banks. J. Bank. Financ. **26**(5), 953–976 (2002). https://doi.org/10.1016/S0378-4266(01)00266-7

Johnson, S., Kaufmann, D., Shleifer, A.: The unofficial economy in transition. Brook. Pap. Econ. Act. **2**, 159–239 (1997). https://doi.org/10.2307/2534688

Jokipii, T., Monnin, P.: The impact of banking sector stability on the real economy. J. Int. Money Financ. **32**(1), 1–16 (2013). https://doi.org/10.1016/J.JIMONFIN.2012.02.008

Kamarudin, F., Sufian, F., Md. Nassir A.: Does country governance foster revenue efficiency of Islamic and conventional banks in GCC countries? EuroMed J. Bus. **11**(2), 181–211 (2016). https://doi.org/10.1108/EMJB-06-2015-0026

Kaufmann, D., Kraay, A., Mastruzzi, M.: The worldwide governance indicators: methodology and analytical issues. Hague J. Rule Law **3**(2), 220–246 (2011). https://doi.org/10.1017/S1876404511200046

Laeven, L., Majnoni, G.: Loan loss provisioning and economic slowdowns: too much, too late? J. Financ. Intermed. **12**(2), 178–197 (2003). https://doi.org/10.1016/S1042-9573(03)00016-0

Lei, A.C.H., Song, Z.: Liquidity creation and bank capital structure in China. Glob. Financ. J. **24**(3), 188–202 (2013). https://doi.org/10.1016/j.gfj.2013.10.004

Levine, R., Loayza, N., Beck, T.: Financial intermediation and growth: causality and causes. J. Monet. Econ. **46**(1), 31–77 (2000). https://doi.org/10.1016/S0304-3932(00)00017-9

Metropolis, N., Rosenbluth, A.W., Rosenbluth, M.N., Teller, A.H., Teller, E.: Equation of state calculations by fast computing machines. J. Chem. Phys. **21**(6), 1087–1092 (1953). https://doi.org/10.1063/1.1699114

Metropolis, N., Ulam, S.: The monte carlo method. J. Am. Stat. Assoc. **44**(247), 335–341 (1949). https://doi.org/10.1080/01621459.1949.10483310

Miller, T., Kim, A.B., Roberts, J.M., Lucia, S.: 2019 index of economic freedom, vol. 25, pp. 474 (2019). https://www.heritage.org/index/about

Nagaraj, P., Zhang, C.: Regulatory quality, financial integration and equity cost of capital. Rev. Int. Econ. **27**(3), 916–935 (2019). https://doi.org/10.1111/roie.12403

Ozili, P.K.: Banking stability determinants in Africa. Int. J. Manag. Financ. **14**(4), 462–483 (2018). https://doi.org/10.1108/IJMF-01-2018-0007

Pliskin, J., Kennedy, P.: A guide to econometrics. J. Am. Stat. Assoc. **82**(399) (1987). https://doi.org/10.2307/2288828

Sarpong-Kumankoma, E., Abor, J., Aboagye, A.Q.Q., Amidu, M.: Freedom, competition and bank profitability in Sub-Saharan Africa. J. Financ. Regul. Compliance **26**(4), 462–481 (2018). https://doi.org/10.1108/JFRC-12-2017-0107

Sufian, F., Habibullah, M.S.: Does economic freedom fosters banks' performance? Panel evidence from Malaysia. J. Contemp. Account. Econ. **6**(2), 77–91 (2010a). https://doi.org/10.1016/j.jcae.2010.09.003

Sufian, F., Habibullah, M.S.: Has economic freedom fostered bank performance? Panel evidence from China. China Econ. J. **3**(3), 255–279 (2010b). https://doi.org/10.1080/17538963.2010.562039

Umar, M., Sun, G., Majeed, M.A.: Bank capital and liquidity creation: evidence of relation from India. J. Asia Bus. Stud. **11**(2), 152–166 (2017). https://doi.org/10.1108/JABS-12-2015-0208

Zhang, D., Cai, J., Dickinson, D.G., Kutan, A.M.: Non-performing loans, moral hazard and regulation of the Chinese commercial banking system. J. Bank. Financ. **63**, 48–60 (2016). https://doi.org/10.1016/j.jbankfin.2015.11.010

Predicting the Relationship Between Influencer Marketing and Purchase Intention: Focusing on Gen Z Consumers

Cuong Nguyen, Tien Nguyen, and Vinh Luu

Abstract The study aims to predict the relationship between influencer marketing and purchase intention among Gen Z consumers in Vietnam. Research using quantitative method with PLS-SEM model based on 250 samples using Smart-PLS 3 software. The results show that Perceived Influencer Credibility, Entertainment Value and Perceived Expertise of Influencer positively impact the Purchase Intention of Vietnamese Gen Z consumers. Besides, Perceived Influencer Credibility mediates the relationship between Entertainment Value and Purchase Intention. Perceived Expertise of influencers also mediates the relationship between Peer Review and Purchase Intention. This study also provides managerial implications for marketers to utilise influencer marketing strategies on Vietnamese Gen Z consumers.

Keywords Influencer Marketing · Purchase Intention · PLS-SEM · Gen Z Consumer · Vietnam

1 Introduction

Vietnam has had a massive advantage in the era of technology when there are more than 68 million Internet users in 2020, increasing by 10% compared to 2019; moreover, 65/68 million are social media users (Kemp 2020). The data collected by GSO of Vietnam (2020), which can explain this statement, is that Vietnam internet penetration accounts for more than 70% of the total 95 million population. Tiamo and Verissimo (2014) stated that the digital is much more developed thanks to much

C. Nguyen (✉)
Faculty of Commerce and Tourism, Industrial University of Ho Chi Minh City, Ho Chi Minh City, Vietnam
e-mail: nguyenquoccuong@iuh.edu.vn

T. Nguyen
Business School, University of Greenwich, London, UK

V. Luu
Faculty of Tourism and Hospitality Management, HUTECH University, Ho Chi Minh City, Vietnam

N. Ngoc Thach et al. (eds.), *Financial Econometrics: Bayesian Analysis, Quantum Uncertainty, and Related Topics*, Studies in Systems, Decision and Control 427,
https://doi.org/10.1007/978-3-030-98689-6_31

more comfortable and cheaper access to the internet. Due to the fast-paced advance in technology, marketing strategies have been changed much (Opreana and Vinerean, 2015), especially marketing campaigns on social media, which are implemented as an affordable but effective way to deliver messages of brands to customers in the 4.0 era (Kirtiş and Karahan 2011). The recent outbreak of information and digital technology led to social media development with social network platforms such as Facebook, Youtube, Tiktok, Instagram, and Twitter. It is witnessing a change in consumer behaviour and how brands had better communicate with their consumers and reach other customer segmentations. Influencer marketing is well-known as the most efficient and trustworthy advertising is by recommendations from family and friends, and 92% of consumers worldwide trust an individual's recommendation rather than traditional advertising (Nielsen, 2012, 2015). We can see that people tend to rely on their friends' recommendations and influencers' advice more than direct messages from brands, so influencer marketing is growing and popularising as a series of brands connect and interact with more customer segmentation. This fact is also the reason why the author would like to research influencer marketing. The effectiveness of people surrounding customers' influence and KOLs' influence have not been compared yet. Hence, this concern is also aimed at dealing with the study. Gen Z takes account of 25% of Vietnam's future workforce, according to Nielsen (2015), equivalent to about 15 million potential consumers. Gen Z will become the most potential customers replacing Millennials in the future since, in 2020, 30% of the total global population belongs to this generation (Sparks and Honey marketing agency 2019). However, there are only a few types of research about influencer marketing's influence on gen Z purchase intention globally and little about that in Vietnam despite this generation's potential. As a result, the author would like to discuss the Vietnam market topic to satisfy this concern. In 2020, the world witnessed the destruction of the COVID-19 pandemic in every aspect, significantly when annual global GDP growth decreased by 0.5% from 2.9% in 2019 to 2.4% in the first half of 2020 (OECD 2020). To survive in the crisis, 69% of brands participated in a survey conducted by Influencer Marketing Hub in March 2020 with 237 brands. Meanwhile, in another survey conducted by Kantar with the participation of 35,000 consumers around the world, 75% of respondents would like to know about brand's activities on media in the COVID-19 period (Kantar 2020); and people in the world generally and in Vietnam mainly spend more time online in the time of "social distancing" in order to get up-to-date information about the pandemic and its impact on economy and society (Ogilvy Social Lab 2020). Therefore, all companies must develop marketing solutions that can effectively cut off marketing costs and approach customers. Therefore, influencer marketing would be applied as a part of the marketing strategy for businesses in the crisis, which was proved by a study conducted by Launchmetrics with 600 professionals in marketing, communication and PR (Launchmetrics 2020). As a result, the author wants to research this topic to develop a marketing solution for businesses in a current situation. This research aims to predict the relationship between influencer marketing and Gen Z purchasing intention in Vietnam. The findings are expected to provide practical managerial recommendations for businesses to improve marketing strategy on social media in the crisis.

2 Literature Review

2.1 Influencer Marketing

Influencer marketing is based on the concept of "trust advisor" and is likely familiar with TV commercials collaborating with celebrities to leverage their reputation to promote the brand (Brown and Hayes 2008). Then, social media is stated as a vital source of communication marketing that helps businesses, researchers, and marketers convey messages to targeted customers, which also means that the value of advertising is valuable, so influencer marketing has been used as an effective form of marketing widely in the current age of technology (Saxena and Khanna 2013; Wong, 2014). Byrne et al. (2017) defined influencer marketing as a type of marketing that focuses on using key leaders to drive a brand's message to the larger market. Li and Du (2011) also believe that an influencer is similar to an opinion leader. Hence, they defined an influencer as an influential person with a solid personal brand. Influencers can be anyone as long as they can impact others in the specific community and industry and encourage them to try brands' products and services according to their recommendations and advice (Brown and Fiorella 2013). influencer marketing uses a person who has built up many followers on a social media platform like Instagram or blogs. Furthermore, the person is also seen as trustable, and brands are the person to spread products and services (De Veirman et al. 2017). There are two main types of influencers: macro influencer and micro-influencer. The former reaches many users on social networks and creates many interactions with them, such as commenting and sharing. However, the connection between the macro influencer and their followers is loose and not much determinate (Brown and Fiorella 2013). On the other hand, micro-influencers have a smaller community, so it is more affordable for business. They can offer unique and exclusive advantages for their followers. Because of micro influencers' small community on social media, businesses must work with more than one micro-influencer in the same period and campaign to expand brand awareness. However, it is likely more effective than using a few macro-influencers because it creates a too high engagement rate.

2.2 Influencer Marketing Models

Figure 1 shows that influencers have large networks on social media and reach many potential customers to improve brand awareness, so influencers are placed at the model's centre. They are expected to guide potential customers within a specific community having the same interest and keyword definitions.

On the other hand, it supposes that customers would be put at the influencer marketing model's centre due to their right to make decisions. Hence, influencer marketing strategy's effectiveness might improve if brands directly transfer the messages to them and use their voice to spread them in their ways, presented as

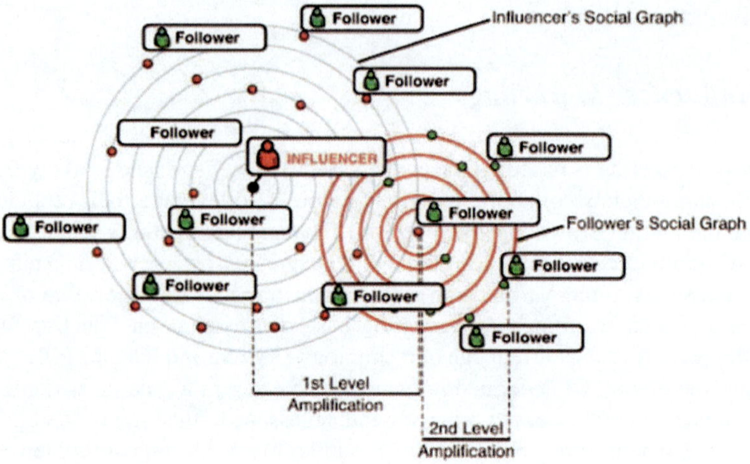

Fig. 1 Fisherman's influencer marketing model (Brown and Fiorella 2013)

Customer-Centric Influencer Marketing Model (in Fig. 2) by Brown and Fiorella (2013). Moreover, it can expand new networks around customers such as offline and online relations when centring customers.

Marketers use both models to express brands' meaningful messages that contain market insights relevant to their products and target customers.

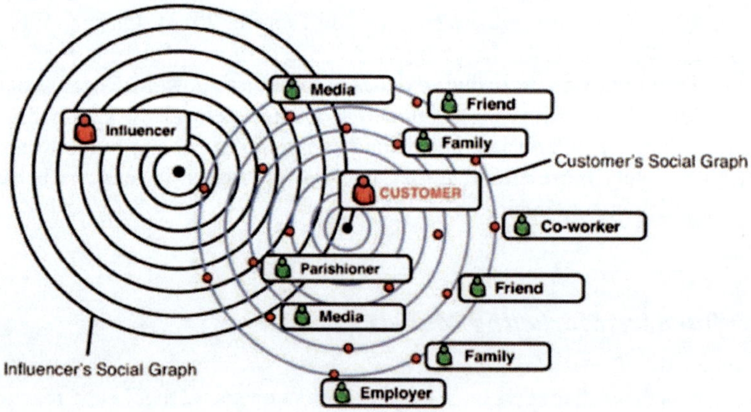

Fig. 2 Customer-centric influence marketing model (Brown and Fiorella 2013)

2.3 Purchase Intention

According to Bawa and Ghosh (1999), understanding consumers purchase intentions is of great importance, as it relates to consumers' behaviour, perception and attitude, and can therefore be used to predict the buying process. Purchase intention can be used to predict actual buying behaviour (Morrison, 1979). Consumers' purchase decisions are complex, and purchase intention is a part of this process (Kotler and Armstrong 2010). Furthermore, Kotler and Armstrong (2010) suggest that the consumer's decision-making process consist of five stages; need recognition, information search, evaluation of alternatives, purchase decision, and post-purchase behaviour. Spears and Singh (2004) define purchase intention as "an individual's conscious plan to make an effort to purchase a brand". Purchase intention can be described as cognitive behaviour regarding buying a particular brand (Shah et al. 2012). Influencer marketing is strongly associated with an influencer's brand endorsements. Influencers are perceived as more credible, believable, knowledgeable, and they are better at explaining how the product works. According to Ajzen (2011), consumer intent consists of beliefs and impulses, directly affecting consumer behaviour. They indicate whether someone is willing to try or attempt to buy and use a product to satisfy their needs. Furthermore, purchase intention can be measured by shopping expectations and consumer reviews of that product (or service) (Laroche et al. 1996). Pavlou (2003) confirmed that when a customer intends to use online transactions to purchase in the social network context, it is called online purchase intent. Paylou (2003) stated that the process of searching, exchanging information and buying goods through the internet is considered as online transactions. Shim et al. (2001) suggested that as long as a Web user intends to perform a given online behaviour, he or she will likely succeed in doing so to the extent that the person is provided with the required opportunities and resources.

2.4 Hypotheses Development

In terms of online marketing, customers can be influenced by source credibility. Perceived credibility, which is said as a valid, impartial and unbiased claim by Hass (1981), influences people to follow influencers on social media by Nam and Dan (2018). In other words, they tend to trust influencers who they already have perceived credibility on them (Hsu et al. 2013; Lee and Koo 2015; Jabr and Zheng 2014). Moreover, perceived credibility was proved as a determinant of customers' subsequent actions and can lead to promising results, such as increasing customer purchasing intention (Chu and Kamal 2008; Dimitrova and Rosenbloom 2014). Previous studies indicated that the congruence between brands and influencers selected for brands' specific products and services could positively impact brand trust and increase purchase intention (Choi and Rifon 2012; Liengpradit et al. 2014; Zietek 2016; Nam and Dan 2018). Furthermore, Xu and Pratt (2018) explained that customers tend to follow influencers who share the same lifestyle, personality traits, and behaviour preferences, linking influencers and followers. As a result, the H1 and H2 hypothesis would be stated as below:

H1. Perceived Influencers Credibility positively affects Gen Z consumer purchasing intention.

H2. Perceived Influencers Credibility mediates the relationship between Entertainment Value and Gen Z consumer purchasing intention.

The customer could be affected positively by high-quality information on the internet, leading to their corresponding actions, that is, intention to purchase products (Cheung et al. 2009; Liengpradit et al. 2014; Dao et al. 2014). Moreover, these informative contents conveyed by influencers would level up followers' trust in branded posts and subsequently turn into the intention to purchase products because influencers are recognised as knowledgeable people in their field so that they can convince their followers with their opinions and experiences about specific products and services relevant to their major (Lou and Yuan 2019). According to and Bergkvist et al. (2016), expert influencers, whose specific expertise determines building up relations and trust between them and followers, can impact customers' attitudes and behaviour towards brands and even kick off their following positive actions. For example, intention to buy when the trust is levelling up thanks to informative value converted through influencer's content on social media. Besides, According to Kitchen and Proctor (2015), gen Z is one of the most vital social platforms users, continuously exchanging information among its peers due to the technology blooming era they are living in. Cruz (2016) said that Gen Z's attitude and decisions are preserved by their peer's thanks to the period they have been sharing information quickly and easily on social media. As a result, he suggested that the brand should focus on stimulating conversations about brands among Gen Z's community and give them personal advice to each other reliably within their community instead of investing in well-known influencers. Previously, peer's recommendations and opinions were stated as the component that can influence gen Z customer purchase intention rather than brands' direct information (Kim 2007; Lu et al. 2014b, Cruz, 2016). As a result, the H3 and H4 hypothesis would be formed as below:

H3. Perceived Expertise of Influencers positively affects Gen Z consumer purchase intention.

H4. Perceived Expertise of Influencers mediates the relationship between Peer Review and Gen Z consumer purchase intention.

Dao et al. (2014) confirmed that one of the three determinants of advertising value and further customer purchase intention is advertising entertainment value. Therefore, entertainment value would positively impact online purchase intention through influencers' content on social media. On the other hand, Lou and Yuan (2019) claimed that influencers' entertainment value has no impact on their followers' trust and subsequent actions relating to purchase intention. As a result, this research cannot form a hypothesis about the positive relationship between the entertainment value of influencers' content and customer purchase intention. Based on all the hypotheses above and research models conducted by Nam and Dan (2018), Chetioui et al. (2020) and Lou and Yuan (2019), the H5 and H6 hypotheses are stated as follows:

H5. Entertainment value is positively associated with Gen Z customer purchase intention.

H6. Perceived Expertise of Influencers mediates the relationship between Entertainment Value and Gen Z customer purchase intention.

3 Research Method

This study was conducted an online survey with 273 Vietnamese Gen Z. However; only 250 valid responses were collected. Therefore, the research would be empirically approached. The hypotheses are based on previous studies (Nam and Dan 2018; Chetioui et al. 2020; Lou and Yuan 2019). The questionnaire comprises the questions adopted from previous studies (Li et al. 2012; Yang et al. 2013; Martins et al. 2019; Xu and Pratt 2018; Bergkvist et al. 2016; and Dao et al. 2014). The questionnaire includes demographic questions and topic's questions scaled by the 5-point Likert scale method having seven levels: (1) Totally disagree, (2) Disagree, (3) Neutral, (4) Agree and (5) Totally Agree. The main objective of this research is to predict the relationship between the factors of influencer marketing on Gen Z customers purchase intention via the 06 hypotheses from H1 to H6. The authors used the Smart-PLS 3.0 software for data analysis. The data analysis method is partial least squares (PLS) with the PLS-SEM model. This technique permits the simultaneous estimation of multiple equations and performs factor analysis, including regression analysis, all in one step (Hair et al. 2019). Testing for the convergence of factors will be performed through the index of outer loading greater than 0.5 and the average variance extracted (AVE) greater than 50%. Hair et al. (2019) state that the factor meets the requirement of the reliability as measuring the path through the observed variables when the composite reliabilities (CR) were greater than 0.7 and Cronbach's Alpha coefficient > 0.7. Moreover, Fornell and Larcker (1981) require that the factors must be completely distinguishable. The discriminant test is based on the quadratic value of AVE, which is larger than the correlation coefficients.

4 Results

4.1 Descriptive Statistics of Respondents

According to Table 1, the age 19–25 years old took 79,2%, the most significant number compared to the other age category. The female participants (72,4%) are more than the male ones (26,8%). They come from provinces around the country like Hanoi, Danang, Binh Dinh, Bien Hoa, Tay Ninh, and Ho Chi Minh, the biggest

Table 1 Demographic description of respondents

Variable	Category	Frequency	Percentage
Gender	Female	181	72,4
	Male	67	26,8
	Other	2	0,8
Age	12–18 years old	29	11,6
	19–25 years old	198	79,2
	26–29 years old	23	9,2
City	Ho Chi Minh	181	72,4
	Da Nang	60	24
	Ha Noi	3	1,2
	Others	6	2,4
Types of influencers	KOLs	170	68
	Celebrity	199	79,6
	Expert	93	37,2
	Peer, family	116	46,4
Total		250	100%

city in Vietnam with 72,4%. In terms of influencer types, respondents can choose more than 1 type of influencer depending on their perspective. The result shows that such famous people as KOLs and celebrities had the most selections, 170 and 199 respectively, besides that 37,2% people said that experts are influencers to them and 46,4% of participants chose their friends, relatives, acquaintances as the people can impact on their purchase intention.

4.2 Measurement Model Test

The reliability and validity analysis will be carried out through the analysis of individual reliability (factorial loads and commonality) and internal consistency (composite reliability (CR), convergent validity through the Average Variance Extracted (AVE) and finally, discriminant validity via Heterotrait–Monotrait ratio). The results show that the convergence of the factors showed that the factors all reached the convergence value with the outer loading coefficients all greater than 0.5 and the AVE values were all greater than 50%. Besides, Cronbach's Alpha coefficient is both greater than 0.7 and CR is also greater than 0.7, indicating that all factors are reliable (Hair et al. 2019). The details are listed in Table 2.

The discriminant validity was tested using the Fornell-Larcker criterion and the Heterotraitmonotrait ratio (Ringle et al. 2015). All the ratios showed good discriminant validity properties (Tables 3 and 4). The result of the measurement model indicates that various validity and reliability criteria were satisfied. Therefore, the constructs and measures could be adequately discriminated against and appropriated to predict relevance for the structural model and associated hypotheses.

Table 2 Validity and reliability of the measurement model

	Cronbach's Alpha	rho_A	Composite reliability	Average variance extracted (AVE)
Entertainment value	0,872	0,880	0,912	0,722
Peer review	0,821	0,823	0,882	0,650
Perceived expertise	0,859	0,865	0,905	0,703
Perceived influencer Credibility	0,899	0,901	0,929	0,767
Purchase intention	0,871	0,873	0,912	0,721

Table 3 Discriminant validity (Fornell-Larcker criterion)

	Entertainment value	Peer review	Perceived expertise	Perceived influencer credibility	Purchase intention
Entertainment value	0,850				
Peer review	0,443	0,807			
Perceived expertise	0,595	0,564	0,839		
Perceived influencer credibility	0,532	0,509	0,547	0,876	
Purchase intention	0,570	0,640	0,596	0,679	0,849

Table 4 Heterotrait-Monotrait Ratio (HTMT)

	Entertainment value	Peer review	Perceived expertise	Perceived influencer credibility	Purchase intention
Entertainment value					
Peer review	0,514				
Perceived expertise	0,676	0,664			
Perceived influencer Credibility	0,594	0,583	0,611		
purchase Intention	0,647	0,756	0,683	0,762	

Table 5 Inner VIF values

	Entertainment value	Peer review	Perceived expertise	Perceived influencer credibility	Purchase intention
Entertainment value			1,244	1,00	1,709
Peer review			1,244		
Perceived expertise					1,750
Perceived influencer Credibility					1,575
purchase intention					

As shown in Table 5, the values of the Heterotrait-Monotrait Ratio of each of the constructs were lower than 0.85. Therefore, the criteria of discriminant validity were established for HTMT.

4.3 Structure Model Assessment

Collinearity assessment is the first step in the structural model analysis. The procedure is necessary to ensure that the path coefficients estimated by regressing endogenous variables on the attached exogenous variables are not biased. Collinearity issues exist between the exogenous and endogenous variables (Lowry and Gaskin 2014). According to Wong (2013), if the variance inflation factor (VIF) value is greater than 5 or lesser than 0.2, collinearity issues could be with the latent variables. The results in Table 5 show that there are no collinearity issues with the latent variables.

4.4 Hypothesis Testing

The structural model path coefficients assessment of the model was carried on utilising the bootstrapping procedure. Hair et al. (2019) suggest that bootstrapping is a resampling technique to estimate the standard error without relaying distributional assumptions. The bootstrap result approximates the normality of data. It is used to calculate the significance of the t statistic associated with path coefficients (Wong 2013). Table 6 and Fig. 3 showed significant values for the path coefficients determined from the bootstrapping process. PLS-SEM shows the results of all hypotheses, and model connections are illustrated in Table 6.

Table 6 Final results of the relationship checking of model's constructs

Hypothesis	Relationship	Regression weight	P value	Results
H1	Perceived influencer credibility->Purchase intention	0,447	0,000	Supported
H2	Entertainment value_ ->Perceived influencer credibility->Purchase intention	0,238	0,000	Supported
H3	Perceived expertise->Purchase intention	0,238	0,002	Supported
H4	Entertainment value_->Perceived expertise->Purchase intention	0,102	0,004	Supported
H5	Entertainment value_->Purchase intention	0,190	0,011	Supported
H6	Peer review->Perceived expertise ->Purchase intention	0,089	0,011	Supported

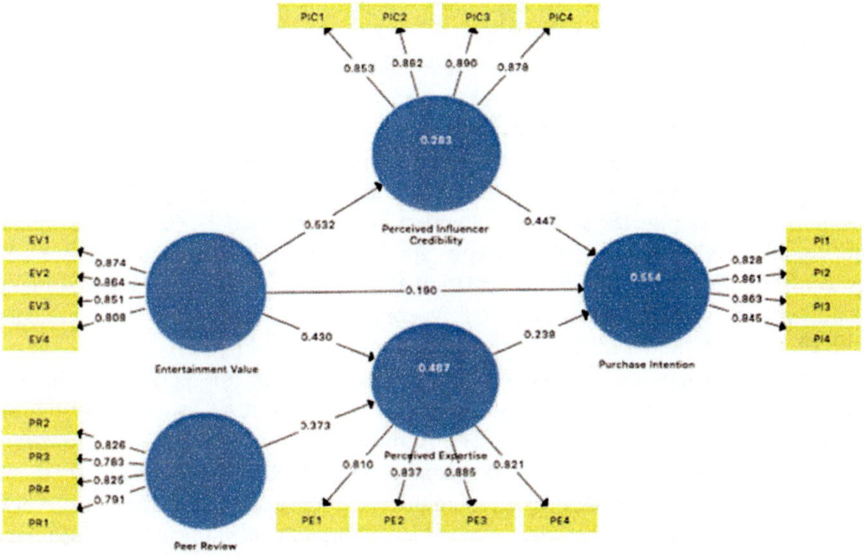

Fig. 3 PLS-SEM result

Next, the R2 value (Coefficient of determination) was evaluated. Entertainment Value represented 28.3% of the explained variance for Perceived Influencer Credibility. Entertainment Value and Peer Review represented 46.7% of the explained variance for Perceived Expertise. Also, Perceived Influencer Credibility, Entertainment Value and Perceived Expertise represented 55.4% of the explained variance for Purchase Intention. Then, Stone-Geisser's Q2 was employed using the blindfolding technique to check the predictive relevance of the model. The Q2 values were above zero (Hair et al. 2019) for Perceived Expertise (0.315), Perceived Influencer Credibility (0,210) and Purchase Intention (0.391), thus, the model's predictive relevance is supported.

5 Conclusion

The research objective is to predict the relationship between influencer marketing and Gen Z consumers' purchase intention in Vietnam. The results concluded that Perceived Influencer Credibility, Entertainment Value and Perceived Expertise of Influencer positively impact the purchase intention of Vietnamese Gen Z consumers. Besides, Perceived Influencer Credibility mediates the relationship between Entertainment Value and Purchase Intention, and this finding is consistent with many previous studies (Cakim 2009; Bennett 2014; Cruz 2016). Perceived Expertise of influencers also mediates the relationship between Peer Review and Purchase Intention. This finding is consistent with the influencer model created by Brown and Fiorella (2013); Customer-Centric Influence Marketing, which places customers at the influencer map's centre, stated the efficiency of consumers' influence on the purchase intention process. For managerial implications, influencer marketing should be applied effectively based on the research's findings and previous results, proving that customers' influencer in their community is the most influential factor in Gen Z consumers' purchase intention. Influencers can have a significant impact on Gen Z to encourage the purchase intention of certain products. Gen Z consumers are young people who have a high ego, are self-centred and want to make decisions after collecting sufficient information from many sources. Gen Z consumers can approach information more efficiently, connect with other customers to consult with them about the products or service, and be persuaded by their accurate opinions rather than famous influencers on social media. From the perspective of the Customer-Centric Influence Marketing Model (Brownll 2013), marketers should utilise influencer marketing on Gen Z consumers and drive their influence towards the Gen ample. Marketers can carry on social media campaigns that can call for their volunteer participation, such as expressing their feelings and perspectives, positively sharing their moments and stories relevant to the brand's meaningful messages but not directly related to products services. All touch the customer's right insights so that the message will be spread out in the small but potential community and help the marketing strategy more efficient. At that time, influencer marketing strategy should be conducted with highly credible influencers to express the messages more widely and improve customers' trust in products and brands.

References

Ajzen, I.: The theory of planned behaviour: reactions and reflections (2011).

Bawa, K., Ghosh, A.: A model of household grocery shopping behavior. Mark. Lett. **10**(2), 149–160 (1999)

Bennett, S.: Social Media Business Statistics, Facts, Figures and Trends 2014 [INFOGRAPHIC] (2014). Adweek.com. http://www.adweek.com/digital/social-business-trends-2014. Accessed 15 June 2020

Bergkvist, L., Hjalmarson, H., Mägi, A.W.: A new model of how celebrity endorsements work: attitude toward the endorsement as a mediator of celebrity source and endorsement effects. Int. J. Advert. **35**(2), 171–184 (2016)

Brown, D., Fiorella, S.: Influence marketing: How to create, manage and measure brand influencers in social media marketing. Que Publishing (2013a)

Brown, D., Hayes, N.: Influencer marketing. Routledge (2008)

Byrne, E., Kearney, J., MacEvilly, C.: The role of influencer marketing and social influencers in public health. In:.Proceedings of the Nutrition Society, vol. 76 (OCE3) (2017)

Cakim, I.M.: Implementing word of mouth marketing: online strategies to identify influencers, craft stories, and draw customers. Wiley (2009)

Chetioui, Y., Benlafqih, H., Lebdaoui, H.: How fashion influencers contribute to consumers' purchase intention. J. Fash. Mark. Manag. Int. J. (2020)

Cheung, C.M., Lee, M.K., Thadani, D.R.: The impact of positive electronic word-of-mouth on consumer online purchasing decision. In: World Summit on Knowledge Society, pp. 501–510. Springer, Berlin, Heidelberg (2009, September)

Choi, S.M., Rifon, N.J.: It is a match: the impact of congruence between celebrity image and consumer ideal self on endorsement effectiveness. Psychol. Mark. **29**(9), 639–650 (2012)

Chu, S.C., Kamal, S.: The effect of perceived blogger credibility and argument quality on message elaboration and brand attitudes: an exploratory study. J. Interact. Advert. **8**(2), 26–37 (2008)

Cruz, M.M.M.D.: Generation Z: influencers of the decision-making process: the influence of WOM and Peer interaction in the decision-making process. Doctoral dissertation (2016)

De Veirman, M., Cauberghe, V., Hudders, L.: Marketing through Instagram influencers: the impact of number of followers and product divergence on brand attitude. Int. J. Advert. **36**(5), 798–828 (2017)

Dimitrova, B.V., Rosenbloom, B.: Retailer brand image building: evidence from two European retailers. J. Euromarketing **23**, 124–143 (2014)

Fornell, C., Larcker, D.F.: Evaluating structural equation models with unobservable variables and measurement error. J. Mark. Res. **18**(1), 39–50 (1981)

gso.gov.vn. 2020. Statistical Data-GSO Of VIETNAM. https://www.gso.gov.vn/default_en.aspx?tabid=774. Accessed 19 July 2020

Hair, J.F., Risher, J.J., Sarstedt, M., Ringle, C.M.: When to use and how to report the results of PLS-SEM. In: European Business Review (2019)

Hass, R.G.: Effects of source characteristics on cognitive responses in persuasion. In: Cognitive Responses in Persuasion, pp.141–172 (1981)

Hsu, C., Chuan-Chuan Lin, J., Chiang, H.: The effects of blogger recommendations on customers' online shopping intentions. Internet Res. **23**(1), 69–88 (2013). https://doi.org/10.1108/10662241311295782

Jabr, W., Zheng, Z.: Know yourself and know your enemy. MIS Q. **38**(3), 635-A10 (2014)

Kemp, S.: Digital 2020: Vietnam—Datareportal–Global Digital Insights (2020). https://datareportal.com/reports/digital-2020-vietnam. Accessed 15 June 2020

Kim, D.K.: Identifying opinion leaders by using social network analysis: a synthesis of opinion leadership data collection methods and instruments. Doctoral dissertation, Ohio University (2007).

Kirtiş, A.K., Karahan, F.: To be or not to be in social media arena as the most cost-efficient marketing strategy after the global recession. Procedia Soc. Behav. Sci. **24**, 260–268 (2011)

Kitchen, P.J., Proctor, T.: Marketing communications in a post-modern world. J. Bus. Strateg. **36**(5), 34–42 (2015). https://doi.org/10.1108/JBS-06-2014-0070
Kotler, P., Armstrong, G.: Principles of Marketing. Pearson Education (2010)
Laroche, M., Kim, C., Zhou, L.: Brand familiarity and confidence as determinants of purchase intention: an empirical test in a multiple brand context. J. Bus. Res. **37**(2), 115–120 (1996)
Lee, Y., Koo, J.: Athlete endorsement, attitudes, and purchase intention: the interaction effect between athlete endorser-product congruence and endorser credibility. J. Sport Manag. **29**(5), 523–538 (2015)
Li, F., Du, T.C.: Who is talking? An ontology-based opinion leader identification framework for word-of-mouth marketing in online social blogs. Decis. Support Syst. **51**(1), 190–197 (2011)
Li, Y.M., Lee, Y.L., Lien, N.J.: Online social advertising via influential endorsers. Int. J. Electron. Commer. **16**(3), 119–154 (2012)
Liengpradit, P., Sinthupinyo, S., Anuntavoranich, P.: A conceptual framework for identify specific influencer on social network. Int. J. Comput. Internet Manag. **22**(2), 33–40 (2014)
Lou, C., Yuan, S.: Influencer marketing: how message value and credibility affect consumer trust of branded content on social media. J. Interact. Advert. **19**(1), 58–73 (2019)
Lowry, P.B., Gaskin, J.: Partial least squares (PLS) structural equation modeling (SEM) for building and testing behavioral causal theory: When to choose it and how to use it. IEEE Trans. Prof. Commun. **57**(2), 123–146 (2014)
Lu, L.C., Chang, W.P., Chang, H.H.: Consumer attitudes toward blogger's sponsored recommendations and purchase intention: the effect of sponsorship type, product type, and brand awareness. Comput. Hum. Behav. **34**, 258–266 (2014b)
Martins, J., Costa, C., Oliveira, T., Gonçalves, R., Branco, F.: How smartphone advertising influences consumers' purchase intention. J. Bus. Res. **94**, 378–387 (2019)
Morrison, D.G.: Purchase intentions and purchase behavior. J. Mark. **43**(2), 65–74 (1979)
Nam, L.G., Dân, H.T.: Impact of social media Influencer marketing on consumer at Ho Chi Minh City. Int. J. Soc. Sci. Humanit. Invent. **5**(5), 4710–4714 (2018)
Nielsen: Consumer Trust In Online, Social And Mobile Advertising Grows (2012). http://www.nielsen.com/us/en/insights/news/2012/consumer-trust-in-online-socialand-mobile-advertising-grows.html. Accessed 15 July 2020.
Nielsen: Recommendations from friends remain most credible form of advertising (2015). http://www.nielsen.com/eu/en/press-room/2015/recommendations-from-friendsremain-most-credible-form-of-advertising.html. Accessed 15 July 2020
Opreana, A., Vinerean, S.: A new development in online marketing: Introducing digital inbound marketing. Expert J. Mark. **3**(1) (2015)
Pavlou, P.A.: Consumer acceptance of electronic commerce: integrating trust and risk with the technology acceptance model. Int. J. Electron. Commer. **7**(3), 101–134 (2003)
Ringle, C.M., Wende, S., Becker, J.M.: SmartPLS 3. SmartPLS GmbH, Boenningstedt (2015)
Shah, S.S.H., Aziz, J., Jaffari, A.R., Waris, S., Ejaz, W., Fatima, M., Sherazi, S.K.: The impact of brands on consumer purchase intentions. Asian J. Bus. Manag. **4**(2), 105–110 (2012)
Shim, S., Eastlick, M.A., Lotz, S.L., Warrington, P.: An online prepurchase intentions model: the role of intention to search: best overall paper award—The Sixth Triennial AMS/ACRA Retailing Conference, 2000☆. J. Retail. **77**(3), 397–416 (2001)
Spears, N., Singh, S.N.: Measuring attitude toward the brand and purchase intentions. J. Current Issues Res. Advert. **26**(2), 53–66 (2004)
Tiago, M.T.P.M.B., Veríssimo, J.M.C.: Digital marketing and social media: Why bother? Bus. Horiz. **57**(6), 703–708 (2014)
Van-Tien Dao, W., Nhat Hanh Le, A., Ming-Sung Cheng, J., Chao Chen, D.: Social media advertising value: the case of transitional economies in Southeast Asia. Int. J. Advert. **33**(2), 271–294 (2014)
Wong, K.K.K.: Partial least squares structural equation modeling (PLS-SEM) techniques using SmartPLS. Mark. Bull. **24**(1), 1–32 (2013)
Wong, K.: The explosive growth of influencer marketing and what it means for you. Diakses pada February, vol. 16, pp. 2018 (2014)

Xu, X., Pratt, S.: Social media influencers as endorsers to promote travel destinations: an application of self-congruence theory to the Chinese Generation Y. J. Travel Tour. Mark. **35**(7), 958–972 (2018)

Yang, B., Kim, Y., Yoo, C.: The integrated mobile advertising model: the effects of technology-and emotion-based evaluations. J. Bus. Res. **66**(9), 1345–1352 (2013)

Zietek, N.: Influencer Marketing : the characteristics and components of fashion influencer marketing. Dissertation (2016). http://urn.kb.se/resolve?urn=urn:nbn:se:hb:diva-10721

The Effect of Exchange Rate Volatility on FDI Inflows: A Bayesian Random-Effect Panel Data Model

Le Thong Tien, Nguyen Chi Duc, and Vo Thi Thuy Kieu

Abstract This article sheds new light on the effect of exchange rate volatility on FDI inflows through the design of a Bayesian random-effect panel-data model. The dataset covers the years from 2002 to 2019 and is divided into six periods for measuring the three-year standard deviation of the exchange rate. The research results provided statistical evidence that the positive effect of exchange rate volatility on FDI, a theoretical proxy of production flexibility, exceeded the negative effect of those in emerging countries. However, the negative effects of exchange rate volatility on FDI, supporting risk aversion perspectives, accounted for over 20%. With the exception of inflation showing a negative correlation, the effect of other determinants of FDI inflows, such as GDP growth per capita, trade openness growth, and monetary supply growth, proved to be significantly positive.

Keywords Bayesian · Exchange rate volatility · Production flexibility · Risk aversion

JEL C11 · F43 · I31 · O15

1 Introduction

Many attempts to quantify determinants of foreign direct investment, namely FDI, have been implemented over the period of recent decades. There was numerous previous materials investigating the effect of exchange rate volatility on FDI. Debates

L. T. Tien · N. C. Duc
Saigon University, 273 An Dương Vương, District 5, Ho Chi Minh City, Vietnam
e-mail: lttien@sgu.edu.vn

N. C. Duc
e-mail: ncduc@sgu.edu.vn

V. T. T. Kieu (✉)
Banking University HCMC, 36 Ton That Dam Street, District 1, Ho Chi Minh City, Vietnam
e-mail: kieuvtt@buh.edu.vn

N. Ngoc Thach et al. (eds.), *Financial Econometrics: Bayesian Analysis, Quantum Uncertainty, and Related Topics*, Studies in Systems, Decision and Control 427,
https://doi.org/10.1007/978-3-030-98689-6_32

on the effect of exchange rate volatility on FDI inflows have come from many different schools of theory, and two prominent views in which have both opposing and complementary arguments, the production flexibility and the risk aversion. While the former expects a possible positive correlation, the latter makes many arguments for negative effects. Most of those revolved around the traditional frequentist approach, concentrating on statistical hypothesis testing by p-value in comparison with the given level of significance (Bernanke 1983; Jayaratnam 2003; Furceri and Borelli 2008; Osinubi and Amaghionyeodiwe 2009; Sharifi-Renani and Mirfatah 2012; Khamis and Hussin 2015; Mansoor and Bibi 2019; Dang et al. 2020).

It is clearly undeniable that the frequentist approach has become familiar to most scientific research in the fields of economics and finance. One of the outstanding advantages of the frequentist approach is the stringency in data science, which appears to find the evidence objective. Furthermore, frequentist approach has huge available resources, which have been regularly supplemented and have full of previous literature and empirical models, including methods for solving model misspecification in regression analysis and inferencing results scientifically. Specifically, the frequentist approach's target is to compare the estimated p_value with given significance levels for the purpose of Null hypothesis significance testing. Nevertheless, this is also a notable limitation because a conclusion about an insignificant effect by frequentist approach can only claim that statistical evidence has not been found for various reasons, instead of no effect.

Recently, the Bayesian approach has attracted a lot of scientific concerns. It must be admitted that the Bayesian approach makes the statistical inference more straightforward and flexible. To be more detailed, prior information and posterior distribution of model parameters in Bayesian analysis produce the reliable intervals more comprehensively and more flexibly than the confidence intervals of frequentist approach. Especially, Bayesian analysis simulates probability of each posterior parameter similar to sensitivity analysis, while the frequentist analysis builds and tests prespecified hypotheses. This facilitates Bayesian analysis in determining probability distributions and generating reliable simulations despite the existence of outliers. Another large advantage of the Bayesian approach is that it does not require a large sample size like that of the frequentist approach, which will limit the ability to access many areas of research whose data are not yet available or insufficient. The Bayesian approach generates parameter estimates using numerous simulations, thus ensuring that the posterior parameters become reliable in case the sample size is not big enough. However, the calculation of complex models by the Bayesian approach proved to be extremely time-consuming and frustrating for specifying prior distributions. The estimation of posterior parameters has been conveniently supported by Stata 17, whose command packages provided default settings or customization.

Our study aims to employ a new Bayesian approach to compare and evaluate the positive and negative correlation probability of exchange rate volatility on FDI, instead of just concluding on the statistical evidence found as frequentist approach. The results of the study also strengthened helpful evidence for further research.

The rest of this paper is organized as follows. Section 2 presents literature review on the theoretical and empirical effect of exchange rate volatility on FDI by frequentist approach. The next section describes the design of an empirical Bayesian panel-data model and the introduction of the collected data. Then, the main empirical results are analyzed and compared in Sect. 4. Last but not least, concluding remarks and implications are further discussed in the final section.

2 Literature Review

2.1 The Effect of Exchange Rate Volatility on FDI Inflows

According to the theory of Irreversible Investment pioneered by Dixit and Pindyck (1994), most foreign investments have two important characteristics. First, the investment is seemingly irreversible. The investment cannot be easily recovered because of the existence of sunk costs. Manufacturers need to commit capital investments in foreign and domestic capabilities before all costs of goods sold are actualized in the future. Second, investments can be delayed, which allows investors to wait for new information on prices, costs, and other market conditions before committing financial resources. This has important implications for the determinants of investment decisions. Because of being irreversible, investments are particularly sensitive to risk, originating from future cash flows, interest rates, or exchange rate movements. Thus, if the goal of policy is to stimulate investment, stability and credibility may be more important than tax or interest rate incentives (Dixit and Pindyck 1994).

Therefore, exchange rate volatility also has implications for FDI activity. In this regard, measures of exchange rate volatility actually provide additional useful information about the impact of exchange rate volatility on macroeconomic performance. This article employed the standard deviation (σ) of the exchange rate over time, ever used by Furceri and Borelli (2008).

However, there has been no clear consensus in the existing literature on the impact of exchange rate volatility on FDI. Some previous studies on this effect found negative, positive and even insignificant effects (Osinubi and Amaghionyeodiwe 2009). The theoretical arguments were primarily divided into two arguments. These two arguments, based on the production flexibility and risk aversion perspectives, make predictions in two different directions about the impact of the exchange rate volatility on FDI.

Based on the production flexibility argument, exchange rate volatility increases foreign investment because firms can adjust the variable production inputs after recognizing nominal or real shocks (Osinubi and Amaghionyeodiwe 2009). Aizenman (1992) demonstrated that the effect of exchange rate volatility on foreign investment depended on sunk costs, or irreversible progress of investment, competitive structure of the industry and the price elasticity of the profit function. It is obvious that the flexibility production argument is based on the assumption which firms can adjust for variable factors, so, this argument will not hold if factors are fixed. Following a production structure with fixed factors, the expected impact of price changes on profits is likely to be reduced. The more convenient the transformation, the more FDI inflows are attracted (Osinubi and Amaghionyeodiwe 2009).

Another possible explanation for the positive effect is that FDI can replace exports. In other words, an increase in exchange rate volatility causes multinationals in the host country to adopt local production instead of exporting, creating isolation from currency risk (Osinubi and Amaghionyeodiwe 2009). Markusen (1995) argued consistency with the export substitution for FDI. He argued that firms would engage in FDI to avoid the costs of international trade, including currency risk. When the exchange rate becomes unstable, many companies choose to meet foreign markets through their domestic production based on FDI rather than exporting.

As regards the risk aversion argument, FDI decreases as exchange rate volatility increases. This view is consistent with the Theory of Investment Under Uncertainty, which implies that uncertainty causes a decline in investment (Bernanke 1983; Henry 1974). This is because higher exchange rate volatility lowers the expected returns in future periods (Goldberg and Kolstad 1995; Cushman 1988). Besides, investors will require an additional premium for the risk of exchange rate volatility on the rate of return on investment. If the exchange rate fluctuates sharply, the expected value of investment projects will be reduced, and then FDI inflows will decrease. Assuming finite foreign direct investments, countries with high levels of currency risk will miss out on FDI opportunities compared to countries with more stable currencies (Foad 2005). The empirically results are confirmed for investment in US's wholesale industries, especially in the case of high entry costs (Campa 1993).

The difference between the risk aversion and production flexibility requires a distinction between short-term and long-term exchange rate volatility (Goldberg and Kolstad 1995; Avdjiev et al. 2017). The risk aversion is more convincing under short-term volatility because it is somewhat difficult to adjust for short-term variable factors due to the fixed contracts. In the short term, factors of production are usually fixed, and as a result, most firms only perceive risk-averse to their future returns. However, the argument for production flexibility seems convincing under long-term volatility because firms can adjust their use of variable factors (Jayaratnam 2003).

By frequent inference, Cushman (1988) and Stokman et al. (1996) found a significant positive relationship between exchange rate volatility and FDI inflows, outflows from the United States and the Netherlands. De Menil (1999) confirmed that a sustained increase in exchange rate volatility of 10% (as measured by the standard deviation of the exchange rate) will eventually increase the level of FDI by 15% across Europe. Pain and Van Welsum (2003), Foad (2005) discovered evidence to support this finding, a positive effect on FDI inflows, for industrialized countries, including the UK, Germany, Canada and the US.

Furceri and Borelli (2008) analyzed the role of exchange rate volatility in explaining the development of FDI inflows by frequent inference. Based on an empirical model of macroeconomic determinants of FDI, their results asserted that exchange rate volatility has a positive or invalid effect on closed economies, but it actually has a negative impact on highly open economies. This result may be consistent with economies in transition, such as some European countries and Commonwealth of Independent States—CIS.

That being said, Darby et al. (1999) found a negative relationship between exchange rate volatility and investment in France, Germany, USA, UK and Italy. The similar evidence was also observed in developing or transition countries by Benassy-Quere et al. (2001), Brozozowski (2006), and Sharifi-Renani and Mirfatah (2012). Currency risk reduces FDI inflows from Germany to developing countries (Hubert and Pain 1999). However, exchange rate volatility may be a symptom of deeper economic structural and institutional problems in developing countries.

Regarding FDI into the US, Gorg and Wakelin (2001) acknowledged that the exchange rate volatility did not have a statistically significant effect on FDI inflows. Osinubi and Amaghionyeodiwe (2009) did not reference any empirical evidence on the impact of exchange rate volatility on foreign direct investment (FDI) in Nigeria from 1970 to year 2004.

2.2 *The Other Determinants of FDI Inflows*

Khamis and Hussin (2015) investigated the effect of inflation rate and GDP per capita on foreign direct investment inflows into United Arab Emirates (UAE) for duration of 1980–2013. The inflation had insignificant impact on foreign direct investment inflows while economic growth was found to have positive effects. Similar results with time-series data from 1980 to 2016 were also obtained by Mansoor and Bibi (2019) in Pakistan.

Omankhanlen (2011) analyzed effects of the inflation and some economic variables on foreign direct investment. Although, the empirical results showed that the inflation has no effect on FDI inflows; trade openness still has a vital role in explaining the macroeconomic linkage with FDI inflows. Jayakumar et al. (2014) stated that liberalization policies increased FDI inflows in the India.

Assessing the determinants of foreign direct investment inflows in Iran by covering the 1980–2006 period, Sharifi-Renani and Mirfatah (2012) discoveried that economic growth and trade openness positively affect on FDI inflows while exchange rate volatility have negatively influence on foreign direct investment. The positive effect of trade openness is reinforced by Delali (2003), Parajuli (2012).

Dang et al. (2020) found that private investment was positively affected by monetary policies, which was strongly supported by Keynesian theory. Accordingly, the expansion of monetary supply (M2) could be seen as attractant for private investment development in Vietnam. Previously, Fu and Liu (2015) examined the impact of monetary policy on the investment adjustments of China's listed firms during 2005–2012. An increase in the monetary supply could promote an increase in investment and a decrease in interest rates, thereby leading to a positive impact on economic activities. As a result, unemployment decreases, and the value of assets in market shares tends to increase. The attractiveness of the domestic market will stimulate an increase in foreign direct investment inflows (Shafiq et al. 2015).

3 Design of a Bayesian Model

3.1 *Bayesian Estimation*

Bayesian estimation is expressed as follows:

$$p(\theta|y) \propto p(y|\theta)p(\theta) \tag{1}$$

The posterior distribution of θ is established as a compromise between the prior distribution and the likelihood function. In other words, the posterior mean depends on the weighted algebraically arrangement between the prior mean, and the data proportion of the observed sample. It is extremely difficult to realize prior distribution and the likelihood function correctly in order to produce estimated parameters without convergence problems. Historically, formulas for prior distributions getting along with the likelihood function were simplified and even defaulted to conveniently generate the goodness of fit of a Bayesian model (Kruschke 2015).

In the recent decades, many numerical approximation methods for doing Bayesian Data analysis have been developed. One of the prominent methods is generally known as Markov chain Monte Carlo (MCMC) method, which randomly samples a large number of representative combinations of parameter values from the posterior distribution. The continuous development of such MCMC method has been strongly supporting Bayesian statistical methods closer to wide practical applications (Kruschke 2015).

There is a assumption in the pooled model that the same regression line was appropriate for all individuals. All individuals in the individual-effect models had regression lines with the same slopes, but possibly different intercepts. In the individual-effect models, posterior inference can be carried out by setting up a Gibbs sampler. In the frequentist econometric, the random-effect and fixed-effect models are two popular types of individual-effect models (Koop 2003).

Unlike the frequentist framework, both fixed-effect and random-effect estimation in the Bayesian framework are treated as random parameters, scattered in the distributions. The dependent variable is distributed around a mean value that depends on certain parameters, together with regressors. These parameters are, in turn, distributed around a mean value determined by other parameters called hyperparameters, which are also random. The random coefficients model allows intercepts for every individual to be different, but their hierarchical prior means are not too different. Such a model might be reasonable in many applications. While a fixed effects estimation updates the distribution of the parameters, a random effects estimation updates the distribution of the hyperparameters. From a Bayesian perspective, there is no distinction between fixed-effect and random-effect estimates (Rendon 2012; Koop 2003).

Nevertheless, Bayesian computation faces a few difficulties in approximating marginal distribution and posterior moments. These problems are solved by simulations and application of the central limit theorem in practice. Markov chain Monte Carlo (MCMC) methods can be considered as effective solutions for approximately sampling from posterior distributions. One of the most prevalent versions of MCMC, the Metropolis–Hastings algorithm, will be used in the research.

3.2 Design of an Empirical Model

This article performs a Bayesian analysis by identifying a posterior model. The posterior model combines given data and prior information to describe the probability distribution of all parameters. Thus the posterior distribution by Bayes rule has two components: a likelihood function that includes information about the model parameters based on observed data and the prior distribution includes the previous information about the model parameters. For large amounts of data, the Bernstein and Von Mises theorem stated that a posterior distribution converges to a multivariate normal distribution centered at the maximum likelihood estimator. In other words, the posterior distribution is relatively independent of the prior distribution, the inferences based on Bayes, and likelihood function, thereby generating fairly similar results (Vaart 1998).

Bayesian normal regression uses Metropolis–Hastings and Gibbs sampling with the simulation of MCMC iterations (Burn-in ones and MCMC ones of sample size). Its Likelihood is assumed to normal prior with variance defined {sigma2}. Bayesian normal regression can be described by:

$$FDIGDP = \theta X + \varepsilon \tag{2}$$

where FDIGDP indicates net FDI inflows as a percent of GDP; X illustrates a set of determinants of net FDI inflows as a percent of GDP, which are exchange rate volatility, economic growth per capita, inflation, monetary supply (M2), trade openness.

We defaulted normal priors for regression coefficients and assumed inverse-gamma priors for the variance parameters. The identification of informative priors was supported by Stata 17 as follows:

Likelihood:

$$\text{FDIGDP} \sim \text{normal}(\text{xb\_FDIGDP}, \{\text{sigma2}\}) \tag{3}$$

Priors:

$$\begin{aligned}
&\theta_{\text{lnVolExr}} \sim \text{normal}(0, 10000)\\
&\theta_{\text{gGDPcap}} \sim \text{normal}(0, 10000)\\
&\theta_{\text{inf}} \sim \text{normal}(0, 10000)\\
&\theta_{\text{gM2}} \sim \text{normal}(0, 10000)\\
&\theta_{\text{gTrade}} \sim \text{normal}(0, 10000)\\
&\theta_{\text{_cons}} \sim \text{normal}(0, 10000)\\
&\{\text{U[individuals]}\} \sim \text{normal}(0, \{\text{var_U}\})\\
&\{\text{sigma2}\} \sim \text{igamma}(0.01, 0.01) \text{ or } \{\text{sigma2}\} \sim \text{igamma}(50, 50) \text{ or } \{\text{sigma2}\} \sim \text{igamma}(100, 100)
\end{aligned} \tag{4}$$

Hyperprior:

$$\{\text{var\_U}\} \sim \text{igamma}(0.01, 0.01);\ \text{or } \{\text{var\_U}\} \sim \text{igamma}(50, 50);\ \text{or } \{\text{var\_U}\} \sim \text{igamma}(100, 100) \tag{5}$$

where θ was assumed as the effect of each determinant on net FDI inflows as a percent of GDP, namely FDIGDP; lnVolExr was the 3-year standard deviation of the logarithm of the exchange rate, called exchange rate volatility; gGDPcap was the economic growth per capita; inf standed for inflation; gM2 represented broad monetary supply growth (M2); trade openness growth was measured by gTrade; _cons was proxy for the mean of random effects; sigma2 was proxy for the error variance; and var_U was proxy for the variance of random effects. The assumption of inverse-gamma priors for the variance parameters was based on the half of degrees of freedom of the model, with a sample size of 174 observations.

3.3 Data

The dataset of 29 emerging countries[1] between 2002 and 2019 was collected and aggregated from the World Development Indicators of World Bank database, accessed at https://data.worldbank.org/. This research dataset was grouped into six periods: 2002–2004, 2005–2007, 2008–2010, 2011–2013, 2014–2016, and 2017–2019 periods. The sample size was relatively large enough to perform the Bayesian estimation after splitting the periods. Excepting for the exchange rate volatility, which was the standard deviation of the logarithm of the exchange rate, the remaining factors were transformed into a 3-year average. In order to clarify the meaning of the research findings, the Bayesian estimation was compared with the previous studies. The panel data was balanced to ensure the convenience of identifying the prior mean and variance in the normal distribution.

4 Empirical Results

This Bayesian estimation had utilized Metropolis–Hastings and Gibbs sampling for the simulation of 25,000 MCMC iterations, including 5000 burn-in ones and 20,000 MCMC ones of sample size. This is to ensure a robust Bayes inference when the MCMC chains converge to a stationary range.

More specifically, the convergence of MCMC chains should be examined before performing Bayesian inferences. Information about efficiency of the simulation was presented in Table 1, providing objective criterion for convergence. Visual inspections are also included in Appendices to further reinforce the examine the convergence of MCMC convergent tests of MCMC chains in terms of data visualization. Appendices 1, 2 and 3 described the visual diagnostics for MCMC convergence of Bayesian estimation with default priors, igammaprior(50,50) and igammaprior(100,100) respectively. The graphical summaries for parameters did not show any convergence problems.

According to visual inspections, the trace plots illustrated almost well-mixing, exhibiting low autocorrelation, with simulation efficiency for reliable estimation. The trace plots looked a good coverage of the domains of the marginal distributions, did not show trends and fluctuated with their averages. This reveal more rapid traversing of the marginal posterior ones. The histogram and kernel density plots for variance parameters {sigma2} and {var_U} almost resembled the expected shape of inverse-gamma distributions. The histograms and density plots for other parameters were similar for all chains, following normal distributions.

[1] Bangladesh, Bulgaria, Brazil, Colombia, Czech, Chile, China, Hungary, Indonesia, India, Israel, Korea, Sri lanka, Morocco, Mexico, Mauritius, Nigeria, Pakistan, Peru, Poland, Philippine, Romania, Sudan, Tunisia, Turkey, Thailand, Ukraine, Vietnam, South Africa.

Table 1 The estimation result for posterior mean

Random-Effect Panel-data model	Bayesian estimation					
	Default		igammaprior (50,50)		igammaprior (100,100)	
	(4a)		(4b)		(4c)	
Estimator	*Mean*	*Efficiency*	*Mean*	*Efficiency*	*Mean*	*Efficiency*
$\theta_{\_cons}$	2.3602	0.1387	2.4323	0.2413	2.3975	0.1828
$\theta_{lnVolExr}$	2.0073	0.4439	2.4821	0.5560	2.2216	0.5057
$\theta_{gGDPcap}$	0.1312	0.3478	0.1240	0.3820	0.1282	0.3529
θ_{inf}	−0.0534	0.2263	-0.0668	0.3971	-0.0596	0.3377
θ_{gM2}	0.0439	0.6219	0.0438	0.7333	0.0440	0.6760
θ_{gTrade}	9.0424	0.7182	9.1834	0.6880	9.0746	0.7834
sigma2	9.8423	0.2734	6.4728	0.4050	4.8105	0.5083
var_U	3.1243	0.0755	1.2050	0.2504	1.1397	0.4044
Acceptance rate	0.8155		0.8170		0.8153	
Number of obs	174		174		174	
Number of groups	29		29		29	
Note:	*Likelihood:* FDIGDP ~ normal(xb_FDIGDP,{sigma2}) *Priors:* {FDIGDP:gGDPcap inf gM2 gTrade lnVolExr _cons} ~ normal(0,10,000) {U[individuals]} ~ normal(0,{var_U})					
	{sigma2 var_U} ~ igammaprior (0.01, 0.01)		{sigma2 var_U} ~ igammaprior (50, 50)		{sigma2 var_U} ~ igammaprior (100, 100)	

As regards scatterplots, there were no significant correlations among the parameters. The sequences of multiple chains for graphical diagnostics did not reveal substantial differences.

As can be seen from the Table 1, acceptance rates of the Model (4b), (4c), (4d) were respectively 0.8155, 0.8170, 0.8153 close to 1. Efficiency of the simulation, which raises convergence and precision of credible interval, ranged [0.0755; 0.7182]; [0.2413; 0.7333]; [0.1828; 0.7834] for Model (4a), (4b), (4c) correspondingly. It is worth noting that the results of posterior means demonstrated little difference in the magnitude of the effect, but there was no difference in the direction of impact among the levels of HDI.

The information about means in the Bayesian method is supplemented by the posterior probability of parameters, thereby further researching results more comprehensively. For further analysis, probability distributions, differed in the direction of

Table 2 The Bayesian posterior probability of parameters

Sign of impact	Default		igammaprior (50,50)		igammaprior (100,100)	
	+	−	+	−	+	−
θ_{lnVolExr}	69.75%	30.25%	78.96%	21.04%	79.64%	20.36%
θ_{gGDPcap}	82.79%	17.21%	87.50%	12.50%	91.40%	08.60%
θ_{inf}	23.46%	76.54%	12.68%	87.32%	12.24%	87.76%
θ_{gM2}	86.80%	13.20%	91.58%	08.42%	94.57%	05.43%
θ_{gTrade}	96.65%	03.35%	99.02%	00.98%	99.61%	00.39%

the effects, were analyzed more deeply in Table 2. For Bayesian framework, the posterior mean of exchange rate volatility (lnVolExr) had a dispersed effect on FDI inflows. The model with default priors simulated the positive effect of exchange rate volatility with a probability of 69.75% compared to 30.25% of the probability of negative effect. In terms of magnitude, the positive influences in the two models with igammaprior(50,50) and igammaprior(100,100) for variance parameters accounted for a higher probability at 78.96%, 79.64%, respectively. Simultaneously, the probability of negative effect also decreased to 21.04% and even 20.36% accordingly. The research results did not invalidate any of the theoretical arguments, the production flexibility and risk aversion perspectives. Even so, this suggests that the production flexibility still prevails in emerging countries. This also implies a floating exchange regime had been gradually dominating to better attract FDI inflows. However, acknowledging this positive effect in reality will easily lead to harsh reactions by the policies of fixed exchange regime. The negative tendency still exists at a lower probability level, but not completely eliminated. Moreover, considering more macroeconomic and political issues, the positive influence of exchange rate volatility needs to be taken seriously into account.

Markedly, the positive effect of GDP growth per capita (gGDPcap) was in the majority and consistent with expectations of the positive effect on FDI inflows, the lowest 82.79% and the highest 91.40%. Moreover, the negative effect accounts for only 17.21% at most in the model with default priors, and even less than 10% in model (4c). This implied that FDI inflows had really been increased by GDP growth per capita (gGDPcap) in recent years.

Similarly, the impact of trade openness growth (gTrade) witnessed a positive trend, more than 95%, and highest proportion at the model with igammaprior (100,100) was approximately 99.61%. However, Bayesian estimation with igammaprior(50,50) and igammaprior(100,100) produced better efficiency of variance parameters {var_U} than Bayesian estimation with default priors. The research results did not deny a negative effect of trade openness growth, which was still occurring with a minor probability density. Or rather, the positive effect of trade openness growth extremely exceeded its negative effect. Strong evidence of the tendency of integration was strengthened over the recent decades. As a component of globalization, it is plausible to explain why trade openness growth is primarily responsible for the changes in FDI.

In addition, the monetary supply growth (gM2) also recorded a predominantly positive trend, with probability ranging from 86.80% to 94.57%. The positive effect of monetary supply growth has principle that the buying power of the consumer and the growth of industries raise when monetary supply increases. This result also confirmed the evidence of Shafiq et al. (2015). It is plausible that emerging countries have been pursuing the need for expansionary monetary policy, addressing possible recession and unemployment.

Nevertheless, the inflation (inf) exposed opposite influences, constituting 76.54%, 87.32% and 87.76% respectively. The decreasing level of inflation increases FDI inflows. When inflation reduces, the request for cost of capital is lower; leading to increases in the FDI inflows. The tightening monetary policies such as raising the interest rate, which is pursued for higher inflation, can weaken the international competition among the countries. This increases the request for cost of capital and make it less attractive to FDI inflows. This positive effect found was consistent with the research result of Mansoor and Bibi (2019).

5 Concluding Remarks and Implications

To conclude, our article has provided new insight into the correlation apparently associated with probable probabilities for the effect of exchange rate volatility on net FDI inflows as a percent of GDP.

While the previous studies used the frequentist framework to find statistical evidence to accept or reject null hypothesis, the Bayesian framework retrieves to provide probabilities of both positive and negative effects and support for the more dominant probability.

Our research on the Bayesian framework acknowledged that the positive effects of exchange rate volatility on FDI exceeded the negative effects of those in emerging countries. In other words, the production flexibility may be more predominant in terms of magnitude of probability, depending on the characteristics of each country and specific period. However, the negative effects of exchange rate volatility on FDI, supporting risk aversion perspectives, accounted for over 20%. From a panorama, a floating exchange rate regime betters attract FDI inflows. Regards controlling factors, GDP growth, trade openness and monetary supply recorded positive effects on FDI inflows; only the inflation (inf) exposed negative influence.

Appendices

Appendix 1. Visual Diagnostics for MCMC Convergence of Bayesian Estimation with Default Priors

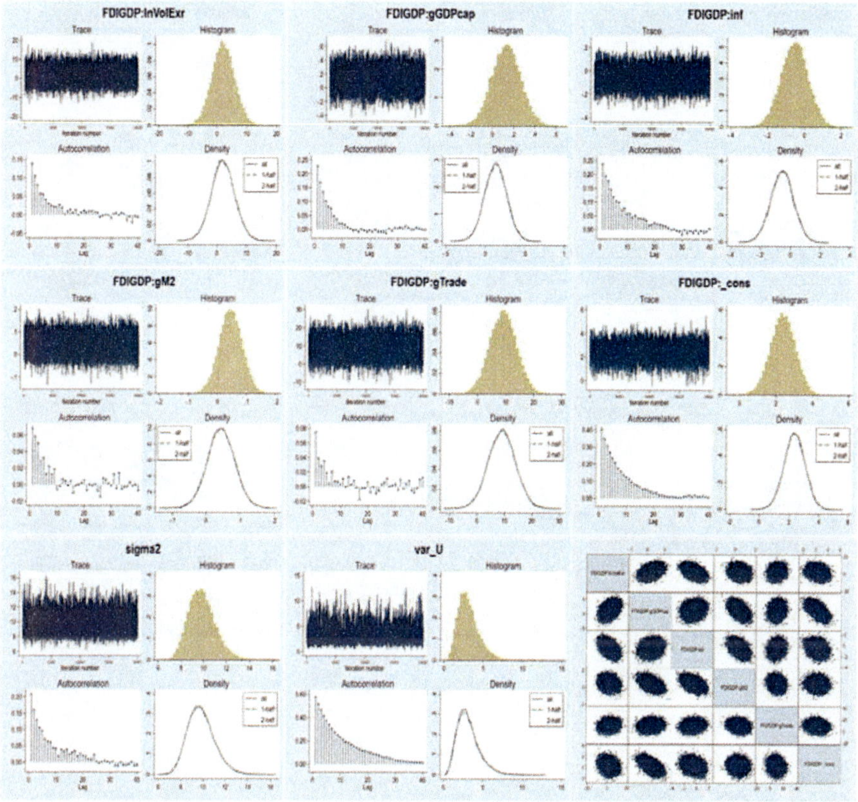

Appendix 2. Visual Diagnostics for MCMC Convergence of Bayesian Estimation with igammaprior (50,50)

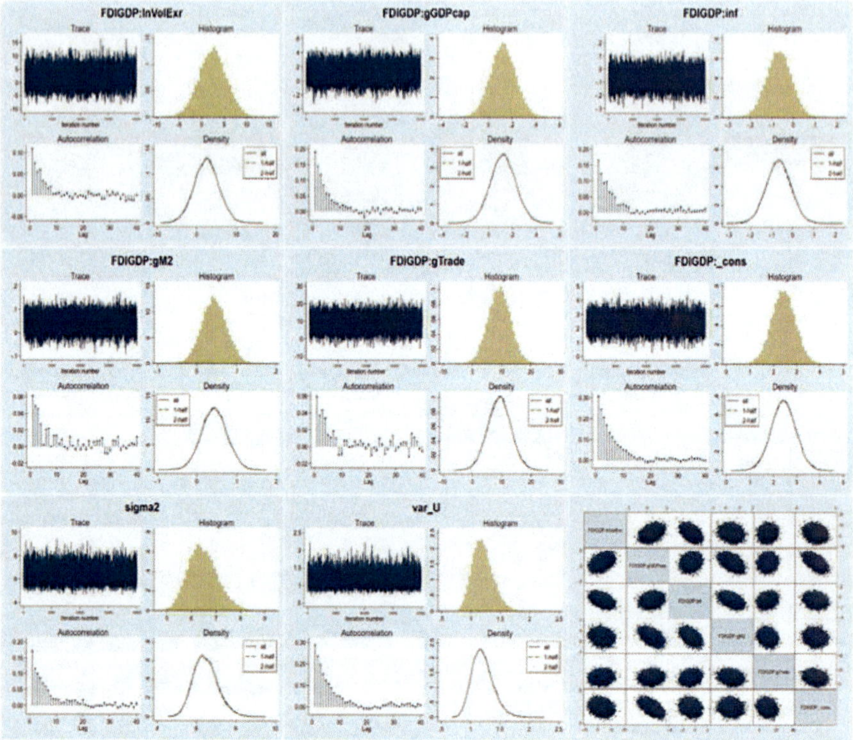

Appendix 3. Visual Diagnostics for MCMC Convergence of Bayesian Estimation with igammaprior (100,100)

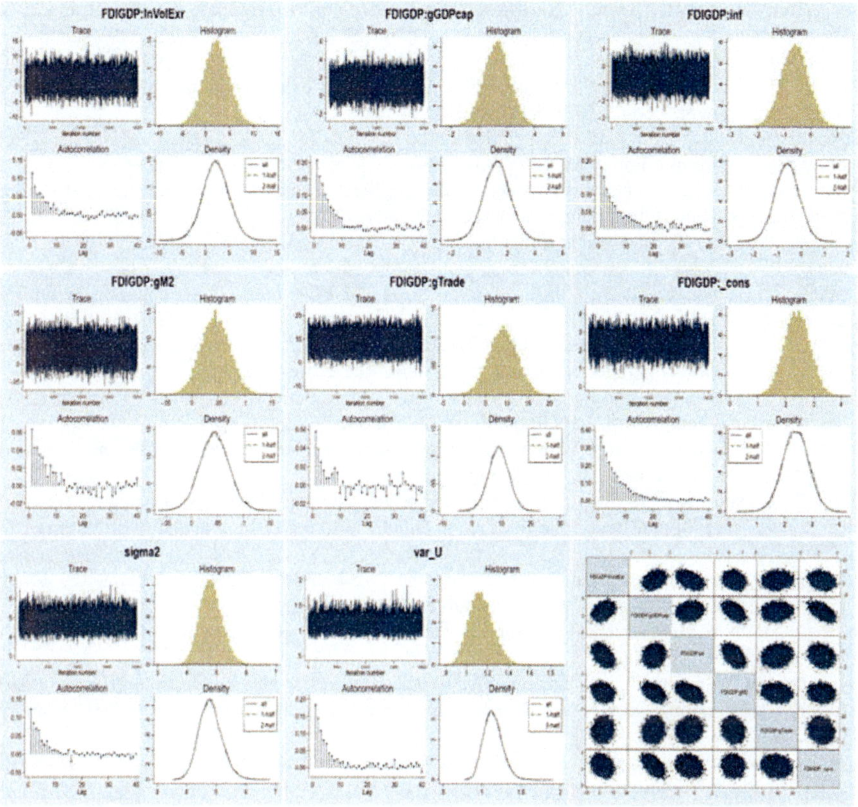

References

Aizenman, J.: Exchange rate flexibility, volatility and patterns of domestic and foreign direct investment. Int. Monetary Fund Staff Papers **39**(4), 890–922 (1992)

Avdjiev, S., Gambacorta, L., Goldberg, L.S., Schiaffi, S.: The shifting drivers of international capital flows. NBER Working Paper No 23565, Washington, DC: National Bureau of Economic Research (2017)

Benassy-Quere, A., Fontagne, L., Lahreche-Revil, A.: Exchange-rate strategies in the competition for attracting Foreign Direct Investment. J. Japan. Int. Econ. **15**, 178–198 (2001)

Bernanke, B.S.: Non-monetary effects of the financial crisis in the propagation of the great depression. Am. Econ. Rev. **73**(3), 257–276 (1983)

Brozozowski, M.: Exchange rate variability and foreign direct investment: consequences of EMU enlargement. Eastern Eur. Econ. **44**(1), 5–24 (2006). https://doi.org/10.2139/ssrn.1441247

Campa, J.M.: Entry by foreign firms in the United States under exchange rate uncertainty. Rev. Econ. Stat. **75**, 614–622 (1993)

Cushman, D.O.: Exchange-rate uncertainty and foreign direct investment in the United States. Weltwirtschaftliches Archiv. **124**, 322–335 (1988). https://www.jstor.org/stable/40439599

Dang, T.T., Pham, A.D., Tran, D.N.: Impact of monetary policy on private investment: evidence from Vietnam's provincial data. Economies **8**(3), 70 (2020). https://doi.org/10.3390/economies8030070

Darby, J., Hallett, A.H., Ireland, J., Piscitelli., L.: The impact of exchange rate uncertainty on the level of investment. Econ. J. **109**, 55–67 (1999)

De Menil, G.: Real capital market integration in the EU: how far has it gone? What will the effect of the Euro be? Econ. Policy **28**, 165–189 (1999)

Delali, A.: The determinants and impacts of foreign direct investment. Munich Personal RePEc Archive, 3084 (2003). https://mpra.ub.uni-muenchen.de/3084/

Dixit, A., Pindyck, R.: Investment under uncertainty. Princeton University Press, Princeton, New York (1994)

Foad, H.S.: Exchange rate volatility and export-oriented FDI, pp. 2–7. A Paper from Emory University, Atlanta, GA (2005)

Fu, Q., Liu, X.: Monetary policy and dynamic adjustment of corporate investment: a policy transmission channel perspective. China J. Account. Res. **8**, 91–109 (2015)

Furceri, D., Borelli, S.: Foreign direct investments and exchange rate volatility in the EMU neighbourhood countries. J. Int. Global Econ. Stud. **1**(1), 42–59 (2008)

Goldberg, L.S., Kolstad, C.D.: Foreign direct investment, exchange rate volatility and demand uncertainty. Int. Econ. Rev. **36**, 855–873 (1995)

Gorg, H., Wakelin K.: the impact of exchange rate volatility on US direct investment. In: GEP Conference on FDI and Economic Integration (June 29–30th, 2001), University of Nottingham (2001)

Henry, C.: Investment decisions under uncertainty: the "irreversibility effect". Am. Econ. Rev. **64**(6), 1006–1012 (1974). https://www.jstor.org/stable/1815248

Hubert, F., Pain, N.: Investment, innovation and the diffusion of technology in Europe. Cambridge University Press (1999)

Jayakumar, A., Kannan, L., Anbalagan, G.: Impact of foreign direct investment, imports and exports. International Review of Research in Emerging Markets and the Global Economy (IRREM), **1** (2014)

Jayaratnam, A.: How does the black market exchange premium affect Foreign Direct Investment (FDI)? A Paper from Stanford University, pp. 1–16 (2003)

Khamis, H.A., Hussin, R.M.: The impact of inflation and GDP per capita on foreign direct investment: the case of United Arab Emirates. Invest. Manag. Finan. Innov. **12**(3), 18–27 (2015)

Koop, G.: Bayesian Econometrics. Wiley, England. ISBN: 978–0–470–84567–7 (2003)

Kruschke, J.K: Doing Bayesian Data Analysis (Second Edition). Inferring a Binomial Probability via Exact Mathematical Analysis (2015)

Mansoor, A., Bibi, T.: Dynamic relationship between inflation, exchange rate, FDI and GDP: evidence from Pakistan. Acta Universitatis Danubius. Œconomica **15**(2), 431–444 (2019)

Markusen, J.R.: The boundaries of multinational enterprises and the theory of international trade. J. Econ. Rev. **9**, 169–189 (1995)

Omankhanlen, A.E.: The effect of exchange rate and inflation on foreign direct investment and its relationship with economic growth in Nigeria. EA1, **1** (2011)

Osinubi, T.S., Amaghionyeodiwe, L.A.: Foreign direct investment and exchange rate volatility in Nigeria. Int. J. Appl. Econ. Quant. Stud. **6**(2) (2009)

Pain, N., Welsum, V.: Untying the Gordian Knot: the multiple links between exchange rates and foreign direct investment. J. Common Market Stud. **41**, 823–846 (2003)

Parajuli, S.: Examining the relationship between the exchange rate, foreign direct investment and trade. LSU Doctoral Dissertations, 2670 (2012). https://digitalcommons.lsu.edu/gradschool_dissertations/2670

Silvio, R.: Rendon: Fixed and Random Effects in Classical and Bayesian Regression, Oxford Bulletin Of Economics And Statistics. (2012). https://doi.org/10.1111/j.1468-0084.2012.00700.x

Shafiq, N., Ahmad, H., Hassan, S.: Examine the effects of monetary supply M1 And GDP on FDI in Pakistan. Int. J. Curr. Res. **7**(03), 13498–13502 (2015)

Sharifi-Renani, H., Mirfatah, M.: The impact of exchange rate volatility on foreign direct investment in Iran. Proc. Econ. Financ. **1**, 365–373 (2012)

Stokman, A.C.J., Vlaar, P.J G.: Volatility, international trade and capital flows. In: Bruni, F., Fair, D.E., O'Brien, R. (eds.) Risk Management in Volatile Financial Markets, Financial and Monetary Policy Studies, vol. 32, Springer, Boston, MA (1996). https://doi.org/10.1007/978-1-4613-1271-0_7

Vaart, A.W., van der: 10.2 Bernstein–von Mises Theorem. Asymptotic Statistics. Cambridge University Press. ISBN 0-521-49603-9 (1998)

COVID-19, Stimulus, Vaccination and Stock Market Performance

Linh D. Nguyen

Abstract Although the COVID-19 outbreak harms economies and causes panic among investors, stock markets keep climbing and few researchers have addressed the problem. This study finds that the negative effect exists only in the short run. When Fed provides stimulus packages to support the economy, especially after vaccine campaigns, this impact changes from negative to positive. The rationale is that the combination of stimulus packages and COVID-19 vaccines makes announcements of confirmed cases to become a buy signal for optimist investors. The investors think that stock markets will soon correct themselves and increase; hence, they buy stocks and then wait for the market to catch up. As a result, the stock market performance increases with the increase in total COVID-19 confirmed cases.

Keywords Bayesian structural time series · COVID-19 · Causal impact analysis · Stimulus · Stock market performance

JEL Classification G01 · G14 · G33

1 Introduction

The COVID-19 pandemic, which could be described as a "Black Swan event", has a strongly adverse impact on economic growth and development of many countries, and causes fear and panic among investors (He et al. 2020). Among losses caused by the novel coronavirus, financial markets all over the world also experienced a sharp decline. In the United States, Baker et al. (2020) show that the volatility level of the U.S. stock market rivals or surpasses that last seen in the global financial crisis in 2008, Black Monday in 1987 and Great depression in 1933.

The strong fluctuations in stock markets due to COVID-19 are increasingly attracting scholarly attention, and an increasing number of research have been

L. D. Nguyen (✉)
Department of Finance, Banking University of Ho Chi Minh City, 36 Tôn Thất Đạm, District 1, Ho Chi Minh City, Vietnam
e-mail: linhnd@buh.edu.vn

N. Ngoc Thach et al. (eds.), *Financial Econometrics: Bayesian Analysis, Quantum Uncertainty, and Related Topics*, Studies in Systems, Decision and Control 427,
https://doi.org/10.1007/978-3-030-98689-6_33

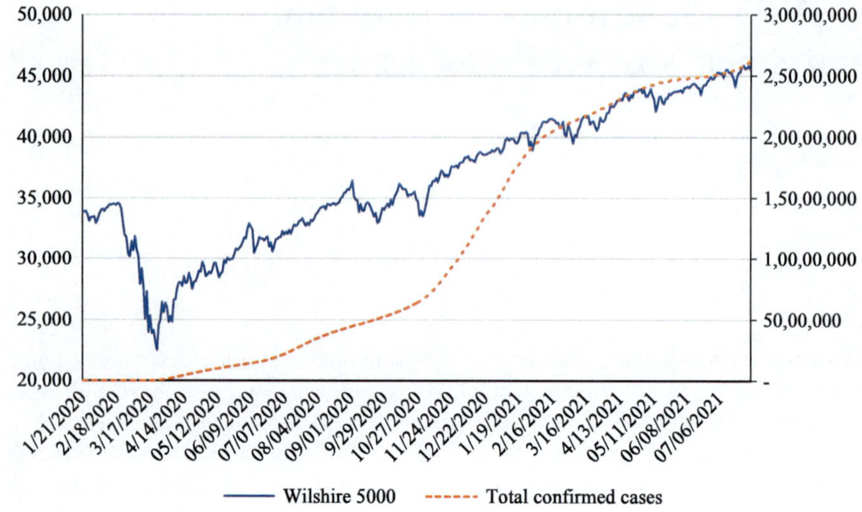

Fig. 1 The evolution of Wilshire 5000 Total Market Index and total confirmed cases. This figure shows the evolution of Wilshire 5000 Total Market Index and total COVID-19 confirmed cases. The data is obtained from Refinitiv Datastream. Federal Reserve announced to provide a $2.3 trillion stimulus package to support the economy on April 9, 2020. December 14, 2020 is the first day the USA implements the vaccination program

performed recently to analyze the impact of the COVID-19 pandemic. The consensus in the growing literature is that the outbreak harms the stock market performance. However, in the real world, stock markets keep climbing despite real-world turmoil, as shown in Fig. 1. Few researchers have addressed this fact. Hence, this study is conducted to fill the gap in the literature and contribute to the discussion on the economic impact of the COVID-19 pandemic on stock market performance.

Using daily data on the growth of total COVID-19 confirmed cases, the change in the Wilshire 5000 Total Market Index and control variables from January 21, 2020 (the first COVID-19 confirmed case in the U.S.) to July 30, 2021, this investigation attempts to find the explanation for the increase in the stock market performance while economies struggle. At a first glance, this study also finds the negative relationship between total confirmed cases and the stock market performance, as shown in the previous literature.

However, with further analyses, the research finds that the negative relationship is short-lived. This relationship changes from negative to positive when Fed provides stimulus packages to support the economy. The stock market is forward-looking and stock prices reflect the expectation of investors. With the stimulus, investors think that they can find opportunities in a crisis, i.e., a sharp decrease in stock price due to COVID-19. Hence, the more confirmed cases there are, the more stock investors buy up and wait for the market to catch up. As a result, the stock market performance has a positive relationship with the total confirmed cases. Furthermore, the positive impact of confirmed cases on the stock market performance becomes even stronger as governments conduct both stimulus packages and vaccination campaigns. The reason is that vaccination appears to be the best solution to reignite the suffering economy and to increase the confidence of investors. The results keep robust when the analyses are replicated with alternative proxies and different approaches, i.e., causal impact analysis using the Bayesian structural time series model.

The rest of this article is organized as follows: Sect. 2 briefly reviews the existing literature. Section 3 describes the data used and the methodology applied. Section 4 reports and discusses results for empirical analyses. Section 5 conducts robustness checks while Sect. 6 concludes the study.

2 Literature Review

The COVID-19 crisis has been bringing considerable human suffering (Baldwin and Mauro 2020a). To slow the worldwide spread of COVID-19, governments around the world have performed a range of strong measures, such as quarantine, travel restrictions, lockdown, social distancing, closure of non-essential businesses and schools (WHO 2020). These strong measures cause manufacturing and consumption almost to a standstill, which leads to supply and demand disruptions (Baldwin and Mauro 2020b). Supply shocks (production falls) and demand shocks (purchases fall) have caused negative impacts on economies as well as the financial markets around the world.

The combination of the pandemic and governments' strong actions results in dramatic fluctuations in financial markets and triggered the interest of many researchers. Therefore, an increasing number of studies have been conducted to examine the impact of the COVID-19 outbreak on stock markets by using different econometric techniques. The techniques have generally followed three tracks: regression, event study and text analytics.

The first group is papers using regression analyses. To examine the effect of contagious infectious diseases on the Chinese stock market, Al-awadhi et al. (2020) use panel data analysis with the two main variables being daily growth in total confirmed cases and daily growth in total cases of deaths. The authors find that both measures have significant negative effects on stock returns across all companies. Bahrini and Filfilan (2020) investigate the impact of the COVID-19 on the daily returns of the stock market indices in the Gulf Cooperation Council countries. Their results show that new cases and total COVID-19 confirmed deaths significantly negatively affect these indices, while the number of COVID-19 confirmed cases is not significant. In a larger scale of sample, Ashraf (2020) investigate the effect of growth in COVID-19 confirmed cases on the stock market returns from 64 countries. The author also finds the negative relationship between stock market returns and the growth in confirmed cases. Other papers, which use the similar method, also find the same results, such as He et al. (2020), Takyi and Bentum-Ennin (2021) and Zhang et al. (2020).

The second track is to use an event study method. Liu et al. (2020) apply this method to examine the impact of the COVID-19 pandemic on leading stock market indices in 21 major affected countries. Findings indicate that the outbreak affects negatively and significantly the stock market returns of all countries. Similarly, Harjoto et al. (2021) use the WHO announcement on 11 March 2020 as the event representing the shock. They find that COVID-19 caused a negative shock to the global stock markets, especially in emerging markets and for small firms. The results are consistent with that of Nguyen (2021), Ramelli and Alexander (2020) and Yan (2020).

The final strand is to use text analytics. Baker et al. (2020) apply a text-based approach to analyze the unprecedented stock market impact of COVID-19. The authors show that the COVID-19 outbreak has forcefully impacted the U.S. stock market, while previous pandemics, including the Spanish Flu, left only mild traces. Similarly, Ahundjanov et al. (2020) examine the relationship between Google Trends search related to COVID-19 and the stock market returns around the world. They find that Coronavirus Search Trends significantly reduced financial indices. Amstad et al. (2020) use information on internet searches on Google and Baidu to measure investors' risk attitudes related to the pandemic. Their findings show that stock markets around the world are sensitive to changes in risk attitude index, especially in more financially developed countries.

In summary, the consensus in the growing literature on the effect of the COVID-19 outbreak on stock market performance is that the pandemic negatively affects the stock market performance. However, few researchers have explained why stock markets keep climbing despite real-world turmoil. Therefore, this study is conducted to fill the gap in the literature and contribute to the discussion and research on the economic impact of the COVID-19 pandemic on the stock market performance.

3 Data and Methodology

3.1 Data

To analyze the impact of the COVID-19 pandemic on the stock market performance of the USA, this research uses models which are composed of the following variables. The main dependent variable is the daily change in the Wilshire 5000 Total Market Index (*WILSHIRE*). Approximately 5,000 capitalization-weighted security returns are used to adjust the index. Hence, *WILSHIRE* measures the performance of all U.S. equity securities with readily available price data and is the most complete measure of the entire U.S. stock market.[1]

With respect to explanatory variables, the main considered variable is the daily change in the number of daily Covid-19 confirmed cases (*CASE*), which represents the impact of the COVID-19 pandemic. Since daily stock market performance could be affected by potential factors, this study includes a vector of other independent

[1] https://www.wilshire.com/Portals/0/analytics/indexes/fact-sheets/wilshire-5000-fact-sheet.pdf.

Table 1 Descriptive statistics

	N	Mean	Std. Dev	Min	Median	Max
WILSHIRE	398	0.090	1.820	−12.290	0.130	9.400
CASE	398	5.110	14.800	0.000	1.110	163.640
VOL	398	2.040	22.500	−55.370	−0.300	198.290
OIL	398	−0.710	17.360	−30.597	0.060	35.020
GOLD	398	0.050	1.110	−4.940	0.090	5.090
EXRATE	398	0.940	0.034	0.893	0.929	1.030
VIX	398	0.285	0.085	0.150	0.274	0.730

The table shows the summary statistics (number of observations, mean, standard deviation, minimum, median and maximum) of the sample. *WILSHIRE* is defined as the daily change in the Wilshire 5000 Total Market Index. *CASE* is estimated as the daily growth of the total confirmed cases. *VOL* is calculated as the daily change in total share volume on the U.S. stock market. *OIL* is defined as the daily change in the spot prices of West Texas Intermediate crude oil. *GOLD* is estimated as the daily change in spot gold prices. *EXRATE* is calculated as the daily change in the U.S. Dollar Index. *VIX* is estimated as the change in the S&P500 volatility index

variables to control the impact. Specifically, *OIL* is used to control the impact of daily fluctuation of oil price, *GOLD* for gold price, *EXRATE* for the U.S. Dollar Index, *VOL* for total share volume and *VIX* for the S&P500 volatility index (*VIX*). All the necessary data are obtained from Refinitiv Datastream.

In terms of the research period, the first confirmed case of 2019-nCoV infection in the U.S. is recorded on January 21, 2020. Hence, the considered period is from January 21, 2020 to July 30, 2021. Since this research includes lagged variables, the first-day observations are lost. Thus, the final sample consists of 398 observations. Summary statistics of the sample and the definition of the variables are shown in Table 1.

Table 1 shows that the U.S. stock market performance is 0.09% per day, while the growth of the total COVID-19 confirmed case is 5.11% per day. During the research period, all considered variables experience an increase in value, except for the oil price. It is not surprising because the COVID-19 crisis has severely affected the level of global economic activities, thereby causing a historic decline in global oil demand.

3.2 Methodology

Al-awadhi et al. (2020) contend that the pandemic lasts for several days and the peak of the event is not the start date, so the event study is an inappropriate method to investigate the effect of the pandemic on stock markets. Hence, the authors do not employ an event study approach, but a regression analysis. Since this paper focuses on only the U.S. stock market, the panel data regression is not eligible. Hence, the ordinary least square (OLS) regressions are implemented to analyze the impact of the outbreak on stock market performance. This technique is also used by Amstad et al. (2020) and Liu et al. (2020). The following is the baseline regression model:

$$WILSHIRE_t = \alpha_0 + \alpha_1 CASE_{t-1} + \alpha_2 CONTROL_{t-1} + \varepsilon_t \tag{1}$$

$CONTROL_{t\text{-}1}$ is a vector of control variables, including *VOL*, *OIL*, *GOLD*, *EXRATE* and *VIX*. $\varepsilon_{i,t}$ consists of the error terms of the model. All explanatory variables are lagged by one year to mitigate concerns regarding reverse causality problems. Standard errors are robust to heteroscedasticity.

4 Results and Discussion

4.1 *The COVID-19 Pandemic and Stock Market Performance*

Table 2 presents the results of this investigation. Model 1 reports the results of analysis whose independent variable is only *CASE*, while Model 2 shows results of Eq. (1).

Consistent with the finding of previous studies, this research indicates the significantly negative relationship between the daily growth of the total confirmed cases and the stock market performance, even after controlling the potential impact of changes in oil price, gold price, exchange rate, trading volume and the market's expectations for volatility on stock market performance. Specifically, in Model 1, the estimated coefficient shows that when the daily growth of the total confirmed cases increases 1%, the stock market performance will decrease 1.323%. The impact decreased by 1.319% as the control variables are included in the model.

However, the results have not explained the fact why the stock market performance increases with the increase in the total confirmed cases, as shown in Fig. 1.

Table 2 Total confirmed cases and stock market performance

	Model 1	Model 2
CASE	−1.323**	−1.319**
	(−2.151)	(−1.999)
CONTROL	No	Yes
No. of Obs	398	398
R-square	0.012	0.111

This table shows results of OLS regressions that examine the effect of total confirmed cases on stock market performance. Model 1 has only one independent variable, while Model 2 adds a vector of control variables, i.e., *CASE, VOL, OIL, GOLD, EXRATE,* and *VIX*. All explanatory variables are lagged by one year. The definition of the variables is shown in Table 1. To conserve space, the coefficients on *CONTROL* are not presented. The *t*-statistics use robust standard errors which are in parentheses. *p < 0.10, **p < 0.05, ***p < 0.01

Therefore, this study digs deeper into the effects of the pandemic on the stock market performance.

4.2 *Fed's Prescription for the Economy*

In response to the COVID-19 crisis, among others, Federal Reserve announced to provide a $2.3 trillion stimulus package to support the economy on April 9, 2020. Using this date in an event study approach, Harjoto et al. (2021) find that the U.S. stock market exposures positive abnormal returns from this Fed's stimulus.

Based on their findings, this study divides the research period into two subperiods. The first subperiod covers the period from January 21, 2020 to April 8, 2020 (hereafter *pre-stimulus period*). The second one is from April 9, 2020 to July 30, 2021 (hereafter *post-stimulus period*). Equation 1 continues to be used for this analysis and results are shown in Table 3.

Model 1 presents the findings for the pre-stimulus period, while Model 2 is for the post-stimulus one. In Model 1, the estimated coefficient of *CASE* is negative, which is associated with the aforementioned finding in Sect. 4.1. However, in Model 2, this coefficient is positive. This result implies that the daily growth of the total confirmed cases is positively related to the stock market performance. This finding appears to be hard to interpret in terms of theory perspective, but to be consistent with the real world.

Takyi and Bentum-Ennin (2021) contend that pandemics could negatively affect stock markets in the short-term period, but stock markets eventually remedy themselves in the long-term period. Hence, Fed's stimulus appears to be a prescription for the recovery of the economy. The stock market is forward-looking and stock prices reflect the expectation of investors. With the stimulus, investors think that they can

Table 3 Fed stimulus and stock market performance

	Model 1	Model 2
CASE	−0.636*	19.838***
	(−0.377)	(3.276)
CONTROL	Yes	Yes
No. of Obs	56	342
R-square	0.316	0.141

This table shows results of OLS regressions that examine the effect of total confirmed cases on stock market performance. Model 1 shows the results of the pre-stimulus period, while Model 2 shows that of the post-stimulus period. *CONTROL* is a vector including *CASE, VOL, OIL, GOLD, EXRATE,* and *VIX*. All explanatory variables are lagged by one year. The definition of the variables is shown in Table 1. To conserve space, the coefficients on *CONTROL* are not presented. The *t*-statistics use robust standard errors which are in parentheses. *p < 0.10, **p < 0.05, ***p < 0.01

find opportunities in a crisis, i.e., a sharp decrease in stock price due to COVID-19. Hence, the more confirmed cases there are, the more stock investors buy up and wait for the market to catch up. As a result, the stock market performance has a positive relationship with the daily growth of the total confirmed cases.

4.3 A Combination of Fed's Stimulus and COVID-19 Vaccine

In addition to Fed's stimulus, the COVID-19 vaccine can also be a potential factor for the positive relationship. The vaccine allows governments to relax the COVID-19 restrictions and lockdowns. Accordingly, economics could gradually reopen and the stock market could rebound. Hence, COVID-19 vaccination accompanied with the stimulus is conjectured to distribute to the positive relationship between total confirmed cases and the stock market performance.

To test the conjecture, this research conducts one more examination. Specifically, the sample is divided into three subsamples. The first subsample covers the period from January 21, 2020 to April 8, 2020 (hereafter *blank period*) to capture the effect of COVID-19 cases on the stock market performance without the Fed's stimulus and vaccine. The second subsample is from April 9, 2020 to December 13, 2020 (hereafter *stimulus period*) to reflect the effect of the appearance of Fed's stimulus. The final one is from December 14, 2020 to July 30, 2021 (hereafter *combination period*) to investigate the effect of the combination of stimulus and vaccine. December 14, 2020 is chosen because this date is recorded as the first day the U.S. implements the vaccination program, following the database of Refinitiv Datastream. The findings of this analysis are shown in Table 4.

Table 4 Fed stimulus and stock market performance

	Model 1	Model 2	Model 3
CASE	−0.636*	15.423**	33.398**
	(−0.377)	(1.989)	(2.331)
CONTROL	Yes	Yes	Yes
No. of Obs	56	177	165
R-square	0.316	0.176	0.188

This table shows results of OLS regressions that examine the effect of total confirmed cases on stock market performance. Model 1 shows the results of the blank period, while Models 2 and 3 show that of the stimulus period and combination period, respectively. *CONTROL* is a vector including *CASE, VOL, OIL, GOLD, EXRATE,* and *VIX*. All explanatory variables are lagged by one year. The definition of the variables is shown in Table 1. To conserve space, the coefficients on *CONTROL* are not presented. The t-statistics use robust standard errors which are in parentheses. *$p < 0.10$, **$p < 0.05$, ***$p < 0.01$

Model 1 shows the results of the blank period, while Models 2 and 3 present the findings of the stimulus and vaccine periods, respectively. The investigation indicates that the daily growth of the total confirmed cases harms the stock market performance in the blank period, as aforementioned in Sect. 4.2. In contrast, this impact is significantly positive in the stimulus and vaccine periods. However, the positive impact of total confirmed cases on the stock market performance is stronger in the combination period than in the stimulus one. In particular, an 1% increase in the daily growth of total confirmed cases increase the stock market performance by 0.15% points in the stimulus period, and by 0.33% points in the combination period.

This finding is not surprising because getting vaccinated can help improve people's health and decrease infection rates. Vaccination appears to be the best solution to reignite the suffering economy, minimize the risks from supply disruptions and return to some semblance of normal life. Accordingly, the more vaccination rate is, the lesser COVID-19 restrictions are. As a result, with the combination of stimulus packages and the COVID-19 vaccine, total confirmed cases are considered as the opportunity to buy stocks at a good price and then wait for the market to catch up. This is the answer to the question of why stock market prices are rising despite the COVID-19 pandemic.

5 Robustness Checks

5.1 Alternative Measures of the Dependent Variable

To verify that the findings are robust, Eq. (1) is re-run with alternative measures of the dependent variables. Specifically, *WILSHIRE* is replaced by the daily change in the three most widely followed indexes in the U.S. They are the S&P 500 (*S&P500*), Nasdaq Composite (*NASDAQ*) and Dow Jones Industrial Average (*DOW_JONES*). Results are shown in Table 5.

The findings suggest that the baseline results remain robust to the different measures of the dependent variable. In particular, the daily growth of the total confirmed cases harms the stock market performance in the blank period but is positively related to the performance in the stimulus and vaccine periods.

5.2 Alternative Measures of COVID-19's Impact

In addition to the daily growth of the total COVID-19 confirmed cases, the literature also uses the daily growth of total death cases caused by COVID-19 to capture the impact of the outbreak, such as Al-awadhi et al. (2020), Bahrini and Filfilan (2020). Therefore, this study uses the daily growth of total death cases (*DEATH*) to replace

Table 5 Alternative measures of the dependent variable

	Model 1	Model 2	Model 3
Panel A: S&P 500			
CASE	−0.633*	14.813*	30.725**
	(−0.375)	(1.950)	(2.273)
CONTROL	Yes	Yes	Yes
No. of Obs	56	177	165
R-square	0.306	0.175	0.211
Panel B: NASDAQ			
CASE	−0.698*	15.252*	51.158**
	(−0.424)	(1.736)	(2.593)
CONTROL	Yes	Yes	Yes
No. of Obs	56	177	165
R-square	0.317	0.182	0.166
Panel C: DOW_JONES			
CASE	−0.186	13.255**	20.150**
	(−0.101)	(1.648)	(1.565)
CONTROL	Yes	Yes	Yes
No. of Obs	56	177	165
R-square	0.292	0.171	0.166

This table shows results of OLS regressions that examine the effect of total confirmed cases on stock market performance. Model 1 shows the results of the blank period, while Models 2 and 3 show that of the stimulus period and combination period, respectively. The dependent variables in Panels A, B and C are *S&P500, NASDAQ* and *DOW_JONES*. *CONTROL* is a vector including *CASE, VOL, OIL, GOLD, EXRATE,* and *VIX*. All explanatory variables are lagged by one year. The definition of the variables is shown in Table 1. To conserve space, the coefficients on *CONTROL* are not presented. The *t*-statistics use robust standard errors which are in parentheses. *$p < 0.10$, **$p < 0.05$, ***$p < 0.01$

CASE in Eq. (1). Table 6 presents findings of this examination that confirm the robustness of the main results.

5.3 *Causal Impact Analysis Using Bayesian Structural Time Series Model*

Bayesian structural time series (BSTS) model is a statistical technique used for time series forecasting, inferring causal impact and other applications. Droste et al. (2018) state that the BSTS model can be used to estimate the post-event difference between the observed times series (a factual time series) and a forecasted time series that

Table 6 Impact of total death cases

	Model 1	Model 2	Model 3
DEATH	−5.092*	7.742*	56.765***
	(−1.933)	(1.962)	(3.191)
CONTROL	(−0.764)	(−2.221)	(−0.581)
No. of Obs	27	177	165
R-square	0.467	0.176	0.234

This table shows results of OLS regressions that examine the effect of total confirmed cases on stock market performance. Model 1 shows the results of the blank period, while Models 2 and 3 show that of the stimulus period and combination period, respectively. *CONTROL* is a vector including *CASE, VOL, OIL, GOLD, EXRATE,* and *VIX*. All explanatory variables are lagged by one year. The definition of the variables is shown in Table 1. To conserve space, the coefficients on *CONTROL* are not presented. The *t*-statistics use robust standard errors which are in parentheses. *p < 0.10, **p < 0.05, ***p < 0.01

would have occurred without the event (a simulated counterfactual time series). Takyi and Bentum-Ennin (2021) use this model to estimate the potential causal impact of COVID-19 on the stock market performance in thirteen African countries.

The following robustness checks are conducted based on the approach of Takyi and Bentum-Ennin (2021). However, there is the main difference. They use the pre-COVID-19 period data to simulate the data on the post-COVID-19 period, while this research uses the dynamic time warping (DTW) algorithm to search for the input data, then uses using 10,000 Markov chain Monte Carlo (MCMC) samples to stimulate data.

5.3.1 Causal Effect of the COVID-19 Pandemic

For this robustness check, this study firstly searches for time series matches that are similar to the most recent 90 days of the first confirmed case (named *pre-actual-COVID-19 period*) in the last 10 *years* of Wilshire 5000 Total Market Index. Then, the period, which is similar to the pre-actual-COVID-19 period, (hereafter *pre-similar-COVID-19 period*) and 79 trading days following the pre-similar-COVID-19 period (hereafter *post-similar-COVID-19 period*), which are corresponding to the blank period, are used to train models.

The DTW algorithm addresses that the period, which is similar to the most recent 90 days of the first confirmed case, is from October 12, 2016 to February 21, 2017. Figure 2 shows the evolution of Wilshire 5000 Total Market Index during the two periods. The holding-period return is 11.3% for pre-actual-COVID-19 period and 11.2% for the pre-similar-COVID-19 period.

After specifying the similar period, this study conducts the causal impact analysis by using the Causal Impact package in R software, which is developed by Brodersen et al. (2015). Results of causal impact analysis of the COVID outbreak are presented in Panel A of Table 7 and visualized in Fig. 3. The mean value of the actual data

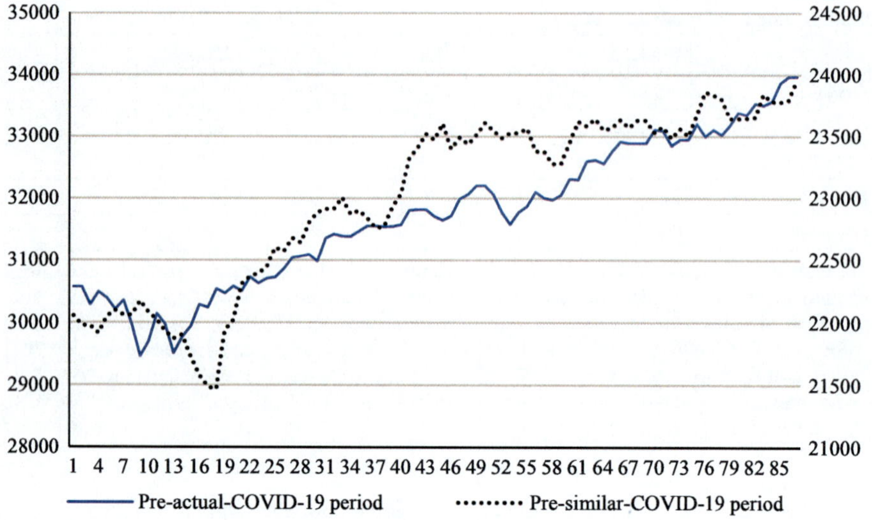

Fig. 2 The evolution of Wilshire 5000 Total Market Index during the two periods. This figure shows the similarity in the evolution of Wilshire 5000 Total Market Index between the two periods. The pre-actual-COVID-19 period is the most recent 90 days of the first confirmed case (20 September, 2019 to January 20, 2020). The pre-similar-COVID-19 period is from October 12, 2016 to February 21, 2017. The similar period is found by the DTW algorithm

Table 7 Causal effect of the COVID-19 pandemic

	Actual	Prediction	Absolute Effect	Relative Effect	Posterior tail-area probability
	(1)	(2)	(3)	(4)	(5)
Panel A: Causal effect of the COVID-19 pandemic	30,121	33,951 [33,735, 34,276]	−3,830 [-4,157, -3,614]	−11% [−12%, −11%]	0.013
Panel B: Causal effect of the combination	42,202	38,228 [37.744; 38690]	3,974 [3.512; 4.458]	10% [9.2%, 12%]	0.001

This table shows results of causal impact analysis. The mean value of the actual data (Wilshire 5000 Total Market Index in the pre-actual-COVID-19 period) is shown in Column 1, while Columns 2, 3 and 4 show the average value of the forecasted data, the absolute effect, the relative impact, respectively. Column 5 presents the posterior tail-area probability. Panel A shows the causal impact of the COVID-19 pandemic on the Wilshire 5000, while Panel shows the causal impact of the combination of Fed's stimulus and vaccination campaigns. The 95% interval is in brackets

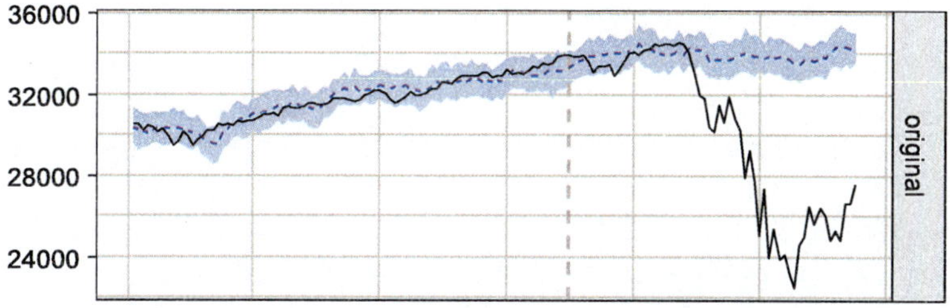

Fig. 3 Bayesian posterior distribution graphs for the causal effect of COVID-19. The figure shows the causal effect of the COVID-19 pandemic. On the original panel, the blue-dotted and the black solid lines horizontal indicate the time path of predicted series and actual series, respectively

(Wilshire 5000 Total Market Index in the pre-actual-COVID-19 period) is shown in Column 1, while Columns 2, 3 and 4 show the average value of the forecasted data, the absolute effect, the relative impact of the COVID-19, respectively. Column 5 presents the posterior tail-area probability.

During the post-intervention period, Wilshire 5000 had an average value of approx. 30,121. By contrast, in the absence of an intervention, Wilshire 5000 is expected to be 33,951. The 95% interval of this counterfactual prediction is [33,765; 34,278]. Subtracting this prediction from the observed response yields an estimate of the causal effect the intervention had on the response variable. This effect is -3,830 with a 95% interval of [−4,157; −3,614].

In relative terms, the response variable showed a decrease of −11%. The 95% interval of this percentage is [−12%, −11%]. This means that the negative effect observed during the intervention period is statistically significant. The probability of obtaining this effect by chance is very small (Bayesian one-sided tail-area probability $p = 0.013$). This means the causal effect can be considered statistically significant.

5.3.2 Causal Effect of the Combination

This subsection examines the causal effect of the combination of Fed's stimulus and vaccination campaigns. Similar to Sect. 5.3.1, the DTW algorithm is used to find the similar period. The period from August 6, 2020 to December 13, 2020 is named the pre-actual-combination period, while the period from August 6, 2020 to December 13, 2020 is the pre-similar-combination period. Figure 4 shows the similarity in the evolution of Wilshire 5000 Total Market Index between the two periods, while Fig. 5 visualizes results of the causal impact analysis.

Panel B of Table 7 shows results of the investigation. During the post-intervention period, Wilshire 5000 had an average value of 42,202. By contrast, in the absence of an intervention, Wilshire 5000 is expected to be 38,229. The causal effect is 3,973 with a 95% interval of [3,512; 4,458]. In relative terms, Wilshire 5000 showed an increase of 10%. The 95% interval of this percentage is [+9%, +12%]. The probability of obtaining this effect by chance is very small (Bayesian one-sided tail-area probability

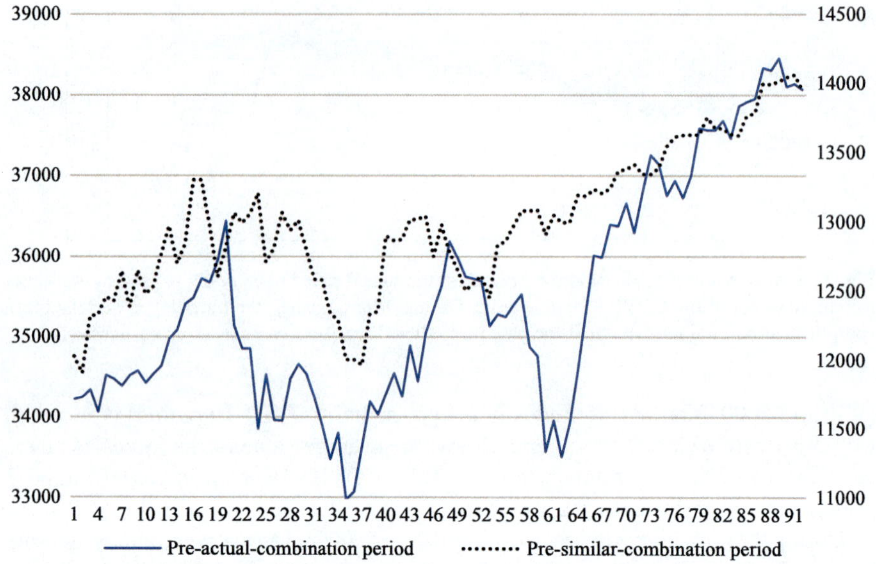

Fig. 4 The evolution of Wilshire 5000 Total Market Index during the two periods. This figure shows the similarity in the evolution of Wilshire 5000 Total Market Index between the two periods. The pre-actual-combination period is August 6, 2020 to December 13, 2020. The pre-similar-combination period is from October 6, 2011 to September 28, 2012. The similar period is found by the DTW algorithm

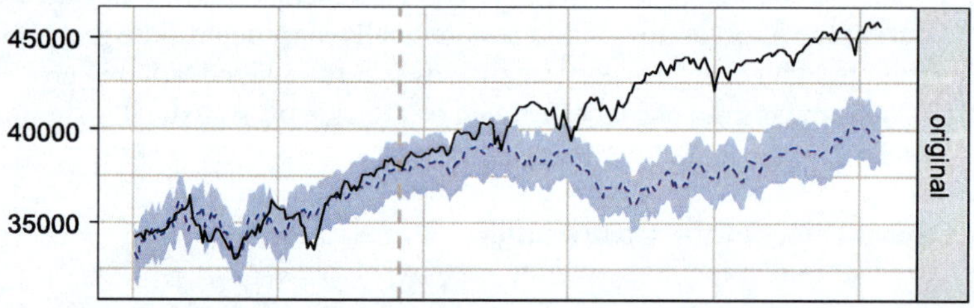

Fig. 5 Bayesian posterior distribution graphs for the causal effect of combination. The figure shows the causal effect of the combination of stimulus and vaccination campaigns on the stock market performance. On the original panel, the blue-dotted and the black solid lines horizontal indicate the time path of predicted series and actual series, respectively

$p = 0.001$). This means that the positive effect observed during the intervention period is statistically significant and unlikely to be due to random fluctuations.

In summary, causal impact analyses provide evidence supporting the fact that the COVID-19 pandemic harms the stock market performance, whereas Fed's stimulus and vaccination campaigns boost the stock market performance.

6 Conclusion

Coronavirus has spread to every corner of the world, brought considerable human suffering as well as badly hit worldwide economies. Its influence has become an important topic to discuss in many research areas, including the stock market performance. Despite an increasing number of papers, few studies have explained the problem of why the stock market keeps climbing while real-world turmoil. This study fills the gap in the literature and thus contributes to the literature on the economic impact of the COVID-19 pandemic on stock market performance.

The following results are drawn from the investigation. At the first glance, this study also provides evidence supporting the negative relationship between the daily growth of total COVID-19 confirmed cases and the stock market performance, as addressed in previous studies. However, further analyses suggest that this negative relationship exists only in the short run. When Fed provides stimulus packages to support the economy, the relationship changes from negative to positive. Especially after governments conduct vaccine campaigns, the positive relationship becomes stronger. This finding is consistent with what is happening in the real world. The rationale is that the combination of stimulus packages and COVID-19 vaccines makes announcements of confirmed cases to become a buy signal for optimist investors. They think that stock markets will soon correct themselves and increase, so they buy stocks at very favorable prices and then wait for the market to catch up. As a result, the stock market performance increases with the increase in total confirmed cases.

Conflict of Interest The authors declare no conflicts of interest in this paper.

References

Ahundjanov, B.B., Akhundjanov, S.B., Okhunjanov, B.B.: Information search and financial markets under COVID-19. Entropy **22**(7), 1–18 (2020)

Al-awadhi, A.M., Alsaifi, K., Al-awadhi, A., Alhammadi, S.: Death and contagious infectious diseases: impact of the COVID-19 virus on stock market returns. J. Behav. Exper. Finan. **27**, 100326 (2020)

Amstad, M., Cornelli, G., Gambacorta, L., Xia, F.D.: Investors' risk attitudes in the pandemic and the stock market: new evidence based on internet searches. BIS Bull. **25** (2020)

Ashraf, B.N.: Stock markets' reaction to COVID-19: cases or fatalities. Res. Int. Business Financ. **54**, 101249 (2020)

Bahrini, R., Filfilan, A.: Impact of the novel coronavirus on stock market returns: evidence from GCC countries. Quant. Financ. Econ. **4**(4), 640–652 (2020)

Baker, S.R., Bloom, N., Davis, S.J., Kost, K., Sammon, M.C., Viratyosin, T.: The unprecedented Stock Market Impact of COVID-19. Rev. Corp. Financ. Stud. **9**(April), 622–655 (2020)

Baldwin, R., di Mauro, B.W.: Economics in the Time of COVID-19. CEPR Press, A VoxEU.org eBook (2020a)

Baldwin, R., di Mauro, B.W.: Mitigating the COVID economic crisis: act fast and do whatever. CEPR Press, A VoxEU.org eBook (2020b)

Brodersen, K.H., Gallusser, F., Koehler, J., Remy, N., Scott, S.L.: Inferring causal impact using Bayesian structural time-series models. Ann. Appl. Stat. **9**(1), 247–274 (2015)

Droste, N., Becker, C., Ring, I., Santos, R.: Decentralization effects in ecological fiscal transfers: a Bayesian structural time series analysis for Portugal. Environ. Resource Econ. **71**(4), 1027–1051 (2018)

Harjoto, M.A., Rossi, F., Paglia, J.K.: COVID-19: stock market reactions to the shock and the stimulus. Appl. Econ. Lett. **28**(10), 795–801 (2021)

He, Q., Liu, J., Wang, S., Yu, J.: The impact of COVID-19 on stock markets. Econ. Political Stud. **8**(3), 275–288 (2020)

Liu, H., Manzoor, A., Wang, C., Zhang, L., Manzoor, Z.: The COVID-19 outbreak and affected countries stock markets response. Int. J. Environ. Res. Public Health **17**(8), 1–19 (2020)

Nguyen, L.D.: Liquidity management and stock price reactions in an economic crisis. Int. Econ. Conf. Vietnam **1**, 197–208 (2021)

Ramelli, S., Alexander, F.W.: Feverish stock price reactions to covid-19. Rev. Corp. Financ. Stud. **9**(3), 622–655 (2020)

Takyi, P.O., Bentum-Ennin, I.: The impact of COVID-19 on stock market performance in Africa: A Bayesian structural time series approach. J. Econ. Bus. **115**, Forthcoming (2021)

WHO. Coronavirus disease situation report-48 (Issue 48). https://www.who.int/emergencies/diseases/novel-coronavirus-2019 (2020)

Yan, C.: COVID-19 Outbreak and Stock Prices: Evidence from China. *Available at SSRN 3574374* (2020).https://doi.org/10.2139/ssrn.3574374

Zhang, D., Hu, M., Ji, Q.: Financial markets under the global pandemic of COVID-19. Financ. Res. Lett. **36**, 101528 (2020)

Determinants of Bank Profitability in Vietnam

Bui Dan Thanh, Nguyen Ngoc Thach, and Tran Anh Tuan

Abstract The article studied the impact of internal and external factors on the profitability of joint-stock commercial banks during 2009–2019. The paper used a data sample of 24 joint-stock commercial banks that account for a large proportion of the total assets of the Vietnamese commercial banking system. Bank profitability is measured by return on assets (ROA) and return on equity (ROE). To be more specific, ROA is a dependent variable with nine independent variables of micro and macro components of the economy. Through Bayesian multivariate regression, the results showed that external factors such as gross domestic product (GDP) and inflation rate (INF) have a positive relationship with ROA and ROE. Internal factors have different effects on ROA & ROE. While size (SIZE) owns a positive relationship to ROA and ROE, credit risk (LLR), operating costs (COSR), income diversification (DIV) are together negatively related to ROA and ROE. The study also reported loan (LOAN) and liquidity (LIQ) both have a positive correlation with ROE yet negative one with ROA, meanwhile equity size (CAP) has a positive relation with ROE and negative relation with ROE.

Keywords Determinants · Bank profitability · Commercial bank

B. D. Thanh (✉) · N. N. Thach
Banking University of Ho Chi Minh City, 36 Ton That Dam Street, District 1, Ho Chi Minh City, Vietnam
e-mail: thanhbd@buh.edu.vn

N. N. Thach
e-mail: thachnn@buh.edu.vn

T. A. Tuan
Department of Planning and Investment of Ho Chi Minh City, 32 Le Thanh Ton, District 1, Ho Chi Minh City, Vietnam

N. Ngoc Thach et al. (eds.), *Financial Econometrics: Bayesian Analysis, Quantum Uncertainty, and Related Topics*, Studies in Systems, Decision and Control 427,
https://doi.org/10.1007/978-3-030-98689-6_34

1 Introduction

In an economy, commercial banks serve the instrumental role of supplying capital for the production needs of its entities, thus they constitute a useful tool to help the state regulate economic activities and help the central bank implement appropriate monetary and financial policies at a specific time period. In recent years, bank profits, expressed in return on assets (ROA) and return on equity (ROE), have been highly volatile in Vietnam. Since 2011, the Vietnam banking system has been facing a lot of challenges. Although interest rates have been maintained at moderate levels, credit growth has been low due to a weak capital absorption capacity in the economy. In addition, the inadequacy of the mechanisms and policies on collateral handling, the provisions of the law on land and real estate leads to many difficulties in handling collateral of non–performing loans. Credit institutions therefore have to increase their provision for credit losses. These factors all affect the profitability of banks. There is a large amount of studies published on bank probability, for example, Ally (2014), Dawood (2014), Wang and Wang (2015), Wahdan and El Leithy (2017), etc. Besides, there are also some works on this issue in Vietnam, for example, Tran (2014), Hien (2017). Nevertheless, most previous studies employed outdated frequency-based methods, which might produce unreliable outcomes. So, this study aims to analyse the effects of the selected factors on bank probability in Vietnam through a Bayesian approach.

2 Research Overview and Research Methodology

2.1 Theoretical Basis

As a matter of fact, while profitability is an effective measurement in terms of money, it alone is not a sufficient condition for maintaining financial balance; profitability assessment must be derived from period of observation. The concept of profitability is widely applied in all economic activities, resulting from using a combination of tangible assets and financial assets, i.e., economic capital held by the business to generate profits for the business.

Profitability is among key metrics in evaluating the financial performance of a commercial bank, which is a combination of business results and resources usage. Profitability is an important foundation for banks to innovate, diversify their products, and thus operate effectively. It is an internationally common practice to measure the profitability of commercial banks by such quantitative indicators as absolute value of profit after tax, profit growth rate, profit structure and especially those that show profitability ratios such as return on equity (ROE), return on total assets (ROA), interest margin (NIM)… The higher its profitability index group is, the more effectively a bank performs. Therefore, profitability ratio is arguably one of the most important factors reflecting banking performance. There are several definitions and concepts of the rate of return used in various economic contexts regarding the amount of money received relative to the capital used to generate interest or the number of expenses consumed to generate interest.

2.2 Research Overview

Ally (2014), with a data set of independent variables and macroeconomics factors, studied the factors affecting the profitability of Tanzanian commercial banks in the period 2009–2013. The former includes asset size, equity size, non–performing loan, operating costs and liquidity; yet the latter covers GDP growth rate, inflation rate and real interest rate. The empirical research shows that most of the intrinsic factors have an effect on banking profitability, while the macroeconomic factors do not significantly affect the profitability of commercial banks.

Dawood (2014) researched into the factors affecting the return on investment (ROA) of 23 commercial banks in Pakistan in the period 2009–2012. The author used the pooled OLS estimation method in this study with independent variables including operating costs, liquidity, equity size, deposit–to–total assets ratio, and bank size. The result has drawn conclusions that while operating costs and liquidity have an inverse relation, the size of equity positively correlated with the banking ROA.

Wang and Wang (2015) identified the determinants of profitability of commercial banks in the US in 2002–2014. Accordingly, the study indicates that banks will have higher profitability when the ratio of loans to total assets is lower, customer deposits are lower than the total ratio of liabilities, lower NPLs to total outstanding loans, lower operating expenses, and more income diversification.

Wahdan and El Leithy (2017) conducted a study on the factors affecting the profitability of commercial banks in Egypt in the period from 2011 to 2015. Accordingly, the authors used a sample set of different internal and external independent variables of the banks. Whereas the internal ones include capital adequacy ratio, credit risk, the growth rate of asset size, operating costs, net interest income, total non–interest income; the external factors contain annual GDP rate, inflation rate. Research results has come up with the assumptions that ROA is positively affected by capital adequacy ratio, total non–interest income and inflation rate. Meanwhile, ROE is positively affected by capital adequacy ratio, credit risk, net interest income, total non–interest income and inflation rate.

Also studying the factors affecting the profitability of Vietnamese commercial banks, Tran (2014) continued to develop a study to determine through a sample of 22 banks in the period from 2006 to 2012. Research result reveals that state ownership, capital structure, and economic growth (GDP) are negatively related to the profitability of commercial banks. Besides, the study also added that the foreign ownership factor has not significantly impacted the banking profitability in the sample. On the contrary, credit risk and the inflation rate negatively affect ROA. Factors such as foreign ownership, asset size, liquidity risk, deposit size, and total debt have not affected the profitability of commercial banks in Vietnam during the study period as mentioned above.

Hien (2017) studied the factors affecting the profitability of 22 commercial banks in Vietnam in the period 2006–2015. Apart from the research on the impact of outstanding factors inside the bank such as loan balance, credit risk, non–performing loans, liquidity, deposit size, interest payment expenses, non–interest income, operating expenses, and asset size, in this study, the author took into account the corporate governance variables through the number of members of the Bank's Board of Directors and the number of female members on the Board. On the contrary, banks with high NPL ratios and poor cost management efficiency have a high correlation with ROA. The study did not find a correlation between the variables of board size and the profitability of commercial banks.

2.3 Research Methodology

In the context of a deepening crisis in classical frequentist statistics (see, for example, Nguyen and Thach 2018; Anh et al. 2018; Nguyen and Thach 2019; Hung et al. 2019a, b; Sriboonchitta et al. 2019; Svitek et al. 2019; Kreinovich et al. 2019; Tuan et al. 2019; Thach et al. 2019; Thach 2020a, b), the current study applies the Bayesian framework to achieve the research purpose (Table 1).

The study was conducted to evaluate the influence of the selected factors on the profitability of 24 Vietnam commercial banks from 2009 to 2019. Referring to earlier studies, to obtain the research goal, the authors propose two research models as follows:

Model 1:

$$\begin{aligned} \mathrm{ROA}_{\mathrm{it}} &= \beta_0 + \beta_1 \mathrm{SIZE}_{\mathrm{it}} + \beta_2 \mathrm{CAP}_{\mathrm{it}} + \beta_3 \mathrm{LOAN}_{\mathrm{it}} + \beta_4 \mathrm{LLR}_{\mathrm{it}} \\ &\quad + \beta_5 \mathrm{COSR}_{\mathrm{it}} + \beta_6 \mathrm{LIQ}_{\mathrm{it}} + \beta_7 \mathrm{DIV}_{\mathrm{it}} + \beta_8 \mathrm{GPD}_{\mathrm{it}} + \beta_9 \mathrm{IFN}_{\mathrm{it}} + \varepsilon_{\mathrm{it}}. \end{aligned}$$

Model 2:

$$\begin{aligned} \mathrm{ROE}_{\mathrm{it}} &= \alpha_0 + \alpha_1 \mathrm{SIZE}_{\mathrm{it}} + \alpha_2 \mathrm{CAP}_{\mathrm{it}} + \alpha_3 \mathrm{LOAN}_{\mathrm{it}} + \alpha_4 \mathrm{LLR}_{\mathrm{it}} \\ &\quad + \alpha_5 \mathrm{COSR}_{\mathrm{it}} + \alpha_6 \mathrm{LIQ}_{\mathrm{it}} + \alpha_7 \mathrm{DIV}_{\mathrm{it}} + \alpha_8 \mathrm{GPD}_{\mathrm{it}} + \alpha_9 \mathrm{IFN}_{\mathrm{it}} + \varepsilon_{\mathrm{it}}. \end{aligned}$$

Since previous studies were often performed using out–of–date frequentist methods, which might provide unreliable and inconsistent results, the Bayes setting is applied. Note that the number of observations in this study is relatively large, so

Table 1 Model variables

Variable		Notation	Formula
Dependent	Return on assets	ROA	$\frac{\text{Net income}}{\text{Total asset}}$
	Return on equity	ROE	$\frac{\text{Net income}}{\text{Shareholder Equity}}$
Independent	Bank size	SIZE	Logarit of total assest
	Size of equity	CAP	$\frac{\text{Equity}}{\text{Total assest}}$
	Loan	LOAN	$\frac{\text{Total loan}}{\text{Total assest}}$
	Credit risk	LLR	$\frac{\text{Provision for credit losses}}{\text{Total loan}}$
	Operating costs	COSR	$\frac{\text{Operating expenses}}{\text{Total operating income}}$
	Liquidity	LIQ	$\frac{\text{Liquid Asset}}{\text{Total assest}}$
	Income diversification	DIV	$DIV = 1 - \left[(\frac{NET}{NETTOP})^2 + (\frac{NON}{NETTOP})^2 \right]$ where *NON* is Non–Interest Income; *NET* is Net Interest Income; *NETOP* is Net Income; NETOP = NON + NET
	Gross domestic product	GDP	Gained from data of WB
	Inflation	INF	Gained from data of WB

the prior information might not overwhelm the data distribution and affect the posterior estimates. In this case, Block et al. proposed specifying the standard Gaussian distributions with different prior distributions and conduct a sensitivity analysis to choose the best among the considered models. For each of the two models, we run six simulations (Table 2).

The authors apply suggested regressions and the obtained results are as recorded in Table 3. After that, a Bayesian factor analysis and a Bayes model test are carried out. The preferred model has the smallest DIC, the largest Log BF and the largest Log ML estimates. As a result, simulation 1 and simulation 6 are the most suitable and chosen for further investigation.

By moving to Bayesian inference, MCMC algorithms require to conduct chain convergence. Several common tests for convergence are practiced. As usual, checks are conducted through trace plots and correlation plots.

Figure 1 demonstrates that the diagnostic graphs for all model parameters are relatively reasonable. The autocorrelation plots indicate low autocorrelation; the trace plots show no trends, depicting good mixing. Hence, it can be concluded that the MCMC chains have converged. We also acquire the similar results for model 2.

Table 2 Model specifications

Model 1	
Likelihood function	ROA ~ N(μ, σ)
Prior distribution	
Simulation 1	β_i ~ N(0; 1) σ^2 ~ Invgamma(0, 01; 0.01)
Simulation 2	β_i ~ N(0; 10) σ^2 ~ Invgamma(0, 01; 0, 01)
Simulation 3	β_i ~ N(0; 100) σ^2 ~ Invgamma(0, 01; 0, 01)
Simulation 4	β_i ~ N(0; 1000) σ^2 ~ Invgamma(0, 01; 0, 01)
Simulation 5	β_i ~ N(0; 10000) σ^2 ~ Invgamma(0, 01; 0, 01)
Model 2	
Likelihood function	
Prior distribution	
Simulation 6	α_i ~ N(0; 1) σ^2 ~ Invgamma(0, 01; 0.01)
Simulation 7	α_i ~ N(0; 10) σ^2 ~ Invgamma(0, 01; 0, 01)
Simulation 8	α_i ~ N(0; 100) σ^2 ~ Invgamma(0, 01; 0, 01)
Simulation 9	α_i ~ N(0; 1000) σ^2 ~ Invgamma(0, 01; 0, 01)
Simulation 10	α_i ~ N(0; 10000) σ^2 ~ Invgamma(0, 01; 0, 01)
i = 1, 2, 3, 4, 5, 6, 7, 8, 9	

Source Authors' synthesizer

3 Discussion

Additionally, we consider some important initial indicators, the rate of acceptance and efficiency. Moreover, Max Gelman–Rubin test is also useful for multiple chains. The results in Tables 4 and 5 point out that the acceptance rates have reached 1 (owing to efficient Gibbs sampling), while minimum efficiency has obtained 0.92 (exceeds the required level of 0.01). The maximum Rc value of the coefficients is 1 (lower that permissible level of 1.2). Besides, the summary tables provide Monte–Carlo standard errors (MCSE), which indicate the accuracy of the posterior estimates. According to Flegal et al. (2008), the closer the MCSEs are to zero, the stronger the MCMC

Table 3 Results of a test via Bayes factor and a Bayes model test

Model 1

	Chains	Avg DIC	Avg log (ML)	Avg Log BF	P(M|y)
simulation1	3	−384.8434	151.5677	1.000	1
simulation2	3	−384.7733	143.5319	−8.0358	0
simulation3	3	−384.7358	135.4192	−16.1485	0
simulation4	3	−384.8788	127.2551	−24.3126	0
simulation5	3	−384.7892	119.2788	−32.2889	0

Model 2

	Chains	Avg DIC	Avg log (ML)	Avg Log BF	P(M|y)
simulation6	3	−814.2425	374.2324	1.000	1
simulation7	3	−813.5492	363.2676	−10.9648	0
simulation8	3	−813.3353	351.8842	−22.3483	0
simulation9	3	−813.4661	340.2077	−34.0247	0
simulation10	3	−813.4377	328.7899	−45.4426	0

Source Authors' calculation

series, claiming that the MCSE values less than 6.5% of the standard deviation are acceptable and smaller 5% of the standard deviation are optimal.

Regression results show that macroeconomic factors, including GDP and inflation, positively affect ROA and ROE. Meanwhile, the internal factors have different effects on ROA and ROE. CAP increases ROA but has a negative impact on ROE. Conversely, LOAN and LIQ improve ROE but harm ROA. SIZE has a positive effect on both ROA and ROE. The remaining factors, including LLR, COSR and DIV, all negatively affect ROA and ROE. We cannot know the probability of a negative effect of the dependent variable on ROA and ROE within the frequentist method. However, we can compute this probability in the Bayesian framework. This will help us have a more comprehensive assessment of the impact of these factors on banking profitability.

Bank size

It is apparent that the SIZE of a Bank has a positive effect on its profitability, the probability of affecting ROA and ROE being 94% and 100%, respectively. This result is consistent with the research by Wang and Wang (2015), Alper and Anbar (2011), on the ground that banks will increase profits thanks to economies of scale. Specifically, as banks increase in size and network expansion, they help increase brand recognition and transaction convenience for customers; Banks with larger asset sizes will also enjoy more opportunities to exploit profits through their other operations in the market on top of the core operations. In fact, large–scale banks have high profitability ratios, especially state–owned joint–stock commercial banks such as Joint Stock Commercial Bank for Foreign Trade of Vietnam (VCB), Vietnam Joint Stock Commercial Bank for Industry and Trade (CTG) and Bank for Investment and Development of Vietnam (BIDV).

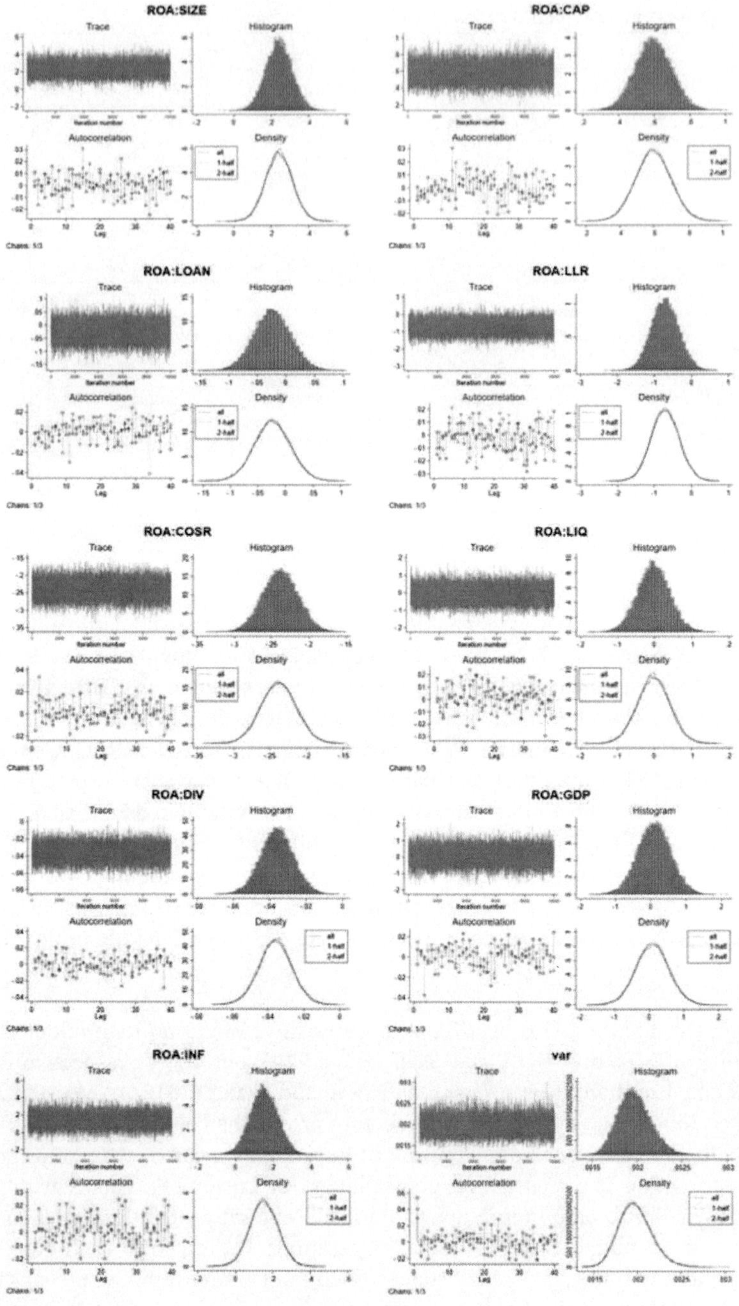

Fig. 1 Graphical checks for MCMC convergence. *Source* Authors' calculation

Table 4 Bayesian simulation outcomes for ROA

ROA	Mean	Std. Dev	MCSE	Median	Equal–tailed	
					[95% Cred. Interval]	
SIZE	0.0257	0.0168	0.0001	0.0257	–0.0077	0.0588
CAP	0.0616	0.0242	0.0001	0.0614	0.0143	0.1094
LOAN	−0.0026	0.0076	0.0000	−0.0026	−0.0175	0.0123
LLR	−0.0829	0.0933	0.0005	−0.0827	−0.2640	0.1003
COSR	−0.0238	0.0056	0.0000	−0.0238	−0.0348	−0.0128
LIQ	−0.0005	0.0102	0.0001	−0.0005	−0.0206	0.0194
DIV	−0.0036	0.0021	0.0000	−0.0036	−0.0078	0.0005
GDP	0.0126	0.1216	0.0007	0.0125	−0.2258	0.2535
INF	0.0148	0.0172	0.0001	0.0148	−0.0189	0.0485
_cons	−0.0367	0.0382	0.0002	−0.0366	−0.1116	0.0388
var	0.0001	0.0000	0.0000	0.0001	0.0001	0.0001
Avg acceptance rate	1					
Avg efficiency: min	0.9177					
Max Gelman–Rubin Rc	1					

Table 5 Bayesian simulation outcomes for ROE

ROE	Mean	Std. dev	MCSE	Median	Equal–tailed	
					[95% Cred. Interval]	
SIZE	0.2956	0.0713	0.0004	0.2952	0.1570	0.4348
CAP	−0.1841	0.1016	0.0006	−0.1845	−0.3831	0.0147
LOAN	0.0161	0.0324	0.0002	0.0162	−0.0469	0.0796
LLR	−0.4448	0.3714	0.0022	−0.4441	−1.1630	0.2848
COSR	−0.2623	0.0239	0.0001	−0.2622	−0.3090	−0.2148
LIQ	0.0671	0.0435	0.0003	0.0671	−0.0185	0.1522
DIV	−0.0443	0.0089	0.0001	−0.0443	−0.0617	−0.0270
GDP	0.1130	0.4673	0.0027	0.1166	−0.8099	1.0201
INF	0.1496	0.0736	0.0004	0.1496	0.0072	0.2948
_cons	−0.3862	0.1614	0.0009	−0.3857	−0.7020	−0.0700
var	0.0020	0.0002	0.0000	0.0020	0.0016	0.0024
Avg acceptance rate	1					
Avg efficiency: min	0.9291					
Max Gelman–Rubin Rc	1					

Size of equity

Equity size improves ROA reach to 99% probability, which shows an apparent impact of CAP on ROA. Consequently, banks with large equity capital have an edge in cost of capital, higher tolerance for market risks, and enhanced customer confidence in using their banking products and services, which inevitably increases their business efficiency. This result is consistent with most of the studies of Syfari, Pasiouras and Kosmidou (2007), Sufian and Chong (2008). However, as for ROE, equity size has a positive effect on ROE of only 0.037, meaning the probability of a negative effect is up to 96%. This signifies that when banks do not utilize and exploit equity effectively, as the size of equity increases, the ROE will decrease. This view is also consistent with the study of Ruochen Wang and Xuan Wang (2015).

Loan

Within the scope of research, loan variable has a very vague effect on bank profitability, which is not statistically significant. The probability of a positive effect of this variable on ROE is only 68%, while for the ROA variable, it is 37%, i.e., that 63% of the Loan variable negatively affects ROA. This result is consistent with the actual situation of joint–stock commercial banks in Vietnam. While lending is still considered as one of the main profitable operations for banks, the growth of outstanding loans together with credit quality control, will benefit the bank in increasing net interest income and improve profitability. This result is also consistent with the studies of Gul et al. (2011) or Syafri (2012).

Operating costs

The negative effect of COSR on both ROA and ROE is virtually 100%. This result is in line with expectations and is completely consistent with the studies of Syfari, Dawood (2014), San and Heng or Wang and Wang (2015). As a result, if the bank administrators control all kinds of operating costs while promoting and increasing the application of technology in its operation, reducing staff costs, which account for the major proportion in its operating cost structure, the profits will be improved, and thereby increasing profitability.

Liquidity

The LIQ variable has an inconspicuous effect on ROA when the probability is only 48%, but its positive effect on ROE reaches 94%. This result is consistent with the research of San and Heng, which suggested that banks with high liquidity are capable of tolerating higher financial risks, minimizing their risk of bankruptcy. At the same time, these banks also significantly reduce borrowing costs from external funding sources, thereby improve profitability for shareholders.

Income diversification

The results in Table 6 show that the variable DIV hurts both ROA and ROE with the probability of reaching almost 100%. This implied that income diversification does not increase income for banks as expected. Diversifying products to diversify income

Table 6 Probability of correlation between independent variables and dependent variable

	ROA			ROE		
	Mean	Std. Dev	MCSE	Mean	Std. Dev	MCSE
SIZE	0.9376	0.2419	0.0014	1.0000	0.0000	0.0000
CAP	0.9945	0.0742	0.0004	0.0371	0.1889	0.0011
LOAN	0.3654	0.4816	0.0028	0.6865	0.4639	0.0027
LLR	0.1872	0.3901	0.0023	0.1168	0.3212	0.0019
COSR	0.0000	0.0000	0.0000	0.0000	0.0000	0.0000
LIQ	0.4804	0.4996	0.0029	0.9361	0.2445	0.0014
DIV	0.0414	0.1992	0.0012	0.0000	0.0000	0.0000
GDP	0.5396	0.4984	0.0029	0.5974	0.4905	0.0028
INF	0.8052	0.3961	0.0023	0.9787	0.1445	0.0008

Source Authors' calculation

could increase the operating expenses, thereby reducing the banking profitability. However, income diversification could reduce the banking operational risk; this issue should be clarified in future studies.

Inflation

The variable of INF has a positive effect on ROA and ROE with the probability of 80% and 99%, respectively. This result is consistent with the study by Gul, Irshad and Zaman, Sufian. This positive correlation shows that when banks predict higher inflation in the future, they have adjusted interest rates to the right level to ensure profitable business.

4 Conclusions and Recommendations

The results of the study have contributed to a better explanation of the impact of both internal and external factors on the profitability of commercial banks; therefore, the authors propose some recommendations to bank managers and investors with a view to better assessing the importance of factors affecting profitability, thereby making strategic plans in business plans to increase business efficiency. For managers, it is necessary to increase capital size through the sale of shares to strategic investors and improve capital use efficiency by having a clear strategy to increase capital. For operating costs, it is advisable to have appropriate control, reduce capital costs, modernize technology, possibly in collaboration with Fintech companies, and set specific norms for expenses for production and business activities for each branch and department. In addition, it is necessary to increase liquidity by diversifying capital mobilization and use operations, improving macroeconomic forecasting to prepare for market fluctuations. For investors, before making an investment decision, it is necessary to consider commercial banks in terms of financial management, focusing on financial indicators such as equity size, ensuring a safe level of liquidity accounts, and diversify income sources to ensure appropriate and stable profitability.

References

Ally, Z.: Determinants of bank's profitability in a developing economy: empirical evidence from Tanzania. Eur. J. Bus. Manag. **6**, 31 (2014). ISSN 2222–1905 (Paper) ISSN 2222–2839 (Online)

Alper, Anbar, A.: Bank specific and macroeconomic determinants of commercial bank profitability: Empirical evidence from Turkey. Bus. Econ. Res. J. (2), 135–152 (2011)

Anh, L.H., Dong, L.S., Kreinovich, V., Thach, N.N. (eds.): ECONVN 2018. SCI, vol. 760. Springer, Cham (2018). https://doi.org/10.1007/978-3-319-73150-6

Chiorazzo, V., Milani, C., Salvini, F.: Income diversification and bank performance: evidence from Italian banks. J. Financ. Serv. Res. **33**(3), 181–203 (2008)

Davydenko, A.: Determinants of bank profitability in Ukraine. Undergraduate Econ. Rev. **7**(1), Article 2 (2011)

Dawood, U.: Factors impacting profitability of commercial banks in Pakistan for the period of (2009–2012). Int. J. Sci. Res. Publ. **4**(3) (2014). ISSN 2250–3153

Flegal, K., Furie, K., et al.: Heart disease and stroke statistics—2008 update: A report from the American Heart Association Statistics Committee and Stroke Statistics Subcommittee. Circulation **117**, 25–146 (2008). http://dx.doi.org/10.1161/CIRCULATIONAHA.107.187998

Hi`ên, N.T.T.: The factors that determine the profitability of Vietnamese Commercial Banks. Ind. Trade Mag. **7** (2017)

Hung, N.T., Songsak, S., Thach, N.N.: On quantum probability calculus for modeling economic decisions. In: Kreinovich, V., Sriboonchitta, S. (eds.) Structural changes and their econometric modeling, TES 2019a. SCI, vol. 808, pp. 18–34. Springer, Cham (2019a). https://doi.org/10.1007/978-3-030-04263-9_15

Hung, N.T., Trung, N.D., Thach, N.N.: Beyond traditional probabilistic methods in econometrics. In: Kreinovich, V., Thach, N.N., Trung, N.D., Van Thanh, D. (eds.) Beyond traditional probabilistic methods in economics, ECONVN 2019b. SCI, vol. 809. Springer, Cham (2019b). https://doi.org/10.1007/978-3-030-04200-4_13

Kreinovich, V., Thach, N.N., Trung, N.D., Van Thanh, D. (eds.): ECONVN 2019. SCI, 'vol. 809. Springer, Cham (2019). https://doi.org/10.1007/978-3-030-04200-4

Lee, C.C., Hsieh, M.F.: The impact of bank capital on profitability and risk in Asian banking. J. Int. Money Financ. **32**, 251–281 (2013)

Nguyen, H.T., Thach, N.N.: A closer look at the modeling of economics data. In: Kreinovich, V., Thach, N.N., Trung, N.D., Van Thanh, D. (eds.) ECONVN 2019. SCI, vol. 809, pp. 100–112. Springer, Cham (2019). https://doi.org/10.1007/978-3-030-04200-4_7

Nguyen, H.T., Nguyen, N.T.: A panorama of applied mathematical problems in economics. Thai J. Math. Spec. Issue Ann. Meet. Math. **17**(1), 1–20 (2018)

Pasiouras, F., Kosmidou, K.: Factors influencing the profitability of domestic and foreign commercial banks in the European Union. Int. Bus. Finan. **21**, 222–237 (2007)

Phan, T.H.N.: The determinants of listed banks' profitability. Bank. Rev. **68**, 20–25 (2011)

Sriboonchitta, S., Nguyen, H.T., Kosheleva, O., Kreinovich, V., Nguyen, T.N.: Quantum approach explains the need for expert knowledge: on the example of econometrics. In: Kreinovich, V., Sriboonchitta, S. (eds.) TES 2019. SCI, vol. 808, pp. 191–199. Springer, Cham (2019). https://doi.org/10.1007/978-3-030-04263-9_15

Sufian, F., Chong, R.R.: Determinants of bank profitability in a developing economy: Empirical evidence from Philippines. Asian Acad. Manag. J. Acc. Finan. **4**, 91–112 (2008)

Svítek, M., Kosheleva,O., Kreinovich,V., Nguyen, T.N.: Why quantum (wave probability) models are a good description of many non-quantum complex systems, and how to go beyond quantum models. In: Kreinovich, V., Thach, N.N., Trung, N.D., Van Thanh, D. (eds.) ECONVN 2019. SCI, vol. 809, pp. 168–175. Springer, Cham (2019). https://doi.org/10.1007/978-3-030-04200-4_13

Syafri: Factors Affecting Bank Profitability in Indonesia, The 2012 International Conference on Business and Management, 6 – 7 September 2012, Phuket – Thailand (2012)

Thach, N.N., Anh, L.H., An, P.T.H.: The effects of public expenditure on economic growth in Asia countries: a Bayesian model averaging approach. Asian J. Econ. Bank. **3**, 126–149 (2019)

Thach, N.N.: How to explain when the ES is Lower than One? A Bayesian nonlinear mixed-effects approach. J. Risk Financ. Manag. **13**, 21 (2020)

Thach, N.N.: The variable elasticity of substitution function and endogenous growth: an empirical evidence from Vietnam. Int. J. Econ. Bus. Admin. VIII, 263–277 (2020b)

Tr`ân V.D.: Determining factors affecting the profitability of Vietnam Commercial Banks. Banking Rev. **16**, 2–11 (2014)

Tuan, T.A., Kreinovich, V., Nguyen, T.N.: Decision making under interval uncertainty: beyond Hurwicz Pessimism-optimism criterion. In: Kreinovich, V., Thach, N.N., Trung, N.D., Van Thanh, D. (eds.) ECONVN 2019. SCI, vol. 809, pp. 176–184. Springer, Cham (2019). https://doi.org/10.1007/978-3-030-04200-4_14

Vinh, V.X., Mai, T.T.P.: Profitability and risk in the case of income diversification of Vietnam Commercial Banks. Econ. Manag. Rev. **26**(8), 54–70 (2015)

Wang, R., Wang, X.: What determines the profitability of banks? Evidence from the US. Simon Fraser University, Fall (2015)

Wahdan, M., El Leithy, W.: Factors affecting the profitability of commercial banks in Egypt over the last 5 year (2011–2015). Int. Bus. Manage. **11**, 342–349 (2017)

The Determinants of Financial Inclusion in Asia—A Bayesian Approach

Nguyen Duc Trung and Nguyen Thi Nhu Quynh

Abstract Financial inclusion has recently become one of the main goals all over the world. Using data extracted from the Global Financial Inclusion Database (Global Findex) of the World Bank, the paper aims to investigate the determinants of financial inclusion in Asia in 2017. The authors use two indicators for proxies representative financial inclusion, including debit and credit card ownership. The study employs Bayesian estimations and finds that the individual characteristics factors (such as gender, age, income, education, and employment status) affect the access and use of the products and services of formal financial institutions. Among these factors, education is the strongest factor influence on financial inclusion while age has a linear relationship and the weakest effect on inclusive finance. Furthermore, individuals with higher income, in the workforce and being a man relate positively in both indicators on financial inclusion. From these findings, the authors propose some important recommendations to develop financial inclusion in Asia.

Keywords Financial inclusion · Bayesian regression · Credit card ownership · Debit card ownership · Asia

1 Introduction

These days, financial inclusion has become one of the matters of the main concern all over the world. According to Worldbank (2017), in 2017, 69% of people owned accounts at official financial institutions worldwide, up 7% from 2014 and 18% from 2011. This implies that the world is growing strongly in financial inclusion, which has important implications for socio-economic development, especially in underdeveloped and developing countries. Indeed according to Bruhn and Love (2014) and

N. D. Trung (✉) · N. T. N. Quynh
Banking University Ho Chi Minh City, 36 Ton That Dam Street, District 1, Ho Chi Minh City 700000, Vietnam
e-mail: trungnd@buh.edu.vn

N. T. N. Quynh
e-mail: quynhntn@buh.edu.vn

N. Ngoc Thach et al. (eds.), *Financial Econometrics: Bayesian Analysis, Quantum Uncertainty, and Related Topics*, Studies in Systems, Decision and Control 427,
https://doi.org/10.1007/978-3-030-98689-6_35

Demirgüç-Kunt and Klapper (2012), financial inclusion leads to economic benefits, it can support deprived and poverty-stricken people enabling them to increase their income and reduce unemployment, because of the accessibility of financial products enabling people to invest in their education as well as financial projects. In contrast, in countries lacking financial services, poverty and unemployment can arise, which is a barrier to economic development.

Similar to all over the world countries, inclusive financial growth is also the most important policy goal in Asia. Although in recent years, Asia has had some success in sustaining economic expansion that has lifted millions out of poverty, however, poverty remains a major challenge in this region. That is a reason why financial inclusion is necessary since increasing the poverty-stricken's access to financial products is usually seen as an effective instrument that can help reduce poverty. Hence, considering the determinants of financial inclusion in Asia is crucial to suggest some recommendations to enhance inclusive finance, in order to reduce poverty and develop socio-economic status.

Concern topic about the factors affecting financial inclusion has received the attention of various scholars such as Demirgüç-Kunt et al. (2013a, b), Demirgüç-Kunt et al. (2013a, b), Soumaré et al. (2016), Zins and Weill (2016), Allen et al. (2016). However, these research results are inconclusive. Furthermore, according to the author's knowledge, these studies mostly use frequency methods such as logit or probit estimators. When employing these methods, the estimations are dependent on only observed data without incorporating prior information. This causes the estimation results to sometimes be biased.

This paper aims to consider the determinants of financial inclusion in Asia. Compared with previous studies, this paper has some important contributions to the existing literature. Firstly, distinguishing from the earlier works, our analysis is carried out within the Bayesian framework. Regarding the Bayesian approach, the Bayesian results are not only based on observed data, but also prior observations. This makes the results of this method more accurate than traditional estimations methods (logit, probit). Secondly, our paper was conducted in Asia—the second largest continent in the world with observational data of 39 countries in 2017.

The remainder of this paper is organized as follows. Section 2 discusses financial inclusion and provides some related literature. Section 3 presents descriptive data, model and methodology. Section 4 analyzes empirical results and finally is the conclusion and some policy recommendations.

2 Literature Review

2.1 The Concepts of Financial Inclusion

According to Worldbank (2018), financial inclusion is the provision of appropriate and economical financial services and products for individuals and businesses, that meets their needs about businesses, payments, credit, savings, and insurance in a manner that is responsible and sustainable. Financial inclusion is determined with three benchmarks (i) accessing financial services, (ii) using the financial services, and (iii) the quality of financial services and products.[1] In agreement with this view, United Nations (UN) also describes financial access as meaning comprehensive access to financial services, delivered with quality, to all who can use financial services, by that increased financial capability.[2] IMF (2015) defined financial inclusion as when households and firms can access and use formal financial services.

From a research perspective, Leyshon and Thrift (1995) introduce the concept of financial exclusion, the authors indicate that financial exclusion is a process that prevents poverty-stricken and disadvantaged social groups from accessing the formal financial system. Hannig and Jansen (2010) suggest financial inclusion motivates the "unbanked" people to use the formal financial system so that they are able to access various financial services such as saving, payments, credit and insurance conveniently. Financial inclusion also is a process to ensure access to financial services and the demand of credit for vulnerable groups such as the poor, low income with an affordable cost. These are reflected in the accessibility through the bank account (for example saving, credit, payment,…) (Khan 2011).

In sum, from these definitions, in this paper, financial inclusion is a process that all the citizens, especially the poor, vulnerable can access and use the financial services, meet their requirements at affordable costs, is provided by reputable official financial services institutions.

Financial inclusion focuses on providing financial solutions for the low income population. It ensures that all of the population has a more positive view of participating in the financial institutions, has better financial literacy, and has more opportunities in accessing quality, effective and secure products and services at an appropriate cost.

2.2 *The Role of Financial Inclusion*

Financial inclusion plays an important role in promoting socio-economic development, making it easier for households and businesses to plan their finances and deal with unexpected difficulties. Thereby it makes improving the quality of life, contributing to economic development and reducing poverty (Worldbank 2018). Park and Mercado (2015) find that financial inclusion is an efficient policy on reducing poverty rates and income inequality in emerging economies in Asia. Hence, inclusive finance has become a priority goal in numerous countries. In several countries, financial inclusion is an instrument for inclusive growth, supporting everyone in improving their future financial well-being and driving economic growth (Chakrabarty 2011).

[1] Access from https://www.worldbank.org/en/topic/financialinclusion/overview.

[2] Access from https://www.un.org/development/desa/socialperspectiveondevelopment/issues/financial-inclusion.html.

In addition, financial inclusion also helps the Governments reduce the cost of social security through payment via bank account. Thereby, it increases transparency, anti-corruption is more efficient. From these results, the social manager is better.

2.3 The Related Literature

Until now, there are various researches on the topic of the factor affecting financial inclusion. However, these results are inconclusive. Kumar (2013) analyzes determinants of financial inclusion in India. The research employs panel data from 1995 to 2008 of 29 major states, by FEM, REM, GMM estimators with the Kendall index. The research finds that the branch network of financial institutions is the strong factor effect positive on financial inclusion in India. Besides, the level of factories and employee base relates positively to penetration indicators. When conducting a sample data of Argentina, Tuesta et al. (2015) employ probit models to examine the three dimensions affecting financial inclusion. The results find that these dimensions include the factors on the supply side (for example branches and the number of ATM), the factors in term use (such as education, income, age), and the factors affecting the perception (income and age) influence financial stability in Argentina. Kabakova and Plaksenkov (2018) use sample data of 43 emerging and low-income economics (based on classifications of the World Bank and Standard and Poor's), the paper indicates that there are three compositions influence on financial inclusion: (i) high socio-demographic and political; (ii) high social, technological and economical and (iii) political and economical.

Demirgüç-Kunt et al. (2013a, b) use the global financial database Global Findex in 2012 from 98 developing countries to examine accessing individual finance. The paper indicates the presence of significant gender gaps in holding of account and usage of credit products and savings even when controlling the several of individual-specific such as education, income, employment status, age and residence. As women, they are more likely to be exclusive finance, they have more challenges in approaching formal credit as well as limited financial literacy. Moreover, the women have various difficulties in business and proof of collateral. However, in the informal financial system in some countries, women can access higher unofficial finance than men. The results also demonstrate that the token of gender norms, for example, the proportion of violence against women and the incidence of early marriage also contribute to explain variation in the use of financial products between women and men. In addition to these results, the research shows that income is the main factor that influences access to individual financial inclusion, in which, the cost associated with owning an account is a barrier to low-income in accessing financial services. Similar to this topic, Demirgüç-Kunt et al. (2013a, b) consider the financial access of Muslim adults with a sample of more than 65,000 adults from 65 Muslim countries. After controlling the country and individual-level characteristics, the researchers find that compared with non-Muslim, Muslims are significantly less likely to own an institution account or saving at a formal financial institution. However, regarding borrowing, compare with non-Muslim there is no evidence that Muslims are less likely to own a borrowing in a formal as well as an informal institution.

When identifying the determinants of financial inclusion in Central and West Africa, Soumaré et al. (2016) use the Global Findex database (the Global Financial Inclusion) of the World Bank. By the probit and logit estimations, the paper indicates that, in Central and West Africa, educated, working-age, full-time employed, and urban residents are significant access to financial inclusion. However, in Central Africa, the factors that affect positively accessing finance is male or/and married, whereas, in West Africa, the factor is income. In addition, household size relates negatively to account ownership in the West and not in Central Africa. When considering the other indicators as proxies for financial inclusion (such as borrowing, saving, or the time of use), the research implies that these above factors remain significant for Africa. Zins and Weill (2016) also use data from 37 countries in Africa to investigate the factors that influence financial inclusion. By the probit estimations, the authors conclude that being a man, better income, higher educated and older prefer inclusive finance, in particular, education and income are strongly positive on using the formal financial institution services. However, age has a nonlinear relationship with financial inclusion. Older adults have a higher level of accessing finance, but at a certain age older adults' access to finance decreases.

Also using the Global Findex in 2012 with over 150,000 adults in over 40 countries, by the probit, OLS, 2SLS estimations, Allen et al. (2016) explores the determinants associated with financial inclusion including individual and country specifics such as female, rural, poor or young individuals. The results find that lower account costs, stronger legal rights, the relationship with financial intermediaries and degree stability of political environments have positive on financial inclusion. In addition to the above studies, Fungáčová and Weill (2015) also use data from Global Index database for 2011 in China to examine financial inclusion. The research finds that in China several determinants such as being a man, higher education, richer, and being older are positive on using formal account and legal credit. Furthermore, income and literacy affect the access to alternative sources of borrowing.

In sum, from the above studies, the determinants affecting financial inclusion can be classified into two groups factor, the first is supply-side (such as density of bank branches, the number of ATMs,...); the second is demand-side (such as gender, education, income, employment status,...). In this paper, we focus on analyzing supply-side factors. Academically, there are several scholars research these factors (see Demirgüç-Kunt et al. (2013a, b), Demirgüç-Kunt et al. (2013a, b), Soumaré et al. (2016), Zins and Weill (2016), Allen et al. (2016)), however, most of these studies apply the dichotomous models (logit, probit). Moreover, the previous studies are conducted in Africa or other economies. This paper declares an in-depth analysis of the determinants of financial inclusion in Asia by the Bayesian approach. From the reliable results, the authors suggest some useful recommendations to develop financial inclusion in Asia.

3 Methodology, Models and Data

3.1 Methodology

In science, there are two schools in the use of statistics, including frequency statistic and Bayesian statistic. These two schools differ in their scientific philosophy and especially in their understanding of the concept of probability. In particular, frequency statistics are based on the current data, without incorporating prior information, whereas, Bayesian statistics are not only based on the current data set but also predicated prior information. In the Bayesian approach, the prior information constitutes the theoretical basis, so the results depend on the known facts combined with observed data (van de Schoot and Depaoli 2014). Hence, the outputs of Bayesian give higher accuracy. For this reason, Bayesian approach is more popular, especially in the branches of medicine and social science.

However, in Bayesian analysis, posterior distribution needs to identify, this is a challenge for the researcher. In which, it can be defined, posterior model is the probability distribution of all models parameter prior depends on the data observed and some priors choice. According to Bayes law, the posterior distribution has the following distribution:

$$\text{Posterir distribution} \propto \text{Likelihood} \times \text{Prior information}$$

In which, posterior probability has two components, a likelihood function, that gives information model parameters based on observed data, whereas prior information includes prior information on the model parameters (this information is inferred from previous knowledge, or expert opinions or belief). In the case the posterior distribution can be derived from in a closet form, we can proceed with the inference of the Bayesian analysis immediately. However, with the exception of several special models, most of posterior distribution needs to be estimated through simulation. In practice, there have some methods to estimate the posterior probability, such as MCMC, Metropolis–Hastings (MH) and Gibbs.

In this paper, the authors perform a Bayesian approach through MH algorithm with Gibbs sampling to identify determinants of financial inclusion in Asia. Metropolis et al. (1953) are the first to propose the Metropolis algorithm, then Hastings (1970) develops a more efficient algorithm. According to Gelfand et al. (1990) indicates that the Gibbs sample method is a special case of the MH algorithm. The reasons that the authors choose this method, the first is most of the previous studies apply frequency approach, so there is no information on the prior distribution of the parameters models; the second is in the case of big data, prior information may cause the model to be bises (Thach 2020). On this basis, the authors use the normal distribution of N(1,10,000) for the variables in the models and igamma distribution (0.01,0.01) for the variance.

3.2 Models

This paper focuses on analyzing demand-side factors affecting financial inclusion, including the individual-characteristic such as age, gender, income, level academic. Based on Demirgüç-Kunt, Klapper, and Singer (2013a, b), Soumaré et al. (2016), Zins and Weill (2016), Allen et al. (2016), the general research models are as follows:

$$\begin{aligned} X_i = & \alpha_0 + \alpha_1 Gender_i + \alpha_2 age_i + \alpha_3 income_i + \alpha_4 education_i \\ & + \alpha_5 employment_t + \varepsilon_i \end{aligned} \quad (1)$$

In which i refer to country, α_0 is constant, α_i (i = 1;...;5) are regression coefficients of the explanatory variables and ε_i is the error. In model (1) X_i are the dependent variables as proxies representative for financial inclusion. In this paper, the authors use 2 variables as dependent variables are owner debit cards and owner credit cards. These variables are dummy variables equal to one if they have a debit card or credit card and zero else. Gender is also a dummy variable, it takes the value one if the surveyor is a man (female) and zero else. Age is as proxy represented with two measures: the first is the number of ages (Age) and the second is its square (Age^2) to control the nonlinear relationship between financial inclusion and age. Regarding income, the authors use five dummy variables, in which the person who is poorest 20%, second 20%, third 20%, four 20% and richest 20%, respectively takes the value one, two, three, four, five, respectively. Concerning the level of academic, we use three dummy variables, in which the individual has completed primary school or less, secondary education and tertiary education, respectively take the value one, two, three, respectively. Finally, with the employment represents employment status of an individual, it is a dummy variable, equal to one if the individual is in the workforce and zero else.

From these bases, the specific research models are as follows:

$$\begin{aligned} Debit\_card_i = & \alpha_0 + \alpha_1 Gender_i + \alpha_2 Age_i + \alpha_3 Age_i^2 + \alpha_4 income_i \\ & + \alpha_5 education_i + \alpha_6 employment_t + \varepsilon_1 \end{aligned}$$

$$\begin{aligned} Credit\_card_i = & \beta_0 + \beta_1 Gender_i + \beta_2 Age_i + \beta_3 Age_i^2 + \beta_4 income_i \\ & + \beta_5 education_i + \beta_6 employment_t + \varepsilon_2 \end{aligned}$$

Table 1 shows the definition of the variables in the research models.

3.3 Data Description

Similar to Zins and Weill (2016), Allen et al. (2016), this paper uses the World Bank's 2017 Global Findex database to perform our analyses. The Global Findex database is

Table 1 Definition of the variables

Variables	Definition
Debit_card	Dummy variable, equal to one if the individual has a debit card, zero elsewise
Credit_card	Dummy variable, equal to one if the individual has a credit card, zero elsewise
Gender	Dummy variable, equal to one if the individual is man, zero else
Age	Age in the number of years
Age^2	The square of age
income	Dummy variable, equal to one, two, three, fourth, five respectively if the individual is in poorest 20%, %, second 20%, third 20%, four 20% and richest 20%, respectively
education	Dummy variable, equal to one, two, three respectively if the individual has completed primary school or less, secondary education and tertiary education, respectively
employment	Dummy variable, equal to one if the individual is in the workforce and zero elsewise

Source Compilation authors

obtained thanks to surveys in 143 countries and covering more than 150,000 people all over the world, in which, each economy presents roughly 1000 person participate the survey. The Global Findex database performs a large number of indicators on financial inclusion, such as account ownership, the use of financial services (saving, debit cards, credit cards, as well as mobile payments,...). In addition, the Global Findex database also provides information about individual-characteristic (such as gender, age, income, academic level).

However, to identify the effect of the factors on financial inclusion in Asia, our sample restricts to thirty-nine (39) countries of Asia included in the database with a total of 44,403 observations. Table 2 presents the descriptive statistics of the variables in the research models. From the results of table 2, in 2017, the percentage of individual who has a debit card and credit card is 44.8% and 17.8%, respectively in Asia. The average age of citizens in Asia is 41 years old.

Table 2 Descriptive statistics

Variable	Obs	Mean	St. Dev
Debit_card	44,403	0.448	0.497
Credit_card	44,403	0.178	0.382
Gender	44,403	0.471	0.499
Age	44,403	41.189	16.667
Age^2	44,403	1974.309	1543.612
income	44,403	3.141	1.430
education	44,403	1.836	0.711
employment	44,403	0.633	0.482

Source The authors' calculation

4 Empirical Results

The paper performs the MH algorithm as a Gibbs sampling method. Hence, before proceeding to Bayesian inference, the study has to employ several tests to ensure Bayesian inference is efficient. In the Bayesian approach, two kinds of tests need to be conducted are autocorrelation histograms and effective sample size (ESS), in which, the basic indicators such as acceptance and efficiency rate impact chain convergence. The acceptance rate implies that the proportion of the proposals model parameters is accepted in total proposals. Whereas the mixing properties of MCMC sequences are indicated by efficiency rate. High efficiency means good mixing in the MCMC sequences; conversely, low-efficiency shows the simulated MCMC sample mix bad. From the regression results, in both models, the rates of acceptance are 1, while the efficiency rates are 0.99, 1 and 1, respectively regarding for the dependent variable model Credit_card, regarding the model with the dependent variable Debit_card, these rates are 0.96, 0.99 and 1, respectively.

Besides, according to Kreinovich et al. (2019); Sriboonchitta et al. (2019); Svítek et al. (2019), Bayesian's advantage over frequency is Bayesian synthesis. For this reason, Bayesian approach does not employ many tests similar to frequency statistical such as c. Especially, when the MCMC chain converges, the Bayesian approach results are reliable, which means that it overcomes all the problems encountered by the frequency method. This is also the advantage of Bayesian approach over frequency method. In order to test for chain convergence, Figs. 1 and 2 presents diagnostic plots for the model the dependent variable debit_card and credit_card, respectively, including trace, autocorrelations, histograms and density plots. As shown in Fig. 1, trace plots for both the model parameters run quickly through a posterior domain, do not show trend and stop the average. While all the autocorrelation plots have no lags, the histogram plot has the shape unimodal and the kernel density plots have graphs that resemble the shape of the posterior distributions of the model parameters. In sum, none of these plots implies any issues in our sample. Besides diagnostic by MCMC chain convergence, the authors also use ESS (chain convergence) benchmark to evaluate the MCMC convergence. The estimated number of ESS is as close to the number of MCMC chains (10,000) as positive. Thus, it is possible to conclude that the parameters of both research models have converged to reasonable values and we can proceed to inference (Figs. 1, 2 and Table 3).

Tables 4 and 5 present the summary for all the model parameters, Monte Carlo chain standard error (MCSE) value is close to zero, this is reasonable for an MCMC algorithm. As shown in Tables 4 and 5, the variables including Gender, Age, income, education, employment relate positively to the debit_card and credit_card variables as proxies for financial inclusion indicators, whereas the variable of Age^2 relates negatively with the level weak on these indicators.

From the empirical results, the paper provides the following discussions.

Firstly, GENDER correlates positively on financial inclusion, that means, comparing the individual is women, the men have the trend to use credit and debit cards more. This finding is consistent with Zins and Weill (2016), which shows that

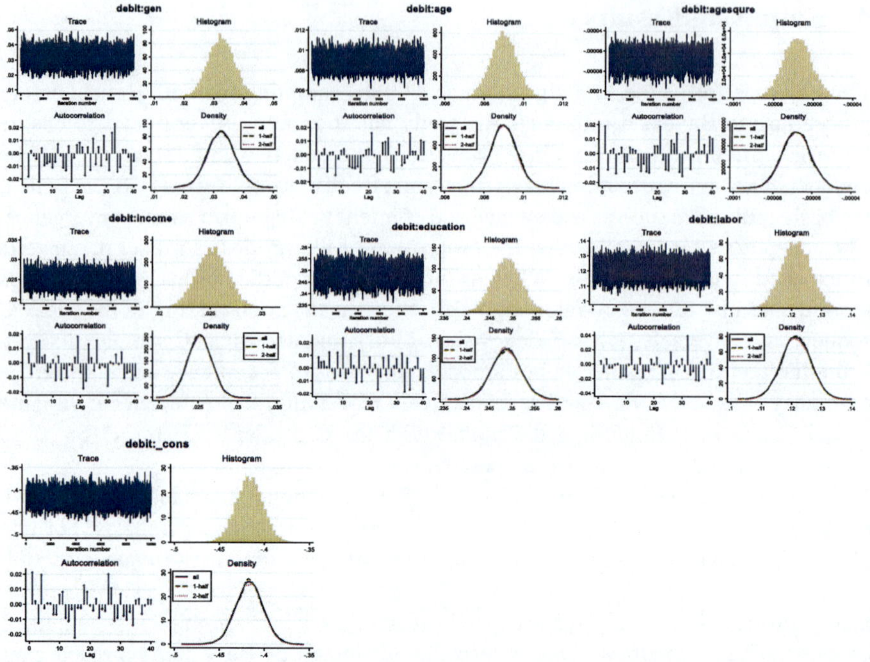

Fig. 1 Convergence diagnostics for the model with dependent debit_card parameters. *Source* The authors' calculations

women's barriers to financial inclusion are higher than men's. This also implies that there is still a distinction between men and women in the use of formal financial products in 2017 in Asia countries. The reason for this character is that, in some Asia countries, gender inequality still exists. According to ADB (2018), in the sub-region, Asia, and the Pacific, men have more access to financial services than women. Besides, in Asia, the average income of women is only 70–90% of mens' income, women are only allowed to work in the lowest salary jobs and informal employment. Moreover, in all countries, women do most of the family care works and unpaid housework, on average 2.5 times more than men.[3] For these reasons, the consumption needs of women are limited than that of men, which is an obstacle to using official services of financial institutions such as debit cards, credit cards.

Secondly, age has a nonlinear relationship with both indicators of financial inclusion, with a positive coefficient for Age and a negative for Age^2 variables to own a debit card or credit card. This result is similar to Allen et al. (2016), Zins and Weill (2016), which is explained that older people are more likely to access and use financial services, but this only affects certain ages. Actually, as the individual matures, jobs and incomes become more to more stable, social relationships become more and

[3] Access from website https://www-cdn.oxfam.org/s3fs-public/file_attachments/ib-inequality-womens-work-asia-310516-vn.pdf, on August 22nd, 2021.

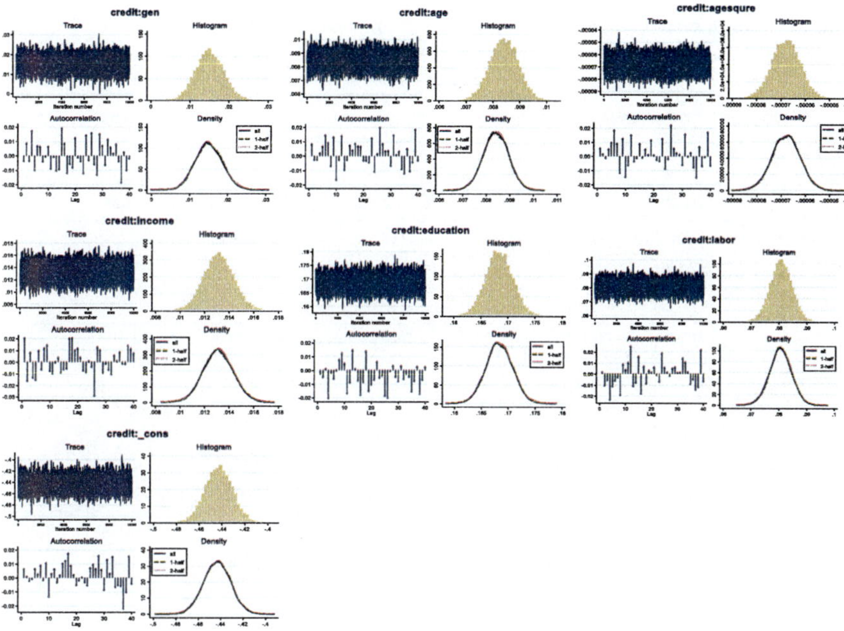

Fig. 2 Convergence diagnostics for the model with dependent credit_card parameters. *Source* The authors' calculations

Table 3 Effective sample size

	Model of dependent variable Debit_card			Model of dependent variable Credit_card		
	ESS	Corr. time	Efficiency	ESS	Corr. time	Efficiency
Gender	10,000.00	1.00	1.000	10,000.00	1.00	1.000
Age	10,000.00	1.00	1.000	10,000.00	1.00	1.000
Age^2	10,000.00	1.00	1.000	10,000.00	1.00	1.000
income	10,000.00	1.00	1.000	10,000.00	1.00	1.000
education	9690.25	1.03	0.969	10,000.00	1.00	1.000
employment	10,000.00	1.00	1.000	9566.15	1.05	0.957
_cons	10,000.00	1.00	1.000	10,000.00	1.00	1.000
sigma2	10,000.00	1.00	1.000	10,000.00	1.00	1.000

Source The authors' calculations

more important. Hence, the greater the demand for consumption, saving and investment, the more the demand for official services of financial institutions increases. However, up to a certain age, elderly people are often afraid to learn and use financial services that they have never used. In addition, these people have demand to use the financial products less than younger ones. So the older people's debit or credit card

Table 4 Posterior model of the dependent variable debit_card summary

Parameters	Mean	Median	Std. Dev	MCSE	Equal-tailed [95% Cred. Interval]
Gender	0.032	0.032	0.004	0.000	0.024, 0.041
Age	0.009	0.009	0.001	0.000	0.008, 0.010
Age^2	0.000	0.000	0.000	0.000	−0.000, −0.000
income	0.025	0.025	0.002	0.000	0.022, 0.028
education	0.249	0.249	0.003	0.000	0.243, 0.255
employment	0.122	0.122	0.005	0.000	0.112, 0.131
_cons	−0.416	−0.416	0.015	0.000	−0.446, −0.387
var	0.203	0.203	0.001	0.000	0.200, 0.205

Source The authors' calculations

Table 5 Posterior model of the dependent variable credit_card summary

Parameters	Mean	Median	Std. Dev	MCSE	Equal-tailed [95% Cred. Interval]
Gender	0.015	0.015	0.004	0.000	0.008, 0.022
Age	0.008	0.008	0.001	0.000	0.007, 0.009
Age^2	0.000	0.000	0.000	0.000	−0.000, −0.000
income	0.013	0.013	0.001	0.000	0.011, 0.016
education	0.168	0.168	0.002	0.000	0.164, 0.173
employment	0.081	0.081	0.004	0.000	0.0729, 0.088
_cons	−0.443	−0.443	0.012	0.000	−0.467, −0.420
var	0.126	0.126	0.001	0.000	0.124, 0.128

Source The authors' calculations

ownership decreases. Regarding Asia, in 2017, the impact of age on the accessing and use of financial services is not large when the regression coefficients of Age and Age^2 variables are approximately zero in all the models.

Thirdly, the results of the research indicate that income is positively associated with the indicators of financial inclusion. This suggests that individuals with higher income have more demand for financial products and services (such as debit cards, credit cards,…), so they have more opportunities to access financial inclusion than those with lower income. Demirgüç-Kunt et al. (2013a, b), Allen et al. (2016), Zins and Weill (2016) also find the same results. In practice, in order to own a debit card or a credit card, the owner must pay an accompanying cost, which becomes a major barrier for the low-income individual. Furthermore, compared to low-income individuals, higher-income individuals have more opportunities to consume, save and invest and trade. Therefore, the demand for official products of financial institutions has also increased. Whereas for low-income individuals, limited consumption, saving and investment activities make them more inclined to use cash. This explains why the higher an individual's income, the easier it is to access financial services.

Fourthly, education has a positive influence on financial inclusion in Asia, which indicates that the people with more education have more opportunities to use financial products and services. Among all the model parameters, the coefficients of the education variable are the highest, suggesting that financial inclusion is most strongly influenced by the level of academics. Zins and Weill (2016), Fungáčová and Weill (2015), Allen et al. (2016) also presents the same results. As a matter of fact, for a group of people with a high level of education and financial literacy, they usually use formal financial products and services at an affordable cost, to solve their demands and difficulties in finance. Whereas, people with low education and no financial literacy, usually pay attention to informal financial products, with various associated risks. In addition, compared with low-educated individuals, those of higher education are more likely to have formal jobs with higher incomes. This makes them easy in conducting online transactions as well as expenditures.

Finally, employment status also relates positively to both indicators of financial inclusion. This is consistent with reality, people in the workforce, usually receive their salary via debit card, so they tend to use products and services of formal financial institutions more. At the same time, income stability helps them afford the accompanying fees. Besides, people in the workforce have a higher credit to be able to access loan products of financial institutions (for example credit cards or bank loans,...) Meanwhile, the people outside the workforce, prefer using cash more, so they have more barriers in accessing financial inclusion.

In short, in the individual-characteristic factors, the paper finds that the determinant of financial inclusion includes gender, age, income, education, and employment status. In particular, being a man, income, and level of academic associates positively on financial inclusion, whereas age has a nonlinear relationship on accessing financial services. These findings are consistent with several previous studies such as Allen et al. (2016), Zins and Weill (2016), Fungáčová and Weill (2015).

5 Conclusions and Policy Recommendations

5.1 Conclusions

Financial inclusion is influenced by large numbers of factors, consisting of both demand-side and supply-side factors. The paper investigates the demand-side determinants of financial inclusion in Asia countries. The paper uses two measures to represent financial inclusion, including debit and credit card ownership. By the Bayesian regression with a Gibbs sampling method, the paper finds that gender, age, income, education, and employment status affect the access and use of the products and services of formal financial institutions. Among these factors, education is the strongest factor influence on financial inclusion while age has a linear relationship and the weakest effect on inclusive finance. Furthermore, individuals with higher income, in the workforce and being a man relate positively in both indicators on financial inclusion. Although the results research is quite interesting, however, the results in the Bayesian approach depend on the distribution for the observed variables and variance in the model. Hence in the next researches, the authors will check the results of research with the other distribution.

5.2 Policy Recommendations

From the above findings, the authors propose some important recommendations as follows.

Firstly, the results indicate that vulnerable groups in society (such as low income, low education, outside of the workforce, and woman) always have more barriers in accessing financial inclusion. So in order to make financial inclusion easily accessible to all, formal financial institutions should develop more kinds of financial services to meet the demand of the citizens, especially vulnerable groups. Moreover, to perform social responsibility, official financial institutions could consider reducing accompanying fees so that individuals (especially low income) have more opportunities to access financial products.

Secondly, the results implied that compared to men, women have more barriers to access financial inclusion in Asia. So, on the one hand, the countries should have policies to reduce gender inequality, at the same time enhancing the role of women in the family and society more as well as developing the services for women taking care of the family and unpaid jobs. On the other hand, to advance motivate the role of men in financial inclusion, financial institutions consider collaborating with market research organizations to develop products or services that are more tailored to mens' requirements.

Thirdly, from the results, education is the strongest influence on financial inclusion. So the governments should encourage young people to attend school at the highest possible level, as well as improve the quality of education, especially higher education. Because higher education is the foundation for students to have great jobs and income. However, certainly, not everyone can attend university, so the countries should organize courses in personal finance, in order to improve financial literacy, especially for groups of teenagers studying at all levels (primary, junior-high and high schools). Moreover, for groups with challenges in accessing financial inclusion services (such as low-income, outside the workforce, and women), mass organizations as well as financial institutions guide, popularize fundamental financial products or services for them.

Fourthly, the individual with high income has tended to use more formal financial services than others. Hence, one of the policy recommendations is that Governments consider several solutions to increase individuals' income (such as improving the quality of public investment, creating an investment environment to attract foreign direct investment,...). Besides, in order to promote a high-income individual to use financial services, it is necessary to have luxurious and classy products designed specifically for high-class customers and enjoy many attractive incentives.

Finally, as a matter of fact, the use of financial services depends on consumption, spending, transaction,... So, in order for financial inclusion to reach all individuals, Governments restrict people from using cash, instead of requiring them to use payment via financial institutions, applying information technology in transactions. To do this, developed countries can assist underdeveloped countries with financial types of equipment (ATM, POS,...) and knowledge of financial services.

References

ADB: Gender equality and the sustainable development goals in Asia and the Pacific: Baseline and Pathways for Transformative change by 2030 (2018). Access from https://www.adb.org/publications/gender-equality-sdgs-asia-pacific August 22nd, 2021

Allen, F., Demirguc-Kunt, A., Klapper, L., Peria, M.S.M.: The foundations of financial inclusion: understanding ownership and use of formal accounts. J. Financ. Intermed. **27**, 1–30 (2016)

Bruhn, M., Love, I.: The real impact of improved access to finance: evidence from Mexico. J. Financ. **69**(3), 1347–1376 (2014)

Chakrabarty, K.: Financial inclusion and banks: Issues and perspectives. RBI Bull. (2011). Access from: https://rbidocs.rbi.org.in/rdocs/Bulletin/PDFs/02SEPC1111FL.pdf

Demirgüç-Kunt, A., Klapper, L.F.: Measuring financial inclusion: the global findex database. World Bank Policy Research Working Paper(6025) (2012)

Demirgüç-Kunt, A., Klapper, L.F., Randall, D.: Islamic finance and financial inclusion: measuring use of and demand for formal financial services among Muslim adults. World Bank Policy Research Working Paper (6642) (2013a)

Demirgüç-Kunt, A., Klapper, L.F., Singer, D.: Financial inclusion and legal discrimination against women: evidence from developing countries. World Bank Policy Research Working Paper(6416) (2013b).

Fungáčová, Z., Weill, L.: Understanding financial inclusion in China. China Econ. Rev. **34**, 196–206 (2015). https://doi.org/10.1016/j.chieco.2014.12.004

Gelfand, A.E., Hills, S.E., Racine-Poon, A., Smith, A.F.: Illustration of Bayesian inference in normal data models using Gibbs sampling. J. Am. Stat. Assoc. **85**(412), 972–985 (1990)

Hannig, A., Jansen, S.: Financial inclusion and financial stability: current policy issues (2010). Access from: https://papers.ssrn.com/sol3/papers.cfm?abstract_id=1729122.

Hastings, W.K.: Monte Carlo sampling methods using Markov chains and their applications. Biometrika **57**(1), 97–109 (1970)

IMF: Financial Inclusion: Can It Meet Multiple Macroeconomic Goals. IMF Staff Discussion Note 15/17, International Monetary Fund Washington, DC (2015)

Kabakova, O., Plaksenkov, E.: Analysis of factors affecting financial inclusion: eosystem view. J. Bus. Res. **89**, 198–205 (2018). https://doi.org/10.1016/j.jbusres.2018.01.066

Khan, H.: Financial inclusion and financial stability: are they two sides of the same coin. Address by Shri HR Khan, Deputy Governor of the Reserve Bank of India, at BANCON (2011)

Kreinovich, V., Thach, N.N., Trung, N.D., Thanh, D.V.: Beyond Traditional Probabilistic Methods in Economics. Springer, Cham. (2019). https://doi.org/10.1007/978-3-030-04200-4

Kumar, N.: Financial inclusion and its determinants: evidence from India. J. Financ. Econ. Policy **5**(1), 4–19 (2013). https://doi.org/10.1108/17576381311317754

Leyshon, A., Thrift, N.: Geographies of financial exclusion: financial abandonment in Britain and the United States. Trans. Inst. Br. Geogr. **20**(3), 312–341 (1995)

Metropolis, N., Rosenbluth, A.W., Rosenbluth, M.N., Teller, A.H., Teller, E.: Equation of state calculations by fast computing machines. J. Chem. Phys. **21**(6), 1087–1092 (1953)

Park, C.-Y., Mercado, R:. Financial inclusion, poverty, and income inequality in developing Asia. Asian Development Bank Economics Working Paper Series(426) (2015)

Soumaré, I., Tchana Tchana, F., Kengne, T.M.: Analysis of the determinants of financial inclusion in Central and West Africa. Trans. Corporat. Rev. **8**(4), 231–249 (2016)

Sriboonchitta, S., Nguyen, H.T., Kosheleva, O., Kreinovich, V., Nguyen, T.N.: Quantum approach explains the need for expert knowledge: on the example of econometrics. In: Kreinovich, V, Sriboonchitta, S. (Eds.), Structural Changes and Their Econometric Modeling, vol. 808, pp. 191–199. Springer (2019). https://doi.org/10.1007/978-3-030-04263-9_15.

Svítek, M., Koshelevac, O., Kreinovich, V., Nguyen, T.N.: Why quantum (wave probability) models are a good description of many non-quantum complex systems, and how to go beyond quantum models. In: Kreinovich, V., Thach, N.N., Trung, N.D., Van Thanh, D. (Eds.), Beyond Traditional Probabilistic Methods in Economics, vol. 809, pp. 168–175. Springer (2019). https://doi.org/10.1007/978-3-030-04200-4_13

Thach, N.N.: How to explain when the ES is lower than one? A Bayesian nonlinear mixed-effects approach. J. Risk Financ. Manag. **13**(2), 21 (2020)

Tuesta, D., Sorensen, G., Haring, A., Camara, N.: Financial inclusion and its determinants: the case of Argentina. Madrid: BBVA Research. Working Paper No.15/03 (2015)

van de Schoot, R., Depaoli, S.: Bayesian analyses: where to start and what to report. Europ. Health Psychol. **16**(2), 75–84 (2014)

Zins, A., Weill, L.: The determinants of financial inclusion in Africa. Rev. Dev. Financ. **6**(1), 46–57 (2016). https://doi.org/10.1016/j.rdf.2016.05.001

IMF—Measured Stock Market Development and Firms' Use of Debt: Evidence from Developing Countries

Bich Loc Tram, Van Thuan Nguyen, Van Tuan Ngo, and Thanh Liem Nguyen

Abstract Studies on the relationship between stock market development and capital structure in developing countries are still limited. In addition, most previous studies applied outdated frequentist methods (such as FEM, REM, GMM, etc.), which cannot provide a straightforward and intuitive probabilistic interpretation of results as compared to Bayesian analysis. Therefore, using Bayesian estimation method, this paper focuses on the impact of stock market development, measured by an index constructed by IMF, on firms' capital structure in five developing countries in ASEAN (Indonesia, Malaysia, Philippines, Thailand and Vietnam) for the period 2010–2019. Research results show that stock market development and corporate profitability have a negative impact on corporate capital structure, which is true for total debt, and both short- and long-term debt. Meanwhile, inflation and GDP per capita growth rate positively affect capital structure. Firm size, asset maturity and TOBINQ have a positive impact on total debt and long-term debt, but negative impact on short-term debt. Based on the research findings, we offer several implications for relevant stakeholders.

Keywords Capital structure · ASEAN · Stock market development

JEL Classification Code C12 · C13 · E44 · F15

B. L. Tram (✉)
Sai Gon University, Ho Chi Minh City, Vietnam

V. T. Nguyen
Ho Chi Minh City, Vietnam

V. T. Ngo
Banking University HCMC, Ho Chi Minh City, Vietnam

T. L. Nguyen
University of Economics and Law, Ho Chi Minh City, Vietnam

N. Ngoc Thach et al. (eds.), *Financial Econometrics: Bayesian Analysis, Quantum Uncertainty, and Related Topics*, Studies in Systems, Decision and Control 427,
https://doi.org/10.1007/978-3-030-98689-6_36

1 Introduction

Capital structure is one of the key factors in increasing firm value. Capital structure can be changed through the issuance of securities on financial markets in general and stock markets in particular. According to Kumar et al. (2017), stock markets in developing countries are often quite small in scale, lack transparency and are not really efficient, so firm investment has typically depended on bank loans. Studies on the determinants of capital structure have not been considered in the context of linkages with stock market. ASEAN is a vibrant area that comprises Brunei, Cambodia, Indonesia, Laos, Malaysia, Myanmar, Philippines, Singapore, Thailand and Vietnam. Although empirical studies have delved more and more in the context of developing countries, there is still very little research done in the ASEAN region (Phooi et al. 2017), even though this area records high economic growth rates that require large volumes of funding. This paper aims to analyze the impact of stock market development on capital structure of listed firms in ASEAN to provide important implications to improve capital access and increase firm value.

2 Relevant Literature and Hypothesis Development

2.1 Concepts and Theories of Capital Structure

Capital structure is the combination of debt and equity that an organization uses to finance its operations and investments (Kumar et al. 2017). Based on a number of assumptions including no taxes, no asymmetric information, no costs of bankruptcy and financial distress, Modigliani and Miller (1958) argue that capital structure does not affect cash flows, so it does not affect firm value. Modigliani and Miller (1963) added the factor of corporate income tax to Modigliani and Miller (1958)'s theory, and state that the value of levered firm (firms that use debt) is equal to value of unlevered firm plus present value of tax shield, and thanks to the tax shield, the average cost of capital decreases as firm increases its debt ratio. In other words, M&M theory implies that it is beneficial for a company to use debt as much as possible.

Nonetheless, if a firm continuously increases debt ratio, its value does not always increase. Myers (1993) constructed Trade-off Theory, in which capital structure would be determined by trade-offs between costs and benefits of borrowed capital. According to Trade-off Theory, value of the levered firm is equal to the value of unlevered firm, plus the benefit from the debt tax shield and the reduction of agency cost associated with free cash flows, minus costs (including agency costs when firms borrow heavily, bankruptcy costs and financial distress). The optimal capital structure is achieved when the marginal benefit of increasing debt equals the marginal cost of debt increase (Miglo 2016).

Contrary to Trade-off Theory, Pecking Order Theory proposed by Myers (1984) does not aim to predict the optimal debt ratio. The Pecking Order Theory states the preferences of funding sources in financing investments. Specifically, when a business needs funding, the manager will prioritize internal capital (such as retained earnings because this capital has little problems of information asymmetry), then borrowed capital and finally stock issuance. The above order is related to cost of using funds, the lower the cost, the higher the priority. Therefore, a firm's capital structure is the result of company's financing requirements over time and effort to minimize the cost of adverse selection resulted from information asymmetry issue.

Market timing theory holds that capital structure is not the outcome of dynamic optimization strategy, but merely an accumulation of past efforts to capitalize on market timing. In other words, firms issue shares depending on stock market conditions, and capital structure changes are due to this strategy, not because the company adjusts their debt to the target ratio (Baker and Wurgler 2002).

2.2 The Impact of Stock Market Development on Capital Structure and the Measure of Stock Market Development

The stock market channels capital directly from investors to issuers, thereby providing medium and long-term capital for the economy. The stock market serves important functions even in economies with a well-developed banking sector, because equity and debt financing are generally not perfect substitutes. Equity capital has a key role in managing conflicts of interest that may arise between different stakeholders (Demirguc-Kunt and Maksimovic 1996). Levine (2002) points out the growth-promoting role of well-performing stock markets, specifically: (i) creating greater incentive to research companies, (ii) strengthening corporate governance by making it easier to tie management responsibilities to firm performance, and (iii) creating risk management opportunities. In other words, stock markets can stimulate search for businesses' information. As the market becomes larger and more liquid, market participants may have more incentives to use resources to analyze firm performance, because it is easier for them to benefit from this information gained from the analysis by trading stocks in large and liquid markets (Grossman and Stiglitz 1980; Holmstrom and Tirole 1993). Consistently, Merton (1987) argues that larger and more liquid markets provide incentives for participants to generate information that facilitate efficient capital allocation.

When a stock market grows, it would be expected to have some direct effects on corporate debt ratios. First, the substitution effect occurs when external equity is substituted for external debt, which reduces the debt ratio. Second, external equity is substituted for internal equity, and this is not supposed to affect the firm's debt ratio. In addition, the development of stock markets may also have an indirect influence on corporate leverage. Stock markets aggregate information that investors obtain about firms, which makes it less expensive for investors and financial intermediaries to monitor firms, resulting in equity and external debt becoming less risky so external financing is more likely to be granted. However, it remains unclear whether equity

Table 1 IMF Financial market development index

<table>
<tr><th colspan="3">Fiancial Markets Index (FMI)</th></tr>
<tr><th>Fiancial markets depth index</th><th>Fiancial markets acess index (FMA)</th><th>Fiancial markets efficiency index (FME)</th></tr>
<tr><td>1. Stock market capitalization to GDP</td><td rowspan="4">1. Percent of market capitalization outside of top 10 largest firms</td><td rowspan="5">Stock market turnover ratio (stocks traded to capitalization)</td></tr>
<tr><td>2. Stock traded to GDP</td></tr>
<tr><td>3. International debt securities of government to GDP</td></tr>
<tr><td>4. Total debt securities of financial corporations to GDP</td></tr>
<tr><td>5. Total debt securities of nonfinancial corporations to GDP</td><td>2. Total number of issuers of debt (domestic and external, nonfinancial and financial corporations)</td></tr>
</table>

Source IMF (2015, 2016)

prevails debt when stock markets become more efficient and grow larger. To the extent that debt is provided by suppliers and by banks, who may already be well informed about the firm's creditworthiness, leverage is expected to decrease as the stock market reduces supervisory costs for investors. The different likely impacts of stock market development on firm leverage suggest that it is not straightforward to determine the influence of stock markets on firms' debt (Demirguc-Kunt and Maksimovic 1996, 1999; Yartey 2009; Bokpin 2010).

To gauge the level of development of financial markets, International Monetary Fund—IMF (2015, 2016) has launched a new set of indicators on three aspects of stock markets: depth, accessibility and effectiveness (Table 1). Rather than analyzing how each aspect of financial market, we use FMI as an overall index to measure the development of stock market.

2.3 Literature Review

Kumar et al. (2017) provide a meta-analysis based on 167 research articles on capital structure determinants. The study offers three findings. First, studies on determinants of corporate capital structure are numerous and quite complete in developed economies, but this issue has not been thoroughly considered in emerging markets. Meanwhile, capital markets in emerging markets are relatively inefficient and inadequate compared with developed markets. Therefore, the results found in studies in developed markets might not be generalized to developing ones. Second, most of the literature studies on the determinants of corporate leverage tend to use firm-level variables, and industry- and country-level factors have not received much attention. In particular, few empirical articles examine the impact of stock market development on

Table 2 Variables measuring stock market development in empirical research

Name	Research used
Stock market capitalization/GDP	Demirguc-Kunt and Maksimovic (1996, 1999), Agarwal and Mohtadi (2004), Deesomsak et al. (2009), Le and Ooi (2012), Le (2017), Zafar et al. (2019)
Shares traded value/GDP	Demirguc-Kunt and Maksimovic (1996), Agarwal and Mohtadi (2004), Bokpin (2010), Doku et al. (2011), Le and Ooi (2012), Le (2017)
Value of shares traded/Stock market capitalization	Demirguc-Kunt and Maksimovic (1996, 1999), Agarwal and Mohtadi (2004), Deesomsak et al. (2004), Doku et al. (2011), Le and Ooi (2012), Le (2017)
INDEX 1 (Average of 3 indices: Stock market capitalization/GDP, Stock value traded/GDP, and Value of shares traded/Stock market capitalization)	Demirguc-Kunt and Maksimovic (1996), Le and Ooi (2012)
INDEX 2 (average of INDEX 1 and cost of equity—CAPM)	Demirguc-Kunt and Maksimovic (1996)
Bond market capitalization to GDP	Bokpin (2010), Zafar et al. (2019)

Source compiled from previous studies

capital structure. Besides, most studies use single variables to evaluate stock market development (Table 2), so it is not possible to fully assess market development.

Demirguc-Kunt and Maksimovic (1996) show that stock market development significantly affects corporate capital structure, but the direction of impact depends on the level of stock market development and corporate size. It is found that stock market development has a positive impact on capital structure only for large firms. However, the result of Demirguc-Kunt and Maksimovic (1996) is inconsistent with Agarwal and Mohtadi (2004). Deesomsak et al. (2004) document similar results with Agarwal and Mohtadi (2004): stock market activity negatively affects capital structure, because firms tend to use equity instead of debt. However, Deesomsak et al. (2009) show that the impact of stock markets on corporate capital structure depends on the level of stock market development. Typically, when stock markets grow, firms in the region where the stock market is relatively developed can raise equity capital more easily, compared to less developed markets. Zafar et al (2019) show that stock market development (measured the ratio of stock market capitalization to GDP) does not have a significant impact on capital structure for high-middle-income countries; however, it has a negative effect on capital structure in low-middle- and high-income countries. Interestingly, Doku et al (2011) show that the direction of impact of stock market development depends on the proxies of the stock market development. The research results of Doku et al (2011) are quite similar to the study of Le (2017), with only one difference being that the variable of the ratio of stock market capitalization to value of traded shares was excluded from the model due to problem of multicollinearity. Lemma and Negash (2012) argue that stock market development has a positive impact on capital structure in the long run, but negative and insignificant impact in the short term.

Due to the multifaceted characteristic of stock markets, using single variables to gauge the development of stock market cannot provide a complete view of the impact of stock market development on capital structure, the study will use the IMF's financial market development indicators to examine the impact of stock market development on capital structure.

2.4 Research Hypothesis

2.4.1 Variables Representing Stock Market Development

Theoretical basis and most empirical research shows that stock market development can affect capital structure in two different ways. First, the expansion of stock markets is likely to generate more funding options for firms and encourage firms to use equity instead of debt, thus having a negative impact on capital structure. Second, the development of stock markets promotes the search for corporate information, facilitates corporate monitoring with lower costs, which positively affects capital structure (Grossman and Stiglitz 1980; Merton 1987; Holmstrom and Tirole 1993; Levine 2002; Antzoulatos et al. 2016). The stock market development is demonstrated through the following factors.

Market depth: One of the most popular measures of depth is market capitalization. The larger the market size, the more investors, creating more favorable conditions for businesses to issue securities to raise capital. Therefore, stock market capitalization to GDP ratio (equal to the total market value of all listed shares divided by GDP) is used to measure the stock market's ability to allocate capital to investment projects and opportunities to diversify risks for investors. Besides, the ratio of bond market capitalization to GDP is also used by many studies to measure the size of bond market.

Accessibility to stock market (stock market concentration): represented by the ratio of stock market capitalization (excluding the 10 largest firms) to total market capitalization (IMF 2015, 2016). A high degree of market concentration implies that the disparity in size between the 10 largest firms and the rest of the market is narrowing, so the likelihood of firms accessing finance will be high. The increase in size of large enterprises can creates difficulties for new issuers in entering the stock market in developing countries, and at the same time have a negative significant impact on investors in the market. Therefore, accessibility to the stock market is expected to have a certain impact on corporate capital structure.

Efficiency: Empirical studies often use total value of stocks traded to market capitalization (turnover ratio), which is a variable that can help to measure liquidity (along with the total value of shares traded to GDP) and market efficiency. A small but dynamic stock market will have a small ratio of total traded shares to GDP and a high market capitalization of shares traded. While the ratio of total traded shares to market capitalization is not a direct measure of liquidity, a high ratio usually indicates low transaction costs. Additionally, the simultaneous use of transaction value and capitalization data overcomes limitations of using just one value. The high index means good market performance and strong liquidity, making the stock market more dynamic and can assist firms in raising capital.

Previous studies suggest that the impact of stock market development on capital structure is inconclusive, and this relationship depends on the way the variable is measured, the country of study and the way the sample is divided. For the 5 developing countries in ASEAN (Indonesia, Malaysia, Philippines, Thailand and Vietnam), the sizes of stock markets are still quite small, and banks are the main source of capital for these economies. Therefore, when the stock markets are more developed, it is expected to make it easier for firms to issue stocks to raise capital, reducing dependence on bank loans. In this paper, rather than analyzing how each aspect of financial market, we use FMI as an overall index to measure the development of stock market.

The research hypothesis of the study is as follows:

Stock market development has a negative impact on corporate use of debt.

2.4.2 Control Variables

The control variables in the model comprise of 2 groups: the group of firm-level characteristics and the group of macroeconomic variables.

Firm-level variables

We choose 4 variables representing the business characteristics, including:

Firm size: According to Trade-off Theory, firm size is expected to have a positive impact on capital structure, but the Pecking Order Theory predicts the opposite effect. Most of the empirical studies use the natural logarithmic of total assets to represent firm size and give positive results.

Profitability: The Trade-off Theory suggests a positive relationship between leverage and profitability (Sbeiti 2010; Darskuviene 2010). This view is also consistent with signaling models of capital structure (Darskuviene 2010). In contrast, the Pecking Order Theory shows an inverse relationship between profitability and firm leverage (Sbeiti 2010; Baker and Martin 2011). Most of the empirical studies have negative effects. In this study, the authors use ROA because this is a widely used indicator of profitability and risks faced by the business.

Tangibility: The Fixed Assets to Total Assets ratio is used to represent tangibility, consistent with Deesomsak et al. (2004), Bokpin (2010), Doku et al. (2011), Lemma and Negash (2013), Le and Ooi (2012), Antzoulatos et al. (2016), and Zafar et al. (2019). According to the Trade-off Theory, this ratio and capital structure are positively related. However, the Pecking Order Theory suggests opposite result.

Growth opportunity: According to Trade-off Theory, the Theory of Free Cash Flow by Jensen (1986) and the Timing Market Theory, growth opportunities have a negative impact on capital structure. However, Pecking Order Theory does not offer a clear prediction and it has to depend on whether the model being considered is a simple or complex model. According to Sbeiti (2010), growth opportunity is measured by the Tobin-Q index because it captures change between future investment opportunities and existing assets. Tobin-Q is also a measure used in the research of Deesomsak et al. (2004), Bokpin (2010), and Le and Ooi (2012). Therefore, the study will use the Tobin-Q index to represent the growth opportunity.

Macroeconomic variables

GDP per capita growth rate: Economic growth will influence financial decisions because economic growth is an indicator of companies' financial needs. If investment opportunities and the economy are interrelated, the faster growth rate an economy has, the more resources firms will need to capitalize on the opportunities. However, the relationship between GDP growth rate and capital structure is elusive, because when funds are needed, firms can increase capital in many ways such as bank loans (increase leverage) or issuing additional shares to raise capital (reduce leverage).

Inflation: Inflation rate is often seen as an indicator to evaluate the ability of the government to manage the economy, it also provides information about the stability of a certain currency in the long-term contracts (Demirguc-Kunt and Maksimovic 1999). According to Trade-off Theory, inflation will have a positive effect on leverage because it not only reduces the real value of debt but also increases the tax shield from debt (Lemma and Negash 2013).

3 Methodology and Data

3.1 *Models*

Based on the theoretical basis, the models of previous studies and research hypotheses set out in Sect. 3.1, the research model is:

$$D_{it} = \alpha_{it} + \alpha_1 CM_{jt} + \alpha_2 BC_{it} + \alpha_3 MA_{jt} + \varepsilon_{it}$$

- D_{it} (Debt) is the ratio of debt to total assets. Debt is measured using total debt—TD_{it}, short-term debt—SD_{it} and long-term debt—LD_{it} of firm i at time t;
- CM_{jt} (Capital market) is the variable representing the capital market of country j at time t, which is Financial Markets Index (FMI) of IMF;
- BC_{it} is variable representing the characteristics of businesses, including: firm size (logarithm of total assets—SIZE), profitability (net profit/total assets—ROA), asset nature (fixed assets to total assets—TANG), and growth opportunities (Tobin-Q index);
- MA_{jt} (Macroeconomics) are variables representing macroeconomic conditions, including inflation rate (INF) and GDP per capita growth rate (GDPGR).
- ε_{it} is the residual.

3.2 Estimation Methods

Previous studies tend to apply traditional estimation methods (frequentist approaches) such as FEM, REM, GMM. The frequentist approach assumes that the observed data are a repeatable random sample and that parameters are unknown but fixed and constant across the repeated samples, which may not always be feasible (StataCorp 2021). Therefore, the authors use the Bayesian approach to bridge the research gap.

Bayesian analysis is a statistical analysis that answers research questions about unknown parameters of statistical models by using probability statements. Bayesian analysis rests on the assumption that all model parameters are random and can incorporate prior knowledge. This assumption is in sharp contrast with the more traditional, also called frequentist, statistical inference where all parameters are considered unknown but fixed. Bayesian analysis follows a simple rule of probability, the Bayes rule, which provides a formalism for combining prior information with evidence from the data at hand. The Bayes rule is used to form the so-called posterior distribution of model parameters. The posterior distribution results from updating the prior knowledge about model parameters with evidence from the observed data. Bayesian analysis uses the posterior distribution to form various summaries for the model parameters including point estimates such as posterior means, medians, percentiles, and interval estimates. Moreover, all statistical tests about model parameters can be expressed as probability statements based on the estimated posterior distribution. Therefore, Bayesian analysis is a powerful analytical tool for statistical modeling, interpretation of results, and prediction of data. It can be used when there are no standard frequentist methods available or the existing frequentist methods fail (StataCorp 2021).

Bayesian analysis starts with the specification of a posterior model. The posterior model describes the probability distribution of all model parameters conditional on the observed data and some prior knowledge. The posterior distribution has two components: a likelihood, which includes information about model parameters based on the observed data, and a prior, which includes prior information (before observing the data) about model parameters.

Likelihood:

$$\text{td} \sim \text{normal}(\text{xb\_td}, \{sigma2\})$$
$$\text{Id} \sim \text{normal}(\text{xb\_Id}, \{sigma2\})$$
$$\text{sd} \sim \text{normal}(\text{xb\_sd}, \{sigma2\})$$

Priors:

$$\{\text{td : fmi size roa tang tobinq inf gdgr\_ cons}\} \sim \text{normal}(0, 10000)$$
$$\{\text{Id : fmi size roa tang tobinq inf gdgr\_ cons}\} \sim \text{normal}(0, 10000)$$

$$\{sd: \text{fmi size roa tang tobinq inf gdgr\_ cons}\} \sim \text{normal}(0, 10000)$$
$$\{\text{sigma2}\} \sim \text{igamma}(0.01, 0.01)$$

where all the model parameters are normally distributed with the zero mean and the same variance of 10,000, the overall variance (sigma2) has an Igamma(0.01, 0.01) prior.

3.3 Research Data

To ensure the sample is large enough and representative of developing countries in ASEAN, the study uses the financial statements of the listed firms in five countries (Indonesia, Malaysia, Philippines, Thailand and Vietnam) with data retrieved from Thomson Reuters for the period 2010–2019.

Research data are collected according to the following specific steps: First, we remove financial institutions (such as banks, financial companies, insurance companies) from the sample. Second, we remove extreme observations, for example firms with negative stock prices or / and negative equity, total debt value is greater than total assets. Finally, the study excluded businesses that only had 1 to 2 years data.

4 Empirical Results and Discussions

The statistical value described in Table 3 shows that the average total debt ratio (TD) of listed firms in 5 countries reached 28.51%, showing that more than a quarter of

Table 3 Descriptive statistics of the variables in the model

Variable	Obs	Mean	Std. Dev	Min	Max
TD	11,917	0.285	0.170	0.010	0.932
LD	11,917	0.137	0.137	$4.83e^{-06}$	0.86
SD	11,917	0.148	0.135	0	0.862
FMI	11,917	0.517	0.153	0.26	0.74
SIZE	11,917	18.707	1.798	13.282	25.154
ROA	11,493	0.045	0.082	−0.925	0.953
TANG	11,779	0.343	0.242	0.000	0.98
TOBINQ	10,562	1.662	2.57	0.02	49.64
INF	11,917	0.035	0.041	−0.007	0.213
GDPGR	11,917	0.041	0.014	0.004	0.07

Source Results from Stata software

the total assets of business are financed by debt. The minimum value of the total debt ratio is about 1% and the maximum is 93.15%, indicating that some firms are safe because they use little debt, but some other businesses are very risky due to the intensive use of debt. The short-term and long-term debt of enterprises are quite similar with the smallest value almost zero, and the highest value about 86%. Stock market development indexes have been standardized, so they have values ranging from 0 to 1. The average value of FMI index reaches 0.517 with values ranging from 0.26 to 0.74.

For business characteristics, the average firm size (SIZE) is 18.71, the smallest value is 13.28 and the largest value is 25.15. Average profitability rate (ROA) is 4.52%, showing that for every 100 dong of assets, 4.52 dong of profit can be generated. The ratio of fixed assets to total assets (TANG) has an average value of 34.28% with values ranging from 1.45e-04 to 98%. Tobin-Q index (equal to market value over book value of total assets) reaches an average value of 1.66 times with a rather large range from 0.02 to 49.64 times.

For macroeconomic variables, the average inflation rate (INF) is 3.5% with values ranging from -0.7% to 21.26%, the standard deviation is quite high (4.06%). Meanwhile, the average GDP per capita growth rate (GDPGR) reached 4.11% with the range from 0.36% to 6.99%.

Concerning the tests for chain convergence, Fig. 1 shows diagnostic graphs that contain trace plots, histograms and density plots for our MCMC sample and autocorrelations. According to the results of the convergence test by the chart, trace plots run quickly through the posterior domain, do not show trends, stop with averages, and the value of variance is directed to a constant. The histogram plots resemble the shape of the posterior distributions of the model parameters; the autocorrelations have no lags; the CUSUM plots are jagged, intercepting the X-axis; the kernel density estimation graph has graphs that resemble the shape of the posterior distributions of the model parameters. None of these plots indicate any problems with our sample. Thus, it is possible to conclude that the parameters of the research model have converged to suitable values (Figs. 2 and 3).

First, the results in Tables 4, 5 and 6 show that the means and medians of all variables are similar. Thus, we can conclude that three models has symmetric posterior distributions. Second, the standard deviation for all model parameters is small. Third, the less the MCSE values, the more precise the mean estimates of the model parameters, and all regression coefficients have the MCSE estimates far below one decimal. The 95% credible intervals of FMI (in case of TD and LD) do not contain the value of zero. Hence, we can make probabilistic statements that FMI has strong negative effects on TD and LD. While the impact of FMI on SD is likely to be moderate, as the zero value falls into its 95% credible interval, ranging from -0.033 to 0.004. However, these results show that stock market development has a negative impact on debts. When the stock market develops in general, it can encourage businesses to reduce debt (both long-term and short-term debt). The result that business reduce their debt ratio is similar to hypothesis Demirguc-Kunt and Maksimovic (1996), Bokpin (2010), and Le and Ooi (2012). The reason is that when the stock market

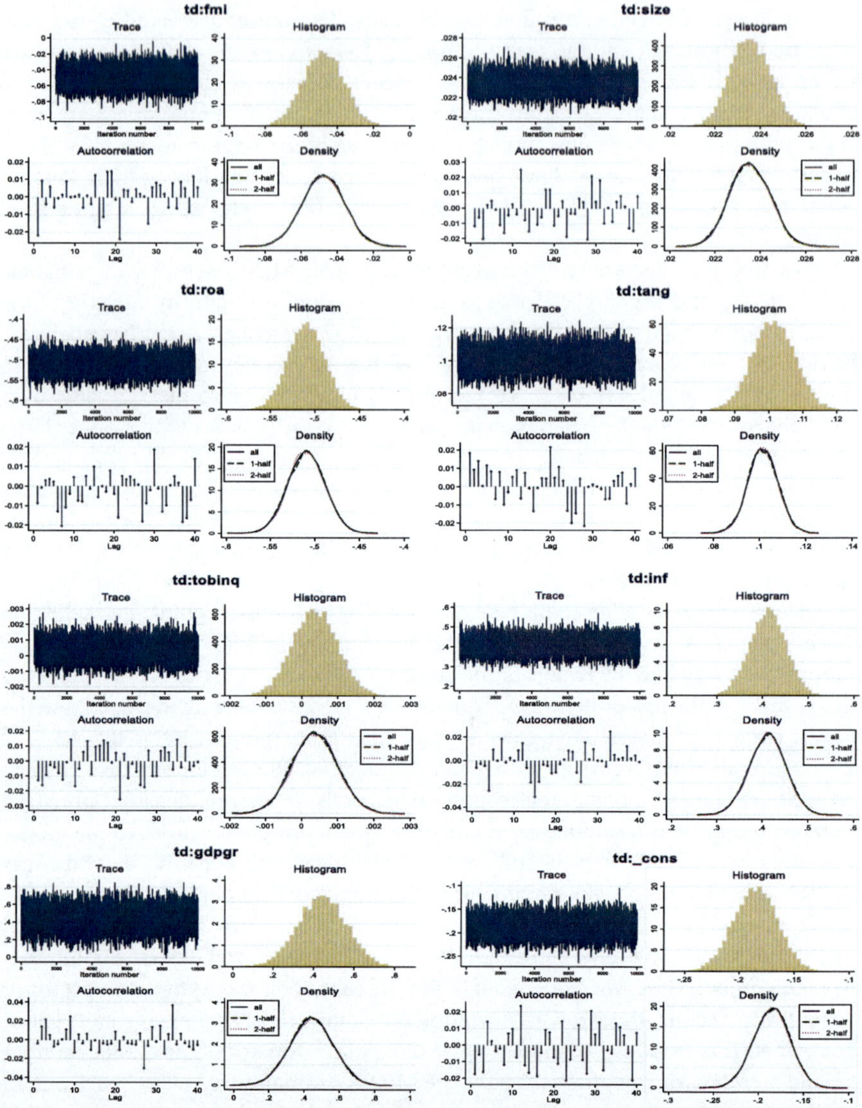

Fig. 1 Diagnostic plots of total debt. *Source* Results from Stata software

develops more, businesses will have more diversified forms of capital mobilization and reduce dependence on bank loans.

The 95% credible intervals of SIZE, ROA, TANG, INF and GDPGR do not contain the value of zero. Hence, we can make probabilistic statements that ROA has strong negative effects, while INF and GDPGR have positive effects on the three dependent variables. In addition, SIZE and TANG have positive effects on TD and LD, but

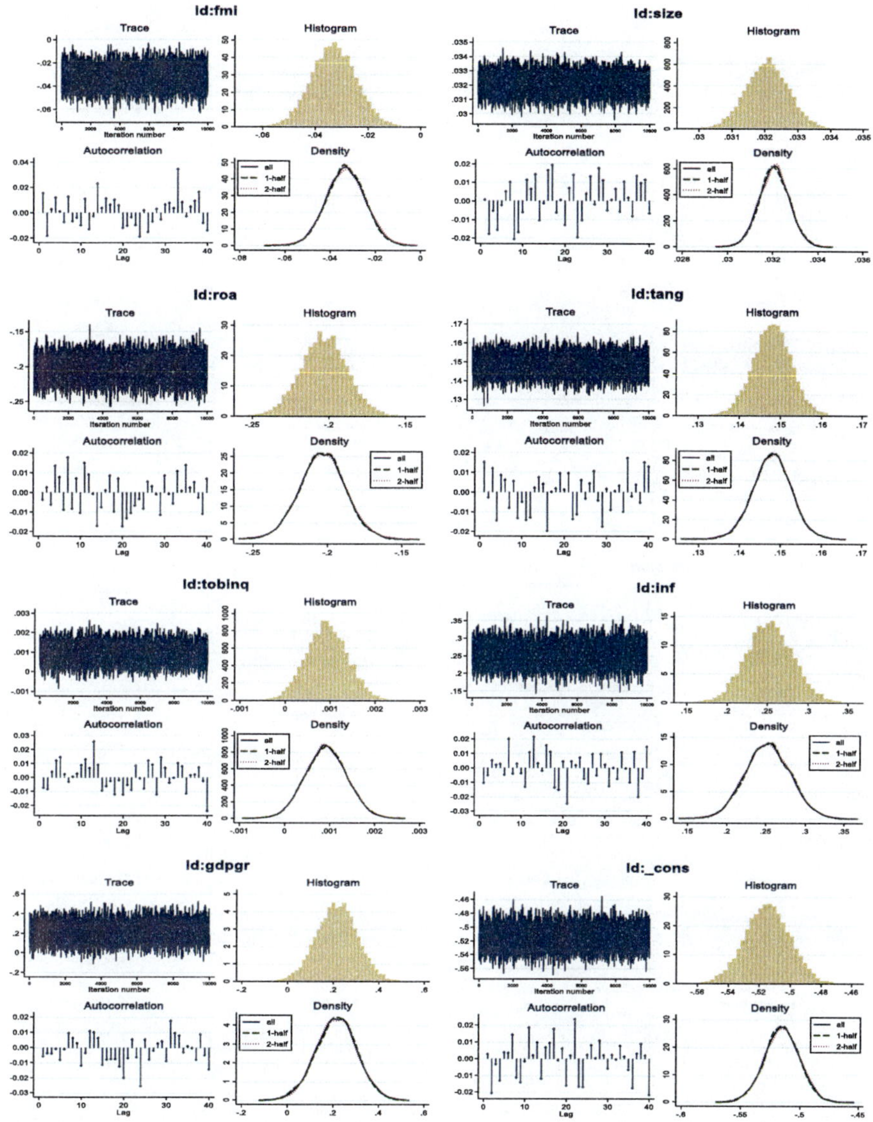

Fig. 2 Diagnostic plots of long term debt. *Source* Results from Stata software

negative effects on SD. The positive effects of SIZE supports the trade-off theory and is similar to studies Deesomsak et al. (2004, 2009), Sbeiti (2010), Lucey and Zhang (2011), Le and Ooi (2012), Antzoulatos et al. (2016), and Le (2017). Meanwhile, TANG has a positive effect on total debt and long-debt supporting the trade-off theory and similar to the results of Deesomsak et al. (2004, 2009), Le and Ooi (2012).

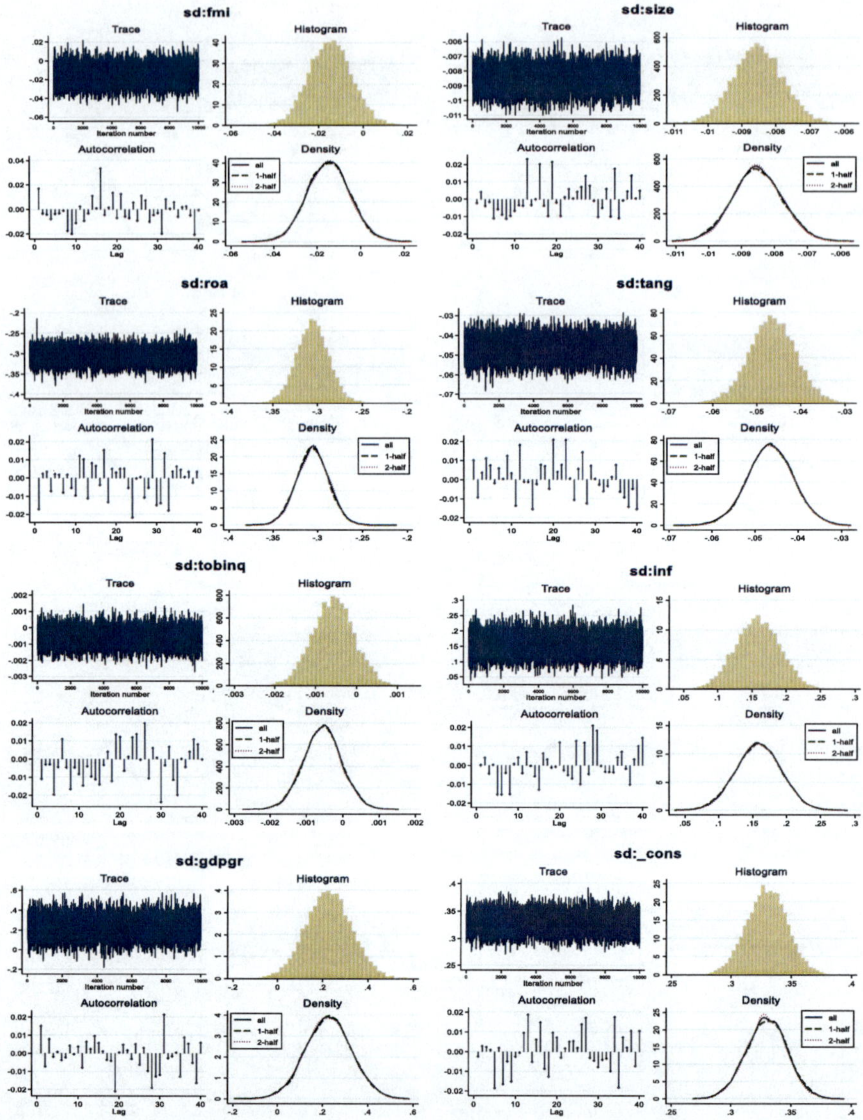

Fig. 3 Diagnostic plots of short term debt. *Source* Results from Stata software

However, SIZE and TANG have the negative effect on short-debt, which support pecking order theory. The results show that enterprises with large scale and many tangible assets will have easier access to loans due to transparent information and collateral for loans; therefore, they tend to borrow more, especially more long-term debt and reduce short-term one.

Table 4 The impact of stock market development on total debt

td	Mean	Std. Dev	MCSE	Median	Equal-tailed [95% Cred. Interval]	
fmi	−0.0476744	0.0115587	0.000113	−0.0476305	− 0.0703158	−0.0249187
size	0.02355	0.0008858	8.8e-06	0.0235358	0.0218514	0.0252917
roa	−0.5095096	0.0207635	0.000208	−0.5095384	−0.5501882	−0.4686515
tang	0.1014818	0.0063338	0.000064	0.1014421	0.088999	0.1138071
tobinq	0.0003657	0.0006245	6.le-06	0.0003604	−0.0008584	0.0015969
inf	0.4115753	0.0396637	0.000397	0.4119666	0.3333923	0.4890132
gdpgr	0.4440425	0.1211576	0.001212	0.4445917	0.2052557	0.6795243
_cons	−0.1852516	0.0198882	0.000195	−0.1847603	−0.2246521	−0.1469364
sigma2	0.0241194	0.000332	3.3e-06	0.0241149	0.0234663	0.0247797

Source Results from Stata software

Table 5 The impact of stock market development on long term debt

Id	Mean	Std. Dev	MCSE	Median	Equal-tailed [95% Cred. Interval]	
fmi	−0.0330568	0.0084669	0.000084	−0.0331404	−0.0493651	−0.0163401
size	0.0320662	0.0006481	6.5e−06	0.0320722	0.0307864	0.0333429
roa	−0.2034426	0.0152077	0.000152	−0.2034785	−0.2334155	−0.1733961
tang	0.1479701	0.0046597	0.000047	0.1480274	0.1388075	0.1572947
tobinq	0.0009216	0.0004532	4.5e−06	0.000917	0.0000514	0.0018121
inf	0.2521762	0.0286513	0.000284	0.2519662	0.1974496	0.3093159
gdpgr	0.2161728	0.0891562	0.000892	0.2163062	0.0401497	0.3914803
_cons	−0.5147797	0.0145572	0.000146	−0.5147703	−0.5436797	−0.4863748
sigma2	0.0127827	0.0001756	1.8e-06	0.0127824	0.0124477	0.0131328

Source Results from Stata software

Table 6 The impact of stock market development on short term debt

sd	Mean	Std. Dev	MCSE	Median	Equal-tailed [95% Cred. Interval]	
fii	−0.0146327	0.009644	0.000098	−0.0146391	−0.0331527	0.0043041
size	−0.008521	0.0007424	7.4e−06	−0.0085303	−0.0099831	−0.0070586
roa	−0.3055024	0.0177273	0.000174	−0.3053295	−0.3405497	−0.2711857
tang	−0.0464495	0.0052617	0.000053	−0.0464471	−0.0570594	−0.0360973
tobinq	−0.0005581	0.0005202	5.le−06	−0.0005524	−0.0015706	0.0004654
inf	0.1583249	0.0330707	0.000331	0.1584482	0.0928996	0.2218695
gdpgr	0.2264653	0.1006244	0.001021	0.2263418	0.0287667	0.425768
_cons	0.3296726	0.01671	0.000167	0.3297646	0.295997	0.3621339
sigma2	0.0166294	0.0002316	2.3e-06	0.0166255	0.0161794	0.0170836

Source Results from Stata software

Contrary to SIZE and TANG, profitability (ROA) has a negative impact on debt, supporting the pecking order theory and being similar to the results of Zafar et al. (2019). This shows that businesses with high profitability will reduce debt use because they already have retained earnings or are more favorable in mobilizing capital by other channels in both short and long term, reducing dependence on bank loans. Firm growth opportunity (TOBINQ) has a positive effect on total debt and long-term debt, which supports the pecking order theory in a simple model and is similar to the research results by Deesomsak et al. (2004, 2009), Zafar et al. (2019). However, TOBINQ impact is quite small. Regarding the credible intervals, we can claim that the mean coefficient for variable TOBINQ belongs to the range between -0.0008584 and 0.0018121 with a probability of 95%. The inflation rate (INF) has a positive impact on debt, which is similar to the study of Lucey and Zhang (2011), Lemma and Negash (2013), Zafar et al. (2019). This implies that when inflation increases, businesses tend to increase their debt ratio (both in the short and long term debt), possibly to take advantage of currency devaluation to reduce interest expenses. The GDP per capita growth rate (GDPGR) has a positive effect on debt, showing that businesses will increase total debt when the economy develops.

5 Conclusions

The study provides empirical evidence that stock market development does indeed have a negative impact on corporate use of debt. When the stock market develops, firms are more likely to reduce their dependence on bank loans (both in long-term and short-term debt). At the same time, the capital structure of firms in the five developing countries supports both the trade-off theory and the pecking order, which means that firms' borrowing will depend on a balance between profitability and costs of using borrowed capital. However, due to the existence of information asymmetry in the economies, firms still prefer to use internal capital. Therefore, development of the stock market is expected to contribute to diversifying capital mobilization channels and depending on the position of enterprises, they will choose the most profitable source of capital to access.

To promote development of stock market, it is necessary to increase the number of goods on the market and improve legal framework as well as infrastructure. This will help improve stock market development in terms of depth, accessibility and efficiency. For developing countries, information transparency is important. Therefore, regulatory bodies need to closely supervise listed firms as well as require them to disclose timely information to investors, avoiding delays and falsification. In addition, the State Securities Commission should continue to study and develop technical criteria to detect market manipulation and insider transactions in order to propose appropriate penalty based on level of damage and adverse effects on stock market.

References

Agarwal, S., Mohtadi, H.: Financial markets and the financing choice of firms: evidence from developing countries. Glob. Financ. J. **15**, 57–70 (2004)

Anderson, T.W., Hsiao, C.: Estimation of dynamic models with error components. J. Am. Stat. Assoc. **76**, 598–606 (1981)

Anderson, T.W., Hsiao, C.: Formulation and estimation of dynamic models using panel data. J. Econ. **18**, 47–82 (1982)

Antzoulatos, A.A., Koufopoulos, K., Lambrinoudakis, C., Tsiritakis, E.: Supply of capital and capital structure: the role of financial development. J. Corp. Finan. **38**, 166–195 (2016)

Arellano, M., Bond, S.: Some tests of specification for panel data: monte Carlo evidence and an application to employment equations. Rev. Econ. Stud. **58**(2), 277–297 (1991)

Arellano, M., Bover, O.: Another look at the instrumental variable estimation of error-components models. J. Econ. **68**(1), 29–51 (1995)

Baker, H.K., Martin, G.S.: Capital Structure and Corporate Financing Decisions—Theory, Evidence and Practice. Wiley, Hoboken, New Jersey (2011)

Baker, M., Wurgler, J.: Market timing and capital structure. J. Financ. **57**(1), 1–32 (2002)

Blundell, R.W., Bond, S.R.: Initial conditions and moment restrictions in dynamic panel data models. J. Econ. **87**, 115–143 (1998)

Bokpin, G.A.: Financial market development and corporate financing evidence from emerging market economies. J. Econ. Stud. **37**(1), 96–116 (2010)

Cuyvers, L., Chen, L., Lombaerde, P.D.: 50 years of regional integration in ASEAN. Asia Pacific Bus. Rev. (2019)

Darskuviene, V.: Financial Markets. Leonardo Da Vinci—Transfer of Innovation, Education and Culture DG—Lifelong Learning Programme (2010)

Deesomsak, R., Paudyal, K., Pescetto, G.: The determinants of capital structure: evidence from the Asia Pacific region. J. Multinatl. Financ. Manag. **14**(2004), 387–405 (2004)

Demirguc-Kunt, A., Maksimovic, V.: Stock market development and financing choices of firms. World Bank Econ. Rev. **10**(2), 341–369 (1996)

Demirguc-Kunt, A., Maksimovic, V.: Institutions, financial markets, and firm debt maturity. J. Financ. Econ. **54**(1999), 295–336 (1999)

Doku, J.N., Adjasi, C.K.D., Sarpong-Kumankuma, E.: Financial market development and capital structure of listed firms—empirical evidence from Ghana. Serbian J. Manag. **6**(2), 155–168 (2011)

Grossman, S.J., Stiglitz, J.: On the impossibility of informationally efficient markets. Amer. Econ. Rev. **70**, 393–408 (1980)

Holmstrom, B., Tirole, J.: Market liquidity and performance monitoring. J. Polit. Econ. **101**, 678–709 (1993)

IMF—Svirydzenka, K.: Introducing a New Broad—based index of Financial Development. International Monetary Fund (IMF Working Paper) (2016)

IMF Staff members—Sahay, R., Cihák, M., N'Diaye, P., Barajas, A., Bi, R., Ayala, D., Gao, Y., Kyobe, A., Nguyen, L., Saborowski, C., Svirydzenka, K., Yousefi, S.R.: Rethinking financial deepening: stability and growth in emerging markets. Int. Monetary Fund (2015)

Jensen, M.C.: Agency costs of free cash flow, corporate finance, and takeovers. Am. Econ. Rev. **76**(2), 323–329 (1986)

Kumar, S., Colombage, S., Rao, P.: Research on capital structure determinants: a review and future directions. Int. J. Manag. Financ. **13**(2), 106–132 (2017)

Le, M.T.: Impact of the financial markets development on capital structure of firms listed on ho chi minh stock exchange. Int. J. Econ. Financ. Issues **7**(3), 510–515 (2017)

Le, T.T.T., Ooi, J.T.L.: Financial structure of property companies and capital market development. J. Property Invest. Financ. **30**(6), 596–611 (2012)

Lemma, T.T., Negash, M.: Institutional, macroeconomic and firm-specific determinants of capital structure the African evidence. Manag. Res. Rev. **36**(11), 1081–1122 (2013)

Levine, R.: Bank-based or market-based financial systems: which is better? J. Financ. Intermed. **11**, 398–428 (2002)
Malarvizhi, C.A.N., Zeynali, Y., Mamun, A.A., Ahmad, G.B.: Financial development and economic growth in ASEAN-5 countries. Glob. Bus. Rev. **20**(1), 57–71 (2019)
Merton, R.C.: A simple model of capital market equilibrium with incomplete information. J. Financ. **42**, 483–510 (1987)
Miglo, A.: Capital Structure in the Modern World. Springer Nature (2016). ISBN 978–3–319–30712–1
Modigliani, F., Miller, M.H.: The cost of capital, corporation finance and the theory of investment. Amer. Econ. Rev. **48**(3), 261–297 (1958)
Modigliani, F., Miller, M.H.: Corporate income taxes and the cost of capital: a correction. Amer. Econ. Rev. **53**(June), 433–443 (1963)
Myers, S.C.: The capital structure puzzle. J. Financ. **39**(3), 574–592 (1984)
Myers, S.C.: Still searching for optimal capital structure. J. Appl. Corporate Financ. 6(1) (1993)
Nickell, S.: Biases in dynamic models with fixed effects. Econometrica **49**, 1417–1426 (1981)
Phooi M'ng J.C., Rahman, M., Sannacy S.: The determinants of capital structure: Evidence from public listed companies in Malaysia, Singapore and Thailand. Cogent Econ. Financ. (2017)
StataCorp: Stata Bayesian Analysis Reference manual: Release 17. Statistical Software. College Station, TX: StataCorp LLC (2021)
Sbeiti, W.: The determinants of capital structure: evidence from the GCC countries. Int. Res. J. Financ. Econ. Issue, 47 (2010)
Yartey, C.A.: The Stock Market and the Financing of Corporate Growth in Africa: The Case of Ghana. Emerging Markets Finance and Trade (2009)
Zafar, Q., Wongsurawat, W., Camino, D.: The determinants of leverage decisions: evidence from Asian emerging markets. Cogent Econ. Financ. **7**(1), 1598836 (2019)

Factors Affecting Cash Dividend Policy of Food Industry Businesses in Vietnam

Bui Dan Thanh, Doan Thanh Ha, and Nguyen Hong Ngoc

Abstract The objective of this study is to measure the factors affecting the cash dividend policy (CDP) of the food companies listed on the Vietnamese stock market. The data in this study is collected from the financial statements of 52 food industry enterprises for the period 2011–2020. The factors included in the research model are earnings per share, revenue growth rate, solvency, debt-to-equity, corporate income tax rate, while the dependent variable is the probability of the occurrence or non-occurrence of the cash dividend payment of a food business in Vietnam. By taking advantage of Bayesian Binary Logistic regression, the results show that cash dividend payout ratio (DPR) is positively related to earnings per share (EPS), solvency (LR), income tax rate corporate income (TAX), but negative impact on growth rate (GR) and debt to equity (DE).

Keywords Dividend policy · Cash dividend · Food business · Binary logistic model

1 Introduction

The existence, operation and development of enterprises are greatly affected by dividend policy (DP). In recent years, the source of income from dividends has begun to be of interest to investors in the Vietnamese stock market. However, most businesses have not properly realized the importance of DP, have not had clear

B. D. Thanh · D. T. Ha
Banking University of Ho Chi Minh City, 36 Ton That Dam Street, District 1, Ho Chi Minh City, Vietnam
e-mail: thanhbd@buh.edu.vn

D. T. Ha
e-mail: hadt@buh.edu.vn

N. H. Ngoc (✉)
Ho Chi Minh City College of Economics, 33 Vinh Vien Street, District 10, Ho Chi Minh City, Vietnam
e-mail: ngocnh@kthcm.edu.vn

N. Ngoc Thach et al. (eds.), *Financial Econometrics: Bayesian Analysis, Quantum Uncertainty, and Related Topics*, Studies in Systems, Decision and Control 427,
https://doi.org/10.1007/978-3-030-98689-6_37

and long-term orientations for this issue. In fact, in Vietnam, there are still not many quantitative studies related to this issue, the reason why listed companies often tend to pay dividends in shares, therefore, has not properly reflected the DP in the past. Because stock dividends are also a form of retaining profits for reinvestment rather than distributing profits to shareholders and thus the research results do not reflect accurately. In addition, in Vietnam, previous studies have mostly studied all companies listed on the Vietnamese stock market. Meanwhile, the difference in industry characteristics, development policies of enterprises, the research sample is not large enough (currently there are more than 1,000 enterprises listed on Vietnam stock exchange),... are the limitations of previous studies. With these arguments, the authors propose to focus only on a specific industry group that is the food industry to evaluate the factors affecting the CDP. The food industry accounts for more than 20% of the total capitalization of the Vietnamese stock market. There are published studies on cash dividend policy, for example, Edmund (2018), Al-Najjar et al. (2018), Megha (2018), Gangil and Nathani (2018), Farman and Nawaz (2017), etc. Besides, In Vietnam there are also some works on this issue, such as Truong et al. (2015). This study aims to analyze the impact of selected factors on the probability of cash dividend payment of food businesses in Vietnam through Bayesian Binary Logistic regression application method.

2 Theoretical Basis

2.1 *Dividend Irrelevance Theory*

Miller and Modigliani (1961), gave the theory of DP independence, that is, DP has no effect on stock prices, this theory is based on the assumption that there are no taxes or transaction costs. The authors believe that no matter how a business distributes profits, the value of the business will be determined by its earning ability and investment policy. In other words, investors calculate the value of a business basing on the value of capitalization of income in the future, and it is not affected whether the business pays dividends.

2.2 *Theory of High Cash Dividend Policy (CDP)*

According to the high CDP theory (Bird in the hand), also known as the bird in the palm theory, Lintner (1956) and Gordon (1963) proposed the theory that investors prefer high dividends. In an imperfect, incomplete and asymmetric market, investors often prefer to choose firms with high DP. This theory states that investors prefer high cash payout ratios and often invest in high-dividend businesses because cash dividends are less risky than future capital gains.

2.3 Agency Cost Theory

According to Jensen and Meckling (1976), the agency cost comes from the separation between ownership and management rights. This cost is incurred when the company's board of directors carries out activities aimed at their own benefit instead of those that benefit the shareholders (who own the business). The benefits from this decision can be direct or indirect. In addition, agency costs can also be incurred by shareholders of a business which there is too much cash and shareholders often require dividends instead of re-investment. The payment of dividends can serve to realign the interests of the parties, and at the same time to minimize disagreements between managers and shareholders because in reality, the business managers always have the conflicts with other shareholders.

2.4 Signal Theory

According to Signaling theory, investors can infer information about the earnings of business through signals coming from dividend announcements. According to this theory, dividends act as a signal of a company's growth prospects when insiders have more information than the market. The researchers believe that, in fact, there is always an information gap between managers of corporation and investors because managers control the operations of business every day and know a lot of inside information than outside investors. Under such asymmetry information condition, dividends are the most effective way to communicate information between firms and shareholders. Lintner (1956) and Gordon (1963) said that investors are very risk averse, do not want to taking the risk which explains why they prefer current dividends and stable growth rather than expecting the stock price increase. Also according to the two authors, dividends provide a signal about the development of the business to investors. Bhattacharyya (1979) said that managers have the confidential information about the distribution of cash flows and they signal to the market through the choice of dividend payout rate.

2.5 Clientele Effects Theory

Clientele Effects Theory said that the various investors will have different needs and tend to prefer buying the shares of companies who have DP that meet the needs of investors. According to Black and Scholes (1974), the investors who have to pay the high tax often prefer businesses to retain the profits to avoid heavy taxes payment, while investors in the low payment group of tax prefer to receive the dividends. In addition, investors have to pay tax rates for both cases of dividend receiving and share price increasing beside payment for the transaction costs while they buy and sell the stock.

2.6 Relevant Studies

Gangil and Nathani (2018), analyzed the determinants of dividend payment of FMCG (fast-moving consumer) industry in India. The study concluded that growth has a negative impact and profitability has a positive impact on DP decision. Truong et al. (2015) when studying the DP of listed companies in Vietnam with the data of 236 non-financial enterprises in the period 2010–2012 also agreed with the idea that there is a positive effect of profit on dividend payout ratio.

Demirgunes (2015) also research the growth factors negatively affecting dividend payout ratio. This author studied at enterprises operating in the non-metallic (cement) industry in Turkey. Similar to the study of Demirgunes (2015), Leon and Putra (2014) also found evidence of a negative impact of growth factors on dividend payout ratio when studying non-financial companies in Indonesia..

Regarding the factor of using financial leverage, Asad and Yousef (2014) using OLS regression method of 44 manufacturing enterprises in Pakistan. The author said that business managers should decide the level of financial leverage and suitable DP to ensure the stability of capital. Agreeing with this opinion, Edmund (2018) have concluded that debt has a negative effect on DP when the author studied the impact of financial leverage on the dividend payout ratio within 10 years for banks listed on the Ghana Stock Exchange.

While previous tax studies only focused on the impact of taxes to economic, Akhlaqul et al. (2013) investigated the relationship between tax shield (TS) and DP of 33 companies listed on the Karachi Exchange Securities. However, the results found no evidence that tax shield was significantly related to dividend payout ratio. Contrary to the study of Akhlaqul et al. (2013), Farman and Nawaz (2017) found evidence that, tax (Taxation), risk P/E ratio (RISK), firm size (FS) and financial leverage (Leverage) had an influence on dividend payment decisions of pharmaceutical companies.

On the other hand, Khoirunnisa et al. (2018) studied the impact of solvency on the DP of listed business and service enterprises in Malaysia. The regression results of fixed effects model (FEM) showed that the solvency has a positive influence on dividend payment policy. Ahmed and Murtaza (2015) also found this result when studying cement enterprises in Pakistan.

Through a review of previous studies, the authors propose to select the dominant factors that are found to have an impact on dividend payment policy such as profitability, liquidity, leverage. Finance, growth and factor have not been found to have a clear relationship that is taxes.

3 Models and Data

3.1 General Model

$$Y_{it} = \propto + \beta_k X_{kit} + u_{it}$$

In there:

Y_it- value of the dependent variable corresponding to enterprise i at time t;

X_kit—the value of the independent variable k representing the factors affecting the dependent variable of enterprise i at time t;

u_it—random error of enterprise i at time t.

3.2 Research Hypotheses

Hypothesis H1: Firms with high profits are likely to pay large dividends. Earnings per share (EPS).

According to the theory of agency costs, high-profit enterprises their shareholders always expect the enterprise to pay more dividends than reinvest, in order to adjust the benefits between managers and managers, corporate managers and shareholders. Most of the previous empirical studies support the view that there is a positive relationship between earnings per share and dividend payout ratio. In other words, profit is the first condition that businesses need to do to have money to pay dividends to shareholders. Typically, the studies of Gangil and Nathani (2018) and Truong et al. (2015) all agree with this argument.

Hypothesis H2: Firms with high growth rate have low dividend payment ability. Revenue growth rate (symbol GR).

According to Signal theory, when a company pays low dividends, it can indicate that the cash flow of the business is focused on investment and business expansion. Research by Demirgunes (2015) also mentioned that when businesses with fast growth rates often have large capital needs to finance new investment projects. Therefore, businesses will tend to limit dividends to retain profits for reinvestment. In the article of Leon and Putra (2014) also supports this view when it is said that enterprises will pay low dividends when there are opportunities to expand business activities.

Hypothesis H3: Firms with high solvency are likely to pay large dividends. Solvency (symbol LR).

According to Khoirunnisa et al. (2018), businesses with good quick payment ability also tend to be proportional to dividend payments because they can convert short-term assets into cash to meet their needs, short-term demand such as dividend payments to shareholders. Ahmed et al. (2015) also said that the enterprise has a good current ratio, which means that the enterprise can meet its short-term payment obligations, including paying dividends to shareholders. This hypothesis also supports the high cash return theory that investors prefer cash over future capital gains.

Hypothesis H4: Firms with high financial leverage have low ability to pay dividends. Debt to equity (symbol DE).

Edmund (2018) and Asad and Yousef (2014) argue that, when firms have large debt ratios, firms have pressure to pay interest and pay principal. As a result, the profit distribution cash flow will be smaller, which means that the dividend payment to shareholders will also be reduced or will not pay dividends but prioritize debt payment. The theory of incompatible DP also shows that there is no relationship between DP and investment policy of enterprises.

Hypothesis H5: Firms that pay a lot of income tax are likely to pay high dividends. Corporate income tax rate (symbol TAX).

In the studies of Farman and Nawaz (2017) and Akhlaqul et al. (2013) did not find the impact of corporate income tax on dividend payout ratio. The theory of customer group effect shows that a group of investors prefer to buy shares of businesses that can implement a dividend payment policy according to their needs. Specifically, when a business has to pay a high income tax, it is usually an efficient and profitable business. Therefore, businesses will tend to pay more dividends to shareholders as a signal about the good financial health of the business.

3.3 Data

In the context of a deepening crisis in classical frequentist statistics (see, for example, Nguyen and Thach 2019; Hung et al. 2019a, b; Sriboonchitta et al. 2019a, b; Kreinovich et al. 2019), the current study applies the Bayesian framework to achieve the research purpose.

The data used in this study is panel data collected mainly from audited financial statements for the period 2011–2020 of 52 listed food companies in the market. Vietnam stock market (364 observations) (Table 1).

The results of Table 2 show that, the correlation coefficients between the independent variables in the research model are low (absolute value <0.5), so the selection of these independent variables for research is complete.

Since previous studies were often performed using out-of-date frequentist methods, which might provide unreliable and inconsistent results, the Bayes setting is applied. Note that the number of observations in this study is relatively large, so the prior information might not overwhelm the data distribution and affect the posterior estimates. In this case, Block et al. (2011) proposed specifying the standard Gaussian distributions with different prior distributions and conduct a sensitivity analysis to

Table 1 Definition of variables

Variable name	Explain	Expectation sign
Dependent variable		
Cash dividend payout ratio (DPR)	As a percentage of the par value of the shares (VND 10,000)	
Independent variables		
Earnings per share (EPS)	Profit after tax/Average number of shares	+
Revenue growth rate (GR)	((This year's revenue/Previous year's revenue) –1)*100 (%)	–
Ability to Pay (LR)	(Current Assets/Current Liabilities)	+
Debt to Equity (DE)	(Debt/Equity)*100 (%)	–
Corporate income tax rate (TAX)	(Corporate income tax expense/Total profit before tax)*100(%)	+

Table 2 Correlation coefficient matrix

	EPS	GR	LR	DE	TAX
EPS	1,0000				
GR	0,0292	1,0000			
LR	−0,1762	−0,0790	1,0000		
DE	−0,0413	−0,0819	−0,0270	1,0000	
TAX	0,2140	0,0071	−0,1023	0,0738	1,0000

Source Result extracted from Stata

choose the best among the considered models. For each of the two models, we run FIVE simulations (Table 3).

In the next step, the authors perform suggested regressions and the obtained results are recorded in Table 4. The simulation selected in the Bayesian factor analysis is the simulation with the largest mean Log BF, largest Log ML, and minimum DIC mean. Thus, the results of Bayesian factor analysis in Table 4 show that simulation 1 has the most advantage. Moreover, the results of Table 4 also show that simulation 1 has superior posterior probability compared to other simulations; therefore, simulation 1 has the most suitable prior information.

To ensure that the Bayesian inference based on the MCMC sample is robust, we use the convergence diagnostic. This diagnosis can be made visually through trace plots, autocorrelation plots, histograms, and kernel density plots.

From Fig. 1, it can be seen that all the parameter estimates in the graphs are quite reasonable: the trace and autocorrelation plots show good mixing, i.e. low autocorrelation. The autocorrelation plot reflects all lags that are within the effective limit. In addition, the histograms in Fig. 1 all show a good fit with the simulated density distribution, the shape of the histogram is uniform (unimodal). Histogram and Density plots both show the simulation of the shape of the normal distribution

Table 3 Model specifications

Model 1	
Likelihood function	DIS $\sim N(\mu, \sigma)$
Prior distribution	
Simulation 1	$\beta_i \sim N(0; 1)$ $\sigma^2 \sim \text{Invgamma}(0, 01; 0.01)$
Simulation 2	$\beta_i \sim N(0; 10)$ $\sigma^2 \sim \text{Invgamma}(0, 01; 0, 01)$
Simulation 3	$\beta_i \sim N(0; 100)$ $\sigma^2 \sim \text{Invgamma}(0, 01; 0, 01)$
Simulation 4	$\beta_i \sim N(0; 1000)$ $\sigma^2 \sim \text{Invgamma}(0, 01; 0, 01)$
Simulation 5	$\beta_i \sim N(0; 10000)$ $\sigma^2 \sim \text{Invgamma}(0, 01; 0, 01)$

Table 4 Results of a test via Bayes factor and a Bayes model test

	DIC	log(ML)	log(BF)	P(Mly)
Simulation 1	446.9683	−232.287	1	0.9966
Simulation 2	447.8905	−237.961	−5.67405	0.0034
Simulation 3	447.95	−244.728	−12.4406	0
Simulation 4	447.9859	−251.602	−19.3146	0
Simulation 5	447.8554	−258.583	−26.2955	0

of the parameters. From the results of testing the convergence of the above model series, it can be concluded that the MCMC series have converged to a stationary distribution, so the Bayesian inference will be stable.

Additionally, we consider some important initial indicators, the rate of acceptance and efficiency. The results in Table 5 point out that the acceptance rates have reached 0.2125, while minimum efficiency has obtained 0.1603 (exceeds the required level of 0.01). Besides, the summary tables provide Monte-Carlo standard errors (MCSE), which indicate the accuracy of the posterior estimates, According to Flegal et al. (2008), the closer the MCSEs are to zero, the stronger the MCMC series, claiming that the MCSE values less than 6.5% of the standard deviation are acceptable and smaller 5% of the standard deviation are optimal.

The results of Table 5 shows the sign of the regression coefficients, according to which the values of EPS and TAX have a positive impact on the variable DPR. In contrast, the remaining value harm DPR. However, to determine the magnitude of the impact, it is necessary to calculate the probability of these effects.

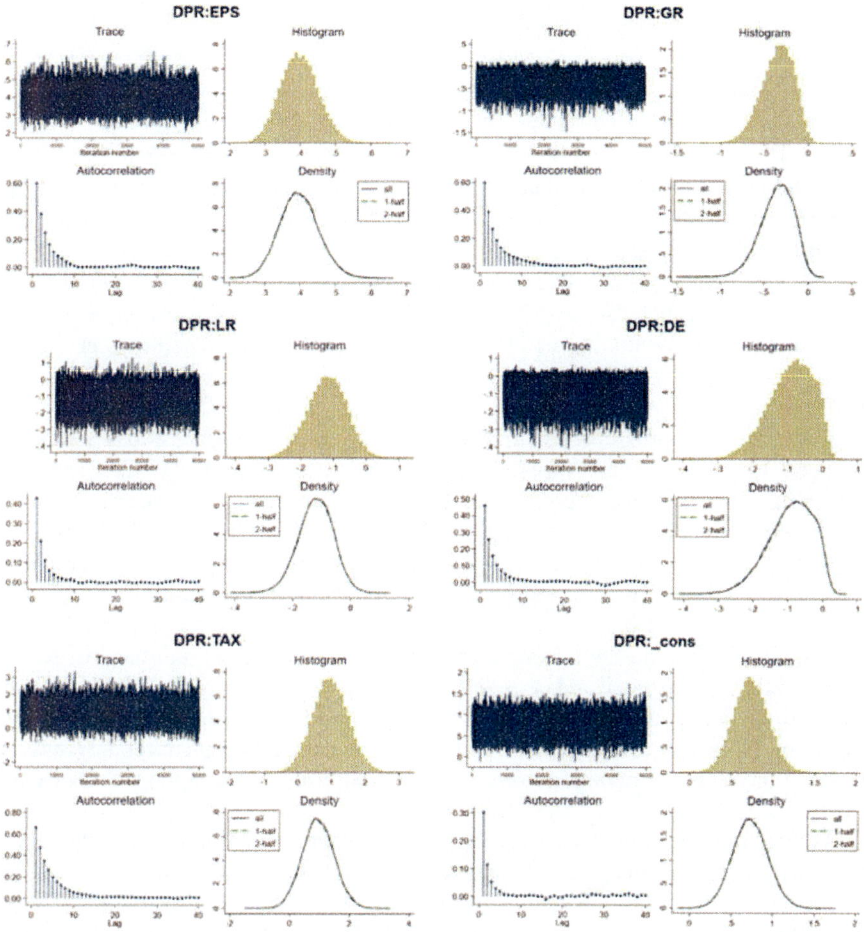

Fig. 1 Graphical convergence

Table 5 Bayesian simulation outcomes

DPR	Mean	Std. Dev	MCSE	Median	Equal-tailed	
					[95% Cred. Interval]	
EPS	0.3979	0.0547	0.0005	0.3965	0.2958	0.5094
GR	−0.3568	0.1858	0.0019	−0.3413	−0.7566	−0.0464
LR	−0.1205	0.0604	0.0005	−0.1189	−0.2436	−0.0070
DE	−0.0925	0.0646	0.0005	−0.0859	−0.2346	0.0086
TAX	0.9644	0.5298	0.0059	0.9542	−0.0559	2.0302
_cons	0.7321	0.2115	0.0013	0.7283	0.3263	1.1557
Acceptance rate	0.2125					
Efficiency: min	0.1603					

3.4 Discussing Research Results

Table 6 shows the probability of impact of the independent variables on DPR. Whereby:

(EPS) This study finds consensus with the research results of Gangil et al. (2018), Asad and Putra (2014), Ahmed and Murtaza (2015) and Truong et al. (2015). That is, earnings per share have a positive impact on dividend payout ratios with the probability that the impact 100%. This is completely consistent based on signal theory, which means that when profits increase, the dividend payout ratio of the company will also increase as a signal of the business efficiency. The analysis results from the model show that in the absence of other factors, when earnings per share increase by 1%, the dividend payout ratio on par value will increase by 0.39% and vice versa.

(GR) Similar to previous research results of Gangil and Nathani (2018), Demirgunes (2015) and Leon and Putra (2014). This study has determined that the revenue growth rate has a negative impact on the dividend payout ratio. It can be affirmed that the impact of this factor on DPR is very clear when the probability of this factor's impact is up to 99.14%. Businesses that have a new business plan or are in a growth phase tend to retain profits for reinvestment. From there, it is possible to save capital costs when borrowing but also reduce the dividend payout ratio for shareholders.

(LR) Different from the studies of Khoirunnisa et al. (2018), Farman Ali Khan and ctg (2017) and Ahmed and ctg (2015) found a positive impact of solvency on the dividend payout ratio, this study gives the opposite result when LR has a negative impact on DPR, which is explained that food businesses in Vietnam tend to reduce dividend payments to maintain liquidity. For businesses, because the specificity of the food industry is the very high demand for working capital, therefore, businesses need to have a large enough cash reserve, which means a reduction in cash dividend payments in order to ensure that business operations remain safe and sustainable.

(DE) Consistent with the studies of Edmund (2018), Khoirunnisa Mohd Nazari et al. (2018) and Asad et al. (2014). This study found a negative relationship with an impact probability of 95.21%, the impact level is very clear, the use of financial leverage has a negative impact on the dividend payout ratio. This is completely consistent with the fact that when businesses have a lot of debt, they tend to prioritize debt obligations before allocating profits. (TAX) Contrary to previous studies by Farman Ali Khan and ctg (2017) and Leon and ctg (2014) did not find statistical

Table 6 An interval test

LnZScore	Mean	Std. Dev	MCSE
Probability (DPR:EPS) > 0	1	0.000	0.0000
Probability (DPR:GR) < 0	0.9914	0.0922	0.00048
Probability (DPR:LR) < 0	0.9818	0.1336	0.00073
Probability (DPR:DE) < 0	0.9521	0.2135	0.00116
Probability (DPR:TAX) > 0	0.9673	0.1779	0.00135

significance. In this study, the authors found that the corporate income tax rate has a positive impact on the dividend payout ratio with a probability of up to 96.73%. The difference is because the sample is different in different countries and the tax policy is not the same, this study is based on the Client Effects Theory, the results show that shareholders tend to prefer cash dividends rather than capital gains. Therefore, when the enterprise pays more income tax, it is because the business operates efficiently, thereby bringing in large profits and tends to pay more dividends to shareholders.

4 Conclusions and Policy Suggestions

4.1 Conclude

The results of data research of 52 food industry enterprises listed on the Vietnamese stock market in the period 2011–2020 show that the average cash dividend payout ratio (calculated on par value) of enterprises reached about 13.69%. Besides, the Bayesian simulation outcomes showed that earnings per share (EPS), corporate income tax rate (TAX) have a positive impact; Ability to Pay (LR), revenue growth rate (GR), financial leverage (DE) negatively affect dividend payout ratio of food industry enterprises.

4.2 Policy Suggestions

The food industry is a serving the people's dietary needs. The capital investment in equipment, production lines, warehouses, manpower, etc. are relatively large. The management and transportation costs are high and dependent many intertwined factors. The time to invest until getting profit is long with many fluctuations and risks, so dividend policy needs to be studied to harmonize the interests of shareholders and the enterprise.

When the long-term investors decide to buy shares of businesses, the first factor they pay attention is the annual dividend payment rate. The plan to distribute profits and dividends are the news that investors look forward to the most. The good, stable dividend policy also shows the potential finance, the effective ability and management ability, especially cash flow management.

The first, basing on the above research results, investors can build a set of indicators including EPS, GR, LR, DE and TAX to chose and buy stocks of food industry businesses. This can be a signal of a business with a stable cash dividend payout ratio and gradual growth over the years. If the factors simultaneously signaled as positive and high EPS, stable GR, stable LR, stable solvency, low DE, low corporate income tax, at this time, investors can consider buying shares of this business and holding for long-term.

The second, the business management board should pay dividends and maintain a stable policy according to the plan. When the enterprise has an increase in the profit expressed through the index (EPS) and the tax expense (TAX) increase, the enterprise should pay the dividend to shareholders. It not only shows the responsibility of the business but also to be a companion to share benefits. When an enterprise has growth opportunities as shown by an increasing index (GR), the enterprise can maintain a low dividend payout ratio to accumulate capital for development plans. When a business prioritizes working capital showing high solvency (LR), the business is reducing its ability to pay dividends. When the enterprise has high financial leverage as shown by the large index (DE), the enterprise should also maintain a low dividend payout ratio to prioritize the performance of obligations to creditors.

The third, the enterprise should maintain a low rate of stable dividend payment and retain a certain standby amount, with this policy, it can be seen that the enterprise has taken precautions when the profit declines not as planned but the dividend payout ratio still remained at the planned level. Or when the business has a profit that exceeds expectations, it will immediately distribute that profit to its shareholders. On the side of investors, there will be a good image when the business maintains its reputation, the market has strong fluctuations, and shareholders are still profitable. On the business side, it will win the trust of shareholders, increase the number of loyal shareholders, and be ready to accompany the business when there are difficulties. This is a very suitable policy in attracting long-term shareholders.

The fourth, setting the target dividend payout ratio, in addition to the policy of paying periodic low-level dividends plus the additional amount if there is a sudden change, the enterprise can still maintain a stable dividend policy equal to how to build a target dividend every years. Basing on the development strategy of the business and the forecast of profit, capital demand for the following years, Businesses should set a long-term goal of dividend payment to income ratio, avoid cutting dividends as much as possible even when the business has good investment opportunities, and businesses should maintain a debt ratio optimized according to the target capital structure.

Fifth, Dividend policy should not stand alone, it must be built on the investment policy and financing policy in a long-term plan. Dividend policy can be according to the project life cycle in the investment stages, there are five stages:

- Beginning stage: low dividend.
- Growth stage: average dividend.
- Development stage: high dividend.
- Saturation stage: good dividend.
- Recession stage: moderate dividend.

For each stage, the dividend policy must be harmonized with the financing policy to ensure the target capital structure. When it comes to a recession stage, businesses are forced to look for new investment policies and new projects. At this time, businesses can maintain a dividend policy at a moderate level to accumulate capital, start implementing new projects, and the end of the old project's lifecycle. Starting a new project at the starting stage with a dividend policy at a low level but higher than that of the previous project's starting stage, in order to show that the business has a ladder of growth stable and sustainable.

References

Ahmed, S., Murtaza, H.: Critical analysis of the factors affecting the dividend payout: evidence from Pakistan. Int. J. Econ. Financ. Manag. Sci. **3**(3), 204–212 (2015)

Ahmed et al.: Liquidity, Profitability and the Dividends Payout Policy. World Review of Business Research Vol. 5. No. 2. April 2015 Issue. pp 73–85 (2015)

Akhlaqul, H., Mubashar, T., Muhammad, S., Muhammad, M.: Tax shield and its impact on corporate dividend policy: evidence from pakistani stock market. Scient. Res. Open Access **5**, 184–188 (2013)

Anh, L.H., Dong, L.S., Kreinovich, V., Thach, N.N. (eds.): ECONVN 2018. SCI, vol. 760. Springer, Cham (2018). https://doi.org/10.1007/978-3-319-73150-6

Asad, M., Yousef, S.: Impact of leverage on dividend payment behavior of Pakistani manufacturing firms. Int. J. Inov. Appl. Stud. **6**(2), 216–221 (2014)

Bhattacharya, S.: Imperfect information, dividend policy and the "Bird-in-Hand" fallacy. Bell J. Econ. **10**, 259–270 (1979)

Black, F., Scholes, M.: The effects of dividend yield and dividend policy on common stock prices and return. J. Financ. Econ. **1**(1), 1–22 (1974)

Demirgunes, K.: Determinants of target dividend payout ratio: a panel autoregressive distributed lag analysis. Int. J. Econ. Financ. Issues **5**(2), 418–426 (2015)

Edmund, N.K.N.: Determinants of dividend policy among banks listed on the Ghana stock exchange. J. Bus. Financ. Affairs **7**(1), 314 (2018)

Farman, A.K., Nawaz, A.: Determinants of dividend payout: an empirical study of pharmaceutical companies of Pakistan stock exchange (PSX). J. Financ. Stud. Res. (2017). https://doi.org/10.5171/2017.538214

Farman, Al.K., Nawaz, A.: Determinants of Dividend Payout: An Empirical Study of Pharmaceutical Companies of Pakistan Stock Exchange (PSX). J. Financ. Stu. Res. (2017) DOI: https://doi.org/10.5171/2017.538214

Gangil, R., Nathani, N.: Determinants of dividend policy: A study of FMCG sector in India. IOSR J. Bus. Manag. (IOSR-JBM) **20**(2), 40–46 (2018)

Gordon, M.J.: Optimal investment and financing Policy. J. Financ. **18**(2), 264–272 (1963)

Hung, N.T., Songsak, S., Thach, N.N.: On quantum probability calculus for modeling economic decisions. In: Kreinovich, V., Sriboonchitta, S. (eds.) Structural Changes and their Econometric Modeling, TES 2019a. SCI. vol. 808, pp. 18–34. Springer, Cham (2019a). https://doi.org/10.1007/978-3-030-04263-9_15

Hung, N.T., Trung, N.D., Thach, N.N.: Beyond traditional probabilistic methods in econometrics. In: In: Kreinovich, V., Thach, N.N., Trung, N.D., Van Thanh, D. (eds.) Beyond Traditional Probabilistic Methods in Economics, ECONVN 2019b. SCI. vol. 809. Springer, Cham (2019b). https://doi.org/10.1007/978-3-030-04200-4_13

Jensen, M.C., Meckling, W.H.: Theory of the firm: Managerial behavior, agency costs and ownership structure. J. Financ. Econ. **3**, 305–360 (1976)

Khoirunnisa, M.N., Salwani, A., Nur, A.M.A., Nabilah, A.S.: The Determinants of dividend policy: evidencefrom trading and services companies in Malaysia. J. Global Bus. Soc. Entrepreneurship (GBSE) **4**(10), 106–113 (2018)

Kreinovich, V., Thach, N.N., Trung, N.D., Van Thanh, D. (eds.): ECONVN 2019. SCI, 'vol. 809. Springer, Cham (2019). https://doi.org/10.1007/978-3-030-04200-4

Leon, F., Putra: The determinant factor of dividend policy at non-finance listed companies. Int. J. Eng. Bus. Enterp. Appl. (IJEBEA) **7**(1), 22–26 (2014)

Leon, F.M., & Putra, P.M.: The determinant factor of dividend policy at non finance listed companies. Int. J. Eng. Bus. Enterp. Appl. **7**(1), 22–26 (2014)

Lintner, J.: Distribution of incomes of corporations among dividends, retained earning and taxes. Am. Econ. Rev. **46**(2), 97–133 (1956)

Narang, M.: Impact of Capital Structure on Financial Performance: A Study of listed firms on National Stock Exchange. Int. J. Adv. Edu. Res. 251–254. (2018)

Miller, M.H., Modiglinani, F.: Dividend policy, growth, and the valuation of shares. J. Bus. **34**(4), 411–433 (1961)

Nguyen, H.T., Nguyen, N.T.: A panorama of applied mathematical problems in economics. Thai J. Math. Spec. Issue Ann. Meet. Math. **17**(1), 1–20 (2018)

Nguyen, H.T., Thach, N.N.: A closer look at the modeling of economics data. In: Kreinovich, V., Thach, N.N., Trung, N.D., Van Thanh, D. (eds.) ECONVN 2019. SCI, vol. 809, pp. 100–112. Springer, Cham (2019). https://doi.org/10.1007/978-3-030-04200-4_7

Truong Dong Loc & Pham Phat Tien: Determinants of dividend policy of listed companies in Ho Chi Minh Stock Exchange. Science J. Can Tho University **38**, 67–74 (2015)

The Financing Decisions—A Case of Firms in Vietnam

Linh D. Nguyen

Abstract The empirical results show that managers exhibit market timing behavior when they make debt and equity issue decisions. Specifically, they have a higher propensity to issue equity when their firm's stocks experience a price increase. Similarly, firms also show market timing behavior in the debt capital market with respect to events of debt issues accompanied by equity repurchases and debt retirements accompanied by equity issues. Meanwhile, securities repurchase decisions are consistent with the adjustment to target capital structure behavior. The adjustment to target ratio behavior, however, appears to play a supplementary role to the market timing behavior in the equity issue decision.

Keywords Capital structure · Market timing · Target leverage

1 Introduction

Trade-off theory is one of the traditional capital structure theories which has been received a lot of attention in corporate finance literature. There are a number of empirical evidence supporting this theory, see, for example, Hovakimian et al. (2001), Jensen and Meckling (1976) and Myers (1977). Being a recent strand of the theories of capital structure, market timing theory also significantly contributes to the explanation of empirically observed capital structure. In the study done by Graham and Harvey (2001), two-thirds of 392 CFOs admitted market timing behavior. In a comprehensive analysis of the financing choices of real estate investment trusts (REITs), Ooi et al. (2010) developed a hybrid hypothesis where REITs time their financing decisions, and simultaneously look for a target leverage ratio in the long run.

This study conducts an extensive investigation to examine how market conditions and firms' existing capital structure impact financing choices each year, then test the

L. D. Nguyen (✉)
Department of Finance, Banking University of Ho Chi Minh City, 36 Tôn Thất Đạm street, District 1, Ho Chi Minh City, Vietnam
e-mail: linhnd@buh.edu.vn

N. Ngoc Thach et al. (eds.), *Financial Econometrics: Bayesian Analysis, Quantum Uncertainty, and Related Topics*, Studies in Systems, Decision and Control 427,
https://doi.org/10.1007/978-3-030-98689-6_38

ability of trade-off theory and market timing theory to explain marginal financing decisions of listed firms on the stock market in Vietnam from 2000 to 2019. Specifically, financing events are firstly divided into nine sub-categories, namely (0) no change in capital structure, (1) pure equity issues, (2) equity repurchases, (3) pure debt issues, (4) debt retirements, (5) dual issues, (6) debt issues accompanied by equity repurchases, (7) equity issues accompanied by deb retirements, and (8) dual repurchases. Next, a multinomial logistic model (MNL) is used with a dependent variable being financing choices and explanatory variables capturing capital market conditions and deviation from target leverage ratio.

The main results suggest that market timing theory provides an adequate explanation for issuing equity. Specifically, managers have a higher propensity to issue equity when their firm's stock experiences a price increase. The aim of this decision exploits the low cost of equity capital relative to other forms of capital. Similarly, firms also show market timing behavior in the debt capital market with respect to events of debt issues accompanied by equity repurchases and debt retirements accompanied by equity issues. Meanwhile, securities repurchase decisions are consistent with the adjustment to target capital structure behavior. This behavior, however, seems to play a supplementary role to the market timing behavior in equity issue decisions.

The remainder of this paper is organized as follows. Section "Literature review" reviews the literature involving firms' marginal financing decisions. Section "Multinomial logistic model" describes the research model and the data description is shown in Sect. 4. Section "Empirical results" discusses the empirical results, and Section "Conclusions" summarises the findings.

2 Literature Review

Trade-off theory is one of the traditional capital structure theories which has been received a lot of attention in corporate finance literature. According to trade-off theory, firms have a target capital structure. The firms make financing decisions based on considering the various benefits and costs of debt financing and adjust their capital structure towards a target (Jensen and Meckling 1976; Stulz 1990). Hovakimian et al. (2001) use a two-stage estimation process to capture the difference between estimated target and actual debt ratios. Their findings propose that firms strive to adjust their current capital structure towards a target as they make financing choices, which support trade-off theory. They also predict that the deviation from the target debt ratio seems to be more important in the equity repurchase and debt retirement decisions than in the issuance decision. Similarly, Kayhan and Titman (2007) perform a two-stage regression to predict the value of the leverage ratio and then examine the effects of financial deficits, market conditions, profitability variables on changes in leverage. Their results support the view that firms have target debt ratios.

Hovakimian et al. (2004) provide evidence supporting the notion that firms have target leverage when they study the behavior of dual debt and equity issuers. Dual issuers passively accumulate earnings and losses to offset the deviation from the target capital structure. Similarly, Leary and Roberts (2005) employ a duration model to study the motivation of financing decisions. They found strong evidence supporting a rebalancing effort of capital structure by firms. They suggest that firms actively readjust their capital structure toward a target ratio or optimal range. Among the papers supporting trade-off theory, Graham and Harvey (2001) provide strong practice-based evidence based on surveying 392 CFOs. They report that 37% of the CFOs have a flexible target, 34% have a somewhat tight target or range, and 10% have a strict target debt ratio.

Nevertheless, market timing behavior still plays an important role in corporate financing decisions. Also, in the study of Graham and Harvey (2001), CFOs said that market condition is one of the most popular factors affecting the decisions to issue common stocks and bonds. More than 60% of CFOs admit that they would issue stock when their firm's stock price has risen and more than 40% of those would issue debt when current interest rates have been low.

Baker and Wurgler (2002) argue that observed capital structure is a cumulative outcome of timing the equity market; managers have market timing behaviors and the attempts to time the market have persistent effects on capital structure. Baker et al. (2003, p. 284) define market timing as "raising finance in whatever form is currently available at the lowest risk adjusted cost" when they test the impact of market timing theory on debt issues. A number of studies provided empirical evidence about practices supporting market timing behavior, such as Jung et al. (1996) and Loughran and Ritter (1995).

In another study, Ooi et al. (2010) analyzed the financing decisions of REITs. They propose that market timing behavior has a more important role than target leverage behavior in the financing decisions of REITs. They develop a hybrid hypothesis where REITs time their financing events and simultaneously seek a target leverage ratio in the long run. Therefore, this paper focuses on analyzing how the effects of market conditions and firms' existing capital structure on financing choices are each year, and then test the ability of trade-off theory and market timing theory to explain marginal financing decisions of listed firms on the Vietnamese stock market.

3 Multinomial Logistic Model

To examine how market conditions and firms' existing capital structure impact financing choices, this research uses the multinomial logistic model (MNL). In statistics, multinomial logistic regression is the linear regression analysis to be conducted when the dependent variable is nominal with more than two levels. This model is used to predict the probabilities of the different possible outcomes of a dependent variable based on multiple independent variables. Several studies conduct MNL regression to examine the financing decisions, i.e., Hovakimian et al. (2001), Hovakimian et al. (2004), Huang and Ritter (2009), and Ooi et al. (2010). Briefly, the MNL model is represented as follows: Let y denotes a random variable taking the values $\{1, 2, \ldots, j\}$, x denotes a set of conditioning variables, and (x_i, y_i) is a random drawn from the population. The primary interest is to explain how changes in the elements of x affect the response probabilities, $P(y = j|x)$, $j = 1, 2, \ldots, n$, all else equal. In other words, the main interest is to specify how x changes the probability the ith firm chooses option j. Since the probabilities must sum to unity, $P(y = 0|x)$ is determined once we know the probabilities for $j = 1, 2, \ldots, n$. Let x be a $1 \times K$ vector, the response probabilities of the MNL model:

$$P(y = j|x) = \frac{\exp(x\beta_j)}{\left[1 + \sum_{h=1}^{n} exp(x\beta_h)\right]}, j = 1, 2, \ldots, n$$

where β_j is $K \times 1, j = 1, \ldots$, n. The MNL model expresses the probability that a specific alternative is chosen is the exponent of the utility of the chosen alternative divided by the exponent of the sum of all alternatives.

This study categorizes financing events into nine groups. Firms that did not experience any changes in their capital structure during the review period are taken as the base option (0), while firms that made changes in capital structure are classified as follows: (1) pure equity issues, (2) pure equity repurchases, (3) debt issues, (4) debt retirements, (5) dual issues, (6) debt issues accompanied by equity repurchases, (7) equity issues accompanied by debt retirements, and (8) dual repurchases. Therefore, MNL identifies how independent variables change the probability the firms choose financing event *i*, which *i* takes the values of {1, 2, …, 8}, relative to the base option.

4 Data and Descriptive Statistics

4.1 Data

Data necessary to empirically employ these models comes from DataStream. Sample construction starts by determining all firms listed in the Hanoi Stock Exchange and the Ho Chi Minh Stock Exchange at any point between 2000 and 2019. Firms in the financial sector with a Fama and French industry code 11 (SIC codes 6000–6999) are not included in the sample because their capital structure is likely to be significantly different from other firms in the sample.

The dependent variable is *Decision*, which captures nine financing events as discussed in Sect. 3. A firm is defined as issuing equity when net equity issue divided by the total assets (e/A) exceeded 5% or as repurchasing equity when e/A is lower –5%. Similarly, a firm is defined as issuing debt when net debt issue divided by the total assets (d/A) exceeded 5% or as retiring debt when d/A is lower –5%. Table 1 shows summary information on nine financing decisions.

To capture market timing behavior in the equity market, the paper includes two variables in MNL model, namely *Stock performance* and *Market performance*. *Stock performance* is a measure of a firm' stock price change over one year. The change of

Table 1 Financing decisions

Financing events	Count	Percent (%)
(0) No change	511	22.0
(1) Pure equity issues	166	7.1
(2) Equity repurchases	99	4.3
(3) Pure debt issues	664	28.5
(4) Debt retirement	407	17.5
(5) Dual issues	232	10.0
(6) Debt issues, equity repurchases	88	3.8
(7) Debt retirement, equity issues	90	3.9
(8) Dual repurchases	69	3.0
Total	**2,326**	**100.0**

VN-Index over one year is used as *Market performance.* To capture market timing behavior in the debt market, *10-year bond* and *Risk premium* are used. *Risk premium* is measured as the interest rate of 10-year government bond minus that of 1-year government bond, and the interest rate on the 10-year bond of Vietnam Government is employed to represent *10-year bond.* Ooi et al. (2010) found evidence supporting the idea that larger-sized firms are more active in the capital structure. Hence, firm size (*Size*) is included in the MNL model.

The last independent variable is *Deviation,* which is used to investigate how deviation from the target capital structure affects financing decisions. To estimate *Deviation,* leverage ratio is regressed on a vector of independent variables. Following Hovakimian et al. (2001), Kayhan and Titman (2007) and Ooi et al. (2010), explanatory variables are market-to-book ratio, profitability and size of the individual firms. In addition, to account for industry effects, firms are grouped according to the 12 Fama and French industries (excluding financial firms) and cross-sectional regressions are performed for each industry. With estimated coefficients taken from these regressions, target capital structure for individual firm-years observations is predicted, and then the discrepancy between the firm's actual leverage and its predicted target leverage is used as a measure of deviation from the target leverage. After dropping observations with inadequate variables, the final sample consists of 2.326 firm years covering the 2000–2019 period.

4.2 *Descriptive Statistics*

Table 2 reports the mean values of explanatory variables according to nine types of financing activities, and the full sample in the last column. In terms of two market timing variables in the equity market, firms show market timing behavior in groups where issuing equity, specifically group of pure equity issues and dual issues. On average, they issued equity in the years of high stock price performance. The mean

Table 2 Descriptive statistics

Predictor variables	No changes	Pure equity issues	Equity repurchases	Pure debt issues	Net debt retirement	Dual issues	Debt issues and equity repurchases	Equity issues and debt retirement	Dual repurchase	Full sample
Stock performance	−2.4	26.6	−20.0	−2.5	−8.8	30.9	−23.2	8.5	4.3	4.2
Market performance	2.6	10.2	2.8	2.3	2.5	16.0	1.8	13.6	11.5	5.6
10-year bond	−3.1	1.9	−11.7	−0.5	3.1	−4.6	−11.9	0.5	−11.7	−2.1
Risk premium	−1.44	−1.57	−1.53	−1.46	1.40	−1.67	−1.72	−1.56	−1.89	−1.50
Size	19.9	20.1	19.7	20.2	19.7	20.5	20.3	19.7	19.4	20.0
Deviation	−5.7	−15.3	−10.0	7.2	0.8	−1.5	1.8	−9.2	−1.8	−1.1
Observations	511	166	99	664	407	232	88	90	69	2,326

This table reports the mean values of explanatory variables by nine types of financing activities and a full sample. *Stock performance* is defined as the appreciation of individual firm's stock price appreciation over one year. *Market performance* is defined as the appreciation of VN-Index over one year. *10-year bond* is defined as the appreciation of interest rate on the 10-year bond of Vietnam Government over one year. *Risk premium* is defined as the appreciation of difference between interest rate of 10-year and 1-year government bond over one year. *Size* is defined as the natural logarithm of total assets. *Deviation* is defined as the actual book leverage minus the target book leverage values of their stock appreciation are 26.6% and 30.9%, respectively, which are higher than other groups. Similarly, these two groups also undergo a bullish period of the stock market when they issue equity.

Similar to equity market timing behavior, firms also show market timing behavior in the debt capital market with respect to their debt issuances. Firms issue pure debt or simultaneously issue debt and equity repurchases when the costs of debt issuances are relatively lower. Specifically, firms issue debt when the interest rate or risk premium is lower than those in the previous year.

Inconsistent with the predictions of target capital structure theory, the firms with pure equity issues, pure debt issues, debt issues accompanied by equity repurchases, or equity issues accompanied by debt retirement move them away from their target capital structure. For example, underleveraged firms issue more equity and over-leveraged firms issue more debt. However, the practice of debt retirement, equity repurchases or dual repurchases is consistent with target capital structure theory.

5 Empirical Results

Two analyses are conducted to investigate how market timing behavior and target ratio behavior change the probability a firm chooses a financing decision. They are multivariate regression analysis and Bayesian multinomial logistic analysis. The consistency of the outcomes of the two different approaches should quantify the robustness of the main result.

5.1 Multivariate Regression Analysis

In this section, the ability of market timing theory and trade-off theory to explain firms' financing decisions are tested. The study regresses a MNL that conducts each of the eight issue and repurchase decisions against a no change transaction. The estimated coefficients from this MNL provide the comparisons about the probability of (1) equity issues, (2) equity repurchases, (3) debt issues, (4) debt retirements, (5) dual issues, (6) debt issues accompanied by equity repurchases, (7) equity issues accompanied by debt retirements, and (8) dual repurchases, relative to the probability of no change event in that year. It should be noted that a significantly positive coefficient in MNL would imply that a higher value of the explanatory variable increases the likelihood of each of financing decisions against a no change transaction, and vice versa. Table 3 reports the results of the MNL regression.

Table 3 provides strong evidence supporting the equity market timing behavior. Firms that issue new equity capital (pure equity issues, dual issues or equity issues

Table 3 Multinomial Logistic Regression comparing firms that issue and repurchase securities to those that do not

	Pure equity issues	Equity repurchases	Pure debt issues	Net debt retirement	Dual issues	Debt Issue and equity repurchases	Equity issues and debt retirement	Dual repurchase
Stock performance	0.007***	0.003	0.002	−0.004	0.007***	0.003	0.005**	−0.003
Market performance	−0.002	−0.001	−0.002	0.003	−0.001	−0.027*	−0.004	−0.007
10-year bond	0.043*	−0.109***	−0.023***	−0.003	0.066	−0.093***	0.030***	−0.113
Risk premium	0.000	0.002**	0.000	0.000	0.001	0.002*	0.000	0.000
Size	0.127**	−0.122	0.168***	−0.064	0.349***	0.131	−0.073	−0.312*
Deviation	−0.024	−0.009**	−0.035	0.016***	0.011**	−0.020**	−0.009	0.011

This table reports results of the MNL regression with the base financing choice to be "No change (0)". The definition of the variables is shown in Table 2

* $p < 0.10$, ** $p < 0.05$, *** $p < 0.01$

accompanied by debt retirements) have significantly positive coefficients of stock performance. This result implies that firms experiencing a high appreciation of stock price would have a higher propensity to increase the likelihood of an equity issue against the no transaction alternative. The purpose of this decision could be to exploit the low cost of equity capital relative to other forms of capital. While stock price performance plays an important role in equity issue decisions, the other proxy for market conditions, market performance, does not significantly influence financing decisions.

The empirical results also provide evidence supporting the debt market timing behavior predicting that an increase in interest rate would reduce events of debt issues because the cost of debt financing is relatively high. The higher value of interest rate (*10-year bond*) decreases the probability of a debt issue against the no transaction event.

Consistent with the adjustment to target capital structure behavior, deviation from target capital structure has an important role in security repurchase decisions. Specifically, firms repurchase equity to increase their leverage when they are underleveraged, or overleveraged firms have a higher tendency to reduce their leverage by retiring only debt. The effect of these practices would move their capital structure toward a target leverage or a target range.

Lastly, the estimated coefficients of *Size* are significantly positive in most groups. This finding implies that larger firms have a higher tendency to be more active in the capital market than smaller firms. The result is consistent with the findings of Ooi et al. (2010).

In summary, the regression results of MNL model confirm that firms exhibit market timing behaviors in both equity and debt capital markets. The adjustment to target capital structure behavior plays an important role in securities repurchase decisions. However, its role appears to be less important than that of the market timing behavior.

5.2 *Bayesian Multinomial Logistic Regression*

To examine the ability of this model to provide unbiased parameter estimates, this section performs the Bayesian multinomial logistic regression that specifies 1,000 Markov chain Monte Carlo (MCMC) samples. The findings are displayed in Table 4.

For each parameter, the posterior mean and associated 95% credible interval in square brackets are presented. The model successfully estimates the variance parameters with mean estimated values all within 5% of the true values used to simulate the datasets. In addition, the estimated means from 1,000 simulations are similar to the findings in Table 3.

Table 4 Bayesian multinomial logistic regression

	Pure equity issues	Equity repurchases	Pure debt issues	Net debt retirement	Dual issues	Debt Issue and equity repurchases	Equity issues and debt retirement	Dual repurchase
Stock performance	0.006	0.003	0.003	–0.002	0.007	0.003	0.005	–0.001
	[0.003; 0.009]	[–0.001; 0.008]	[0.000; 0.006]	[–0.006; 0.001]	[0.003; 0.011]	[–0.003; 0.007]	[0; 0.01]	[–0.007; 0.005]
Market performance	–0.002	–0.006	–0.001	0.003	0.000	0.027	–0.005	–0.013
	[–0.008; 0.003]	[–0.033; 0.013]	[–0.006; 0.003]	[–0.001; 0.008]	[–0.005; 0.004]	[0.009; 0.047]	[–0.012; 0.002]	[–0.038; 0.009]
10-year bond	0.041	–0.114	0.026	–0.002	0.065	–0.100	0.027	–0.128
	[0.028; 0.057]	[–0.165; –0.079]	[0.018; 0.036]	[–0.013; 0.013]	[0.051; 0.078]	[–0.136; –0.07]	[0.008; 0.047]	[–0.191; –0.082]
Risk premium	0.000	0.002	0.000	0.001	0.001	0.002	–0.001	0.001
	[–0.002; 0.001]	[0.000; 0.003]	[–0.001; 0.001]	[0.000; 0.001]	[0.000; 0.002]	[0.000; 0.003]	[–0.003; 0.001]	[–0.001; 0.002]
Size	0.129	–0.125	0.170	–0.065	0.333	0.127	–0.055	–0.369
	[0.039; 0.212]	[–0.237; –0.011]	[0.163; 0.178]	[–0.077; –0.048]	[0.318; 0.352]	[0.052; 0.212]	[–0.072; –0.036]	[–0.476; –0.308]
Deviation	–0.029	–0.012	0.025	0.014	0.002	0.011	–0.015	0.010
	[–0.037; –0.02]	[–0.019; –0.002]	[0.019; 0.032]	[0.009; 0.021]	[–0.004; 0.010]	[–0.001; 0.022]	[–0.025; –0.004]	[–0.002; 0.02]

This table shows results of the Bayesian multinomial logistic regression which include mean and the 95% credible intervals in square brackets. The definition of the variables is shown in Table 2

6 Conclusion

The existing empirical evidence on target leverage and market timing behavior is mixed. Therefore, to comprehensively examine how market timing behavior and the adjustment to target capital structure behavior affect marginal financing decisions, this study categorizes financing events into nine subsamples, namely (0) no changes in capital structure, (1) equity issues, (2) equity repurchases, (3) debt issues, (4) debt retirements, (5) dual issues, (6) debt issues accompanied by equity repurchases, (7) equity issues accompanied by debt retirements, and (8) dual repurchases.

A MNL model is conducted with the dependent variable being financing choices. The option (0) no changes in capital structure is chosen as a basic option. The regression results of MNL model confirm that firms exhibit market timing behaviors in both debt and equity capital markets. The adjustment to target capital structure behavior plays an important role in securities repurchase decisions. In general, its role appears to be less important than that of market timing behavior.

References

Baker, M., Greenwood, R.M., Wurgler, J.: The maturity of debt issues and predictable variation in bond returns. J. Financ. Econ. **70**(2), 261–291 (2003)

Baker, M., Wurgler, J.: Market timing and capital structure. J. Financ. **57**(1), 1–32 (2002)

Graham, J.R., Harvey, C.R.: The theory and practice of corporate finance: evidence from the field. J. Financ. Econ. **60**(2–3), 187–243 (2001)

Hovakimian, A., Hovakimian, G., Tehranian, H.: Determinants of target capital structure: the case of dual debt and equity issues. J. Financ. Econ. **71**(3), 517–540 (2004)

Hovakimian, A., Opler, T., Titman, S.: The debt-equity choice. J. Financ Quant. Anal. **36**(1), 1–24 (2001)

Huang, R., Ritter, J.R.: Testing theories of capital structure and estimating the speed of adjustment. J. Financ. Quant. Anal. **44**(2), 237–271 (2009)

Jensen, M.C., Meckling, W.H.: Theory of the firm: Managerial behavior, agency costs and ownership structure. J. Financ. Econ. **3**(4), 305–360 (1976)

Jung, K., Kim, Y.-C., René, S.: Timing, investment opportunities, managerial discretion, and the security issue decision. J. Financ. Econ. **42**(2), 159–185 (1996)

Kayhan, A., Titman, S.: Firms' histories and their capital structure. J. Financ. Econ. **83**(1), 1–32 (2007)

Leary, M.T., Roberts, M.C.: Do firms rebalance their capital structures? J. Financ. **60**(6), 2575–2619 (2005)

Loughran, T., Ritter, J.R.: The new issues puzzle. J. Financ. **50**(1), 23–51 (1995)

Myers, S.C.: Determinants of corporate borrowing. J. Financ. Econ. **5**(2), 147–175 (1977)

Ooi, J.T.L., Ong, Seow-Eng, Lin, L.: An analysis of the financing decisions of reits: the role of market timing and target leverage. J Real Estate Financ. Econ. **40**(2), 130–160 (2010)

Stulz, R.: Managerial discretion and optimal financing policies. J. Financ. Econ. **26**(1), 3–27 (1990)

A Cointegration Analysis of Vietnamese Bond Yields

Nguyen Thanh Ha and Bui Huy Tung

Abstract The key objective in the current paper is to analyze the cointegration relationship in the Vietnam bond market. Employing two different methods regarding linear cointegration test on Error Correction Model and nonlinear cointegration framework relied on the smooth transition regression error correction model, the study finds that the one-year and two-year yields or one-year and three-year yields are empirically proved to be linear cointegrated while the one-year and five-year yields have the nonlinear cointegration relationship and the correction error follows smooth transition regression dynamics. From then the crucial implications will be drawn for the investors and monetary policymakers.

1 Introduction

The relationship among the yields on default-free securities that differ only in their term to maturity is called the term structure of interest rates (TSIR). The investors are interested in the TSIR because it contains information about the rate of returns. The information of the TSIR is also crucial for policymakers in assessing the impacts of macroeconomic policies on the economy. For those reasons, the TSIR is of interest to both market participants and policy authorities.

Some theories try to explain the TSIR such as the Market Segmentation Theory, the Preferred Habitat Theory, The Liquidity Preference Theory, the Expectations Hypothesis While the Market Segmentation Theory holds that the long and short term interest rates are not related to each other, the others such as the Expectations Hypothesis posit that the short-term and long-term interest rates have an equilibrium relationship in the long run, meaning that they are cointegrated.

N. T. Ha (✉) · B. H. Tung
Banking University of Ho Chi Minh City, Ho Chi Minh City, Vietnam
e-mail: hant_tkt@buh.edu.vn

B. H. Tung
e-mail: tungbh@buh.edu.vn

N. Ngoc Thach et al. (eds.), *Financial Econometrics: Bayesian Analysis, Quantum Uncertainty, and Related Topics*, Studies in Systems, Decision and Control 427,
https://doi.org/10.1007/978-3-030-98689-6_39

Most studies such as Campbell and Shiller (1987), Hall et al. (1992), Cuthbertson (1996), Domlnguez and Novales (2000), Cooray (2003), Michelis and Koukouritakis (2007), Beechey et al. (2009), among many others assume a linear cointegration relationship. However, some recent studies such as Clements and GalVao (2003), Clarida et al (2006), Mili et al. (2012), Arac and Yalta (2015), Guidolin and Thornton (2018) applied the nonlinear method and found evidence of nonlinear cointegration relationship between interest rates.

In the Vietnam bond market, it is observed that the long-term bond yields with different maturities tended to fluctuate in the same trend during the sample period. Therefore, we suspect whether the bond yields have a cointegration relationship and which model is suitable to measure the relationship between the bond yields with different maturities. Such questions have been researched in many previous studies but have never been studied in the Vietnam bond market. This paper attempts to fill this gap in the literature.

The critical objective in this study is to analyze the TSIR using a cointegration approach. Unlike the aforementioned empirical studies, we employ both linear and nonlinear cointegration to find an appropriate model measuring the relationship between Vietnamese bond yields.

The study of the cointegration relationship between interest rates through error correction models helps analyze the short-run and long-run relationship between interest rates. This problem is essential not only for investors but also for policymakers.

The rest of this paper is constructed as follows. Section 2 explains the cointegration test with Nonlinear Smooth Transtion Error Correction Model introduced by Kapetanios et al. (2006), Arac and Yalta (2015). The method and data are described in Sect. 3, while the main results are presented and discussed in Sect. 4. Finally, the conclusion is drawn in Sect. 5.

2 Testing for Linear and Nonlinear Cointegration

If the series are non-stationary and the first order differences are stationary then the series are integrated of order 1, denoted by I(1). If y_t and x_t are I(1) series but a linear combination of this collection is stationary, then the series are said to be co-integrated. To test the cointegration relationship between two series y_t and x_t, Engle and Granger (1987) proposed a procedure that the residual obtained from regression is tested using Augmented Dickey Fuller (ADF) unit root test. They also show that if the variables have a cointegration relationship, an error correction mechanism (ECM) exists to respond to short run imbalance and ensure equilibrium in the long run. Conversely, if the variables are I(1) and there is an ECM, they have a cointegration relationship. Therefore, one method to test the linear cointegration relationship between interest rates is to estimate ECM. The linear approach has been highlighted by a series of papers, including, Cuthbertson (1996), Cooray (2003).

However, some authors indicate that the conventional linear cointegration test may lead to erroneous results. Recent studies have applied nonlinear method to test the EH, such as Clements and GalVao (2003), Clarida et al (2006), Mili et al. (2012), Arac and Yalta (2015), Guidolin and Thornton (2018).

Based on Engle and Granger (1987) with ECM, Kapetanios et al. (2006) propose a new testing procedure to detect the presence of a cointegration relationship that the error correction follows a stationary smooth transition process, called the Conditional exponential Smooth Transtion Regression Error Correction Model (STR ECM). We then remind this procedure as follows.

Assume y_t and x_t are I(1) processes. The general nonlinear vector error correction model has the form:

$$\Delta y_t = \phi u_{t-1} + \gamma u_{t-1} g(u_{t-1}) + \sum_{i=1}^{p} \phi_i \Delta y_{t-i} + \sum_{j=0}^{q} \psi_j \Delta x_{t-j} + v_t, \quad (1)$$

where u_t is the residual when y_t is regressed on x_t, v_t is a white noise.

Consider g(.) as the exponential smooth transition function:

$$g(u_{t-1}) = 1 - e^{-\theta (u_{t-1} - c)^2}, \quad (2)$$

where it is assumed that $\theta \geqslant 0$ for identification purposes and c is a transition parameter.

Equations (1) and (2) give the STR ECM:

$$\Delta y_t = \phi u_{t-1} + \gamma u_{t-1} \left(1 - e^{-\theta (u_{t-1} - c)^2}\right) + \sum_{i=1}^{p} \phi_i \Delta y_{t-i} + \sum_{j=0}^{q} \psi_j \Delta x_{t-j} + v_t, \quad (3)$$

If $\theta > 0$ then y_t and x_t have a nonlinear cointegration relationship. Therefore, to check whether the nonlinear cointegration relationship is consistent such that the error follows STR ECM, it is necessary to test the null hypothesis of no cointegration as $H_0 : \theta = 0$ against the alternative of nonlinear STR ECM cointegration of $H_1 : \theta > 0$. However, it is difficult to test H_0 because θ is not identified.

Performing Taylor expansion with the series $1 - e^{-\theta (u_{t-1} - c)^2}$, the model (3) is approximated to a new regression model as follows:

$$\Delta y_t = \delta_1 u_{t-1} + \delta_2 u_{t-1}^2 + \delta_3 u_{t-1}^3 + \delta x_t + \sum_{i=1}^{p} \phi_i \Delta y_{t-i} + \sum_{j=1}^{q} \psi_j \Delta x_{t-j} + v_t, \quad (4)$$

Kapetanios et al. (2006) propose a t-type statistic for $\delta = 0$ (no cointegration) against $\delta < 0$ (nonlinear cointegration). They use a two-step regression procedure to test the nonlinear cointegration between y_t and x_t. In the first step, y_t is regressed on x_t and we obtain the residuals e_t. In the second step, the Eq. (4) is estimated and the t-type test is employed.

We can extend to consider the regression model with trending time series.

Based on the procedure proposed by Kapetanios et al. (2006), Arac and Yalta (2015) consider the particular case when c = 0 in the model (3). In addition, to study the cointegration relationship between the interest rates with k periods to maturity (R_t^k) and the interest rate with one period (R_t^1) applying STR ECM, the authors reduce the model (4) to the form:

$$\Delta R_t^k = \delta e_{t-1}^3 + \omega \Delta R_t^1 + \sum_{i=1}^{p} \phi_i \Delta R_{t-i}^k + \sum_{j=1}^{q} \psi_j \Delta R_{t-j}^1 + v_t, \tag{5}$$

where e_t is the residual obtained from the regression R_t^k on R_t^1, Δ denotes the difference of the lagged series of R_t^k and R_t^1, v_t is a white noise.

To test $\delta = 0$ against $\delta < 0$, the t-statistics is compared to the critical value calculated based on Monte Carlo simulation. This critical value depends on whether the time series are trending or non-trending. Specifically, with two trending time series, the critical values at the significance levels of 10%, 5%, 1% calculated by Kapetanios et al. (2006) are: −3.3; −3.59; −4.17, respectively.

3 Methodology and Data

Methodology

It isn't easy to know that the relationship between interest rates with different maturities in the Vietnam bond market is linear or nonlinear. Therefore, in this study, we examine the cointegration relationship between R_t^i and R_t^j, i, j = 1, 2, 3, 5 in both linear and nonlinear approaches.

First of all, the necessary condition for a cointegration relationship is that the series must be I(1). Therefore, in the first step, the stationarity of the time series is tested by the ADF unit root test. In the second step, we regress on and obtain the residual. We pretest whether is stationary using the standard ADF test without trend is employed in STR ECM framework:

$$\Delta R_t^i = \gamma e_{t-1}^3 + \sum_{m=1}^{p} \phi_m \Delta R_{t-m}^i + \sum_{n=0}^{q} \psi_n \Delta R_{t-n}^j + v_t, \tag{6}$$

or ECM framework:

$$\Delta R_t^i = \gamma e_{t-1} + \sum_{m=1}^{p} \phi_m \Delta R_{t-m}^i + \sum_{n=0}^{q} \psi_n \Delta R_{t-n}^j + v_t, \tag{7}$$

The order of autoregressive terms included in the relevant models is selected by the coefficients' significance and the Akaike Information Criterion or the Schwarz Criterion. To examine the presence of autocorrelation, the Breusch—Godfrey test (BG test) is applied and to examine the issue of heteroscedasticity, Engle (1982) LM test for first-order autoregressive conditional heteroscedasticity (ARCH(1)) is used. The cointegration relationship between R_t^i and R_t^j is tested through examining the null hypothesis $H_0 : \gamma = 0$ of no cointegration against $H_1 : \gamma < 0$.

Instead of using p-values for significance test, Bayesian alternatives to null hypothesis significance test are employed. The study uses Bayes factor (BF) to transform a p-value-based explanation into a more reliable Bayesian one. BF compares the likelihood of the data y under the null hypothesis H_0 to the likelihood under the alternative hypothesis H_1. The value of the BF reflects the strength of evidence against H_0 (See Held and Ott (2016)).

Data

Our data set covers 1-year, 2-year, 3-year, 5-year Vietnam government bond yields collected from Bloomberg. The starting date is August 6, 2009, to December 31, 2019, when the data is available. We use a weekly average to avoid some missing daily values. The total number of observations in the study is 552. All the variables are expressed in percentage.

Let R_t^k be the k-year government bond yields at time t (k = 1, 2, 3, 5).

4 Empirical Results

Data description

Figure 1 shows that the bond yields with different maturities tend to fluctuate with each other. Therefore, we speculate that the bond markets with different maturities are not independent and separate but interconnected. The variation of the bond yields is quite considerable, ranging from 1.4% to 13.9%. They reached the highest value in 2011 when Vietnam's economy faced many uncertainties after the global economic crisis in 2009. The inflation rate increased sharply to 18.58% due to the consequences of previously ineffective and outrageous stimulus packages. The high budget deficit and debt repayment obligation also caused government bond issuance to skyrocket during this period. Since 2013, thanks to the gradual lowering of the inflation rate and better control of money supply growth, the bond market has remained stable, investors' faith has been maintained and interest rates have tended to decrease. By the time of the study, the yields on long-term government bonds are only about 1.5% (Table 1).

In general, the average of the yields on long-term bonds are higher than the shorter-term ones. As stated another way, the mean of bond yields increase with the term. As a result, the yield curve is upward sloped.

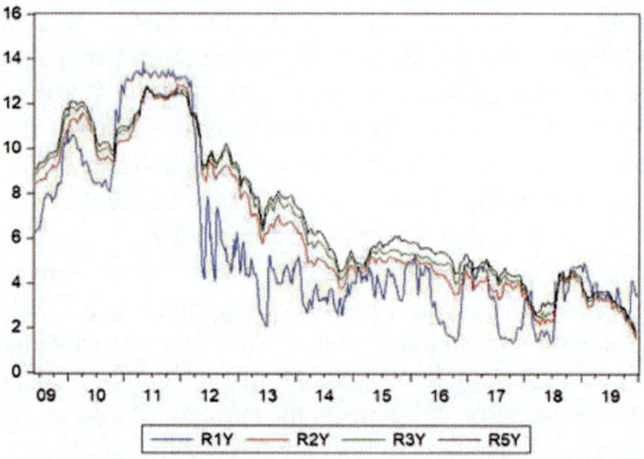

Fig. 1 Bond yields with different maturities. *Source(s)* Bloomberg. *Note(s)* R1Y, R2Y, R3Y, R5Y are the 1-year, 2-year, 3-year, and 5-year bond yields, respectively

Table 1 Data description

Yield	Mean	Standard deviation
One-year	5.6734	3.5591
Two-year	6.5221	3.1984
Three-year	6.8780	3.1609
Five-year	7.1172	3.1078

Source(s) Bloomberg

Cointegration tests

The ADF results indicate that for the series in levels, there is a weak evidence on rejecting the null hypothesis of a unit root while there is a strong evidence on rejecting the null hypothesis of a unit root for the first differenced series. This result suggests that all series are I(1). Since the series display a trending behavior, we regress R_t^i on R_t^j with an intercept and a linear time trend. When testing the unit root tests of the residuals e_t, it appears that i = 1 and j = 2, 3 and 5, e_t are stationary and used to estimate Eqs. (6) and (7) (Table 2).

The BFs are from 1/30 to 1/10. This result provides substantial or strong evidence against the null hypothesis of the non-significant coefficients.

When testing $H_0 : \gamma = 0$ against $H_1 : \gamma < 0$ in model (6) with 1-year yield and 5-year yield data, we find that t-statistics of the adjustment coefficient is smaller than the critical value at the significance level of 5% (−3.59). Bayesian t-test also shows the strong evidence against the null hypothesis. This implies that we find enough evidence to reject H0 at the significance level of 5%. With two pairs of interest rates with maturities one year and two years or one year and three years, it is found that t-statistics of the adjustment coefficient is larger than the critical value at conventional significance levels, meaning that we cannot reject H0. These results

Table 2 Estimate STR ECM $\Delta R_t^1 = \gamma e_{t-1}^3 + \phi \Delta R_t^i + \sum_{n=1}^{q} \psi_n \Delta R_{t-n}^1 + v_t$

	i = 2		i = 3		i = 5	
	Coefficient	BF	Coefficient	BF	Coefficient	BF
e_{t-1}^3	−0.0045	0.0113	−0.0045	0.0425	−0.005	0.0238
	(0.0017)		(0.0015)		(0.0014)	
ΔR_t^i	0.3107	0.0524	0.4385	0.0179	0.4714	0.0789
	(0.1298)		(0.1308)		(0.1322)	
ΔR_{t-1}^1	0.2685	0.0474	0.2696	0.0519	0.2726	0.0529
	(0.0466)		(0.045)		(0.0442)	
ΔR_{t-16}^1	0.0965	0.0185	0.0959	0.0825	0.0989	0.0562
	(0.0422)		(0.0414)		(0.0416)	
ΔR_{t-20}^1	−0,0870**	0.0256	−0,081*	0.0123		
	(0,0422)		(0,041)			
ΔR_{t-28}^1	0.073	0.0245	0.0749	0.0397		
	(0.0421)		(0.0415)			
ΔR_{t-30}^1			−0.109	0.0925		
			(0.0414)			

Note(s) Standard errors of coefficients are in parentheses

imply the presence of the nonlinear cointegration between the one-year yield and five-year yield. However, the same inclusion does not hold for the pairs of one-year yield and two-year yield or one-year yield and three-year yield.

The fact that the interest rates with maturities of one year and five years have a nonlinear cointegration relationship in STR ECM framework indicates that these interest rates tend to move together in the long run. In the relation between one-year yield and five-year yield, five-year yield plays a role as a leading one. Once one-year yield is separated from the common trend, the imbalance adjusts five-year yield in the previous period. The smooth transition regression model allows correction process to take place continuously, smoothly for each different value of the exponential transition function in the interval (0,1) and convert two mechanisms by two extreme values 0 and 1.

The coefficient of the variable ΔR_t^5 is statistically significant and positive, showing that in the short run, the change of the five-year yield has a positive effect on the change of the one-year yield. If the change of the five-year yield increases 1% (absolute), the change of the one-year yield will increase 0.471%.

We continue to consider the linear cointegration relationship between the one-year and two-year yields or the one-year and three-year yields.

When regressing R_t^1 on a trending time and R_t^2 or R_t^3, ADF test statistic of the residual is less than the critical value at 5% significance level (-3.5). Therefore, according to the Engle-Granger causality test, there exists a linear cointegration relationship between these pairs of interest rates. Thus, there exists a mechanism to correct errors to ensure their balance in the long run, that is, ECM exists (Table 3).

Table 3 Estimate ECM $\Delta R_t^1 = \gamma e_{t-1} + \phi \Delta R_t^i + \sum_{n=1}^{q} \psi_n \Delta R_{t-n}^1 + v_t$

	i = 2		i = 3	
	Coefficient	BF	Coefficient	BF
e_{t-1}	−0.0318	0.0124	−0.0244	0.0231
	(0.0135)		(0.0123)	
ΔR_t^i	0.2649	0.0235	0.4513	0.0316
	(0.1256)		(0.1299)	
ΔR_{t-1}^1	0.2549	0.0285	0.2558	0.0341
	(0.0456)		(0.045)	
ΔR_{t-2}^1			−0.1052	0.0442
			(0.0443)	
ΔR_{t-16}^1	0.1026	0.0275	0.0935	0.0527
	(0.042)		(0.0419)	
ΔR_{t-5}^i	−0.2561	0.0613		
	(0.1151)			
ΔR_{t-7}^i	0.3163	0.0394		
	(0.115)			

Note(s) Standard errors of coefficients are in parentheses

The linear cointegration test $H_0 : \gamma = 0, H_1 : \gamma < 0$ in model (7) is also produced. Bayesian t test shows the strong evidence against the null hypothesis. Besides, we find that t-statistics of the adjustment coefficient is smaller than the critical value at the conventional significance level. Thus, the ECM model is suitable to represent the linear cointegration relationship between one-year yield and two-year yield or one-year yield and three-year yield. This reflects that they have a long run relationship and move together in a common trend. The two-year yield and three-year yield become the leading interest rate. However, the adjustment coefficients in two cases are quite small, showing that the adjustment speed is slow. The results also indicate that the changes in two-year yield and three-year yield affect positively to the changes in 1-year yield.

Discussion and policy recommendations

The empirical results suggest that the one-year yield and five-year yield have a nonlinear cointegration relationship. Besides, the STR ECM is suitable to represent the short-run and long-run relationship between these series. Meanwhile, the cointegration relationship between the one-year yield and the two-year yield or the three-year yield are linear and a more appropriate model to reflect the short-run and long-run relation between the interest rates may be the ECM.

A general result from the ECM and STR ECM estimation is that the longer-term yields play a leading role to keep long run equilibrium relationship between the one-year yield and the longer term yield. This result is not similar to that found in Arac and Yalta (2015) who found that the one-year interest rate plays a leading role.

The relation between the interest rates in the Vietnam bond market indicates that the Market Segmentation Theory is rejected.

The relationship between the one-year yield and two-year or three-year yields is empirically found to be consistent with the linear framework while the appropriate model to measure the term structure of the one-year yield and five-year yield is nonlinear. This fact leads us to doubt that the nonlinear cointegration could happen when the gap between the maturities is large. This result is found in Bachmeier (2002).

The empirical results have important implications for investors and for policymakers in setting monetary policy. First, long term bond yields with different maturities move together in a common trend. When the policymakers directly regulate one-year interest rate, the long-term interest rates will also move in the same direction and affect aggregate demand. This is the monetary policy transmission mechanism to the production activities of the macroeconomy. A change in short-term interest rates could affect long-term interest rates and pass on to the real activity. Second, the relations between the one-year interest rate and longer-term interest rates suggest that the investors could observe the one-year interest rate to understand and predict the future long term interest rate.

5 Conclusion

Our study contributes to the literature evidence on the cointegration relationship between Vietnam government bond yields with maturities of one year to five years in two frameworks: linear and nonlinear cointegration. Empirical evidence shows a nonlinear cointegration relationship in the framework of the STR ECM between the one-year and five-year yields. In contrast, the cointegration relationship between the one-year and two-year yields or one-year and three-year yields are still linear, consistent with the ECM model. These results also suggest that the one-year interest rate could be used to reference the future long term interest rates. Besides, the cointegration between the short term and long term interest rates reflects the monetary policy mechanism channel to the real economy.

References

Arac, A., Yalta, A.Y.: Testing the expectations hypothesis for the Eurozone: a nonlinear cointegration analysis. Financ. Res. Lett., 41–48 (2015)

Bachmeier, L.: Is the term structure nonlinear? A semiparametric investigation. Appl. Econ. Lett. **9**(3), 151–153 (2002)

Beechey, M., Hjalmarsson, E., Österholm, P.: Testing the expectations hypothesis when interest rates are near integrated. J. Banking Financ. **33**(5), 934–943 (2009)

Bekaert, G., Hodrick, R.J., Marshall, D.A.: On biases in tests of the expectations hypothesis of the term structure of interest rates. J. Financ. Econ. **44**(3), 309–348 (1997)

Campbell, J.Y., Shiller, R.J.: Cointegration and tests of present value models. J. Political Econ. **95**(5), 1062–1088 (1987)

Clarida, R.H., Sarno, L., Taylor, M.P., Valente, G.: The role of asymmetries and regime shifts in the term structure of interest rates. J. Bus. **79**(3), 1193–1224 (2006)

Clements, M.P., GalVao, A.B.C.: Testing the expectations theory of the term structure of interest rates in threshold models. Macroeconomic Dyn. **7**(4), 567–585 (2003)

Cooray, A.: A test of the expectations hypothesis of the term structure of interest rates for Sri Lanka. Appl. Econ. **35**(17), 1819–1827 (2003)

Cuthbertson, K.: The expectations hypothesis of the term structure: the UK interbank market. Econ. J. **106**(436), 578–592 (1996)

Domlnguez, E., Novales, A.: Testing the expectations hypothesis in eurodeposits. J. Int. Money Financ. **19**(5), 713–736 (2000)

Fama, E.F.: Term-structure forecasts of interest rates, inflation and real returns. J. Monetary Econ. **25**(1), 59–76 (1990)

Guidolin, M., Thornton, D.L.: Predictions of short-term rates and the expectations hypothesis. Int. J. Forecast. **34**(4), 636–664 (2018)

Hall, A.D., Anderson, H.M., Granger, C.W.: A cointegration analysis of treasury bill yields. Rev. Econ. Stat., 116–126 (1992)

Held, L., Ott, M.: How the maximal evidence of p-values against point null hypotheses depends on sample size. Amer. Stat. **70**(4), 335–341 (2016)

Kapetanios, G., Shin, Y., Snell, A.: Testing for cointegration in nonlinear smooth transition error correction models. Econometri. Theory, 279–303 (2006)

Mankiw, N.G., Miron, J.A.: The changing behavior of the term structure of interest rates. Quart. J. Econ. **101**(2), 211–228 (1986)

Michelis, L., Koukouritakis, M.: Enlargement and the EMU. J. Econ. Integr., 156–180 (2007)

Mili, M., Sahut, J.M., Teulon, F.: New evidence of the expectation hypothesis of interest rates: a flexible nonlinear approach. Appl. Financ. Econ. **22**(2), 165–176 (2012)

How Do Macroprudential Policy and Institutions Matter for Financial Stability? New Evidence from Eagles

Nguyen Tran Xuan Linh, Nguyen Ngoc Thach, and Vu Tien Duc

Abstract The study was conducted to assess the impact of macroprudential policy and institutional quality on Emerging and growth-leading economies in 2008–2018. Through Bayesian multivariate regression via MCMC simulations, research has confirmed the important role of institutional quality in financial stability. In addition, the research results revealed that when countries loosen credit growth would lead to the erosion of the stability of the financial system. An astonishing finding is that loan-to-value instrument adversely influences financial stability. Limits on foreign currency lending and capital surcharges for systemically important institutions obviously improve financial stability. The remaining instruments are Limits on Interbank Exposures and FX and/or Countercyclical Reserve Requirements, which tend to strengthen the stability of the financial system, but their impact is ambiguous.

Keywords Macroprudential policy—Institutional quality · Financial stability · EAGLEs group

1 Introduction

The financial system is a vital pillar in a country's economy. Therefore, any financial system malfunction often results in serious consequences for the real economy; the 2007–2008 global financial system proves this statement. This crisis also indicated

N. T. X. Linh (✉)
Ho Chi Minh Industry and Trade College, 20 Tang Nhon Phu, District 9, Ho Chi Minh City, Vietnam
e-mail: xuanlinh86@gmail.com

N. N. Thach
Institute for Research Science and Banking Technology, Banking University HCMC, 36 Ton That Dam, District 1, Ho Chi Minh City, Vietnam
e-mail: thachnn@buh.edu.vn

V. T. Duc
Banking University HCMC, 36 Ton That Dam, District 1, Ho Chi Minh City, Vietnam
e-mail: ducvt@buh.edu.vn

N. Ngoc Thach et al. (eds.), *Financial Econometrics: Bayesian Analysis, Quantum Uncertainty, and Related Topics*, Studies in Systems, Decision and Control 427,
https://doi.org/10.1007/978-3-030-98689-6_40

that the established supervisory regulations before this event were insufficient to keep the financial system stable. In order to control the risks that could erode the soundness of the financial system as well as increase the resistance of the financial system, besides traditional tools such as monetary and fiscal policy, the financial system needs to be managed and monitored as a whole and its interconnection with the entire economy through the implementation of macroprudential policy (Freixas et al. 2016).

After the crisis, many studies were conducted to assess the role of macroprudential policy in maintaining financial system stability. These studies, however, have only analyzed the impact of macroprudential policy on some aspects of the financial system, such as loan loss provisions, relative interest income (Foos Norden et al. 2010); credit growth and asset prices, real estate (Pan and Wang 2013; Cerutti et al. 2017a, b) or the credit cycle (Claessens et al. 2013; Galati and Moessner 2018a, b; Carreras et al. 2018). These factors do not yet cover in all respects such a multidimensional economic phenomenon as the financial system.

Another issue that needs to be considered in maintaining the stability of the financial system in a country is the quality of institutions. Nonetheless, there are not many studies on the relationship between institutional quality and financial stability. According to the authors' literature review, only a few studies are trying to address this relationship, such as Klomp and Haan (2014), Bermpei et al. (2018) assessed the impact of Quality of Institutions on bank risk.

In summary, financial system stability is an important research issue, especially after the global financial crisis of 2007–2008. However, current studies only consider the influence of macroeconomic policies on some aspects of the financial system, which may lead to biased policy implications. In addition, another important factor in maintaining financial stability is institutional quality, but research on this issue is relatively scanty. In this study, the authors used the financial stability index to comprehensively assess the impact of macroprudential policy and institutional quality on the financial system in Emerging and growth-leading economies (EAGLEs). By adopting Bayesian multivariate regression via MCMC simulations, the study confirmed the importance of institutional quality in maintaining financial stability. In addition, the study also exerted the clear improvement of two macroprudential tools, namely **limits on foreign currency lending** and **capital surcharges for systemically important institutions** for financial stability; while the results also reveal that **loan-to-value** adversely influence financial stability; The role of the other two tools, **Limits on interbank exposures** and **FX and/or countercyclical reserve requirements** for the stability of the financial system, is vague.

The paper is divided into 4 sections: Sect. 1 is the introduction. In Sect. 2, we will clarify the views on financial stability and methods of measuring financial stability. We then review the impact of macroprudential policy and institutional quality on financial stability. Section 3 will propose a set of indicators to measure financial stability for the group of EAGLEs and the research model. Section 4 is the results of the research and discussion. Finally, Sect. 5 is the conclusion and policy implications.

2 Literature Review

This section discusses definitions and empirical studies on financial stability and the role of macroprudential policy in financial stability.

Financial stability

There are two approaches to financial stability. *The first* approach describes what situations are not considered instability financially, that is, to give a concept of financial instability. According to Mishkin (1999), financial instability occurs when shocks to the financial system interfere with information, causing the financial system to fail to mobilize capital and allocate capital. Rosengren (2011a, b) argued that financial instability is when the real economy is severely affected by instability from credit providers. From this point of view, several researchers have developed methods to measure the stability of the financial system.

Kibritçioğlu (2002) proposed the "financial stress index," which includes indicators of bank credit to the private sector, external debt of the banking system, and bank deposits. An increase in this index is a sign of instability in the banking system and vice versa. Nelson and Perli (2005) constructed the "financial fragility index". In this study, the authors focused on the US financial system, and they demonstrated that this composite index could contribute to forecasting the probability that a turbulent period is likely to occur out. Van den End (2006) built the "financial stability conditions index" for the Netherlands. This index is built based on indicators that characterize monetary conditions: interest rates, real effective exchange rates, stock prices, the solvency of financial institutions, and stock index volatility. Low indices mean increased volatility, while too high values could lead to the accumulation of financial imbalances. Although the above authors have tried to develop a set of indicators to measure financial stability, the above indicators only show some aspects of the financial system; still, they could not fully reflect a multidimensional economic phenomenon such as financial stability.

According to the *second approach*, financial stability is whenever the financial system is in the range of stability that is able to facilitate (rather than inhibit) the functioning of the economic system and disperse financial asymmetries originating from within the economy or unintended negative impacts from external shocks (Schinasi 2004). Foot (2003) stated that a financial system reaches financial stability when it consists of four main components: stable currencies; the employment rate in the real economy should be close to full employment; confidence in the operation of financial institutions; and stability in real and financial asset prices. Freedman and Goodlet (2007), Rosengren (2011a, b) emphasized that a stable financial system is not only capable of allocating savings, transmitting information, performing well the payment system function, and responding to credit demand promptly to ensure growth for the real economy but also the ability of the financial system to withstand shocks.

According to Davis (2003), the first approach is easier to measure because of its focus on certain components and more convenient to observe because it is defined through financial instability. However, Borio and Drehmann (2008) argue that this approach focuses on few aspects of the financial system such as asset prices, the efficiency of financial services, availability of credit. In addition, designing policies towards avoiding financial instability and crises could lead to biased policy decisions (Schinasi 2004). For this reason, the authors approach and measure financial stability from the second perspective.

The set of indicators often mentioned when measuring financial stability according to the second approach is the "Financial soundness indicator—FSI" developed by the IMF after the global economic crisis in the 1980s and 1990s. This set of indicators covers six main sectors: real economy, household sector, external sector, corporate sector, financial sector, and financial markets. Besides the FSI index of the IMF, the European Central Bank (ECB) also initiated the "Macro-Prudential Indicators—MPI," which includes 303 indexes covering the entire financial and banking sector, the macroeconomic sector, non-financial sector, and the interconnection factor to monitor the stability of the European banking system (Mörttinen et al. 2005).

Although constructed very comprehensively, the sets of indicators of the IMF and ECB require too large data, which is difficult to collect and too complicated to interpret and compare. Albulescu (2008) proposed the Aggregate Financial Stability Index to measure the stability of the Romanian financial system. The set of financial stability indicators constructed by Albulescu (2008) includes 20 indicators divided into 4 sub-index groups, including Financial Development Index, Financial Vulnerability Index, Financial Soundness Index, World Economic Climate Index. The measurement method is relatively simple; nonetheless, this index proved quite effective in measuring the health of the Romanian financial system. Based on his research background, Albulescu (2010) used the Stochastic Simulation Model to conduct forecasting of Romanian financial stability. Besides identifying the shape of the financial stability trend, this financial stability index has performed quite well its role in providing early warning signs when the shape of this index represents a downward trend during the Romanian banking crises of 1998, the capital market crash, the Argentina crisis of 2001 and the subprime lending crisis of 2007.

With these advantages, the financial stability index developed by Albulescu (2008) has been inherited by many researchers and made appropriate adjustments to create a set of indicators to measure the financial stability of the country or the country group of countries they observe (see Morris 2010; Goran and Karanovic 2015; Azka and Sahara 2018). Based on Albulescu's (2008) research, Linh (2021) supplemented public debt to GDP and foreign exchange reserves to foreign debt to make this index suit the characteristics of the EAGLEs group. Besides, Linh (2021) also replaced the Financial Institutions Index and the Financial Market Index developed by the IMF for Total credit to GDP and Market capitalization to GDP proposed by Abulescu (2008) to measure the development of the financial institution system financial institutions, and financial markets, as the measure of IMF is more comprehensive. In this article, the authors used the Financial stability Index proposed by Linh (2021) to measure the stability of the financial system of the EAGLEs group.

Macroprudential policy

The macroprudential policy uses policy instruments to improve the stability of the financial system to reduce the likelihood of financial system failure and handle common risks to avoid serious consequences for the real economy (Clement 2010). The framework of analyzing the impact of macroprudential policy on financial stability is expressed through the time and cross-sectional of systemic risk.

The time dimension is expressed through macroprudential policy control of the financial boom and bust cycle (Borio et al. 2011). The financial boom and bust cycles could be derived from the demand for credit or the behavior of financial institutions. Through the data of 49 emerging and developed countries, Nabar and Ahuja (2011) demonstrated that the Loan-to-value (LTV) instrument restrains the growth in house prices and mortgage lending while the Debt-to-income (DTI) tool reduces growth in property lending. For the credit supply side, the results also show that the LTV ratio improves credit quality by reducing non-performing loans, but the DTI ratio is not statistically significant. With the data of 49 countries in the period 2000–2010 collected from the IMF survey, Lim et al. (2011) also found out similar results, besides the study also showed the instruments such as Limits on Domestic Currency Loans (CG); FX, and/or Countercyclical Reserve Requirements (RR) also curb financial cycle volatility.

Claessens et al. (2013), applying panel data GMM regression, Claessens et al. (2013) analyzed the change in balance sheets of 2,800 commercial banks in 23 developed and 25 emerging economies in the period 2000–2010. The regression results showed that the LTV, DTI, CG, and Limits on Foreign Currency Loans (FC) effectively control financial leverage abuse and excessive lending growth during the boom period. While capital buffering tools such as RR, Time-Varying/Dynamic Loan-Loss Provisioning (DP) mitigate adverse effects during recessions.

The cross-sectional dimension is exhibited through the provisions of macroprudential policy for the interaction between financial institutions. The 2007–2008 financial crisis spread globally due to the amplified impact of the collapse of several important credit institutions in the financial system. Hence, after this crisis, the feature of the cross-sectional dimension became the focus of policy discussions. In order to reduce the risk of similar crises in the future, the Basel III framework tends to focus on the supervision of financial institutions in the financial system. By imposing capital surcharges on important financial institutions, the Basel III framework is expected to minimize adverse externalities stemming from interactions between credit institutions (BCBS 2011). By employing data from 64 countries from Q1.2000 to Q4.2014, Cerutti et al. (2017a, b) confirmed that Concentration Limits (CONC), Limits on Interbank Exposures (INTER), improve the soundness of the banking system. In this study, Cerutti et al. (2017a, b) also suggested that LTV and RR are more efficient when executed in a countercyclical trend. Carreras et al. (2018) concluded that, along with LTV and DTI, the Levy/Tax on Financial Institutions (TAX) and INTER proved to be more effective than other macroprudential tools in maintaining financial stability at 18 OECD countries Q1.2000 to Q4.2014.

In summary, macroprudential instruments could reduce the risk-taking behavior of credit institutions. For example, capital-based instruments are seen as a capital preservation buffer, improving credit institutions to accumulate capital during the flourishing period. This capital will become a buffer to support credit institutions absorb and minimize financial stress shocks. Macroprudential tools such as LTV and DTI can restrain the boom spiral in asset prices, thereby reducing the volatility of financial cycles. In addition, after the 2007–2008 financial crisis, many macroprudential policy tools were established targeting important financial institutions such as Concentration Limits, Limits on Interbank Exposures, Capital Surcharges on Important Financial Institutions, Levy/Tax on Financial Institutions are expected to reduce systemic risks arising from interactions between credit institutions.

Institutional quality

Institutional quality is defined as that captures law, individual rights, and high quality government regulation and services (Bruinshoofd 2016). Many research confirmed the interrelationship between institutional quality and economic performance (Bruinshoofd 2016). However, studies on the role of institutional quality on financial stability are relatively few. Empirical studies showed that countries with a sufficiently strong institutional environment would have more sound banking systems. Klomp and Haan (2014) argued that institutional quality could affect bank regulation and supervision, thus affecting the soundness of the banking system. Through data analysis of 371 banks from nonindustrial countries from 2002–2008, Klomp and Haan (2014) demonstrated that stricter regulation and supervision, which depend on Institutional quality, reduce banking risk. With a similar analytical framework, with data from approximately 1,050 commercial banks from 69 emerging and developing economies from 2004–2013, Bermpei et al. (2018) clarified the impact of institutional quality on bank regulations and supervision regarding the effect of the latter on bank stability.

Thus, it can be seen that studies on financial stability and influencing factors received a lot of attention from researchers. However, current studies have only examined the impact of macroeconomic policies on one aspect of the financial system, such as credit growth and asset prices, real estate values (see Borio et al. 2011; Cerutti et al. 2017a, b) Credit Cycle (see Lim et al. 2011; Claessens et al. 2013; Galati and Moessner 2018a, b). To assess the impact of macroprudential policy and institutional quality on financial stability in the EAGLEs group more comprehensively, in this study, the authors use a set of financial measurement indicators proposed by Linh (2021).

3 Research Method

The study was conducted to examine the impact of macroprudential policy and institutional quality on financial stability from 2008 to 2018 in Emerging and growth-leading economies (EAGLEs), including 15 countries: Bangladesh, Brazil, China,

Egypt, India, Indonesia, Iran, Malaysia, Mexico, Nigeria, Pakistan, Philippines, Turkey, Russia, and Vietnam. This is a group of countries created by Banco Bilbao Vizcaya Argentaria (BBVA research) to identify emerging economies expected to significantly contribute to global economic growth over the next decade. To achieve high economic growth, countries had strong support from the financial system. Data from World Development Indicators of the World Bank (World Bank - WB 2020) showed that in order to boost the economy, many countries in the EAGLEs group had boosted the supply of credit to the economy, in some countries has Domestic credit growth to the private sector by banks exceeded 100% of GDP such as China is 161%; Malaysia 120% and Vietnam hit 133% in 2018. Khan et al. (2019) asserted that EAGLEs are encountering a challenging period. The financial system in EAGLEs is not really strong, especially credit institutions in countries including India, Egypt, Bangladesh, Nigeria, Pakistan, and Vietnam are relatively weak and vulnerable to the effects of shock, which could lead to fierce consequences for the economy. Hence, it is necessary to research and find solutions to maintain the stability of the financial system to create a foundation for economic growth as expected.

The literature review reveals that the research on macroprudential policy and financial stability is mostly conducted through the frequentist approach. Still, this method has the main disadvantage; the accuracy estimation depends greatly on the sample size; however, it is challenging to obtain a large number of observations at the national level. The Bayesian approach, through interpreting the results as a probability distribution of parameter values, regardless of the sample size, can thus overcome the small sample weakness in the studies (Mariëlle et al. 2017). In fact, there have been numerous studies detailing the advantages of Bayesian approaches (see, for instance, Hung and Thach 2018; Anh et al. 2018; Hung and Thach 2019; Hung et al. 2019; Sriboonchitta et al. 2019; Svitek et al. 2019; Thach et al. 2020). The main disadvantage of the Bayesian approach is that the algorithm is complicated, the computational cost is high, but McNeish (2016) emphasizes that with the strong development of computer science, the Bayesian method has become accessible and popular. Based on a comprehensive review of 15 years, van de Schoot (2016) notes that the number of empirical studies using the Bayesian method has increased nearly fivefold between 2010 and 2015. For this reason, the authors applied the Bayesian method to assess the impact of macroprudential policy and institutional quality on financial stability in EAGLEs.

From the literature review, the authors proposed the following research model:

$$FSTI_{it} = \beta_1 + \beta_2 CRE_{it} + \beta_3 MP_{it} + \beta_4 INS_{it} + \varepsilon_{it}$$

i = 1, 2, …, 15.
t = 2008, 2009, …, 2018.

where i is the countries in the EAGLEs group, respectively; t is time; β is the regression coefficients; FSTI represents financial stability index; CRE is Domestic credit growth to the private sector by banks; it represents the macroprudential policy view in a period; if this policy expands, CRE will increase and vice versa. MP is a vector of macroprudential instruments. INS is Institutional quality (Table 1).

Table 1 Model variables

Notation		Variables of model	Expected sign	Data source
Dependent variable				
FSTI		Financial stability Index		Linh (2021)
Independent variable				
CRE		Domestic credit growth to private sector by banks	−	WB
MP	FC	Limits on Foreign Currency Loans	+	Eugenio dataset (2018)
	LTV	Loan-to-value	+	Eugenio dataset (2018)
	INTER	Limits on Interbank Exposures	+	Eugenio dataset (2018)
	RR	FX and/or Countercyclical Reserve Requirements	+	Eugenio dataset (2018)
	SIFI	Capital surcharges for systemically important institutions	+	Eugenio dataset (2018)
INS		Institutional quality	+	WGI

Source Authors' synthesizer

The FSTI index proposed by Linh (2021) includes four sub-indices: Financial Development Index (FDI); Financial Vulnerability Index (FVI); Financial Soundness Index (FSI); World Economic Climate Index (WECI) (Table 2).

In order to be considered for inclusion in a composite index, individual indicators must be normalized. The standard values for the normalization procedure are expressed as the worst and best indicator for the time period analyzed. To facilitate the synthesis and analysis of financial stability indicators, standardized indicator values are in the range [0; 1]. The value "0" indicates an instability situation and is equal to the worst recorded value and the value "0" reflects the opposite case.

The normalization process has the followed formula:

$$I_{qc}^{t} = \frac{X_{qc}^{t} - \min X_{q}^{t0}}{\max X_{q}^{t0} - \min X_{q}^{t0}} \tag{1}$$

where I_{qc}^{t} is the normalization of the q-index for country c in period t.

$\min X_{q}^{t0}$ is worst value for index q.

$\max X_{q}^{t0}$ is the best value for the index q.

After the normalization process, Financial Development Index (FDI), Financial Vulnerability Index (FVI), Financial Soundness Index (FSI), World Economic Climate Index (WECI) and Financial Stability Index (FSTI) are calculated through the formula:

Table 2 Factors that constitute the financial stability index

Individual indicators	Notation	Data source
Financial Development Index—FDI		
Financial Institution index	FD1	Financial Development, IMF
Financial Markets index	FD2	Financial Development, IMF
Interest spread	FD3	Global Financial Development WB
Financial Vulnerability Index—FVI		
Inflation rate	FV1	Global Financial Development, WB
General budget deficit (% GDP)	FV2	Global Financial Development, WB
Current account deficit (% GDP)	FV3	Global Financial Development, WB
REER excessive depreciation or appreciation	FV4	Bruegel datasets
Credit to private sector/Total credit	FV5	Global Financial Development, WB
Loans/deposits	FV6	Global Financial Development, WB
Deposits/M2	FV7	Global Financial Development, WB
Public debt (% GDP)	FV8	Global Financial Development, WB
Financial Soundness Index—FSI		
Non-performing loans/Total loans	FS1	Financial Soundness Index, IMF
Regulatory capital/Risk-weighted assets	FS2	Financial Soundness Index, IMF
Bank capital/Asset	FS3	Global Financial Development, WB
Liquid reserves/Bank assets	FS4	Global Financial Development, WB
Z-score	FS5	Global Financial Development, WB
ROA of bank	FS6	Global Financial Development, WB
Total reserves (% of total external debt)	FS7	Economic Climate Index—CESifo
World Economic Climate Index—WECI		
Economic Climate Index—CESifo	WE1	Economic Climate Index, CESifo
World Inflation	WE2	Global Financial Development, WB
World Economic Growth Rate	WE3	Global Financial Development, WB

Linh (2021)

$$FDI_t = \frac{\sum_{i=1}^{3} FD_{it}}{3} \tag{2}$$

$$FVI_t = \frac{\sum_{i=1}^{8} FV_{it}}{8} \tag{3}$$

$$FSI_t = \frac{\sum_{i=1}^{7} FS_{it}}{7} \tag{4}$$

$$WECI_t = \frac{\sum_{i=1}^{3} WE_{it}}{3} \tag{5}$$

$$FSTI_t = \frac{3 \times FDI_t + 8 \times FVI_t + 7 \times FSI_t + 3 \times WECI_t}{21}$$

Although the financial system consists of two main pillars, namely the credit institution system and the financial market, in reality, the financial market of the EAGLEs has not really developed. According to the data of the World Bank (2020), the average stock trading volume of the EAGLEs group in 2018 was only 25.74% of GDP, of which there are countries with very low stock trading volume, such as Iran, which is 5.78% GDP, Egypt 5.8% GDP, Bangladesh 5.89% GDP, Mexico 7.67% GDP, Philippines 8.42% GDP, Russia 8.80% GDP, Pakistan 9.88% GDP, Indonesia is 10.04% GDP. This Fig. 1 in developed countries such as the OECD group of countries was 114.03%, even in Hong Kong, this figure was 626.74% of GDP in 2018. Meanwhile, bank credit to the private sector of EAGLEs is 65% of GDP; for OECD countries, this figure is 146.39% of GDP and 223.1% of GDP for Hong Kong. Thus, the banking system occupies a crucial position in the financial system of the EAGLEs group. Hence, the Financial Stability Index developed by Linh (2021) from the composite index of Albulescu (2008) is very consistent with the practice of the EAGLEs group of countries when it focuses on the banking system.

Macroprudential policy instruments include Loan-to-value; limits on interbank exposures; capital surcharges for systemically important institutions; limits on foreign currency loans; FX and/or countercyclical reserve requirements. These macroprudential policy instruments are used commonly in EAGLEs.

The institutional quality is calculated by averaging the six dimensions of governance quality of the Worldwide Governance Indicators (WGI) collected and calculated by the World Bank. These six dimensions include voice and accountability; political stability and absence of violence; government effectiveness; regulatory quality; rule of law; and control of corruption.

Previous studies primarily applied the frequentist approach; therefore, the prior information for this study is not available. In this case, Block et al. (2011) proposed specifying the standard Gaussian distributions. To conduct this process, five simulations are performed corresponding to specified priors that decrease from the strongest to the weakest (Table 3).

4 Simulation Results and Discussion

According to Bayes factor analysis, the chosen prior information simulation having with Maximum Log BF, Maximum Log ML and minimum DIC. Table 4 showed that simulation 1 conforms to the criteria to become simulation having the most appropriate prior information. Moreover, the Bayes model results also demonstrate

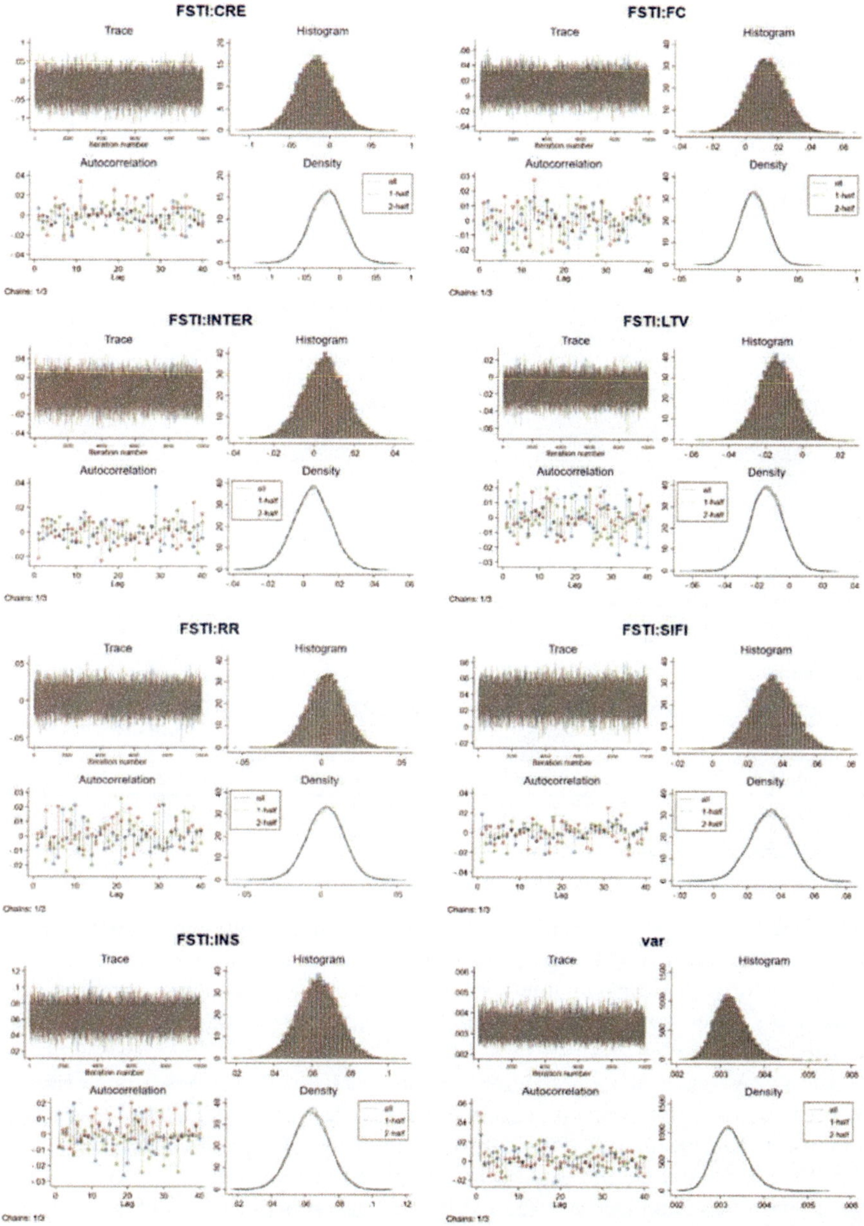

Fig. 1 Graphical tests for MCMC convergence. *Source* Authors' calculation

Table 3 Simulations summary

Likelihood	FSTI $\sim$ N (μ, δ)
Prior distributions	
Simulation 1	$\alpha_i \sim$ N (0, 1) $\delta^2 \sim$ Invgamma $(0.01, 0.01)$
Simulation 2	$\alpha_i \sim$ N (0, 10) $\delta^2 \sim$ Invgamma $(0.01, 0.01)$
Simulation 3	$\alpha_i \sim$ N (0, 100) $\delta^2 \sim$ Invgamma $(0.01, 0.01)$
Simulation 4	$\alpha_i \sim$ N (0, 1000) $\delta^2 \sim$ Invgamma $(0.01, 0.01)$
Simulation 5	$\alpha_i \sim$ N (0, 10000) $\delta^2 \sim$ Invgamma $(0.01, 0.01)$
i = 1, 2, 3, 4, 5, 6, 7	

Source Authors' synthesizer

Table 4 A Bayesian factor test and a model test

	Chain	Avg DIC	Avg log(ML)	log(BF)	P(M|y)
Simulation 1	3	−469.8564	198.2841	1	0.9999
Simulation 2	3	−469.9375	189.1588	−9.1253	0.0001
Simulation 3	3	−470.0221	179.9421	−18.3419	0
Simulation 4	3	−469.9692	170.7931	−27.4909	0
Simulation 5	3	−469.9286	161.6151	−36.6690	0

Source Authors' calculation

that simulation 1 has an advantage over the other simulation; hence, simulation 1 with a prior information N (0, 1) is selected for further analysis.

To ensure that Bayesian inference based on MCMC samples is reasonable, the authors tested convergence of MCMC of parameter estimates through visual diagnosis by the graph.

The graph visual diagnosis shows that all graphs of the model parameters are reasonable, trace plots, and correlation plots show low correlation. The chart shape is uniform and appearance a normal distribution. The above graphs also show a good mix; the correlation coefficients in the graphs fluctuate around 0.02, showing the consistency with simulated density distribution and reflecting lags within efficiency limits.

Convergent diagnostic analysis by images and Grubin only shows the characteristics of the MCMC chains; the fit of the model with the observed data has not been shown. To perform the Bayesian model fit test, Gelman et al. (2014) proposed to

Table 5 The posterior predictive p-value

	Mean	Std. dev.	E(T_obs)	P(T > = T_obs)
minsl	0.409194	0.0274225	0.4159595	0.4399

Source Authors' calculation

use the smallest observation to measure the difference between observed and simulated data for the estimated model parameters. We could use the posterior predictive p-value to evaluate the difference between the observed and simulated data.

Table 5 shows the mean and the standard deviation of the minimum FSTI value (minsl) in the simulation data (T) as 0.4 and 0.02 respectively. The minimum FSTI value in the observed sample (T_obs) is 0.416. The last column of Table 5 P(T > = T_obs) shows the probability of predicting the mean value of the minimum FSTI value in the simulations greater than or equal to the minimum FSTI value of the observed sample. The P(T > = T_obs) value, ideally, should be close to 0.5, however in experimental studies, this value is between 0.05 and 0.95, the simulations of the Bayesian regression model are considered appropriate (Gelman et al. 2014). In this study, the P(T > = T_obs) value is 0.44, satisfying the above condition.

Table 5 shows that the average acceptance rate of the model is reached 1; the minimum efficiency of the model is 0.89, far exceeding the required level of 0.01. Therefore, the above model meets the convergence requirement. In addition, the Monte-Carlo Standard Error (MCSE) of all the parameters is very small, according to Flegal et al. (2008); the closer the MCSE is to zero, the more valid the MCMC chains. Flegal et al. (2008) also consider that MCSE values less than 6.5% of the standard deviation are acceptable, and less than 5% are optimal. Besides, according to Gelman and Rubin (1992), Brooks and Gelman (1998), the diagnostic value Rc of any parameters greater than 1.2 is considered non-convergent. In practice, Rc must follow a stricter rule that this value must be less than 1.1 to be considered convergent. The analysis results in Table 6 show that all the maximum Rc values of the coefficients are 1, so we could conclude that the MCMC chains satisfy the convergence requirements.

Regression results in Table 6 indicate that most macroprudential policy instruments, except LTV, improve financial stability in EAGLEs. Meanwhile, credit growth hurts financial stability in the EAGLEs group. The regression result also reveals that the institutional environment encourages the soundness of the financial system. The loan-to-value (LTV) variable that hurts the financial stability of EAGLEs is stunning and needs to be further analyzed by calculating the probability of its impact. A major advantage of the Bayesian over the frequency method is that the Bayesian method allows us to determine the probability of the impact factors (Table 7).

The results of Table 6 show the probability that when implementing the easing of bank credit for the private sector, the financial stability tends to decrease with a probability of 78%. Theoretically, when credit growth is eased, commercial banks tend to reduce loan standards, and non-performing loans may increase, eroding the

Table 6 Bayesian simulation outcomes

FSTI	Mean	Std. dev.	MCSE	Median	Equal-tailed [95% Cred. Interval]	
CRE	−0.0184	0.0242	0.0001	−0.0184	−0.0657	0.0286
FC	0.0128	0.0119	0.0001	0.0127	−0.0107	0.0358
INTER	0.0051	0.0106	0.0001	0.0050	−0.0157	0.0257
LTV	−0.0142	0.0101	0.0001	−0.0143	−0.0338	0.0057
RR	0.0028	0.0119	0.0001	0.0028	−0.0207	0.0262
SIFI	0.0345	0.0121	0.0001	0.0345	0.0106	0.0585
INS	0.0631	0.0109	0.0001	0.0631	0.0418	0.0847
_cons	0.6097	0.0096	0.0001	0.6098	0.5911	0.6284
var	0.0032	0.0004	0.0000	0.0032	0.0026	0.0041
Avg. acceptance rate	1.0000					
Avg. efficiency: min	0.8899					
Max Gelman-Rubin Rc	1.0000					

Source Authors' calculation

Table 7 A probabilistic test

	Mean	Std. Dev	MCSE
Probability {FSTI: CRE} < 0	0.7791	0.4149	0.0024
Probability {FSTI: FC} > 0	0.8572	0.3498	0.0020
Probability {FSTI: INTER} > 0	0.6857	0.4643	0.0027
Probability {FSTI: LTV} < 0	0.9225	0.2674	0.0015
Probability {FSTI: RR} > 0	0.6016	0.4896	0.0028
Probability {FSTI: SIFI} > 0	0.9978	0.0469	0.0003
Probability {FSTI: INS} > 0	1.0000	0.0000	0.0000

Source Authors' calculation

financial stability. In addition, if credit growth accelerates, there is also the risk of asset bubbles, threatening the safety of the entire system. This result is similar to the study of Foos et al. (2010), Adrian and Shin (2012).

An astonishing finding in the macroprudential policy instrumental variable regression model is that the loan-to-value ratio (LTV) harms financial stability with more than 90% probability. This result can be explained by the fact that EAGLEs are characterized by very rapid growth; hence asset prices of these countries increase rapidly during the credit boom because of the large amount of credit flowing into an investment property, such as real estate. The increase in asset prices causes the collateral value to increase, so investors meet the LTV ratio standard. During periods of the boom asset price, banks also tend to make exaggerated assessments of asset values. Still, favorable conditions in the housing market along with the high economic growth of the period have erased these misjudgments, leading to the overconfidence of banks.

When an economic slowdown, the large loan value has become a financial burden for borrowers when they have to bear large interest rates, result in non-performing loans increase. Besides, when the value of collateral decreases, the loans no longer guarantee the prescribed LTV ratio, then more credit institutions try to reduce the leverage ratio by speeding up liquidating their investment assets, leading to a downward spiral in asset market prices and the inevitable consequence of a credit crunch (Dell'Ariccia et al. 2012).

The rest of the macroprudential policy instruments tend to improve the financial stability of EAGLEs groups. However, the probability of impact of FX and/or Countercyclical Reserve Requirements (RR) is only 60%, and Limits on Interbank Exposures (INTER) is 68% is relatively low, so we do not have enough evidence to confirm the role of these two tools in maintaining financial stability at EAGLEs.

Meanwhile, the probability that the impact of Capital surcharges for systemically important institutions (SIFI) is up to 99.8% has confirmed its role in supporting the stability of the financial system. This is a tool recommended by the Financial Stability Board to reduce the spatial risk of the financial system. Similar to SIFI, the probability of a positive impact on the financial stability of Limits on Foreign Currency Loans (FC) reached 86%, which also shows the effectiveness of FC in ensuring the stability of the financial system in EAGLEs group. According to the IMF (2011), the exchange rate rises, the borrower incurs additional costs due to the exchange rate, which can lead to failure to repay the loan; Conversely, a decrease in the exchange rate could cause losses to credit institutions, so the FC tool is expected to reduce the risk of exchange rate fluctuations for both borrowers and lenders, thereby maintaining financial stability. This result is consistent with the study of Tovar et al. (2012), Hahm et al. (2012).

Finally, the research results also asserted the crucial role of institutional quality in maintaining financial stability as the probability of impact of this factor reaches 100%. This result is similar to the study of Kunt and Detragiache (1998), Klomp and Haan (2014), Bermpei et al. (2018).

5 Conclusion and Policy Implications

Research results show that bank credit growth for the private sector harms bank stability; when credit growth is eased, banks tend to reduce loan standards, and non-performing loans may increase; hence the stability of the banking system is also reduced. Besides, if credit growth accelerates, there is also the risk of asset bubbles, threatening the safety of the entire system. Therefore, when expanding bank credit to promote economic growth, these countries should proceed with caution and control to avoid causing the accumulation of risks to the banking system.

In addition, the study evaluated the impact of 5 macroprudential policy instruments that are relatively used in the EAGLEs group: Limits on Foreign Currency Loans (FC), Loan-to-value (LTV), Limits on Interbank Exposures (INTER), FX and/or Countercyclical Reserve Requirements (RR), Capital surcharges for systemically important institutions (SIFI). The results show that the role of INTER and RR in maintaining financial stability is vague when the probability of impact is only 60% and 69%, respectively.

An astonishing finding is the LTV has a negative impact on the stability of the financial system with a probability of more than 92%, this implies that these countries should employ more flexibly this instrument, it should be tightened as economic growth is high, avoid creating asset bubbles. However, when the economy is in shock, this tool should be eased to relieve pressure on both borrowers and lenders; only then, this tool plays its function as a buffer to absorb the risks shock and maintain the financial system stable.

The remaining instruments, including FC and SIFI, have an obvious impact on improving the stability of the financial system with the probability of 92% and more than 99%, respectively. This implies that EAGLEs group should deploy these instruments to improve the soundness of their financial systems, creating a solid foundation to maintain sustainable growth for these countries.

Finally, research continues to confirm the important role of institutions in maintaining financial stability. Institutional quality has supported regulation and supervision become stricter, and as a result, the financial system is more stable.

References

Albulescu, C.T.: Assessing Romanian financial sector stability: The importance of the international economic climate. MPRA Paper, No. 16581 (2008). http://mpra.ub.uni-muenchen.de/16581/1/MPRA_paper_16581.pdf. Accessed 20 Sept 2020

Albulescu, C.T.: Forecasting the Romanian financial system stability using a stochastic simulation model. Romanian J. Econ. Forecast. **13**(1), 81–98 (2010). http://mail.ipe.ro/rjef/rjef1_10/rjef1_10_6.pdf. Accessed 20 Sept 2020

Adrian, T., and Shin, H.S. Procyclical leverage and value-at-risk. Federal Reserve Bank of New York Staff Report, **338**, New York (2012)

Anh, L.H., Kreinovich, V., Thach, N.N. (eds.): Econometrics for Financial Applications. Springer, Cham (2018)

Azka, A.D., Sahara, L.A.: Impact of financial inclusion on financial stability based on income group countries. Bull. Monetary Econ. Banking **20**(4), 1–14 (2018)

Basel Committee on Banking Supervision (BCBS) (2011) Basel III: a global regulatory framework for more resilient banks and banking systems. Basel: Bank for International Settlements

Bermpei, T., Kalyvas, A., Nguyen, T.C.: Does institutional quality condition the effect of bank regulations and supervision on bank stability? Evidence from emerging and developing economies. Int. Rev. Financ. Anal. **59**, 255–275 (2018)

Block, J.H., Jaskiewicz, P., Miller, D.: Ownership versus management effects on performance in family and founder companies: a Bayesian reconciliation. J. Family Bus. Strategy **2**, 232–45 (2011)

Borio, C., Drehmann, M., Tsatsaronis, K.: Anchoring countercyclical capital buffers: the role of credit aggregates. Int. J. Cent. Bank. **7**(4), 189–240 (2011)

Borio, C., Drehmann, M.: Towards an operational framework for financial stability: "fuzzy" measurement and its consequences. Paper presented at the 12th Annual Conference of the Banco de Chile—Financial stability, monetary policy and central banking, Santiago, 6–7 November (2008)

Brooks, S., Gelman, A.: General methods for monitoring convergence of iterative simulations. J. Comput. Graph. Stat. **7**(4), 434–455 (1998)

Bruinshoofd, A.: Institutional quality and economic performance. Rabobank Research Economic Report (2016). https://economics.rabobank.com/publications/2016/january/institutional-quality-and-economic-performance/

Carreras, O., Davis, E.P., Piggott, R.: Assessing macroprudential tools in OECD countries within a cointegration framework. J. Financ. Stab. **37**, 112–130 (2018)

Cerutti, E., Claessens, S., Laeven, L.: The use and effectiveness of macroprudential policies: new evidence. J. Financ. Stab. **28**, 203–224 (2017a)

Cerutti, E., Correa, R., Fiorentino, E., Segalla, E.: Changes in prudential policy instruments—a new cross-country database. Int. J. Cent. Bank. **13**(2), 477–503 (2017b)

Claessens, S., Ghosh, S.R., Mihet, R.: Macroprudential policies to mitigate financial system vulnerabilities. J. Int. Money Financ. **39**, 153–185 (2013)

Clement, P.: The term "Macroprudential": origins and evolution. BIS. Quarterly Rev. **1**, 59–67 (2010)

Davis, E.P.: Towards a typology for systemic financial instability. Public Policy Discussion Papers 03–20, Economics and Finance Section, School of Social Sciences, Brunel University (2003)

Dell'Ariccia, G., Igan, D., Laeven, L., Tong, H.: Credit booms and lending standards: evidence from the subprime mortgage market. J. Money, Credit Bank. **44**(2–3), 367–384 (2012)

Demirgüç-Kunt, A., Detragiache, E.: The determinants of banking crises in developing and developed countries. IMF Staff. Pap. **45**, 81–109 (1998)

Demirgüç-Kunt, A., Detragiache, E.: The Determinants of Banking Crises in Developing and Developed Countries. IMF Staff Papers **45**, pp. 81–109, (1997)

Eugenio, E.: Macro-prudential policies dataset (2018). Available from. https://www.eugeniocerutti.com/Datasets [20 October 2020]

Flegal, J.M., Haran, M., Jones, G.L.: Markov chain Monte Carlo: can we trust the third significant figure? Stat. Sci. **23**(2), 250–260 (2008)

Foos, D., Norden, L., Weber, M.: Loan growth and riskiness of banks. J. Bank. Financ. **34**(12), 2929–2940 (2010)

Foot, M.: What Is "Financial Stability" and How Do We Get It? The Roy Bridge Memorial Lecture (United Kingdom: Financial Services Authority) (2003)

Freedman, C., Goodlet, C.: Financial stability: what it is and why it matters. C.D. Howe Institute Commentary, C.D. Howe Institute, issue 256, November (2007)

Freixas, X., Laeven, L., Peydró, J.-L.: Systemic Risk, Crises, and Macroprudential Regulation. The MIT Press (2016)

Galati, G., Moessner, R.: What do we know about the effects of macroprudential policy? Economica **85**(340), 735–770 (2018a)

Galati, G., Moessner, R.: What do we know about the effects of macroprudential policy? Economica **85**(340), 735–770 (2018b)

Gelman, A., Rubin, D.: Inference from iterative simulation using multiple sequences. Stat. Sci. **7**, 457–511 (1992)

Gelman, A., Hwang, J., Vehtari, A.: Understanding predictive information criteria for Bayesian models. Stat. Comput. **24**, 997–1016 (2014)

Hahm, J.H., Mishkin, F., Shin, H.S., Shin, K.: Macroprudential policies in open emerging economies. NBER Working Paper Series 17780, Cambridge, National Bureau of Economic Research, Massachusetts (2012)

Nguyen, H.T., Sriboonchitta, S., Thach, N.N.: On quantum probability calculus for modeling economic decisions. In: Kreinovich, V., Sriboonchitta, S. (eds.) Structural changes and their econometric modeling, TES 2019, Studies in Computational Intelligence. Springer, Cham, vol. 808, pp. 18–34 (2019)

Hung, N.T., Thach, N.N.: A closer look at the modeling of economics data. In: Kreinovich, V., Thach, N., Trung, N., Van Thanh, D. (eds.) Beyond traditional probabilistic methods in economics, ECONVN 2019, Studies in Computational Intelligence. Springer, Cham, p. 809 (2019)

Hung, N.T., Thach, N.N.: A panorama of applied mathematical problems in economics. Thai J. Math. Special Issue Ann. Meeting Math. **17**(1), 1–20 (2018)

IMF: Housing Finance And Financial Stability—Back to Basics. Chapter 3, Global Financial Stability Report, April 2011

Karanovic, G., Karanovic, B.: Developing an aggregate index for measuring financial stability in the Balkans. Proc. Econ. Financ. **33**, 3–17 (2015)

Khan, M.A., Kong, D., Xiang, J., Zhang, J.: Impact of institutional quality on financial development: cross-country evidence based on emerging and growth-leading economies. Emerg. Mark. Financ. Trade **56**(4), 1–17 (2019)

Kibritçioğlu, A.: Excessive risk-taking, banking sector fragility, and banking crises. U of Illinois, Commerce and Bus. Admin. Working Paper No. 02–0114 (2002)

Klomp, J., Haan, J.: Bank regulation, the quality of institutions, and banking risk in emerging and developing countries: an empirical analysis. Emerg. Mark. Financ. Trade **50**(6), 19–40 (2014)

Lim, C.H., Columba, F., Costa, A., Kongsamut, P., Otani, A., Saiyid, M., Wezel, T., Wu, X. Macroprudential policy: what instruments and how to use them? IMF Working Paper 11/238 (2011)

Mariëlle, Z., Peeters, M., Depaoli, S., Schoot, R.V.: Where do priors come from? Applying guidelines to construct informative priors in small sample research. Res. Hum. Dev. **14**(4), 305–320 (2017)

McNeish, D.M.: Using data-dependent priors to mitigate small sample bias in latent growth models: a discussion and illustration using Mplus. J. Educ. Behav. Statis. **41**, 27–56 (2016)

Mishkin, F.S.: Global financial instability: Framework, events, issues. J. Econ. Perspect. **13**(4), 3–20 (1999)

Morris, V.C.: Measuring and forecasting financial stability: the composition of an aggregate financial stability index for Jamaica. Bank of Jamaica (2010). http://boj.org.jm/uploads/pdf/papers_pamphlets/papers_pamphlets_Measuring_and_Forecasting_Financial_Stability__The_Composition_of_an_Aggregate_Financial_Stability_Index_for_Jamaica.pdf. Accessed 20 Sept 2020

Mörttinen, L., Poloni, P., Sandars, P., Vesala, J.: Analysing banking sector conditions. How to use Macro-prudential indicators. ECB Occasional Paper No 26 (2005)

Nabar, M., Ahuja, A.: Safeguarding banks and containing property booms; Cross-country evidence on macroprudential policies and lessons from Hong Kong SAR. IMF Working Papers 11/284 (2011)

Nelson, W.R. Perli, R.: Selected indicators of financial stability. In: 4th Joint central bank research conference on "risk measurement and systemic risk", ECB Frankfurt am Main, November 2005

Linh, N.T.X.: The impact of monetary policy and macroprudential policy on financial stability: the case of emerging and growth-leading economies. Doctoral Dissertation, Ho Chi Minh University of Banking (2021)

Pan, H., Wang, C.: House prices, bank instability, and economic growth: evidence from the threshold model. J. Bank. Financ. **37**(5), 1720–1732 (2013)

Rosengren, E.: Defining financial stability, and some policy implications of applying the definition. Speech from Federal Reserve Bank of Boston vol. 46 (2011a). https://www.bostonfed.org/-/media/Documents/Speeches/PDF/060311.pdf

Rosengren, E.: Defining financial stability, and some policy implications of applying the definition. Speech (June 3) Federal Reserve Bank of Boston (2011b). https://www.bostonfed.org/-/media/Documents/Speeches/PDF/060311.pdf. 8 Oct 2020

Schinasi, G.J.: Defining financial stability. IMF Working Papers 04/187, International Monetary Fund (2004)

Sriboonchitta, S., Nguyen, H.T., Kosheleva, O., Kreinovich, V., Nguyen, T.N.: Quantum approach explains the need for expert knowledge: on the example of econometrics. In: Kreinovich, V., Sriboonchitta, S. (eds.) Structural changes and their econometric modeling, TES 2019, Studies in Computational Intelligence, vol. 808. Springer, Cham (2019)

Svítek, M., Kosheleva, O., Kreinovich, V., Nguyen, T.N.: Why quantum (wave probability) models are a good description of many non-quantum complex systems, and how to go beyond quantum models. In: Kreinovich, V., Thach, N., Trung, N., Van Thanh, D. (eds.) Beyond traditional probabilistic methods in economics, ECONVN 2019, Studies in Computational Intelligence, vol. 809. Springer, Cham (2019)

Thach, Nguyen Ngoc.: The variable elasticity of substitution function and endogenous growth: an empirical evidence from Vietnam. International J. Econ. Bus. Admin. VIII: 263–77 (2020)

Tovar, C., Mercedes, E., Garcia-Escribano, Martin, M.V.: Credit growth and the effectiveness of reserves requirements and other macroprudential instruments in Latin America. IMF Working Paper 12/142 (2012)

van de Schoot, R.: 25 years of Bayes in psychology. Paper presented at the 7th Mplus Users' Meeting, Utrecht, The Netherlands (2016). http://mplus.fss.uu.nl/wp-content/uploads/sites/24/2012/07/opening-review-short.pptx. 21 Oct 2020

Van den End, J.W.: Indicator and boundaries of financial stability. DNB Working Papers 097, Netherlands Central Bank, Research Department (2006)

World Bank. World Development Indicators (2020). http://datatopics.worldbank.org/worlddevelopment-indicators/ . Accessed 20 Sept 2020

Fintech Credit, Digital Payments, and Income Inequality: Ridge and Bayesian Ridge Approach

Pham Thi Thanh Xuan and Nguyen Duc Trung

Abstract This paper investigates how fintech credit and financial inclusion affect income inequality for a panel of 60 countries. Ridge and Bayesian Ridge Regression provided conclusive results highlighting financial inclusion as driving factors reducing income inequality. Whereby the more usages of financial services, the less income inequality. We also found that extending digital payment contributes more than the other variables to significantly reducing income inequality across countries. Interestingly, the poorest group exerts a more substantial effect than the richest group concerning reducing income inequality. By contrast, credit supply via FinTech (i.e., FinTech credit) remains negligible indiscernible effect on income inequality.

1 Introduction

Nowadays, the multidimensional relationships among financial inclusion, fintech credit, and income inequality have emerged as an exciting topic promising to gain much attraction from researchers and practitioners. Indeed, fintech, the short name of financial technologic services, ie... the financial services offered through digital platforms, particularly fintech credit and financial inclusion, is considered important for a country's sustainable development.

Fintech credit, the credit supply through the fintech platform, has rapidly grown worldwide in recent years. It is considered an alternative source of financing for businesses and households and could improve access to credit for some underserved

P. T. T. Xuan (✉)
Faculty of Finance–Banking, University of Economics and Law, Vietnam National University, Ho Chi Minh City, Vietnam
e-mail: xuanptt@uel.edu.vn

Vietnam National University, Ho Chi Minh City, Vietnam

N. D. Trung
Banking University, Ho Chi Minh, Vietnam
e-mail: trungnd@buh.edu.vn

N. Ngoc Thach et al. (eds.), *Financial Econometrics: Bayesian Analysis, Quantum Uncertainty, and Related Topics*, Studies in Systems, Decision and Control 427,
https://doi.org/10.1007/978-3-030-98689-6_41

segments (a part of people and businesses are still excluded from traditional financial sources. It could also enhance the efficiency of financial intermediation. Thus, Fintech credit could help stimulate the economy and reduce income distribution inequality among different income classes. Surprisingly, to the best of our knowledge, no previous studies reveal the potential impact of fintech credit on income inequality. This, thus, motivates this study.

Another motivation of this study comes from the controversy on the impact of financial inclusion on income inequality. No theory has been legitimated yet due to this controversy. Numerous cross-country studies pointed out that higher levels of financial inclusion are associated with lower levels of income inequality (Allen et al. 2016; Mookerjee and Kalipioni 2010; Sahay et al. 2015). Demir et al. (2020), via quantile regression analysis, also finds evidence that financial inclusion is a key channel, which reduces income inequality. Meanwhile, other studies showed the inverse (Kochar 2011; Dimova and Adebowale 2018). Even others demonstrated that this relationship might change over time (Huang and Zhang 2020). A possible explanation for this controversy could be due to the different research methodologies applied across studies. Another reason for that is that each study uses a distinguished proxy for financial inclusion. The questionable point is which one should be used as a presentative variable for financial inclusion while there are over 60 sub-indicators for financial inclusion measurement. This questionable issue strongly motivates this study.

This study is carried out on a sample of 60 countries. The key point of this study is a cross-country analysis using advanced data analysis techniques, including Ridge and Bayesian Ridge regressions instead of traditional OLS regressions. With the support of these two models, this study allows us to simultaneously evaluate the effects of all three proxies of financial inclusion (financial account, saving account, and digital payment adoption) on income inequality. We could say that no study has ever reached this aspect so far.

Our study aims to answer the following question: How do fintech credit and financial inclusion affect income inequality? By clarifying it, this paper seeks to make two main contributions. First, to the best of our knowledge, this is the first study to provide evidence of a link between fintech credit and income inequality at the cross-country level. Second, this study revisits the impact of financial inclusion on income inequality, thus reaching a unified conclusion for the literature. Besides, this study makes it possible to clarify the role of digital payment adoption, which has also not been studied before. This work helps enrich the knowledge in the field.

The paper is structured as follows. Section 2 presents a literature review on fintech credit, financial inclusion, and income inequality. In Sect. 3, database and econometric methodology are presented. Section 4 reports the results. Finally, Sect. 5 concludes.

2 Literature Review

Empirical evidence on the impact of financial inclusion on income inequality is numerous but controversial.

Numerous studies showed that financial inclusion reduces income inequality by providing broader and easier access to different financial resources for low-income households and small businesses. Mookerjee and Kalipioni (2010) argued that the higher the percentage of bank branches in the country, the lower the income inequality. Supported by Honohan (2008), there is a significant negative relationship between the share of the adult population owning an account at a financial institution and income inequality. The higher the financial inclusion, the lower the income inequality level. This fact is also confirmed by Demir et al. (2020), who investigates the interrelationship between fintech, financial inclusion, and income inequality for a panel of 140 countries. Demir et al. (2020) pointed out that Fintech is a key channel through which financial inclusion reduces income inequality, at all quantile levels, primarily among higher-income countries.

However, other empirical studies demonstrated the inverse (Kochar 2011; Dimova and Adebowale 2018). Kochar (2011), working on cross-sectional data of households in Uttar Pradesh, India, indicated that not all households benefit equally from expanding the banking services system. This asymmetric is evident, especially in rural areas. The expansion of rural bank branches has not increased access to formal financial services by the poor. Further, it improves access to credit for the rich. As a result, income inequality continues to be severe. Huang and Zhang (2020) found an inverse effect from the short run to the long run. Financial inclusion increases urban–rural income inequality in the short run. Still, it reduces it in the long run by the grace of upgrading financial infrastructure and financial awareness in rural areas over time.

This controversial issue is still a big question for researchers in this field. A possible explanation for this could be due to the different research methodology applied. Indeed, financial inclusion has many representative variables, and each author chooses a proxy for financial inclusion. While Demir et al. (2020) use the proportion of the population with an account or a saving account at a formal financial institution as a proxy, Huang and Zhang (2020) use sub-dimensional indexes of financial inclusion, including financial accessibility and availability.

To clarify more the literature, we decide to revisit the nexus between financial inclusion and inequality by taking a new approach and new proxies that might best represent financial inclusion. Furthermore, the impact of fintech credit on income inequality has been taken into account. We believe that this study is the very first one focusing on this relationship.

3 Model Design, Database, and Econometric Approach

3.1 Model Design

This study aims to examine two main hypotheses as below:

- H1: FinTech Credit has a significant impact on reducing income inequality.
- H2: Extension of financial inclusion helps reduce income inequality.

The proof of these hypotheses is the core of this study compared to previous studies. For that purpose, to examine the impact of Fintech Credit and Financial Inclusion on income inequality within-country, we develop the following model:

$$Inequality_i = \alpha_0 + \alpha_1 FinCredit_i + \alpha_2 Financial\,inclusion_i + \sum_{k=1}^{K} \beta_k X_{ki} + u_i \quad (1)$$

where,

Inequality represents income inequality as measured by the Gini index (Demir et al. 2020), having values ranging from 0 (perfect equality) to 100 (perfect inequality). Gini index is the most widely used measure of income inequality in the literature on the relationship between finance and inequality (Beck et al. 2007; Haan and Sturm 2016; Jauch and Watzka 2015).

Fintech credit (Fincredit) is the credit supplied through fintech platforms. Fintech credit is an emerging term that has appeared and been rapidly developing due to bringing in Fintech into the financial industry. According to Frost and Turner (2018), Fintech credit was defined to "include all credit activity facilitated by electronic (online) platforms that commercial banks do not operate." In this study, Fintech credit plays as an independent variable to test the impact of Fintech on income inequality by promoting digital credit activity.

Financial inclusion is broadly defined as access to and use of formal financial services by households and firms (Honohan 2008; Mookerjee and Kalipioni 2010; Hermes 2014; Sahay et al. 2015; Kim 2016; Aslan et al. 2017; Park and Mercado 2018; Turegano and Herrero 2018).

Following the database of Demir et al. (2020), 60 subindexes are measuring financial inclusion. This study takes 3 of them as proxies based on the following argument:

Financial account (Finaccount) is measured by the share of the adult population (aged 15 years and older) owning an account at a formal financial institution, following Allen et al. (2016), Demir et al. (2020), Khera et al. (2021).

Saving account (Saved) is measured by the proportion of adults owning a saving account at a formal financial institution, following Allen et al. (2016), Demir et al. (2020), Khera et al. (2021).

Digital payment adoption (Digitalpayment) is measured by the proportion of adults who made or received digital payments in the past year (%). It refers to the percentage of respondents who reported using digital payment methods in the past 12 months. Including mobile money, debit or credit card, a mobile payment, paying for bills through the internet, or buying something online. This variable also includes respondents who reported using the following financial services in the past 12 months. Including paying bills, sending, or receiving remittances, receiving payments for agricultural products, receiving government transfers, receiving wages, or receiving a public sector pension directly from or into a financial institution account or through a mobile money account in the past 12 months. ***Digital payment adoption*** is our new proxy first recommended in this model. There has been no official research yet on the relationship between digital payment and income inequality. However, numerous studies have proved that digital payment adoption helps to stimulate economic growth. Liu et al. (2020) found that the rapidly developed digital payment platforms have greatly reduced financial services transaction costs, improving payment and transfer efficiency for household consumption.

Besides, we include in the model the set of control variables X, including population growth (Popgrowth) (annual %) (Demir et al. 2020); GDP per capita growth (GDPCapitagrowth) (annual %) (Demir et al. 2020); financial development (FinDevelopment) (Demir et al. 2020); trade (% of GDP) (Demir et al. 2020); regulatory quality (Hermes 2014; Lacalle-Calderon et al. 2019; Mookerjee and Kalipioni 2010); fixed broadband subscriptions per 100 people (BroadAccess) (Andrianaivo and Kpodar 2012; Demirgüç-Kunt et al. 2018; Ghosh 2016), and general government final consumption expenditure (% of GDP) (GovConsumption) (Demir et al. 2020).

It's important to note that, different from Demir et al. (2020), in this study, we do not use lagged variables for two reasons. Firstly, we only use the dataset of 2017, which is the most completed; hence the lagged variable cannot be applied. Second, even if the data are available for all three years (2011, 2014, and 2017), the lagged value for 2017 is 2014; hence, adding lagged variables does not make much significance, according to Minges (2015).

3.2 Database

Data were collected from three main sources: Gini indices representing income inequality from the Standardized World Income Inequality Database (SWIID), financial inclusion from the Global Financial Inclusion Database (Findex), and FinTech Credit (FinCredit) from Cornelli et al. (2020). FinCredit was chosen as a proxy for FinTech in this first investigation into FinCredit's relevance to income inequality.

Data for the control variables were collected from the World Development Indicators Database and the Worldwide Governance Indicators. Data were collected from 60 countries in 2017, representing the most extensive sample currently available.[1]

3.3 Econometric Approach

We first applied correlation tests between the variables in the model. We found that multicollinearity occurs because the three proxies of financial inclusion (financial account, saving account, and digital payment) are highly correlated. Therefore, to include these three variables simultaneously, traditional regressions such as Panel OLS are not available. We have tested a battery of regressions; two algorithms are adopted: Ridge and Bayesian Ridge Regression.

Ridge and Bayesian ridge regression were adopted for three reasons: (1) to solve the potential multicollinearity problem existing in the main interest variable—financial inclusion and FinTech (Assaf et al. 2019; Bager et al. 2017); (2) to allow many proxies of financial inclusion to be simultaneously included in the model, rather than one variable at a time, as in Demir et al. (2020), or the use of a composite index, like Sarma (2012). Moreover, these models, which provide the regression coefficients, make it possible to identify the effects of Fintech Credit and financial inclusion on income inequality. Finally, Ridge and Bayesian ridge regression perform well even in small sample settings.

Ridge Regression

Ridge regression uses regularization, which looks for β to minimize the sum of squares of residuals and a penalty term:

$$\min L = \sum_{i=1}^{n} \left(y_i - \sum_{j=0}^{m} \beta_j x_{ij}\right)^2 + \lambda \sum_{j=0}^{m} \beta_j^2 \quad (2)$$

$\lambda \sum_{j=0}^{m} \beta_j^2$ is a penalty function with $\lambda (\lambda \geq 0)$ is a parameter (penalty coefficient) that controls the impact of the penalty.

Ridge Regression is a technique used when the data suffers from multicollinearity (independent variables are highly correlated). Even though the least squares estimates (OLS) are unbiased in multicollinearity, their variances are significant, deviating the observed value far from the true value. By adding a degree of bias to the regression estimates, ridge regression reduces the standard errors.

[1] While data for more than 60 countries are included in each of database individually, only 60 countries have complete data across all three sources for the year 2017. Hence, this is the largest sample currently possible.

Bayesian Ridge Regression

While Ridge regression uses the regularization method, Bayesian regression is defined in probabilistic terms, with explicit priors on the parameters.

Model and Assumptions: Let observations $y = (y_1,...y_n)^T \in R^n$. Let features $X = [1_n, x_1, \ldots, x_m] \in R^{n(m+1)}$, where x_i are column vector in R^n, $i = 1, \ldots, m$, and $1_n = (1, \ldots, 1)^T \in R^n$.

We assume that each y_i has likelihood.

$$p(y_i|\omega, \alpha) = N(y_i| \sum\nolimits_{j=1}^{m} \omega_i X_j, \alpha) \tag{3}$$

Where $\omega = (\omega_0, \omega_1, \ldots, \omega_m)^{T \in} R^m$ weights and $\alpha \in R$ the variance (standing for the noise).

We assume that each ω_j is drawn from.

$p(\omega_j|\lambda) = N(\omega_j|0, \lambda^{-1})$.

Furthermore, we assume that α and λ are Gamma distributed, i.e.

$p(\alpha) = \Gamma(\alpha|\alpha_1, \alpha_2)$.

and.

$p(\lambda) = \Gamma(\lambda|\lambda_1, \lambda_2)$.

where we choose $\alpha_1 = \alpha_2 = \lambda_1 = \lambda_2 = 10^{-6}$ as default value, in sklearn.

The posterior in Bayesian Ridge Regression. Based on Bayesian inference, the posterior is.

$p(\omega, \alpha, \lambda)\, p(y|\omega, \alpha, \lambda)\, p(\omega, \alpha, \lambda)$.

where $p(y|\omega, \alpha, \lambda)$ and $p(\omega, \alpha, \lambda)$ can be computed explicitly as follows,

$p(y|\omega, \alpha, \lambda) = p(y|\omega, \alpha) = \prod_{i=1}^{n} p(y_i|\omega, \alpha)$.

and.

$p(\omega, \alpha, \lambda) = p(\alpha)p(\omega,\lambda) = p(\alpha)p(\omega|\lambda)p(\lambda) = p(\alpha)p(\lambda)\prod_{j=1}^{m} p(\omega_j|\lambda)$.

$p(y_i|\omega, \alpha)$, $p(\omega_j|\lambda)$, $p(\alpha)$ and $p(\lambda)$ have been defined as above.

Estimation. The estimated weighted $\widehat{\omega} = (\widehat{\omega}_0, \widehat{\omega}_1, \ldots, \widehat{\omega}_n)$ and estimated parameter $\widehat{\alpha}, \widehat{\lambda}$ are defined as $(\widehat{\omega}, \widehat{\alpha}, \widehat{\lambda}) = \arg\max p(\omega, \alpha, \lambda)$.

Bayesian Regression can be very useful when we have insufficient data in the dataset or the data is poorly distributed. The output of a Bayesian Regression model is obtained from a probability distribution compared to traditional regression techniques where the output is just received from a single value of each attribute. The output, 'y', is generated from a normal distribution (where mean and variance are normalized). Bayesian Linear Regression aims not to find the model parameters but rather the "posterior" distribution for the model parameters. Thus for Bayesian Ridge Regression, a large amount of training data is needed to make the model accurate.

In this study, estimation of this model using ridge and Bayesian ridge regression helps determine whether and how the growth of FinTech and financial inclusion affects income inequality. In the interest of a comparison, we applied all tests for the richest and poorest groups and the full sample.

4 Empirical Analysis

4.1 Feature Selection

First, we conducted two preliminary methods to identify which variable could be included in the model, including Pearson's correlation coefficient and Feature importance.

Table 1 presents the correlation coefficients for income inequality's predictors in the full sample and two subsamples of the richest and the poorest. Feature importance tests for income inequality's predictors are shown in Figs. 1, 2, and 3.

Table 1 Correlation

	Feature	Full sample	Richest	Poorest
1	Digital payments	−0.6543	−0.6123	−0.6973
2	Financial account	−0.6410	−0.6014	−0.6791
3	Saved Account	−0.6461	−0.6178	−0.6889
4	BroadAccess	−0.5891		
5	GovConsumption	−0.2087		
6	FinCredit	0.1847		
7	RegulatoryQuality	−0.3338		
8	GDPCapita growth	−0.1851		
9	Population growth	0.4353		
10	FinDevelopment	−0.4126		
11	Trade	−0.4752		

Source Author's estimation

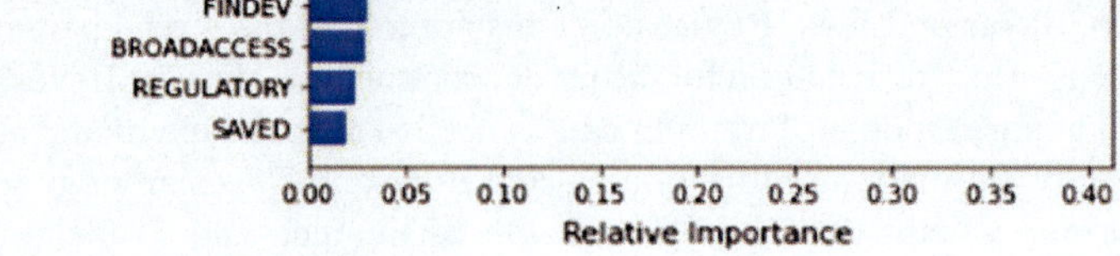

Fig. 1 Feature importance_Full Sample

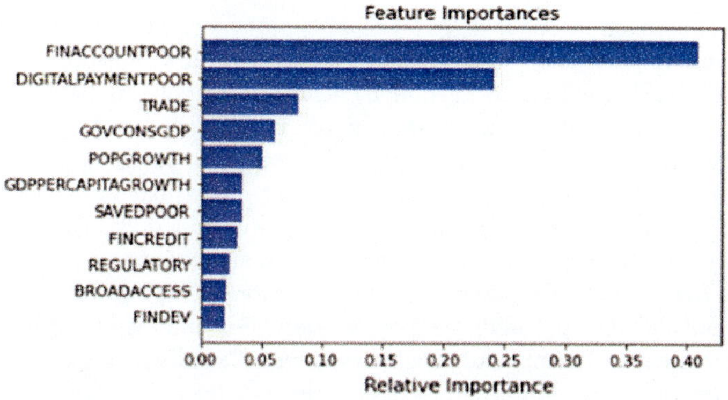

Fig. 2 Feature importance_Poorest

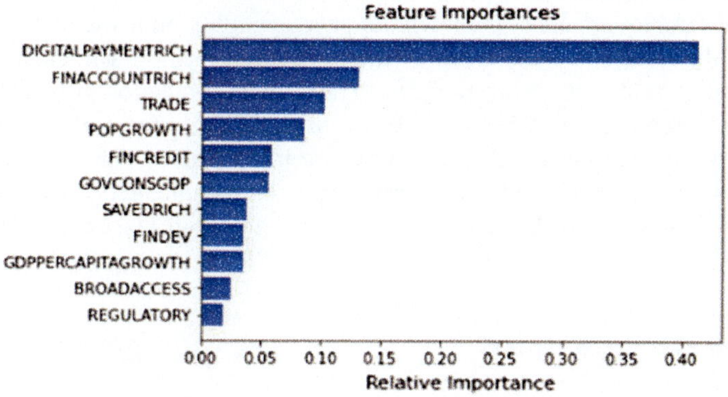

Fig. 3 Feature importance_Richest

Of 60 Findex indicators, the correlation test highlights the three most highly correlated with income inequality, with the coefficient (in absolute value) higher than 0.6 among all three samples. So, they are selected as proxies for financial inclusion. The sign of these coefficients indicates that financial inclusion and income inequality are negatively correlated. Next, it is impossible not to mention another essential predictor for income inequality which is "Broad Access," representing Fixed broadband subscriptions per 100 people. Besides, we also select into the model other indicators summarizing perspectives on government concern and governance quality (GovConsumption, RegulatoryQuality); development of the economy (GDPcapita growth, population growth, trade) and development of Fintech (FinDevelopment and FinCredit). Table 1 demonstrates that all variables have significant relationships with income inequality. However, the variable with the lowest correlation with income inequality is FinCredit (0.1847), suggesting something entirely unexpected and exciting.

Another approach to better determines the significance of an independent variable is feature importance. We represent three figures below illustrating the feature importance of each predictor for income inequality corresponding to the full sample and two subsamples of the richest and the poorest.

The most striking result to emerge from the feature importance tests is that the two most important features are 'digital payment' and 'financial account.' They significantly impact income inequality (with significant scores of 0.38 and 0.23, respectively), consistently occupy the top positions, and maintain a precise distance from the others across techniques and samples, as visualized in Figs. 1, 2, and 3. In other words, an increase in digital payment adoption and the use of banking and financial services significantly contribute to income inequality reduction in the rich or the poor sample. This reinforces the model design and the correlation results above. In line with the previous studies, this is entirely reliable and confirms that financial inclusion clearly impacts income inequality. However, the feature importance tests do not prove the significant effect of 'Saving account.' A slight difference between the two subsamples appeared here. Digital payment adoption is the most important feature in the rich, while the share of the population owning a financial account plays a leading role in the poor. Besides, 'Trade' and 'Population growth' emerged as more essential than the other variables, with significant scores of 0.09 and 0.08, respectively. Concerning FinCredit, the feature importance yields the same result as predictive power. FinCredit has a weak impact on income inequality, both in the lowest and richest income households, with significant scores close to 0.05.

4.2 Regressions Results

Table 2 presents the Ridge and Bayesian Ridge regression analysis results for income inequality, respectively, on the entire sample, the richest and the poorest income households.

Ridge regression and Bayesian Ridge regression provide consistent results.

In all cases, all three proxies of financial inclusion, 'digital payments,' 'saved account,' and 'financial account,' are strongly linked to income inequality with negative coefficients. This indicates that the extension of financial inclusion reduces income inequality significantly. This result provides a robust answer for the first hypothesis. This is also consistent with those of earlier studies (Demir et al. 2020) and sheds new light on the debate of financial inclusion on income inequality.

The negative sign of the coefficients for 'financial account' indicates that an increase in the share of the population owning an account at a formal financial institution is associated with a reduction in income inequality. In Bayesian Ridge Regression, a 1-percentage-point increase in formal account ownership results in a 0.0241– and 0.0376 percentage-point reduction in the Gini coefficient in the richest and the poorest, respectively. The same increase in formal savings leads to a 0.047– and 0.051-percentage-point decrease in the Gini index for the rich and poor, respectively.

Table 2 Regression coefficients

	Feature	Ridge regression			Bayesian Ridge regression		
		Full sample	Richest	Poorest	Full sample	Richest	Poorest
1	Digital payments	−0.0283	−0.0244	−0.0328	−0.0477	−0.0376	−0.0591
2	Financial account	−0.0213	−0.0177	−0.0254	−0.0313	−0.0241	−0.0376
3	Saved Account	−0.0243	−0.0260	−0.0219	−0.0488	−0.0470	−0.0517
4	BroadAccess	−0.0128	−0.0137	−0.0114	−0.0274	−0.0280	−0.0238
5	GovConsumption	−0.0014	−0.0015	−0.0011	0.0013	−0.0004	0.0034
6	FinCredit	3.410E-05	3.523E-05	3.219E-05	0.0000	0.0000	0.0000
7	RegulatoryQuality	−0.0056	−0.0059	−0.0052	−0.0007	−0.0020	0.0004
8	GDPCapita growth	−0.0037	−0.0036	−0.0038	−0.0037	−0.0037	−0.0037
9	Population growth	0.0120	0.0126	0.0113	0.0091	0.0101	0.0086
10	FinDevelopment	−0.0097	−0.0109	−0.0079	0.0032	−0.0047	0.0153
11	Trade	−0.0175	−0.0180	−0.0167	−0.0218	−0.0225	−0.0203

Source Author's estimation

Noticeably, the increase in digital payment adoption has the most substantial impact on reducing income inequality, especially in the poorest. Bayesian Ridge estimates showed that a 1-percentage-point increase in formal account ownership results in a 0.0376– and 0.0591 percentage-point reduction in the Gini coefficient in high-income and low-income households, respectively. The extension of digital payment benefits individuals, households, and small businesses, directly and indirectly, reduce income inequality.

Comparison of these effects across the richest and poorest groups yields an exciting finding. The poorest has a more substantial impact than the richest group on reducing income inequality, with higher absolute coefficients in all cases. To reduce income inequality, the behavior of the poor plays a critical role.

An equally striking finding is that "digital payments" take the most important influence on income inequality with a significant gap with the followed variables 'saved account' and 'financial account'. As such, 'digital payment' contains more valuable information that helps accurately predict income inequality than 'financial account' contains. This point represents the new and convincing evidence that we anticipated.

On the other hand, unfortunately, the results rejected the second hypothesis, showing that Fintech Credit has no noticeable impact on inequality. Regression results show that Fintech credit has very weak regression coefficients (close to zero). A possible explanation for this is that credit through financial technology platforms was still too modest, so its impact on the economy and income inequality are not easy to capture.

Next, two other important variables are broad access and trade. More available broad access to the internet encourages e-commerce, digital payments. These help boost consumption, saving, investment, and then economic growth and can reduce poverty. Trade development both improves income and reduces inequality. Countries with higher trade openness will have higher living standards and better income redistribution between the rich and the poor. Then, population growth and economic growth directly affect income redistribution. The higher the country's population, the greater the income inequality, typically in India and China.

Furthermore, it is easier for countries with stronger economies to distribute income between the rich and the poor. Besides, government policy also plays a crucial role in income redistribution. Government acts through two channels: consumption and regulation. Any government that cares about this income inequality problem will spend much budget to have the right strategies. Also, better regulation quality leads to better social security, helps to lower income inequality. Finally, it is impossible not to mention the impact of fintech development. Fintech, through the economic and financial development channel, encourages e-commerce, investment, saving, and business growth, contributing to poverty reduction.

5 Conclusion

Our study aims to find the answer to an essential question of the literature: How do fintech credit and financial inclusion, including digital payment adoption, affect income inequality? To respond to this question, we conducted an empirical investigation on a panel of 60 countries in 2017. To the best of our knowledge, this is the first study to investigate the impact of fintech credit and digital payment adoption on income inequality at the cross-country level.

As expected, Ridge and Bayesian Ridge estimation provided new empirical evidence but conclusive results, which helps to shed more light on the relationship of financial inclusion to income inequality. By enlarging access to financial services, financial inclusion helps reduce income inequality in the richest and the poorest, corroborating Demir et al. (2020) findings. Primarily, our study outlines a new finding that digital payment adoption is one of the most effective channels to reduce inequality. No study has ever touched this aspect so far. Interestingly, the poorest group exerts a more substantial effect than the richest group concerning reducing income inequality. This exciting finding proves that their behavior helps them to escape poverty.

Unfortunately, this study did not capture the impact of Fintech Credit on income inequality, with the regressive coefficient estimated close to zero. A possible explanation for this might be the amount of credit supplied through this new technology channel is still small. This is still an open question for future research which should significantly pay more attention to the hidden impact of this new credit supply.

Our research results not only fill the research gap in the literature but also provide important policy implications. First, financial sector policies to reduce income inequality should emphasize developing digital payment services and building the banking infrastructure system that benefits the poor and low-income groups. Second, to strengthen the impact of financial inclusion on reducing inequality, it is vital to improve technological infrastructure and regulatory infrastructure for digital payments. Finally, although this study does not capture the influence of fintech credit with success, policymakers and financial institutions should still promote this new credit supply channel. We believe this could improve access to credit for some underserved segments, contributing to economic growth and poverty reduction in the future, particularly in the COVID-19 environment.

Acknowledgments This research is funded by University of Economics and Law (VNU-HCMC) under grant number: CS/2021-11.

References

Allen, F., Demirguc-Kunt, A., Klapper, L., Martinez Peria, M.S.: The foundations of financial inclusion: Understanding ownership and use of formal accounts. J. Financ. Intermed. **27**, 1–30 (2016)

Andrianaivo, M., Kpodar, K.: Mobile phones, financial inclusion, and growth. Rev. Econom. Inst. **3**(2), 30 (2012)

Aslan, G., Deléchat, C., Newiak, M.M., Yang, M.F.: Inequality in financial inclusion and income inequality. In: International Monetary Fund (2017)

Assaf, A.G., Tsionas, M., Tasiopoulos, A.: Diagnosing and correcting the effects of multicollinearity: Bayesian implications of ridge regression. Tourism Manag. **71**, 1–8 (2019)

Bager, A., Roman, M., Algedih, M., Mohammed, B.: Addressing multicollinearity in regression models: a ridge regression application (2017)

Beck, T., Demirgüç-Kunt, A., Levine, R.: Finance, inequality and the poor. J. Econ. Growth **12**(1) (2007). https://doi.org/10.1007/s10887-007-9010-6

Cornelli, G., Frost, J., Gambacorta, L., Rau, P.R., Wardrop, R., Ziegler, T.: Fintech and big tech credit: a new database, No. 887. BIS Working Paper (2020)

Demir, A., Pesqué-Cela, V., Altunbas, Y., Murinde, V.: Fintech, financial inclusion and income inequality: a quantile regression approach. Eur. J. Financ. 1–22 (2020). https://doi.org/10.1080/1351847X.2020.1772335

Demirgüç-Kunt, A., Klapper, L., Singer, D., Ansar, S., Hess, J.: The Global Findex Database 2017: Measuring Financial Inclusion and the Fintech Revolution. Washington, DC: World Bank. Digital Innovations and Societal Transformation: M-Pesa in Kenya, 81 (2018)

Dimova, R., Adebowale, O.: "sa", J. Develop. Stud. Taylor & Francis **54**(9), 1534–1550 (2018)

Frost, J., Turner, G.: Fintech credit markets around the world: size, drivers and policy issues. In: BIS Quarterly Review, September, pp. 29–49 (2018)

Ghosh, S.: Does mobile telephony spur growth? Evidence from Indian states. Telecommun. Policy **40**(10–11), 1020–1031 (2016)

De Haan, J., Sturm, J.: Finance and Income Inequality: A Review and New Evidence (CESifo Working Paper, No. 6079 Provided) (2016)

Hermes, N.: Does microfinance affect income inequality? Appl. Econ. **46**(9), 1021–1034 (2014)

Honohan, P.: Cross-country variation in household access to financial services. J. Bank. Finance **32**(11), 2493–2500 (2008)

Huang, Y., Zhang, Y.: Financial inclusion and urban–rural income inequality: Long-run and short-run relationships. Emerg. Mark. Financ. Trade **56**(2), 457–471 (2020)

Jauch, S., Watzka, S.: Financial development and income inequality: a panel data approach. Empiric. Econ. (2015). https://doi.org/10.1007/s00181-015-1008-x

Khera, P., Ng, S., Ogawa, S., Sahay, R.: Measuring Digital Financial Inclusion in Emerging Market and Developing Economies: A New Index. IMF Publications **2021**(090) (2021)

Kim, J.-H.: A study on the effect of financial inclusion on the relationship between income inequality and economic growth. Emerg. Mark. Financ. Trade **52**(2), 498–512 (2016)

Kochar, A.: The distributive consequences of social banking: a microempirical analysis of the indian experience. Econ. Dev. Cult. Change **59**(2), 251–280 (2011). https://doi.org/10.1086/657122

Lacalle-Calderon, M., Larrú, J.M., Garrido, S.R., Perez-Trujillo, M.: Microfinance and income inequality: new macrolevel evidence. Rev. Dev. Econ. **23**(2), 860–876 (2019)

Liu, D., Jin, Y., Pray, C., Liu, S.: The effects of digital inclusive finance on household income and income inequality in China? (2020). https://ageconsearch.umn.edu/record/304238/files/18011.pdf

Minges, M.: Exploring the relationship between broadband and economic growth, background paper: digital dividends', World Development Report, pp. 1–21 (2015)

Mookerjee, R., Kalipioni, P.: Availability of financial services and income inequality: the evidence from many countries. Emerg. Mark. Rev. **11**(4), 404–408 (2010)

Park, C.-Y., Mercado JR, R.: Financial inclusion, poverty, and income inequality. Singapore Econ. Rev. **63**(01), 185–206 (2018)

Sahay, R., Cihák, M., N'Diaye, P., Barajas, A.: Rethinking financial deepening: Stability and growth in emerging markets | Repensar La Profundización Financiera: Estabilidad Y Crecimiento En Los Mercados Emergentes. Revista de Economia Institucional (2015)

Sarma, M.: Index of Financial Inclusion–A Measure of Financial Sector Inclusiveness, Centre for International Trade and Development, School of International Studies Working Paper Jawaharlal Nehru University. Delhi (2012)

Turegano, D.M., Herrero, A.G.: Financial inclusion, rather than size, is the key to tackling income inequality. Singapore Econ. Rev. **63**(01), 167–184 (2018)

Factors Influencing the Financial Distress Probability of Vietnam Enterprises

Nguyen Duc Trung, Bui Dan Thanh, Bui Ngoc Mai Phuong, and Le Thi Lan

Abstract The purpose of the research is to identify the impact factors on the likelihood of financial distress of Vietnamese companies in the period from 2014 to 2019. Accordingly, the paper has used several research models and estimation methodologies with regards to a data set of 623 Vietnamese enterprises. The research applied regression model to pinpoint macro and micro factors that influence an enterprise financial distress. These elements are financial leverage, company size, net working capital to current assets ratio, retained earnings to total assets ratio and pre-tax profit ratio and interest on total assets. The dependent variable is the probability of occurrence or non-occurrence of financial distress of Vietnamese companies. By taking advantage of Bayesian Binary Logistic regression, the results reveal that financial distress (DIS) level increases as leverage (LEV) increases. While company size (SIZE), the ratio of net working capital to short-term asset (WC), the ratio of retained earnings to total assets (RETA) as well as the ratio of profit before interest and taxes to total assets (NITA) together have a negative impact on financial distress (DIS).

Keywords Financial distress · Listed companies · Binary logistic model

1 Introduction

The economy in Viet Nam went through multiple ups and downs from 2014 to 2019. Upon results of the General Statistics Office in 2019, there were 91,282 enterprises that went bankrupt or stopped operating, of which 92% were small and medium

N. D. Trung · B. D. Thanh · B. N. M. Phuong (✉) · L. T. Lan
Banking University of Ho Chi Minh City, 36 Ton That Dam Street, District 1, Ho Chi Minh City, Vietnam
e-mail: phuongbnm@buh.edu.vn

N. D. Trung
e-mail: trungnd@buh.edu.vn

B. D. Thanh
e-mail: thanhbd@buh.edu.vn

N. Ngoc Thach et al. (eds.), *Financial Econometrics: Bayesian Analysis, Quantum Uncertainty, and Related Topics*, Studies in Systems, Decision and Control 427,
https://doi.org/10.1007/978-3-030-98689-6_42

sized enterprises (business with charter capital under 10 billion VND). However, in the same period of the year, Vietnam recorded a tremendous number with officially reported indicator as of 138,140 newly established firms. Thus, on average, the number of enterprises went bankrupt, dissolved or ceased to operate accounted for 66% of the number of newly established enterprises. If we look back into history, such as in 2014, this figure is even more impressive, every 9 out of 10 newly established enterprises declared bankruptcy (67,823 bankrupt enterprises, and 74.842 established enterprises). The former only covers public data on bankruptcy and dissolution of businesses, have not covered businesses that were in financial distress, having a high debt situation and unable to pay, or in danger of bankruptcy but still trying to manage their own business without publicly disclosing information transparently. Financial distress is certainly not a main cause of bankruptcy, but a bankrupt business is bound to experience financial distress. This leads to the need of forecasting the possibility of financial distress, which has been increasingly attracting the attention of investors, creditors and managers. There are published studies on corporate financial distress, for example, Altman, Graham et al. (2011), Altman (2000), Mario Hernandez Tinoco, Tinoco and Wilson (2013), Platt, HD and Platt, MB (2006), etc. Besides, in Vietnam there are also some works on this topic, such as Pham Thi Hong Van (2015). As a result, our group decided to conduct this research with an aim to point out the determinants on the probability of financial distress occurrence of Vietnamese enterprises, assessing the level of impact, identifying shortcomings and recommendations for improving the effectiveness through Bayesian Binary Logistic regression.

2 Research Overview and Research Methodology

2.1 Theoretical Basis

On the first approach, a company is considered as financially distressed when it has missed interest payments or breached its debt covenants. This definition clearly distinguishes financial distress from bankruptcy: A company can go bankrupt without defaulting to bankruptcy. Nevertheless, bankruptcy is unlikely to happen without an earlier period of financial distress.

The second aspect is to restructure debt in order to prevent default, then: Purnanandam agreed with Gilbert's (1990) study that financial distress has different financial characteristics than bankruptcy: Financial distress is characterized by negative cumulative earnings for at least one or a few continuous years, losses, and poor performance. Bankruptcy is one of the possible outcomes of financial distress.

In conclusion, financial distress is a status when a business fails to pay its debts or has difficulty in debt settlement. Sometimes financial distress leads to bankruptcy, sometimes it just simply means the company is in financial trouble. However, a company that has gone bankrupt certainly experienced financial distress. In that sense, the states of financial distress include: failure, insolvency, default, bankruptcy and dissolution. Nonetheles, in common practice these states are very difficult to observe (Hashi 1997). Due to the limited availability of data, views on financial distress are different among researchers.

As mentioned earlier, due to the limitations in availability of data as well as the variety and complexity of distress patterns, the absence of clear boundaries between periods of financial distress makes perspectives about financial distress vary among researchers (Wruck 1990). Therefore, it is necessary to clarify the financial distress opinions to find the most suitable financial distress view. According to Atlman and Hotckis (2005), firms in financial distress can be identified by the difficult periods they have gone through. There are four general stages discovered in the literature on financial distress to distinguish a company as a financial distress: Failure, Insolvency, Default and Bankruptcy.

As regulated in Decree 58/2012/NĐ-CP of Vietnam government, there are detailed regulations when an enterprise is warned, controlled, delisted (negative profit after tax on financial statements or negative undistributed profit after tax), an enterprise has the similarity with the signs mentioned above or is delisted (an enterprise has suffered losses for three consecutive years or the total accumulated loss exceeds the amount of actually authorized capital in the audited financial statements of the most recent year; the enterprise is not accepted by the audit organization or refuses to comment on the last year's financial statements; the Stock Exchange and the Securities Commission detect that the enterprise has falsified listing documents, or listing documents contain serious false information that affects investors' decisions). These signs are inarguably one of the most important factors reflecting businesses in a state of financial distress.

2.2 Research Overview

Altman researched to evaluate the possibility of bankruptcy, as an illustrative case, the method of analyzing financial ratios and comparing it with a specific number to show the current financial status of enterprise. This research provided a discriminant model in which an observed sample of 60 companies was classified into two groups: bankrupt and non-bankrupt companies. As already stated, the author divided sample into two groups 30–30: (1) Bankruptcy group. (2) Companies which are not bankrupt. A total of 22 variables were statistically counted as potentially useful variables, which were then grouped into 5 standard ratios: liquidity, leverage, solvency, activity ratio. The model testing was conducted and showed that there are 5 variables having an impact on the bankruptcy of the business: Working capital/Total assets, Retained income/Total assets, Earnings before interest and taxes/Total assets, Equity/Book value of total debt, Sales/Total assets. The study has proposed a set of financial ratios combined with an analytical approach to the problem of predicting corporate bankruptcy. The model results proved to be extremely 94 percent accurate in predicting bankruptcy. One limitation of the study is that the companies examined are all publicly available for comprehensive production, these data companies can be easily obtained, including market value. The study will extend the scope of analysis to companies with relatively smaller assets and unincorporated companies.

Graham et al. (2011) conducted a research on the factors affecting insolvency and leading to financial distress during 1926–1938 (the crisis period) and 2008–2009 (normal period). With the sample of 443 non-financial companies, by means of logistic regression, the author has determined that the factors affecting financial distress are mostly focused on the micro level and the company's financial ratios including: Market value to book value; Return on equity; Price volatility; Operating profit; Enterprise size; debt to total assets ratio; Corporate credit rating; Investment of the enterprise; Liquidity through funding rates; The age of the business; get the vallue 1 if the company is in the heavy industry sector (1 character in SIC code is 1,3,4) otherwise get the value 0. Thus, the above study has drawn conclusion to identify companies using high capital debt and earning low credit rating. These firms have a high possibility of financial distress during the recession (period 1926–1938) and in the period 2008–2009 as well.

Edward I. Altman (2000) proposed two models, Z-Score (1968) and ZETA (1977). The research specified the characteristics of financially distressed firms, quantified the variables and predicted the firm's financial distress. In the Z-score model, by using the MDA method (multiple discriminant analysis), from the initial list of 22 variables, the author selected 5 variables to build a prediction model on financial distress: Working capital/Total assets, retained earnings/Total assets, Earnings before interest and taxes/Total assets, Market value of Equity/Book value of Total Liabilities, Sales/Total Assets. As concluded from the model results, the author revealed that the accuracy in predicting failure two years before financial distress and decreased as the lead time increased. In 1977, the author developed the ZETA model (an improvement to the Z-core model, which not only forecast for large enterprises but also for small and medium enterprises) on two samples of 53 bankrupt companies and 58 non-bankrupt companies based on seven following variables: Return on assets, stability of earnings, earnings before interest and taxes/total interest payments, retained earnings of the company (balance sheet)/total assets, operating capital/total assets, common capital/total capital, total assets of the company. The ZETA model's bankruptcy classification accuracy ranges is from over 96% a period before bankruptcy to 70% before five years of annual report.

Mario Hernandez Tinoco, Tinoco and Wilson (2013) with a data set of 23,218 companies from 1980 to 2011, studied the benefit of combining accounting data, market-based and macroeconomic data. This research thus developed risk models for listed companies to predict bankruptcy. The author used the logistic model to test the proposed hypotheses and had the final results for the variables that have an impact on the financial distress: TFOTL: Total amount from operations to total liabilities pay; TLTA: Total Liabilities to Total Assets; NOCREDINT: No credit period.

Platt and Platt (2006) the author pointed out the differences that many researchers often confused between financial distress and bankruptcy previously. A business falling into financial distress is not necessarily going to bankrupt, but a business that goes bankrupt will inevitably fall into financial distress. The sample of research conducted in 1999 and 2000 consisted of 1,403 companies, of which 276 were financially distressed operating in 14 industries and 1127 were not distressed. Firms are catergorized into two groups of financial distress and non-financial distress. The author used logistic regression model to measure financial distress since it is a statistically significant and flexible model. The results show that the factors affecting financial distress: Cash flow/Revenue; interest and depreciation on total assets (EBITDA/TA); long-term debt to total assets; interest coverage ratio (EBIT/interest expense); fast payout ratio.

2.3 Research Methodology

In the context of a deepening crisis in classical frequentist statistics (see, for example Nguyen and Thach 2019; Tuan et al. 2019; Thach et al. 2019; Thach 2020), the current study applies the Bayesian Binary Logistic framework to achieve the research purpose.

The study was conducted to evaluate the influence of the selected factors on the profitability of 632 Vietnamese enterprises from 2014 to 2019. Referring to earlier studies, to obtain the research goal, the authors propose research model as follows:

$$\text{DIS} = \text{Ln}\left(\frac{pi}{1-pi}\right) = \beta o + \Sigma\, \beta_i X_{it} + u_{it}$$

In there:

P_i: the probability of the event occur, in this case the probability of the firm's financial distress.

β_0: The slope of the model.

β_i: Regression coefficient of variable Xit.

Xi_t: Independent variables of enterprise i at time t.

t: time: 1,2,3...6 years.

u_{it}: model residuals.

The authors used Binary Logistic regression with the dependent variable Y as the research model to analyze the factors affecting the possibility of financial distress of Vietnamese enterprises. The variable Y takes the value 1 if the company occurs financially distressed at any time in 2014–2019, otherwise Y takes 0:

$$\text{Ln}\,(\frac{pi}{1-pi}) = \alpha + \beta_1 \text{LEV} + \beta_2 \text{SIZE} + \beta_3 \text{WC} + \beta_4 \text{RETA} + \beta_5 \text{NITA} + \text{uit}$$

$$\text{Pi} = \frac{1}{\alpha + \beta_1 \text{LEV} + \beta_2 \text{SIZE} + \beta_3 \text{WC} + \beta_4 \text{RETA} + \beta_5 \text{NITA} + \text{uit}}$$

Measure the Dependent Variable

With reference to research in the same field in Vietnam by Pham Thi Hong Van, the author team provides data sources to determine a company in financial distress includes:

+ Firstly, companies were delisted on HOSE, HNX (except in the case of mergers and exchanges) in the period from 2014–2019 due to collection from Viet Stock company or bankruptcy declaration or securities are controlled and warned.

+ Secondly, companies were control or warning listed on HOSE, HNX (only considering 2 indicators: profit after tax and undistributed profit after tax) in the period from 2014-2019.

Measure the Independent Variable:

The independent variables are measured as follows:

- LEV: Financial leverage = (Total debt)/(Total assets)
- SIZE: Size of the company = Decimal logarithm of total assets
- WC: Net working capital to current assets = (Net working capital)/(Current assets)
- RETA: Ratio of retained earnings to total assets = (Retained profit)/(Total assets)
- NITA: Ratio of profit before tax and interest divided by total assets = (Profit before tax and interest)/(Total assets)

The independent variables of the authors attempt to affect financial distress likelihood of Vietnamese enterprises in the period 2014–2019. In this research, the hypothesis is made as follows:

Hypothesis H1: Firms with a higher financial leverage are more likely to fall into financial distress

Financial leverage: total debt divided by total assets, where as total debt includes all types of debt of the Balance Sheet. According to M&M's postulate I, in the perfect market situation, capital structure is not related to firm value while the firm's investment decisions are already in place. In the author's trade-off theory mentioned above, firms are more concerned about capital structure as using a lot of debt can be a capital structure aimed at maximizing firm value due to tax shield from debt. However, this is also associated with the enterprise risks, especially when the enterprise does not use debt with regards to its characteristics, the problem of choosing an inappropriate time and amount of debt is a premise leading to the possibility of high financial distress for the business in the future. A company with a high debt ratio is sometimes too cautious making investment decisions, consequently easily ignoring good investment opportunities, thus not maximizing business value. In addition, the debt ratio also reflects how much debt each dollar of assets is financed, the larger the index is, the more clearly it shows the low financial autonomy of the business and vice versa.

From the background theory: the trade-off theory represents that there exists a cost of financial distress when enterprise takes too much debt to total assets; or from the pecking order theory has also shown that debt is only second priority (this theory adds clearer clarification to the trade-off theory), because debt always contains a cost and a certain risk.

According to Graham et al. (2011), Ohlson (1980), Tinoco and Wilson (2013), Luong Trong Duc (2012), the author has proposed the financial leverage indicator in this research, maybe this basis indicator is proposed in almost every research.

Hypothesis H2: Larger firms are less likely to fall into financial distress

The size of the company, in fact, small companies have many difficulties in accessing and receiving financial support from outside such as banks, stock market because they have less assets than large-sized companies, which means it is difficult for them to borrow. On a broader scale, small companies are less able to convince investors to contribute capital. Besides, small companies do not have a strong apparatus to run or make timely decisions at important time. Small firms have a simpler management structure than large companies or even do not have specialized teams performing professional market monitoring and control to make timely forecasts. The nefarious behaviors that easily emerge in small companies cause asymmetric information to customers, creditors or even their managers. Since between these subjects, there is always a potential conflict of interest (agent cost theory). As a consequence, large companies will have an advantage in controlling these costs. According to Ohlson (1980), Luong Trong Duc (2012), the author included the company size variable in their researches.

Hypothesis H3: Firms with a higher net working capital to current assets ratio are less likely to fall into financial distress

Net working capital to current assets ratio: Net working capital is the long-term source of fund to finance short-term assets. Hence the positive net working capital means that long-term capital is not only finance long-term assets but also finance short-term assets. Enterprises do not trade short-term capital for long-term investment or all long-term assets of the enterprise are fully financed by stable capital sources, i.e. long-term capital. Thus then leads to the question is how a funding structure is appropriate and minimize the possibility of enterprises falling into financial distress. According to Solani Hazarika and Narasimhan (2011), Luong Trong Duc (2012), Tran Thi Bich Ngoc (2013), the author team suggests a ratio of net working capital to short-term assets to study the factors affecting financial distress in their research.

This indicator conveys what percentage of short-term assets is financed by long-term capital. If this ratio is high, the enterprise will have a safe financial structure and low payment risk, whereas if the enterprise uses short-term capital to finance short-term assets, the payment risk will be high but at the same time maintain a low cost of capital. Yet as the initial assessment of financial distress is "insolvency", the author will revolve around the theory to provide the basis for the assessment. In some researches related to financial distress, the authors even consider this as a similar indicator to the liquidity of a business. From the former and latter ideas, the author will hypothesize for this index.

Hypothesis H4: Firms with lower retained earnings to total assets are more likely to fall into financial distress

Retained earnings divided by total assets: Retained earnings are the net (after-tax) income left over to businesses after companies have paid dividends to shareholders. Retained earnings helps: Firstly, for the hedging motive, firms maintain liquidity to react to unforeseen contingencies. Secondly, for the transactional motive, it withholds profits to pay for the regular operations. Thirdly, firms may retain cash internally to take advantage of investment opportunities since the existence of information asymmetries can increase the cost of external budget, which may even lead to ignore positive NPV investment opportunities. Therefore, managers hold high liquid assets such as cash to reduce the cost of external budget. Finally, for tax incentives, firms may prefer to retain profits rather than pay dividends to shareholders to avoid paying taxes on dividends.

As stated above, the author proposes a ratio of retained earnings to total assets and based on pecking order theory, the recent investors are more informed than new investors. This makes new funding less optimal than existing funding. Therefore, enterprises using finance from retained earnings will be given more priority and reduce the possibility of enterprises falling into financial distress.

The author relies on research by Altman (2000), Pham Thi Hong Van proposed, to include this factor in the model and hypothesized.

Hypothesis H5: The higher the ratio of profit before tax and interest on total assets, the more difficult it is for firms to fall into financial distress

Profit before tax and interest divided by total assets, as mentioned above, the author takes into account EAT to classify an enterprise is likely fall into financial distress or not. It is very necessary to consider whether the profit before tax and interest has an impact on financial distress or not. Initially, the author mentioned the cash flow perspective in Chap. 2: a company, investors love the profit index, and managers love cash flow. Also in this chapter, the author also mentioned the benefits of revenue and expense management with OCF.

In order to show the comprehensiveness of the research, in addition to promoting the ability of a business to generate money, high profits also has its own attractive characteristics: for instance, the loose sales policy is very sufficient. Therefore, the higher a business has a high profit index (removing the impact of taxes and interest, as the interest rate has been previously measured through EAT) showing the sales policy is very sufficient (high revenue) and adequate cost management (low cost). That is, the higher this index, the lower the likelihood of financial distress occurence.

As per research by Altman (2000), Ohlson (1980), Platt and Platt (2006), the author has included this ratio in the research model (Table 1).

Table 1 Independent variables used in the research model

Name of variables	Explain	Expected
Financial leverage (LEV)	Total assets/Total debt	+
Company Size (SIZE)	Decimal Logarithm of Total Assets	–
Net Working Capital (WC) Ratio	Net Working Capital/Total Assets	–
Retained Earnings Ratio (RETA)	Retained Earnings/Total Assets	–
Profit before tax (NITA)	Profit before tax and interest/total assets	–

3 Results and Discussion

The results of Table 2 show that the correlation coefficients between the independent variables in the research model are low (absolute value <0.5), so the selection of these independent variables for the research is completely appropriate.

Since previous studies often used out-of-date frequentist methods, which might provide unreliable and inconsistent results, the Bayes setting is then applied. Note that the number of observations in this study is relatively large, so the prior information might not overwhelm the data distribution and affect the posterior estimates. In this case, Block et al. (2011) proposed specifying the standard Gaussian distributions with different prior distributions and conduct a sensitivity analysis to choose the best among the considered models. For each of the two models, we run six simulations (Table 2).

In the next step, the authors perform suggested regressions and the obtained results are recorded in Table 3.

The simulation selected in the Bayesian factor analysis is the simulation with the largest mean Log BF, largest Log ML, and minimum DIC mean. Thus, the results of Bayesian factor analysis in Table 3 show that simulation 2 has an advantage over other candidates in terms of log(BF) and log(ML), but DIC is not as good as simulations 3, 4, 5. Therefore, the authors continue to conduct Bayes test model (Table 4).

The results in Table 5 show that Simulation 2 has a higher P(M|y) than the rest of the candidates, so it can be concluded that Simulation 2 is the most suitable. This model can be used for inference and prediction.

Table 2 Correlation matrix analysis between variables in the model

	DIS	LEV	SIZE	WC	RETA	NITA
DIS	1.000					
LEV	0.3142	1.000				
SIZE	−0.0497	0.3206	1.000			
WC	−0.4995	−0.45	−0.1409	1.000		
RETA	−0.0353	0.1363	0.0713	−0.0626	1.000	
NITA	−0.2616	−0.339	−0.0154	0.2116	0.04	1.000

Source Result from STATA

Table 3 Model specifications

Model 1	
Likelihood function	DIS ~ N(μ, σ)
Prior distribution	
Simulation 1	$\beta_i \sim$ N(0; 1) $\sigma^2 \sim$ Invgamma(0, 01; 0.01)
Simulation 2	$\beta_i \sim$ N(0; 10) $\sigma^2 \sim$ Invgamma(0, 01; 0, 01)
Simulation 3	$\beta_i \sim$ N(0; 100) $\sigma^2 \sim$ Invgamma(0, 01; 0, 01)
Simulation 4	$\beta_i \sim$ N(0; 1000) $\sigma^2 \sim$ Invgamma(0, 01; 0, 01)
Simulation 5	$\beta_i \sim$ N(0; 10000) $\sigma^2 \sim$ Invgamma(0, 01; 0, 01)

Table 4 Results of bayes factor

	DIC	log(ML)	log(BF)
Simulation 1	1714.947	−896.137	
Simulation 2	1691.146	−862.006	34.13045
Simulation 3	1689.569	−862.315	33.822
Simulation 4	1690.085	−868.304	27.83241
Simulation 5	1690.39	−875.064	21.07318

Table 5 Bayesian model tests

	log(ML)	P(M)	P(M|y)
Simulation 1	−896.137	0.2	0.0000
Simulation 2	−862.006	0.2	0.5759
Simulation 3	−862.315	0.2	0.4230
Simulation 4	−868.304	0.2	0.0011
Simulation 5	−875.064	0.2	0.0000

To check for MCMC convergence, the authors use additionally a visual tool via cusum plots. Cusum plots are a useful tool to detect convergence of MCMC chains. If cusum lines do not cross the X-axis, then that is a sign of poorly-mixing sequences. However, Fig. 1 demonstrates a fairly good mixing of all the MCMC chains.

Additionally, we consider some important initial indicators, the rate of acceptance and efficiency. The results in Table 6 point out that the acceptance rates have reached 0.2994, while minimum efficiency has obtained 0.02 (exceeds the required level of 0.01). Besides, the summary tables provide Monte-Carlo standard errors (MCSE), which indicate the accuracy of the posterior estimates. In the opinion of Flegal et al. (2008), the closer the MCSEs are to zero, the stronger the MCMC series, claiming

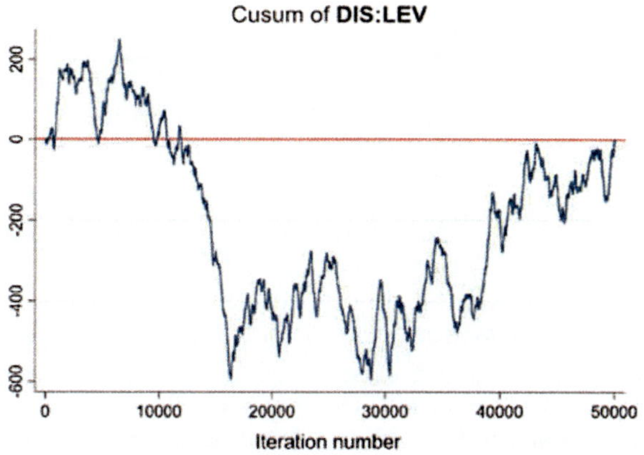

Fig. 1 Graphical checks for MCMC convergence. *Source* Authors' calculation

Table 6 Bayesian simulation outcomes

DIS	Mean	Std. Dev	MCSE	Median	Equal-tailed	
					[95% Cred. Interval]	
LEV	1.6641	0.3713	0.0032	1.6628	0.9428	2.3985
SIZE	−0.2154	0.0654	0.0019	−0.2155	−0.3433	−0.0872
WC	−5.7940	0.2689	0.0029	−5.7892	−6.3377	−5.2833
RETA	−0.6217	0.2459	0.0027	−0.6183	−1.1070	−0.1427
NITA	−4.7909	0.7139	0.0185	−4.7891	−6.2113	−3.4055
_cons	0.9827	0.7487	0.0233	0.9866	−0.4908	2.4478
Acceptance rate	0.2994					
Efficiency: min	0.02074					

that the MCSE values less than 6.5% of the standard deviation are acceptable and smaller 5% of the standard deviation are optimal.

The results of Table 6 show the sign of the regression coefficients, according to which the values of LEV have a positive impact on the variable DIS. In contrast, the remaining value harm DIS. However, to determine the magnitude of the impact, it is necessary to calculate the probability of these effects (Table 7).

The results of the posterior probabilities analysis showed that LEV, NITA increased DIS with the probability to reach 100%, showing a very clear impact. The remaining factors all tend to reduce DIS very clearly with the probability of SIZE is 99.95%; WC hits 100% and RETA is 99.43%.

Varible LEV

Companies with high financial leverage will face the risk of financial distress. This expression is also on the ground of the Empirical Research by Graham et al. (2011),

Table 7 Posterior probabilities

LnZScore	Mean	Std. Dev	MCSE
Probability (DIS:LEV) > 0	1	0.0000	0.0000
Probability (DIS:SIZE) < 0	0.9995	0.0232	0.0001
Probability (DIS:WC) < 0	1.0000	0.0000	0.0000
Probability (DIS:RETA) < 0	0.9943	0.0750	0.00037
Probability (DIS:NITA) < 0	1.0000	0.0000	0.0000

Ohlson (1980), Tinoco and Wilson (2013), Tran Thi Bich Ngoc (2013), Luong Trong Duc (2012).

The trade-off theory (Kraus and Litezeberger 1973) considers financial distress in order to select the appropriate level of debt use towards the goal of achieving the highest firm value. The trade-off theory provides the maximum constraint on the level of debt use in the capital structure of the enterprise, which is the risk of financial distress. Next, the authors present the finite order theory (Myers and Majluf 1984), which is supplementary statement for the trade-off theory. However, the author will prefer to use debt to explain the assumptions of the topic. In this theory, loans are in the second order of priority, after internal sources of capital and prior to direct contributions from owners. Besides the statements about the usefulness of debt, the pecking order theory also indicates the limitations of borrowing, which are: debt is a fixed financial cost, no matter how active your business is, earns a lot or is even at a loss, the business still have to make sure to pay the loan interest on time and repay the principal when it is due, thus an increase in the debt of the business is linked to an increase financial risk. Accordingly, businesses always need to carefully consider when to use debt and with a reasonable ratio.

Research results show that the risk of increasing financial distress when using financial leverage is almost 100%; This result is consistent with the original hypothesis. LEV means that the greater the financial leverage is the higher financial distress probability. Besides, this index also signifies the proportion of assets financed by debt, which can also assess the financial autonomy of the company. Once this ratio is too low, it means that the company has not fully exploited the benefits of debt (financial leverage). On the contrary, if this ratio is too high, it can be recognized that the company does not have strong financial autonomy, capital is mainly composed of borrowing (However, the benefit of leverage is not denied). Therefore, the risk of falling into financial distress of the enterprise will increase (If the enterprise relies on this tool too much).

Variable SIZE

It is apparent that the probability of reducing the risk of financial distress by firm size is more than 99%, which shows that the impact of this variable on financial distress is very significant. This result is consistent with the original hypothesis and also similar to the research of Ohlson (1980), Luong Trong Duc (2012), company with large size, the less likely it is to face financial distress.

Agency cost theory (Jensen and Meckling 1976) has mentioned the conflict of interest groups of 3 groups: shareholders, managers, creditors. Owing to the immediate interests, managers can do harm to the benefit of shareholders and creditors. Consequently, large size enterprises often have a controlling department, so that decrease the risk of financial distress.

Variable WC

The probability that the variable WC reducing the risk of financial distress virtually hits 100%. Utilizing high long-term debt to finance short-term assets (at a reasonable level) implies that enterprises having stable financial structure and are less likely to fall into financial distress. There are researches with similar conclusions: by Altman (2000), Pham Thi Hong Van.

At the same time, this is considered a conservative model. Enterprises that proactively meet most of their working capital needs with long-term capital sources, including regular and temporary needs, will ensure high liquidation and the level of financial safety in the enterprise, creating perfect conditions for business activities to take place continuously and stably. This model makes the company's finances more solid and reduces the risk of financial distress.

Variable RETA, NITA

The probability that RETA and NITA variables reducing the risk of financial distress reaches 99.43% and 100%, respectively. Enterprises with high profit ratios show that enterprises have stable financial structure; therefore, less likely to fall into financial distress. This result is in line with expectations and is completely consistent with the similar studies: Graham et al. (2011), Luong Trong Duc (2012), Tran Thi Bich Ngoc (2013), Altman (2000); Pham Thi Hong Van proposed.

This view is consistent with pecking order theory, which is existing investors have more information than new investors, causing the cost of new capital to be valued higher than for existing capital. Accordingly, enterprises taking good advantage from the finance of retained earnings will be prioritized first and can reduce the possibility of falling into financial distress.

4 Conclusions and Recommendations

The results of the research have contributed to a better explanation of the impact of factors on financial distress of Vietnamese enterprises; therefore, the authors propose some recommendations to business managers and investors with a view to better assesing the importance of factors affecting the possibility of financial distress, thereby developing strategic plans in the business plan to increase business efficiency.

Firstly: Adjusting the debt structure in the direction to reduce the level of debt usage, focusing mainly on reducing high-interest loans will contribute a positive change in the capital structure of the enterprise by saving debt usage costs and reducing the pressure to pay interest expenses.

Secondly: the corporate financial managers can identify the influence of each factor on the capital structure, thereby orienting the establishment of capital structure according to the objectives set for each specific development period of the enterprise. Furthermore, firms can actively plan their capital structure for the future use, by substituting different combinations of the target values of the independent variables to determine the expected capital structure, and from there, the financial managers continue to study the relationship between capital structure and the conditions of profit constraints for owners and financial distress to make the final better choice of capital structure, as a basis for establishing implementation measures, and orienting the organization of capital mobilization plan.

Finally: If viewed from the perspective of an enterprise that has not yet experienced financial distress, it is possible to prevent the negative impact of the SIZE variable on the probability of distress occurence (not by increasing the size of the firm), which is to increase the creditworthiness of the company in the eyes of the capital suppliers (typically what the authors want to mention is debt from credit institutions).

References

Altman, E.I.: Predicting financial distress of companies: revisiting the Z-score and zeta. J. Financ. (2000)

Andrade, G., Kaplan, S.N.: How Costly is Financial (Not Economic) Distress? Envidence from Highly Leveraged Transactions that Became Distressed (1998)

Gilbert, L.R., Menon, K., Schwartz, K.B.: (n.d.) Predicting bankruptcy for firms in financial distress

Graham, J.R., Hazarika, S., Narasimhan, K.: Financial distress in the great depression. Financ. Manage. **40**(4), 821–844 (2011)

Hashi.: The Economics of Bankruptcy, Reorganization, and Liquidation: Lessons for East European Transition Economies (1997)

Lương Trọng Đức.: Reseacrh on factors affecting financial distress and prediction model to joint stock companies in Ho Chi Minh City (2012)

Ohlson, J.A.: Financial ratios and the probabilistic prediction of bankruptcy. J. Account. Res. **18**(1), 109–131 (1980)

Platt, H.D., Platt, M.B.: Understanding differences between financial distress and bankruptcy. Rev. Appl. Econ. Lincoln University, Department of Financial and Business Systems **2**(2), 1–17 (2006)

Phạm Thị Hồng Vân.: Measuring financial distress of listed firms (2015)

Purnanandam, A.: Financial distress and corporate risk management: Theory and evidence. J. Fin. Econo. **87**(3), 706–739 (2008)

Svítek, M., Kosheleva, O., Kreinovich, V., Nguyen, T.N.: Why quantum (wave probability) models are a good description of many non-quantum complex systems, and how to go beyond quantum models. In: Kreinovich, V., Thach, N.N., Trung, N.D., Van Thanh, D. (eds.) ECONVN 2019. SCI, vol. 809, pp. 168–175. Springer, Cham (2019). https://doi.org/10.1007/978-3-030-04200-4_13

Trần Thị Bích Ngọc.: Factors affecting financial distress (2013)

Võ Thị Thu Nguyệt.: Factors affecting the financial distress profitability of listed companies on Ho Chi Minh City Stock Exchange (2017)

Thach, N.N.: How to explain when the ES is lower than one? A Bayesian nonlinear mixed-effects approach. J. Risk Financ. Manage. **13**, 21 (2020)
Thach, N.N., Anh, L.H., An, P.T.H.: The effects of public expenditure on economic growth in Asia countries: a Baytesian model averaging approach. Asian J. Econ. Bank. **3**, 126–149 (2019)
Thach, N.N.: The variable elasticity of substitution function and endogenous growth: an empirical evidence from Vietnam. Int. J. Econ. Bus. Admin. **VIII**, 263–277 (2020b).
Tinoco, M.H., Wilson, N.: Financial distress and bankruptcy prediction among listed companies using accounting, market and macroeconomic variables. Int. Rev. Financ. Anal. **30**(C), 394–419 (2013)
Tuan, T. A., Kreinovich, V., & Nguyen, T. N.: Decision making under interval uncertainty: beyond hurwicz pessimism-optimism criterion. In International Econometric Conference of Vietnam (pp. 176–184). Springer, Cham. (2019, January)
Whitaker, R.B.: The early stages of financial distress. J. Econ. Financ. **23**(2), 123–132 (1999)
Wruck, K.H.: Financial distress, reorganization, and organizational efficiency (1990)

Intention to Buy Organic Food to Keep Healthy: Evidence from Vietnam

Bui Huy Khoi and Ngo Van Tuan

Abstract The paper was conducted to understand the factors affecting the intention to buy organic food from Vietnamese consumers to keep healthy. The official study carried out an online survey through Google forms using a questionnaire with the participation of 200 consumers who have experience in shopping at stores and supermarkets in Ho Chi Minh City, Vietnam. The results of the study show that 6 factors all affect the intention to buy organic food for consumers. This study uses the optimal choice by the BIC Algorithm for the intention to buy organic food for Vietnam consumers to keep healthy. The paper gives managerial implications for the factors that help organic food producers and distributors improve sales efficiency, make production and distribution policies suitable to consumers in Vietnam and the international market.

Keywords BIC algorithm · Organic food safety · Health consciousness · Environmental consciousness · Quality · Price · Keep health

1 Introduction

The consumption of organic food is gaining ground globally due to consumers' concerns for personal health and food safety. Several countries, such as Japan, are turning their focus to promoting organic food consumption, but research is scarce on Japan's organic food market (Talwar et al. 2021). Organic food is an issue many people are interested in and have discussed for a long time. A few years ago, this type of food was not popular and was usually only sold in stores that specialized in healthy, safe foods. About three years ago, organic food has gradually become familiar to consumers. The global market is valued at 96.7 billion USD, wherein

B. H. Khoi (✉)
Industrial University of Ho Chi Minh City, Ho Chi Minh City, Vietnam
e-mail: buihuykhoi@iuh.edu.vn

N. Van Tuan
Banking University of Ho Chi Minh City, Ho Chi Minh City, Vietnam
e-mail: tuannv@buh.edu.vn

N. Ngoc Thach et al. (eds.), *Financial Econometrics: Bayesian Analysis, Quantum Uncertainty, and Related Topics*, Studies in Systems, Decision and Control 427,
https://doi.org/10.1007/978-3-030-98689-6_43

more than 100 countries have already adopted distinct regulations for the cultivation of organic food (Talwar et al. 2021). While spending on the health of Vietnamese consumers is showing signs of increasing, averaging over 6% of income, especially up to 47% of Vietnamese consumers, is increasingly interested in natural food. Thus, the potential for the development of safe organic food in Vietnam is still very large (Nielsen 2020). Especially now, green consumption and safe food consumption are being human consumption trends when the environment becomes a major concern of many countries around the world. As consumers are increasingly concerned about the environment, they place more importance on eco-friendly buying behavior. The recent number of people willing to pay more for eco-friendly products shows that the market for eco-friendly products is expanding. The rapid development of industries is always accompanied by many environmental consequences that need our attention. Global warming, water pollution, dusty, toxic air, the environment is increasingly polluted. In addition, the application of new scientific and technical achievements in animal husbandry, cultivation, production, and food processing also increases the risk of food being contaminated because of residues of pesticides including pesticides, herbicides, growth stimulants, chemical residues in preserving vegetables, tubers, and fruits; residues of growth substances in animal meat; the percentage of controlled livestock and poultry processing and slaughter establishments is still very low. According to statistics of the Vietnamese Ministry of Health in 2020, the whole country recorded 48 cases of food poisoning, causing over 870 people to get sick, 824 people to be hospitalized for treatment, and 22 cases of death. In the first 6 months of 2020, functional forces in 63 provinces and cities inspected nearly 282,000 establishments, detecting over 38,100 food safety violations. When looking at theories related to organic food consumption, many studies have examined the motivation to buy organic food such as (Agarwal 2019; Secapramana and Ang 2019; Waqas and Hong 2019; Teng and Wang 2015; Švecová and Odehnalová 2019; Rahman and Hossain 2019). The study by (Rahman and Hossain 2019) concluded that the intention to buy organic food includes concerns about health, environment, food safety, social influence, and price sensitivity. However, previous studies have focused little on factors that can help consumers create positive intentions and attitudes towards organic food, or explore combinations of factors that can increase the consumption of organic food (Teng and Wang 2015). The authors have argued that consumer concern about safe and healthy foods is a prerequisite for a successful business because consumers are often hesitant to take action. buying, unless they trust the safety as well as the health values that organic food brings to the seller (Waqas and Hong 2019). In addition, beliefs about a food may be more important in a consumer's decision to buy organic than conventional food. Belief about a food is essential to organic food buying behavior (Teng and Wang 2015). This study uses the optimal choice by the BIC Algorithm for the intention to buy organic food for Vietnam consumers to keep health, aiming to fill the gaps identified above while providing meaning and a basis for producers, distributors, and traders in the organic food sector to have effective strategies for organic food market development in Vietnam.

2 Literature Review

2.1 *Organic Food Purchasing Intention (OFPI)*

Rashid (2009) states that the intention to buy organic food is an individual's ability and will to give his preference to organic food over conventional food in shopping considerations. Ramayah et al. (2010) argue that the intention to buy organic food is one of the specific manifestations of buying action. In the consumer purchase journey, the intention to purchase a product does not always lead to the actual purchase afterward (Pham et al. 2019; Tandon et al. 2021). There have been many statistical studies that have demonstrated that there is a certain disagreement between intention and actual buying behavior (Hidalgo-Baz et al. 2017; Lian 2017; Sultan et al. 2020). And Vietnamese consumers, especially consumers in southern Vietnam, are not exempt from such a situation (Nguyen et al. 2019). On the other hand, in studies conducted during the last decade, intentions towards buying organic food have positively and significantly influenced consumers' quick buying behavior (Carfora et al. 2019; Ham et al. 2018; Lee 2016; Ogorevc et al. 2020; Švecová and Odehnalová 2019). In summary, (Ramayah et al. 2010) intend to buy organic food in the next shopping time, showing that organic food has great value for consumers. Because using organic food brings good health and can minimize adverse effects on the environment (Rashid 2009; Yadav and Pathak 2016). Those are the factors that conventional food cannot meet, so consumers tend to use organic food more than conventional food. They will actively seek it in the market (Saleki and Seyedsaleki 2012) and may encourage friends, neighbors, and family to buy organic food (Teng and Lu 2016; De Toni et al. 2018).

2.2 *Organic Food Safety (OFS)*

Today, food safety is receiving more attention than ever before by governments and policymakers, health professionals, the food industry, the biomedical community, and last but not least. An important part is a public (Crutchfield and Roberts 2000), safety has become one of the most important attributes when it comes to food. Food safety is the degree to which it is safe from eating foods. These safety levels are risks assessed from the time the product is grown, packaged, distributed, and processed (Pambudi and Ekawati 2021). According to (Bauer et al. 2013), food safety is a term that refers to consumers using a product without causing harm because it does not contain chemical ingredients, so it ensures clean natural food and is not harmful to health. In summary, food safety is of great concern to consumers when deciding to buy organic food (Nguyen et al. 2019). They are always interested in the origin and attributes of the product (Shaharudin et al. 2010). Consumers have always believed that organic food does not contain harmful chemicals such as pesticides, fertilizers, and genetically modified ingredients (Slamet et al. 2016). Foods with additives and processed foods that do not ensure food safety are rejected by consumers (Chen 2009).

2.3 Health Consciousness (HS)

Health consciousness is defined as consumers who are interested in a reasonable diet with adequate nutrients and a healthy lifestyle (Newsom et al. 2005). Many studies demonstrate that health consciousness has a relationship with organic food consumption as a study conducted by (Kriwy and Mecking 2012) revealed that consumers who believe that organic food brings good health, will consume it without any fear and doubt. In addition, healthy organic food has become a driving force for Australian consumers to buy organic food (Lockie et al. 2002). Health-conscious consumers are "Consumers who are well aware of their health status and care about their health benefits. They are willing to do things to maintain good health, improve health and enhance the quality of life" (Kraft and Goodell 1993). These people tend to prevent illness by engaging in healthy activities. They are knowledgeable about nutrition and participate in sports activities. Consumers don't pay attention to marketing factors, but they care a lot about service quality and health consciousness (TRAN et al. 2020). As the economy develops, people's lives are improved, consumers pay more and more for consumption activities and they pay attention to health issues in the process of choosing their products in general and food products in particular. Organic food is considered to be healthy, so it will influence consumers' decisions when buying food for daily meals. Concluding that today, when people's living standards are increasingly improved, they prioritize health as the biggest concern when making food purchase intentions (Lee et al. 2019; Thøgersen et al. 2015). Most people believe that organic food helps maintain good health. They are always thinking about problems related to their health. Always looking for ways to have good health, know how to consume healthily. On the other hand, skeptical consumers are always interested in whether organic food is good for health or not (Shaharudin et al. 2010; Gracia Royo and de-Magistris 2007)?

2.4 Environmental Consciousness (EC)

Kim and Chung (2011) show environmental awareness means that consumers are awake and aware of the threatened environment and the depletion of natural resources. Awareness of the negative effects that common food chemicals can have on the environment (either through sewers or from manufacturing plants) is increasing (Boxall et al. 2012). Besides, compared to other products, organic products are known to be more environmentally friendly because absolutely no chemicals, pesticides, etc. are used in the farming process. At the same time, according to Winter and Davis (2006), "Organic food is a food that helps protect the environment because the production and business process does not use chemicals and technologies that pollute the environment". Therefore, environmental consciousness is considered as one of the factors affecting whether consumers use organic products or not. Research by Magnusson et al. (2001) has found that up to 89% of respondents agree that they always pay attention to the consequences of using products and food related to environmental problems. It can be seen that at present, environmental issues are the concern of all mankind. Consumers in Vietnam are aware of the importance of protecting the living environment, so it affects their purchasing decision process (Chauke and Duh 2019; Lee 2016; Lian 2017; Nguyen et al. 2019). Consumers here prefer to use recycled or reusable products (Lee 2016). They firmly believe that, with the current environmental pollution problems, we need to act immediately (Kushwah et al. 2019b; Yadav and Pathak 2016; Chen 2009). Most people recommend using organic food to protect the environment (Voon et al. 2011).

2.5 Food Quality (FQ)

The definition of food quality is related to sensory factors such as the taste of food, food consumption experience, food enjoyment (Magnusson et al. 2001). In addition, the quality of organic food is described as the appearance of the product; this may include insect infestation and a lack of uniformity in size (Thompson 1998). Kushwah et al. (2019a) consider food quality as a relevant attribute in purchasing organic food. It is the relationship between expected quality and perceived quality that determines consumer satisfaction. The sticker ensures that the consumer is buying an organic product. Especially in the case of certified organic foods, asymmetry of information can be harmful, since consumers cannot determine the absence of chemical components in the food—a basic requirement in organic farming. In summary, for consumers in Vietnam, the quality factor is as important as other factors when choosing to buy food. They prefer to choose good food, the quality of both the appearance and the inside of the food. They believe that organic food will be of high quality because this food has a lot of nutritional content and is grown using advanced, natural farming methods. Therefore, consumers find that organic food will be of better quality than conventional food (Nguyen et al. 2019; Thøgersen et al. 2015).

2.6 Price (PE)

Price is the amount a buyer must pay to obtain a product or service. Consumers often have the mentality that high prices are indicative of high-quality products (Kotler and Armstrong 2010). It is also recognized that the price of organic food is higher than that of conventional food (Magnusson et al. 2001). There is a lot of research on the willingness to pay for organic food. The price of organic food plays a major role in generating purchase intention and buying behavior of consumers. Often the price is a deterrent to purchase because the price of organic food is often higher than that of conventional food (Boccaletti and Nardella 2000; Magnusson et al. 2001; Fotopoulos and Chryssochoidis 2001; Zanoli and Naspetti 2002; Padel and Foster 2005; Hughner et al. 2007). In summary, price is a necessary condition to influence consumers' intention to buy organic food in Vietnam. Most consumers are always concerned about the price when buying organic food (Wang et al. 2020). Due to the complex production process of organic food requiring advanced and modern equipment, the cost of this food is always medium and high. This means that consumers are always aware that the price of organic food will be high (Lacaze 2009). If there is a reasonable price for this type of food, it would be easier to choose to buy this food (Canavari et al. 2003). On the other hand, for some consumers, as long as organic food meets the requirements as well as satisfies their aspirations, they are willing to pay more for this food even though it is more expensive than real food products (Wang et al. 2020; Canavari et al. 2003; Gil and Soler 2006).

2.7 Knowledge of Organic Food (KOF)

According to Çabuk et al. (2014), organic food knowledge refers to the consumer's understanding of the structure and nature of organic food. Gracia and De Magistris (2008) have found that organic knowledge can not only increase the attitude and likelihood of buying organic food but can also enhance the spending level of existing consumers. Besides, cognitive knowledge about organic food is recognized as an important premise of consumer confidence in organic food (O'Fallon et al. 2007). Ghali (2019) also pointed out that insufficient knowledge and awareness about organic food is considered an important barrier to the decision to buy organic food. In summary, if there is little knowledge of organic food, the decision-making process to buy organic food will be easier (Çabuk et al. 2014; Yadav and Pathak 2016; Chen 2009). Consumers also expect that they have little knowledge about organic food to be able to choose to buy; distinguishing between organic food and conventional food (Edenbrandt 2018) can realistically identify packaging and labels before the difficulty to identify "counterfeit goods" (Effendi et al. 2015).

3 Methodology

3.1 Sample Approach

The female sex accounted for the highest proportion at 62.2% while the number of males-only accounted for 37.8%. The data shows that the survey respondents are mainly female. Table 1 describes statistics of sample characteristics.

Table 1 Statistics of Sample

Characteristics		Amount	Percent (%)
Sex and Age	Male	71	37.8
	Female	117	62.2
	Below 18	9	4.8
	18–35	86	45.7
	36–55	57	30.3
	Above 55	36	19.1
Education	Certificate	20	10.6
	Diploma	28	14.9
	Degree	100	53.2
	Master	40	21.3
Career	Student	38	20.2
	Officials	33	17.6
	Homemaker	71	37.8
	Officer	40	21.3
	Other	6	3.2
Income/Month	Below 5 million VND	36	19.1
	5–10 million VND	44	23.4
	11–15 million VND	70	37.2
	16–20 million VND	35	18.6
	Over 20 million VND	3	1.6
Organic food awareness	Use	188	100.0
	No Use	0	0

In the age group, the age group from 18 to under 35 years old had the highest rate at 45.1%, followed by the group from 36 to under 55 years old accounted for 30.3%, the group over 55 years old accounted for 19.1% and the group under 18 years old accounted for 30.9% accounted for the lowest rate 4.8%. This shows that the respondents to the survey are mainly 18 years of age or older. In terms of education level, most survey respondents are with diplomas, degrees, and masters, accounting for 14.9%, 53.2%, and 21.3% respectively. For the lower Certificate level, the lowest rate is 10.6%. Regarding Career, Table 1 shows that the survey respondents are mainly homemakers, accounting for 37.8% of the total survey respondents, the second is officers accounting for 21.3%, the third is students accounted for 20.2%, the fourth was employees and officials accounted for 17.6% and the fifth was people with other professions accounting for 3.2%. Through the data, it is shown that the subjects taking part in the survey are of working age and have income. It shows that the educational level of the surveyed subjects is relatively high. Regarding income, Table 1 shows that the surveyed subjects with income from 11 to 15 million/month are the dominant group, accounting for the highest proportion of 37.2%. Next, the group with income from 5 to less than 10 million has the rate of 23.4%. The group with income below 5 million accounted for 19.1%. The group with income from 16 to 20 million accounted for 18.6% and the last group with income over 20 million accounted for the lowest proportion of 1.6%. This shows that the survey respondents are mainly in the middle-income segment. Corresponding to the target group are officers and homemakers. Through the data, it shows the match between qualifications and occupation and the income level of the survey subjects as described above. The level of awareness about organic food, Table 1 displays that most survey respondents know about organic food, accounting for the highest proportion of 100% of the total survey subjects.

3.2 Reliability Test

Cronbach's Alpha coefficient of 0.6 or more is acceptable (Nunnally 1978; Peterson 1994; Slater 1995) in case the concept being studied is new or new to the subject with respondents in the research context. However, according to Nunnally et al. (1994), Cronbach's Alpha (α) does not show which variables should be discarded and which should be kept. Therefore, besides Cronbach's Alpha coefficient, one also uses Corrected item-total Correlation (CITC) and those variables with Corrected item-total Correlation greater than 0.3 will be kept.

4 Results

Factors and items are in table 2, Cronbach's Alpha coefficient of the greater than 0.6 and Corrected Item—Total Correlation is higher than 0.3 reliable enough to carry out further analysis and lower than 0.3 it is not reliable as EC4, FQ4, KOF4, OFPI3, and OFPI3. There arc somc new items in table 2.

4.1 BIC Algorithm

BIC (Bayesian Information Criteria) was used to choose the best model for R software. BIC has been used in the theoretical context for model selection. As a regression model, BIC can be applied, estimating one or more dependent variables from one or more independent variables (Raftery et al. 1997). An essential and useful measurement for deciding a complete and straightforward model is the BIC. Based on the BIC information standard, a model with a lower BIC is selected. The best model will stop when the minimum BIC value (Kaplan 2021; Raftery et al. 1997; Raftery 1995). R report shows every step of searching for the optimal model. BIC selects the best 2 models as table 3.

Table 2 Factor and Item

Factor	α	CITC	Item	Source
OFS	0.867	0.769	I always care about food hygiene and safety	New
		0.639	I think that organic food does not contain harmful chemicals such as pesticides, fertilizers	Slamet et al. (2016)
		0.771	I avoid foods with additives	Shaharudin et al. (2010)
		0.692	I avoid eating processed foods	Chen (2009)
HS	0.863	0.761	I always take care of my health	Lee et al. (2019), Lian (2017), Thøgersen et al. (2015)
		0.635	I think organic food helps me maintain good health	Lian (2017)
		0.689	I think it's important to know how to eat healthily	Shaharudin et al. (2010)
		0.762	I care if organic food is healthy or not	Shaharudin et al. (2010)
EC	0.745	0.686	I always pay attention to environmental issues when shopping	Chauke and Duh (2019), Lian (2017), Nguyen et al. (2019)
		0.683	I like to use recycled or reusable products	Lee (2016)
		0.664	People advise me to use organic food to protect the environment	Voon et al. (2011)
		0.224	Environmental pollution will improve if we act	Kushwah et al. (2019a, b)
FQ	0.780	0.683	I always prioritize choosing quality products when shopping	Nguyen et al. (2019)
		0.656	Organic food is better than conventional food	New
		0.800	I think organic food is of high quality	Thøgersen et al. (2015)
		0.239	I believe that organic food is of excellent quality because of its advanced farming methods	New
PE	0.662	0.607	I always care about the price when buying organic food	Wang et al. (2020)
		0.518	I think the price of organic food is high	Lacaze (2009)
		0.285	I don't mind paying more for organic food	Canavari et al. (2003)
		0.454	I will buy organic food even though the price is higher than conventional food	Wang et al. (2020), Gil and Soler (2006)

(continued)

Table 2 (continued)

Factor	α	CITC	Item	Source
KOF	0.700	0.633	I think I'm knowledgeable about organic food	Edenbrandt (2018)
		0.620	I think I have enough knowledge to distinguish organic food from conventional food	Edenbrandt (2018)
		0.693	I can recognize organic food packaging and labels	Effendi et al. (2015)
		0.010	I want to have more knowledge about organic food before shopping	Chen (2009), Yadav and Pathak (2016)
OFPI	0.605	0.512	I plan to buy organic food next time	Ramayah et al. (2010)
		0.340	I will buy organic food instead of conventional food for better health	Rashid (2009)
		0.256	I plan to buy organic food to reduce bad environmental problems	Yadav and Pathak (2016)
		0.268	I will actively look for organic food	Saleki and Seyedsaleki (2012)
		0.595	I will encourage friends, neighbors, and family to buy organic food	Teng and Lu (2016), De Toni et al. (2018)

$\alpha = \frac{k}{k-1}\left[1 - \frac{\sum \sigma^2(x_i)}{\sigma_x^2}\right]$

Table 3 BIC model selection

OFPI	Probability (%)	SD	Model 1	Model 2
Intercept	100.0	0.18438	1.61821	1.48077
OFS	100.0	0.02031	0.13664	0.13155
HS	100.0	0.02523	0.15171	0.14906
EC	100.0	0.01936	0.13178	0.13320
FQ	100.0	0.01726	0.07150	0.06840
PE	100.0	0.01687	0.10234	0.10307
KOF	42.2	0.02459		0.04231

There are six independent and one dependent variable. Organic food safety (OFS), Health consciousness (HS), Environmental consciousness (EC), Food Quality (FQ), and Price (PE) influence Organic food purchasing intention (OFPI) with Probability is 100% and Knowledge of organic food (KOF) influence Organic food purchasing intention (OFPI) with Probability is 42.2%.

Table 4 Model test

Model	nVar	R^2	BIC	post prob
model 1	5	0.553	−125.06443	0.578
model 2	6	0.564	−124.43574	0.422
BIC = -2 * LL + log(N) * k				

4.2 Model Evaluation

According to the results from Table 4, BIC shows model 1 is the optimal selection because BIC (-125.06443) is minimum. Organic food safety (OFS), Health consciousness (HS), Environmental consciousness (EC), Food Quality (FQ), and Price (PE) impact Organic food purchasing intention (OFPI) is 55.3% in Table 4. BIC finds model 1 is the optimal choice and five variables have a Probability of 57.8%. The above analysis shows the regression equation below is statistically significant.

$$\mathbf{OFPI = 1.61821 + 0.13664OFS + 0.15171HS + 0.13178EC + 0.07150FQ + 0.10234PE}$$

5 Conclusions

The aim of the study shows the optimal choice by the BIC Algorithm for the Organic food purchasing intention of consumers. The processed regression results have 5 factors affecting Organic food purchasing intention (OFPI), namely Health consciousness (HS) has the strongest impact ($\beta = 0.15171$), followed by Organic food safety ($\beta = 0.13664$), Environment ($\beta = 0.13178$), Price ($\beta = 0.10234$), Food Quality ($\beta = 0.07150$) and the weakest impact was Knowledge of Organic Food ($\beta =$ 0.116). The results are quite similar to the results of some previous studies. Factors such as food safety, price, and environmental consciousness have a prime influence on the intention to buy organic food that studies by Bagher et al. (2018) and Nguyen et al. (2020) have been published. This shows that the paper's research results are consistent.

Limitations

Although the research results have certain scientific and practical contributions to the research field of safe food—organic food. However, the study still has some limitations that need to be considered: first, because of limitations in various aspects such as implementation time and cost, the study was only conducted with a sample size of 200 and used a non-probability sampling method, so it cannot be representative for the general people who have shopping experience in the food sector. Therefore, the author proposes the next research direction that can be with a larger sample size and can use the probability sampling method to improve the generalization and representativeness of the study population. Second, because of the influence of the SARS-COVID-19 epidemic, people also had limited contact, so it was not possible to conduct in-depth interviews with all consumers to find out more information that the questions in the table could not be used a survey that had not been performed. The author also wishes that the following research topics can be interviewed directly for more specific and accurate information. Third, regarding the application of the results, this study is only limited to food stores in Ho Chi Minh City, Vietnam but not widely available to locations in big cities such as Hanoi, Hai Phong, Da Nang, Nha Trang, Can Tho… This limitation comes from many reasons, ranging from time, the cost to shopping culture reasons of consumers in each region. Therefore, in the next research direction, the author will expand the research scope to other provinces in Vietnam. Fourth, the results of the regression analysis with the R^2 value of 0.553 shows that the model explains 55.3% of the change of the dependent variable. Other observed variables also affect the intention to buy organic food of consumers in Ho Chi Minh City but have not been condensed in this research model. Thus, there are still 49.3% of other factors that affect consumers' intention to buy organic food for further research directions. Finally, in terms of research methods, the study only tested the theoretical model by multiple linear regressions but did not use other methods, such as the linear structural modeling method (SEM) to test it. The hypothesis has just more clearly defined the cause-and-effect relationship between the research concepts. All the above limitations open up a lot of directions for future research, especially for researchers in the field of safe food—organic food in Vietnam.

Acknowledgements This research is funded by the Industrial University of Ho Chi Minh City, Vietnam.

References

Agarwal, P.: Theory of reasoned action and organic food buying in India. Srusti Manag. Rev. **12**, 28–37 (2019)

Bagher, A.N., Salati, F., Ghaffari, M.: Factors affecting intention to purchase organic food products among Iranian consumers. Acad. Market. Stud. J. **22**, 1–23 (2018)

Bauer, H.H., Heinrich, D., Schäfer, D.B.: The effects of organic labels on global, local, and private brands: more hype than substance? J. Bus. Res. **66**, 1035–1043 (2013)

Boccaletti, S., Nardella, M.: Consumer willingness to pay for pesticide-free fresh fruit and vegetables in Italy. Int. Food Agribus. Manag. Rev. **3**, 297–310 (2000)

Boxall, A.B., Rudd, M.A., Brooks, B.W., Caldwell, D.J., Choi, K., Hickmann, S., Innes, E., Ostapyk, K., Staveley, J.P., Verslycke, T.: Pharmaceuticals and personal care products in the environment: what are the big questions? Environ. Health Perspect. **120**, 1221–1229 (2012)

Çabuk, S., Tanrikulu, C., Gelibolu, L.: Understanding organic food consumption: attitude as a mediator. Int. J. Consum. Stud. **38**, 337–345 (2014)

Canavari, M., Nocella, G., Scarpa, R.: Stated willingness to pay for environment-friendly production of apples and peaches: web-based versus in-person surveys. 83rd EAAE Seminar, Chania (2003)

Carfora, V., Cavallo, C., Caso, D., Del Giudice, T., De Devitiis, B., Viscecchia, R., Nardone, G., Cicia, G.: Explaining consumer purchase behavior for organic milk: including trust and green self-identity within the theory of planned behavior. Food Qual. Prefer. **76**, 1–9 (2019)

Chauke, D.X., Duh, H.I.: Marketing and socio-psychological factors influencing organic food purchase and post-purchase outcomes. J. Food Products Market. **25**, 896–920 (2019)

Chen, M.F.: Attitude toward organic foods among Taiwanese as related to health consciousness, environmental attitudes, and the mediating effects of a healthy lifestyle. British Food J. (2009)

Crutchfield, S.R., Roberts, T.: Food safety efforts accelerate in the 1990's. Food Rev. National Food Rev. **23**, 44–49 (2000)

De Toni, D., Eberle, L., Larentis, F., Milan, G.S.: Antecedents of perceived value and repurchase intention of organic food. J. Food Products Market. **24**, 456–475 (2018)

Edenbrandt, A.K.: Demand for pesticide-free, cisgenic food? Exploring differences between consumers of organic and conventional food. British Food J. (2018)

Effendi, I., Ginting, P., Lubis, A.N., Fachruddin, K.A.: Analysis of consumer behavior of organic food in North Sumatra Province, Indonesia. J. Bus. Manag. **4**, 44–58 (2015)

Fotopoulos, C., Chryssochoidis, G.: Factors affecting the decision to purchase organic food. J. Euromarket. **9**, 45–66 (2001)

Ghali, Z.: Motives of willingness to buy organic food under the moderating role of consumer awareness. J. Scient. Res. Reports, 1–11 (2019)

Gil, J.M., Soler, F.: Knowledge and willingness to pay for organic food in Spain: evidence from experimental auctions. Acta Agric. Scand Sect. C **3**, 109–124 (2006)

Gracia, A., De Magistris, T.: The demand for organic foods in the South of Italy: a discrete choice model. Food Policy **33**, 386–396 (2008)

Gracia Royo, A., de-Magistris, T.: Organic food product purchase behaviour: a pilot study for urban consumers in the South of Italy. Span. J. Agric. Res. **5**, 439–451 (2007)

Ham, M., Pap, A., Stanic, M.: What drives organic food purchasing?–evidence from Croatia. British Food J. (2018)

Hidalgo-Baz, M., Martos-Partal, M., González-Benito, Ó.: Attitudes vs. purchase behaviors as experienced dissonance: the roles of knowledge and consumer orientations in organic market. Front. Psychol. **8**, 248 (2017)

Hughner, R.S., McDonagh, P., Prothero, A., Shultz, C.J., Stanton, J.: Who are organic food consumers? A compilation and review of why people purchase organic food. J. Consumer Behav. Int. Res. Rev. **6**, 94–110 (2007)

Kaplan, D.: On the quantification of model uncertainty: a bayesian perspective. Psychometrika **86**, 215–238 (2021)

Kim, H.Y., Chung, J.E.: Consumer purchase intention for organic personal care products. J. Consum. Market. (2011)

Kotler, P., Armstrong, G.: Principles of marketing, Pearson education (2010)

Kraft, F.B., Goodell, P.W.: Identifying the health conscious consumer. Mark. Health Serv. **13**, 18 (1993)

Kriwy, P., Mecking, R.A.: Health and environmental consciousness, costs of behaviour and the purchase of organic food. Int. J. Consum. Stud. **36**, 30–37 (2012)

Kushwah, S., Dhir, A., Sagar, M.: Understanding consumer resistance to the consumption of organic food. A study of ethical consumption, purchasing, and choice behaviour. Food Qual. Prefer. **77**, 1–14 (2019a)

Kushwah, S., Dhir, A., Sagar, M., Gupta, B.: Determinants of organic food consumption. A systematic literature review on motives and barriers. Appetite **143**, 104402 (2019b)

Lacaze, V.: Sustainable food consumption in Argentine: an estimation of willingness to pay for fresh and processed organic food for consumers in the case Buenos Aires's consumers. Food Sci. Technol. Abstracts Revista Agroalimentaria **15**, 87–100 (2009)

Lee, H.-J.: Individual and situational determinants of US consumers' buying behavior of organic foods. J. Int. Food Agribusiness Market. **28**, 117–131 (2016)

Lee, T.H., Fu, C.-J., Chen, Y.Y.: Trust factors for organic foods: consumer buying behavior. British Food J. (2019)
Lian, S.B.: What motivates consumers to purchase organic food in Malaysia. Asian Soc. Sci. **13**, 100–109 (2017)
Lockie, S., Lyons, K., Lawrence, G., Mummery, K.: Eating 'green': motivations behind organic food consumption in Australia. Sociol. Rural. **42**, 23–40 (2002)
Magnusson, M.K., Arvola, A., Hursti, U.K.K., Åberg, L., Sjödén, P.O.: Attitudes towards organic foods among Swedish consumers. British Food J. (2001)
Newsom, J.T., McFarland, B.H., Kaplan, M.S., Huguet, N., Zani, B.: The health consciousness myth: implications of the near independence of major health behaviors in the North American population. Soc. Sci. Med. **60**, 433–437 (2005)
Nguyen, H.V., Nguyen, N., Nguyen, B.K., Lobo, A., Vu, P.A.: Organic food purchases in an emerging market: the influence of consumers' personal factors and green marketing practices of food stores. Int. J. Environ. Res. Public Health **16**, 1037 (2019)
Nguyen, T.M.H., Nguyen, N.T., Nguyen, H.T.: Factors affecting voluntary information disclosure on annual reports: listed companies in Ho Chi Minh City stock exchange. J. Asian Financ. Econ. Bus. **7**, 53–62 (2020)
Nielsen: Successful passing of firstcovid-19 Vietnam become the second optimistic country in the world (Vietnamese) (2020). https://www.nielsen.com/wp-content/uploads/sites/3/2020/08/CCI-Q2-2020-VI.pdf?cid=socSprinklr-Nielsen+Vietnam
Nunnally, J.C.: Psychometric Theory, 2d edn. McGraw-Hill (1978)
O'Fallon, M.J., Gursoy, D., Swanger, N.: To buy or not to buy: impact of labeling on purchasing intentions of genetically modified foods. Int. J. Hosp. Manag. **26**, 117–130 (2007)
Ogorevc, M., Primc, K., Slabe-Erker, R., Kalar, B., Dominko, M., Murovec, N., Bartolj, T.: Social feedback loop in the organic food purchase decision-making process. Sustainability **12**, 4174 (2020)
Padel, S., Foster, C.: Exploring the gap between attitudes and behaviour: understanding why consumers buy or do not buy organic food. British Food J. (2005)
Pambudi, D.B., Ekawati, R.: Juridical Assessment of Food Safety in Packaged Processed Food Product. KnE Life Sci., 129–133–129–133 (2021)
Peterson, R.A.: A meta-analysis of Cronbach's coefficient alpha. J. Consumer Res. **21**, 381–391 (1994)
Pham, T.H., Nguyen, T.N., Phan, T.T.H., Nguyen, N.T.: Evaluating the purchase behaviour of organic food by young consumers in an emerging market economy. J. Strateg. Mark. **27**, 540–556 (2019)
Raftery, A.E.: Bayesian model selection in social research. Soc. Methodol, 111–163 (1995)
Raftery, A.E., Madigan, D., Hoeting, J.A.: Bayesian model averaging for linear regression models. J. Am. Stat. Assoc. **92**, 179–191 (1997)
Rahman, T., Hossain, M.A.: Organic food buying intention among young people. Barishal Univ. J. **6**, 162–178 (2019)
Ramayah, T., Lee, J.W.C., Mohamad, O.: Green product purchase intention: some insights from a developing country. Resour. Conserv. Recycl. **54**, 1419–1427 (2010)
Rashid, N.: Awareness of eco-label in Malaysia's green marketing initiative. Int. J. Bus. Manag. **4**, 132–141 (2009)
Saleki, Z.S., Seyedsaleki, S.M.: The main factors influencing purchase behaviour of organic products in Malaysia. Interdisciplinary J. Contemp. Res. Bus. **4**, 98–116 (2012)
Secapramana, L.V.H., Ang, L.G.K.: Antecedents affecting organic food purchase intentions. Int. J. Organ. Innov. **12**, 140–150 (2019)
Shaharudin, M.R., Pani, J.J., Mansor, S.W., Elias, S.J.: Factors affecting purchase intention of organic food in Malaysia's Kedah state. Cross-Cultural Commun. **6**, 105–116 (2010)
Slamet, A.S., Nakayasu, A., Bai, H.: The determinants of organic vegetable purchasing in Jabodetabek region. Indonesia. Foods **5**, 85 (2016)
Slater, S.F.: Issues in conducting marketing strategy research. J. Strateg. Mark. **3**, 257–270 (1995)

Sultan, P., Tarafder, T., Pearson, D., Henryks, J.: Intention-behaviour gap and perceived behavioural control-behaviour gap in theory of planned behaviour: moderating roles of communication, satisfaction and trust in organic food consumption. Food Qual. Preference **81,** 103838 (2020)

Švecová, J., Odehnalová, P.: The determinants of consumer behaviour of students from Brno when purchasing organic food. Rev. Econ. Perspect. **19**, 49–64 (2019)

Talwar, S., Jabeen, F., Tandon, A., Sakashita, M., Dhir, A.: What drives willingness to purchase and stated buying behavior toward organic food? A Stimulus–Organism–Behavior–Consequence (SOBC) perspective. J. Cleaner Prod. **293,** 125882 (2021)

Tandon, A., Jabeen, F., Talwar, S., Sakashita, M., Dhir, A.: Facilitators and inhibitors of organic food buying behavior. Food Qual. Preference **88,** 104077 (2021)

Teng, C.-C., Lu, C.-H.: Organic food consumption in Taiwan: motives, involvement, and purchase intention under the moderating role of uncertainty. Appetite **105**, 95–105 (2016)

Teng, C.-C., Wang, Y.-M.: Decisional factors driving organic food consumption: Generation of consumer purchase intentions. British Food J. (2015)

Thøgersen, J., de Barcellos, M.D., Perin, M.G., Zhou, Y.: Consumer buying motives and attitudes towards organic food in two emerging markets: China and Brazil. Int. Market. Rev. (2015)

Thompson, G.D.: Consumer demand for organic foods: what we know and what we need to know. Am. J. Agr. Econ. **80**, 1113–1118 (1998)

Tran, T.A., Pham, N.T., Pham, K.V., Nguyen, L.C.T.: The roles of health consciousness and service quality toward customer purchase decision. J. Asian Financ. Econ. Bus. **7**, 345–351 (2020)

Voon, T.J.P., Ngui, K.S., Agrawal, A.: Determinants of willingness to purchase organic food: an exploratory study using structural equation modeling. Int. Food Agribus. Manag. Rev. **14**, 103–120 (2011)

Wang, J., Pham, T.L., Dang, V.T.: Environmental consciousness and organic food purchase intention: a moderated mediation model of perceived food quality and price sensitivity. Int. J. Environ. Res. Public Health **17**, 850 (2020)

Waqas, A., Hong, C.: Study on consumer behaviour and food safety of organic products in Pakistan. E3S Web of Conferences, 2019. EDP Sciences, 02021

Winter, C., Davis, S.: Organic foods. J. Food Sci. (2006)

Yadav, R., Pathak, G.S.: Intention to purchase organic food among young consumers: evidences from a developing nation. Appetite **96**, 122–128 (2016)

Zanoli, R., Naspetti, S.: Consumer motivations in the purchase of organic food: a means-end approach. British Food J. (2002)

How the Exchange Rate Reacts to Google Trends During the COVID-19 Pandemic

Chaiwat Klinlampu, Pichayakone Rakpho, Supareuk Tarapituxwong, and Woraphon Yamaka

Abstract This study investigates the nonlinear impact of the COVID-19 pandemic on the exchange rates of Great Britain Pound, European Euro, and Chinese Yuan against the US$ (USD) which become the most tradeable currency pairs in recent years. Various COVID-19 indicators, namely the number of COVID-19 cases and deaths and Google Trends are considered in our analysis. Google Trends is relevant as it allows us to evaluate the panic and fear of investors during this pandemic. We utilize the Markov Switching regression model to divide the foreign exchange markets into the depreciation and the appreciation regimes. The results show that although there exists a similar sign of the COVID-19's impact on the foreign exchange markets under two different regimes, the magnitude of the impact of COVID-19 in the depreciation regime is greater than the appreciation regime. Moreover, we found that the number of COVID-19 cases and that of COVID-19 deaths can positively affect the exchange rates(depreciation), while Google Trends is likely to exert a negative effect on the foreign exchange markets(appreciation).

Keywords COVID-19 · Google trends · Markov Switching regression model · Exchange rates

1 Introduction

The spread of COVID-19 has had a massive impact on various constituents of the economy. Virtually all countries worldwide have faced the problems of employment reduction, business failure, and the resuled lower GDP growth Bartik et al. (2020).

C. Klinlampu (✉) · P. Rakpho · W. Yamaka
Center of Excellence in Econometrics, Chiang Mai University, Chiang Mai 50200, Thailand
e-mail: woraphon.econ@gmail.com

W. Yamaka
Faculty of Economics, Chiang Mai University, Chiang Mai 50200, Thailand

S. Tarapituxwong
Faculty of Management Sciences, Chiang Mai Rajabhat University, Chiang Mai 50200, Thailand

N. Ngoc Thach et al. (eds.), *Financial Econometrics: Bayesian Analysis, Quantum Uncertainty, and Related Topics*, Studies in Systems, Decision and Control 427,
https://doi.org/10.1007/978-3-030-98689-6_44

However, the impacts of COVID-19 can differ across different economic domains and groups as reflected by the growth rate approximately 3% falling of the global GDP compared to the 4–6.5% GDP fall in developint countires due to the pandemic Mou (2020). Lan et al. (2020) also revealed that the COVID-19 pandemic also influences the financial markets' movement. The motivation of our study comes from several papers that show how COVID-19 has impacted financial markets. In the literature, Al-Awadhi et al. (2020) found that the daily increase in total confirmed COVID-19 cases and deaths has a considerable negative effect on stock returns across all businesses. Ahmed et al. (2021) showed that the COVID-19 caused lockdown policy is a key factor affecting oil prices. Not only the oil and stock markets but the foreign exchange markets have also been significantly and negatively affected by such pandemic. Devpura (2021) confirmed that the COVID-19 shows some effect on the exchange rate during March 2020. Besides, in terms of fundamentals, the epidemic can affect the currency market through global investment flows, as investment flows often depend on economic growth expectations in the country. High volatility of investment flows, which will result in high volatility in the exchange rates.

From an academic standpoint, the effect of the COVID-19 crisis on the behavior and dynamics of exchange rates has been showing through an adjustment from panic. Feng et al. (2021) noted that this pandemic had caused panic in all sectors of the global economy, leading to an adjustment in the operation of businesses and financial institutions to avoid the economic disaster. The adjustment results in exchange rate variation Pasiouras and Daglis (2020). However, some research findings revealed that the spread of COVID-19 can reduce the volatility of exchange rate. For example, Narayan (2020) reported that, prior to the COVID-19 era, the Japanese Yen was nonstationary; but it becomes more stable during the COVID-19 period. We would like to note that the fluctuation of the exchange rate has been found to contribute a negative consequence to export demand Camba and Camba (2020), stock market returns Adjasi et al. (2008), and FDI inflow Kyereboah-Coleman and Agyire-Tettey (2008).

In the analysis of the impacts of COVID-19, most of the studies considered the number of confirmed cases and deaths as the representative of the COVID-19 indicators. Unlike the previous studies, in this study, we aim to show how major foreign exchange markets reacted to the fear of the coronavirus. To measure fear, we rely on internet searches of COVID-19-related terms. Bento et al. (2020) showed that the general public's response to news about local COVID-19 cases is to search for more information on the internet. Mahfuza et al. (2020) showed that Google Trends is the popular and potential alternative tool for researching and monitoring the influence of COVID-19. Google Trends (GT) creates indexes based on a topic's relative web-search volume over time. These indexes can be retrieved for selected geographic areas or on a worldwide scale. The meaning of Google Trends indices is simple: the higher the value of a given GT, the greater the public interest in that topic. Recently, GT has been successfully used for predicting and explaining the movement of economics and finance. Moreover, previous studies have used GT to assess the impact of COVID-19 on a wide range of variables such as well-being Brodeur et al. (2021) and stock markets resilience Costola et al. (2021), Ding et al. (2020).

Although there were numerous studies of the relationship between COVID-19 and financial markets, few studies have examined the relationship between COVID-19 and the exchange rate. In this respect, our study retrieves the GT country index for the topic of COVID-19 as the indicator of country-level public attention and investigates its impact on the foreign exchange markets, including the British Pound, European Euro, and Chinese Yuan during the outbreak of this recent crisis.

The method we adopt to investigate the impact of COVID-19 is the Markov Switching regression model Hamilton (1989). The appealing aspect of this model is that it handles the nonlinear causal effect of predictor variables on response variables and captures more complex dynamic patterns of time series data Maneejuk et al. (2018), Pastpipatkul et al. (2016), Tansuchat et al. (2016). For this reason, we suggest using the Markov Switching regression model to investigate the switching effects of the COVID-19 pandemic on the foreign exchange markets. We believe that the study results can be used to inform investors' currency investment decisions or policymaker on the COVID-19 epidemic in each country.

The rest of this paper is organized as follows: Sect. 2 briefly describes the Markov Switching regression model, Sect. 3 provides data description, Sect. 4 discusses the results of the COVID-19 pandemic effect on the exchange rates, and Sect. 5 concludes this study.

2 Methodology

Before explaining the approach of the Markov Switching regression model, let us provide the concept of the linear regression model.

2.1 *Linear Regression Model*

The linear regression model captures the relationship between two or more predictor variables and a response variable by fitting a linear equation to observed data as illustrated below

$$Y_t = \alpha + \beta_1 X_{1,t} + \beta_2 X_{2,t} + \cdots + \beta_P X_{P,t} + \varepsilon_t, \tag{1}$$

where Y_t is the dependent variable, $X_{1,t}$, ..., $X_{P,t}$ are independent variables, α, $\beta_1, \ldots, \beta_p$ are the estimated parameters, and ε_t is the residuals.

2.2 *Markov Switching Regression Model*

The Markov Switching regression model was presented by Hamilton (1989). Its parameters are time-dependent based on the types of regimes (states) and the transition probabilities between them. The following equations can describe the Markov Switching regression model

$$Y_t = \alpha(s_t) + \beta_1 X_1(s_t) + \beta_2 X_2(s_t) + \cdots + \beta_P X_P(s_t) + \varepsilon_t(s_t), \tag{2}$$

where s_t denotes the regime or state variable, which is assumed to have two regimes, namely depreciation and appreciation regimes. The model assumes that the unobservable realization of the regime s_t is governed by the first-order discrete state of the Markov stochastic process which is defined by

$$p_{ij} = \Pr(s_{t+1} = j\,|s_t = i)\ ,\ \sum_{j=1}^{2} p_{ij} = 1, ; i, j = 1, 2. \tag{3}$$

The probability of switching from one regime to the other is captured in matrix Q , which is known as the transition matrix

$$Q = \begin{bmatrix} p_{11} & p_{12} \\ p_{21} & p_{22} \end{bmatrix}. \tag{4}$$

3 Data Description

In this study, we investigate the impacts of COVID-19 pandemic on three exchange rates, including the Great Britain Pound (GBP), European Euro (EUR), and Chinese Yuan (CNY), against the US$ (USD). We consider these exchange rates because they have become the most tradeable currency pairs in recent years. For the COVID-19 indicators, we consider the intensity of the Covid-19 (represented by confirmed cases (CASES), and the number of deaths (DEATHS) in each country), and the fear of the coronavirus (represented by GT indexes relating to COVID-19). We have collected GT related to COVID-19 by using two English words: "coronavirus" and "covid19" Pan et al. (2020).

All the considered variables are collected on daily frequencies covering the period from January 1, 2020, to July 30, 2021. From the beginning, we convert all the series into growth values. First, Google Trends data are obtained from https://trends.google.co.th, while COVID-19 cases and deaths are obtained from https://ourworldindata.org. Next, all exchange rates data are obtained from https://investing.com. The description, normality test, and unit root test for each variable are presented in Table 1.

Table 1 Descriptive statistics

Variable	Mean	Median	Max	Min	Std.Dev.	Skewness	Kurtosis	JB test	ADF test
Great Britain Pound									
GBP	0.275	0.270	0.352	0.139	0.046	−0.232	2.364	14.127[0.000]	−3.338[0.004]
GT _coro	1.377	1.322	4.605	−2.408	1.026	0.115	4.772	72.766[0.000]	−2.810[0.019]
GT _cov19	1.058	0.971	4.605	−4.605	1.326	−0.436	6.280	262.552[0.000]	−2.725[0.024]
I	7.439	8.113	11.130	0.000	2.924	−1.543	4.580	273.839[0.000]	−2.580[0.036]
DTH	3.807	3.951	7.510	0.000	2.260	−0.282	1.916	34.063[0.000]	−3.414[0.003]
European Euro									
EUR	0.150	0.166	0.209	0.067	0.042	−0.478	1.761	55.790[0.000]	−3.480[0.002]
GT _coro	0.980	0.770	4.387	−2.996	1.179	0.651	3.724	50.638[0.000]	−3.163[0.007]
GT _cov19	1.544	1.569	4.605	−2.813	1.131	−0.067	4.132	29.620[0.000]	−3.307[0.004]
I	7.159	7.720	10.800	0.000	2.873	−1.397	4.244	213.224[0.000]	−2.697[0.026]
DTH	3.512	4.043	7.458	0.000	2.240	−0.296	1.775	42.173[0.000]	−3.442[0.003]
Chinese Yuan									
CNY	−1.910	−1.915	−1.852	−1.970	0.038	0.019	1.357	61.537[0.000]	−2.569[0.037]
GT _coro	0.343	0.000	4.554	−3.507	1.307	0.881	3.802	85.397[0.000]	−2.897[0.015]
GT _cov19	0.561	0.000	4.533	0.000	1.110	1.805	4.920	380.919[0.000]	−4.599[0.000]
C	2.940	2.773	9.625	0.000	1.595	1.038	5.633	256.222[0.000]	−3.438[0.003]
DTH	0.431	0.000	7.162	0.000	1.194	2.840	10.208	1919.759[0.000]	−6.971[0.000]

Note [ ] denotes Minimum Bayes factor (MBF) computed by $e^{p \log p}$, where p is the p-value. (See detail in Maneejuk and Yamaka (2021))

From Table 1, GBP has the highest average growth while CNY has the lowest. We also observe that the volatilities of the two GTs are similar for all countries. Moreover, the growth of Chinese CASES and DEATHS are low, whereas the other two countries tend to have high CASES and DEATHS growth. Furthermore, we observe that exchange rates, Google Trends, cases, and deaths are not normally distributed as the Minimum Bayes factor (MBF) values indicate that the growth data are decisive, not normally distributed (see, Jarque-Bera test). The unit root test is also conducted to investigate the stationarity of the data, and the results show that all growth data are stationary as the MBF values are close to zero. Obviously, these variables can be used to estimate the Markov Switching (MS) regression model in the next step. Thus, our empirical model can be expressed as follows:

$$\begin{aligned} Y_{GBP} &= \alpha_0(s_t) + \beta_0 GT\_\text{cov}\,19(s_t) + \beta_1 GT\_\text{coro}(s_t) + \beta_2 C(s_t) + \beta_3 DTH(s_t) + \varepsilon_t(s_t) \\ Y_{EUR} &= \alpha_1(s_t) + \beta_4 GT\_\text{cov}\,19(s_t) + \beta_5 GT\_coro(s_t) + \beta_6 C(s_t) + \beta_7 DTH(s_t) + \varepsilon_t(s_t) \\ Y_{CNY} &= \alpha_3(s_t) + \beta_8 GT\_\text{cov}\,19(s_t) + \beta_9 GT\_coro(s_t) + \beta_{10} C(s_t) + \beta_{11} DTH(s_t) + \varepsilon_t(s_t) \end{aligned} \quad (5)$$

where GT_cov19 refer to Google Trend Search "covid19", GT_coro is Google Trend Search "coronavirus", C is new confirmed cases, and DTH is new confirmed death.

4 Results

In this section, we report the impacts of COVID-19 on GBP, EU, and CNY in Tables 2, 3 and 4. The present study considers two different regimes in the foreign exchange markets, the depreciation market (Regime 1) and the appreciation market (Regime 2).

From Table 2, it can be seen that the signs of the coefficients under both regimes are similar, indicating the same effect of COVID-19 on GBP in the depreciation and

Table 2 The impact of COVID-19 on GBP

	Regime 1 ($s_t = 1$)			Regime 2 ($s_t = 2$)		
Parameter	Estimate	Std. Error	MBF	Estimate	Std. Error	MBF
Constant	0.271	0.003	0.000	0.190	0.006	0.000
GT_cov19	−0.033	0.002	0.000	−0.019	0.002	0.000
GT_coro	0.002	0.003	1.000	−0.007	0.004	0.499
C	0.009	0.001	0.000	0.018	0.001	0.000
DTH	0.003	0.001	0.002	0.012	0.001	0.000
p_{11}	0.601					
p_{22}	0.859					
AIC-linear regression	−4.262					
AIC-MS regression	−4.439					

Table 3 The impact of COVID-19 on EUR

	Regime 1 ($s_t = 1$)			Regime 2 ($s_t = 2$)		
Parameter	Estimate	Std. Error	MBF	Estimate	Std. Error	MBF
Constant	0.1025	0.0024	0.000	0.0240	0.0077	0.032
GT_cov19	0.0003	0.0019	1.000	−0.0177	0.0027	0.000
GT_coro	−0.0050	0.0012	0.002	−0.0101	0.0028	0.002
C	0.0061	0.0009	0.000	0.0247	0.0013	0.000
DTH	0.0058	0.0011	0.000	0.0080	0.0015	0.000
p_{11}	0.5914					
p_{22}	0.8511					
AIC-linear regression	−4.5556					
AIC-MS regression	−4.8764					

Table 4 The impact of COVID-19 on CNY

	Regime 1 ($s_t = 1$)			Regime 2 ($s_t = 2$)		
Parameter	Estimate	Std. Error	MBF	Estimate	Std. Error	MBF
Constant	−1.8822	0.0040	0.000	−1.9456	0.0018	0.000
GT_cov19	−0.0045	0.0011	0.000	−0.0020	0.0009	0.226
GT_coro	−0.0027	0.0015	0.469	−0.0030	0.0009	0.017
C	0.0032	0.0013	0.154	0.0020	0.0007	0.103
DTH	0.0182	0.0014	0.000	0.0008	0.0010	1.000
p_{11}	0.8620					
p_{22}	0.9024					
AIC-linear regression	−4.1085					
AIC-MS regression	−4.5472					

appreciation periods. We find that new confirmed cases and deaths due to COVID-19 have a significantly positive effect on GBP with decisive evidence. However, GT_covid19 has a negative and significant effect on GBP with decisive evidence. The positive impact of CASES and DEATHS on GBP can be supported by the likelihood that investors were worried about the problematic situation in Britain and accordingly adjusted downward their expectations about policies and the economy, leading to capital outflow from Britain. Our finding is consistent with Feng et al. (2021), who found that the rise in the number of COVID-19 cases and deaths would increase exchange rate growth (depreciation). In terms of GT_covid19, the pandemic effect is favorable for GBP in both regimes. The possible reason is that public information access can effectively reduce the uncertainty and panic caused by COVID-19 by sending positive signals to the markets and investors whose decisions and activities helped decrease the exchange rate's growth (depreciation). Lyócsa et al. (2020) confirmed that social media could reduce investors' panic and stabilize the market as social media can improve investors' confidence during this pandemic.

The results for EUR reported in Table 3 show that new confirmed cases and deaths due to COVID-19 have a significantly positive effect on EUR with decisive evidence, while GT_coronavirus has a significantly negative impact on EUR with decisive evidence in both regimes. We now turn attention to the CNY case. Similar to the first two currencies, and we also find that Internet Searches on the two COVID-19 related terms have a negative impact on CNY, while the COVID-19 cases and deaths remain giving a positive impact on CNY.

According to the above results, we find that the report of COVID-19 cases and deaths has a positive effect on the foreign exchange markets in both regimes but in terms of deaths in appreciation regimes not significant (see the MBF). However, the appreciation regime's coefficient values are greater than those under the depreciation regime, indicating that COVID-19 has a greater positive effect on foreign exchange rates in the appreciation regime. Nevertheless, the effects of GT related to COVID-19 are found to be negative in both regimes, but more pronounced in the appreciation regime.

A perception variable represented by the Google Trends search shows whether the currency is depreciating or appreciating as more searches in "covid19" and "coronavirus" would result in a fall in the exchange rates (currency appreciation). One of the reasons may be that as people become more knowledgeable of the COVID-19, their anxiety and worry will decrease due to an increase in perceived effective self-defense techniques to protect themselves from the disease Liu et al. (2020), and as a result, consumers can begin to continue their normal spending and daily-life activities, thus increasing the country's demand for currency. While the variable of COVID-19's impact shown by new cases and deaths, it was found that the worse the coronavirus crisis, the higher the exchange rate (currency depreciation). This is partly due to investor distrust in the economy Teresiene et al. (2021), which reduces demand for a country's currency, or the implementation of lockdown policies, which restrict economic activity and lead to lower household expenditure Coffey et al. (2020).

To further assist in interpreting the different regimes, we illustrate the filtered probabilities plots of being in the depreciation and appreciation regimes in Figs. 1, 2 and 3. We can see that the probability of staying in the depreciation regime during this COVID-19 crisis is high for GBP and EUR. While the result of Fig. 3 shows that the probability of staying in the depreciation regime of CNY is high in the first half of the COVID-19 pandemic period. In addition, the Chinese yuan performs persistent in staying in the appreciation regime after October 2020. This corresponds to the strong economy of China during this COVID-19 period. Not surprisingly, although the COVID-19 originated from China, the recovery rate and the number of COVID-19 cases in China are pretty low after the successful lockdown in Wuhan province.

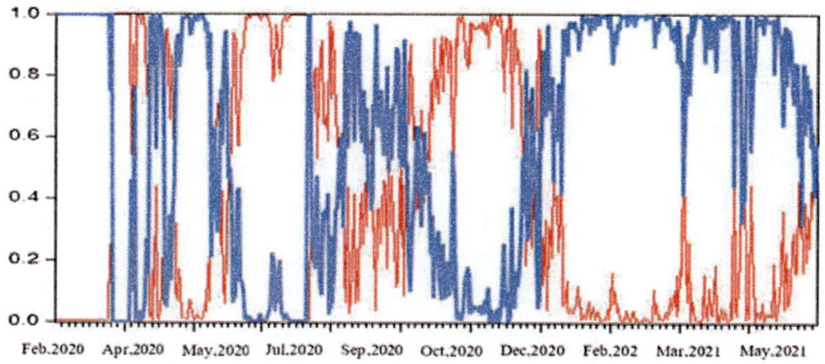

Fig. 1 Filtered probabilities of GBP market in depreciation (blue line) and appreciation (red line) regimes

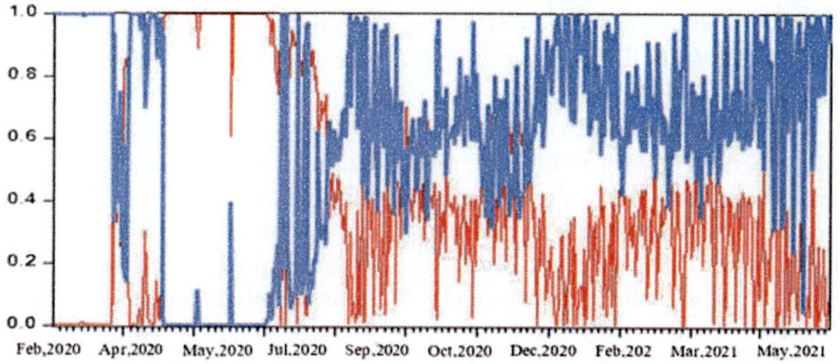

Fig. 2 Filtered probabilities of EUR market in depreciation (blue line) and appreciation (red line) regimes

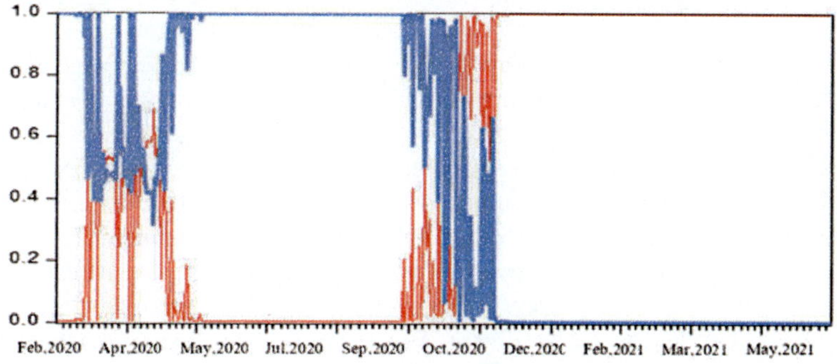

Fig. 3 Filtered probabilities of CNY market in depreciation (blue line) and appreciation (red line) regimes

Finally, to confirm the robustness of the MS regression model, we compare the performance of the model with the linear regression (Sect. 1 in "On The Skill of Influential Predictions") using the Akaike Information Criterion(AIC). The comparison results are reported in the last two rows of Tables 2, 3 and 4, and it is evident that the MS regression performs better than the linear regression as the lower AIC values are obtained for all markets.

5 Conclusion

In this study, we examine the effect of the COVID-19 pandemic on three exchange rates: Great Britain Pound, European Euro, and Chinese Yuan. The conventional indicators of COVID-19 (i.e., cases and deaths) and the new indicator based on the fear of the coronavirus are also considered. This study uses the internet searches of Google Trends of COVID-19-related terms to represent the fear of the people toward COVID-19.

We then conduct the Markov Switching regression analysis on various COVID-19 indicators and discuss these indicators' switching impact (nonlinear) on the exchange rates. Based on such analysis, there is substantial evidence that COVID-19 contributes a nonlinear impact on the exchange rates of Great Britain Pound, European Euro, and Chinese Yuan. Given that the number of new confirmed cases and deaths due to the COVID-19 pandemic have risen, COVID-19 has demonstrate to positively affect the exchange rates in both regimes. However, Google Trends revealed it can have a role to play as more searches in words "covid19" and "coronavirus" would result in currency appreciation due to public access to COVID-19 related information (such as emergency medical input and vaccine research) can improve investors' confidence in pandemic prevention and control, thus curbing exchange rate volatility caused by any panic Béjaoui et al. (2021). In other words, when people become more knowledgeable of COVID-19, the panic will decrease because of the increase in perceived effective self-defense techniques, and consumers can begin to continue regular spending and increase the country's demand for currency. Meanwhile, the variables of new cases and deaths are also crucial as their higher number will lower investors' trust in the performance of the local economy, thereby moving their investments to other countries and resulting in currency depreciation in their home country.

However, we cannot say definitively whether the resulting appreciation or depreciation will positively or negatively affect each country. The exchange rate changes have both positive and negative effects. Therefore, the conclusion of whether the currency is strengthening or weakening will have a positive effect or vise versa on that country must compare the net effect on various domestic sectors derived from exchange rate changes Kandil (2015).

Acknowledgements The authors are grateful to the Centre of Excellence in Econometrics, Chiang Mai University, for financial support. The authors are grateful to Dr.Laxmi Worachai for her helpful comments and suggestions.

References

Adjasi, C., Harvey, S.K., Agyapong, D.A.: Effect of exchange rate volatility on the Ghana stock exchange. African J. Account. Econ. Financ. Banking Res. **3**(3) (2008)

Ahmed, F., Syed, A.A., Kamal, M.A., de las Nieves López-García, M., Ramos-Requena, J.P., Gupta, S.: Assessing the impact of COVID-19 pandemic on the stock and commodity markets performance and sustainability: a comparative analysis of South Asian countries. Sustainability **13**(10), 5669 (2021)

Al-Awadhi, A.M., Alsaifi, K., Al-Awadhi, A., Alhammadi, S.: Death and contagious infectious diseases: impact of the COVID-19 virus on stock market returns. J. Behav. Exp. Financ. **27**, 100326 (2020)

Bartik, A.W., Bertrand, M., Cullen, Z., Glaeser, E.L., Luca, M., Stanton, C.: The impact of COVID-19 on small business outcomes and expectations. Proc. Natl. Acad Sci. **117**(30), 17656–17666 (2020)

Bento, A.I., Nguyen, T., Wing, C., Lozano-Rojas, F., Ahn, Y.Y., Simon, K.: Evidence from internet search data shows information-seeking responses to news of local COVID-19 cases. Proc. Natl. Acad. Sci. **117**(21), 11220–11222 (2020)

Béjaoui, A., Mgadmi, N., Moussa, W., Sadraoui, T.: A short-and long-term analysis of the nexus between Bitcoin, social media and Covid-19 outbreak. Heliyon **7**(7), e07539 (2021)

Brodeur, A., Clark, A.E., Fleche, S., Powdthavee, N.: COVID-19, lockdowns and well-being: evidence from google trends. J. Public Econ. **193**, 104346 (2021)

Camba, A.L., Camba, A.C., Jr.: The effect of Covid-19 pandemic on the Philippine stock Exchange, Peso-Dollar rate and retail price of diesel. J. Asian Financ. Econ. Bus. **7**(10), 543–553 (2020)

Coffey, C., Doorley, K., O'Toole, C., & Roantree, B.: The effect of the COVID-19 pandemic on consumption and indirect tax in Ireland (No. 2021/3). Budget Perspect. (2020)

Costola, M., Iacopini, M., Santagiustina, C.R.: Google search volumes and the financial markets during the COVID-19 outbreak. Financ. Res. Lett. **42**, 101884 (2021)

Ding, D., Guan, C., Chan, C.M., Liu, W.: Building stock market resilience through digital transformation: using Google trends to analyze the impact of COVID-19 pandemic. Front. Bus. Res. China **14**(1), 1–21 (2020)

Devpura, N.: Effect of COVID-19 on the relationship between Euro/USD exchange rate and oil price. MethodsX **8**, 101262 (2021)

Feng, G.F., Yang, H.C., Gong, Q., Chang, C.P.: What is the exchange rate volatility response to COVID-19 and government interventions? Econ. Anal. Policy **69**, 705–719 (2021)

Hamilton, J.D.: A new approach to the economic analysis of nonstationary time series and the business cycle. Econometrica: J. Econ. Soc., 357–384 (1989)

ILO: Impact of the COVID-19 crisis on loss of jobs and hours among domestic workers (2020). Retrieved June 20 2021 from https://www.ilo.org/wcmsp5/groups/public/—ed_protect/—protrav/-travail/documents /publication/wcms_747961.pdf

Kandil, M.: On the benefits of nominal appreciations: contrasting evidence across developed and developing countries. Borsa Istanbul Rev. **15**(4), 223–236 (2015)

Kyereboah-Coleman, A., Agyire-Tettey, K.F.: Effect of exchange-rate volatility on foreign direct investment in Sub-Saharan Africa: the case of Ghana. J. Risk Financ. (2008)

Lan, C., Huang, Z., Huang, W.: Systemic risk in China's financial industry due to the COVID-19 pandemic. Asian Econ. Lett. **1**(3), 18070 (2020)

Liu, B., Lin, S., Wang, Q., Chen, Y., Zhang, J.: Can local governments' disclosure of pandemic information decrease residents' panic when facing COVID-19 in China? Int. Public Manag. J. **24**(2), 203–221 (2020)

Lyócsa, Š, Baumöhl, E., Výrost, T., Molnár, P.: Fear of the coronavirus and the stock markets. Financ. Res. Lett. **36**, 101735 (2020)

Mahfuza, N., Syakurah, R.A., Citra, R.: Analysis and potential use of google trends as a monitoring tool for risk communication during covid-19 pandemic in Indonesia. Int. J. Public Health Sci. (Ijphs) (2020)

Maneejuk, P., Yamaka, W., Sriboonchitta, S.: A Markov-switching model with mixture distribution regimes. In: International Symposium on Integrated Uncertainty in Knowledge Modelling and Decision Making, pp. 312–323. Springer, Cham (2018)

Maneejuk, P., Yamaka, W.: Significance test for linear regression: how to test without P-values? J. Appl. Stat. **48**(5), 827–845 (2021)

Mou, J.: Research on the impact of COVID19 on global economy. In: IOP Conference Series: Earth and Environmental Science, Vol. 546, No. 3, p. 032043. IOP Publishing (2020)

Narayan, P.K.: Has COVID-19 Changed exchange rate resistance to shocks?. Asian Econ. Lett. **1**(1) (2020). https://doi.org/10.46557/001c.17389

Pan, Z., Nguyen, H. L., Abu-gellban, H., Zhang, Y.: Google trends analysis of COVID-19 pandemic. In: 2020 IEEE International Conference on Big Data (Big Data), pp. 3438–3446. IEEE (2020)

Pasiouras, A., Daglis, T.: The dollar exchange rates in the Covid-19 Era: evidence from 5 currencies. Europ. Res. Studies **23**(2), 352–361 (2020)

Pastpipatkul, P., Panthamit, N., Yamaka, W., Sriboochitta, S.: A copula-based Markov switching seemingly unrelated regression approach for analysis the demand and supply on sugar market. In: International Symposium on Integrated Uncertainty in Knowledge Modelling and Decision Making, pp. 481–492. Springer, Cham (2016)

Tansuchat, R., Maneejuk, P., Wiboonpongse, A., Sriboonchitta, S.: Price transmission mechanism in the Thai rice market. In: Causal Inference in Econometrics, pp. 451–461. Springer, Cham (2016)

Teresiene, D., Keliuotyte-Staniuleniene, G., Liao, Y., Kanapickiene, R., Pu, R., Hu, S., Yue, X.G.: The impact of the COVID-19 pandemic on consumer and business confidence indicators. J. Risk Financ. Manag. **14**(4), 159 (2021)

Impact of Financial Institutions Development on Capital Structure of Listed Firms in Asean Developing Countries

Bich Loc Tram, Van Thuan Nguyen, Van Tuan Ngo, and Thanh Liem Nguyen

Abstract Using Bayesian approach, this research aims to analyze the impact of financial institutions development on corporate capital structure in 5 developing countries in ASEAN region from 2010 to 2019. We find that financial institutions development and profitability have a negative impact on firms 'capital structure (including total debt, long-term debt and short-term debt), while inflation positively affects capital structure. National institutional quality and GDP per capita growth rate can have a bidirectional effect on capital structure. Firm size and asset properties have a positive impact on total debt and long-term debt, but negative impact on short-term debt. Meanwhile, TOBINQ has a positive impact on long-term debt, but bidirectional effect on total debt and short-term debt. Our research findings on the capital structure of firms in the five ASEAN developing countries supports both the trade-off theory and the pecking order theory.

Keywords Capital structure · ASEAN · Financial institutions development

JEL Classification Code C12 · C13 · E44 · F15

1 Introduction

Financial intermediaries (financial institutions) act as a "bridge" between those who have excess capital and those who need capital. However, studies on the determinants of capital structure have not been considered in the context of linkages with

B. L. Tram (✉)
Sai Gon University, Ho Chi Minh City, Vietnam

V. T. Nguyen
University of Finance - Marketing, Ho Chi Minh City, Vietnam

V. T. Ngo
Banking University HCMC, Ho Chi Minh City, Vietnam

T. L. Nguyen
University of Economics and Law, Ho Chi Minh City, Vietnam

N. Ngoc Thach et al. (eds.), *Financial Econometrics: Bayesian Analysis, Quantum Uncertainty, and Related Topics*, Studies in Systems, Decision and Control 427,
https://doi.org/10.1007/978-3-030-98689-6_45

financial institutions. Previous studies present mixed results and no single theory seems to be adequate in explaining leverage dynamics. Furthermore, the legal and institutional environments are significantly different in emerging markets, which helps explain the inconsistencies in findings from emerging countries (Wald 1999; Yarba and Guner 2019). ASEAN is an economically vibrant region with more than 600 million residents. Although empirical studies have delved more and more in the context of developing countries, there is still very little research done in the ASEAN region (Phooi et al. 2017). Therefore, examining the impact of financial development on capital structure of listed companies in developing countries in ASEAN is necessary to help businesses improve capital efficiency and increase value.

2 Relevant Literature and Hypothesis Development

2.1 Concepts and Theories of Capital Structure

Capital structure is the combination of debt and equity that an organization uses to finance its operations and investments (Kumar et al. 2017). Based on a number of assumptions including no taxes, no asymmetric information, no costs of bankruptcy and financial distress, Modigliani and Miller (1958) argue that capital structure does not affect cash flows, so it does not affect firm value. Modigliani and Miller (1963) added the factor of corporate income tax to Modigliani and Miller (1958)'s theory, and state that the value of levered firm is equal to value of unlevered firm plus present value of tax shield, and thanks to the tax shield, the average cost of capital decreases as firm increases its debt ratio. In other words, M&M theory implies that it is beneficial for a company to use debt as much as possible.

Nonetheless, if a firm continuously increases debt ratio, its value does not always increase. Myers (1993) constructed Trade-off Theory, in which capital structure would be determined by trade-offs between costs and benefits of borrowed capital. According to Trade-off Theory, value of the levered firm is equal to the value of unlevered firm, plus the benefit from the debt tax shield and the reduction of agency cost associated with free cash flows, minus costs (including agency costs when firms borrow heavily, bankruptcy costs and financial distress). The optimal capital structure is achieved when the marginal benefit of increasing debt equals the marginal cost of debt increase (Miglo 2016).

Contrary to Trade-off Theory, Pecking Order Theory proposed by Myers (1984) does not aim to predict the optimal debt ratio. The Pecking Order Theory states the preferences of funding sources in financing investments. Specifically, when a business needs funding, the manager will prioritize internal capital (such as retained earnings because this capital has little problems of information asymmetry), then borrowed capital and finally stock issuance. The above order is related to cost of using funds, the lower the cost, the higher the priority. Therefore, a firm's capital structure is the result of company's financing requirements over time and effort to minimize the cost of adverse selection resulted from information asymmetry issue.

Market timing theory holds that capital structure is not the outcome of dynamic optimization strategy, but merely an accumulation of past efforts to capitalize on market timing. In other words, firms issue shares depending on stock market conditions, and capital structure changes are due to this strategy, not because the company adjusts their debt to the target ratio (Baker and Wurgler 2002).

The impact of financial institutions development on capital structure and the measure of financial institutions development

Asymmetric information has long been identified as a pervasive problem in financial markets and it leads to problems of negative selection and moral hazard. Therefore, intermediary financial institutions were established to partially overcome the information asymmetry issue in the economy (Mishkin 2016). In addition, information asymmetry and agency costs are identified as the main frictions (Jensen and Meckling 1976; Myers 1977, 1984). In mitigating these frictions as well as in facilitating firms' access to capital, financial intermediaries play an important role (Leland and Pyle 1977; Diamond 1984). Hence, corporate debt is expected to increase with financial institutions development (Yarba and Guner 2019).

Financial development is the improvement of the financial system with regard to its five functions, including: (i) Providing information before making investment decisions; (ii) Monitoring investments and implementing corporate governance; (iii) Diversification and risk management; (iv) Accumulation of savings; and (v) Exchange of goods and services (Levine 2005). The above definition shows that financial institution development is part of overall financial development.

The main function of financial intermediaries is to monitor borrowers. As Diamond (1984) argues, intermediaries are more likely to use the information gathered to discipline borrowers, compared to individual investors. Therefore, a developed banking sector is expected to facilitate access to external finance. However, the impact of financial institutions development on corporate debt maturity is ambiguous. A developed banking sector leads to an increase in the availability of short-term financing, as this form of financing allows intermediaries to use their comparative advantage in monitoring firms. However, the economies of scale of banks and their ability to monitor firms also allow them to provide long-term loans. Therefore, which of these trends prevails is an empirical question (Diamond 1984; Demirguc-Kunt and Maksimovic 1999).

In addition, the institutional quality of a country also plays a very important role and will have a certain impact on the capital structure of enterprises. When the legal system is inefficient or costly to use, short-term debt is more likely to be employed than long-term debt. As Diamond (1991, 1993) and Rajan (1992) argue, short-term financing makes it more difficult for borrowers to defraud creditors. The creditors can review the firm's decisions more frequently and, if necessary, adjust the terms of the financing to protect their interests. Thus, we expect an inverse relation between the inefficiency of a country's legal system and the use of long-term debt. To the extent that there are fixed litigation costs in enforcing contracts, long-term debt is

Table 1 IMF financial institutions development index

Financial Institutions Index (FII)		
Financial Institutions Depth Index (FID)	Financial Institutions Access Index (FIA)	Financial Institutions Efficiency Index (FIE)
1. Private sector credit to GDP 2. Pension fund assets to GDP 3. Mutual fund assets to GDP 4. Insurance premiums (life + non-life) to GDP	1. Bank branches per 100,000 adults 2. ATMs per 100,000 adults	1. Net Interest Margin (NIM) 2. Lending—deposits spread 3. Non-interest income to total income 4. Overhead costs to total assets 5. ROA 6. ROE

Source IMF (2015, 2016)

likely to be used most heavily by large firms. The fixed costs also make the use of long-term debt, particularly by small firms, less responsive to small year-to year changes in the economic environment. This implies a trade-off between the use of long-term and short-term debt (Demirguc-Kunt and Maksimovic 1999). In addition, the law and finance view holds that finance is a set of contracts. These contracts are defined and made more or less effective by legal rights and enforcement mechanisms. Therefore, a well-functioning legal system facilitates the operation of both markets and intermediaries.

To gauge the level of development of financial markets, International Monetary Fund—IMF (2015, 2016) has launched a new set of indicators on three aspects of stock markets: depth, accessibility and effectiveness (Table 1). Rather than analyzing how each aspect of financial market, we use FMI as an overall index to measure the development of stock market.

Note Net interest margin (NIM) is the accounting value of bank's net interest revenue as a share of its average interest-bearing assets. ROA (Return on assets) is an indicator of how profitable a firm is relative to its total assets. ROE (Return on equity) is a measure of a company's financial performance, calculated by dividing net income by shareholders' equity. ROA and ROE is displayed as a percentage; the higher the ROA and ROE is, the better.

2.2 Empirical Research Related to the Topic

Kumar et al. (2017) provide a meta-analysis and find that studies on determinants of corporate capital structure are numerous and quite complete in developed economies, but this issue has not been thoroughly considered in emerging markets. Capital markets in emerging markets are relatively inefficient and inadequate. Most of the literature studies only firm level factors, but less attention is given for country-level determinants. Although there have been many studies on the factors affecting capital structure, each work studies a different aspect and very few works focus on the impact of financial institutions development and national institutional quality on capital structure. Besides, most studies use single variables/dimensions to represent financial institutions development (Table 2), and this can be inadequate.

The results of previous studies are also extremely diverse. Agarwal and Mohtadi (2004) show that the ratio of total bank deposits to GDP has a positive effect on capital structure, while Palacin-Sanchez and Pietro (2015) show the opposite result. Deesomsak et al. (2009) show that the ratio of bank assets to GDP affects capital structure, but this link depends on level of country's development. Le (2017) shows that size of banking system fluctuates inversely with capital structure. However, Demirguc-Kunt and Maksimovic (1996) show that in countries with large banking sector, small firms are less likely to borrow short-term, and the size of banking system has no impact on the capital structure of large firms. Bokpin (2009) suggests that the impact of banking system credit to GDP on capital structure depends on how firm capital structure is measured. Palacin-Sanchez and Pietro (2015) show that the easier it is to access finance, the more loans businesses use.

Kirch et al. (2012) show that the level of financial development does not influence debt maturity, whereas the institutional quality of a country has a significant positive effect on the level of long-term debt. The level of investor protection determines their disposition toward providing funding to firms. Demirguc-Kunt and Maksimovic (1998) suggest that an effective legal system is important to improve long-term financing for firms, because firms must be able to commit credibly to controlling opportunistic behavior by corporate insiders, and because long-term creditors commonly use debt covenants to control for opportunistic behavior. Since short-term debt forces managers to deal with creditors more frequently, we expect that this type of financing prevails in countries with poor creditor protections (Kirch et al. 2012). Therefore, corporate financial decisions may be affected by the legal framework and the quality of legal enforcement (La Porta et al. 1998). La Porta et al. (2000) show that more developed financial markets may be an outcome of better investor protection. According to Levine (1999), the development of financial intermediaries may depend on the quality of legal systems and of accounting standards, since financial activities are based on contractual arrangements and information about corporations. If financial development is simply an outcome of better investor protection, it should not have a first-order effect on corporate financial decisions, after controlling for the quality of investor protection.

Therefore, in order to accurately assess the impact of the development of financial institutions on capital structure, this study will include institutional quality factor in the models. In addition, a review of previous studies also shows that there has not been any research on the 5 countries in ASEAN (including Indonesia, Malaysia, Philippines, Thailand and Vietnam) in the period 2010–2019. Therefore, this study will use the IMF's financial institutions development indicators to examine the impact of financial institutions development on capital structure.

Table 2 Variables measuring financial institutions development in empirical research

Name	Research used
M3/GDP (M3 is a collection of the money supply that includes M2 money as well as large time deposits, institutional money market funds, short-term repurchase agreements, and larger liquid funds)	Demirguc-Kunt and Maksimovic (1996), Le and Ooi (2012), Agarwal and Mohtadi (2004)
M2/GDP (M2 is a measure of the money supply that includes cash, checking deposits, and easily convertible near money)	Bokpin (2010), Doku et al. (2011), Dekle et al. (2016)
Domestic assets of depository banks/GDP	Demirguc-Kunt and Maksimovic (1996, 1999), Deesomsak et al. (2009), Dorrucci et al. (2009), Doku et al. (2011), Le and Ooi (2012), Le (2017)
FINDEX1 (average of the ratio M3/GDP and the ratio of domestic credit granted to the private sector/GDP)	Demirguc-Kunt and Maksimovic (1996)
FINDEX2 (average ratio of total assets of depository banks, financial assets of private non-banking institutions, assets of private insurance and pension companies to GDP)	Demirguc-Kunt and Maksimovic (1996)
Lerner Index	Palacin-Sanchez and Pietro (2015)
Bank branches per 100,000 adults	Dorrucci et al. (2009), Palacin-Sanchez and Pietro (2015), Dekle et al. (2016)
Bank deposits/GDP	Rajan et al. (2003), Agarwal and Mohtadi (2004), Palacin-Sanchez and Pietro (2015)
Bank assets to GDP ratio divided by market capitalization to GDP	Bokpin (2010)
$\frac{\left[\frac{\frac{M3}{GDP}+\frac{PRIV}{GDP}+\frac{BANK}{GDP}}{3}+\frac{BOND}{GDP}\right]}{2}$ With PRIV: the domestic credit granted to the private sector, BANK: the domestic asset of depository banks, BOND: the capitalization of the bond market	Le and Ooi (2012)
Credit growth rate of the banking system	Le (2017)
Credit granted by banks/GDP	Bokpin (2009)
Average debt to the domestic non-financial sector by banks per GDP	Lemma and Negash (2013)
Domestic credit to the private sector/GDP	Levine (2002), Rajan et al. (2003), Bokpin (2010), Lucey and Zhang (2011), Le and Ooi (2012), Dekle et al. (2016), Zafar et al. (2019)
Bank loan interest	Deesomsak et al. (2004), Le (2017)

Source Co-authors compiled from previous studies

2.3 *Research Hypothesis*

2.3.1 Variables Representing Financial Institutions Development

In general, the variables used in previous empirical studies mainly measure the development of banking system, so they do not provide an overview of impact of financial institutions development on capital structure. Therefore, this study uses the development indicators constructed by the IMF (2015, 2016) to evaluate the impact of financial institutions development on corporate capital structure. In addition, for the 5 developing countries in ASEAN (Indonesia, Malaysia, Philippines, Thailand and Vietnam), banks are still the main source of capital for the firms. Therefore, based on the theoretical basis and experimental research results, the hypothesis is as follows:

H1: Financial institutions index (FII) has a positive impact on corporate use of debt.

2.3.2 Control Variables

The control variables in the model comprise of 2 groups: the group of firm-level characteristics and the group of macroeconomic variables.

Business Characteristic Variables

Firm-level characteristics are one of the most important determinants of capital structure. Therefore, this is one of the variable groups that cannot be ignored to ensure the reliability of the model. We choose 4 variables representing the business characteristics, including:

Firm size: According to Trade-off Theory, firm size has a positive impact on capital structure, but the Pecking Order Theory gives the opposite effect. However, most of the empirical studies use the Logarithmic of total asset value to represent firm size and suggest positive relationships [Deesomsak et al. (2004), Doku et al. (2011), Le and Ooi (2012), Le (2017)].

Profitability: The Trade-off Theory suggests a positive relationship between leverage and profitability (Sbeiti 2010; Darskuviene 2010), consistent with signaling models of capital structure (Darskuviene 2010). In contrast, the Pecking Order Theory shows an inverse relationship between profitability and firm leverage (Sbeiti 2010; Baker and Martin 2011). However, most of the empirical studies suggest negative effects. In this study, the authors use the ROA because this is a widely adopted indicator of profitability [Demirguc-Kunt and Maksimovic (1996), Bokpin (2010), Doku et al. (2011), Kirch et al. (2012)].

Tangibility: The Fixed Assets to Total Assets ratio is used to represent tangibility, in line with Deesomsak et al. (2004), Bokpin (2010), Doku et al. (2011), Le and Ooi (2012), Kirch et al. (2012), Tresierra et al. (2017), and Zafar et al. (2019). According to the Trade-off Theory, this ratio and capital structure are positively related. However, the Pecking Order Theory implies an opposite result.

Growth opportunity: According to Trade-off Theory, the Theory of Free Cash Flow by Jensen (1986) and the Market Timing Theory, growth opportunities have a negative impact on capital structure. However, Pecking Order Theory offers an unclear prediction. Tobin-Q is a measure in the research of Deesomsak et al. (2004), Bokpin (2010), Sbeiti (2010), Le and Ooi (2012), Kirch et al. (2012), and Tresierra et al. (2017).

In addition, based on arguments related to the slow adjustment of corporate leverage to the optimal level due to adjustment costs, the leverage of the previous period needs to be controlled (Agarwal and Mohtadi 2004; Doku et al. 2011; Flannery and Hankins 2013; Yarba and Guner 2019). Therefore, a dynamic panel data model is used in this study to control for leverage of the previous period in addition to firm heterogeneity and time invariant difference (not observed) among businesses.

Macroeconomic Variables

GDP per capita growth rate: Economic growth can influence financial decisions because economic growth can increase the requirement for capital. However, the relationship between GDP growth rate and capital structure is difficult to predict, because when firms need capital, they can increase bank loans (increase leverage) or issuing additional shares to raise capital (reduce leverage).

Inflation: According to Trade-off Theory, inflation will have a positive effect on leverage because it not only reduces the real value of debt but also increases the tax shield from debt (Lemma and Negash 2013).

In addition, this study will use the institutional quality variable [Kirch et al. (2012), and Tresierra et al. (2017)]. National institutional quality consists of six components: Control of Corruption, Government Effectiveness, political stability and absence of Violence/Terrorism, Regulatory Quality, Rule of Law, Voice and Accountability. The component variables have values ranging from −2.5 to 2.5 points. The quality of national institutions is to ensure fair treatment of market participants, and to prevent fraudulent practices. Higher institutional quality facilitates the functioning of both markets and intermediaries. The level and quality of financial services improve resource allocation (Levine 2002; Beck and Levin 2002).

3 Methodology and Data

3.1 Models

Based on the theoretical and empirical review, the research model is:

$$D_{it} = \alpha_{it} + \alpha_1 FI_{jt} + \alpha_2 BC_{it} + \alpha_3 MA_{jt} + \alpha_4 GOV_{jt} + \varepsilon_{it}$$

- D_{it} (Debt) is the ratio of debt to total assets calculated by book value (total debt—TDit, short-term debt—SDit and long-term debt—LDit) of firm i at time t;

- FI_{jt} is the variable representing the financial institutions development of country j at time t, which is Financial Institutions Index (FII);
- BC_{it} is a vector representing the characteristics of businesses, including: Enterprise size (logarithm of total assets—SIZE), profitability (net profit/total assets—ROA), asset nature (fixed assets to total assets—TANG), and growth opportunities (Tobin-Q index);
- MA_{jt} (Macroeconomics) are variables representing macroeconomic, including: inflation rate (INF) and GDP per capita growth rate (GDPGR);
- GOV_{jt} is the variable representing the institutional quality of country j at time t, this study will use the factor rotation method (PCA) to merge the six component variables into a single representative variable;
- ε_{it} is the residual.

3.2 Estimation Methods

Concerning previous studies related to this topic, most of them applied traditional estimation methods (frequentist approaches) such as FEM, REM, GMM. The frequentist approach assumes that the observed data are a repeatable random sample and that parameters are unknown but fixed and constant across the repeated samples, which may not always be feasible (StataCorp 2021). Therefore, the authors use the Bayesian approach to supplement the research gap.

Bayesian analysis is a statistical analysis that answers research questions about unknown parameters of statistical models by using probability statements. Bayesian analysis rests on the assumption that all model parameters are random quantities and thus can incorporate prior knowledge. This assumption is in sharp contrast with the more traditional, also called frequentist, statistical inference where all parameters are considered unknown but fixed quantities. Bayesian analysis follows a simple rule of probability, the Bayes rule, which provides a formalism for combining prior information with evidence from the data at hand. The Bayes rule is used to form the so-called posterior distribution of model parameters. The posterior distribution results from updating the prior knowledge about model parameters with evidence from the observed data. Bayesian analysis uses the posterior distribution to form various summaries for the model parameters including point estimates such as posterior means, medians, percentiles, and interval estimates such as credible intervals. Moreover, all statistical tests about model parameters can be expressed as probability statements based on the estimated posterior distribution. Therefore, Bayesian analysis is a powerful analytical tool for statistical modeling, interpretation of results, and prediction of data. It can be used when there are no standard frequentist methods available or the existing frequentist methods fail (StataCorp 2021).

Bayesian analysis starts with the specification of a posterior model. The posterior model describes the probability distribution of all model parameters conditional on the observed data and some prior knowledge. The posterior distribution has two components: a likelihood, which includes information about model parameters based on the observed data, and a prior, which includes prior information (before observing the data) about model parameters.

Likelihood:

$$\text{td} \sim \text{normal}(\text{xb\_td}, \{\text{sigma2}\})$$
$$\text{Id} \sim \text{normal}(\text{xb\_Id}, \{\text{sigma2}\})$$
$$\text{sd} \sim \text{normal}(\text{xb\_sd}, \{\text{sigma2}\})$$

Priors:

$$\{\text{td : fii gov size roa tang tobinq inf gdpgr\_cons}\} \sim normal(0, 10000)$$
$$\{\text{Id : fii gov size roa tang tobinq inf gdpgr\_cons}\} \sim normal(0, 10000)$$
$$\{\text{sd : fii gov size roa tang tobinq inf gdpgr\_cons}\} \sim normal(0, 10000)$$
$$\{\text{sigma2}\} \sim \text{igamma}(0.01, 10000)$$

where all the model parameters are normally distributed with the zero mean and the same variance of 10,000, the overall variance (sigma2) has an Igamma(0.01, 0.01) prior.

3.3 Research Data

To ensure the sample is large enough and representative of developing countries in ASEAN, the study uses the financial statements of the listed companies in five countries (including Indonesia, Malaysia, Philippines, Thailand and Vietnam) with data retrieved from Thomson Reuters for the period 2010–2019.

Research data are collected according to the following specific steps: First, we remove financial institutions from the sample. We remove extreme observations, for example, firms with negative stock prices or/and negative equity, total debt value is greater than total assets. Finally, the study excludes businesses that only had 1 to 2 years data.

4 Empirical Results and Discussions

The statistical value described in Table 3 shows that the average total debt ratio (TD) of listed companies in 5 countries reached 28.51%, showing that more than a quarter of the total assets are financed by debt. The short-term and long-term debt of enterprises are quite similar with the smallest value almost zero, and the highest

Table 3 Descriptive statistics of the variables in the model

Variable	Obs	Mean	Std. Dev	Min	Max
TD	11,917	0.285	0.170	0.010	0.932
LD	11,917	0.137	0.137	$4.83e^{-06}$	0.86
SD	11,917	0.148	0.135	0	0.862
FII	11,917	0.545	0.158	0.28	0.72
GOV	11.917	0.009	0.923	−1.938	1.965
SIZE	11,917	18.707	1.798	13.282	25.154
ROA	11,493	0.045	0.082	−0.925	0.953
TANG	11,779	0.343	0.242	0.000	0.98
TOBINQ	10,562	1.662	2.57	0.02	49.64
INF	11,917	0.035	0.041	−0.007	0.213
GDPGR	11,917	0.041	0.014	0.004	0.07

Source Results from Stata software

value about 86%. The average value of FII index reaches 0.545 with values ranging from 0.28 to 0.72. The national institutional quality (GOV) has a mean value close to zero (reaching 0.009) with a range ranging from −1.94 to 1.96 points. The average value of business size (SIZE) is 18.71, and the mean ROA is 4.52%. The ratio of fixed assets to total assets (TANG) has an average value of 34.28%. Tobin-Q index reaches an average value of 1.66. The average inflation rate (INF) is 3.5%, and the average GDP per capita growth rate (GDPGR) reached 4.11%.

Concerning the tests for chain convergence, Fig. 1 shows diagnostic graphs that contain trace plots, histograms and density plots for our MCMC sample and autocorrelations. According to the results of the convergence test by the chart, trace plots run quickly through the posterior domain, do not show trends, stop with averages, and the value of variance is directed to a constant. The histogram plots resemble the shape of the posterior distributions of the model parameters; the autocorrelations have no lags; the CUSUM plots are jagged, intercepting the X-axis; the kernel density estimation graph has graphs that resemble the shape of the posterior distributions of the model parameters. None of these plots indicate any problems with our sample. Thus, it is possible to conclude that the parameters of the research model have converged to some suitable values (Figs. 2 and 3).

First, the results in Tables 4, 5 and 6 show that the means and medians of all variables are similar. Thus, we can conclude that three models have symmetrical posterior distributions. Second, the standard deviation for all model parameters is small. Third, the less the MCSE values, the more precise the mean estimates of the model parameters, and all regression coefficients have the MCSE estimates far below one decimal. The 95% credible intervals of FII do not contain the value of zero. Hence, we can make probabilistic statements that FII has strong negative effects on corporate's capital structure (including total debt, long-term debt and short-term

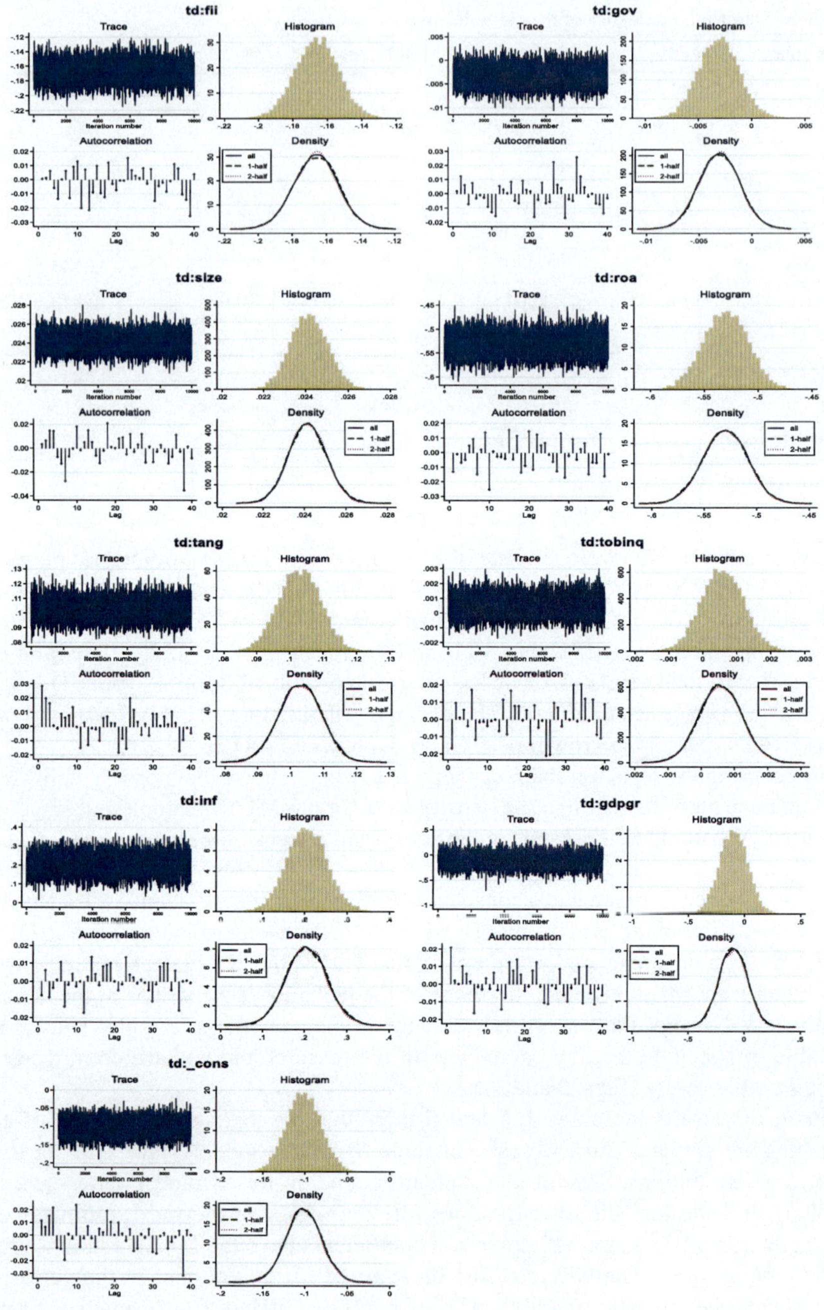

Fig. 1 Diagnostic plots of total debt. *Source* Results from Stata software

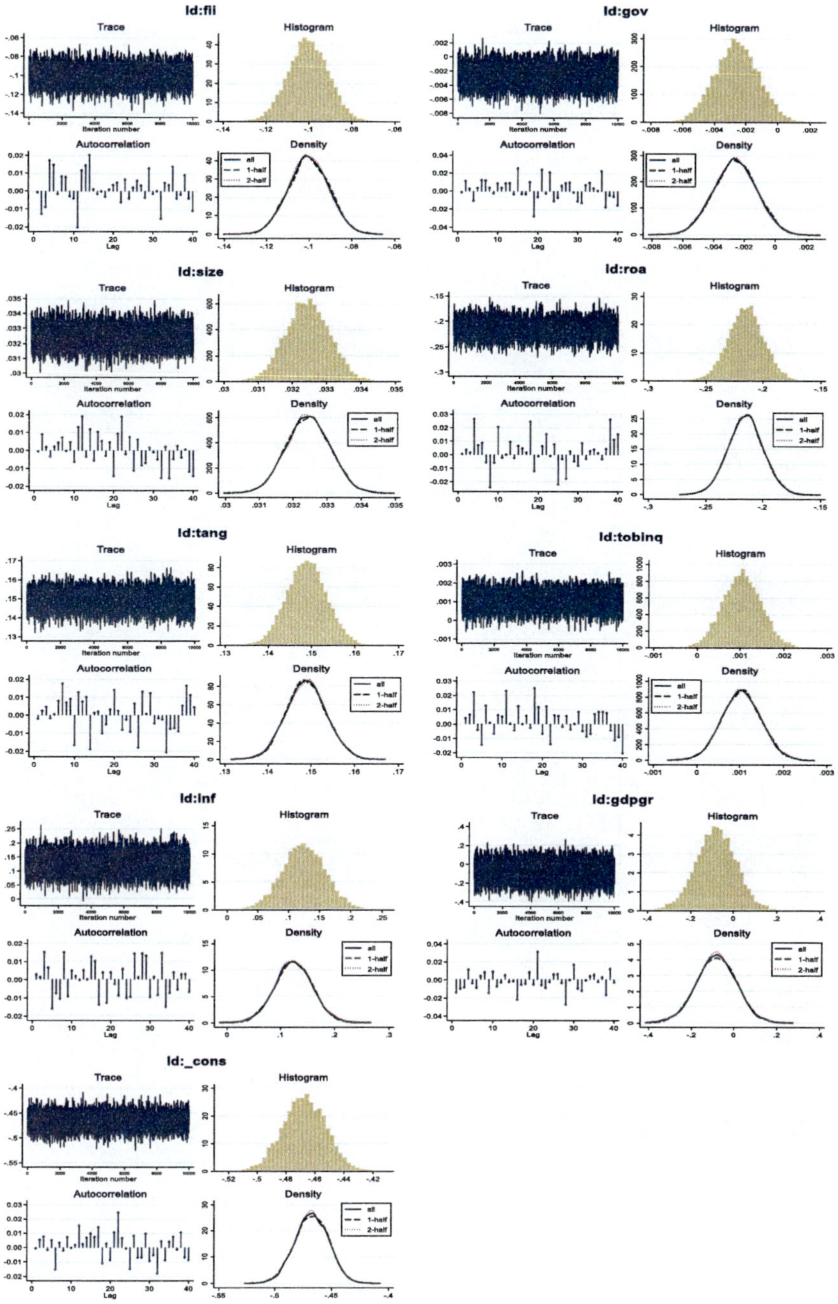

Fig. 2 Diagnostic plots of long-term debt. *Source* Results from Stata software

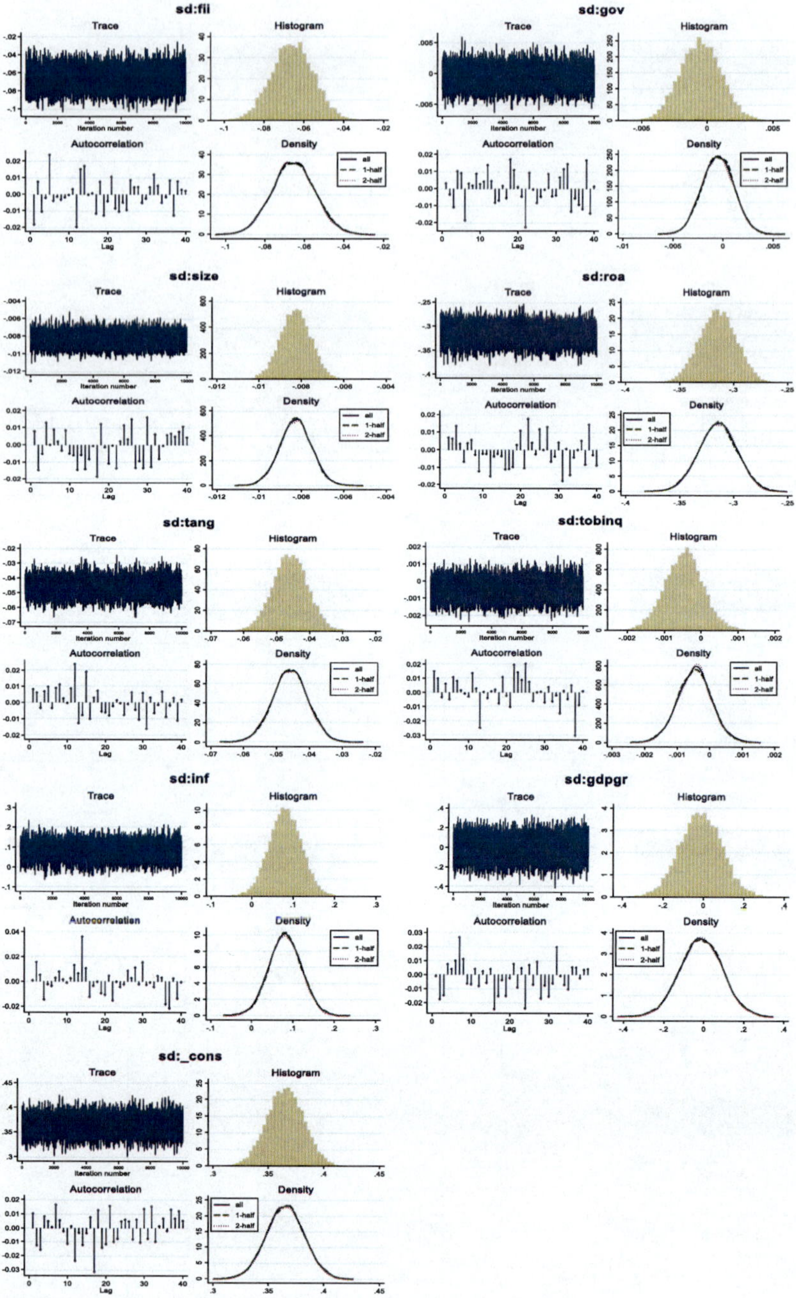

Fig. 3 Diagnostic plots of short-term debt. *Source* Results from Stata software

Table 4 The impact of financial institutions development on total debt

td	Mean	Std. Dev	MCSE	Median	Equal- [95% Cred	tailed Interval]
fii	−0.16606	0.012789	0.000128	−0.16597	−0.19118	−0.141
gov	−0.0031	0.001902	0.000019	−0.00309	−0.00679	0.000661
size	0.024169	0.000895	3.00E-06	0.024162	0.022429	0.02594
roa	−0.52997	0.021024	0.000207	−0.5298	−0.57129	−0.48907
tang	0.10302	0.006361	0.000068	0.103019	0.090542	0.115279
tobinq	0.000579	0.000632	5.20E-06	0.000575	−0.00066	0.00183
inf	0.205241	0.047736	0.000477	0.204988	0.113007	0.300573
gdpgr	−0.09814	0.125312	0.001279	−0.09788	−0.34384	0.144218
_cons	−0.10144	0.020447	0.000207	−0.10149	−0.14157	−0.06142
sigma2	0.023772	0.000331	3.30E-06	0.023769	0.023138	0.0244212

Source Author calculation from research data

Table 5 The impact of financial institutions development on long term debt

Id	Mean	Std. Dev	MCSE	Median	Equal- [95% Cred	tailed Interval]
fii	−0.10029	0.009347	0.000093	0.100298	−0.118669	−0.08205
gov	−0.00261	0.001375	−0.000014	0.002602	−0.005285	4.82E-05
size	0.032442	0.000644	−6.40E-06	0.032442	0.0312042	0.033687
roa	−0.21515	0.015386	0.000154	0.215052	−0.245316	−0.18505
tang	0.148854	0.004601	−0.000046	0.148848	0.1399853	0.157906
tobinq	0.001029	0.000449	4.50E-06	0.001034	0.0001521	0.001908
inf	0.12408	0.034188	0.000342	0.123789	0.0568164	0.191978
gdpgr	−0.08226	0.091959	0.000897	0.081893	−0.26648	0.095267
cons	−0.46743	0.014701	−0.000147	0.467415	−0.495893	−0.43946
sigma2	0.012663	0.000176	1.80E-06	0.012661	0.012324	0.013016

Source Author calculation from research data

debt). When financial institutions develop, businesses tend to reduce the use of debt, which is contrary to hypothesis H1. The reason is that financial depth is measured based on four variables (relative to GDP), including: Domestic credit to the private sector, pension/mutual fund assets, life and non-life insurance premiums longevity; so this is a composite indicator. When domestic credit to the private sector increases, it shows that enterprises have easier access to bank capital, which leads to an increase in total debt, but this variable only accounts for a smaller proportion 40% in FID and less than 25% in FII (IMF 2015, 2016). While the value of the other 3 variables has improved, this is an indirect investment in the stock market, making it more convenient for businesses to mobilize by equity instead of debt. Therefore, it is reasonable that FII has a negative impact on capital structure. The impact of GOV on

Table 6 The impact of financial institutions development on short term debt

sd	Mean	Std. Dev	MCSE	Median	Equal- [95% Cned	tailed Intenval]
fii	−0.0655907	0.0107575	0.000106	−0.0656888	−0.0863945	−0.0442852
gov	−0.0005033	0.001578	0.00016	−0.0005139	−0.0035479	0.0025886
size	−0.0082628	0.0007398	7.40E-06	−0.0082618	−0.0097294	−0.0068075
noa	−0.3145348	0.0.0175941	0.000176	−0.3142898	−0.3492252	−0.2808432
tang	−0.0458152	0.0053117	0.000053	−0.0458084	−0.0562156	−0.0354746
tobinq	−0.000462	0.000517	5.20e-06	−0.0004532	−0.0014913	0.000576
inf	0.0814209	0.0391127	0.000391	0.0811293	0.0047358	0.1581423
gdpgr	−0.0128381	0.1038871	0.001039	−0.0128837	−0.2158092	0.1912404
_cons	0.3655951	0.016873	0.000166	0.3656801	0.3321587	0.3987313
sigma2	0.0165766	0.0002278	2.30E-06	0.0165748	0.0161361	0.0170306

Source Author calculation from research data

capital structure is likely to be moderate, as the zero value falls into its 95% credible interval, ranging from −0.0068 to 0.0026.

The 95% credible intervals of SIZE, ROA, TANG and INF do not contain the value of zero. Hence, we can make probabilistic statements that ROA has strong negative effects, while INF has positive effects on three dependent variables. In addition, SIZE and TANG have positive effects on TD and LD, but negative effects on SD. The direction of SIZE impact supports the trade-off theory and is similar to studies Deesomsak et al. (2004, 2009), Sbeiti (2010), Lucey and Zhang (2011), Le and Ooi (2012), Antzoulatos et al. (2016), and Le (2017). Meanwhile, TANG has a positive effect on total debt and long-debt supporting the trade-off theory and similar to the results of Deesomsak et al. (2004, 2009), Le and Ooi (2012). However, SIZE and TANG have the negative effect on short-term debt, which supports pecking order theory. The results show that firms with large scale and many tangible assets will have easier access to loans due to transparent information and collateral for loans. Therefore, they tend to borrow more, especially long-term debt and reduce short-term one. Contrary to SIZE and TANG, profitability (ROA) has a negative impact on the capital structure, supporting the pecking order theory and being similar to the results of Zafar et al. (2019). This shows that profitability will reduce debt. Firm growth opportunity (TOBINQ) has a positive effect on long-term debt, which supports the pecking order theory in a simple model, consistent with Deesomsak et al. (2004, 2009), Zafar et al. (2019). Regarding the credible intervals, we can claim that the mean coefficient for variable TOBINQ belongs to the range between −0.00149 and 0.00183 with a probability of 95%. The inflation rate (INF) has a positive impact on debt, which is similar to the study of Lucey and Zhang (2011), Lemma and Negash (2013), Zafar et al. (2019). This implies that when inflation increases, businesses tend to increase their debt ratio (both in the short and long term debt), possibly to take advantage of currency devaluation to reduce interest expenses. The impact of GDP per capita growth rate (GDPGR) on capital structure is likely to be moderate, as the zero value falls into its 95% credible interval, ranging from −0.34384 to 0.19124.

5 Conclusion and Policy Implications

The study has provided empirical evidence that financial institutions development really has an impact on corporate capital structure. When financial institutions are developed, enterprises tend to reduce their dependence on loans. The impact direction is different from previous studies because the measurement of the variable is more integrated, generalizing the development of financial institutions in general, not just the banking system. In addition, the capital structure of firms in the five ASEAN developing countries supports both the trade-off theory and the pecking order theory. Because problem of information asymmetry still exists in the economy, enterprises still prefer to use internal capital.

The growth of financial institutions is currently hampered by poor financial literacy and inaccessibility of banking services, as services are not designed for ASEAN's new middle-class customers. The lack of financial awareness and understanding has limited the use of financial services by emerging middle classes across ASEAN, so it is important for banks to improve accessibility and usage in the era of digital financial services and Fintech. Awareness and education about financial services should go hand in hand with raising awareness about adaptation and how to use them; therefore, these programs should be coordinated. One possible approach is the agency business model, which can be used to understand customer needs and to educate those on the last mile who may not be able to reach affiliates. Business agents are retail agents engaged by the bank to provide banking services in remote locations with no branches or ATMs. Agents perform many functions such as identifying borrowers, collecting small deposits, disbursing low-value credits, collecting principal/interest, selling insurance, and transferring low-value funds. These agents also educate clients and raise financial awareness, as well as work to understand their clients' needs and challenges.

Similarly, developing diversified market segments such as venture capital, factoring and leasing markets would be a good way to improve the accessibility of businesses' assets. In addition, domestic market development and contract savings are important to finance long-term investments. However, the transition from short-term to medium- and long-term funding requires certain policy initiatives with a long-term perspective, such as gradually extending the maturities of government and corporate bonds and accelerate the pension reform process. Developing the pension and value-added insurance sector is to strengthen the country's social security system. The development of liquid and well-functioning local currency bond markets provides a long-term resource for infrastructure investment. A key factor to this growth is the strong presence of domestic institutional investors with long maturities—such as national-funded pension schemes, mutual funds and domestic insurance company. Institutional investors not only mobilize long-term savings for long-term investment, but also reinforce market disciplines as they exercise creditor's right to oversee business operations. In addition, the well-developed pension and insurance sectors will also have an impact on the desired demand for the long-term segment of the local currency bond market in the future as the maturity of the instruments gradually elongated. Several middle-income countries including Indonesia, Malaysia, the Philippines and Thailand have promoted investment funds for the development of the pension and insurance sectors, but the size of assets held by institutional investors in management area is still much lower than in Europe and America.

References

Agarwal, S., Mohtadi, H.: Financial markets and the financing choice of firms: evidence from developing countries. Glob. Financ. J. **15**, 57–70 (2004)

Anderson, T.W., Hsiao, C.: Estimation of dynamic models with error components. J. Am. Stat. Assoc. **76**, 598–606 (1981)

Anderson, T.W., Hsiao, C.: Formulation and estimation of dynamic models using panel data. J. Econ. **18**, 47–82 (1982)

Antzoulatos, A.A., Koufopoulos, K., Lambrinoudakis, C., Tsiritakis, E.: Supply of capital and capital structure: the role of financial development. J. Corp. Finan. **38**, 166–195 (2016)

Arellano, M., Bond, S.: Some tests of specification for panel data: monte Carlo evidence and an application to employment equations. Rev. Econ. Stud. **58**(2), 277–297 (1991)

Arellano, M., Bover, O.: Another look at the instrumental variable estimation of error-components models. J. Econ. **68**(1), 29–51 (1995)

Baker, H.K., Martin, G.S.: Capital structure and corporate financing decisions—Theory, evidence and practice. John Wiley & Sons Inc., Hoboken, New Jersey (2011)

Baker, M., Wurgler, J.: Market timing and capital structure. J. Financ. **57**(1), 1–32 (2002)

Beck, T., Levine, R.: Industry growth and capital allocation: does having a market-or bank-based system matter?. J. Finan. Econ. (2002)

Blundell, R.W., Bond, S.R.: Initial conditions and moment restrictions in dynamic panel data models. J. Econ. **87**, 115–143 (1998)

Bokpin, G.A.: Macroeconomic development and capital structure decisions of firms—evidence from emerging market economies. Stud. Econ. Financ. (2009)

Bokpin, G.A.: Financial market development and corporate financing Evidence from emerging market economies. J. Econ. Stud. **37**(1), 96–116 (2010)

Cuyvers, L., Chen, L., Lombaerde, P.D.: 50 years of regional integration in ASEAN. Asia Pacific Bus. Rev. (2019)

Darskuviene, V.: Financial markets. Leonardo Da Vinci—Transfer of Innovation, Education and Culture DG—Lifelong Learning Programme (2010)

Deesomsak, R., Paudyal, K., Pescetto, G.: Debt maturity structure and the 1997 Asian financial crisis. J. Multinatl. Financ. Manag. **19**(2009), 26–42 (2009)

Deesomsak, R., Paudyal, K., Pescetto, G.: The determinants of capital structure: evidence from the Asia Pacific region. J. Multinatl. Financ. Manag. **14**(2004), 387–405 (2004)

Dekle, R., và Pundit, M.: The recent convergence of financial development in asia, emerging markets finance and trade. Emerg. Mark. Financ. Trade **52**(5), 1106–1120 (2016)

Demirguc-Kunt, A., Maksimovic, V.: Stock market development and financing choices of firms. World Bank Econ. Rev. **10**(2), 341–369 (1996)

Demirguc-Kunt, A., Maksimovic, V.: Law, finance, and firm growth. J. Financ. **53**, 2107–2137 (1998)

Demirguc-Kunt, A., Maksimovic, V.: Institutions, financial markets, and firm debt maturity. J. Financ. Econ. **54**(1999), 295–336 (1999)
Diamond, D.: Financial intermediation and delegated monitoring. Rev. Econ. Stud. **51**, 393–414 (1984)
Diamond, D.W., Dybvig, P.H.: Bank runs, deposit insurance, and liquidity. J. Polit. Econ. **91**, 401–419 (1983)
Diamond, D.W.: Monitoring and reputation: The choice between bank loans and directly placed debt. J. Polit. Econ. **99**, 689–721 (1991)
Doku, J.N., Adjasi, C.K.D., Sarpong-Kumankuma, E.: Financial market development and capital structure of listed firms—empirical evidence from Ghana. Serbian J. Manag. **6**(2), 155–168 (2011)
Dorrucci, E., Meyer-Cirkel, A., và Santabárbara, D.: Domestic Financial Development in Emerging economies—Evidence and Implications. European Central Bank (Eurosystem) (2009)
Flannery, M., Hankins, K.W.: Estimating dynamic panel models in corporate finance. J. Corp. Finan. **19**, 1–19 (2013)
IMF—Svirydzenka, K.: Introducing a New Broad—based index of Financial Development. International Monetary Fund (IMF Working Paper) (2016)
IMF Staff members–Sahay, R., Cihák, M., N'Diaye, P., Barajas, A., Bi, R., Ayala, D., Gao, Y., Kyobe, A., Nguyen, L., Saborowski, C., Svirydzenka, K., Yousefi, S.R.: Rethinking financial deepening: stability and growth in emerging markets. Int. Monetary Fund (2015)
Jensen, M., Meckling, W.: A theory of the firm: managerial behavior, agency costs, and ownership structure. J. Financ. Econ. **3**, 305–360 (1976)
Jensen, M.C.: Agency costs of free cash flow, corporate finance, and takeovers. Am. Econ. Rev. **76**(2), 323–329 (1986)
Kirch, G., Terra, R.R.S.: Determinants of corporate debt maturity in South America: do institutional quality and financial development matter? J. Corp. Finan. **18**(2012), 980–993 (2012)
Kumar, S., Colombage, S., Rao, P.: Research on capital structure determinants: a review and future directions. Int. J. Manag.Financ. **13**(2), 106–132 (2017)
Le, M.T.: Impact of the financial markets development on capital structure of firms listed on ho chi minh stock exchange. Int. J. Econ. Financ. Issues **7**(3), 510–515 (2017)
Le, T.T.T., Ooi, J.T.L.: Financial structure of property companies and capital market development. J. Property Invest. Financ. **30**(6), 596–611 (2012)
Leland, H., Pyle, D.: Informational asymmetries, financial structure, and financial intermediation. J. Financ. **32**, 371–387 (1977)
Lemma, T.T., Negash, M.: Institutional, macroeconomic and firm-specific determinants of capital structure The African evidence. Manag. Res. Rev. **36**(11), 1081–1122 (2013)
Levine R.: Finance and Growth: Theory and Evidence, Handbook of Economic Growth, Volume 1A. Edited by Aghion P. and Steven N. D (2005)
Levine, R.: Law, finance, and economic growth. J. Financ. Intermed. **8**, 8–35 (1999)
Levine, R.: Bank-based or market-based financial systems: which is better? J. Financ. Intermed. **11**, 398–428 (2002)
Lucey, B.M. và, Zhang, Q.: Financial integration and emerging markets capital structure. J. Banking Financ. (2011)
Malarvizhi, C.A.N., Zeynali, Y., Mamun, A.A., Ahmad, G.B.: Financial development and economic growth in ASEAN-5 countries. Glob. Bus. Rev. **20**(1), 57–71 (2019)
Miglo, A.: Capital Structure in the Modern World. Springer Nature (2016). ISBN 978-3-319-30712-1
Mishkin, F.S.: The Economics of Money, Banking, and Financial Markets, 11th, Pearson (2016)
Modigliani, F., Miller, M.H.: The cost of capital, corporation fnance and the theory of investment. Amer. Econ. Rev. **48**(3), 261–297 (1958)
Modigliani, F., Miller, M.H.: Corporate income taxes and the cost of capital: a correction. Amer. Econ. Rev. **53**(June), 433–443 (1963)
Myers, S.C.: Determinants of corporate borrowing. J. Financ. Econ. **5**, 147–175 (1977)
Myers, S.C.: The capital structure puzzle. J. Financ. **39**(3), 574–592 (1984)

Myers, S.C.: Still searching for optimal capital structure. J. Appl. Corporate Financ. **6**(1) (1993)
Nickell, S.: Biases in dynamic models with fixed effects. Econometrica **49**, 1417–1426 (1981)
Palacin-Sanchez, M., Pietro, F.D.: The Role of the regional financial sector in the capital structure of small and medium-sized enterprises (SMEs). Regional Studies (2015)
Phooi M'ng, J.C., Rahman, M., Sannacy S.: The determinants of capital structure: evidence from public listed companies in Malaysia, Singapore and Thailand. Cogent Econ. Financ. (2017)
Rajan, R.G.: Insiders and outsiders: The choice between informed and arm's length debt. J. Financ. **47**, 1367–1400 (1992)
Rajan, R.G., và Zingales, L.: The great reversals: the politics of financial development in the twentieth century. J. Financ. Econ. (2003)
Sbeiti, W.: The determinants of capital structure: evidence from the GCC Countries. Int. Res. J. Financ. Econ. Issue 47 (2010)
StataCorp: Stata Bayesian Analysis Reference manual: Release 17. Statistical Software. College Station, TX: StataCorp LLC (2021)
Tresierra, A.E., Reyes, S.D.: Effects of institutional quality and the development of the banking system on corporate debt. J. Econ. Financ. Administr. Sci. **23**(44), 113–124 (2018)
Wald, J.K.: How firm characteristics affect capital structure: an international comparison. J. Financ. Res. **22**(2), 161–187 (1999)
Yarba, I., Guner, Z.N.: Leverage dynamics: do financial development and government leverage matter? Evidence from a major developing economy. Central Bank of the Republic of Turkey, Working Paper No: 19/15 (2019)
Zafar, Q., Wongsurawat, W., Camino, D.: The determinants of leverage decisions: Evidence from Asian emerging markets. Cogent Econ. Financ. **7**(1), 1598836 (2019)

The Nonlinear Connectedness Among Cryptocurrencies Using Markov-Switching VAR Model

Namchok Chimprang, Rungrapee Phadkantha, and Woraphon Yamaka

Abstract This paper aims to examine the regime-dependent dynamic relations among the leading cryptocurrencies using the Markov Switching Vector Autoregressive model. Our findings suggest evidence in favor of regime-switching properties in the cryptocurrency market. Furthermore, these findings provide strong evidence in favor of nonlinear connectedness among cryptocurrencies; and, thus, it is necessary to employ the MS-VAR model to determine the dynamic nonlinear connectedness between cryptocurrencies. Moreover, we find that the degree of connectedness and volatility spillover is different in both regimes and, based on the transition probability matrix, the low volatility regime is more lengthy than the high volatility regime.

Keywords Cryptocurrency · Connectedness · Markov switching VAR · Spillover effects

1 Introduction

Understanding the interconnection among cryptocurrencies has become important for risk management, forecasting, and asset allocation purposes. The cryptocurrency market has emerged as a new and hot venue for many institutions and investors. However, it has gradually become more complex and performs more volatility than other financial markets (Ibrahim et al. 2020). Therefore, those investors who invest in cryptocurrencies should examine the risks and benefits involved. To achieve the investment goal, investor needs to understand the system of connectedness among leading cryptocurrencies such as Bitcoin (BTC), Dogecoin (DODGE), Ethereum (ETH), Binance (BNB), Cardano (ADA), Ripple (XRP), THETA and Chainlink (LINK).

N. Chimprang · R. Phadkantha · W. Yamaka
Center of Excellence in Econometrics, Chiang Mai University, Chiang Mai 50200, Thailand

W. Yamaka (✉)
Faculty of Economics, Chiang Mai University, Chiang Mai 50200, Thailand
e-mail: woraphon.econ@gmail.com

N. Ngoc Thach et al. (eds.), *Financial Econometrics: Bayesian Analysis, Quantum Uncertainty, and Related Topics*, Studies in Systems, Decision and Control 427,
https://doi.org/10.1007/978-3-030-98689-6_46

Analysis of the linkages among cryptocurrencies has become a hot topic in recent research. Previous studies have made tremendous attempts to measure the direction and strength of connectedness or spillover among different cryptocurrencies in the market. This is mainly due to the complexity of the connectedness in this controversial and highly volatile market (Kosc et al. 2019). Various methods have been proposed to analyze the connectedness of cryptocurrencies. Lin (2021) analyzed the causal relationship between the performance of cryptocurrencies and investor attention using the Granger Causality test and the Vector Autoregression (VAR) model and found that the past cryptocurrency returns present a significant effect on future attention and weak reverse results. Griffin and Shams (2020) examined whether Tether influenced Bitcoin and other cryptocurrency prices and found that the purchasing with Tether was timed following market downturns and resulted in significant increases in the price of Bitcoin. Huynh et al. (2020) proposed an empirical analysis on cryptocurrency market spillover risk based on VAR, SVAR, Granger causality, and Student's-t Copulas and discovered that the movement of Ethereum is independent while Bitcoin is likely in contributing a large contagion effect to the other currencies. Using the rolling-window Bayesian Vector Autoregressive Model, Moratis (2021) quantified the spillover effects in the cryptocurrency market and revealed the strong interconnection and shock transmission among currencies. Huynh et al. (2020) applied transfer entropy to analyze the spillover effects among cryptocurrencies. They found that Bitcoin remains the most suitable asset for hedging, whilst Tether (USDT), which has a substantial association with the US dollar, performs high volatility.

The above literature revealed essential aspects of the connectedness across leading cryptocurrencies during normal periods; however, it overlooks the existence of nonlinear correlations among the cryptocurrencies. For example, as we can see in Fig. 1, the daily prices of top cryptocurrencies from January 18, 2018 to June 30, 2021 are moving in the same direction but the degree of co-movement is not constant over time, particularly during the first half of 2021. Therefore, the linear model presented in the literature may produce an unreliable result.

In this study, we consider using the Markov switching VAR model to analyze the nonlinear interconnection among cryptocurrencies. The MS-VAR model is part of a larger class of models that characterize a nonlinear data generation process as piecewise linear by confining the process to be linear in each regime, where the regime is unobservable and only a small number of regimes are possible. The assumptions made by this models about the stochastic process that generates the regime differ. In addition, this model is one of the most extensively used stochastic process models for analyzing multivariate time series interdependencies. It has been proven to be a valuable model for describing the behavior of economic and financial time series and forecasting. Recently, it has also been proved to have a higher performance than the VAR model in measuring the spillover effects and connectedness among the economic and financial variables in several studies (Pastpipatkul et al. 2015; Maneejuk et al. 2019; Tansuchat et al. 2016).

This study adds and contributes to the emerging literature on cryptocurrencies by exploring the pattern of shock transmission and return spillovers in different market regimes. Our main empirical results have implications regarding risk management,

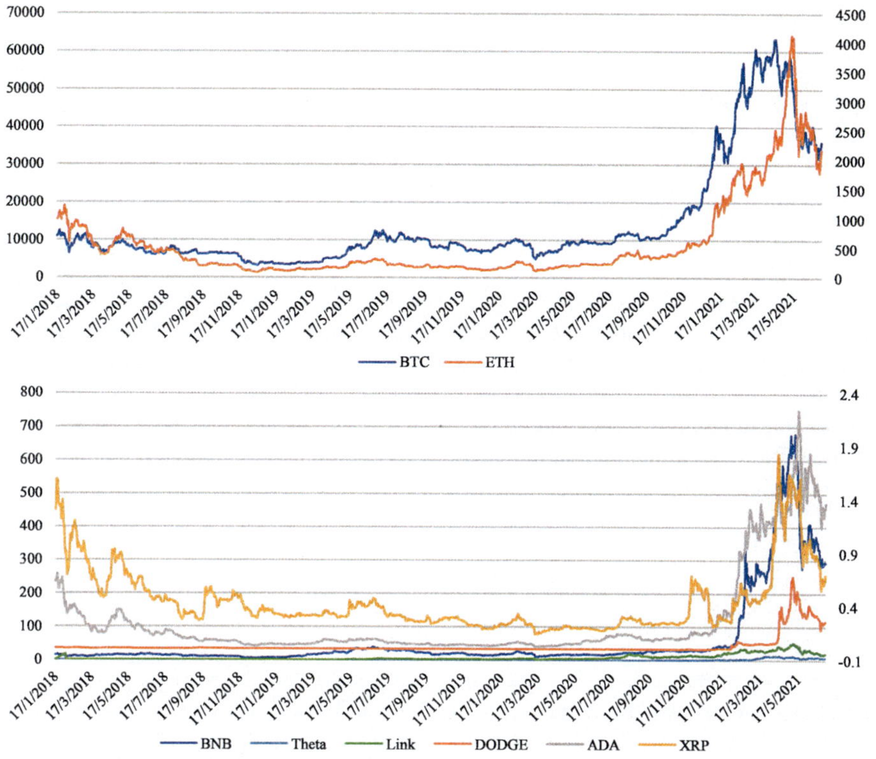

Fig. 1 Price movement of cryptocurrencies

forecasting and asset allocation and the return connectedness within the cryptocurrency market during normal and extreme periods.

The remainder of the paper is laid out as follows. In Sect. 2, we will brief about the Markov switching VAR model. Section 3 introduces variables and data. In Sect. 4, we discuss our empirical results. The last section summarizes our results and concludes the paper.

2 Methodology

The application of this econometric framework is new because of two key properties of the MS-VAR model. First, the MS-VAR model does not limit the number of structural breaks; and second, the model is able to analyze the dynamic behavior of variables under different regimes or states of the economy (Sims et al. 2008).

2.1 Markov-Switching VAR

The model is extended from the single-regime VAR model, which can be written as

$$Y_t = A + \sum_{p=1}^{P} \beta_p Y_{t-p} + U_t, \tag{1}$$

where Y_t is the vector of endogenous variables and U_t is the vector of residuals. A is the intercepts or mean in each regime and β_p is the matrix of coefficient of the lagged endogenous variables. It is worth noting that Eq. (1) only pertains to one regime. Then, Krolzig (1997), Roca et al. (2008), (2011) extended the parameter of the model to be regime dependent. Suppose we consider the two-regime MS-VAR model. The model takes form as

$$Y_t = \begin{cases} A(s_t) + \sum_{p=1}^{P} \beta_p(s_t) Y_{t-p} + U_t(s_t), \ if \ s_t = 1 \\ A(s_t) + \sum_{p=1}^{P} \beta_p(s_t) Y_{t-p} + U_t(s_t), \ if \ s_t = 2, \end{cases} \tag{2}$$

where $A(s_t)$ is the regime-dependent intercepts and $\beta_p(s_t)$ is the regime-dependent coefficient matrix of the lagged endogenous variables. $U_t(s_t)$ vector of normally distributed regime-dependent error terms and are not correlated.

This model assumes that the unobservable regime s_t is regulated by the first-order Markov chain and that the state in the period $t+1$ depends only on the state in the period t. Thus, we can compute the transition probability between two regimes as

$$P_{ij} = \Pr(S_t = j \mid S_t = i) \text{ and } \sum_{j=1}^{2} p_{ij} = 1, \ \ i, j = 1, 2 \tag{3}$$

where p_{ij} is the probability of switching from regime i to regime j, and these transition probabilities can be written in the form of a transition matrix as Eq. (4). It is obvious that there were also different periods characterized by high or low uncertainty (high and low-stress regimes). It should be noticed that the term regime refers to the level of uncertainty in the data series.

$$P = \begin{bmatrix} p_{11} & p_{12} = 1 - p_{11} \\ p_{21} = 1 - p_{22} & p_{22} \end{bmatrix} \tag{4}$$

3 Data Description

The daily prices of the seven leading cryptocurrencies (Bitcoin (BTC) and Dogecoin (DODGE), Ethereum (ETH), Binance (BNB), Cardano (ADA), Ripple (XRP), THETA, and Chainlink (LINK)) are considered in this study. All prices from January 2018 to June 2021 were collected from the Bloomberg database and transformed into natural logarithmic returns.

Table 1 shows the data description of cryptocurrency returns. All mean returns (except for XRP) are positive, with DODGE and THETA scoring the highest mean returns, while XRP has the lowest mean returns. According to the standard deviations, the riskiest cryptocurrency is LINK, while the least risky is BTC. Excess kurtosis is omnipresent, especially for LINK and DODGE. Skewness is positive, except for DODGE, XRP, and THETA. Since both negative and positive skewness exist in the returns of cryptocurrencies, we conduct the Jarque-Bera normality test. Note that the Minimum Bayes factor (MBF) is used in this study to check for significant results. The MBF value indicates that all data series are steady. The MBF can be used as a substitute for the p-value (Held and Ott 2016; Maneejuk and Yamaka 2021). In the interpretation of Bayes Factor values, if $0.33 < MBF < 1$, $0.1 < MBF < 0.33$, $0.033 < MBF < 0.1, 0.01 < MBF < 0.03, 0.003 < MBF < 0.01$, and $MBF < 0.003$ there are a chance that the MBF favors the weak evidence, moderate or substantial evidence, and strong or decisive evidence for $H_1 : \beta \neq 0$ respectively. According to the Jarque-Bera (JB) normality test, it is evident that cryptocurrency returns decisively reject the null hypothesis of a normal distribution. Furthermore, we employ the Augmented Dickey-Fuller (ADF) test, and the results reveal that all data series are stationary with decisive evidence.

Table 1 Data description

	BTC	ETH	BNB	DODGE	ADA	XRP	THETA	LINK
Mean	0.0009	0.0006	0.0024	0.0029	0.0006	−0.0005	0.0029	0.0026
Median	0.0015	0.0014	0.0012	−0.0012	0.0007	−0.0017	0.0012	0.0012
Maximum	0.161	0.2361	0.5298	1.7658	0.2794	0.4448	0.5105	0.8517
Minimum	−0.3159	−0.4236	−0.5515	−0.5069	−0.5036	−0.5505	−0.6039	−3.6519
Std. Dev.	0.0404	0.0522	0.0588	0.0859	0.0605	0.0616	0.0793	0.1285
Skewness	−0.623	−0.8204	−0.0312	7.744	−0.2076	0.1962	0.0219	−18.0395
Kurtosis	8.7965	9.5798	16.6104	151.6623	8.3091	16.2456	10.326	522.9991
Jarque-Bera	1845	2414	9725	1172868	1489	9219	2818	14264288
Probability	0.0000	0.0000	0.0000	0.0000	0.0000	0.0000	0.0000	0.0000
Sum	1.1960	0.7930	3.0656	3.6567	0.7792	−0.6254	3.6301	3.3319
Sum Sq. Dev.	2.0551	3.4327	4.3536	9.2818	4.614	4.7801	7.9119	20.7824
ADF-test	0.0000	0.0000	0.0000	0.0000	0.0000	0.0000	0.0000	0.0000
JB-test	0.0000	0.0000	0.0000	0.0000	0.0000	0.0000	0.0000	0.0000

4 Results

4.1 Model Selection

As the nonlinear connectedness is of interest in this study, the VAR and MS-VAR models are compared to confirm the nonlinear connectedness among cryptocurrencies. In this subsection, we compare the performance of the VAR and MS-VAR models, and select the better model by AIC and BIC criteria. The results reported in Table 2 resolve that, between the VAR model and the MS-VAR model, the appropriate model for our data is the MS-VAR model for its having the lower AIC and BIC. This indicates that the nonlinear connectedness has occurred in the cryptocurrency market.

4.2 Estimation Results

The empirical analyses are based on a lag order of 1, chosen according to AIC, and the estimated results of the MS-VAR model are presented in Tables 3 and 4. The estimated two-regime autoregressive parameters are reported in Table 3, while Table 4 presents the two-regime variance-covariance matrix. Considering the diagonal matrix, the variances of the cryptocurrencies in state 2 are mostly larger than those in state 1, indicating that state 2 is the high volatility regime, while the low volatility regime is state 1. Besides, considering scalar in the lower or upper triangular matrix, the covariance estimates demonstrate that in the low volatility regime, almost all cryptocurrency pairs exhibit a low co-movement, while the potency of this co-movement increases in the high volatility regime. This result is consistent with the idea that the degree of dependency among these cryptocurrencies is strong in a state of high uncertainty.

Considering the results in Table 3, the estimation results demonstrate that the change in the lag of DODGE and XRP returns shows a decisive impact on BTC returns in regimes 1 and 2. The effect of lagged ETH and lagged BTC on BTC returns follows the same pattern in both states with weak evidence and substantial evidence, respectively. Moreover, BTC returns are decisively influenced by BNB, ADA, and LINK lag in both regimes, whereas THETA lag shows a weak effect in regime 1, but decisive evidence in regime 2. Furthermore, the increase in BTC lag

Table 2 Model Selection

	VAR	MS-VAR
AIC	−24.2059	−27.8637
BIC	−23.912	−26.7966

Table 3 MS-VAR with lag 1

State 1

	BTC	ETH	BNB	DODGE	ADA	XRP	THETA	LINK
$A(s_t)$	0.0027	0.0013	0.0014	−0.0031	−0.0033	−0.0043	−0.0017	0.0033
	(−0.0006)	(−0.0283)	(−0.022)	(−0.0136)	(−0.008)	(−0.0145)	(−0.0136)	(−0.009)
	[0.0001]	[0.0933]	[0.1968]	[0.0000]	[0.0273]	[0.0001]	[0.6846]	[0.8377]
BTC(t-1)	−0.0006	−0.0283	−0.022	−0.0136	−0.008	−0.0145	−0.0136	−0.009
	(−0.0522)	(−0.1117)	(−0.0331)	(−0.0207)	(−0.1039)	(−0.0441)	(−0.0955)	(−0.1731)
	[0.1720]	[0.0002]	[0.7750]	[0.8596]	[0.1789]	[0.6599]	[0.4837]	[0.729]
ETH(t-1)	−0.0522	−0.1117	−0.0331	−0.0207	−0.1039	−0.0441	−0.0955	−0.1731
	(−0.0053)	(−0.0006)	(−0.0281)	(−0.0218)	(−0.0135)	(−0.008)	(−0.0144)	(−0.0135)
	[0.0240]	[0.0149]	[1.0000]	[0.9248]	[0.5682]	[1.0000]	[0.0075]	[0.0094]
BNB(t-1)	−0.0053	−0.0006	−0.0281	−0.0218	−0.0135	−0.008	−0.0144	−0.0135
	(−0.0575)	(−0.0604)	(−0.0021)	−0.0141	−0.0518	−0.0144	−0.1726	−0.4029
	[0.0001]	[0.0001]	[1.0000]	[0.0049]	[0.1550]	[0.0921]	[0.1590]	[0.0001]
DODGE(t-1)	−0.0575	−0.0604	−0.0021	0.0141	0.0518	0.0144	0.1726	0.4029
	(−0.009)	(−0.0053)	(−0.0008)	(−0.0361)	(−0.0280)	(−0.0173)	(−0.0103)	(−0.0185)
	[0.0001]	[0.0001]	[0.0001]	[0.0001]	[0.0010]	[0.1661]	[0.4442]	[0.0001]
ADA(t-1)	−0.009	−0.0053	−0.0008	−0.0361	−0.028	−0.0173	−0.0103	−0.0185
	−0.0629	−0.0919	−0.008	−0.0386	−0.0517	−0.0423	(−0.0671)	(−0.3723)
	[0.0001]	[0.0001]	[0.0001]	[0.0001]	[0.0015]	[0.0001]	[1.0000]	[0.0000]
XRP(t-1)	0.0629	0.0919	0.008	0.0386	0.0517	0.0423	−0.0671	−0.3723
	(−0.0173)	(−0.0115)	(−0.0068)	(−0.0006)	(−0.0258)	(−0.0200)	(−0.0124)	(−0.0073)
	[0.0001]	[0.0001]	[0.0001]	[0.0001]	[0.0001]	[0.0001]	[0.0001]	[0.0075]
THETA(t-1)	−0.0173	−0.0115	−0.0068	−0.0006	−0.0258	−0.02	−0.0124	−0.0073
	−0.0547	−0.0917	−0.0506	(−0.1004)	−0.0577	−0.022	(−0.0283)	−0.4947
	[0.3442]	[0.0001]	[0.0001]	[0.0001]	[0.5389]	[1.0000]	[0.0001]	[0.1330]
LINK(t-1)	0.0547	0.0917	0.0506	−0.1004	0.0577	0.022	−0.0283	0.4947
	(−0.0132)	(−0.0124)	(−0.0082)	(−0.0049)	(−0.0013)	(−0.0570)	(−0.0442)	(−0.0274)
	[0.0024]	[0.0105]	[0.0020]	[0.0014]	[0.0017]	[0.0001]	[0.0003]	[0.0003]
MSE	0.6288	0.6177	1.0214	0.5206	2.5469	1.3608	4.3374	24.758
RMSE	0.793	0.7859	1.0106	0.7215	1.5959	1.1665	2.0827	4.9757

State 2

	BTC	ETH	BNB	DODGE	ADA	XRP	THETA	LINK
$A(s_t)$	−0.0027	−0.0006	0.0043	0.014	0.0066	0.0035	0.013	0.0038
	(−0.0011)	(−0.0486)	(−0.0377)	(−0.0233)	(−0.0138)	(−0.0249)	(−0.0233)	(−0.0155)
	[0.0336]	[1.0000]	[0.0002]	[0.0001]	[0.0001]	[0.0417]	[0.0001]	[0.0222]
BTC(t-1)	−0.0011	−0.0486	−0.0377	−0.0233	−0.0138	−0.0249	−0.0233	−0.0155
	(−0.0407)	(−0.1192)	(−0.1576)	(−0.085)	(−0.1019)	(−0.0716)	−0.3173	(−0.1046)
	[0.8356]	[0.1660]	[0.0032]	[0.4533]	[0.0957]	[0.6147]	[0.0001]	[0.2656]
ETH(t-1)	−0.0407	−0.1192	−0.1576	−0.085	−0.1019	−0.0716	0.3173	−0.1046
	(−0.0092)	(−0.0014)	(−0.0642)	(−0.0498)	(−0.0308)	(−0.0182)	(−0.0329)	(−0.0308)
	[0.0404]	[0.2612]	[0.5151]	[0.0001]	[0.8311]	[0.2330]	[0.0613]	[0.7299]
BNB(t-1)	−0.0092	−0.0014	−0.0642	−0.0498	−0.0308	−0.0182	−0.0329	−0.0308
	(−0.0916)	(−0.0815)	(−0.0468)	(−0.2059)	(−0.0318)	(−0.0844)	(−0.0857)	−0.0484
	[0.0001]	[0.0001]	[1.0000]	[0.0001]	[0.0001]	[0.0001]	[0.0028]	[0.0001]

(continued)

Table 3 (continued)

State 1

	BTC	ETH	BNB	DODGE	ADA	XRP	THETA	LINK
DODGE(t-1)	−0.0916	−0.0815	−0.0468	−0.2059	−0.0318	−0.0844	−0.0857	0.0484
	(−0.0205)	(−0.0121)	(−0.0011)	(−0.0486)	(−0.0377)	(−0.0233)	(−0.0138)	(−0.0249)
	[0.0298]	[0.0952]	[0.0334]	[0.4236]	[0.1521]	[0.0001]	[0.1130]	[0.9609]
ADA(t-1)	−0.0205	−0.0121	−0.0011	−0.0486	−0.0377	−0.0233	−0.0138	−0.0249
	−0.0977	−0.1254	−0.0073	−0.2078	−0.1422	−0.1839	(−0.0767)	−0.1453
	[0.0003]	[0.0001]	[0.0001]	[0.0001]	[0.2404]	[0.8837]	[0.1913]	[0.0010]
XRP(t-1)	0.0977	0.1254	0.0073	0.2078	0.1422	0.1839	−0.0767	0.1453
	(−0.0233)	(−0.0155)	(−0.0092)	(−0.0014)	(−0.0642)	(−0.0498)	(−0.0308)	(−0.0182)
	[0.0172]	[0.0557]	[0.0001]	[1.0000]	[1.0000]	[0.0085]	[0.0144]	[0.9813]
THETA(t-1)	−0.0233	−0.0155	−0.0092	−0.0014	−0.0642	−0.0498	−0.0308	−0.0182
	−0.0351	−0.0383	−0.0345	−0.0249	−0.0262	−0.132	(−0.0280)	−0.0116
	[0.0001]	[0.0001]	[0.0001]	[0.0001]	[0.0604]	[0.8538]	[0.2637]	[0.0001]
LINK(t-1)	0.0351	0.0383	0.0345	0.0249	0.0262	0.1320	−0.028	0.0116
	(−0.0329)	(−0.0308)	(−0.0205)	(−0.0121)	(−0.0011)	−0.0486)	(−0.0377)	(−0.0233)
	[0.0001]	[0.0001]	[0.0001]	[0.0001]	[0.3280]	[0.0003]	[0.0001]	[0.8400]
MSE	1.8504	3.23	6.1757	23.3188	8.6871	11.3794	14.9929	8.4636
RMSE	1.3603	1.7972	2.4851	4.8289	2.9474	3.3733	3.8721	2.9092

Note () is standard error, [] is MBF, by 0.0001–0.01 MBF is decisive evidence, 0.01–0.1 is moderate evidence, and 0.1–1 is weak evidence

increases the returns of ETH in regime 1 and BNB and THETA in regime 2. These results indicate that BTC contributes a larger spillover effect to other currencies in the high volatility regime when compared to the low volatility regime.

Regarding the other cryptocurrencies, there is evidence that, in the low volatility regime, ETH has been decisively affected by all the other cryptocurrencies except LINK and its own lag. Meanwhile, in the high volatility regime, a half of the digital assets show a weak impact on ETH. For the case of BNB, it is decisively influenced by the lag of ADA, XRP. THETA, and LINK in both two regimes. We find that oscillations in BTC and ETH lags have a weak impact on BNB returns only in regime 1. Moreover, in state 1, the changes of DODGE, ADA, XRP, and LINK have a decisive significant impact on the dynamics of ADA returns, but these effects fade in state 2, whereas there is only a decisive effect of BNB and substantial evidence of BTC and THETA on ADA. For the XRP returns, LINK and its own lag show a decisive impact in both regimes (Table 5).

For the THETA case, a half of the cryptocurrencies, consisting of ETH, XRP, and LINK, reveal a decisive effect on THETA returns in regime 1, while BTC, BNB, and LINK provide a strong linkage in regime 2. Finally, we consider the linkages of LINK with other currencies. In regime 1, ETH, DODGE, XRP, and its own lag have a decisive influence on LINK returns, while THETA lag has a decisive impact on LINK returns in regime 2.

Figure 2 depicts the smoothed probabilities of being in regimes 1 and 2, computed based on information of all the samples. It can be seen that the smoothed probabilities

Table 4 Estimates of variance-covariance for the MS-VAR model

	BTC	ETH	BNB	DODGE	ADA	XRP	Theta	Link
State 1								
BTC	5.00E-04							
ETH	3.10E-04	4.91E-04						
BNB	1.56E-04	1.62E-04	8.12E-04					
DODGE	1.54E-04	8.26E-05	1.06E-04	4.14E-04				
ADA	1.99E-05	1.79E-05	–9.09E-05	2.23E-05	2.02E-03			
XRP	1.33E-05	3.86E-06	–5.51E-05	1.38E-05	1.22E-03	1.08E-03		
Theta	5.53E-05	3.52E-05	4.63E-06	7.69E-05	1.45E-03	1.04E-03	3.45E-03	
Link	1.93E-04	1.08E-05	8.62E-05	3.53E-04	1.95E-03	1.40E-03	2.01E-03	1.97E-02
State 2								
	BTC	ETH	BNB	DODGE	ADA	XRP	Theta	Link
BTC	1.47E-03							
ETH	1.21E-03	2.57E-03						
BNB	8.11E-04	7.00E-04	4.91E-03					
DODGE	9.03E-04	8.27E-04	5.28E-04	1.85E-02				
ADA	3.43E-04	6.10E-04	1.94E-04	1.08E-03	6.91E-03			
XRP	3.67E-04	6.87E-04	4.75E-04	6.58E-04	4.55E-03	9.05E-03		
Theta	5.00E-05	3.32E-04	8.35E-04	1.24E-04	3.31E-03	2.95E-03	1.19E-02	
Link	2.19E-04	6.21E-04	7.08E-04	6.81E-04	4.29E-03	4.35E-03	3.03E-03	6.73E-03

Table 5 Estimates of transition probabilities for the MS-VAR model

	State 1	State 2
State 1	0.8928	0.1072
State 2	0.2897	0.7103

of being in the first regime are higher than in the second regime (approximately 69.9%), especially in 2019/01–2020/12, while for the rest, regime 2 is outstanding. Thus, it can be said that the low volatility cryptocurrency market regime reflects the financial distress period during the sample period, such as the COVID-19 crisis. The cryptocurrency market appears to have transitioned to a high volatility market environment by the end of 2020.

5 Conclustions

The recent quadruple rise in the market capitalization of cryptocurrencies has sparked a wave of investor interest in alternative asset investments as well as widespread public awareness of this viral new asset class. Unfortunately, at the start of the second

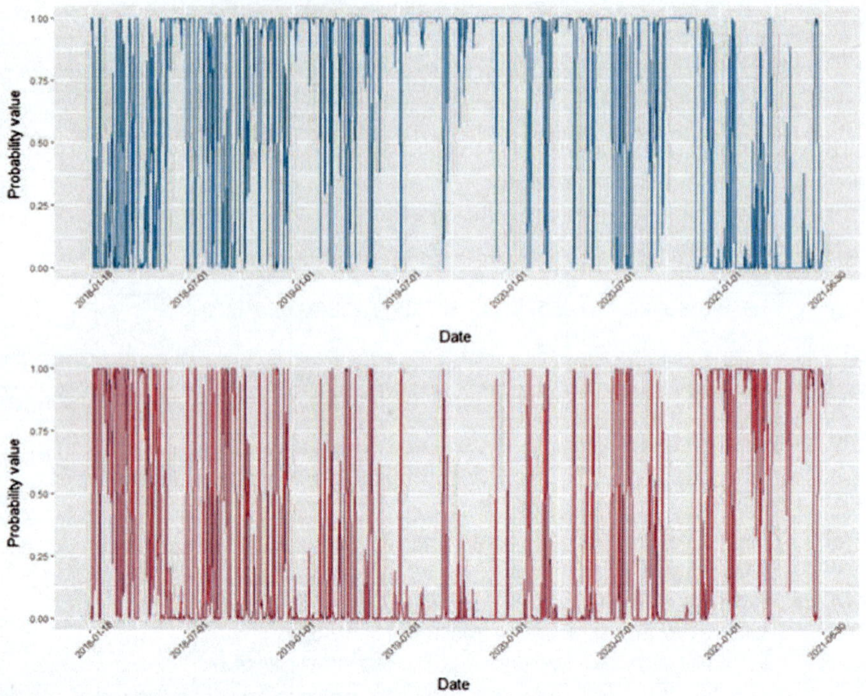

Fig. 2 The smoothed probability plots of a high-volatility regime (top) and a low-volatility regime (bottom)

quarter, practically all cryptocurrency prices collapse in half. Therefore, it can be seen that this tendency reflects the fluctuations in cryptocurrency prices. This paper analyzes the nonlinear relationship of eight cryptocurrencies, consisting of Bitcoin (BTC), Dogecoin (DODGE), Ethereum (ETH), Binance (BNB), Cardano (ADA), Ripple (XRP), THETA, and Chainlink (LINK). We show that the nonlinear connectedness among cryptocurrencies is confirmed as the MS-VAR model presents a better performance compared to the VAR model according to AIC and BIC criteria. The estimated results show that the MS-VAR model can capture the returns' movement of cryptocurrencies and enables us to understand the variations of dependence in different regime. Moreover, this result is consistent with the idea that in a state of high uncertainty, the degree of dependency between these cryptocurrencies varies tremendously.

In this study, cryptocurrencies can be categorized into three groups. First is the group in which a cryptocurrency behaves like a digital gold asset, such as BTC and DODGE, because of its stock-to-flow model, which stabilizes the cryptocurrency inflation rate close to the pace of physical gold production. We notice that under low volatility regimes, BTC and DODGE tend to have decisive evidence of positive effect by a lagged return of less decentralized cryptocurrencies, such as BNB, ADA and XRP. An increase in the returns of Bitcoin and DODGE will increase the returns of BNB, ADA and XRP in the next trading day. Thus, BTC and DODGE are not the good hedge assets for BNB, ADA and XRP.

Second is the ecosystem platform currency consisting of ETH, BNB, ADA, and XRP. The results for the low volatility regime demonstrate that BTC has a weak negative effect on these ecosystem platform cryptocurrency returns while DODGE has decisive evidence and positive impact on these cryptocurrencies. On the contrary, in the high volatility regime, BTC tends to have strong evidence only in BNB with the same direction effect, while DODGE shows a strong impact only on XRP. The difference in the effects of BTC and DODGE may be due to the high investor's reliability on the essence, history's proving, and credibility of BTC, giving it significant independence from other cryptocurrencies, whereas the meme-like DODGE only reflects a rise in viral confidence in cryptocurrencies investment. Moreover, we also find that small capital platform cryptocurrencies (XRP and ADA) have a positive, decisive effect on large platform cryptocurrencies (ETH and BNB) in both high and low volatility regimes. Furthermore, the software company cryptocurrencies, especially the LINK, has a decisive positive impact on XRP and ADA in both regimes.

Finally, it is the software company cryptocurrency group, consisting of LINK and THETA. It is found that cryptocurrencies have a weak effect on software company cryptocurrencies for both regimes except for ecosystem platform cryptocurrencies in the low volatility regime. In future research, we suggest analyzing the herding behavior in the cryptocurrency market. Moreover, it would be challenging to optimize the portfolio investment in the cryptocurrency market for investors' interest.

Acknowledgements The authors are grateful to the Centre of Excellence in Econometrics, Chiang Mai University, for financial support. The authors are grateful to Dr. Laxmi Worachai for their helpful comments and suggestions.

References

Griffin, J.M., Shams, A.: Is Bitcoin really untethered? J. Financ. **75**(4), 1913–1964 (2020)

Held, L., Ott, M.: How the maximal evidence of p-values against point null hypotheses depends on sample size. Amer. Stat. **70**(4), 335–341 (2016)

Huynh, T.L.D., Nasir, M.A., Vo, X.V., Nguyen, T.T.: Small things matter most: the spillover effects in the cryptocurrency market and gold as a silver bullet. North Amer. J. Econ. Financ. **54**, 101277 (2020)

Ibrahim, A., Kashef, R., Li, M., Valencia, E., Huang, E.: Bitcoin network mechanics: forecasting the BTC closing price using vector auto-regression models based on endogenous and exogenous feature variables. J. Risk and Financ. Manag. **13**(9), 189 (2020)

Kosc, K., Sakowski, P., Ślepaczuk, R.: Momentum and contrarian effects on the cryptocurrency market. .Physica A: Stat. Mech. Appl. **523**, 691–701 (2019)

Krolzig, H.M.: The markov-switching vector autoregressive model. In: Markov-Switching Vector Autoregressions, pp. 6–28. Springer, Berlin, Heidelberg (1997)

Lin, Z.Y.: Investor attention and cryptocurrency performance. Financ. Res. Lett. **40**, 101702 (2021)

Maneejuk, P., Yamaka, W., Sriboonchitta, S.: Measuring US business cycle using markov-switching model: a comparison between empirical likelihood estimation and parametric estimations. In: International Conference of the Thailand Econometrics Society, pp. 596–606. Springer, Cham (2019)

Maneejuk, P., Yamaka, W.: Significance test for linear regression: how to test without P-values? J. Appl. Stat. **48**(5), 827–845 (2021)

Moratis, G.: Quantifying the spillover effect in the cryptocurrency market. Financ. Res. Lett. **38**, 101534 (2021)

Pastpipatkul, P., Yamaka, W., Wiboonpongse, A., Sriboonchitta, S.: Spillovers of quantitative easing on financial markets of Thailand, Indonesia, and the Philippines. In: International Symposium on Integrated Uncertainty in Knowledge Modelling and Decision Making, pp. 374–388. Springer, Cham (2015)

Roca, E.D., Wong, V.S.: An analysis of the sensitivity of Australian superannuation funds to market movements: a Markov regime switching approach. Appl. Financ. Econ. **18**(7), 583–597 (2008)

Roca, E.D., Tularam, G.A., Wong, V.S.H.: Markov regime switching modelling and analysis of socially responsible investment funds. J. Math. Stat. **7**(4), 302–313 (2011)

Sims, C.A., Waggoner, D.F., Zha, T.: Methods for inference in large multiple-equation Markov-switching models. J. Econ. **146**(2), 255–274 (2008)

Tansuchat, R., Maneejuk, P., Wiboonpongse, A., Sriboonchitta, S.: Price transmission mechanism in the Thai rice market. In: Causal Inference in Econometrics, pp. 451–461. Springer, Cham (2016)

How Credit Growth and Political Connection Affect Net Interest Margin of Commercial Bank in Vietnam: A Bayesian Approach

Duong Dang Khoa, Phan Thi Thanh Phuong, Nguyen Ngoc Thach, and Nguyen Van Diep

Abstract This paper aims to analyze the impact of credit growth rate and political connection on commercial banks' net interest return (NIM) in Vietnam in 2003–2020. We employ the Bayesian linear regression method through the Gibbs sampling algorithm to overcome the asymptotic, a property that can hinder when using frequentist methods in small-sample contexts. The empirical results indicate that the lending growth rate and political connection have a robust negative effect on NIM. In addition, we also figure out that higher bank funding diversity lowers the spread of commercial banks in Vietnam while higher bank liquidity creation pushes higher NIM. Moreover, the gross domestic product (GDP) growth, the growth of money supply M2 (M2GR) impact positively on NIM. Finally, our study figures out that Covid-19 has adverse impacts on NIM. Our findings are helpful for bank managers and market supervisors to maintain the sustainable growth of the financial market.

Keywords Bayesian linear regression · NIM · Political connection · Credit growth rate · GDP growth · M2 · Banks

D. D. Khoa · P. T. T. Phuong
Faculty of Finance and Banking, Ton Duc Thang University, Ho Chi Minh City, Vietnam
e-mail: duongdangkhoa@tdtu.edu.vn

P. T. T. Phuong
e-mail: phanthithanhphuong1@tdtu.edu.vn

N. N. Thach
Asian Journal of Economics and Banking, Banking University of Ho Chi Minh City, Ho Chi Minh City, Vietnam
e-mail: thachnn@buh.edu.vn

N. Van Diep (✉)
Faculty of Finance and Banking, Ho Chi Minh City Open University, Ho Chi Minh City, Vietnam
e-mail: diep.nv@ou.edu.vn

N. Ngoc Thach et al. (eds.), *Financial Econometrics: Bayesian Analysis, Quantum Uncertainty, and Related Topics*, Studies in Systems, Decision and Control 427,
https://doi.org/10.1007/978-3-030-98689-6_47

1 Introduction

Banks play a vital role in the economy, as it is one of the foundations for sustainable development. Competitions drive their performance, which is why banks are more likely to sustain themselves and grow. Angbazo (1997) showed that banks that have higher interest rate risk typically seek higher margins by choosing to select deposit and loan rates. According to Berger et al. (2009), political connections can affect the performance of commercial banks. They argue that the pressure on banks to provide loans for political purposes can limit their profitability. Thus, testing the impact of political connection on the performance of commercial banks is open for research.

Due to the rapid growth of domestic banks after Vietnam joined WTO in 2007, the competition has increased. To avoid the impact of the economic recession, the State Bank of Vietnam launched various measures to stimulate demand, thus encouraged commercial banks to provide credit to domestic enterprises. Credit growth rate is the top concern of the banking industry, from over 30% in 2003 the government has controlled around 10% in the period 2003–2020. Especially in 2020, despite of the effect of the Covid-19 pandemic, the credit growth keeps the rise of 7.42%. Therefore, we focus on the credit growth rate in our research.

The Covid-19 andemic has become a global concern. Simoens and Vennet (2021), Demirgüç-Kunt et al. (2021), Elnahass et al. (2021) indicate a negative impact of the Covid-19 pandemic on the banking performance across countries. However, there is lack of studies focusing the impacts of Covid-19 on the profitability of commercial banks in Vietnam, especially in 2020 Vietnam economy performed the growth of GDP + 2.91% which was the better than most of other Asian economy.

In this study, we employ the Bayesian approach because of the following reasons. Although there have been many studies on the relationship between bank's NIM, the empirical results from the classical estimation method are mixed and controversial (Andrade 2019; Ho et al. 2019; Wasserstein et al. 2019). Secondly, the current research models become more complex, which requires computational estimations. However, the small sample size poses challenges to classical estimation methods which rely on asymptotic theory. The development of Bayesian estimation techniques helps overcome the shortcomings of the classical estimation method because Bayesian methods do not rely on asymptotic, a property that can hinder when employing frequentist methods in small-sample contexts (Holtmann et al. 2016; McNeish 2016; Nguyen et al. 2019). Also, the Bayesian approach allows incorporating additional information external to the sample by prior distributions. Grzenda (2015) indicates that additional information may improve the accuracy and credibility of estimations. The credible regions incorporate this prior information, while frequentist confidence intervals are based only on the sample data (Grzenda 2015). Therefore, we employ the Bayesian approach to estimate credit growth rate and political connection impacts on Vietnamese banks' performance. This study also contributes to literature and policy debates on commercial banks' performance in Vietnam.

This study is structured as follows: Part 2 provides a review of the literature. Part 3 presents the research model, data, and methodology. In Sect. 4, we are analyzing the estimation results. Section 5 concludes the results and provides some relevant policy implications.

2 Literature Reviews

2.1 *Net Interest Margin*

Net interest margin (NIM) reflects the gap between interest income and interest expense over the total assets. Thomas and Saunders (1981) argue that the degree of competition in the markets and the risk associated with the bank's interest rate exposure are two critical components of the interest margin. Allen (1988) broadened the dealership model by allowing banks to offer various loan and deposit products. Angbazo (1997) showed that banks with higher interest rate risk would typically seek higher margins by selecting deposit and loan rates. Kusi et al. (2020) argue that financial sector transparency led by the private sector is more effective in reducing bank interest margins than those led by the public sector.

In contrast, Borio et al. (2017) found that the relationship between the slope of the yield curve and the level of short-term rates is positive. It suggests that the interest rate structure's positive impact on net interest income is more significant than the negative impact on loan loss provisions and non-interest income. The above studies focus mainly on developed markets; there is a lack of research focus on frontier markets. Recently, Khanh and Tra (2015), Suu et al. (2020) resulted that the level of risk aversion, quality of management impact negatively on NIM. Nguyen et al. (2020) studied Vietnamese banks and found that excess liquidity causes banks to reduce their lending interest rates to expand credit supply. This effect is detrimental to the effectiveness of monetary policy transmission. Fewer studies focus mainly on the impact of lending growth rate and political connection on NIMs of Vietnamese banks before and after Covid-19.

2.2 *Credit Growth*

The loan growth ratio (LGR) represents bank loans' growth and underpins interest income growth. Kundid and Rogosic (2012) examine the determinants of bank profitability in Croatia from 2003 to 2008. It is found that the loan growth positively influences bank profitability while its insignificant effect impacts loan loss provisions. The loan ratio that influences the profitability of European banks is not apparent and varies by measure, according to Menicucci and Paolucci (2016) research of 35 European banks from 2009 to 2013.

Meanwhile, variable loans have a favourable but not substantial association with return on assets (ROA) and return on equity (ROE). In the other multi-country studies, Kosmidou (2008), Foos et al. (2010) examines banks' performance determinants. Kosmidou (2008) investigates profitability for a sample of 23 EU commercial banks from 1990 to 2002. They show that the ratio of net loans to total assets of European banks harms profitability. Foos et al. (2010) report that loan growth is significantly negatively related to bank solvency which is statistically significant at the 1% level.

2.3 *Political Connection*

Political connection or state ownership is a variable used to examine the relationship and impact of political factors on the bank's performance. Prior studies report the controversial relationship between political connection and bank performance. Micco et al. (2007) report that the state-owned banks in developing nations have poorer profitability and higher costs than private and foreign-owned banks. Furthermore, Berger et al. (2009) argue that the Big Four Chinese banks with government ownership generate the worst performance. They explain that the Big Four Chinese banks are constantly under pressure to grant loans for political purposes instead of maximizing profits for the banks. Hung et al. (2017) analyze 70 Chinese banks from 2007 to 2014, and they report that banks with political connections generate higher ROA and lower credit risk. Meanwhile, Altunbas et al. (2001) and La Porta et al. (2002) suggest an indifferent efficiency between state-owned and private banks. These studies indicate that developed countries have better techniques to reduce costs and "distortions" with state-owned banks.

2.4 *Other Determinants of Bank Risk-Taking Behaviours*

2.4.1 Capital Structure

Siddik et al. (2017) indicate the impact of capital structure on the performance of 30 listed banks in Bangladesh over the period 2005–2014. Specifically, they figure out that the capital structure negatively affects ROE, ROA, and earnings per share (EPS). Moreover, the total debt ratio inversely affects ROA and ROE while it positively influences EPS. Yahya et al. (2017), Sivalingam and Kengatharan (2018), Jadah et al. (2020) also figure out the relationship negative relationship between capital structure and bank efficiency. Sivalingam and Kengatharan (2018) and Siddik et al. (2017) suggest that bank managers utilize their internal funding sources instead of relying heavily on external financing to maximize performance.

Almaqtari et al. (2019) reveal that a higher leverage capital structure adversely affects ROA, while this study finds an insignificant leverage impact on the ROE in Indian banks. El-Chaarani and El-Abiad (2019) examine the relationship between banks' capital structure and performance in Middle East countries for 2011–2016. They also use ROA and ROE to measure the performance of banks. El-Chaarani and El-Abiad (2019) indicate that the ratio of total debt and the ratio of short-term debt have a negative impact on the ROA while the long-term debt ratio positively affects ROA. The study reports the positive effect of the total debt ratio and the short-term debt ratio on ROE, but no evidence of the long-term debt ratio's impact on the ROE.

2.4.2 Bank Funding Diversity

In the current competitive context, diversification is a vital strategy to help banks increase their competitiveness. With the bank's income diversification strategy, studies by Vinh and Mai (2016), Doumpos et al. (2016) and Moudud-Ul-Huq et al. (2018) all pointed out that income diversification helps increase the operational efficiency of banks. Doumpos et al. (2016) believe that revenue diversification can assist banks to enhance their financial resilience and mitigate the adverse effects of the financial crisis. In less-developed nations, income diversification has a more significant influence. Meanwhile, Batten and Vo (2016) state that banks that diversify their revenue with non-interest earning operations are at a higher risk.

Regarding the diversification of funding sources, According to King (2013), diversifying financing sources raises expenses, lowering bank profits. Notably, there is research by Vo (2018) showing that diversification of funding sources has a positive impact on bank earnings.

2.4.3 Bank Size

The size of a bank and its performance are negatively connected, as size is a proxy for reduced risk. Sufian and Chong (2008) and Kasman et al. (2010) found the impact of log of total assets to have a negative impact on bank performance in the Philippines and European Union countries before 2006. Meanwhile, the studies of Anbar and Alper (2011), Menicucci and Paolucci (2016) and Siddik et al. (2017) showed a positive impact of size on bank performance. Notably, the study of Menicucci and Paolucci (2016) with 35 leading European banks from 2009 to 2013 showed the positive effect of size, significant in all cases. The findings demonstrate that large banks are more effective than smaller banks in generating greater and NIM. According to Menicucci and Paolucci (2016), this conclusion stems from the fact that larger banks generally have a substantial part of the market, allowing them to enhance profitability through cost allocation.

Meanwhile, large banks' costs may be cheaper than those of small banks. According to Hauner (2005), economies of scale has a beneficial impact on bank costs through two channels: market power and lower input prices. Shih et al. (2007) demonstrated that bank size is unrelated to efficiency in China. According to Shih et al. (2007), the explanation stems from the fact that banks in China are under regulatory pressure and are supervised by central regulatory agencies.

2.4.4 Bank Liquidity

Liquidity (LIQ) is also an essential factor in influencing a bank's performance. Liquidity has a positive and negative impact on profitability. If the bank effectively holds liquidity, it has a favourable influence on profitability; otherwise, it negatively impacts it. Liquidity has a detrimental influence on bank performance, according to Arif and Nauman Anees (2012), Tran et al. (2016), and Siddik et al. (2017). They feel that the more liquid a bank is, the less efficient it is. According to Arif and Nauman Anees (2012), an increase in the liquidity gap and NPLs has a detrimental impact on a bank's profitability. Liquidity risk is exacerbated by two factors: liquidity gap and non-performing assets. According to Tran et al. (2016), the research findings support the anticipated bankruptcy cost theory, which states that as liquidity rises, so does liquidity risk, raising the probability of bankruptcy and lowering bank performance. Meanwhile, Abbas et al. (2019) used dynamic panel data estimators and the two-step technique of GMM to evaluate the influence on profitability in the post-crisis era 2011–2017 between Asian developed nations and the US banking industry. The findings demonstrate that liquidity and bank capital positively impact bank profitability, whereas credit risk has a negative impact.

2.4.5 Required Reserves

Required reserves ratio (RRR) is the key to risk management in banks. Baltensperger (1982) indicated that reserve requirements are similar to a tax on economic activity. They are designed to have efficiency effects. According to Brock and Rojas Suarez (2000), NIMs are expected to be higher in the context of a transition from a financial system resulting from higher reserve requirements in several Latin American Countries. Glocker and Towbin (2012) show that reserve requirements can only support a price stability objective if the financial market is efficient. Recently, Wei and Han (2020) studied that a reduction in the reserve requirement ratio helps banks lend more to small and micro enterprises, which helps minimize the output gap and reduce the rise in unemployment.

2.4.6 Macroeconomic Indicators (GDP and Money Supply M2)

The macroeconomic factors have an evident impact on the performance of the bank. The study by Tan and Floros (2012) was carried out with 101 Chinese banks from 2003 to 2009 to examine the impact of GDP on banks' profit growth. Using GMM estimation, their empirical results pointed out a negative relationship between the GDP and bank performance and supported the view that the higher the increased banks' entry, the more damp the banks' profitability. Besides, Samad (2015) studied a sample of 43 commercial banks in Bangladesh from 2009 to 2011. Samad (2015) used Panel Ordinary least square (OLS) and reported that GDP is insignificant for banking profitability. Using the OLS multiple regression method, Alkali et al. (2018)

showed GDP growth had a negative effect on the profitability of commercial banks in Nigeria.

An exciting study by Wu et al. (2007) examines the impact of financial development on Chinese banks' operational performance. It is found that the value of M2/GDP is higher, the performance of the banking sector in terms of ROA is better. The results of the OLS method provide evidence that the coefficient of M2/GDP has a positive sign and is statistically significant at the 5% level. Besides, this study contributes that the banks play an essential role in allocating funds; banks' ROA performance will tend to increase if the allocation of funds is effective.

2.5 Covid-19 and Bank Performance

In recent times, the Covid-19 pandemic has become a particular issue, especially its severe impact on the economy as countries implement lockdown, social distancing to limit the spread of the virus. Several studies with the banking sector also show exciting results. Research by Demirgüç-Kunt et al. (2020) conducted with 211 banks in 53 countries reports that the adverse impact of Covid-19 on commercial banks is more pronounced than other financial institutions. Meanwhile, Li et al. (2020) find that the Covid-19 pandemic reduces bank interest margin. According to Yanti et al. (2021), the Covid-19 pandemic impacts the performance of commercial banks in Indonesia. Meanwhile, Almonifi et al. (2021) report that the pandemic has only a modest impact on the Islamic banking sector.

However, Li et al. (2021) document that commercial banks generate higher Non-interest income during the pandemic. The social distance period decreases the bank loans, so the banks introduce payments services to boost the performance. Li et al. (2021) demonstrate that revenue diversification is helpful for commercial banks to sustain during a pandemic.

3 Model and Methodology

3.1 Data

The sample of this study includes 38 joint-stock commercial banks in Vietnam from 2003 to 2020. We collect data from the annual financial reports of the banks. The data of the variables representing the macroeconomy of Vietnam are collected from the World Bank database for the sampling period. We remove observations that do not have sufficient data to calculate relevant variables. We also winsorize data at 1 and 99 percentiles. The final sample contains 491 bank-year observations.

3.2 Methodology

We employ Bayesian linear regressions to analyze the effect of credit growth rate and political connection on commercial banks' NIM in Vietnam. We construct a linear regression model using a probability distribution based on the Bayesian perspective, where y is not estimated as a unique value but is assumed to be drawn from the probability distribution. A Bayesian linear regression model with y is sampled from a normal distribution as follows:

$$y \sim N(\beta^T X, \sigma^2 I)$$

where y is characterized by mean and variance. While Bayesian linear regression does not aim to find the single "best" value of the model parameters, this regression methodology focuses on estimating the posterior distribution for model parameters. The posterior probability of conditional model parameters based on inputs and outputs:

$$P(\beta|y, X) = \frac{P(y|\beta, X)xP(\beta|X)}{P(y|X)}$$

In which $P(\beta|X)$ is prior probability, $P(y|\beta, X)$ is likelihood probability, and $P(\beta|y, X)$ is posterior probability, $P(\beta|X)$ is the normalization constant.

The Bayesian theorem is rewritten as a ratio: posterior probability $\propto$ prior probability $\times$ likelihood probability.

Lemoine (2019) recommends using the normal distribution of N(0, 1) for the observed variables and the distribution $Igamma(0.01, 0.01)$ for the variances in the model. The a priori distribution can be rewritten as follows: $\beta \sim (0, 1)$ and $\sigma^2 \sim (0.01, 0.01)$. Meanwhile, we employ the Markov chain Monte Carlo (MCMC) method with the Gibbs sampler to estimate the posterior distribution. Lynch (2007) and Roy (2020) suggest that Markov chains must support convergence to ensure robust Bayes analysis.

3.3 Regression Models

To examine the relationship between credit growth rate and political connection and NIM of commercial banks in Vietnam, the model is specified as follows:

$$NIM_{it} = \alpha + \beta_1 LGR_{it} + \beta_2 POC_{it} + \delta X_{it} + \varepsilon_{it}$$

where NIM: Net interest margin; LGR: Loan growth rate; POC: Political connection; X_{it} indicates a k $\times$ 1 vector of control variables and ε_{it}: the error term.

First, we estimate the impact of credit growth rate and political connection on NIM of commercial banks after controlling for bank characteristics. King's study (2013), diversifying funding sources is costly and reduces bank profitability, while Vo (2018) argues that diversification helps increase bank income. Besides, Siddik et al. (2017) point out the negative impact of liquidity on bank performance. Therefore, we want to clarify the effect of these two factors on the bank's performance.

Besides, we follow Ozili (2019) to measure loan growth (LGR) and follow Hauner (2005), Shih et al. (2007), Kasman et al. (2010), Anbar and Alper (2011) to measure bank size (LogSIZE). We then examine the effects of loan growth (LGR) and bank size (LogSIZE) on bank performance. Concerning loan growth (LGR), studies by Kosmidou (2008), Ozili (2019) indicate a negative effect, while Kundid and Rogosic (2012), Menicucci and Paolucci (2016) again show a positive impact on bank performance. Besides, with bank size (LogSIZE), there are also mixed results between the size and profitability of commercial banks. As the study of Hauner (2005), Anbar and Alper (2011), Menicucci and Paolucci (2016) gave positive results, while the study of Sufian and Chong (2008), Kasman et al. (2010) again gave negative results to the bank's performance. For that reason, we want to clarify the effects of loan growth (LGR) and bank size (LogSIZE) on the NIM of commercial banks in Vietnam.

In particular, we followed Altunbas et al. (2001), La Porta et al. (2002), Micco et al. (2007), Berger et al. (2009), and Hung et al. (2017) to add the political connection (POC) variable to our model. This is a place in the paper that we believe is interesting. Micco et al. (2007) and Berger et al. (2009) found that political ties had a detrimental influence on bank performance. Hung et al. (2017) discovered a positive correlation, but Altunbas et al. (2001) and La Porta et al. (2002) found no correlation. What drew our attention was that whereas Berger et al. (2009) found a negative association in Chinese banks from 1994 to 2003, Hung et al. (2017) found a positive relationship from 2007 to 2014. Vietnam and China share some similar political and economic characteristics. Therefore, we add the political connection (POC) variable to the model to clarify the influence of POC on the NIM of Vietnamese banks.

Besides, Doku et al. (2019), Wu et al. (2007), Tan and Floros (2012), Al-Qudah and Jaradat (2013), and Samad (2015) argued that the macroeconomic indicators affect bank performance because they influence customer demand for banking products and services. Therefore, we will additionally test the impact of economic growth (RGDP) and money supply M2 growth rate (M2GR) on the performance of commercial banks in Vietnam.

Finally, Demirgüç-Kunt et al. (2020), Li et al. (2020), and Elnahass et al. (2021) concluded that the Covid-19 pandemic had a detrimental influence on the bank's performance. In addition, Li et al. (2020) also believe that the Covid-19 pandemic reduces the optimal interest margin of banks and the government's capital injection. Therefore, we will look at how credit growth rate and political connection, and macroeconomic indicators affect the NIM of commercial banks before and during the Covid-19 pandemic. Table 1 shows the details of variable definitions.

Table 1 Variable definitions

Variables	Notation	Definition
Dependent variables		
Net interest margin	NIM	Apply Thomas and Saunders (1981) model: Gap between interest income and interest expense over the total assets
Independent variables		
Loan growth rate	LGR	Follow Ozili (2019), the LGR is the rate at which a customer's bank loan grows follow this formula: (Total bank's loan of the current year – Total bank's loan of previous year)/(Total bank's loan of the previous year)
Political connection	POC	Follow Rehman et al. (2016), we give POC is the dummy variable measured by: If a member of the Board of Directors is a Party member, it is 1; otherwise, it is recorded as 0
Total debt to total asset	TDTA	Follow Al-Qudah and Jaradat (2013), the TDTA is calculated by this formula: Total debt/Total asset
Bank funding diversity	BFD	Follow bank funding diversity fomular of Vo (2020): $BFD = 1 - \left[\left(\frac{EQU}{Fund}\right)^2 + \left(\frac{GOV}{Fund}\right)^2 + \left(\frac{IBD}{Fund}\right)^2 + \left(\frac{CD}{Fund}\right)^2 + \left(\frac{DER}{Fund}\right)^2 + \left(\frac{LTF}{Fund}\right)^2 + \left(\frac{OTH}{Fund}\right)^2 \right]$
Bank size	LogSIZE	Follow Anbar and Alper (2011), calculate bank size by: Natural logarithm of total assets
Liquidity	LIQ	Follow Al-Qudah and Jaradat (2013), the LIQ variable that measures a bank's liquidity calculated as: Total loan debt/Total customer deposits
Required reserves ratio	RRR	Bank reserve ratio is measured with the formula: Reserve maintained with Central bank/Customer deposits
Economic growth	RGDP	Annual real GDP growth
Money supply growth	M2GR	Annual money supply M2 growth
Covid-19 pandemic	COVID19	Dummy variable taking the value of 1 if the year of 2020, 0 otherwise

4 Empirical Results and Discussion

4.1 Descriptive Statistics

Table 2 provides descriptive statistics of our sample. The average NIM of the banking industry in the sample is 3.28%, indicate that banks have an average spread of the cost of funding and borrowers' charge is 3.28% over the total assets. The highest value is 29.311%, and the lowest is −0.8%, with a standard deviation of 2.012%. This data is similar to Khanh and Tra (2015), Suu et al. (2020), and Nguyen et al. (2020), fully match the NIM indicator during the research period.

The LGR values the average of 0.43499, with a max of 11.31726 and a min of −0.41573. It shows the competitive loan competition between banks in Viet Nam every year. The TDTA has a mean of 0.89529, and the standard deviation is 0.0426, which implies that Vietnamese commercial banks have a high leverage ratio. It is due to the nature that the banking industry uses savings deposits as the primary funding source. This statistic is similar to the study of Le and Nguyen (2020). The bank funding diversity (BFD) has an average value of 0.55888, consistent with Vo (2020). The bank liquidity (LIQ) has an average value of 0.93513. The required reserved ratio (RRR) averages at 0.04678 with the max of 0.55354 and min of 0.0010. For the macroeconomic situation variables, real GDP growth (RGDP) has an average value of 6.144%, and the average yearly growth rate of money supply M2 (M2GR) is 23.16%. Finally, Table 2 explicitly report the descriptive statistics of other control variables.

The sample of the paper covers 38 commercial banks for the period 2003–2020. Of these, 19 banks are completely politically connected, with 50% of the sample's entire banks. There are 11 banks with absolutely no political connection (29%). There are eight banks with political connection and without political connection, depending on the sample's observation moment (Table 3).

Table 2 Descriptive statistics

Variables	Obs	Mean	Std. dev	Min	Max
NIM	491	0.03280	0.02012	−0.00888	0.29311
TDTA	491	0.89529	0.06690	0.33925	1.08915
BFD	491	0.55888	0.10030	0.18758	0.75455
LGR	491	0.43499	0.95533	−0.41573	11.31726
Size	491	31.73785	1.54144	26.09084	34.95530
DC	491	0.02262	0.02384	0.00000	0.33135
LIQ	491	0.93513	0.40543	0.37187	7.29587
RRR	491	0.04678	0.04195	0.00101	0.55354
RGDP	491	0.06144	0.01039	0.02906	0.07547
M2GR	463	0.23136	0.10554	0.04400	0.46100

Table 3 Descriptive statistics for political connection variable (POC)

Political connection	Banks	Percent (%)
With connection	19	50
Without connection	11	29
With and without connection	8	21
Total	38	100

4.2 *Correlation Matrix*

Table 4 examines the correlation matrix between the independent variables. We follow Lee and Wagenmakers (2014) to apply Bayesian factors (BF) to examine the correlation between variables. Table 4 reports the weak correlations between independent variables because all correlation coefficients are <0.4. Moreover, Adepoju and Ojo (2018) indicate that the Bayes estimation method mitigates multicollinearity, so multicollinearity is not a potential issue in our study.

4.3 *MCMC Diagnostics*

When performing Bayes analysis, convergence testing is required. This section performs MCMC diagnostics to consider the convergence of Markov chains or stopping MCMC sampling through effective sample size (ESS) and Gelman-Rubin diagnostic. ESS is the number of independent samples equivalent to a set of correlated Markov chain samples. Gelman-Rubin diagnostic is the most common method to evaluate samples obtained from running MCMC algorithms.

We perform the MCMC convergence test and ESS for all three models, in which NIM is a dependent variable. The results in Table 5 report that the lowest efficiency is 0.9505, and the Rc values of the Gelman-Rubin diagnostic are all <1.1. The acceptance rate is 1.0000 with the Gibbs sampling algorithm, so the regression result has reached the acceptance rate. Thus, the above tests have shown that Markov chains have converged, so Bayesian inference is stable (Gelman et al. 2013; Lynch 2007).

4.4 *Posterior Simulation and Discussion*

Table 6 reports the impacts of credit growth rate (LGR) and political connection (POC) on the NIM. Credit growth also negatively affects NIM. The growth in lending at banks contributes to the growth of interest income. However, increasing lending forces banks to increase provisions for loan losses, which in turn reduces profits. However, to guarantee the stability of the banking system in Vietnam, the implementation of BASEL I standards has been in place since 2005, according to

Table 4 Bayesian correlation matrix

	NIM	TDTA	BFD	LGR	LogSIZE	DC	LIQ	RRR	POC	RGDP	M2GR
NIM	–										
TDTA	−0.327*** [5.315e + 10]	–									
BFD	0.011 [0.058]	−0.322*** [2.181e + 10]	–								
LGR	0.039 [0.082]	−0.144 [9.094]	0.197*** [889.509]	–							
LogSIZE	−0.300*** [5.322e + 8]	0.661*** [3.197e + 59]	−0.316*** [7.573e + 9]	−0.278*** [1.840e + 7]	–						
DC	−0.057 [0.124]	0.011 [0.058]	0.014 [0.059]	−0.089 [0.394]	0.011 [0.058]	–					
LIQ	0.249*** [330,986]	−0.294*** [2.086e + 8]	0.089 [0.392]	0.103 [0.756]	−0.266*** [3.329e + 6]	0.044 [0.091]	–				
RRR	0.046 [0.095]	−0.075 [0.224]	0.071 [0.192]	0.078 [0.251]	−0.085 [0.331]	0.077 [0.244]	0.258*** [1.116e + 6]	–			
POC	−0.054 [0.116]	0.011 [0.058]	0.009 [0.058]	−0.049 [0.101]	0.151* [16.002]	0.098 [0.596]	−0.017 [0.061]	0.060 [0.136]	–		
RGDP	0.091 [0.429]	0.000 [0.057]	0.062 [0.144]	0.132 [3.961]	−0.201*** [1,313]	0.019 [0.062]	0.071 [0.196]	0.029 [0.069]	0.025 [0.065]	–	

(continued)

Table 4 (continued)

	NIM	TDTA	BFD	LGR	LogSIZE	DC	LIQ	RRR	POC	RGDP	M2GR
M2GR	−0.109 [0.904]	0.146 [8.443]	−0.038 [0.081]	−0.270*** [2.147e + 6]	0.275*** [3.776e + 6]	0.034 [0.076]	−0.183*** [140.348]	−0.185*** [173.684]	0.033 [0.075]	−0.319*** [3.011e + 9]	–

Note $^{*}BF_{10} > 10$, $^{**}BF_{10} > 30$, $^{***}BF_{10} > 100$
Source The author's calculation

Table 5 Convergence test and effective sample size for the models

Models	Model 1		Model 2		Model 3	
Variables	Efficiency	Rc	Efficiency	Rc	Efficiency	Rc
TDTA	1.00000	1.00006	1.00000	1.00031	1.00000	1.00013
BFD	1.00000	1.00002	1.00000	1.00001	1.00000	0.99999
LGR	1.00000	1.00005	1.00000	1.00004	1.00000	1.00004
LogSIZE	0.99320	0.99998	1.00000	1.00019	1.00000	1.00003
DC	1.00000	1.00001	0.97710	1.00009	0.98360	1.00004
LIQ	0.96490	0.99998	0.98480	1.00000	0.98480	1.00001
RRR	0.98670	1.00002	1.00000	0.99996	0.98640	1.00000
POC	1.00000	1.00003	0.99230	1.00002	1.00000	1.00000
RGDP	–	–	1.00000	1.00002	0.97240	1.00000
M2GR	–	–	1.00000	0.99999	0.97670	0.99997
COVID19	–	–	–	–	0.99260	1.00004
_cons	0.99030	0.99999	1.00000	1.00021	1.00000	1.00002
var	0.97110	1.00010	0.95050	1.00025	0.96120	1.00005

Source The author's calculation

the State Bank's rules. Banks will test the application of BASEL II standards in 2014, and banks in Vietnam will be required to comply with BASEL II requirements by 2020. Applying the BASEL standard forces banks to reduce or not to lend to doubtful debt groups or customers with bad credit history, strengthen customer credit quality appraisal, and reduce loan loss provisions. Therefore, bank lending activities become more regulated, and it adversely affects the NIM. Our findings are consistent with Kosmidou (2008), but they are inconsistent with Kundid and Rogosic (2012), Menicucci and Paolucci (2016).

In Vietnam, most of the large commercial banks are formed with the support of the state (for example, the Big 4 has state ownership). These banks play a leading role in the banking industry and the stability of the banking system. Although political connection helps banks access new information and policies to adjust and exploit the bank's activities accordingly. These banks must serve the social and economic programs directed by the government. Moreover, politically connected banks are constantly under pressure to grant loans for political purposes instead of maximizing profits for the banks. Therefore, their business objectives deviate from maximizing shareholder wealth, and their NIM reduce accordingly. Our findings are consistent with Micco et al. (2007) and Berger et al. (2009). However, the results of our study are consistent with the study of Hung et al. (2017).

Our results indicate that TDTA significantly affects the performance of Vietnamese commercial banks. For instance, TDTA adversely affects NIM. This result is consistent with the study of Al-Qudah and Jaradat (2013), Siddik et al. (2017), Yahya et al. (2017), Sivalingam and Kengatharan (2018), Almaqtari et al. (2019). According to Siddik et al. (2017), commercial banks prefer using external financing to

Table 6 Posterior simulation results of the models

Models	Model 1		Model 2		Model 3	
Independent variables	Mean	Probability of mean	Mean	Probability of mean	Mean	Probability of mean
LGR	−0.00073 [−0.00266; 0.00120]	0.7730*	−0.00096 [−0.00296; 0.00105]	0.8270*	−0.00097 [−0.00294; 0.00104]	0.8323*
POC	−0.00084 [−0.00458; 0.00289]	0.6723*	−0.00066 [−0.00461; 0.00323]	0.6283*	−0.00069 [−0.00463; 0.00319]	0.6316*
BFD	−0.02293 [−0.04158; −0.00429]	0.9916*	−0.02394 [−0.04339; −0.00461]	0.9918*	−0.02369 [−0.04311; −0.00413]	0.9912*
LIQ	0.00804 [0.00339; 0.01280]	0.9996**	0.00697 [0.00215; 0.01174]	0.9980**	0.00699 [0.00221; 0.01174]	0.9976**
TDTA	−0.06583 [−0.10147; −0.02977]	0.9999*	−0.06661 [−0.10434; −0.02832]	0.9996*	−0.06644 [−0.10399; −0.02849]	0.9997*
LogSIZE	−0.06299 [−0.11278; −0.01256]	0.9928*	−0.06948 [−0.12300; −0.01557]	0.9938*	−0.06934 [−0.12308; −0.01551]	0.9946*
DC	−0.04890 [−0.12340; 0.02498]	0.9021*	−0.04786 [−0.12291; 0.02663]	0.8934*	−0.04793 [−0.12339; 0.02778]	0.8935*
RRR	−0.00405 [−0.04690; 0.03915]	0.5744*	−0.00376 [−0.04878; 0.04063]	0.5648*	−0.00383 [−0.04889; 0.04129]	0.5651*
RGDP	–	–	0.30201 [0.01638; 0.58921]	0.9811**	0.30248 [0.01462; 0.58718]	1.0000**
M2GR	–	–	0.00232 [−0.01730; 0.02153]	0.5939**	0.00227 [−0.01733; 0.02149]	0.5920**
COVID19	–	–	–	–	−0.00226 [−1.96613; 1.94483]	0.5002*
_cons	0.31689 [0.15736; 0.47424]	1.0000**	0.32185 [0.14981; 0.49282]	1.0000**	0.32107 [0.14890; 0.49171]	1.0000**
var	0.00039 [0.00034; 0.00044]	–	0.00040 [0.00035; 0.00045]	–	0.00040 [0.00035; 0.00045]	–

Note *probability of mean <0; **probability of mean >0; credible interval in brackets
Source The author's calculation

equity financing because the Vietnamese capital markets are undeveloped in Vietnam. Therefore, high financing costs erode the NIM of commercial banks.

Table 6 also reports a negative relationship between bank funding diversity (BFD) and NIM. Diversification of funding sources helps maintain stability, but it is costly to maintain diversified funding sources. Therefore, higher funding diversity reduces the NIM of a commercial bank. Our results are consistent with King (2013), while they are consistent with Abbas et al. (2019), and Vo (2018).

On the other hand, bank liquidity (LIQ) has a favourable impact on the profitability of commercial banks in Vietnam. In other words, when banks raise the loan-to-customer ratio (LIQ), it will boost the bank's profit growth. More loans will help banks increase interest income from loans. Our findings align with Arif and Nauman Anees (2012), Tran et al. (2016), and Siddik et al. (2017).

Specifically, the bank size has a strong negative effect on NIM. Larger banks have higher operational costs due to agency problems. The results of this study are consistent with those of Hauner (2005), Anbar and Alper (2011), Menicucci and Paolucci (2016) and Siddik et al. (2017).

Table 6 reports the positive relationship between RGDP and NIM. According to Wu et al. (2007), when the economy grows at a rate that contributes to an increase in GDP per capita, individuals have more money to save and use other banking services. Savings deposits are a low-cost, reliable, and low-risk source of funds. Furthermore, using additional services allows the bank to raise non-interest revenue, improving, enhancing, and developing new goods and services. The bank's performance will then improve. This result is consistent with Wu et al. (2007), Sufian and Habibullah (2009) and Samad (2015).

The M2GR also has favourable impacts on the performance of commercial banks in Vietnam. The increasing money supply means more opportunities for commercial banks to allocate this money to the economy via credit activities. Moreover, the higher money supply reduces the funding costs of commercial banks, so it has positive impacts on NIM. Our results aligns with Al-Qudah and Jaradat (2013) while they are inconsitent with Sufian and Habibullah (2009), and Alkali et al. (2018).

Finally, we examine the impacts of the Covid-19 pandemic on banking performance in Vietnam. Our result indicates that the Covid-19 has negative impacts on the NIM of commercial banks in Vietnam. This negative impact, in our opinion, is the result of the government imposing a blockade and social distance for a long time in order to contain the proliferation of Covid-19, which has a bad influence on customers. Borrowers are unable to engage in manufacturing or business operations. As a result, there's a good chance the client will default. As a result, banks are being forced to boost loan loss provisions, lowering profits. This study's findings are consistent with those of Demirgüç-Kunt et al. (2020), Li et al. (2020), and Elnahass et al. (2021).

5 Conclusions

This paper enriches the literature by adding evidence of credit growth and political connection impacts on bank performance in Vietnam. While prior literature utilizes traditional regression approaches, we employ the Bayesian regressions to analyze a sample of 38 commercial banks from 2003 to 2020. The empirical results indicate that the lending growth rate and political connection have negative impacts on NIM. On the other hand, our findings suggest a positive relationship between bank liquidity, GDP rate, and money supply M2 growth on NIM. In addition, our findings suggest a negative relationship between bank funding diversity, total debt to total asset, bank size and debt credit ratio have negative impacts on NIM. Finally, our study figures out that Covid-19 has adverse impacts on NIM. Our findings are helpful for bank managers and market supervisors to maintain the sustainable growth of the financial market.

References

Abbas, F., Iqbal, S., Aziz, B.: The impact of bank capital, bank liquidity and credit risk on profitability in postcrisis period: a comparative study of US and Asia. Cogent Econ. Financ. **7**(1), 1605683 (2019). https://doi.org/10.1080/23322039.2019.1605683

Adepoju, A.A., Ojo, O.O.: Bayesian method for solving the problem of multicollinearity in regression. Afr. Stat. **13**(3), 1823–1834 (2018)

Alkali, M.A., Sipan, I., Razali, M.N.: An overview of macro-economic determinants of real estate price in Nigeria. Int. J. Eng. Technol. **7**(3.30), 484–488 (2018). https://doi.org/10.14419/ijet.v7i3.30.18416

Allen, L.: The Determinants of Bank Interest Margins: a Note. J. Financ. Quant. Anal. **23**(2), 231–235 (1988). https://doi.org/10.2307/2330883

Almaqtari, F.A., Al-Homaidi, E.A., Tabash, M.I., Farhan, N.H.: The determinants of profitability of Indian commercial banks: a panel data approach. Int. J. Financ. Econ. **24**(1), 168–185 (2019). https://doi.org/10.1002/ijfe.1655

Almonifi, Y.S.A., Rehman, S.U., Gulzar, R.: The COVID-19 pandemic effect on the performance of the islamic banking sector in KSA: an empirical study of Al Rajhi bank. Int. J. Manag. Educ. **12**(4), 533–547 (2021). https://doi.org/10.34218/IJM.12.4.2021.045

Al-Qudah, A.M., Jaradat, M.A.: The impact of macroeconomic variables and banks characteristics on Jordanian Islamic banks profitability: empirical evidence. Int. Bus. Res. **6**(10), 153 (2013). https://doi.org/10.5539/ibr.v6n10p153

Altunbas, Y., Evans, L., Molyneux, P.: Bank ownership and efficiency. J. Money, Credit, Bank. **33**(4), 926–954 (2001). https://doi.org/10.2307/2673929

Anbar, A., Alper, D.: Bank specific and macroeconomic determinants of commercial bank profitability: empirical evidence from Turkey. Bus. Econ. Res. J. **2**(2), 139–152 (2011)

Andrade, C.: The P value and statistical significance: misunderstandings, explanations, challenges, and alternatives. Indian J. Psychol. Med. **41**(3), 210–215 (2019). https://doi.org/10.4103/ijpsym.ijpsym_193_19

Angbazo, L.: Commercial bank net interest margins, default risk, interest-rate risk, and off-balance sheet banking. J. Bank. Financ. **21**(1), 55–87 (1997). https://doi.org/10.1016/S0378-4266(96)00025-8

Arif, A., Nauman Anees, A.: Liquidity risk and performance of banking system. J. Financ. Regulat. Compl. **20**(2), 182–195 (2012). https://doi.org/10.1108/13581981211218342

Baltensperger, E.: Reserve requirements and economic stability. J. Money Credit Bank **14**(2), 205–215 (1982). https://doi.org/10.2307/1991639

Batten, J.A., Vo, X.V.: Bank risk shifting and diversification in an emerging market. Risk Manag. **18**(4), 217–235 (2016). http://www.jstor.org/stable/26624312

Berger, A.N., Hasan, I., Zhou, M.: Bank ownership and efficiency in China: What will happen in the world's largest nation? J. Bank. Financ. **33**(1), 113–130 (2009). https://doi.org/10.1016/j.jbankfin.2007.05.016

Borio, C., Gambacorta, L., Hofmann, B.: The influence of monetary policy on bank profitability. Int. Financ. **20**(1), 48–63 (2017). https://doi.org/10.1111/infi.12104

Brock, P.L., Rojas Suarez, L.: Understanding the behavior of bank spreads in Latin America. J. Dev. Econ. **63**(1), 113–134 (2000). https://doi.org/10.1016/S0304-3878(00)00102-4

Demirgüç-Kunt, A., Martinez Peria, M.S., Tressel, T.: The global financial crisis and the capital structure of firms: Was the impact more severe among SMEs and non-listed firms? J. Corp. Finan. **60**, 101514 (2020). https://doi.org/10.1016/j.jcorpfin.2019.101514

Demirgüç-Kunt, A., Pedraza, A., Ruiz-Ortega, C.: Banking sector performance during the COVID-19 crisis. J. Bank. Financ. **133**, 106305 (2021). https://doi.org/10.1016/j.jbankfin.2021.106305

Doku, J.N., Kpekpena, F.A., Boateng, P.Y.: Capital structure and bank performance: empirical evidence from Ghana. Afr. Dev. Rev. **31**(1), 15–27 (2019). https://doi.org/10.1111/1467-8268.12360

Doumpos, M., Gaganis, C., Pasiouras, F.: Bank diversification and overall financial strength: international evidence. Financ. Mark. Inst. Instrum. **25**(3), 169–213 (2016). https://doi.org/10.1111/fmii.12069

El-Chaarani, H., El-Abiad, Z.: Exploring the financial dimensions of Lebanese SMEs: comparative study between family and non-family business. Int. Rev. Manag. Mark. **9**(3), 19–30 (2019). https://doi.org/10.32479/irmm.7976

Elnahass, M., Trinh, V.Q., Li, T.: Global banking stability in the shadow of Covid-19 outbreak. J. Int. Finan. Markets. Inst. Money **72**, 101322 (2021). https://doi.org/10.1016/j.intfin.2021.101322

Foos, D., Norden, L., Weber, M.: Loan growth and riskiness of banks. J. Bank. Financ. **34**(12), 2929–2940 (2010). https://doi.org/10.1016/j.jbankfin.2010.06.007

Gelman, A., Carlin, J.B., Stern, H.S., Dunson, D.B., Vehtari, A., Rubin, D.B.: Bayesian Data Analysis. CRC Press, New York (2013)

Glocker, C., Towbin, P.: Reserve requirements for price and financial stability—When are they effective? Int. J. Cent. Bank. **8**(1), 65–113 (2012)

Grzenda, W.: The advantages of Bayesian methods over classical methods in the context of credible intervals. Inform. Syst. Manag. **4**(1), 53–63 (2015)

Hauner, D.: Explaining efficiency differences among large German and Austrian banks. Appl. Econ. **37**(9), 969–980 (2005). https://doi.org/10.1080/00036840500081820

Ho, J., Tumkaya, T., Aryal, S., Choi, H., Claridge-Chang, A.: Moving beyond P values: data analysis with estimation graphics. Nat. Methods **16**(7), 565–566 (2019). https://doi.org/10.1038/s41592-019-0470-3

Holtmann, J., Koch, T., Lochner, K., Eid, M.: A Comparison of ML, WLSMV, and Bayesian methods for multilevel structural equation models in small samples: a simulation study. Multivar. Behav. Res. **51**(5), 661–680 (2016). https://doi.org/10.1080/00273171.2016.1208074

Hung, C.-H.D., Jiang, Y., Liu, F.H., Tu, H., Wang, S.: Bank political connections and performance in China. J. Financ. Stab. **32**, 57–69 (2017). https://doi.org/10.1016/j.jfs.2017.09.003

Jadah, H.M., Hameed, T.M., Al-Husainy, N.H.M.: The impact of the capital structure on Iraqi banks' performance. Invest. Manag. Financ. Innov. **17**(3), 122 (2020). https://doi.org/10.21511/imfi.17(3).2020.10

Kasman, A., Tunc, G., Vardar, G., Okan, B.: Consolidation and commercial bank net interest margins: evidence from the old and new European Union members and candidate countries. Econ. Model. **27**(3), 648–655 (2010). https://doi.org/10.1016/j.econmod.2010.01.004

Khanh, H.T., Tra, V.T.: Determinants of net interest margin of commercial banks in Vietnam. J. Econ. Dev. **17**(2), 69–82 (2015)

King, M.: Fisheries biology, assessment and management. Wiley, Hoboken, New Jersey (2013)

Kosmidou, K.: The determinants of banks' profits in Greece during the period of EU financial integration. Manag. Financ. **34**(3), 146–159 (2008). https://doi.org/10.1108/03074350810848036

Kundid, A., Rogosic, A.: E-transparency Of Croatian banks: determinants and disclosure contents. Econ. Res.-Ekonomska Istraživanja **25**(sup1), 86–116 (2012). https://doi.org/10.1080/1331677X.2012.11517558

Kusi, B.A., Agbloyor, E.K., Gyeke-Dako, A., Asongu, S.A.: Financial sector transparency and net interest margins: Should the private or public sector lead financial sector transparency? Res. Int. Bus. Financ. **54**, 101260 (2020). https://doi.org/10.1016/j.ribaf.2020.101260

La Porta, R., Lopez-De-Silanes, F., Shleifer, A.: Government ownership of banks. J. Financ. **57**(1), 265–301 (2002). https://doi.org/10.1111/1540-6261.00422

Le, T.D.Q., Nguyen, D.T.: Capital structure and bank profitability in Vietnam: a quantile regression approach. J. Risk Financ. Manag. **13**(8), 168 (2020). https://doi.org/10.3390/jrfm13080168

Lee, M.D., Wagenmakers, E.-J.: Bayesian Cognitive Modeling: A Practical Course. Cambridge University Press, New York, NY (2014)

Lemoine, N.P.: Moving beyond noninformative priors: Why and how to choose weakly informative priors in Bayesian analyses. Oikos **128**(7), 912–928 (2019). https://doi.org/10.1111/oik.05985

Li, L., Strahan, P.E., Zhang, S.: Banks as lenders of first resort: evidence from the COVID-19 crisis. Rev. Corp. Financ. Stud. **9**(3), 472–500 (2020). https://doi.org/10.1093/rcfs/cfaa009

Li, X., Feng, H., Zhao, S., Carter, D.A.: The effect of revenue diversification on bank profitability and risk during the COVID-19 pandemic. Financ. Res. Lett. **43**, 101957 (2021). https://doi.org/10.1016/j.frl.2021.101957

Lynch, S.M.: Introduction to applied Bayesian statistics and estimation for social scientists. Springer New York, New York, NY (2007)

McNeish, D.: On using Bayesian methods to address small sample problems. Struct. Equ. Model. **23**(5), 750–773 (2016). https://doi.org/10.1080/10705511.2016.1186549

Menicucci, E., Paolucci, G.: The determinants of bank profitability: empirical evidence from European banking sector. J. Financ. Report. Account. **14**(1), 86–115 (2016). https://doi.org/10.1108/JFRA-05-2015-0060

Micco, A., Panizza, U., & Yañez, M.: Bank ownership and performance. Does politics matter? J. Bank. Financ. **31**(1), 219–241 (2007). https://doi.org/10.1016/j.jbankfin.2006.02.007

Moudud-Ul-Huq, S., Ashraf, B.N., Gupta, A.D., Zheng, C.: Does bank diversification heterogeneously affect performance and risk-taking in ASEAN emerging economies? Res. Int. Bus. Financ. **46**, 342–362 (2018). https://doi.org/10.1016/j.ribaf.2018.04.007

Nguyen, H.T., Trung, N.D., Thach, N.N.: Beyond traditional probabilistic methods in econometrics. In: Kreinovich, V., Thach, N.N., Trung, N.D., Van Thanh, D. (eds.) Beyond Traditional Probabilistic Methods in Economics, pp. 3–21. Springer International Publishing, Cham (2019)

Nguyen, T.V.H., Pham, T.T.T., Nguyen, C.P., Nguyen, T.C., Nguyen, B.T.: Excess liquidity and net interest margins: evidence from Vietnamese banks. J. Econ. Bus. **110**, 105893 (2020). https://doi.org/10.1016/j.jeconbus.2020.105893

Ozili, P.K.: Non-performing loans and financial development: new evidence. J. Risk Financ. **20**(1), 59–81 (2019). https://doi.org/10.1108/JRF-07-2017-0112

Rehman, R.U., Zhang, J., Ahmad, M.I.: Political system of a country and its non-performing loans: a case of emerging markets. Int. J. Bus. Perform. Manag. **17**(3), 241–265 (2016)

Roy, V.: Convergence diagnostics for Markov Chain Monte Carlo. Ann. Rev. Stat. Its Appl. **7**(1), 387–412 (2020). https://doi.org/10.1146/annurev-statistics-031219-041300

Samad, A.: Determinants bank profitability: empirical evidence from Bangladesh commercial banks. Int. J. Financ. Res. **6**(3), 173–179 (2015). https://doi.org/10.5430/ijfr.v6n3p173

Shih, V., Zhang, Q., Liu, M.: Comparing the performance of Chinese banks: a principal component approach. China Econ. Rev. **18**(1), 15–34 (2007). https://doi.org/10.1016/j.chieco.2006.11.001

Siddik, M., Alam, N., Kabiraj, S., Joghee, S.: Impacts of capital structure on performance of banks in a developing economy: evidence from Bangladesh. Int. J. Financ. Stud. **5**(2), 13 (2017). https://doi.org/10.3390/ijfs5020013

Simoens, M., Vennet, R.V.: Bank performance in Europe and the US: a divergence in market-to-book ratios. Financ. Res. Lett. **40**, 101672 (2021). https://doi.org/10.1016/j.frl.2020.101672

Sivalingam, L., Kengatharan, L.: Capital structure and financial performance: a study on commercial banks in Sri Lanka. Asian Econ. Financ. Rev. **8**(5), 586–598 (2018)

Sufian, F., Chong, R.R.: Determinants of bank profitability in a developing economy: empirical evidence from Philippines. Asian Acad. Manag. J. Account. Financ. **4**(2) (2008)

Sufian, F., Habibullah, M.S.: Bank specific and macroeconomic determinants of bank profitability: empirical evidence from the China banking sector. Front. Econ. China **4**(2), 274–291 (2009)

Suu, N.D., Luu, T.-Q., Pho, K.-H., McAleer, M.: Net interest margin of commercial banks in Vietnam. Adv. Decis. Sci. **24**(1), 1–27 (2020)

Tan, Y., Floros, C.: Bank profitability and inflation: the case of China. J. Econ. Stud. **39**(6), 675–696 (2012). https://doi.org/10.1108/01443581211274610

Thomas, S.Y.H., Saunders, A.: The determinants of bank interest margins: theory and empirical evidence. J. Financ. Quant. Anal. **16**(4), 581–600 (1981). https://doi.org/10.2307/2330377

Tran, V.T., Lin, C.T., Nguyen, H.: Liquidity creation, regulatory capital, and bank profitability. Int. Rev. Financ. Anal. **48**, 98–109 (2016). https://doi.org/10.1016/j.irfa.2016.09.010

Vinh, V.X., Mai, T.T.P.: Profitability and risk in relation to income diversification of Vietnamese commercial banking system. J. Econ. Dev. **23**(2), 61–76 (2016)

Vo, X.V.: Do firms with state ownership in transitional economies take more risk? Evidence from Vietnam. Res. Int. Bus. Financ. **46**, 251–256 (2018). https://doi.org/10.1016/j.ribaf.2018.03.002

Vo, X.V.: The role of bank funding diversity: evidence from Vietnam. Int. Rev. Financ. **20**(2), 529–536 (2020). https://doi.org/10.1111/irfi.12215

Wasserstein, R.L., Schirm, A.L., Lazar, N.A.: Moving to a world beyond "$p < 0.05$". Am. Stat. **73**(Suppl 1), 1–19 (2019). https://doi.org/10.1080/00031305.2019.1583913

Wei, X., Han, L.: Targeted reduction in reserve requirement ratio and optimal monetary policy in China. Int. Rev. Econ. Financ. **69**, 209–230 (2020). https://doi.org/10.1016/j.iref.2020.04.002

Wu, H.L., Chen, C.H., Shiu, F.Y.: The impact of financial development and bank characteristics on the operational performance of commercial banks in the Chinese transitional economy. J. Econ. Stud. **34**(5), 401–414 (2007). https://doi.org/10.1108/01443580710823211

Yahya, A.T., Akhtar, A., Tabash, M.I.: The impact of political instability, macroeconomic and bank-specific factors on the profitability of Islamic banks: an empirical evidence. Invest. Manag. Financ. Innov. **14**(4), 30–39 (2017). https://doi.org/10.21511/imfi.14(4).2017.04

Yanti, N.I., Komalasari, A., Prasetyo, T.J.: Does Covid-19 have an impact on bank performance in Indonesia? A comparative analysis based on BUKU. J. Dimensie Manag. Pub. Sect. **2**(2), 9–18 (2021). https://doi.org/10.48173/jdmps.v2i2.90

Hyperparameter Tuning with Different Objective Functions in Financial Market Modeling

Minh Tran, Duc Pham-Hi, and Marc Bui

Abstract In this paper, we propose different objective functions to tune hyperparameters in an agent-based simulation of the stock market. To reproduce the stylized facts of the real market, Bayesian optimization is introduced to find the calibrated set of parameters. The experimental results of Bayesian calibration have provided a stable and low-cost hyperparameter tuning method in financial market modeling. With empirical data of the VN Index, statistical analysis of the simulated data showed that the model was able to replicate some of the important stylized facts in real markets such as random walk price dynamics, absence of autocorrelations, heavy tail in distribution, and volatility clustering of the returns.

1 Introduction

Hyperparameter tuning is the problem of choosing a set of optimal hyperparameters for a learning algorithm. A hyperparameter is a parameter of which value is used to control the learning process. Thus, in order to achieve maximal performance, it is important to understand how to optimize them Claesen and Moor (2015). One prominent application of hyperparameter tuning in financial markets is to help models reproduce some empirical statistical regularities, known as "stylized facts", of the real market Challet et al. (2001). In this study, we provide results to answer the

M. Tran (✉)
John von Neumann Institute, Vietnam National University, Ho Chi Minh, Vietnam
e-mail: minh.tran@jvn.edu.vn

PSL Research University, Paris, France

D. Pham-Hi
Financial Engineering Department, ECE Paris, Graduate school of Engineering, Paris, France
e-mail: phamhi@ece.fr

M. Bui
CHArt Laboratory EA 4004 EPHE, PSL Research University, Paris, France
e-mail: marc.bui@ephe.psl.eu

N. Ngoc Thach et al. (eds.), *Financial Econometrics: Bayesian Analysis, Quantum Uncertainty, and Related Topics*, Studies in Systems, Decision and Control 427,
https://doi.org/10.1007/978-3-030-98689-6_48

question of how to have a stable and low-cost hyperparameter tuning method in financial market modeling.

Many models use large parameter sets and run simulations for long periods. However, even for small models, exploring the stylized facts through all parameter combinations is not possible or costly Snoek et al. (2012). Grid search is a method to list all combinations of hyperparameters within a given range, then perform a model test with this list Larochelle et al. (2007). Random search only randomly selects a finite number of hyperparameters from the list to conduct a model test Bergstra and Bengio (2012). Theoretically, Grid Search can find the global value, however, as the number of parameters increases, this becomes more and more impossible due to the time and cost of implementation. Random search does not run enough cases like Grid search, so it will be significantly faster. However, depending on the corresponding random selection, we don't know if there exists a better combination of hyperparameters. For neural networks, it is possible to compute the gradient for hyperparameters and then optimize the hyperparameters using gradient descent Larsen et al. (1996). Evolutionary optimization is used in hyperparameter optimization for statistical machine learning algorithms Kousiouris et al. (2011).

Bayesian optimization is an adaptive approach to parameter optimization, trading off between exploring new areas of the parameter space, and exploiting historical information to find the parameters that maximize the function quickly Snoek et al. (2012). The optimization function is composed of multiple hyperparameters that are set before the learning process and affect how the learning algorithm fits the model to data. Thereby, Bayesian optimization is used to tune hyperparameters for the optimization function in our model. We propose two objective functions based on return and price. The return-based objective function considers the Kolmogorov-Smirnov (KS) test where the statistic in the KS test quantifies a distance between the empirical distribution functions of two samples Marsaglia et al. (2003). Besides, Dynamic Time Warping (DTW), which measures similarity between two series, is introduced to construct the price-based objective function Berndt and Clifford (1994).

We synthesize the key components of our previous work to build a simple agent-based model (ABM) for the stock market Tran et al. (2020). In which, the objective functions are compared based on cost and stability over many simulations. In addition, we also analyze the efficiency of each objective function through its ability to reproduce the properties of the stock market in reality.

The results in this research show that Bayesian optimization can propose a optimal set of parameters with high stability and low cost. The return-based objective function with the KS test shows better results on convergence rate and execution time. In addition, the ABM with different objective functions can represent some stylized facts of the real market. The DTW algorithm can reproduce dynamics of the prices in the real market but is not good for the absence of autocorrelations in returns. In contrast, the KS test does not show a good simulation of prices but successfully explains some important stylized facts of returns, such as the absence of autocorrelations, fat tails in stock return distribution, and volatility clustering. This result is consistent with the definition of the objective function.

The remainder of this paper is organized as follows. The definition of the objective functions are introduced in Sect. 2. Section 3 presents the characteristics of our ABM. Section 4 describes the data and the calibration of the model. Statistical analysis of the model's results and the stylized facts are given in Sect. 5. Finally, Sect. 6 concludes and proposes some extensions of our work.

2 Objective Function

The main goal of ABM is to reproduce stylized facts and to identify sufficient conditions for their creations. In financial markets, stylized facts emerge from statistical analysis of prices and returns. Therefore, the objective function in this study is defined based on the above two variables.

2.1 Return-Based Objective Function

The objective function is constructed by comparing simulated returns and actual returns. The KS test statistic measures the largest distance between the empirical distribution function of simulated returns, $F_s(x)$ and the empirical distribution function of actual returns, $F_a(x)$. The formula for the test statistic is given by:

$$D_{KS} = \sup_x |F_{s,n}(x) - F_{a,n}(x)|, \tag{1}$$

where n is the data size and sup is the supremum function.

2.2 Price-Based Objective Function

The objective function uses the DTW algorithm to calculate an optimal match between simulated prices and actual prices with certain restrictions and rules. The cost is computed as the sum of absolute differences, for each matched pair of indices, among their values. Let x and y be two vectors of lengths m and n, respectively, the algorithm is summarized in the following three steps:

- Optimum-value function: Define $D(i, j)$ as the DTW distance between $(x_1, ..., x_i)$ and $(y_1, ..., y_j)$, with the mapping path starting from $(1, 1)$ to (i, j).
- Recursion:

$$D(i, j) = |x(i) - y(j)| + min\Big\{D(i-1, j), D(i-1, j-1), D(i, j-1)\Big\}, \tag{2}$$

  with the initial condition $D(0, 0) = 0$, $D(0, j) = \infty$, $D(0, j) = \infty$.
- Final answer: $D(m, n)$.

3 Financial Market Model

In this work, we recombine key ingredients of our previous works to come up with a simple agent-based model that can match the stylized facts of financial markets. For more details, see Tran et al. (2020).

3.1 The Agent

The agents in the model are assumed to gather into groups when making investment decisions. Sornette describes the group of traders introduced by is a square lattice and each trader is connected to his four nearest neighbors Sornette and Zhou (2006). In our model, we define the traders' group as a multi-dimensional lattice that allows each agent to be connected to his neighbors in different size groups. This makes the trading behaviors of agents more diverse. Accordingly, the propensity in decision-making behavior of agents is introduced. The term "propensity of agents' decision" defines the probability that one agent will follow the other agents' opinion and that agents will change their decisions on their own.

Let's define the sign function, $sign(x)$, as

$$sign(x) := \begin{cases} 1, & \text{if } x > 0; \\ 0, & \text{if } x = 0; \\ -1, & \text{if } x < 0. \end{cases} \tag{3}$$

At each time step t, the final decision of the agent i , denoted as $s_i(t)$, is given by the following equation:

$$s_i(t) = \text{sign}\Big[\eta_i(t)Q_i(t) + \theta_i(t)\sum_{j\in\Omega_i} s_j(t) + \epsilon_i(t)\Big]. \tag{4}$$

If the outcome of this decision is 1, the agent will be bullish and expect the asset price to rise. As such, he will buy one share of the asset. If the outcome of an agent's decision is -1, the agent will be bearish and expect the asset price to fall. In this case, he will sell the one share of the asset. Otherwise, he will hold his assets. The main components in the Eq. (4) are introduced in detail below.

- News reaction term $\eta_i(t)Q_i(t)$—The news reaction of an agent i implies the sensitivity on the news he get, $\eta_i(t)$, and the impact of news on future prices, $Q_i(t)$. We define the inflow of news arriving in the market as a signal $I(t)$ with $I(t) \sim N(0, \sigma_I^2)$. $I(t)$ is a forecast of the future return and each agent has to decide whether the news is significant or not. To describe the heterogeneity, the news sensitivity $\eta_i(t)$ is allowed to float randomly on a uniform distribution,

$$\eta_i(t) = \sigma_I[1 + U(0, 1)]. \quad (5)$$

At each time step, each agent i has a probability $0 \leq p_i \leq 1$ of updating $\eta_i(t)$. When an agent updates his news sensitivity, he sets it to be equal to the recently observed absolute return, $|r(t)| = |ln\frac{S(t)}{S(t-1)}|$, where $S(t)$ and $r(t)$ are the market price and market return of stock, respectively. Introducing I.I.D random variables $u_i(t), i \in N$, uniformly distributed on [0, 1], the updating scheme is represented as

$$\eta_i(t) = \begin{cases} |r(t)|, & \text{if } u_i(t) < \mathrm{p}_i; \\ \eta_i(t-1), & \text{otherwise.} \end{cases} \quad (6)$$

The news-meaning, $Q_i(t)$, is assumed that news about the stock can either be good or bad. Specifically, the news value presented to the market that is >0 is attributed a value of 1, otherwise, it is set to −1. The formula of $Q_i(t)$ is represented by

$$Q_i(t) = \mathbb{1}_{I(t)>0} - \mathbb{1}_{I(t)<=0}. \quad (7)$$

- Propensity term $\theta_i(t)\sum_{j\in\Omega_i} s_j(t)$—The propensity of agent i implies the trust (decision sensitivity), $\theta_i(t)$, and the sum of decisions of agent i's neighbors, $\sum_{j\in\Omega_i} s_j(t)$. We assume that the agent is sensitive to the decisions of his neighbors in the agents' group. In this way, the decision of each agent influences, and is influenced by the decisions of other agents. The trust variable is initially set to base trust σ_θ, $\theta_i(t) \sim U(0, \sigma_\theta)$. At each time step t, each agent i also has a probability $0 \leq p_i \leq 1$ of updating $\theta_i(t)$

$$\theta_i(t) = \begin{cases} |r(t)|, & \text{if } u_i(t) < p_i; \\ \theta_i(t-1), & \text{otherwise.} \end{cases} \quad (8)$$

  Moreover, when an agent updates his trust, he decides to switch to another group. For each agent i, we denote his neighborhood which contains all other agents that directly connect to him as Ω_i. The sum $\sum_{j\in\Omega_i} s_j(t)$ is the sentiments of all agents j who are in direct contact with agent i.
- Idiosyncratic term $\epsilon(t)$—This variable is included to aid in the representation of the heterogeneous preferences held by agents. In other words, this variable is the outlook that an agent brings to the market, whether he is inherently bullish or bearish, optimistic, or pessimistic, regardless of market behavior Gonçalves (2003). We take it as being normally distributed around zero, $\epsilon_i \sim N(0, \sigma_\epsilon^2)$.

In summary, the main characteristics of the agents are described as follows. At each time step t:

- Agents receive a public information as a signal $I(t)$.
- Each agent i compares the signal to his news sensitivity $\eta_i(t)$ to make his news reaction.
- The agent then looks at the propensity of neighbors' decisions.

- He make decision by combining the news reaction, the propensity, and his own interpretation.
- After calculating the market return, he updated his sensitivities with a probability p_i. If the agent updates his trust, he decides to switch to another group.

3.2 The Environment

The model considers a fixed amount of agents, denoted as N, trading a single stock. At each time step t, a given trader i can choose between three actions: buys/sells one unit of the stock or remains inactive, $s_i(t) \in \{1, -1, 0\}$, respectively. We assume that each agent always has enough wealth and shares. In other words, all agents can always buy or sell.

At each time step t, the excess demand $D(t)$ is calculated as the number of buyers minus the number of sellers, which is formed as

$$D(t) = \sum_{i=1}^{N} s_i(t). \tag{9}$$

The market price and market return of stock are then calculated. The market price $S(t)$ is updated according to

$$S(t) = S(t-1)[1 + r(t)], \tag{10}$$

and return $r(t)$ is determined according to

$$r(t) = g\left(\frac{D(t)}{N}\right), \tag{11}$$

where the price impact function $g : \mathbb{R} \mapsto \mathbb{R}$ is an increasing function with $g(0) = 0$. We follow Cont (2007) to define the market depth λ by: $g(0) = 1/\lambda$. In our model, the chosen price impact function is $g(z) = arctan(z/\lambda)$.

4 Hyperparameter Tuning

We construct an artificial stock market of 1000 traders. In order to interpret the trading periods as days and compare the results with properties of daily returns, we choose $\lambda \in [5, 20]$ which allows a range of daily returns between 5 and 20%. Denote $1/p_i$ as the typical time period during which an agent i will hold a his news sensitivity $\eta_i(t)$. p is chosen in $[0.001, 0.1]$ to be interpreted as days. For example, $p = 0.1$ means the average updating period of agents is 10 days. The standard deviation of the signal,

Table 1 Description of parameters

Parameters	Description	Value
N	Number of agents	1000
λ	Market depth	$[5, 20]$
p	Fraction of agents updating their news reaction	$[0.001, 0.1]$
σ_θ	Trust base	$[0.1, 1]$
σ_I	Standard deviation of the inflow of news	$[0.001, 0.01]$
σ_ϵ	Standard deviation of idiosyncratic term	$[0.01, 1]$

σ_I, can be chosen in the range $[0.001, 0.01]$ to reproduce a realistic range of values for the annualized volatility Cont (2007). The value of the trust base is chosen in the range $[0.1, 1]$. The value of standard deviation in idiosyncratic term is chosen in the range $[0.01, 1]$. The parameters and ranges are summarized in Table 1.

We formulate the optimization problem using '*hyperopt package*' of Python with three main ingredients

- Domain space: Each trader is assigned with initial values derived from random uniform distributions in within the range of Table 1, respectively. Our approach is to start the domain from a wide range and then focus on specific areas around the optimal parameters calculated from the previous run.
- Objective function: return-based and price-based objective functions are used to find the best value of hyperparameters that minimize the loss in 100 iterations.
- Surrogate function and selection function: we follow Snoek et al. (2012) to use Tree Parzen Estimator as the surrogate function and Expected Improvement criterion as the selection function.

5 Experiments

As an empirical basis, we use a data set that covers 4 years of daily quotes of the VN Index, from January 2017 to January 2021. This period contains 1000 daily data, of which the main statistical properties are given in Table 4. All numerical simulation and graphical visualization are done by using Python 3.6.5 version on Intel Core i7-4790 3.6 GHz. The focus of the analysis is on statistical properties of the random walk price dynamics, distribution of stock returns, and the volatility clustering in the market. The visualization of simulations is displayed in Fig. 1.

In this work, we do not compare values of the objective functions due to differences in measurement definitions. The better objective function is picked out by investigating their cost and stability over many simulations. Both KS test and DTW

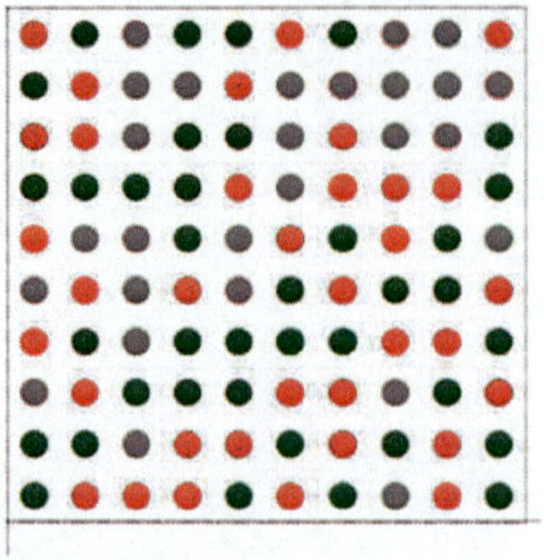

Visualization of agents in 2-Dimensions. Green: buy, Red: sell, Grey: hold.

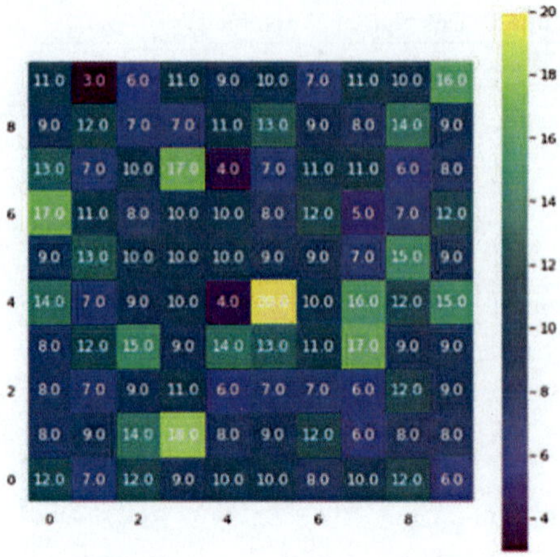

Distribution of number of agents.

Fig. 1 Visualization results

Table 2 Optimized parameters

Method	Optima value	Parameters				
		λ	p	σ_θ	σ_I	σ_ϵ
KS	0.05	12	0.088	0.1	0.001	0.14
DTW	77.74	17	0.037	0.1	0.004	0.34

Table 3 Comparision of cost and stability

Method	Confidence interval	Running time (h)
KS	0.05–0.07	2.78
DTW	77–100	3.19

algorithm can find the optimal set of parameters with a small simulation error as shown in Table 2.

After 100 simulations, the DTW algorithm represents the instability, the optimal value can range from 77 to 100 (see Table 3). In contrast, the KS test shows stability with small fluctuations in the range of 0.05–0.07. Moreover, the execution time of KS is shorter than that of DTW, which means that the KS test costs less than DTW algorithm. Figure 2 shows that the KS convergence is faster than the DTW, and generally, the KS is more stable than the DTW algorithm, because the standard deviation of optimal observed objective is smaller after 20 iterations.

In addition, both algorithms can reproduce some stylized facts of the real market. The results are summarized as follows.

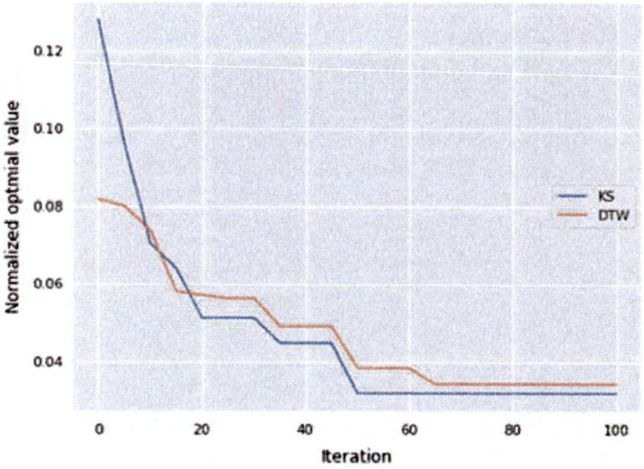

Fig. 2 Comparison of convergence rate

5.1 *Dynamics of the Prices*

First, we investigate the random walk price dynamics in financial markets. The simulated prices are computed from price-based optimization algorithm. Figure 3 shows the daily movement of real prices and simulated prices. Both series have an upward trend, the mean and variance change over time, and prices do not tend to come back towards their average.

The autocorrelation function (ACF) plots in Fig. 4 demonstrate that the correlation between the time series and its lags fall outside the 95% of the confidence interval of no autocorrelation for 50 lags. Therefore, the results confirm that both prices are random walk and the simulated prices are similar to the dynamics of prices in the real market.

5.2 *Distribution of Returns*

The simulated returns are computed from a return-based optimization algorithm. The statistical analysis of the actual returns and simulated returns are given in Tables 4 and 5. The simulated returns have similar properties to the actual returns. We consider

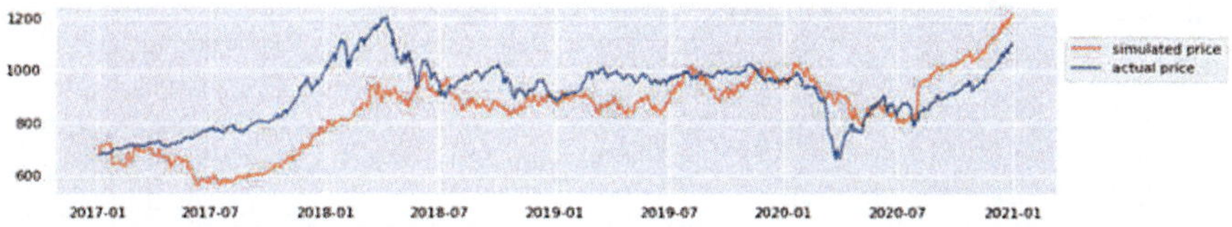

Fig. 3 Simulation of prices

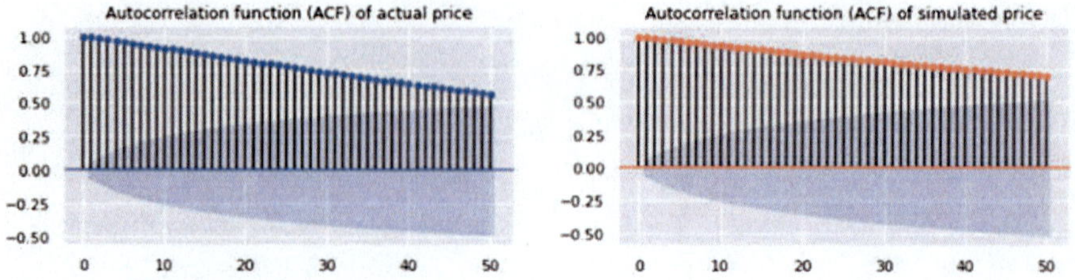

Fig. 4 Autocorrelations of prices

a significance test if there is evidence for a difference θ between two returns, with the null hypothesis $H_0 : \theta = 0$ and the alternative hypothesis $H_1 : \theta \neq 0$. Let $D := r_{actual} - r_{simulated}$, where r_{actual} and $r_{simulated}$ are the actual returns and simulated returns, respectively. The Bayes factor $BF_{10}(D)$ stands for "H_1 versus H_0", is given by:

$$BF_{10}(D) = \frac{P(D|H_1)}{P(D|H_0)}.$$

The result is computed from the R packages "BayestestR", which gives: $BF_{10} = 0.036$. So, $BF_{01} = 1/0.036 = 27.03$. A value of $BF_{01} > 1$ means that H_0 is more strongly supported by the data under consideration than H_1. Using the scale for interpreting Bayes factors by Jeffreys (1998), we obtain a strong evidence that there is no difference between the actual returns and simulated returns.

In financial data, the excess kurtosis of return is larger than 0, indicating that the distribution of returns displays a heavy tail Lux and Alfarano (2016). The quantile-quantile plots in Fig. 5 show that the actual returns and simulated returns are not normally distributed. The returns produced by the model are consistently leptokurtically distributed with fat tails as shown in Fig. 6.

Table 4 Comparision of statistical properties

Data	Mean (%)	Standard deviation (%)	Kurtosis	Skewness
Real returns	0.05	1.10	5.62	−1.04
Simulation	0.03	1.22	5.17	−1.16

Table 5 Comparing statistical properties (cont.)

Data	Min (%)	Quantiles (%)			Max (%)
		25	*50*	*75*	
Real returns	−6.27	−0.34	0.14	0.59	4.98
Simulation	−6.19	−0.43	0.00	0.33	6.60

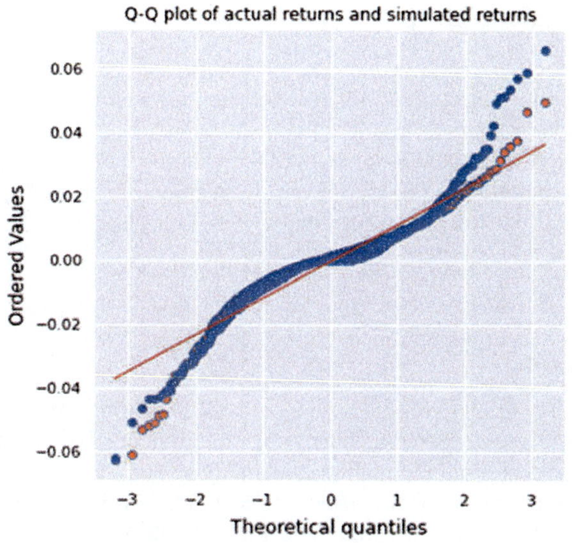

Fig. 5 Quantile-quantile plot of returns. Blue: actual returns, Coral: simulated returns

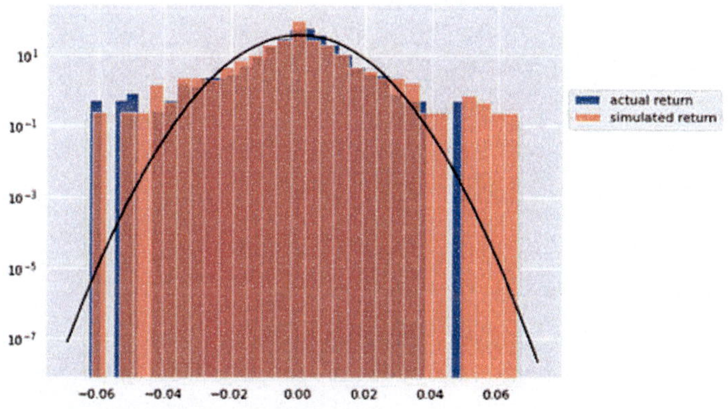

Fig. 6 Distribution of returns with logarithmic scaling on vertical axis

5.3 *Absence of Autocorrelations in Returns*

Figure 7 shows the plots of stock returns for both actual data and simulated data. ACF plots in Fig. 8 demonstrate that there is no correlation between the simulated returns and its lags after the first few lags.

5.4 *Volatility Clustering*

The next stylized fact we study is volatility clustering, which describes the fact that the tendency of large changes in prices to cluster together. A slow decay in the ACF of absolute values of returns is a signature of the volatility clustering that can be seen in Fig. 9. Notice that for both cases, the autocorrelation of absolute returns remains significantly positive over many time lags. This is a clear evidence of volatility clustering.

6 Conclusion

In this paper, we have proposed different objective functions to tune hyperparameters in an agent-based simulation of the stock market. The Bayesian calibration with Kolmogorov-Smirnov test has proposed an optimal set of parameter with high stability and low cost. The model with a return-based objective algorithm was able to

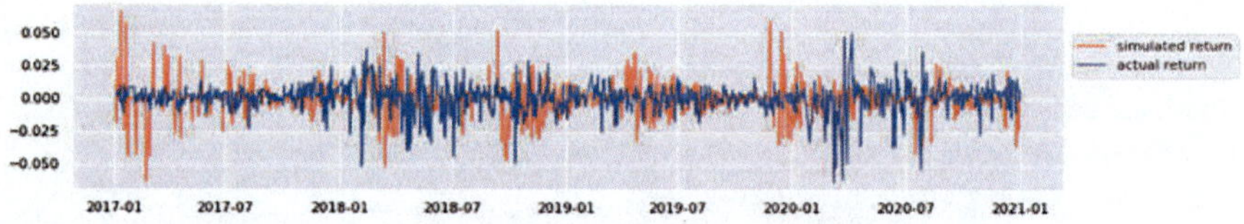

Fig. 7 Simulation of returns

Fig. 8 Autocorrelations of simulated returns

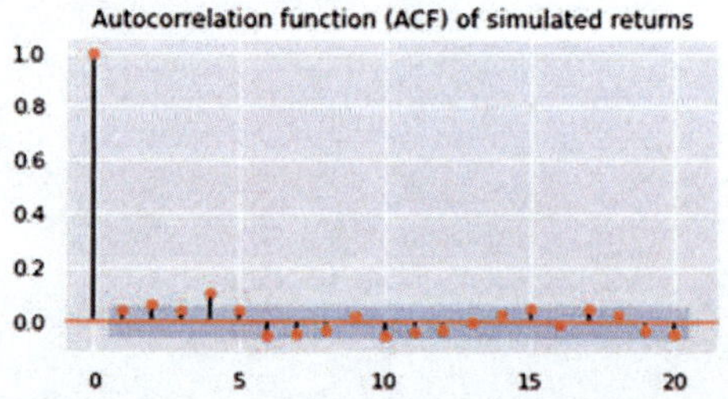

Fig. 9 Autocorrelations of absolute value of simulated returns

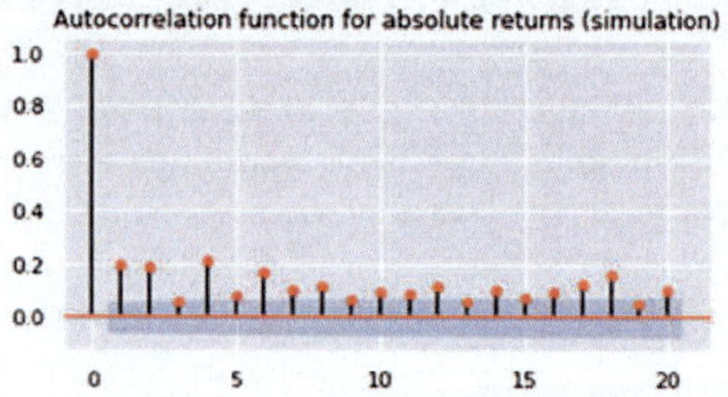

highlight some of the important stylized facts of the stock returns such as the absence of autocorrelations, the heavy tail of the returns, and volatility clustering. Besides, the dynamics of stock prices in the real market are reproduced when simulated with the price-based objective algorithm. In the future, we can adjust to combine the two objective functions to get results for both prices and returns.

Acknowledgements We thank Prof. Vladik Kreinovich for the Bayes factor approach. This work was supported by the John von Neumann Institute—Vietnam National University, Ho Chi Minh, Vietnam.

References

Bergstra, J., Bengio, Y.: Random search for hyper-parameter optimization. J. Mach. Learn. Res. **13**(2) (2012)

Berndt, D.J., Clifford, J.: Using dynamic time warping to find patterns in time series. In: KDD Workshop, vol. 10, no. 16, Seattle, WA, USA, pp. 359–370 (1994)

Challet, D., Marsili, M., Zhang, Y.-C.: Stylized facts of financial markets and market crashes in minority games. Physica A: Stat. Mech. Its Appl. **294**(3–4), 514–524 (2001)

Claesen, M., De Moor, B.: Hyperparameter search in machine learning. arXiv preprint arXiv:1502.02127 (2015)

Cont, R.: Volatility clustering in financial markets: empirical facts and agent-based models. In: Long Memory in Economics, pp. 289–309. Springer (2007)

Gonçalves, C.: Artificial financial market model (2003). http://www.ccl.northwestern.edu/netlogo/models/community/ArtificialFinancialMarketModel

Jeffreys, H.: The Theory of Probability. OUP, Oxford (1998)

Kousiouris, G., Cucinotta, T., Varvarigou, T.: The effects of scheduling, workload type and consolidation scenarios on virtual machine performance and their prediction through optimized artificial neural networks. J. Syst. Softw. **84**(8), 1270–1291 (2011)

Larochelle, H., Erhan, D., Courville, A., Bergstra, J., Bengio, Y.: An empirical evaluation of deep architectures on problems with many factors of variation. In: Proceedings of the 24th International Conference on Machine Learning, pp. 473–480 (2007)

Larsen, J., Hansen, L.K., Svarer, C., Ohlsson, M.: Design and regularization of neural networks: the optimal use of a validation set. In: Neural Networks for Signal Processing VI. Proceedings of the 1996 IEEE Signal Processing Society Workshop, pp. 62–71. IEEE (1996)

Lux, T., Alfarano, S.: Financial power laws: empirical evidence, models, and mechanisms. Chaos Solit. Fractals **88**, 3–18 (2016)

Marsaglia, G., Tsang, W.W., Wang, J., et al.: Evaluating Kolmogorov's distribution. J. Stat. Softw. **8**(18), 1–4 (2003)

Snoek, J., Larochelle, H., Adams, R.P.: Practical Bayesian optimization of machine learning algorithms. Adv. Neural Inform. Process. Syst. **25** (2012)

Sornette, D., Zhou, W.-X.: Importance of positive feedbacks and overconfidence in a self-fulfilling Ising model of financial markets. Physica A: Stat. Mech. Its Appl. **370**(2), 704–726 (2006)

Tran, M., Duong, T., Pham-Hi, D., Bui, M.: Detecting the proportion of traders in the stock market: an agent-based approach. Mathematics **8**(2), 198 (2020)

Shadow Economy, Corruption, and Economic Growth: A Bayesian Analysis

My-Linh Thi Nguyen, Toan Ngoc Bui, Tung Duy Thai, Thuong Thi Nguyen, and Hung Tuan Nguyen

Abstract The objective of this paper is to analyze the impact of shadow economy and corruption on economic growth. The data were collected from 10 ASEAN countries in the period 2002–2019, including: Brunei Darussalam, Indonesia, Cambodia, Lao PDR, Myanmar, Malaysia, the Philippines, Singapore, Thailand, and Vietnam. The Bayesian method is used to estimate the research model. This method is an approach with many outstanding advantages but it is rarely utilized in empirical studies. The results show that economic growth is negatively impacted by shadow economy and is positively impacted by corruption control. In other words, the existence of corruption impacts negatively on economic growth. The probabilities of these impacts are 96.36 and 100%, respectively. Accordingly, shadow economy hinders economic growth, while less corruption can significantly improve economic growth. In particular, this study found a negative but rather low impact of the interaction variable between shadow economy and corruption control on economic growth, with the impact probability of 78.32%. This indicates that improving the level of corruption control can

M.-L. T. Nguyen
Faculty of Finance and Banking, University of Finance – Marketing (UFM), Ho Chi Minh City, Vietnam
e-mail: ntmylinh@ufm.edu.vn

T. N. Bui · T. D. Thai · T. T. Nguyen (✉)
Faculty of Finance and Banking, Industrial University of Ho Chi Minh City (IUH), Ho Chi Minh City, Vietnam
e-mail: nguyenthithuong@iuh.edu.vn

T. N. Bui
e-mail: buingoctoan@iuh.edu.vn

T. D. Thai
e-mail: thaiduytung@iuh.edu.vn

H. T. Nguyen
School of Business and Management, RMIT University, Ho Chi Minh City, Vietnam
e-mail: s3806456@rmit.edu.vn

N. Ngoc Thach et al. (eds.), *Financial Econometrics: Bayesian Analysis, Quantum Uncertainty, and Related Topics*, Studies in Systems, Decision and Control 427,
https://doi.org/10.1007/978-3-030-98689-6_49

help limit the negative impact of shadow economy on economic growth in ASEAN countries, which is an interesting finding of this study. In addition, this study also found a significant impact of the control variables of domestic credit, government expenditure, foreign direct investment, population growth, and inflation on economic growth.

Keywords ASEAN · Bayesian · Corruption · Economic growth · Shadow economy

1 Introduction

In recent years, the nations in Southeast Asia (ASEAN) have experienced rapid economic growth. ASEAN countries overcame the Asian financial crisis of 1997 and the global economic crisis of 2008–2009 to become one of the top 5 economies in the world.

According to the study by Medina and Schneider (2018), the long-term average growth of the shadow economy in Southeast Asia since the early 1990s has been recorded to be generally at a high level. Specifically, in the period 1991–2015, the activity of shadow economy reached 33.4% of regional GDP. This figure was much higher than the 21.2% average of the broader East Asian region including China, South Korea and Japan, and also higher than the global average of 31.9%. According to the Corruption Perception Index (CPI), except for Singapore, the remaining 9 countries in the ASEAN region are at an average level; especially, this situation is quite serious in Cambodia and Myanmar.

It can be said that shadow economy and corruption are two factors that always exist and have more and more impacts on economic growth in ASEAN countries. Therefore, in order to stimulate economic growth effectively, it is essential to consider the impact of shadow economy and corruption on economic growth. This impact has been found in many empirical studies. For example, Schneider and Bajada (2003), Schneider and Enste (2013), Vo and Pham (2014) have stated that shadow economy has an impact on economic growth, but this relationship is not consistent. Alm and Embaye (2013) indicated that shadow economy has a negative impact on economic growth. Some studies such as Schneider et al. (2010), Buehn and Schneider (2008), Williams and Schneider (2016), Buehn and Farzanegan (2012) as well as Dell'Anno et al. (2007) have concluded that there is a negative relationship between the size of the shadow economy and the formal economy.

Shadow economy and corruption can interact. Enterprises will be more motivated to participate in the corruption-based shadow economy. This means increased corruption leads to an increase in the size of the shadow economy (Friedman et al. 2000). In contrast, Choi and Thum (2005), Dreher et al. (2009) have provided a model showing that corruption will decrease when enterprises participate in shadow economy because in this sector, enterprises will not need to give bribes in economic transactions.

In general, shadow economy and corruption may be interrelated. If this happens, the impact mechanism of these two factors on economic growth may be more complicated. Although this is possible in reality, most previous studies have not mentioned this issue. Therefore, if this relationship is identified, there will be a reliable basis for proposing practical and synchronous solutions to stimulate economic growth in a stable and sustainable way. Accordingly, ASEAN countries will limit the negative impacts of shadow economy and corruption on economic growth.

Overall, it can be seen that most of previous studies examine the impact of shadow economy on economic growth. However, the impact of both shadow economy and corruption on economic growth has not been investigated, and the impact of the interaction between shadow economy and corruption on economic growth has not been explored. Finding the limitations in empirical studies and the urgency of this issue both in theory and in practice, based on the results of the research paper "Shadow economy, Corruption and Economic Growth: A Bayesian Analysis" in 10 ASEAN countries from 2002 to 2019, the authors expect to provide many interesting findings compared to previous studies. In order to achieve the research objectives, the authors use the Bayesian Method to estimate the research model. This method is an approach with many outstanding advantages but it is rarely utilized in empirical studies. This will be a reliable basis for the authors to propose solutions in a synchronous manner to limit the negative impacts of the shadow economy and corruption on economic growth in Vietnam.

2 Literature Review

2.1 Impacts of Shadow Economy on Economic Growth

The negative impacts of shadow economy on economic growth can be found in numerous previous studies. First of all, the development of shadow economy influences resource allocation, making income distribution in the society become uncontrollable, and lowering tax revenue (Alm and Embaye 2013). Loayza (1996) found the evidence supporting the decrease in quality of public goods and services, and living standard as the result of araising shadow economy. Moreover, the expansion of shadow economy in terms of size would reduce tax revenue (Kodila-Tedika and Mutascu 2013), and the decrease in tax revenue would result in lower quality of public goods provided by the government (Broms 2011). Second, shadow economy may indirectly show a negative impact on the official economy by discouraging innovation. Shadow economy is described as low productive activities (Friedman et al. 2000; Ihrig and Moe 2004; and La Porta and Shleifer 2014) because this sector allows outdated technology, or most of the shadow economic activities related to consumer service industries, in essence, do not require updating and innovating technology.

The works of Dell'Anno et al. (2007) in France, Greece and Spain, Buehn and Schneider (2008), Schneider et al. (2010) taking sample of 114 countries, Buehn and

Farzanegan (2012), as well as Williams and Schneider (2016), concluded that there were a negative relationship between the size of the shadow economy and the official economy. The increase of the shadow economy results in the decline of the official economy, shown as the fall of GDP growth, because resources and productivity factors taken by the shadow economy from the official sector lead to the decrease of the latter (Alanon and Go'mez-Antonio 2005; Schneider and Enste 2013). In addition, De Paula and Scheinkman (2011) employing the direct survey method on 50,000 small business owners in Brazil, deduced that most small business owners joined the shadow economy to avoid tax payments at the expense of higher cost of capital, less ability of expansion, higher use of human labour, and less capability of innovation than the official side.

2.2 *Impacts of Corruption on Economic Growth*

Impacts of corruption on economic growth are analysed in terms of positiveness and negativeness, that Méon and Sekkat (2005) called "grease the wheels" and "sand the wheels" hypotheses. From the viewpoint of "grease the wheels" hypothesis which was suggested by the previous works of Leys (1965), Bayley (1966), Huntington (1968), Lui (1985), and Beck and Maher (1986), for instance, corruption enhances economic growth by reducing time for waiting to complete administative procedure, limiting the negative effects of bureaucracy, also raising government officials and agents's income to a more adequate level. On the other hand, a remarkable number of studies which followed the "sand the wheels" hypothesis argued that corruption might be harmful to the country at the cost of economic growth. According to Méon and Sekkat (2005), so-called "grease the wheels" of corruption could be diminished in a centralized decision making process. In the other words, corrupted agents might decrease the effectiveness of the system. The fact that too many government official get involved in an investment project might end up with higher cost and demotivate new profitable plans (Shleifer and Vishny 1993). In addition, Mauro (1995) found that corruption reduced investment in private sector, as well as misallocate effective resources, resulting in the slower growth rate under the assumption of endogenous growth theory proposed by Romer (1990). Also, Méon and Sekkat (2005) concluded that more public investment could be distributed into ineffective projects as the result of corruption. Another negative and indirect impact of corruption on economic growth lies in the less spending on education, according to Mauro (1995) and Pellegrini and Gerlagh (2004). Under the endogeneous growth theory, human capital is considered to be one of the main internal factors of economic growth (Romer 1990; Shleifer and Vishny 1993). Therefore, the less investment in human capital is made, the less economic growth can be achived.

Evidence of negative impacts of corruption on economic growth can be found in the previous papers of Méon and Sekkat (2005), d'Agostino et al. (2016), Cieślik and Goczek (2018), Gründler and Potrafke (2019), and Son et al. (2020). The findings of Méon and Sekkat (2005) supported the "sand the wheels" hypothesis and concluded that negative impact of corruption on economic growth might be even stronger in an unstable economy, in line with the statement of Mauro (1995). d'Agostino et al. (2016) used the sample of African countries to provide evidence of corruption's negative effect on economic growth within military burdens. Later, Cieślik and Goczek (2018) with the sample of 142 countries under the period between 1994 and 2014 concluded that uncertainty caused by corruption discouraged investment, and thus reduced economic growth in researched countries. The authors also inferred that the control of corruption ultimately improved investment and economic growth. Gründler and Potrafke (2019) estimated a decrease of 17% in economic growth for each change in standard deviation of corruption perception index. Son et al. (2020) found a negative impact of corruption on economic growth by means of the banking system in 120 countries under the study.

2.3 *Interactive Relationship of Shadow Economy, Corruption and Economic Growth*

Previous studies about impacts of shadow economy and corruption on economic growth mainly showed the same and negative direction. In addition, the statistics made by Eilat and Zinnes (2000) described that most countries of which the size of shadow economy higher than 40% of GDP were developing countries in Asia and Africa, with low GDP per capita. The fact was supported by empirical evidences in the works of Borlea et al. (2017) and Hoinaru et al. (2020). As Chong and Calderon (2000) stated, low institutional quality may be among the reasons of low economic growth. Therefore, it raised a concern that there might be a interaction relationship of shadow economy and corruption which influenced economic growth. Friedman et al. (2000) gave an exlanation for this combined impact by using the function of tax payers motivation. Accordingly, tax payers in high corrupted countries would pay less tax and take advantage of corruption situation to misreport their business activities, resulting in a larger unofficial economic sector.

Baklouti and Boujelbene (2020) believed that shadow economy and corruption complemented each other to make negative impact on economic growth. Their research result confirm previous finding of Dreher and Schneider (2010), and Cooray et al. (2017) who gave evidences of combined impacts of corruption and shadow economy in low-income countries.

3 Econometric Methodology and Data

3.1 Econometric Methodology

The results of previous studies show that economic growth is significantly impacted by shadow economy (Dell'Anno et al. 2007; Buehn and Schneider 2008; Schneider et al. 2010; Buehn and Farzanegan 2012; Shahid and Khan 2020; Esaku 2021; Özgür et al. 2021) and corruption control (Mauro 1995; d'Agostino et al. 2016; Cieślik and Goczek 2018; Gründler and Potrafke 2019; Son et al. 2020). In particular, Baklouti and Boujelbene (2020) argue that the interaction between shadow economy and corruption control can have a significant impact on economic growth. In addition, economic growth can also be impacted by the control variables of domestic credit (Beck and Levine, 2004; Arcand et al. 2015; Ruiz 2018), government expenditure (d'Agostino et al. 2016; Gründler and Potrafke 2019; Son et al. 2020), foreign direct investment (Cieślik and Goczek 2018, and Gründler and Potrafke 2019), population growth (Méon and Sekkat 2005; Ruiz 2018; Son et al. 2020), and inflation (Ruiz 2018; Ehigiamusoe et al. 2019). Based on this, the authors build a research model with the following equation:

$$\begin{aligned}\mathrm{Y_{it}} = \beta_0 + \beta_1\,\mathrm{SE_{it}} + \beta_2\,\mathrm{CC_{it}} + \beta_3\,\mathrm{SE_{it}} \times \mathrm{CC_{it}} + \beta_4\,\mathrm{DC_{it}} + \beta_5\,\mathrm{GOV_{it}} + \beta_6\,\mathrm{FDI_{it}} \\ + \beta_7\,\mathrm{PG_{it}} + \beta_8\,\mathrm{INF_{it}} + \mathrm{e_{it}}\end{aligned}$$

where Y is economic growth, measured through the logarithm of GDP per capita according to the approach of Cieślik and Goczek (2018), Gründler and Potrafke (2019). Shadow economy (SE) is measured by the ratio of the size of the shadow economy to official GDP as used by Elgin and Birinci (2016), Borlear et al. (2017), Wu & Schneider (2019). Corruption control (CC) shows the level of corruption control of each country, ranging from 0 (lowest) to 100 (highest) in accordance with the work of d'Agostino et al. (2016) and Cieślik and Goczek (2018). Domestic credit (DC) is domestic credit to the private sector (% of GDP) (Beck and Levine, 2004; Arcand et al. 2015; Ruiz 2018). Government expenditure (GOV) is the general government final consumption expenditure (% of GDP) as defined in the studies of d'Agostino et al. (2016), Ruiz 2018, Gründler and Potrafke (2019). FDI is foreign direct investment, defined as net capital input (% of GDP) (Cieślik and Goczek 2018; Gründler and Potrafke 2019). PG is the annual growth of population growth (Mauro 1995; Méon and Sekkat 2005; Ruiz 2018; Son et al. 2020). INF is inflation, determined by the annual growth of consumer prices (Ruiz 2018; Ehigiamusoe et al. 2019).

In this study, the authors use the Bayesian method to estimate the research model. Accordingly, the posterior distribution is a combination of the prior distribution information and the collected data, which increases the robustness of the estimation results and can overcome limitations in small data samples (McNeish 2016; Thach 2020a, b, 2021). Although there are many outstanding advantages, the Bayesian method is still rarely used in previous studies on the impact of shadow economy and corruption control on economic growth. This resulted in the authors not having information about the prior distribution to use in this study. Therefore, the authors used a normal distribution and an inverse-gamma distribution based on Lemoine's (2019) point of view, with the default size of 10,000.

3.2 Data

For the research data, the authors used the data sample of 10 ASEAN countries (Brunei Darussalam, Indonesia, Cambodia, Lao PDR, Myanmar, Malaysia, the Philippines, Singapore, Thailand, and Vietnam) in the period 2002–2019. The corruption control index published by the World Bank contains full annual data since 2002; therefore, the authors were only able to collect the data sample in this period. The authors collected the data on shadow economy from the International Monetary Fund (IMF), while the data on the remaining variables were from the World Bank source.

4 Results and Discussion

4.1 Results

Table 1 shows the description results of the data sample of 10 ASEAN countries in the period 2002–2019.

Table 1 shows that SE has the average value of 30.9428%, the lowest value (9.4%) belongs to Singapore in 2012, the highest value (54.1%) belongs to Cambodia in 2002. CC has the average value of 40.4068%, the lowest value (0.47%) belongs to Myanmar in 2011, the highest value (99.52%) belongs to Singapore in 2019. Regarding Y, the average value of this indicator is 8.1234 (equivalent to 10,134.50

Table 1 Description of the data sample

Variable	Mean	Standard deviation	Min	Max
Y	8.1234	1.4911	4.9564	11.1003
SE	30.9428	12.4445	9.4000	54.1000
CC	40.4068	28.3140	0.4700	99.5200
DC	61.8151	45.4892	3.1200	149.3700
GOV	11.8505	5.1520	3.4600	27.1700
FDI	5.5088	5.9953	-1.3200	28.6000
PG	1.3167	0.6466	-1.4700	5.3200
INF	4.7939	6.7554	-2.3100	57.0700

Table 2 The estimation results of the research model through the Bayesian method

Variable	Mean	Std. Dev	MCSE	Median	Equal-tailed [95% Cred. Interval]	
SE	−0.0110	0.0062	0.0000	−0.0110	−0.0231	0.0011
CC	0.0354	0.0046	0.0000	0.0354	0.0264	0.0445
SE × CC	−0.0001	0.0002	0.0000	-0.0001	-0.0005	0.0002
DC	0.0026	0.0013	0.0000	0.0026	0.0001	0.0051
GOV	0.0837	0.0109	0.0001	0.0836	0.0624	0.1053
FDI	0.0180	0.0112	0.0001	0.0180	-0.0042	0.0399
PG	0.1189	0.0652	0.0004	0.1191	-0.0102	0.2454
INF	−0.0330	0.0065	0.0000	-0.0330	-0.0459	-0.0202
_cons	5.9314	0.2419	0.0014	5.9313	5.4582	6.4109
var	0.2277	0.0248	0.0002	0.2260	0.1839	0.2807
Avg acceptance rate	1					
Avg efficiency: min	0.908					
Max Gelman-Rubin Rc	1					

USD), Singapore reached the highest value (11.1003, equivalent to 66,188.78 USD) in 2018, whereas Myanmar reached the lowest value (4.9564, equivalent to 142.08 USD) in 2002.

The estimation results of the research model through the Bayesian method are presented in Table 2.

Table 2 shows that the Avg acceptance rate is 1, exceeding the allowable minimum of 0.1 (Roberts and Rosenthal 2001). Regarding the Max Gelman-Rubin Rc, this is the largest statistical value based on all the model parameters, and the estimation result is guaranteed to converge when the convergence statistics of all the parameters in the model are less than 1.1 (Gelman and Rubin 1992). The estimation results indicate that Max Gelman-Rubin Rc is equal to 1; therefore, the convergence in the model is satisfied (satisfied). In particular, the Avg efficiency (min) reached a value of 0.908, exceeding the allowable minimum of 0.01. The higher efficiency means the smaller Monte Carlo Standard Error (MCSE), resulting in a more accurate posterior mean estimate. The estimation results reveal that the MCSE of all variables is very small, much smaller than the allowable maximum of 0.05, which ensures the robustness of the model (Flegal et al. 2008).

On the other hand, Fig. 1 shows a relatively perfect trace plot. The Autocorrelation histogram demonstrates low autocorrelation, mainly around the sub-0.02 level. At the same time, the Histogram and Density plots also simulate the shape of the normal distribution. Overall, the estimation results have met the criteria for the model acceptance. In other words, the model estimation results are suitable and usable.

Based on Table 2, it is seen that Y is negatively impacted (−0.0110) by SE and is positively impacted (0.0354) by CC. In particular, the interaction variable SE × CC has a negative impact on Y at a rather low level (−0.0001). This shows that shadow

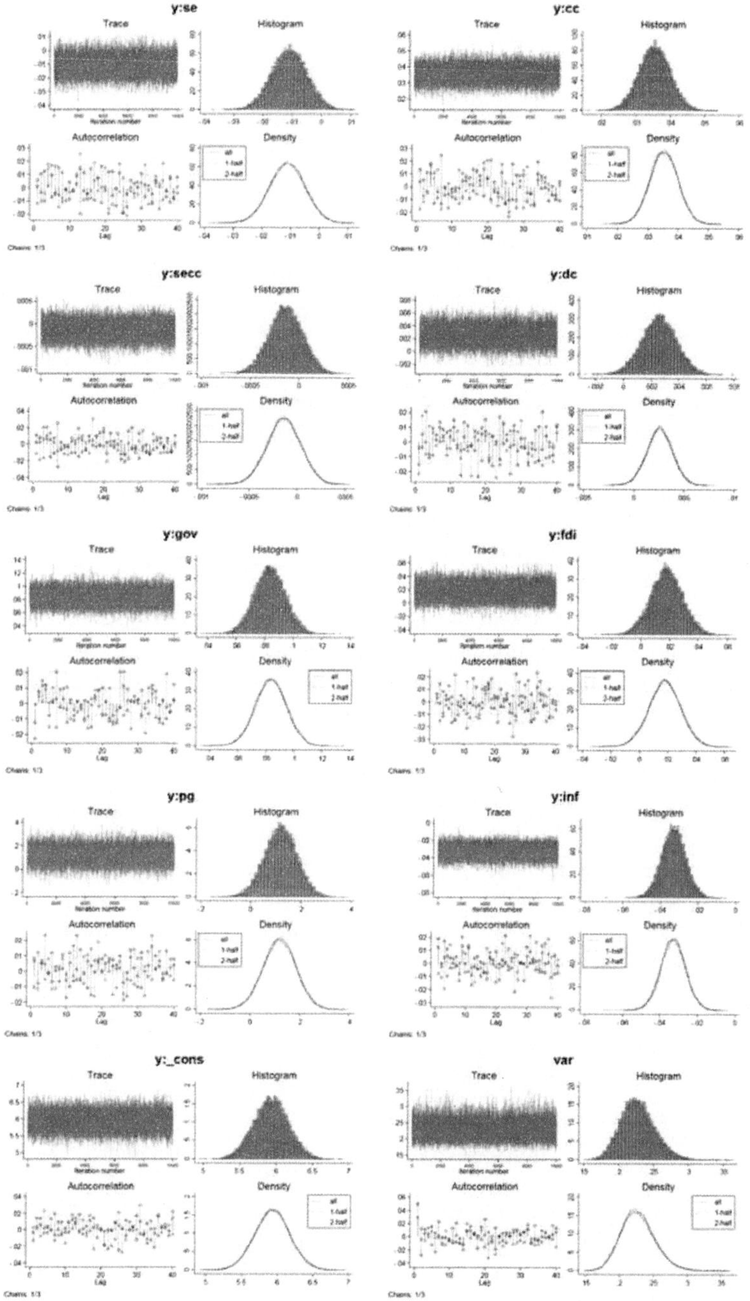

Fig. 1 The results of the Bayesgraph diagnostics

Table 3 The results of the Bayestest interval

Interval tests	Mean	Std. Dev	MCSE
SE	0.9636	0.1874	0.0011
CC	1.0000	0.0000	0.0000
SE × CC	0.7832	0.4121	0.0024
DC	0.9777	0.1477	0.0009
GOV	1.0000	0.0000	0.0000
FDI	0.9434	0.2310	0.0013
PG	0.9656	0.1823	0.0011
INF	1.0000	0.0000	0.0000

economy hinders economic growth. Meanwhile, corruption control can help improve economic growth. Furthermore, the combination of shadow economy and corruption control can significantly reduce the level of the negative impact of shadow economy on economic growth. It can be said that this is the interesting finding of this study. In addition, Y is positively impacted by the control variables of DC (0.0026), GOV (0.0837), FDI (0.0180), and PG (0.1189). Meanwhile, the control variable of INF has a negative impact (−0.0330) on Y.

Table 3 shows the impact probability of independent variables and control variables on the dependent variable. In other words, we can see the probabilities of events occurring, which is also a major advantage of the Bayesian method. Accordingly, the probability that SE has a negative impact on Y is 96.36%. Meanwhile, the probability that CC has a positive impact on Y is 100%. The probability that the interaction variable SE × CC has a negative impact on Y is 78.32%. Although this value is lower than the probability that SE or CC has a separate impact on Y, the value of 78.32% is a considerable number. In addition, the impact probabilities of the control variables on the dependent variable Y are relatively high. Specifically, the probabilities that Y is positively impacted by the control variables of DC, GOV, FDI, and PG are 97.77, 100, 94.34, and 96.56%, respectively. Simultaneously, the probability that INF has a negative impact on Y is 100%.

4.2 *Discussion*

4.2.1 The Impact of Shadow Economy on Economic Growth

Based on the research results, our research results are consistent with Dell'Anno et al. (2007), Buehn and Schneider (2008), Schneider et al. (2010), Buehn and Farzanegan (2012), Williams and Schneider (2016), as well as recent studies by Borlea et al (2017), Shahid and Khan (2020), Esaku (2021), Özgür et al. (2021), that the size of the shadow economy has a negative impact on a nation's economic growth, as measured by GDP per capita growth rate.

The negative impact of shadow economy on economic growth is explained through the existence and development of the informal economy, which reduces government tax revenues, decreases public investment resources, and directly affects the quality of public goods and services. Based on the analysis of Eilat and Zinnes (2000), an increase in tax rates will lead to a decrease in tax revenues according to the Laffer Curve effect because a part of the formal economy will participate in the shadow economy. As a result, the government raises taxes, but it pushes many enterprises to take part in informal activities. This lessens the number of companies that receive funding from the budget, and they are more motivated to operate in the shadow economy. Companies in the informal sector also have difficulties in accessing funding allocated through the banking system due to the inability to disclose their financial status transparently, as well as the lack of collateral according to the standards of the bank. Therefore, the capital of the society will be concentrated on enterprises in the formal sector although these enterprises cannot be more efficient. Moreover, the existence of shadow economy creates unfair competitive advantage. Companies in the informal sector can pay higher wages (because they do not have to pay personal income taxes for employees), lower the selling price of their products and compete directly with formally operated enterprises.

The existence of the shadow economy also has a negative impact on economic growth because it creates related costs that are borne by direct consumers. Specifically, according to Kaufmann and Kaliberda (1996), in order to avoid tax and social obligations, organizations in the shadow economy incur costs to maintain this anonymity, including the time spent on bribes, avoiding the checks for administrative procedures and operating licenses, as well as seeking brokerage services brokerage services as an alternative to public services that they cannot access. Overall, the existence of the shadow economy incurs additional costs for the society, wastes public services and utilities, and reduces revenues to finance these purposes, causing the government to be unable to optimize its policies to support economic growth.

For the countries in Southeast Asia, one of the main drivers of the process of promoting economic growth and catching up with developed countries is to apply technological innovation thoroughly and improve productivity. However, this process can be slowed down by the impact of the size of the shadow economy. As analyzed above, the initial purpose of enterprises when actively participating in the shadow economy is to avoid tax obligations and ever-increasing welfare requirements for employees. Nevertheless, participation in underground economic activities also makes them lose the opportunity to access low-cost funding sources from official distribution channels of the government because they do not meet the requirements of information transparency. As a result, they tend to reduce asset capitalization, and focus on labor-intensive industries. This negatively affects the overall productivity of the economy, and decreases the growth rate.

4.2.2 The Impact of Corruption Control on Economic Growth

Empirical research results show that corruption control improves economic growth. In other words, corruption will hinder economic growth in ASEAN countries. This result is also consistent with the studies of Hoinaru et al. (2020), Gründler and Potrafke (2019), Faber and Gerritse (2012), Jetter and Parmeter (2018), Cieślik and Goczek (2018), and Borsky and Kalkschmied (2019).

First of all, corruption directly affects inequality in business, reducing the competitiveness of enterprises. Corruption distorts the government's behavior by allowing officials to interfere in areas where they should not be, reducing the government's ability to enforce laws, and undermining business confidence. Companies are willing to offer bribes for more development opportunities, whereas those that are not will fall behind. Corruption wastes public resources, disrupting the market and the linear trajectory of an economy (Saha et al. 2009). In the long term, the lack of fairness in business makes it difficult for the overall economy to achieve the desired growth.

In another aspect, corruption affects the quality of public investment, especially investment in basic construction, which causes the quality of infrastructure in countries to deteriorate. Public investment projects, once proven to be inefficient and lack transparency, will be cut down, leading to the decrease in the size of public investment (Tanzi and Davoodi 1998; Baliamoune-Lutz and Ndikumana 2008). Furthermore, the manifestations of corruption are often monopoly, concealment of information, and lack of accountability. In other words, corruption is accompanied by the lack of transparency in information. This makes it impossible for the government to grasp the real needs of the society for a certain kind of goods and service and the government still continue to expand investment, leading to mistakes in public investment.

Corruption creates barriers to attracting investment (Cieślik and Goczek 2018). Companies may refuse to invest in a country because they think that operating costs can increase due to corruption. These expenditures also contribute significantly to total costs, and even in some cases, this makes investment projects unfeasible, as pointed out by Shleifer and Vishny (1993). Corruption also damages national budget revenues through reduced taxes due to bribery (Méon and Sekkat 2005). Good corruption control is regarded as a proof of good institutional quality and public governance. Therefore, in the eyes of foreign investors, this is the improvement in the quality of governance and the demonstration of democracy.

In particular, this study found a negative but rather low impact of the interaction variable between shadow economy and corruption control on economic growth, with the impact probability of 78.323%. This shows that improving the level of corruption control can help limit the negative impact of shadow economy on economic growth in ASEAN countries. In contrast, the increase in corruption reduces the rate of investment, associated with tax evasion, expands the size of the shadow economy, and hinders economic growth (Borlea et al. 2017; Hoinaru et al. 2020; Baklouti and Boujelbene 2020).

In addition, the research results also indicate that domestic credit to the private sector, population growth, foreign direct investment and government expenditure have a positive impact on economic growth, whereas inflation has a negative impact on economic growth. This is also consistent with previous studies of Barro (2003), Moral-Benito (2012).

5 Conclusion

In this study, the authors focus on examining the impact of shadow economy and corruption control on economic growth in 10 ASEAN countries over the period 2002–2019. Regarding the method of analysis, the authors used the Bayesian method to estimate the research model. The estimation results show that economic growth is negatively impacted by shadow economy and is positively impacted by corruption control. This indicates that shadow economy hinders economic growth in ASEAN countries, whereas corruption control can significantly improve economic growth in these countries. In particular, the authors found a negative but rather low impact of the interaction variable between shadow economy and corruption control on economic growth. Accordingly, the combination of shadow economy and corruption control can significantly reduce the level of the negative impact of shadow economy on economic growth. In other words, improving the level of corruption control can help limit the negative impact of the shadow economy on economic growth in ASEAN countries. It can be said that this is an interesting finding of this study. In addition, the estimation results reveal that economic growth is significantly impacted by the control variables of domestic credit, government expenditure, foreign direct investment, population growth, and inflation.

References

Arcand, J.L., Berkes, E., Panizza, U.: Too much finance? J. Econ. Growth **20**(2), 105–148 (2015)

Alañón, A., Gómez-Antonio, M.: Estimating the size of the shadow economy in Spain: a structural model with latent variables. Appl. Econ. **37**(9), 1011–1025 (2005)

Alm, J., Embaye, A.: Using dynamic panel methods to estimate shadow economies around the world, 1984–2006. Public Finance Rev. **41**(5), 510–543 (2013)

Baklouti, N., Boujelbene, Y.: Shadow economy, corruption, and economic growth: an empirical analysis. Rev. Black Polit. Econ. **47**(3), 276–294 (2020)

Baliamoune-Lutz, M., Ndikumana, L.: (2008). Corruption and growth: exploring the investment channel. In: Economics Department Working Paper Series, p. 33. https://doi.org/10.7275/1068835

Barro, R.J.: Determinants of economic growth in a panel of countries. Ann Econ Finance **4**, 231–274 (2003)

Bayley, D.H.: The effects of corruption in a developing nation. Western Polit Quart **19**(4), 719–732 (1966)

Beck, T., Levine, R.: Stock markets, banks, and growth: panel evidence. J. Bank. Finance **28**(3), 423–442 (2004)

Beck, P.J., Maher, M.W.: A comparison of bribery and bidding in thin markets. Econ. Lett. **20**(1), 1–5 (1986)

Becker, S.O., Egger, P.H., Seidel, T.: Common political culture: evidence on regional corruption contagion. Eur. J. Polit. Econ. **25**(3), 300–310 (2009)

Borlea, S.N., Achim, M.V., Miron, M.G.: Corruption, shadow economy and economic growth: an empirical survey across the European Union countries. Studia Universitatis Vasile Goldiș Arad, Seria Științe Econ. **27**(2), 19–32 (2017)

Borsky, S., Kalkschmied, K.: Corruption in space: a closer look at the world's subnations. Eur. J. Polit. Econ. **59**, 400–422 (2019)

Broms, R.: Taxation and government quality. QoG Working Paper Series, No. 16 (2011).Retrieved from https://core.ac.uk/reader/43558930

Buehn, A., Farzanegan, M.R.: Smuggling around the world: evidence from a structural equation model. Appl. Econ. **44**(23), 3047–3064 (2012)

Buehn, A., Schneider, F.: MIMIC models, cointegration and error correction: an application to the French shadow economy. IZA DP No. 3306. (2008) Retrieved from https://core.ac.uk/download/pdf/7141573.pdf

Cieślik, A., Goczek, Ł: Control of corruption, international investment, and economic growth–Evidence from panel data. World Dev. **103**, 323–335 (2018)

Choi, J.P., Thum, M.: Corruption and the shadow economy. Int. Econ. Rev. **46**(3), 817–836 (2005)

Chong, A., Calderon, C.: Causality and feedback between institutional measures and economic growth. Econ. Polit. **12**(1), 69–81 (2000)

Cooray, A., Dzhumashev, R., Schneider, F.: How does corruption affect public debt? An empirical analysis. World Dev. **90**, 115–127 (2017)

d'Agostino, G., Dunne, J.P., Pieroni, L.: Government spending, corruption and economic growth. World Dev. **84**, 190–205 (2016)

De Paula, A., Scheinkman, J.A.: The informal sector: an equilibrium model and some empirical evidence from Brazil. Rev. Income Wealth **57**, S8–S26 (2011)

Dell'Anno, R., Gómez-Antonio, M., Pardo, A.: The shadow economy in three Mediterranean countries: France, Spain and Greece A MIMIC Approach. Empir. Econ. **33**(1), 51–84 (2007)

Dreher, A., Kotsogiannis, C., McCorriston, S.: How do institutions affect corruption and the shadow economy? Int. Tax Public Financ. **16**(6), 773–796 (2009)

Dreher, A., Schneider, F.: Corruption and the shadow economy: an empirical analysis. Public Choice **144**(1), 215–238 (2010)

Eilat, Y., Zinnes, C.: The evolution of the shadow economy in transition countries: consequences for economic growth and donor assistance. Harvard Institute for International Development, CAER II Discussion Paper No. 83 (2000). Retrieved from http://citeseerx.ist.psu.edu/viewdoc/download?doi=10.1.1.470.2018&rep=rep1&type=pdf

Ehigiamusoe, K.U., Lean, H.H., Lee, C.C.: Moderating effect of inflation on the finance–growth nexus: insights from West African countries. Empir. Econ. **57**(2), 399–422 (2019)

Elgin, C., Birinci, S.: Growth and informality: a comprehensive panel data analysis. J. Appl. Econ. **19**(2), 271–292 (2016)

Esaku, S.: Is informality a barrier to economic growth in Uganda? Empricial analysis. Dev. Stud. Res. **8**(1), 109–121 (2021)

Faber, G., Gerritse, M.: Foreign determinants of local institutions: Spatial dependence and openness. Eur. J. Polit. Econ. **28**(1), 54–63 (2012)

Flegal, J.M., Haran, M., Jones, G.L.: Markov chain Monte Carlo: can we trust the third significant figure? Stat. Sci. 250–260 (2008)

Friedman, E., Johnson, S., Kaufmann, D., Zoido-Lobaton, P.: Dodging the grabbing hand: the determinants of unofficial activity in 69 countries. J. Public Econ. **76**(3), 459–493 (2000)

Gelman, A., Rubin, D.B.: Inference from iterative simulation using multiple sequences. Stat. Sci. **7**(4), 457–472 (1992)

Gründler, K., Potrafke, N.: Corruption and economic growth: new empirical evidence. Eur. J. Polit. Econ. 60, 101810 (2019)

Hoinaru, R., Buda, D., Borlea, S.N., Văidean, V.L., Achim, M.V.: The impact of corruption and shadow economy on the economic and sustainable development. Do they "sand the wheels" or "grease the wheels"? Sustainability 12(2), 481 (2020)

Huntington, S.P.: Political order in changing societies. Yale University Press, New Haven (1968)

Ihrig, J., Moe, K.S.: Lurking in the shadows: the informal sector and government policy. J. Dev. Econ. **73**(2), 541–557 (2004)

Jetter, M., Parmeter, C.F.: Sorting through global corruption determinants: institutions and education matter–Not culture. World Dev. **109**, 279–294 (2018)

Kaufmann, D., Kaliberda, A.: Integrating the unofficial economy into the dynamics of post socialist economies: A framework of analyses and evidence. In: Economic Transition in Russia and the New States of Eurasia, pp. 81–120. ME Sharpe, London (1996). https://doi.org/10.1596/1813-9450-1691

Kodila-Tedika, O., Mutascu, M.: Shadow economy and tax revenue in Africa. (2013) Retrieved from https://mpra.ub.uni-muenchen.de/50812/

La Porta, R., Shleifer, A.: Informality and development. J. Econ. Perspect. **28**(3), 109–126 (2014)

Lemoine, N.P.: Moving beyond noninformative priors: why and how to choose weakly informative priors in Bayesian analyses. Oikos **128**(7), 912–928 (2019)

Leys, C.: What is the Problem about Corruption? J. Mod. Afr. Stud. **3**(2), 215–230 (1965)

Loayza, N.V.: The economics of the informal sector: a simple model and some empirical evidence from Latin America. Carn.-Roch. Conf. Ser. Public Policy **45**, 129–162 (1996)

Lui, F.T.: An equilibrium queuing model of bribery. J. Polit. Econ. **93**(4), 760–781 (1985)

McNeish, D.M.: Using data-dependent priors to mitigate small sample bias in latent growth models: a discussion and illustration using M plus. J. Educ. Behav. Stat. **41**(1), 27–56 (2016)

Medina, L., Schneider, M.F.: Shadow economies around the world: what did we learn over the last 20 years? Working Paper No. 18/17 (2018). Retrieved from https://www.imf.org/en/Publications/WP/Issues/2018/01/25/Shadow-Economies-Around-the-World-What-Did-We-Learn-Over-the-Last-20-Years-45583

Méon, P.G., Sekkat, K.: Does corruption grease or sand the wheels of growth? Public Choice **122**(1), 69–97 (2005)

Moral-Benito, E.: Determinants of economic growth: a Bayesian panel data approach. Rev. Econ. Stat. **94**(2), 566–579 (2012)

Mauro, P.: Corruption and growth. Q. J. Econ. **110**(3), 681–712 (1995)

Nguyen, D.V., Duong, M.T.H.: Shadow economy, corruption and economic growth: an analysis of BRICS Countries. J. Asian Finance Econ. Bus. **8**(4), 665–672 (2021)

Özgür, G., Elgin, C., Elveren, A.Y.: Is informality a barrier to sustainable development? Sustain. Dev. **29**(1), 45–65 (2021)

Pellegrini, L., Gerlagh, R.: Corruption's effect on growth and its transmission channels. Kyklos **57**(3), 429–456 (2004)

Roberts, G.O., Rosenthal, J.S.: Optimal scaling for various metropolis-hastings algorithms. Stat. Sci. **16**(4), 351–367 (2001)

Romer, P.M.: Endogenous technological change. J. Polit. Econ. 98(5, Pt 2), S71-S102 (1990)

Ruiz, J.L.: Financial development, institutional investors, and economic growth. Int. Rev. Econ. Financ. **54**, 218–224 (2018)

Saha, S., Gounder, R., Su, J.J.: The interaction effect of economic freedom and democracy on corruption: a panel cross-country analysis. Econ. Lett. **105**(2), 173–176 (2009)

Schneider, F., Bajada, C.: The size and development of the shadow economies in the Asia-Pacific. Working Paper No. 0301 (2003). Retrieved from http://www.economics.uni-linz.ac.at/papers/2003/wp0301.pdf

Schneider, F., Buehn, A., Montenegro, C.E.: New estimates for the shadow economies all over the world. Int. Econ. J. **24**(4), 443–461 (2010)

Schneider, F., Enste, D.H.: The shadow economy: an international survey. Cambridge University Press (2013)

Shahid, S., Khan, R.E.A.: Informal sector economy, child labor and economic growth in developing economies: exploring the interlinkages. Asian Dev. Policy Rev. **8**(4), 277–287 (2020)

Shleifer, A., Vishny, R.W.: Corruption. Q. J. Econ. **108**(3), 599–617 (1993)

Son, T.H., Liem, N.T., Khuong, N.V.: Corruption, nonperforming loans, and economic growth: international evidence. Cogent Bus. Manag. **7**(1), 1735691 (2020)

Tanzi, V., Davoodi, H.: Corruption, public investment, and growth. In: In the Welfare State, Public Investment, and Growth, pp. 41–60. Springer, Tokyo (1998). Retrieved from http://www1.worldbank.org/publicsector/LearningProgram/anticorrupt/TanziDavoodi.pdf

Thach, N.N.: Endogenous economic growth: the arrow-romer theory and a test on vietnamese economy. WSEAS Trans. Bus. Econ. **17**, 374–386 (2020a)

Thach, N.N.: The variable elasticity of substitution function and endogenous growth: an empirical evidence from Vietnam. Int. J. Econ. Bus. Adm. **8**(1), 263–277 (2020b)

Thach, N.N.: How have NESTs grown? Explanations based on endogenous growth theory. Cogent Econ. Finance **9**(1), 1913847 (2021)

Vo, D. H., Pham, T. M., & Authority, E. R: Any link between unofficial economy and official economy? An empirical evidence from the ASEAN. Int. J. Econ. Financ. **6**(11), 139–148 (2014)

Williams, C.C., Schneider, F.: Measuring the global shadow economy: the prevalence of informal work and labour. Edward Elgar Publishing Limited (2016). https://doi.org/10.4337/9781784717995.00002

The Impact of Foreign Direct Investment on Financial Development: A Bayesian Approach

Vo Thi Thuy Kieu, Le Thong Tien, and Nguyen Ngoc Thach

Abstract Our article has provided new insight into the effect of foreign direct investment inflows and the other macroeconomic factors on financial development. The panel dataset was collected in 31 countries over the 2009–2019 period. The frequentist inference, such as GLS estimation, did not provide evidence of statistical significance of the foreign direct investment. By Bayesian estimation, research findings acknowledged both positive and negative effects of foreign direct investment inflows. Simultaneously, the positive effect of foreign direct investment inflows exceeded its possible negative effect on financial development in some emerging and developing countries. In terms of control variables (such as exchange rate, market capitalization and trade openness), they had prominently positive influences on the financial development by Bayesian estimation. Inflation (inf) was the sole determinant of prevailed negative impact on financial development.

Keywords Bayesian · Financial development · Foreign direct investment inflows · Frequentist

JEL C11 · F21 · F43 · F65

1 Introduction

Socio-economic development is always the top concern of countries in the world including developed countries, developing countries, or emerging countries. Nowadays, foreign direct investment (fdi) and financial development (fd) are two important

---

V. T. T. Kieu · N. N. Thach
Banking University HCMC, 36 Ton That Dam Street, District 1, Ho Chi Minh City, Vietnam
e-mail: kieuvtt@buh.edu.vn

N. N. Thach
e-mail: thachnn@buh.edu.vn

L. T. Tien (✉)
Saigon University, 273 An Dương Vương, District 5, Ho Chi Minh City, Vietnam
e-mail: lttien@sgu.edu.vn

N. Ngoc Thach et al. (eds.), *Financial Econometrics: Bayesian Analysis, Quantum Uncertainty, and Related Topics*, Studies in Systems, Decision and Control 427,
https://doi.org/10.1007/978-3-030-98689-6_50

factors that contribute to changes in economic activities (Lee et al. 2017, 2013; Shen et al. 2010). In general, financial development is part of the private sector development strategy to promote growth and decrease poverty (Levine 2003; Eryılmaz et al. 2015). Fundamentally, financial sector development concerns overcoming costs incurred in the financial system (Eryılmaz et al. 2015). Foreign direct investment considered an essential source of capital investment for economies, is considered a crucial tool for boosting economic growth.

Many previous studies have examined the relationship between financial development and foreign direct investment, but the research results are conflicting (Claessens et al. 2001; Dutta and Roy 2009, Nasser and Gomez 2009; Abzari et al. 2011; Korgaonkar 2012; Agbloyor et al. 2011; Bayar and Ozel 2014; Sahina and Ege 2015; Fauzel 2016, and Enisan 2017). Many studies have proven that foreign direct investment promoted the development of the financial sector (Seetanah et al. 2010 for Mauritius; Nasser and Gomez 2009 for 15 Latin American countries; Desbordes and Wei 2014 for 67 developed and developing countries). Besides, the studies by Sahina and Ege (2015) for Greece and neighbouring countries (Bulgaria, Macedonia, and Turkey), Fauzel (2016) for small island economies, Saidi (2018) for low–income Countries, Alsmadi & Oudat (2019) for Bahrai, Ibrahim and Acquah (2020) for 45 African countries revealed that foreign direct investment has an impact on financial development and vice versa. However, Dutta and Roy (2011) found that foreign direct investment boosts financial development to a particular level of foreign direct investment inflows in 97 developed and developing countries; but when exceeds this threshold, foreign direct investment impedes financial development in host countries. On the other hand, foreign direct investment inflows might not affect the financial development of a country (Bayar and Gavriletea 2018).

As far as the methodology is concerned, within the frequentist framework, various datasets and methods were applied in numerous studies on the relationship between financial development and foreign direct investment. For example, among others, Sahina and Ege (2015) performed bootstrap causality between financial development and foreign direct investment in panel data whereas Shahbaz et al. (2020) and Seetanah et al. (2010) utilized a ARDL cointegration to test this relationship. Some other previous studies used traditional estimation methods such as Fauzel (2016) with a Panel Vector Autoregressive Regression (PVAR); Saidi (2018) with an Error Correction Model (ECM); Ibrahim and Acquah (2020) with Granger non-causality.

Therefore, the topic of financial development and foreign direct investment with the new Bayesian estimation method is of interest to researchers. In particular, the development of many countries has now formed emerging economies, whether the impact of foreign direct investment on financial development is different from previous studies. Mainly for these reasons, this work will apply the Bayesian approach to analyze the impact of foreign direct investment on financial development in emerging and developing countries.

The paper is divided into five sections. The second section discusses a literature review; Sects. 3 and 4 present the research method and the empirical results. Section 5 gives a discussion and conclusion.

2 Literature Review

The financial development of a country is also considered as one of the crucial factors for the investment activities of multinational corporations. Campos and Kinoshita (2010) showed an important relationship between structural reform and foreign direct investment. Not only was a better financial system and environment attractive to foreign direct investment, but management knowledge and the spillover effects of technology introduced by foreign direct investment reinforce the positive relationship with growth through a good financial system (Durham 2004; Alfaro et al. 2009, 2011).

In 2010, Seetanah et al. evaluated the factors affecting financial development in Mauritius. With the time series dataset from 1970 to 2008 and the ARDL model, the results showed that foreign direct investment has a positive correlation with financial development. Besides, trade openness, market capitalization, and GDP per capita were also considered to be significant factors in promoting financial development while inflation had a negative influence on financial development even in the short and long run. In their paper, financial development measured the value of credits by financial intermediaries to the private sector divided by the GDP was used to measure financial development.

Analyzing panel data from 97 developed and developing countries, Dutta and Roy (2011) used the ratio of private credit by deposit money bank to GDP to measure financial development. In addition, the authors also used the control variables as annual growth of GDP, inflation, exchange rate, trade openness to examine the relationship between foreign direct investment and financial development. Ordinary least squares (OLS) and Feasible generalized least squares (FGLS) regression results showed that foreign direct investment boosts financial development to a particular level of foreign direct investment inflows; but when exceeds this level, foreign direct investment impedes financial development in host countries. Also studying in developed and developing countries, Desbordes and Wei (2014) found similar results to Dutta and Roy (2011). They suggested that there was a conditional relationship between development finance and direct investment foreign. Specifically, foreign direct investment stimulated financial development in financially vulnerable sectors.

In 2015, Sahina & Ege examined the relationship between foreign direct investment and financial development in Greece and neighboring countries (Bulgaria, Macedonia, and Turkey) over the period 1996–2012. Through Bootstrap causality analysis, the authors found that foreign direct investment was able to predict financial development in all countries, except Macedonia. In addition, the findings indicated that there was a causal relationship between these two variables in Turkey. Domestic credit to private sector as a percentage of GDP was used to measure financial development.

Fauzel (2016) is one of the few authors who has studied small island economies, using the PVAR model to examine the relationship between financial development and foreign direct investment during the period 1990–2013. Research results have shown that foreign direct investment can be an important component for the development of financial sectors in small island economies. While GDP per capita and trade openness had a positive effect on financial development, inflation had a negative effect on financial development. Saidi (2018) studied the relationship between financial development (fd), foreign direct investment (fdi), and economic growth in Low–income countries. The results showed that financial development; foreign direct investment, and GDP growth are linked, which indicated the pursuit of a long-run equilibrium relationship between them. Regression results showed that foreign direct investment has an impact on financial development.

Another study by Shabbir et al. (2018) examined the factors affecting financial development in Pakistan during the period of 1995–2015. Through the method of regression analysis and correlation analysis, the research results showed that inflation, trade openness, market capitalization have a significant impact on financial development. However, while trade openness and market capitalization had a positive effect on financial development, inflation negatively affected financial development. The authors pointed out that the paper's limitation was to consider only five determinants of financial development. Besides, the impact of foreign direct investment on financial sector development was found in the study of Alsmadi and Oudat (2019) and Ibrahim and Acquah (2020). Nevertheless, Alsmadi and Oudat (2019) examined the relationship between foreign direct investment and financial sector development in Bahrai from 1978 to 2015 while Ibrahim and Acquah (2020) used data from 45 African countries over the period 1980 to 2016.

Most recently Majeed et al. (2021) evaluated the impact of foreign direct investment on financial development through cointegration and causality analysis. Regression results from the dataset of 102 Belt and Road Initiative countries on four continents (Asia, Europe, Africa, and Latin America) have shown that foreign direct investment, trade openness, and inflation were statistically significant for financial development. Foreign direct investment, openness to trade increased financial development in Asia, Europe, and Latin America but decreased in Africa. Furthermore, inflation had a negative effect on financial development in all continents.

3 Methodology

3.1 Research Method

In the context of the effect of foreign direct investment and other economic factors on financial development, most previous studies employed the Frequentist approach. In recent years, we have seen a burst of activity in the use of the Bayesian approach (see, for instance, Anh et al. 2018; Nguyen and Thach 2018, 2019; Nguyen et al. 2019a, b; Svítek et al. 2019; Tuan et al. 2019; Sriboonchitta et al. 2019; Kreinovich et al. 2019; Thach 2020; Thach et al. 2019, 2021; Kieu and Tien 2021).

Based on the prevalent rule of probability, the Bayes rule, the benefit of Bayesian analysis is to find answers to unknown parameters using statistical simulations and probability distributions. In other words, Bayesian inferences give straightforward probabilistic interpretations compared to classical econometrics. The prior information and posterior distribution of model parameters in Bayesian analysis also produce more comprehensive and more flexible reliable intervals than the frequentist confidence intervals.

Bayesian estimation proves to be extremely beneficial in generating reliable simulations despite the existence of outliers. Due to perform simulations, Bayesian analysis does not require a large sample size and thus makes it easier to access some fields of research concepts whose data are not available or insufficient. Fortunately, the command packages, provided by Stata 17, have recently facilitated the extremely time-consuming calculation of complex models by the Bayesian approach. The following paragraphs strive to modelize the effects of foreign direct investment on financial development by both frequentist and Bayesian estimation.

The frequentist model for panel data was built as follows:

$$
\begin{aligned}
fd_{i,t} = {} & \beta_0 + \beta_1 \mathrm{fdi}_{i,t} + \beta_2 \mathrm{gdpcap}_{i,t} + \beta_3 \mathrm{ex}_{i,t} + \beta_4 \mathrm{market}_{i,t} \\
& + \beta_5 \mathrm{inf}_{i,t} + \beta_6 \mathrm{open}_{i,t} + \alpha_i + \varepsilon_{i,t}
\end{aligned} \quad (1)
$$

where fd denotes financial development proxied by domestic credit to the private sector, market is calculated by market capitalization to GDP; gdpcap is natural log of GDP per capita; ex represents official exchange rate; open is the proportion of trade volume (the sum of export and import of goods and services); α_i is an unobserved individual-specific random effect; and $\varepsilon_{i,t}$ is an unobserved random error term. In the condition that α_i and $\varepsilon_{i,t}$ are independent, $\mu_i = \alpha_i + \varepsilon_{i,t}$ might be a substitution, setting a random-effect model. Instead of random-effect estimation, Research results using Generalized Least Square estimation (GLS) for settling possible heteroscedasticity and autocorrelation problems.

For comparition, this article continues to perform Bayesian analysis for panel data. Its likelihood is assumed to be normal prior with variance defined var_0. We assumed normal priors for regression coefficients and inverse-gamma priors for the variance parameters. Metropolis–Hastings and Gibbs sampling was also included. The Bayesian random-effect estimation for panel data, supported by Stata 17, can be described by:

Likelihood:

$$
\mathrm{fd} \quad \mathrm{normal}\left(\mathrm{xb\_fd},\ \sigma^2\right) \quad (2)
$$

Priors were set as follows:

$\Theta_{\mathrm{fdi}} \sim$ normal(0,10,000).

$\Theta_{\mathrm{gdpcap}} \sim$ normal(0,10,000).

$\Theta_{\mathrm{ex}} \sim$ normal(0,10,000).

Θ_{market} ~ normal(0,10,000).

Θ_{inf} ~ normal(0,10,000).

Θ_{open} ~ normal(0,10,000).

$\theta_{\_cons}$ ~ normal(0,10,000).

{U[individuals]} ~ normal(0,σ_U^2)

σ^2 ~ igamma(φ,φ)

Hyperprior: σ_U^2 ~ igamma(φ,φ)

where θ is assumed as the effect of each determinant on financial development; _cons is proxy for the mean of random effects; σ^2 was proxy for the error variance; σ_U^2 was proxy for the variance of random effects; φ is varied from 100 to 1000 for checking robustness.

For Bayesian estimation, the selection of priors can more or less affect posterior distributions. However, the Bernstein & Von Mises theorem states that a posterior distribution converges to a multivariate normal distribution centered at the maximum likelihood estimator. Or rather, the posterior distribution can be relatively independent of the prior distribution, the inferences based on Bayesian estimation, and likelihood function, thereby generate similar results (Vaart 1998).

3.2 Data Description

According to the MSCI Market Classification Framework (2014), emerging economies were economies that were in the transition from a developing economy to a developed economy. However, the Economist (2011) had its own criteria based on investment information. Then, Cox (2017) indicated that an emerging market (or an emerging country or an emerging economy) has some characteristics of a developed market but does not fully meet its standards. This included markets that may become developed countries in the future or were in the past. However, there were no clear and universal criteria for determining whether an economy is an emerging economy or not. The Duttagupta[1] and Pazarbasioglu[2] (2021) classified emerging and developing countries in the same group in its world economy documents, without dividing them into two separate groups.

The Table 1 gives information about the description of the variables employed in the current analysis. The financial development (fd) was considered as a dependent variable; whereas the independent variables consist of GDP per capita (gdpcap), official exchange rate (ex), market capitalization to GDP (market), inflation (inf), trade openness (open). The whole dataset was collected from WGI (World Governance

[1] RUPA DUTTAGUPTA is a division chief in the IMF's Strategy, Policy, and Review Department, where CEYLA.

[2] PAZARBASIOGLU is director.

Table 1 Dependent and independent variables used in the study

Variables	Description	Sources
fd (Dependent variables)	Financial development is domestic credit to the private sector that describes the sources of financing provided to the private sector by financial institutions measured as a share of GDP	Seetanah et al. (2010), Dutta and Roy (2011), Desbordes and Wei (2014), Sahina and Ege (2015), Fauzel (2016), Saidi (2018), Alsmadi and Oudat (2019), Ibrahim and Acquah (2020), Majeed et al. (2021)
fdi (Independent variable)	Foreign direct investment is the net inflows of investment to acquire a lasting management interest (10 percent or more of voting stock) in an enterprise operating in an economy other than that of the investor. It is the sum of equity capital, reinvestment of earnings, other long-term capital, and short-term capital as shown in the balance of payments. This series shows net inflows (new investment inflows less disinvestment) in the reporting economy from foreign investors, and is divided by GDP	Kinoshita (2010), Seetanah et al. (2010), Dutta and Roy (2011), Desbordes and Wei (2014), Sahina and Ege (2015), Fauzel (2016), Saidi (2018), Alsmadi and Oudat (2019), Ibrahim and Acquah (2020), Majeed et al. (2021)
gdpcap (Independent variable)	GDP per capita is GDP divided by mid-year population. Data are in constant 2010 U.S. dollars	Seetanah et al. (2010), Fauzel (2016), Majeed et al. (2021)
ex (Independent variable)	EX represents the official exchange rate refers to the exchange rate determined by national authorities or to the rate determined in the legally sanctioned exchange market. It is calculated as an annual average based on monthly averages (local currency units relative to the U.S. dollar)	Dutta and Roy (2011), Shabbir et al. (2018)
market (Independent variable)	Market capitalization to GDP (also known as market value) is the share price multiplied by the number of shares outstanding (including some of their categories) for listed domestic companies	Seetanah et al. (2010), Shabbir et al. (2018), Majeed et al. (2021)
inf (Independent variable)	Inflation, as measured by the consumer price index reflects the annual percentage change in the cost to the average consumer of acquiring a basket of goods and services that may be fixed or changed at specified intervals, such as yearly	Seetanah et al. (2010), Dutta and Roy (2011), Fauzel (2016), Shabbir et al. (2018), Majeed et al. (2021)
open (Independent variable)	Trade openness is the sum of exports and imports of goods and services measured as a share of GDP	Seetanah et al. (2010), Dutta and Roy (2011), Fauzel (2016), Shabbir et al. (2018), Majeed et al. (2021)

Source The authors' computations

Indicators) of World Bank in 31[3] emerging and developing countries, classified by Duttagupta and Pazarbasioglu (2021), between 2009 and 2019.

4 Bayesian Simulation Results

The Table 2 showed the Frequentist estimation and Bayesian estimation for the posterior mean of determinants of financial development. Metropolis–Hastings and Gibbs sampling were included for the simulation of 20,000 MCMC iterations and 500 burn-in ones of a sample size to ensure robust Bayesian inferences when the MCMC chains converge to a stationary range. The acceptance rates were ranged from [0.8094; 0.8126] close to 1. The average efficiency of the simulation measures convergence and precision of reliable interval, ranged [0.5999; 0.8603]. The detailed efficiency was analyzed more deeply in Appendix A.

As can be seen from the Table 2, Generalized Least Square estimation (GLS), instead of the random-effect, was performed to settle heteroscedasticity and autocorrelation problems. According to GLS estimation, there was statistical significance in the exchange rate (ex), market capitalization (market), and trade openness (open). Their coefficients were positively correlated with financial development (fd). The positive effect of exchange rate indicated that financial development in strong-currency countries is gradually maturing, while the rapid growth of weak-currency regions has a greater impetus for financial development. It is clearly understandable for the positive effect of market capitalization (market) as it is almost a component of financial development. Besides, deeper trade integration also encourages the growth of financial development, which explains the positive effect of trade openness (open). However, the estimated results from the GLS model found no evidence of the influence of foreign direct investment, which is of interest in the content of the article. This does not mean that foreign direct investment does not affect financial development. Tablc 2 also showed that the posterior mean by Bayesian estimation is similar to the estimated coefficients by GLS, but the statistical inferences presented in the Table 3 provided a more specific and comprehensive view.

Compared with GLS estimation, the posterior means demonstrated little difference in the magnitude of the effects. The Bayesian estimates, especially the model with igammaprior (100,100), were tangent to those from frequentist model. This will assist in checking the robustness of the Bayesian estimates found. The Bayesian posterior mean needs testing for the convergence of MCMC chains before performing statistical inferences. Appendix A provided the efficiency of the simulation for various variance priors to demonstrate the objective criterion for convergence. This is also additionally illustrated by Visual inspections, which are shown in Appendix B, for the purpose of further reinforce the convergence of MCMC chains.

[3] Bangladesh, Bulgaria, Brazil, Colombia, Czech, Chile, China, Hungary, Indonesia, India, Iran, Israel, Korea, Kuwait, Sri lanka, Morocco, Mexico, Mauritius, Malaysia, Nigeria, Pakistan, Peru, Poland, Philippine, Romania, Tunisia, Turkey, Thailand, Ukraine, Vietnam, South Africa.

Table 2 The research results by GLS estimation and Bayesian random-effect estimation with various variance priors

Dependent variable: Financial development	Frequentist estimation (GLS)	Bayesian random-effect estimation									
		$\sigma^2,\sigma_U^2 \sim$ igamma prior (100, 100)	$\sigma^2,\sigma_U^2 \sim$ igamma prior (200, 200)	$\sigma^2,\sigma_U^2 \sim$ igamma prior (300, 300)	$\sigma^2,\sigma_U^2 \sim$ igamma prior (400, 400)	$\sigma^2,\sigma_U^2 \sim$ igamma prior (500, 500)	$\sigma^2,\sigma_U^2 \sim$ igamma prior (600, 600)	$\sigma^2,\sigma_U^2 \sim$ igamma prior (700, 700)	$\sigma^2,\sigma_U^2 \sim$ igamma prior (800, 800)	$\sigma^2,\sigma_U^2 \sim$ igamma prior (900, 900)	$\sigma^2,\sigma_U^2 \sim$ igamma prior (1000, 1000)
θ_{fdi}	0.1451	0.1450	0.1361	0.1340	0.1348	0.1305	0.1266	0.1280	0.1219	0.1186	0.1186
θ_{gdpcap}	0.6570	0.6259	0.6105	0.6048	0.5884	0.5707	0.5529	0.5419	0.5261	0.5067	0.4894
θ_{ex}	0.9798*	0.9820	0.9851	0.9889	0 0.9918	0.9951	0.9977	0.9973	1.0043	1.0033	1.009
θ_{market}	0.4330***	0.4320	0.4317	0.4311	0.4308	0.4301	0.4297	0.4291	0.4287	0.4280	0.4274
θ_{inf}	−0.2658	−0.2705	−0.2736	−0.2748	−0.2765	−0.2833	−0.2844	−0.2882	-0.2917	-0.2934	-0.2964
θ_{open}	0.2888***	0.2890	0.2883	0.2884	0.2879	0.2877	0.2877	0.2874	0.2872	0.2870	0.2867
Acceptance rate	–	0.8106	0.8094	0.8133	0.8124	0.8096	0.8124	0.8124	0.8104	0.8126	0.8113
Average efficiency	–	0.8603	0.8366	0.7933	0.7748	0.7419	0.7145	0.6818	0.6456	0.6343	0.5999
Number of obs	341										
Number of groups	31										
Likelihood:	–	fd ~ normal(xb_fd,σ^2)									
Priors:	–	{fd: fdi gdpcap ex market inf open} ~ normal(0,10,000) {U[individuals]} ~ normal(0,σ_U^2)									

Source Authors' calculations

Table 3 The posterior probability of parameters

Bayesian random-effect estimation																				
The posterior probability (%)	$\sigma^2, \sigma_U^2 \sim$ igamma prior (100,100)		$\sigma^2, \sigma_U^2 \sim$ igamma prior (200,200)		$\sigma^2, \sigma_U^2 \sim$ igamma prior (300,300)		$\sigma^2, \sigma_U^2 \sim$ igamma prior (400,400)		$\sigma^2, \sigma_U^2 \sim$ igamma prior (500,500)		$\sigma^2, \sigma_U^2 \sim$ igamma prior (600,600)		$\sigma^2, \sigma_U^2 \sim$ igamma prior (700,700)		$\sigma^2, \sigma_U^2 \sim$ igamma prior (800,800)		$\sigma^2, \sigma_U^2 \sim$ igamma prior (900,900)		$\sigma^2, \sigma_U^2 \sim$ igamma prior (1000,1000)	
Effect	+	−	+	−	+	−	+	−	+	−	+	−	+	−	+	−	+	−	+	−
θ_{fdi}	70	30	72	28	74	26	77	23	78	22	79	21	81	19	82	18	82	18	84	16
θ_{gdpcap}	95	05	97	03	98	02	98	02	99	01	99	01	99	01	99	01	99	01	99	01
θ_{ex}	98	02	99	01	99	01	99	01	99	01	100	0	100	0	100	0	100	0	100	0
θ_{market}	100	0	100	0	100	0	100	0	100	0	100	0	100	0	100	0	100	0	100	0
θ_{inf}	15	85	11	89	08	92	06	94	04	96	03	97	02	98	02	98	01	99	01	99
θ_{open}	100	0	100	0	100	0	100	0	100	0	100	0	100	0	100	0	100	0	100	0

Source Authors' calculations

Some of the visual inspections were reduced to optimize the design of article, Appendix B only presents the imaging diagnoses for Bayesian estimation with igammaprior (100,100). The trace plots looked no trends, fluctuated with their averages, and revealed more rapid traversing of the marginal posterior ones. The visual marks indicated almost well-mixing in the trace plots, and thus the autocorrelation seems low enough to be ignored. The histograms and density plots for posterior means and variance parameters were resembled the expected shapes, following the former normal distributions and the latter inverse-gamma distributions. As can be seen from the scatterplots, there were no significant correlations among the parameters. The graphical diagnostics in Appendix B did not show any convergence problems. Hence, the simulation efficiency for all chains becomes reliable for Bayesian estimation.

From an overall perspective, the Table 3 showed the results of the sensitivity analysis according to various variance priors from igammaprior (100,100) to igammaprior (1000,1000). There were no differences in the prevailing direction of impacts on financial development (fd) among the determinants by various variance priors.

According to bayesian estimation, the difference between the two-dimensional effects of foreign direct investment depends on different probability levels, instead of statistically insignificant by GLS estimation. As it is shown, the frequentist model only shows results on statistical evidence at the p_value for null hypothesis acceptance or rejection, whereas bayesian estimation does simulate with probability distributions. In other words, this empirical results from bayesian estimation illustrated that the influence of foreign direct investment has fragmented directions on financial development (fd). Or rather, Bayesian estimation did not deny the existence of a negative effect of foreign direct investment, which was still occurring with a lower probability density.

In terms of robustness, the posterior mean of the model with igammaprior (100,100) was extremely close to coefficients estimated by GLS, this variance prior could be utilized for statistical inferences. More specifically, the positive effect of foreign direct investment (fdi) exceeded its negative effect for the financial development (fd), 70% and 30% respectively, for the model with igammaprior (100,100). Therefore, this suggests that the effects of foreign direct investment inflows on financial development acknowledged both positive and negative direction, but more supporting the previous results reported by Desbordes and Wei (2014); Dutta and Roy (2011); Seetanah et al. (2010); and Nasser and Gomez (2009).

Regards the other control variables, most of the factors, excluding inflation, had prominently positive influences on the financial development. To be more detailed, the positive effect of GDP per capita (gdpcap), the exchange rate (ex), market capitalization (market), and trade openness (open) occupied the lowest probability of 95%.

For further analysis, the posterior means of parameters of exchange rate (ex), market capitalization (market) and trade openness (open) asserted that there were no differences in the direction of influences being relatively consistent with positive correlation regardless of frequentist or bayesian estimation. As a component of financial development, it is plausible to explain why market capitalization (market) is responsible for the climb in financial development. What remains of interest is the positive impact of trade openness strongly supporting international integration rather than trade protectionism.

Besides, the empirical result for Bayesian estimation was different from the GLS estimation for GDP per capita (gdpcap). While GLS estimation makes no claim of statistical significance in GDP per capita (gdpcap), whose effect on the financial development was recorded between 95 and 99% by Bayesian estimation. This was inlined with the results in the previous literature.

Finally, the negative effect of inflation (inf) was in the majority with the lowest figure of 85%. The decline in the inflation rate might lead to an upward trend in financial development. This implied financial development had really been improved by better inflation control in recent years.

5 Concluding Remarks and Implications

In conclusion, the frequentist inference does not provide evidence of statistical significance of foreign direct investment, Bayesian inference can be supplemented by the posterior probability of parameters, thereby assessing researching results more comprehensively. Instead of efforts to seek statistical evidence to the possible correlation as the frequentist estimation, Bayesian method strives to provide probabilities of both positive and negative effects, and support for dominated probability.

In general, our article has provided new insight into dominated probability for the effect of foreign direct investment on financial development by Bayesian estimation. Research findings indicated that the positive effect of foreign direct investment inflows seemingly exceeded its possible negative effect on financial development in some emerging and developing countries.

While GLS estimation found no statistical significance in foreign direct investment inflows, GDP per capita and inflation, it is plausible that exchange rate, market capitalization and trade openness had prominently positive influences on the financial development by Bayesian estimation. The positive impact of trade openness strongly supporting international integration rather than trade protectionism. Apart from that, the rapid growth of weak-currency regions has a greater impetus for financial development. Finally, the inflation rate clearly showed prevailed negative effects on financial development.

Appendices

Appendix A: The Detailed Efficiency of the Simulations

The efficiency of simulations	Bayesian estimation									
	$\sigma^2,\sigma_U^2 \sim$ igamma prior (100,100)	$\sigma^2,\sigma_U^2 \sim$ igamma prior (200,200)	$\sigma^2,\sigma_U^2 \sim$ igamma prior (300,300)	$\sigma^2,\sigma_U^2 \sim$ igamma prior (400,400)	$\sigma^2,\sigma_U^2 \sim$ igamma prior (500,500)	$\sigma^2,\sigma_U^2 \sim$ igamma prior (600,600)	$\sigma^2,\sigma_U^2 \sim$ igamma prior (700,700)	$\sigma^2,\sigma_U^2 \sim$ igamma prior (800,800)	$\sigma^2,\sigma_U^2 \sim$ igamma prior (900,900)	$\sigma^2,\sigma_U^2 \sim$ igamma prior (1000,1000)
θ_{fdi}	1.0000	1.0000	1.0000	1.0000	0.9641	1.0000	1.0000	0.9451	1.0000	1.0000
θ_{gdpcap}	1.0000	1.0000	0.9561	1.0000	0.8740	0.9433	0.8951	0.8144	0.8933	0.7906
θ_{ex}	0.8933	0.8276	0.8067	0.7728	0.7055	0.6984	0.6802	0.6451	0.5929	0.5911
θ_{market}	0.9701	0.8227	0.7827	0.7427	0.7153	0.6938	0.6367	0.6347	0.6239	0.5617
θ_{inf}	0.9550	1.0000	0.9267	0.9455	0.8658	0.8104	0.8280	0.8145	0.7506	0.7826
θ_{open}	0.8483	0.9541	0.8244	0.9455	0.7896	0.7153	0.6779	0.6485	0.6444	0.5815
$\theta_{\_cons}$	0.9654	0.9646	0.8236	0.7756	0.8350	0.7888	0.6962	0.6940	0.6497	0.6149
σ^2	0.7865	0.6114	0.6033	0.5496	0.5040	0.4638	0.3901	0.3405	0.3023	0.2515
σ_U^2	0.3239	0.3489	0.4158	0.4172	0.4243	0.3164	0.3318	0.2739	0.2518	0.2253

Source Authors' calculations

Appendix B: Visual Diagnostics For MCMC Convergence of Bayesian Estimation With σ^2, $\sigma_U^2 \sim$ igammaprior(100,100)

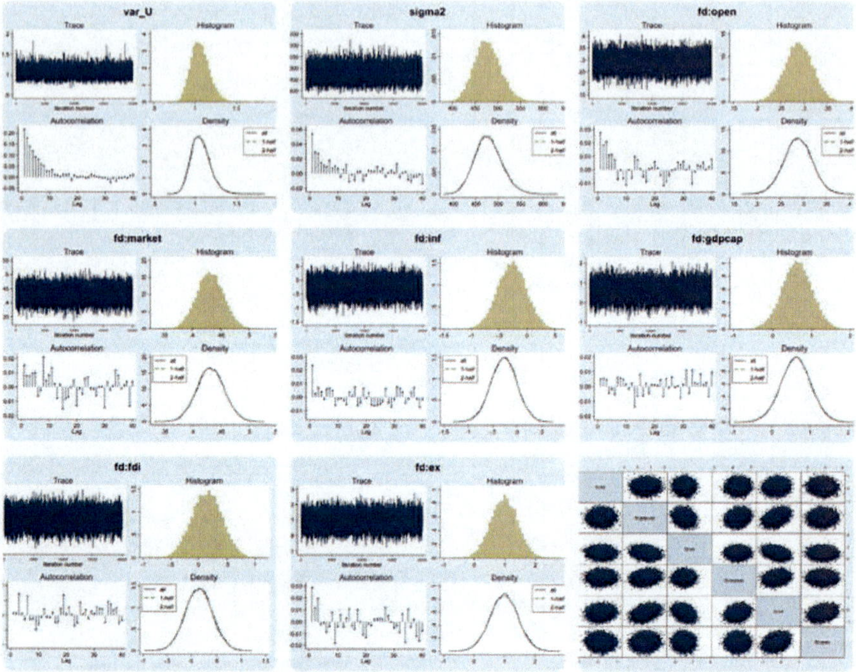

Source Authors' calculations

References

Abzari, M., Zarei, F., Esfahani, S.S.: Analyzing the link between financial development and foreign direct investment among D-8 group of countries. Int. J. Econ. Financ. **3**(6), 148–156 (2011)

Agbloyor, E.K., Abor, J., Adjasi, C.K.D., Yawson, A.: Exploring the causality links between financial markets and foreign direct investment in Africa. Res. Int. Bus. Financ. **28**, 118–134 (2011)

Alfaro, L., Kalemli-Ozcan, S., Sayek, S.: FDI, productivity and financial development. World Econ. **32**(1), 111–135 (2009). https://doi.org/10.1111/j.1467-9701.2009.01159.x

Alfaro, L., Chanda, A., Kalemli-Ozcan, S., Sayek, S.: Does foreign direct investment promote growth? Exploring the role of financial markets on linkages. J. Dev. Econ. **91**(2), 242–256 (2011). https://doi.org/10.1016/j.jdeveco.2009.09.004

Alsmadi, A.A., Oudat, M.S.: The effect of foreign direct investment on financial development: empirical evidence from bahrain. Ekonomski Pregled Hrvatsko Društvo Ekonomista **70**(1), 22–40 (2019)

Anh, L.H., Kreinovich, V., Thach, N.N. (eds.): Econometrics for financial applications. Springer, Cham (2018)

Bayar, Y., Gavriletea, M.D.: Foreign direct investment inflows and financial development in Central and Eastern European Union countries: a panel cointegration and causality. Int. J. Finan. Stud. **6**(2), 55 (2018). https://doi.org/10.3390/ijfs6020055

Bayar, Y., Ozel, H.A.: Determinants of foreign direct investment inflows in the transition economies of European Union. Int. J. Res. Comm. Econ. Manag. **4**(10), 49–53 (2014)

Campos, N.F., Kinoshita, Y.: Structural reforms, financial liberalization, and foreign direct investment. IMF Staff. Pap. **57**(2), 326–365 (2010). https://doi.org/10.1057/imfsp.2009.17

Claessens, S., Demirgüç-Kunt, A., Harry, H.: How does foreign entry affect the domestic banking system. J. Bank. Finance **25**, 891–911 (2001)

Cox, S.: Defining emerging markets. Economist (2017). https://www.economist.com/news/special-report/21729866-self-fulfilling-prophecy-defining-emerging-markets

Desbordes, R., Wei, S.J.: The effects of financial development on foreign direct investment. World Bank Policy Research Working Paper, No. 7065 (2014)

Durham, J.B.: Absorptive capacity and the effects of foreign direct investment and equity foreign portfolio investment on economic growth. Eur. Econ. Rev. **48**(2), 285–306 (2004). https://doi.org/10.1016/S0014-2921(02)00264-7

Dutta, N., Roy, S.: Foreign direct investment, financial development and political risks. MPRA Working Paper No. 10186. https://mpra.ub.unimuenchen.de/10186/ (2009)

Dutta, N., Roy, S.: Foreign direct investment, financial development and political risks. J. Dev. Areas **44**(2), 303–327 (2011)

Duttagupta, R., Pazarbasioglu: Miles to go: the future of emerging markets. In: International Monetary Fund, Finance & Development (IMF F&D), 1–9. https://www.imf.org/external/pubs/ft/fandd/2021/06/pdf/the-future-of-emerging-markets-duttagupta-and-pazarbasioglu.pdf (2021)

Enisan, A.A.: Determinants of foreign direct investment in Nigeria: a Markov regime-switching approach. Rev. Innov. Compet. **3**, 21–48 (2017)

Eryılmaz F, Bakır H, Mercan M.: Financial development and economic growth: panel data analysis. In: Olgu, Ö., Dinçer, H., Hacıoğlu, Ü. (eds.) Handbook of Research on Strategic Developments and Regulatory Practice in Global Finance (2015). https://doi.org/10.4018/978-1-4666-7288-8

Fauzel, S.: Modeling the relationship between FDI and financial development in small island economies: a PVAR approach. Theor. Econ. Lett. **6**, 367–375 (2016)

Ibrahim, M., Acquah, A.M.: Re-examining the causal relationships among FDI, economic growth and financial sector development in Africa. Int. Rev. Appl. Econ. 1–19 (2020). https://doi.org/10.1080/02692171.2020.1822299

Kieu, V.T.T., Tien, L.T.: Determinants of variation in human development index before and after the financial crisis: a Bayesian analysis for panel data model. In: Thach, N.N., Ha, D.T., Trung, N.D., Kreinovich, V. (eds.) Prediction and Causality in Econometrics and Related Topics. ECONVN 2021. Stud. Comput. Intell. **983**, 586–608 (2021). Springer. https://link.springer.com/book/10.1007%2F978-3-030-77094-5

Korgaonkar, C.: Analysis of the impact of financial development on foreign direct investment: a data mining approach. J. Econ. Sustain. Dev. **3**(6), 70–79 (2012)

Kreinovich, V., Thach, N.N., Trung, N.D., Thanh, D.V. (eds.): Beyond Traditional Probabilistic Methods in Economics. Cham, Springer (2019). https://doi.org/10.1007/978-3-030-04200-4

Lee, C.C., Lee, C.C., Chiu, Y.B.: The link between life insurance activities and economic growth: Some new evidence. J. Int. Money Financ. **32**, 405–427 (2013). https://doi.org/10.1016/j.jimonfin.2012.05.001

Lee, C.C., Lee, C.C., Chiou, Y.Y.: Insurance activities, globalization, and economic growth: new methods, new evidence. J. Int. Finan. Markets. Inst. Money **51**, 155–170 (2017). https://doi.org/10.1016/j.intfin.2017.05.006

Levine, R.: More on finance and growth: more finance, more growth? Fed. Res. Bank St. Louis Rev. **85**(4), 31–46 (2003)

Majeed, A., Jiang, P., Ahmad, M., Khan, M.A., Olah, J.: The impact of foreign direct investment on financial development: new evidence from panel cointegration and causality analysis. J. Compet. **13**(1), 95–112 (2021). https://doi.org/10.7441/joc.2021.01.06

MSCI.: Market Classification Framework (2014). https://www.msci.com/documents/1296102/1330218/MSCI_Market_Classification_Framework.pdf/d93e536f-cee1-4e12-9b69-ec3886ab8cc8

Nasser, O.M., Gomez, X.G.: Do well-functioning financial systems affect the FDI flows to Latin America. Int. Res. J. Financ. Econ. **29**, 60–75 (2009)

Nguyen, T.H., Thach N.N.: A panorama of applied mathematical problems in economics. thai journal of mathematics. special issue: Annu. Meet. Math. 1–20 (2018)

Nguyen, T.H., Thach, N.N.: A closer look at the modeling of economics data. In: Kreinovich V., Thach N.N., Trung N.D., Van, T.D. (eds) Beyond Traditional Probabilistic Methods in Economics. ECONVN 2019. Studies in Computational Intelligence, vol. 809. Springer, Cham (2019). https://doi.org/10.1007/978-3-030-04200-4_7

Nguyen, T.H., Sriboonchitta, S., Thach, N.N.: On quantum probability calculus for modeling economic decisions. In: Kreinovich V., Sriboonchitta S. (eds.) Structural changes and their econometric modeling. TES 2019a. Studies in Computational Intelligence, vol. 808, pp. 18–34. Springer, Cham (2019a). https://doi.org/10.1007/978-3-030-04263-9_2

Nguyen, T.H., Trung, N.D., Thach, N.N.: Beyond traditional probabilistic methods in econometrics. In: Kreinovich, V., Thach, N.N., Trung, N.D., Van, T.D. (eds.) Beyond traditional probabilistic methods in economics. ECONVN 2019b. Studies in Computational Intelligence, vol. 809. Springer, Cham (2019b). https://doi.org/10.1007/978-3-030-04200-4_1

Sahina, S., Ege, I.: Financial development and FDI in Greece and neighbouring countries: a panel data analysis. Proc. Econ. Fin. **24**, 583–588 (2015)

Saidi, K.: Foreign direct investment, financial development and their impact on the GDP growth in low-income countries. Int. Econ. J. 1–15 (2018). https://doi.org/10.1080/10168737.2018.1529813

Seetanah, B., Padachi, K., Hosany, J., Seetanah, B.: Determinants of financial development: the case of mauritius. SSRN Electron. J. (2010). https://doi.org/10.2139/ssrn.1724404

Shabbir, B., Jamil, L., Bashir, S., Aslam, N., Hussain, M.: Determinants of financial development. A case study of Pakistan. SSRN Electron. J. (2018). https://doi.org/10.2139/ssrn.3122911

Shahbaz, M., Mateev, M., Abosedra, S., Nasir, M.A., Jiao, Z.: Determinants of FDI in France: Role of transport infrastructure, education, financial development and energy consumption. Int. J. Fin. Econ. **26**(1), 1351–1374 (2020). https://doi.org/10.1002/ijfe.1853

Shen, C.H., Lee, C.C., Lee, C.C.: What makes international capital flows promote economic growth? An international crosscountry analysis. Scottish J. Polit. Econ. **57**(5), 515–546 (2010). https://doi.org/10.1111/j.1467-9485.2010.00529.x

Sriboonchitta, S., Nguyen, H.T., Kosheleva, O., Kreinovich, V., Nguyen, T.N.: Quantum approach explains the need for expert knowledge: on the example of econometrics. In: Kreinovich, V., Sriboonchitta, S. (eds.) Structural Changes and their Econometric Modeling. TES 2019. Studies in Computational Intelligence, vol. 808. Springer, Cham (2019). https://doi.org/10.1007/978-3-030-04263-9_15

Svítek, M., Kosheleva, O., Kreinovich, V., Nguyen, T.N.: Why quantum (wave probability) models are a good description of many non-quantum complex systems, and how to go beyond quantum models. In: Kreinovich, V., Thach, N., Trung, N., Van Thanh, D. (eds) Beyond Traditional Probabilistic Methods in Economics. ECONVN 2019. Studies in Computational Intelligence, vol. 809. Springer, Cham (2019). https://doi.org/10.1007/978-3-030-04200-4_13

Thach, N.N., Anh, L.H., An, P.T.H.: The effects of public expenditure on economic growth in asia countries: a bayesian model averaging approach. Asian J. Econ. Bank. **3**(1), 126–149 (2019)

Thach, N.N., Kieu, V.T.T., An, D.T.T.: Impact of financial development on international trade in ASEAN-6 countries: a bayesian approach. In: Thach, N.N., Ha, D.T., Trung, N.D., Kreinovich, V. (eds) Prediction and Causality in Econometrics and Related Topics. ECONVN 2021. Studies in Computational Intelligence, vol. 983, pp. 169–184. Springer, Cham (2021). https://link.springer.com/book/10.1007%2F978-3-030-77094-5

Thach, N.N.: How to explain when the ES is lower than one? A Bayesian nonlinear mixed-effects approach. J. Risk Fin. Manag. **13**(2) (2020). https://doi.org/10.3390/jrfm13020021

The Economist.: Acronyms BRIC out all over. (2011)

Tuan, T.A., Kreinovich, V., Nguyen, T.N.: Decision making under interval uncertainty: beyond Hurwicz Pessimism-optimism criterion. In: Kreinovich, V., Thach, N., Trung, N., Van Thanh, D. (eds.) Beyond Traditional Probabilistic Methods in Economics. ECONVN 2019. Studies in Computational Intelligence, vol. 809. Springer, Cham (2019). https://doi.org/10.1007/978-3-030-04200-4_14

A Markov Chain Model for Predicting Brand Switching Behavior Toward Online Food Delivery Services

Dinh Hai Dung, Nguyen Minh Quang, and Bui Huu Chi

Abstract One food trend that has been emerging among Vietnamese nowadays is ordering food home through an online service. In Vietnam, the online food delivery service industry is currently growing strong with an annual growth rate of the market at 11%. This new online method of food delivery has started to replace the traditional food delivery method due to numerous factors. Furthermore, due to the COVID-19 pandemic, the way people live, work and consume has been changed and probably changes would remain permanent. As a result, this study seeks to discuss the customers' perception of online food delivery service in Vietnam and also aims to predict the customers' brand switching behavior between the five online food delivery companies Grabfood, Now, Gofood, Baemin, and Loship. This goal will be achieved by using the Markov chain model. An online survey was conducted to analyze data from 244 respondents. The result shows that the most crucial factors influencing the use of online food delivery service are convenience (87.01%), followed by price value (70.13%) and travel time saving (65.3%). On the other hand, the majority of non-user prefers homemade meals (76.92%). High shipping cost is the next reason that hinders Vietnamese to order food online. Regarding market shares, Now (35.06%) is the most used online food delivery service company in Vietnam in 2020, followed closely by Grabfood (33.77%) and Baemin (25.97%). However, according to the transition probability matrix of the Markov chain, in the long run Baemin will have the most significant market share among the companies, with 67.7% of Vietnamese using it, followed by 20.9% to Grabfood and 8.9% to Now.

Keywords Markov chain · Prediction · Switching behavior · Brand loyalty · Online food delivery

D. H. Dung (✉) · N. M. Quang · B. H. Chi
Vietnamese-German University, Thu Dau Mot, Binh Duong, Vietnam
e-mail: dung.dh@vgu.edu.vn

N. Ngoc Thach et al. (eds.), *Financial Econometrics: Bayesian Analysis, Quantum Uncertainty, and Related Topics*, Studies in Systems, Decision and Control 427,
https://doi.org/10.1007/978-3-030-98689-6_51

1 Introduction

In 2015, there were up to 540,000 food and beverage stores across Vietnam. 430,000 of them were small-sized restaurant businesses, 7000 were fast-food chains, 22,000 were coffee shops, and more than 80,000 belonged to other well-known brands. Even until now, small-sized family-owned restaurants cover nearly 80% of food and beverage stores in Vietnam. These proprietors offer great varieties of cheap but delicious food for foreign tourists as well as Vietnamese. It is due to this characteristic, Vietnam has been known for good dishes no matter the size of the restaurants. People living in Vietnam have formed a food culture to wander around and seek out quality food from the smallest food stalls to the biggest food chains. However, in recent years, the number of Vietnamese following this culture seems to have dwindled, according to one research studying the reasons for Vietnamese when choosing a restaurant to eat out. Nowadays, the factor of cleanliness and food quality only stands in second place, with only 34% of respondents saying this is the factor that they value. Overshadowing this factor is the factor of distance and range, with 63% of Vietnamese people saying they value places that are closer to their location. This result was recorded during a time when most restaurants in Vietnam did not have a delivery service. So, in order to enjoy restaurant-quality food, most times, people had to travel the distance. Unfortunately, commuting in Vietnam can cause numerous inconveniences due to the growing urban population and traffic congestions. Hence, it may defer customers from choosing a restaurant if it is too far away, despite the quality of food and service.

However, since the appearance of the third party online food delivery mobile service like Delivery Now in July 2015, followed by the arrival of both domestic and foreign services like Grabfood, Gofood, Baemin, and Loship, the online food delivery market in Vietnam has grown exponentially. Since then, many small-sized restaurants in Vietnam have had access to food delivery services by partnering with these online food delivery services. Thanks to this, customers no longer consider the location to be the main issue when choosing a restaurant. Users can also easily access these online food delivery services through their apps or websites due to the rapid increase in mobile phone ownership and internet penetration. From having little to no existence in the food delivery service, online ordering now accounts for almost half of the food delivery orders in major cities.

Retail food delivery is a courier service used by restaurants, stores, or independent food delivery companies to deliver food to customers. The order is usually recognized through phone calls, websites, or through a food ordering company. Traditional food delivery in Vietnam is generally placed through phone calls. The delivery products can vary from entrees, sides, snacks, drinks, desserts to grocery items and are usually contained in bags or boxes. The means of transportation can be cars, motorbikes, or bikes, but mainly by motorbikes because of its flexibility in Vietnam's crowded streets.

The customers can choose to pay online or in-person, through cash or card. The payment usually includes the food price listed on the restaurant menu and a small shipping fee depending on the distance between the restaurant and the received place. Sometimes depending on the promotion from the restaurant or the delivery company, shipping costs may be excluded.

These online food delivery services are also suitable for the Vietnamese lifestyle in many aspects. Due to the increasingly fast urbanization and digitalization in big cities, Vietnamese citizens became too busy. They do not want to waste much time and energy to have good quality food, thus increasing the demand for online food delivery service.

In this paper we set three objectives of research:

1. Identify factors that attract and hinder Vietnamese toward online food delivery service.
2. Discover the most preferred online food delivery service in Vietnam.
3. Build a Markov Chain to model and predict customers' brand switching behavior and predict the future market share of online food delivery service in Vietnam.

2 Literature Review on Online Food Ordering

Alagoz and Hekimoglu (2012) examined the impact of perceived ease of use, perceived usefulness, trust, innovativeness, and external influences on the attitude towards online food ordering. Using data from 231 questionnaires from undergraduate and graduate students in Turkey, the result revealed a significant positive relationship between perceived ease of use, perceived usefulness, trust, innovativeness, external influences, and university students' attitude toward online food ordering.

Tran and Tran (2020) stated that through combining the technology acceptance model (TAM) and Technology, Personal, and Environment (TPE) framework, they have found out the reasons behind the acceptance of online food delivery service in Vietnam. The above-combined model consists of three technical factors, three personal factors, and two environmental factors. In the technology context, the price value factor has the most significant influence on intention to use, perceived ease of use, and perceived usefulness, followed by convenience and information quality. In the personal context, compatibility influences the most, while the effect of prior purchase experience and perceived innovativeness is not confirmed. In the environment context, electronic Word of Mouth (eWOM) and subjective norm have a significant impact on perceived ease of use, perceived usefulness as well as intention to use.

Lee et al. (2017) studied the relation between user-generated information, firm-generated information, design quality, system quality, perceived usefulness, ease of use, attitude, and the intention to use food delivery apps. Primary data from 350 questionnaires were used to test the research model using Structural Equation Modeling (SEM). The result shows that user-generated information, firm-generated information, and system quality significantly affected perceived usefulness. System quality and design quality of the food delivery apps greatly influenced the perceived ease of use, which improved perceived usefulness. In turn, perceived ease of use and perceived usefulness both affected customers' attitudes toward the use of food delivery mobile apps.

Prabhash (2020) identified the factors that attract Kerala youths toward online food delivery service and analyzed the advantages and disadvantages of the online food delivery system. Primary data is collected from 250 respondents' surveys by using random sampling as the sampling technique. The study reveals that offers and discounts are the most significant factors attracting youths toward online food delivery service, followed by convenience and ease of payment. On the other hand, unawareness, fear of online payment, and trust are some factors that prevent customers from using online food delivery services.

Chai and Yat (2019) studied the relationships of several factors like perceived ease of use, time-saving orientation, convenience motivation, privacy, and security with the behavioral intention toward online food delivery services among Malaysian urban dwellers. The authors used Performing Partial Least Square–Structural Equation Modeling (PLS–SEM) as a data analysis technique to analyze data from 302 respondents' surveys. The result shows that time-saving orientation, convenience motivation, privacy, and security factors have a positive effect on behavioral intention toward adopting online food delivery services. In contrast, the perceived ease of use factor was not significant enough to affect behavioral intention.

Nguyen (2019) studied what factors keep Vietnamese Millennials and Generation Z continue using Delivery Now, an online food delivery service. The study used both primary and secondary data. Preliminary data are gathered from interviews with Delivery Now users and online surveys distributed to Vietnamese. Secondary data are collected from books and peer-reviewed articles. The result shows that food delivery services should focus on using social media to communicate, connect, and display the brands' personalities to young Vietnamese. The main factors that convince the target group to pick one food delivery service over the others are convenience, speed, and quality. However, to ensure the target group stays loyal to one online food delivery service, the company needs to consider the factor of trust.

Kim Dang (2018) examined how the Internet has changed consumers' food-buying behavior and identifies its associated factors and consumers' concern about online food products' safety information. To collect data, researchers gathered 1736 customers from Hanoi to participate in face-to-face interviews using structured questionnaires. The study shows that 81.3% of the respondents use the Internet to search for food products. Convenience (69.1%) and price (59.3%) are the most crucial factors influencing participants to purchase food products through the Internet. However, only one-third of the participants select products based on accurate information about food safety certification. 51.6% of the respondents concerned about the expiration date, while brand and food licensing information were often neglected.

Doan (2013) discovered factors that influence the demand of customers in using online food delivery service in Vietnam. The study stated that, to increase the demand of customers, online food delivery services need to understand and satisfy customers' expectations and increase customers' intention to recommend the service. Intention to recommend is derived from customer loyalty and repurchase intention. However, customer expectation is influenced by many factors like medium characteristics, culture, shopping behavior, and merchant characteristics. The benefit of the service must be higher than the customer's expectation to attract and retain customers.

Chandra and Cassandra (2019) identified stimulus factors that make people interested in online ordering for food delivery. Data from 187 respondents was used to answer the research question using the Stimulus Organism Response Model (SOR Model). The result showed that the privacy of customers and informativeness both had a significant effect on the customer's value in terms of ordering food via an online application, while perceived ease of use, perceived usefulness and facilitating state are the less significant stimulus factors.

Ray et al. (2019) applied the uses and gratifications (U&G) theory to determine eight main gratifications of using online food delivery apps. This study applied a mixed-method research approach costing 125 open-ended essays and 395 online cross-sectional surveys. The result showed that customer experience (coupon discounts loyalty program, advertisement), ease of use (order placement process, a feature to track order progress and filter options), and ability to search for restaurants easily are positively associated with the intention to use food delivery apps. On the other hand, societal pressure, delivery experience, quality control, and convenience do not significantly associate with the use of food delivery apps. Especially, the listing factor had a significantly negative association with said intention.

3 Current Situation of Food Delivery Service in Vietnam

Seeing the numerous disadvantages of traditional food delivery and the rising demand for online food ordering, both foreign and local companies had emerged to seize this investment opportunity in Vietnam. Online food delivery service was introduced to Vietnamese to fill in the vacant market of food delivery services.

Online food delivery is a courier service where the food ordering process is done entirely through a website or a mobile application. After an order is registered, food gets delivered by the restaurant or a third-party delivery company. With the rise of technology, many food delivery models aim to catch customers' needs and simplify their shopping life. There are three main models for food delivery services (Table 1).

The logistics-concentrated business model provides a website based or mobile-based platform, with menus provided by the company's partnered restaurants. This model helps the delivery and order process easier for both restaurants and customers.

Table 1 Differences in online prepared food delivery models (Nguyen 2019)

	Ordering	Cooking	Delivery
Logistics-concentrated	Yes	No	Yes
Aggregator service	Yes	No	No
Full-service on-demand	Yes	Yes	Yes

The company supplies restaurants with external delivery staff. Since restaurants do not have to worry about delivery, they can focus more on improving their food. These companies earn revenue by charging restaurants with commission and customers with delivery fees. In Vietnam, companies like Now, Grabfood, Gofood, Baemin, and Loship are currently following this business model.

The aggregator business model plays the role of a mediator, connecting between customers and restaurants. The ordering process is usually done through a website or application. Because this business model doesn't supply delivery staff, the restaurants have to rely on their own employees to deliver the food. Nonetheless, the aggregator model helps increase the visibility of its partnered restaurants on the internet while also facilitating customers' ordering process. In Vietnam, Vietnammm is one food delivery company following this business model.

The vertically integrated or full-service on-demand delivery business model is mainly adopted by cloud kitchens that do not have a brick-and-mortar location and do not support dine-in experience. These restaurants handle all the ordering, delivery, and cooking process on their own. Smartmeal, Fitfood, and Flavorbox are examples of this model in Vietnam. These companies focus on providing healthy meals to customers.

According to a report done in 2016, the most common way of food ordering is still the traditional food ordering method, with Vietnamese sitting next to a phone with a menu of the local KFC or Dominos in their hands to order food. Ordering through phone still makes up 71% over the other ordering methods, but mostly done by people from 30 to 39 years old. Younger people from 18 to 29 years old prefer a more technological ordering method, which is through mobile phone applications.

However, in July 2015, Foody Corporation, famously known for its platform as a restaurants' review website written by members, established a delivery subsidiary company. Delivery Now is a third-party online food delivery company that follows the logistics-concentrated model, which provides the partnered restaurants with logistics, technology, and marketing to efficiently run their delivery service.

With the advantages of being the first mover in the third-party food delivery service and established under the well-known Foody Corporation, Delivery Now has the highest penetration rate at its appearance, with 76.8% of Vietnamese knowing about it. The application is reported to have up to 10,000 orders a day in 2016. However, Delivery Now was only in a monopoly online food ordering market until 2018, with the appearance of both domestic companies like Lozi's Loship, Vietnammm, Lixibox, Ahamove's Lala together with other international brands like Grab's Grabfood and Gojek's Gofood.

With the arrival of other competitors, Vietnam's online food delivery market had experienced some rearrangement in its market position. Grabfood was gradually replacing Delivery Now to become the market leader in the third quarter of 2019. However, other newcomers did not have as bright an outcome as Grabfood due to the fiercely competitive market. Ahamove's Lala attempted to penetrate this new market segment, only to depart after one year of operation since December 2018. Another example is Vietnammm, a delivery system famous in Vietnam's ex-pat community. It was one of the market leaders for many years and was even able to acquire its competitor FoodPanda in late 2015. Despite this, Vietnammm could not secure its position and was purchased by South Korea's new venture Woowa Brothers Corp. This acquisition opened a unique opportunity for their startup Baemin to enter Vietnam's market. The others, Loship and Lixibox, had been inactive for months.

At the start of 2020, with the COVID-19 pandemic spreading rapidly worldwide, customer demand for food delivery has sky-rocketed. To minimize the risks of transmitting the virus, Vietnam's government had ordered the Vietnamese to effectuate social isolation. People were instructed to stay at home, avoid going out, and limit social interaction unless absolute necessity. Vietnam's government also limited non-essential services, including restaurants, and prohibits crowds of more than 30 people. Vietnamese relied on third-party food delivery applications to order food in the safety of their homes. Without dining-in customers, most Vietnamese restaurants and coffee shops had to convert to using a delivery model to continue operating in this challenging period. Those restaurants and shops also benefit significantly from third party delivery applications.

4 Methodology

Due to the fierce competition among online food delivery companies in Vietnam, this study is established to analyze the factors that attract and hinder Vietnamese toward online food delivery services, in order to help these companies gain market shares. Moreover, the forecast of future market share and customers' brand switching behavior can provide meaningful guidelines for each company to adjust their strategies in order to retain old customers and attract potential ones.

The area of study is in Vietnam, with the respondents from Ho Chi Minh City, Binh Duong Province, Ha Noi, Can Tho, Buon Me Thuot, Bien Hoa, Long An, Quy Nhon, Tay Ninh, Vung Tau. For this study, inductive reasoning is executed. Inductive reasoning is a method of reasoning in which general principles are derived from observed data. This paper employs quantitative research methods, which are used for interpreting data from surveys. Both primary and secondary data are used. Primary data is collected through online surveys conducted by the authors. Secondary data consists of information from trustworthy Internet-based articles, scientific journals and books. 244 Vietnamese were selected as samples through the method of convenience sampling. The collected data is analyzed with appropriate statistical tools. The Homogeneous Markov chain model is also used to predict future market share and customers' brand switching behavior between online food delivery companies in Vietnam.

Suppose X_n with $n \geq 0$ denotes a random variable on discrete space S. The sequence $\{X_n, n \geq 0\}$ is called a stochastic process. This stochastic process is said to be a Markov process if for all state value $i_0, i_1, i_2, \ldots\ldots, i_n \in I$, it possesses the following property:

$$P\{X_{n+1} = j/X_0 = i_0, X_1 = i_1, \ldots\ldots\ldots\ldots\ldots\ldots, X_n = 1\} = P\{X_{n+1} = j/X_n = i\}$$

where, i_0, i_1, i_2, ……, in are the states of state space I. This type of probability is called the Markov Chain probability. The probability of transitioning to state j at time (n + 1) depends only on the current state i at time n, and not on its history prior to time n. This means that the probability of moving to the next state depends only on the present state, and not on the previous states.

The transition probability is the probability associated with the transition from state i to state j. This is also called the one-step transition probability:

$$P\{X_{n+1} = j/X_n = i\} = p_{ij}$$

The Markov chain is homogenous if the transition probabilities above are independent of time (n)

$$P\{X_{n+1} = j/X_n = i\} = P\{X_1 = j/X_0 = i\} = p_{ij} \text{ which } 0 \leq p_{ij} \leq 1 \text{ and } \sum p_{ij} = 1$$

However, if a Markov chain has initial probability vector X_0 and transition matrix P_{ij}, the probability vector after n repetition X_n is calculated by multiplying the initial probability vector X_0 with the transition matrix P_{ij} after nth step P_{ij}^n, which defines the stable state probability vector.

$$X_n = X_0 * P_{ij}^n$$

In order to find out the values for computing the initial vector, we need to conduct the survey and collect data of the market shandings & Interpretatiores of the five players in the market. Then, by setting up the transition probability matrix we can compute the stationary distribution and probability of switching behavior in order to get to the state of the next time step. It is also important to notice that we make an assumption for the market size to stay stable, i.e. we consider neither new customers nor loss of current customers.

5 Findings and Interpretation of Results

5.1 The Current State According to Survey Results

The data shows that, out of 244 respondents, 52.46% of the respondents are male and 47.54% are female. 49.18% of the sample size is from 18 to 22 years old, 48.36% is from 22 to 30 years old, with only a small percentage of 2.46% is over 30 years old. More than 75% of the participants are currently living in Ho Chi Minh City, with the rest of the participants being from the previously mentioned areas. A total of 231 people, covering almost 95% of respondents, have used online food delivery services, and only 13 have not used any kinds of online food delivery services (Fig. 1).

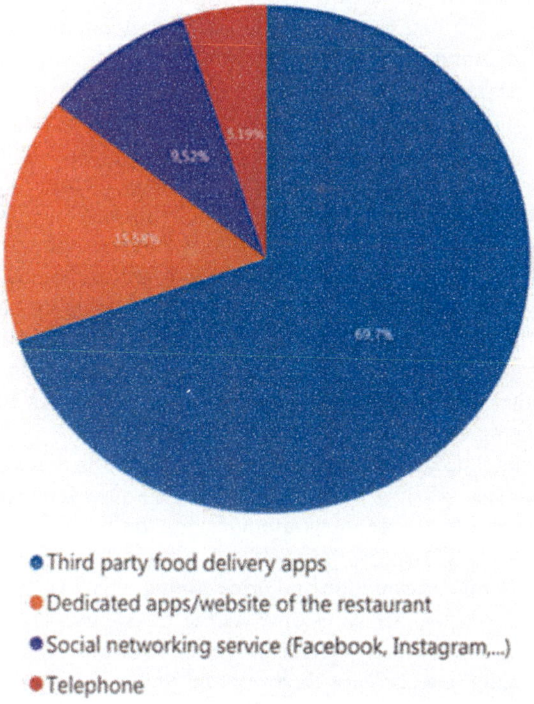

Fig. 1 Survey result for online food ordering methods

Nearly 70% of respondents use third-party food delivery apps to order food and drink online. Using these apps can allow consumers to search and choose between a variety of restaurants, without going to different dedicated restaurant apps and websites to compare information and prices. However, using a dedicated ordering app from restaurants like Dominos, KFC, Lotteria can give customers a closer look into the restaurant, offer a more exclusive experience to customers. With 62 million active social media users in Vietnam that spend an average of 2 h 32 min a day using social media, restaurants can use social network pages like Facebook or Instagram as an alternative ordering channel. Finally, ordering through phone calls has a small percentage of usage by the participants.

Convenience is the most notable factor that attracts youth in using an online food delivery service. After an 8 h working shift, commute time, and curricular activities, it can be exhausting to come home and start cooking dinner. Even getting take-outs is so energy-consuming. Furthermore, people are reluctant to go outside and get a chance of exposure to the virus during this COVID-19 pandemic. This is when the convenience of apps like Delivery Now or Grabfood comes to help. These apps allow users to rest in the comfort and safety of their homes while waiting for hot food to be delivered right to the door. This way, customers can enjoy food without the inconvenience, time, and effort of making it or picking food up at the restaurant. Moreover, these apps only require smartphones with 3G or 4G subscription to fully access and use anytime, anywhere.

Price value also holds an important place for youngsters in Vietnam. Most of the participants are still students or just joining the workforce with little to no money. Being offered with promotions, discounts, or free shipping codes by these online food delivery services can significantly increase the intention to use online food delivery services. According to a Q&Me survey, 67% of Vietnamese tend to look for restaurants with coupons, and 85% favor restaurants that offer free shipping.

With Vietnam having the second-largest motorbike ownership in the world, and having small, low-quality streets; the number of bicycles, motorbikes, cars, buses commuting simultaneously on the road is tremendous. This can cause a light 30-min traffic jam up to 3-h long traffic jams during rush hours. With the honking from other drivers and the smoke from various vehicles, going to a restaurant to take food can be very time consuming and frustrating for the customers. With the COVID-19 pandemic also happening in Vietnam, the incentives for consumers to go out also reduced. 65.37% of the respondents are attracted to the time-saving travel aspect and the ability to get food without leaving the house, which these online delivery services provide.

While information and price comparison, ease of payments, safety from COVID-19, and quick delivery time are the secondary factors that also engage consumers in using online food delivery services as shown in Fig. 2b, the majority of non-users usually can cook for themself and prefer home-made meals more than meals from restaurants. They do not see the hassle of preparing and cooking food. High shipping costs are the next reason that hinders the intention for consumers to use the services. Cold and unappetizing food after slow delivery time, complicated ordering process, and trust issues are the other problems that these online food delivery services should solve in order to attract more customers, see Fig. 3a and b.

Regarding the most important insight we want to gain, 95% of the market share for online food delivery services belongs to the three big players Now, Grabfood and Baemin, with Now having the most market share at 35.06%, followed closely by Grabfood with 33.77%, next is the new start-up Baemin covering almost 26%. This is really impressive due to Baemin only joined the Vietnam online food delivery service a year ago at the end of 2019. Only 5% of the survey participants use GoFood and Loship. These values are the input as an initial state for the Markov model (Fig. 4).

5.2 The Future State—Markov Model to Predict the Future Market Shares

The basic characteristic of the Markov chain is that the future state only depends on the current state. This offers a unique prediction power, since we do not need any further information because all previous information has been embedded in

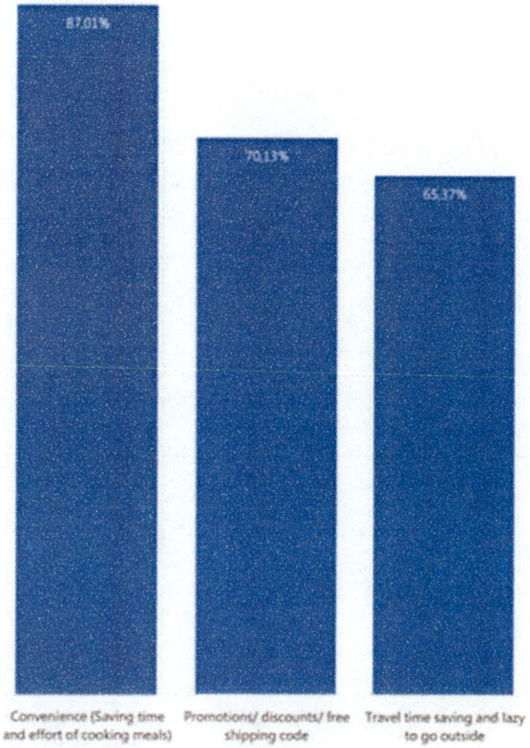

a

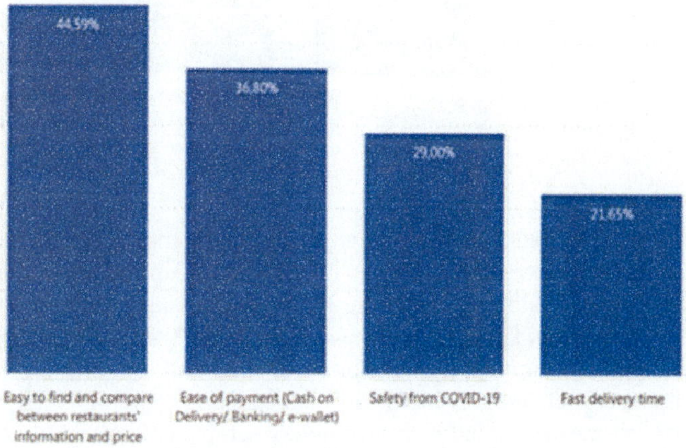

b

Fig. 2 **a** Primary factors that attract customer to online food ordering. **b** Secondary factors that attract customer to online food ordering

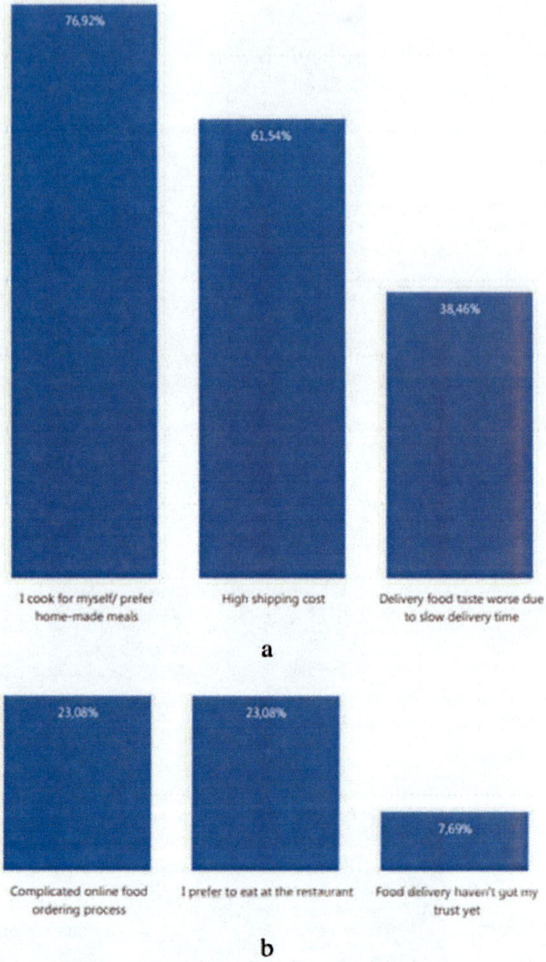

Fig. 3 **a** Primary factors hindering customers from using online food ordering. **b** Secondary factors hindering customers from using online food ordering

the current state. If a consumer's intention to switch is known, the market can be modeled to indicate future market shares. Thus, Markovian models can be applied in examining customer's brand-switching behavior to predict future market share based on the current brand switching behavior. We show the future relative ranking and market shares of the competing brands in the market.

As mentioned earlier, a survey has been conducted on 244 participants in Vietnam asking about their most-used online food delivery service in 2019 and 2020 to study the brand switching behavior. For setting up the Markov chain, we consider six states of transition for customers: Grabfood, Now, Gofood, Baemin, Loship, and Do not use. From this, a transition matrix of customers' brand switching is formed, which has 36 state spaces. Next, the initial probability vector in 2019 is multiplied with the transition matrix to calculate the probability vector in 2020. The market share

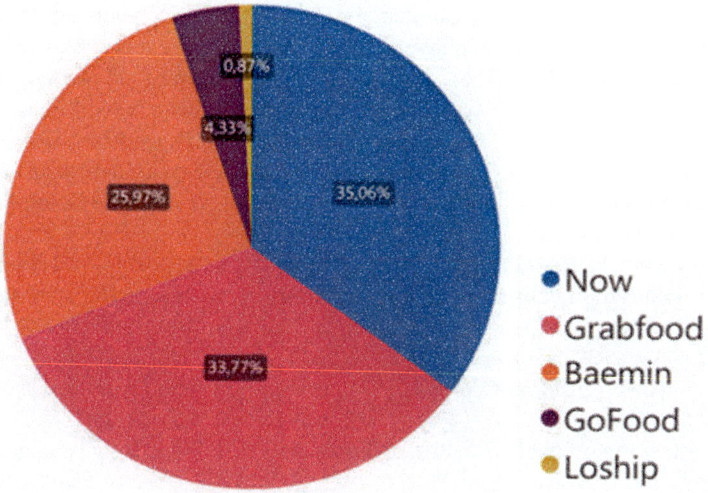

Fig. 4 Current market shares among competitors

in 2021 and in the long run can then be predicted. Table 2 provides survey result as basis for computing the transition probability matrix of brand switching behavior on online food delivery service.

Table 2 Survey result of customers' brand switching behavior

	Grabfood 2020	Now 2020	Gofood 2020	Baemin 2020	Loship 2020	Do not use 2020	Total customers 2019
Grabfood in 2019	54	13	3	23	1	2	**96**
Now in 2019	14	60	0	20	0	1	**95**
Gofood in 2019	0	2	6	3	0	0	**11**
Baemin in 2019	1	0	0	8	0	0	**9**
Loship in 2019	2	1	0	1	1	0	**5**
Did not use in 2019	7	5	1	5	0	10	**28**
Total customers in 2020	**78**	**81**	**10**	**60**	**2**	**13**	**244**

We briefly explain what the numbers in Table 2 mean. Each number explains how many users of an online food delivery service in 2019 switch to others in 2020. For example, in the Grabfood's row, the total demand for Grabfood is 96; there are 54 participants still using Grabfood in 2020; 13 people switch to Now; 3 goes to Gofood from Grabfood; 23 respondents move to Baemin; 1 person transfer to Loship, and 2 people change from using Grabfood in 2019 to do not use any food delivery service in 2020. The same description can be done on the rows of Now, Gofood, Baemin, Loship and Did not use.

There are six states (Grabfood, Now, Gofood, Baemin, Loship, and people do not use any online food delivery services) in this study; therefore, the state space is of the form below:

$$S = \{Grabfood, Now, Gofood, Baemin, Loship, Do\ not\ use\}$$

The transition probability matrix of customers' brand switching behavior can also be constructed as 6×6 matrix P_{ij}:

$$\mathbf{P}_{ij} = \begin{bmatrix} 0.563\ 0.135\ 0.031\ 0.240\ 0.010\ 0.021 \\ 0.147\ 0.632\ 0.000\ 0.211\ 0.000\ 0.011 \\ 0.000\ 0.182\ 0.545\ 0.273\ 0.000\ 0.000 \\ 0.111\ 0.000\ 0.000\ 0.889\ 0.000\ 0.000 \\ 0.400\ 0.200\ 0.000\ 0.200\ 0.200\ 0.000 \\ 0.250\ 0.179\ 0.036\ 0.179\ 0.000\ 0.357 \end{bmatrix}$$

Next, the initial probability state vector X_0 was the initial market share of online food delivery services in 2019. X_0 is calculated by dividing the total number of respondents of each state to the total respondents. The initial state vector X_0 in 2019 is

$$X_0 = (0.3930\ 0.389\ 0.045\ 0.037\ 0.020\ 0.115)$$

The Markov chain model suggests that the state probability for various periods can be obtained by multiplying the initial probability state vector and the transition probability matrix $X_1 = X_0 * P_{ij}$. So, the state probability for online food delivery service in 2020 will be:

$$X_1 = X_o \times P_{ij} = (0.393\ 0.389\ 0.045\ 0.037\ 0.020\ 0.115)$$

$$\times \begin{bmatrix} 0.563\ 0.135\ 0.031\ 0.240\ 0.010\ 0.021 \\ 0.147\ 0.632\ 0.000\ 0.211\ 0.000\ 0.011 \\ 0.000\ 0.182\ 0.545\ 0.273\ 0.000\ 0.000 \\ 0.111\ 0.000\ 0.000\ 0.889\ 0.000\ 0.000 \\ 0.400\ 0.200\ 0.000\ 0.200\ 0.200\ 0.000 \\ 0.250\ 0.179\ 0.036\ 0.179\ 0.000\ 0.357 \end{bmatrix}$$

$$X_1 = (0.320\ 0.332\ 0.041\ 0.246\ 0.008\ 0.053)$$

The above result is the market share in 2020 for Grabfood, Now, Gofood, Baemin, Loship, and people do not use any services, respectively. The above result shows that in 2020, 33.2% of the market share belong to Now, next is 32% of market share by Grabfood, 24.6% to Baemin, 4.1% to Gofood, and 0.8% to Loship. There is also a 6.2% decrease in the number of non-users. The result implies that Now will overtake Grabfood to have the most significant market share in 2020, what actually happens according to the survey result itself. Moreover, Baemin is also expected to make a substantial net gain of 20.9% in only a year. Using the predicted market share in 2020 X_1, the expected market share for 2021 X_2 can also be calculated by:

$$X_2 = X_1 * P_{ij} = (0.273\ 0.271\ 0.034\ 0.388\ 0.005\ 0.029)$$

The forecasted market share in 2021 is experiencing changes in ranking. With a 38.8% market share, Baemin is expected to be the market leader in 2021. Grabfood is in second place with 27.3% of market share, followed closely by Now with 27.1%. Gofood will cover 3.4% and in the last position is Loship with only 0.5% of market share. The people that will not use any online food delivery service in 2021 make up for 2.9%.

5.3 *The Steady State*

The same process can be repeated to predict the market share of online food delivery services in 2022, 2023, 2024, etc. i.e. in the long term. If the process keeps going on, the change between the $X_3, X_4, X_5, \ldots, X_n$ becomes smaller and smaller; and the state probability vector converges to a particular set of values. When this happens, the state probability vector is stable and can be used to forecast the probable future state of the market in the long run. This stable state probability vector X_n can be determined using the nth step transition probability matrix.

$$X_n = X_0 * P_{ij}^n = (0.209\ 0.089\ 0.015\ 0.676\ 0.003\ 0.008)$$

The above stable state probability vector X_n shows that in the long run, there will be 20.9, 8.9, 1.5, 67.6, 0.3% Vietnamese use Grabfood, Now, Gofood, Baemin and Loship, respectively; and only 0.8% Vietnamese will not use any online food delivery services. The implication is that in the long run, Baemin is expected to have the highest share of the Vietnam online food delivery market, followed by Grabfood, Now, Gofood, and lastly, Loship. Given the initial market share in 2019, Now is expected to concede a net loss of 30% of its market share in the long run while Baemin is expected to make a net gain of about 63.9% market shares. All other online food delivery companies are expected to make net losses.

6 Conclusions and Implications

The paper reveals that most of the participants have used online food delivery services. Convenience is the most influential factor in attracting the use of online food delivery services. Price values, saving time and effort to go outside, information and price comparison, ease of payments, safety from COVID-19 and quick delivery time are the other advantages enjoyed by the participants. On the other hand, homemade meals preference, high shipping cost, cold food, complicated ordering process, and trust issues are some factors that hinder customers from using online delivery service. Third-party food delivery apps are also the most used method of online food ordering. Applying the Markov chain, we found out that Now is the most preferred online food delivery service in Vietnam in 2020, followed closely by Grabfood and Baemin. However, also by using the Markov chain, the study predicts that in the long run, Baemin will hold the majority of the food delivery service market share.

There are limitations and potentials for future studies. Firstly, the scope of the survey is small and it is mainly done by young Vietnamese from Generation Z and Millennials, mainly from 18 to 30 years old, which do not cover the other generations like Gen X or Baby Boomers. So, the result may not completely represent the whole Vietnamese population. Moreover, due to our choice for the time range between the change in market share being quite large, future studies can try to reduce the time range to 6 or 3 months, in order to have a more accurate future market share forecast by observing the change in states much more frequently. Yet, the paper offers a good insight into one of the megatrends in the coming year which is online food delivery and important managerial implications can be drawn from the achieved results.

References

Alagoz, S.M., Hekimoglu, H.: A study on Tam: analysis of customer attitudes in online food ordering system. Procedia—Soc. Behav. Sci. **62**, 1138–1143 (2012)

Chai, L.T., Yat, D.N.C.: Online food delivery services: making food delivery the new normal (2019)

Chandra, Y.U., Cassandra, C.: Stimulus factors of order online food delivery. In: 2019 International Conference on Information Management and Technology (ICIMTech), no. 1, pp. 330–333 (2019)

Doan N. H. (2013). Demand creation of online services for B2B and consumer market-Food delivery in Vietnam (Master's thesis)

Kim Dang, A., Xuan Tran, B., Tat Nguyen, C., Le Thi, H., Thi Do, H., Duc Nguyen, H., Ngo, C.: Consumer preference and attitude regarding online food products in Hanoi, Vietnam. Int. J. Environ. Res. Pub. Health **15**(5), 981 (2018)

Lee, E.-Y., Lee, S.-B., Jeon, Y.J.J.: Factors influencing the behavioral intention to use food delivery apps. Soc. Behav. Personal.: Int. J. **45**(9), 1461–1473 (2017). https://doi.org/10.2224/sbp.6185

Nguyen, M.: How Food Delivery Services in Vietnam Accommodate Millennials and Generation Z: Case company: Delivery Now (2019)

Prabhash, M.A.: The consumer perception on online food delivery system among youth in Kerala. EPRA Int. J. Multidisc. Res. (IJMR) **6**(2), 96 (2020)

Ray, A., Dhir, A., Bala, P.K., Kaur, P.: Why do people use food delivery apps (FDA)? A uses and gratification theory perspective. J. Retail. Consum. Serv. **51**, 221–230 (2019)
Tran, T.A., Tran, Y.V.T.: An empirical analysis of the factors affecting consumers adoption of online food delivery in Vietnam. Econ. Manag. Bus. **1374** (2020)

Credit Rating Models for Firms in Vietnam Using Artificial Neural Networks (ANN)

Quoc Hai Pham, Diep Ho, and Sarod Khandaker

Abstract This study investigates the factors that affect the credit rating model for small, medium and large firms in Vietnam in the period 2015–2018. Since Altman's Z-score model (1968), the accuracy of credit rating models fluctuates significantly based on country and period (Altman and Sabato 2007; Tsai et al. 2009; Altman et al. 2017; Jones et al. 2017; Pham et al. 2018). This study uses the dataset including more than 30,000 firms in Vietnam both small, medium and large firms from the Orbis database. In addition, the dependent variable in this study includes ten classes (AAA, AA, … to D) that uses Artificial Neural Networks (ANN) to maximise the model's predictive power (Jones et al. 2017; Sigrist and Hirnschall 2019). The result shows that the significant predictors of the ANN model for the dataset 2015–2018 are NITA (Net Income to Total Assets), ROE (Returns on Equity), the current ratio, the solvency ratio and GDPG (GDP growth rate). The accurate rate of the ANN model is 61.11%, which is the highest accuracy rate compare to the OLR of 55.43% and MDA of 53.1%. This result indicates that ANN model is the most robust model which use less inputs with higher accurate rate.

Keywords Credit ratings · Credit classification · Artificial Intelligence (AI) · Artificial Neural Networks (ANN) · Firms in Vietnam

---

Q. H. Pham (✉) · D. Ho
International University – Vietnam National University, Ho Chi Minh City, Vietnam
e-mail: haipq@uef.edu.vn

D. Ho
e-mail: hdiep@hcmiu.edu.vn

Q. H. Pham
University of Economics and Finance Ho Chi Minh City, Ho Chi Minh City, Vietnam

S. Khandaker
Swinburne University of Technology, Melbourne, VIC, Australia
e-mail: skhandaker@swin.edu.au

N. Ngoc Thach et al. (eds.), *Financial Econometrics: Bayesian Analysis, Quantum Uncertainty, and Related Topics*, Studies in Systems, Decision and Control 427,
https://doi.org/10.1007/978-3-030-98689-6_52

1 Introduction

The forces of globalisation, financial deregulation and innovation have not reduced the importance of credit risk, even though the market and off-balance-sheet risks have received much more attention during recent periods of chaos in international financial markets (Psillaki et al. 2010). However, credit risk remains the most significant risk to Financial Institutions as the overwhelming effects of the credit crisis confirms. The New Basel Capital Accord explicitly requires banks to apply the recommended internal credit risk management practices to evaluate their capital adequacy needs. Through effective credit risk management, banks not only support the sustainability and growth of their businesses but also contribute to systemic stability and efficient capital allocation in the economy (Psillaki et al. 2010).

In 2019, the number of newly-established firms in Vietnam is 138,139 according to the General Statistics Office of Vietnam (GSO), but 28,731 firms declare bankruptcy which increases of 5.9% compared to 2018 (GSO 2020). Moreover, 43,700 firms suspended operations to wait for the default process which also climbs 41.7% compared to 2018. Finally, 16,840 firms complete default procedures which increases 3.2% compared to 2018 (GSO 2020). These numbers confirm the crucial role of a reliable credit rating system that can evaluate the firms' financial health to classify good and bad borrowers effectively. As a result, these credit classifications could help lending decisions for banks, investors and related stakeholders.

In the literature over time, there are two main ways to increase the accuracy of the credit rating model including improving the independent variables and improving the prediction methodology (Altman and Sabato 2007; Psillaki et al. 2010; Wu et al. 2010; Duan et al. 2012; Giordani et al. 2014; Jones et al. 2017; Cornée 2019; Muñoz-Izquierdo et al. 2019; Sigrist and Hirnschall 2019). This research aims to build a credit rating model for small, medium and large Vietnamese firms in the period 2015–2018 and investigates the effects of the local accounting system to the credit rating model. This research significantly contributes to the existing literature for several reasons.

First, this research is significant because most of the influential studies analysed data before 2016 and mostly in US and European contexts (Altman and Sabato 2007; Psillaki et al. 2010; Wu et al. 2010; Duan et al. 2012; Giordani et al. 2014; Jones et al. 2017; Sigrist and Hirnschall 2019; Cornée 2019; Muñoz-Izquierdo et al. 2019). Since Altman's Z-score model (1968), the accuracy of credit rating models have needed updating and rebuilding because it varies significantly based on country and period (Altman and Sabato 2007; Tsai et al. 2009; Altman et al. 2017; Jones et al. 2017; Pham et l. 2018a, b). This study builds a credit rating model for Vietnamese small, medium and large firms in 2015–2018. The dataset includes 39,162 both small, medium and large firms in Vietnam in the period 2015–2018 from the Orbis database.

In addition, Pham et al. (2018a, b) built a default prediction model for Vietnamese listed firms in the period 2003–2016 which only focused on default and non-default (two classes of dependent variable) prediction. This research investigates a credit rating model with ten classes of dependent variable (DV) which is more informative and powerful in predicting the default probability of Vietnamese firms in advance as well as supporting lending decisions. Also, this research uses small, medium and large Vietnamese firms in the period 2015–2018, which better represents a generalisation of Vietnamese firms. As a result, the dataset of this research had significant advantages compared to the study of Pham et al. (2018a, b) by using (1) a more informative DV with ten categories; (2) a bigger dataset including small, medium and large Vietnamese firms; and (3) a current period 2015–2018.

Another significant contribution to the literature is the set of Independent Variables (IV). In the literature, many possible ratios have been identified as helpful in predicting the credit rating of firms. IVs are usually accounting ratios derived from financial statements and include measurements of profitability, liquidity and leverage in a credit rating model (Altman and Sabato 2007; Wu et al. 2010; Giordani et al. 2014; Pham et al. 2018a, b; Sigrist and Hirnschall 2019; Muñoz-Izquierdo et al. 2019). Some studies also include market-based variables and macroeconomic indicators such as stock volatility and past excess returns, GDP or annual interest rates (Wu et l. 2010; Jones et al. 2017; Pham et al. 2018a, b). The recent literature concludes that quantitative variables are not enough to predict business default and that qualitative variables can improve the predictive power of a model (such as firm size and firm age) (Altman and Sabato 2007; Psillaki et al. 2010; Wu et al. 2010; Duan et al. 2012; Giordani et al. 2014; Muñoz-Izquierdo et al. 2019; Sigrist and Hirnschall 2019; Cornée 2019). In this research, the IVs include four main categories: financial indicators, market indicators, firm characteristic indicators and macroeconomic indicators. This research study also ranks each IV based on their impacts on each model, so that this research can identify which IV is the most significant and which is less significant to the credit rating model.

The third significant contribution to the literature is the research methodology. To illustrate, several researchers in the credit rating field have used two categories of DV, 'default' and 'non-default' (two classes) (Altman and Sabato 2007; Psillaki et al. 2010; Wu et al. 2010; Duan et al. 2012; Giordani et al. 2014; Muñoz-Izquierdo et al. 2019; Sigrist and Hirnschall 2019; Cornée 2019). This research classifies credit ratings into ten classes (from AAA to D), which are more informative and better credit ratings compared to two-class DVs. However, the challenge in building more than two classes of DV in a credit rating model is the significantly larger amount of data required to build the model. Machine learning models such as NNs and SVMs are the most robust in term of predictive accuracy, but these models are hard to interpret and apply because they have a 'black box' in the analysis process (Jones et al. 2017; Sigrist and Hirnschall 2019). This research opens the 'black box' of ANN using the Simulink tool of MATLAB R2019a to maximise the predicting power of the credit rating model as well as increasing the interpretive power of the model.

2 Literature Review and Hypothesis Development

2.1 Studies on Factors Affecting the Credit Rating Model

There are many different version of the definitions of credit rating and financial distress because the credit health of a firm has many stages from healthy to bankruptcy (Argenti 1976; Lukason and Hoffman 2014; and Muñoz-Izquierdo et al. 2019). In this research, the credit rating model is an analytical technique that uses the customer's credit information to classify whether the credit customers can pay their loan on time (Altman and Sabato 2007; and Tsai et al. 2009). The dependent variable (DV) is usually a dichotomous variable where the default is 1 and the non-default is zero (Altman and Sabato 2007). The DV can be changed from 'default' or 'non-default' to the probability of default or bankruptcy, depending on the statistical methodology used (Altman and Sabato 2007; Psillaki et al. 2010; Wu et al. 2010; Duan et al. 2012; Giordani et al. 2014; Jones et al. 2017; Pham et al. 2018a, b; Jiang et al. 2018; Muñoz-Izquierdo et al. 2019; Sigrist and Hirnschall 2019; Cornée 2019).

In the literature, many possible ratios are identified as helpful in predicting the default of companies. The independent variables (IVs) are often accounting ratios derived from financial statements and include measurements of profitability, liquidity and leverage in the credit rating models (Altman and Sabato 2007; Wu et al. 2010; Giordani et al. 2014; Pham et al. 2018a, b; Sigrist and Hirnschall 2019; Muñoz-Izquierdo et al. 2019). Some studies also include market-based variables and macroeconomic indicators such as stock volatility and past excess returns, GDP or annual interest rates (Wu et al. 2010). Recent literature has concluded that quantitative variables are not enough to predict business default and that qualitative variable can improve the predictive power of the model (such as firm size and firm age) (Altman and Sabato 2007; Psillaki et al. 2010; Wu et al. 2010; Duan et al. 2012; and Giordani et al. 2014; Muñoz-Izquierdo et al. 2019; Sigrist and Hirnschall 2019; Cornée 2019). This research divides the IVs into four main categories that are significant in the literature, including financial ratios, macroeconomic indicators, market indicators and firm characteristics (Table 1).

2.2 Financial Ratios

This research uses three most common financial ratio categories in the literature including profitability, liquidity and leverage (Altman and Sabato 2007; Psillaki et al. 2010; Wu et al. 2010; Duan et al. 2012; Giordani et al. 2014; Jones et al. 2017; Pham et al. 2018a, b; Jiang et al. 2018; Muñoz-Izquierdo et al. 2019; Chai et al. 2019; Sigrist and Hirnschall 2019; Cornée 2019; Chai et al. 2019).

Table 1 The independent variables in the research

Categories		Variables	Explanation	Previous Studies
Financial ratios	Profitability	EBITTA	EBIT/total assets	Altman (1968), Hillegeist et al. (2004) Wu et al. (2010), Giordani et al. (2014), Jones et al. (2017), Pham et l. (2018a, b)
		NITA	Net income/total assets	Ohlson (1980), Zmijewski (1984), Shumway (2001), Psillaki et al. (2010), Wu et al. (2010), Duan et al. (2012), Jones et al. (2017)
		ORTA	Operating revenues/total assets	(Altman 1968; Wu et al. 2010; Psillaki et al. 2010; Jones et al. 2017)
		ProMar	Profit margin	Altman (1968), Wu et al. (2010), Jones et al. (2017)
		ROE	ROE using EBT	Altman (1968), Wu et al. (2010), Jones et al. (2017)
		ROCE	ROCE using EBT	Altman (1968), Wu et al. (2010), Jones et al. (2017)
		LogEBITDATA	Log (1—EBITDA/total assets)	Altman and Sabato (2007)
		LogRETS	Log (1—retained earnings/total assets)	Altman and Sabato (2007)
		LogEBITDAIE	Log (EBITDA/interest expenses)	Altman and Sabato (2007)
	Liquidity	WCTA	Working capital/total assets	Altman (1968), Ohlson (1980), Hillegeist et al. (2004), Psillaki et al. (2010), Wu et al. (2010), Jones et al. (2017), Pham et al. (2018a, b), Muñoz-Izquierdo et al. (2019)

(continued)

Table 1 (continued)

Categories		Variables	Explanation	Previous Studies
		WCSE	Working capital/shareholders equity	Psillaki et al. (2010), Jones et al. (2017)
		CurrentR	Current ratio	Altman (1968), Wu et al. (2010), Jones et al. (2017)
		SolvencyR	Solvency ratio	Altman (1968), Wu et al. (2010), Jones et al. (2017)
		ORSE	Operating revenue/shareholders equity	Psillaki et al. (2010), Jones et al. (2017)
		CLCA	Current liabilities/current assets	Altman (1968), Zmijewski (1984), Wu et al. (2010), Jones et al. (2017), Sigrist and Hirnschall (2019)
		TRDTL	(Total revenue—depreciation)/total Liabilities	Altman (1968), Wu e tal. (2010), Jones et al. (2017)
Financial ratios	Leverage	TLTS	Total liabilities/total assets	Altman (1968), Hillegeist et al. (2004), Wu et al. (2010), Giordani et al. (2014), Jones et al. (2017)
		TLSE	Total liabilities/shareholders equity	Psillaki et al. (2010), Jones et al. (2017)
		RETA	Retained earnings/total assets	Altman (1968), Wu et al. (2010), Jones et al. (2017) Pham et al. (2018a, b), Sigrist and Hirnschall (2019), Muñoz-Izquierdo et al. (2019)
Macroeconomics indicators		GDPG	GDP growth rate for the year	Wu et al. (2010), Giordani et al. (2014), Pham et al. (2018a, b)

(continued)

Table 1 (continued)

Categories	Variables	Explanation	Previous Studies
	INF	The inflation rate for the year	Wu et al. (2010), Giordani et al. (2014), Pham et al. (2018a, b)
	RIR	The real interest rate for the year	Wu et al. (2010), Giordani et al. (2014), Pham et al. (2018a, b)
Market indicators	LogPL	Log (Closing Price at the end of last year)	Hillegeist et al. (2004), Wu et al. Gaunt and Gray (2010)
	TLMTA	Total liabilities/market value of total assets	Campbell et al. (2008), Wu et al. (2010)
	MBTA	Market/book total assets	Wu et al. (2010)
Firm characteristics	SIZEOS	Log10 (Operating Revenue)	Psillaki et al. (2010), Giordani et al. (2014), Jones et al. (2017)
	SIZENE	Log10 (Number of employees)	Giordani et al. (2014), Jones et al. (2017)
	InTaATS	Intangible assets/total assets	Psillaki et al. (2010)
	InTaATE	Intangible assets/total equity	Psillaki et al. (2010)

2.2.1 Profitability

Wu et al. (2010), Psillaki et al. (2010), Duan et al. (2012) and Doumpos et al. (2019) all define that profitability ratios measure the ability of a firm to use its assets and capital efficiently to generate positive profits or at least remain in business. Some popular profitability ratios include return on assets (ROA), return on equity (ROE) and profit margin (PM). Earnings before interest and taxes (EBIT) divided by total assets measures return on investment (ROI), which does not consider any tax or leverage factors, and is also an essential indicator of a company's wealth (Zavgren 1985; Wu et al. 2010). Altman (1968) mentioned that a firm's existence is based on the earning power of its assets. As a result, there is a negative relationship between profitability and the firm's credit rating (Psillaki et al. 2010; Pham et al. 2018a, b). EBIT to total assets (EBITTA) and net income to total assets (NITA) both calculate the relationship between firm assets and firm profitability. Following Ohlson's (1980) research, Wu et al. (2010) and Jones et al. (2017) measured change in net income as CHIN = $(NI_t - NI_t - 1)/(|NI_t| + |NI_t - 1|)$ where NI_t represents net income in year t. If net income was negative in the previous year but positive in the current year, this variable takes the value 1. If the opposite occurs, the variable will take the value negative one.

Another significant profitability ratio is Total Sales to Total Assets (TSTA) as used by Altman (1968), and a firm with a higher value of TSTA will be less likely to experience bankruptcy. However, Wu et al. (2010) make a contradictory statement. To illustrate, their analysis showed that the book value of total assets for the mean bankrupt firm fell by 9.8% over the year prior to bankruptcy (Wu et al. 2010). Consequently, the ratio of sales to total assets at the end of the year can be used to explain the revenue generation ability of the firm. Finally, Altman's Z-score model (2007) used logged profitability variables (details in Table 2) to eliminate the effect of outlier values for the variables as well as minimizing both type I and type II errors and increasing the accuracy of the model (Altman and Sabato 2007; Muñoz-Izquierdo et al. 2019). The results stated that the logged predictors increased the AR by approximately 12% compared with the unlogged predictors at 75–87% (Altman and Sabato 2007). Consequently, this research also applies Altman and Sabato (2007) logged predictors to build the credit rating model.

2.2.2 Liquidity

Liquidity provides a measure of the cash and cash equivalent resources for meeting payment obligations (Wu et al. 2010; Doumpos et al. 2019). However, firms also need to manage the benefits between (1) high levels of liquidity to minimise default risk but with high cash opportunity costs; and (2) low levels of liquidity to maximise investment but with high default risk (Saunders and Cornett 2007). Liquidity is measured by six main variables: (1) the working capital (current assets minus current liabilities) to total assets ratio; or (2) by the working capital to equity ratio (Altman and Sabato 2007; Psillaki et al. 2010; Wu et al. 2010; Duan et al. 2012; Giordani et al. 2014; Jones et al. 2017; Pham et al. 2018a, b; Sigrist and Hirnschall 2019; Muñoz-Izquierdo et al. 2019). A firm will have revenues or losses that shrink assets in relation to (3) total assets; or (4) equity (Altman 1968; and Psillaki et al. 2010). Bankrupt firms also have (5) higher current liabilities relative to current assets (CLCA) and are more likely to have (6) negative net income over the prior two years (INTWO) (Altman 1968; and Wu et al. 2010). Bankrupt firms are also likely to have (7) a lower ratio of funds from operations minus depreciation to total liabilities (TORDTL) (Ohlson 1980; Wu et al. 2010). Finally, Duan et al. (2012), Giordani et al. (2014) and Jones et al. (2017) stated that firms with (8) higher total cash and other liquid assets to total assets had reduced risk of default.

2.2.3 Leverage

The leverage variables calculate the amount of debt and other obligations of a firm including six main ratios from the literature: (1) the firm's debt to assets; or (2) debt to equity ratio (Altman 1968; Zmijewski 1984; Shumway 2001; Hillegeist et al. 2004; Psillaki et al. 2010; Giordani et al. 2014; Jones et al. 2017; Pham et al. 2018a, b; Sigrist and Hirnschall 2019; Muñoz-Izquierdo et al. 2019). The leverage ratios are often used as indicators of financial ability to repay long-term and short-term debt obligations. This provides a relationship with the borrower's earnings and assets values. Wu et al. (2010) illustrate, the more debts a firm has, the more credit risk the firm incurs. In other words, bankrupt firms are usually carrying more debt than equity in their capital structure (Zavgren 1985). However, decreasing the debt to equity ratio of borrowers may not always be beneficial to banks because of the lower return on equity (ROE). Bankrupt firms have: (3) lower retained earnings relative to total assets (RETA) (Altman 1968; Wu et al. 2010; Jones et al. 2017; Pham et al. 2018a, b; Sigrist and Hirnschall 2019; Muñoz-Izquierdo et al. 2019); (4) lower market value of equity to total liabilities (METL); or (5) lower market value of equity to total assets (Altman 1968 and Wu et al. 2010), and higher total liabilities to total assets (TLTA) (Wu et al. 2010). Bankrupt firms are also more likely to have: (6) total liabilities more significant than total assets (DTLTS) (Ohlson 1980; and Wu et al. 2010) and (7) higher liabilities to market value of total assets (TLMTA) (Hillegeist et al. 2004; Wu et al. 2010; Jones et al. 2017).

H1: The proposed financial ratios including profitability ratios, liquidity ratios and leverage ratios positively affects the credit rating of Vietnamese small, medium and large firms in the period 2015–2018.

2.3 *Macroeconomic Indicators*

Jacobson et al. (2013), Giordani et al. (2014), Pham et al. (2018a, b), Le (2018) state that macroeconomic factors including: (1) GDP annual growth rate; (2) annual real interest rate and (3) inflation rate and shift the mean of the default risk distribution over time and, thereby, are three of the most seminal determinants of the average level of firm failure. To illustrate this in general terms, all else being equal, the Central Bank decreases their real interest rate to encourage organisations to borrow to expand their businesses and to increase the aggregate demands of a country (Mankiw 2020). As the result, the economy grows rapidly; and as GPD growth rate climbs most companies increase their profitability, which helps them to service their debt. Therefore, the default risk will decrease overtimes. But if the economy is on the downturn, GDP growth rate (annual percentage) decreases, profitabilities of organisations decreases and default risk increases. In other words, GDP growth rate has a negative relationship with the credit rating; and the real interest rate has a positive relationship with the credit rating (Jacobson et al. 2013; Giordani et al. 2014; Washington 2014; Pham et al. 2018; Le 2018).

Another macroeconomic factor that affects the credit rating quality is the annual inflation rate. Based on the literature it is not apparent what the effects of the annual inflation rate is on credit quality. To illustrate, all else being equal, in the expansionary period of the economy, when the aggregate demands of a country increases can lead to the rise of GPD growth rate and increase of inflation rate (Mankiw 2020). It means that the inflation rate has a negative relationship with the credit rating since most of the companies can earn more profit and increase their credit rating in the expansion period (Mankiw 2020). However, Schechtman and Gaglianone (2012) and Le (2018) state that inflation rate has a positive relationship with the credit rating. To illustrate, when the inflation rate increases, the central bank also raise the interest rate which leads to the increase in loan interest. Consequently, the firms needs to pay more interest expenses. In another point of view, Washington (2014) concludes that in the short term, the inflation rate does not affect the credit rating because if the firms can plan for the predictable inflation rate, the negative effect of inflation can be ignored. In short, in addition to the set of financial ratios, this research includes three macroeconomic variables, which are annual GPD growth rate (annual percentage), annual real interest rate, and the inflation rate to capture the essential time-varying mean of the failure risk distribution.

H2: The proposed macroeconomic indicators including GDP annual growth rate, annual real interest rate and inflation rate positively affects the credit rating of Vietnamese small, medium and large firms in the period 2015–2018.

2.4 Market Indicators

Market indicators include three main variables that used most in the literature (Doumpos et al. 2019). Campbell et al. (2008) and Wu et al. (2010) used a market-based leverage measure, Total Liabilities to the Market value of Total Assets (TLMTA). These authors state that a market-based measure of leverage performs better than the traditional book value to leverage ratio. If the firm has a high total liabilities to the market value of total assets, mean that they can pay their debt better and has a better credit rating or vice versa. Consequently, TLMTA positively affects the credit rating of the firm.

The second variable of the market indicators is the log of a firm's share price at the end of the previous year (LogPL), as share prices tend to have a positive relationship with firm size. For example, a more prominent firm has a higher share price and vice versa. The share price may have a positive relationship with liquidity. To explain, firms with very low-priced stocks usually have lower liquidity on average, so that they cannot pay their debt and has a risky credit rating also (Wu et al. 2010). Consequently, LogPL also positively affects the credit rating of the firm.

Finally, the third variable is the Market-to-Book Total Assets ratio (MBTA), a combined measure of market miss-valuation and future growth opportunities (Wu et al. 2010) research. The forward default intensities in the market-to-book assets ratio will increase if the market miss-valuation effect increases. Otherwise, there should be negative signs on the coefficients. Their results showed that the estimated forward intensities in the market-to-book assets ratio for most of the forecast horizons were consistent with the increasing control of another covariate (Campbell et al. 2008). To explain, the rise MBTA means that the firm assets have a better revaluation, then the firm can pay their debt better as well as have a safer credit score. Consequently, LogPL also positively affects the credit rating of the firm.

H3: The proposed market indicators including market-based leverage measure, firm's share price at the end of the previous year and market-to-book total asset ratio *positively affects the credit rating of Vietnamese small, medium and large firms in the period 2015–2018.*

2.5 *Firm Characteristics*

2.5.1 Firm Size and Age

Jacobson et al. (2013), Giordani et al. (2014), Jones et al. (2017), Pham et al. (2018a, b) and Doumpos et al. (2019) stated the two important firm-specific control variables are firm size and firm age, as smaller and younger firms are more likely to be riskier than larger and older firms. These two aggregate variables are found to be essential determinants of average default rates. Giordani et al. (2014) and Psillaki et al. (2010) measured firm size (SIZELTS) by the logarithm of the firm's sales. Larger firms are more able to overcome difficult times and less likely to default. Vice versa, bankrupt firms, on average, are usually smaller (Wu et al. 2010). Another result showed that larger firms are more diversified, have better management system and have better organisational and financial structures, so they tend to fail less often than smaller ones. Wu et al. (2010) also state that larger companies are stronger than smaller firms in calling on additional equity or securing external finance during adverse situations. Finally, large firms are also more likely to benefit from a bailout by the government simply because they may be 'too big to fail'. Consequently, firm size is negatively related to business failure. Finally, based on different measures to define the size of a company, the most common measures are the total assets and revenues of firms or the market capitalisation for listed companies (Doumpos et al. 2019).

2.5.2 Tangible Collateral

Psillaki et al. (2010) measured 'the tangible collateral as the ratio of tangible assets divided by either total assets or by the firm's equity'. Collateral is one of the most prevalent conditions used by commercial banks to evaluate borrowers so as to minimise credit risk (Scott 1977; and Psillaki et al. 2010). The relationship between tangible collateral and the credit rating is negative. Firms with strong tangible collateral are less likely to default. Based on banking standards to minimise loan losses, a bank usually requires more collateral if it believes the prospective customer is at high risk. Psillaki et al. (2010) state that large corporations with proven credibility and efficient business operations are less likely to pledge collateral. These large prime borrowers tend to have stronger equity, more cash flows and more certain investment opportunities, so these companies are more likely to get non-collateralised loans. Moreover, the authors argue that collateral might decrease the credit rating efforts of commercial banks. To illustrate, bank officers usually believe that with high collateral, loans are more protected in case of customer default, and hence they are less precise in screening potential customers. Banking services are not a collateral pawn service. What protects banks from loan loss are companies' cash flows from their primary operations, rather than collateral. Jiménez and Saurina (2004) claim that collateralised loans have higher credit ratings and close bank–customer relationships enhance the willingness to take risk.

2.5.3 Growth Opportunities

Growth opportunities are 'measured as the ratio of intangible assets divided either by total assets or by the equity of the firm' (Psillaki et al. 2010). Opposite to tangible assets, intangible assets can be considered the future growth of a firm and value-added of a firm, but cannot be used as collateral (Titman and Wessels 1988; and Psillaki et al. 2010). If the borrowers are companies with more growth opportunities, but these opportunities are the result of high-risk investments, commercial banks usually conclude that, on balance, firms with lots of growth opportunities may be viewed as having higher credit risk (Psillaki et al. 2010).

H4: The proposed firm characteristics including firm size and firm age, tangible collateral and growth opportunities positively affects the credit rating of Vietnamese small, medium and large firms in the period 2015–2018.

3 Research Methodology and Data Analysis

3.1 Artificial Neural Networks (ANN)

ANNs, also often known as Neural Networks (NN), are an Artificial Intelligence (AI) system in which the information processes build the models by mimicking the learning ability of biological systems in understanding an unknown behaviour. In other words, ANNs try to learn and copy the functionality of the nerve cells in the human brain. Parker (2006) states that ANNs use the biological concepts of learning and memory creation by the interconnected nodes. ANNs build a set of connected inputs and outputs like a network in which each connection has specific weights. In the learning process, ANNs adjust the weights of the nodes so that it can fit the input to the most reliable model to produce accurate output.

ANNs can learn, understand and solve financial problems, especially credit rating and default prediction. Goonatilake and Treleaven (1995) and Farhadieh (2011) explain five extraordinary abilities of ANNs to solve financial problems, including learning, adaption, flexibility, explanation and discovery. (1) Learning is one of the most critical abilities of ANNs in learning decisions and tasks from an extensive historical database. (2) Adaptation is the ability to reflect and reform to changes in the data and the environment, for example, changes in policies, regulations or economic conditions. (3) Flexibility is the ability to perform analysis or help to make decisions even with incomplete data and non-normally distributed data. (4) Explanation is the ability to explain how decisions are made based on data processing and data analysis. (5) Discovery is the ability to investigate unknown relationships or unknown findings based on data analysis. However, each AI system has its unique advantages and limitations. As a result, each can only be used for suitable and specific problems (Goonatilake and Treleaven 1995; Farhadieh 2011).

The network builds weights based on the paths between the nodes. The model of ANNs is a learning algorithm to calculate these weight values. The most common ANN is a multilayer perceptron (MLP), which uses the learning algorithm as a 'backpropagation rule' (Rumelhart 1986). To illustrate, this backpropagation learning algorithm adjusts these weights to minimise the difference between the estimated outputs and actual outputs. When input comes into the network, the network will produce an output based on the first version of its weights. Then this output is compared with the actual output by using the mean squared error (MSE). Then, the error value propagates backward through the network so that the ANN makes some small changes to the weights in each layer. This repeated process will become a cycle until the model's MSE minimises and reaches an acceptable value based on the research purposes, the computational capacity or the time consumed. At the end of the learning process, the ANN has learned the problem 'well enough' and is ready to build the model and test the model. The backpropagation learning algorithm is very efficient for credit rating models (Mitchell 1997, Desai et al. 1996; Lee and Chen 2005; Malhotra and Malhotra 2003).

Although ANNs can create models with high predictive accuracy, they have very poor interpretability (Desai et al. 1996; Lee and Chen 2005; Malhotra and Malhotra 2003). It is challenging to understand and analyse the full meaning of the model weights. ANNs also often require a long training time with multiple tests and significant big data. These factors are some of the main limitations of the application of ANNs in credit rating research (Chung and Gray 1999; Craven and Shavlik 1997). This is why many FIs and researchers still choose the traditional statistical classifications although many studies state that ANNs perform with significantly higher predictive accuracy than some conventional classification models including MDA and LR (Desai et al. 1996; Lee and Chen 2005; Malhotra and Malhotra 2003). Thus, the ANN is one of the essential models used in this research to predict the credit rating for Vietnamese firms.

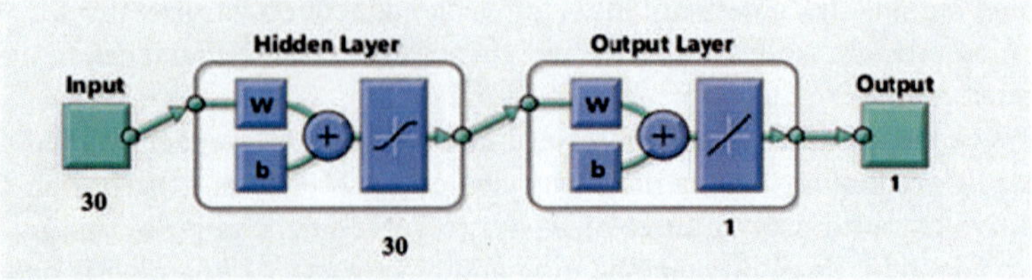

Fig. 1 The artificial neural network (ANN) research model (MATLAB 2019)

Figure 1 shows a model summary of the ANN in the research model that consists of three stages including training, validating and testing. After collecting the data and pre-processing the data, the training stage is the first stage of building the ANN model in which the new dataset is trained by MATLAB R2019a with 70% of the dataset. The network's weight and bias values are adjusted based on its errors. Based on the recommendations of MATLAB R2019a and the testing results, the research use 30 hidden layers to predict the credit rating to produce the most accurate model.

This research chooses 30 hidden layers for the ANN models as the optimal number of hidden layers because this maximises the ANN performance and effectively uses computer and time resources. To test for the ANN performance when changing the number of hidden layers, this research builds another five testing models that apply 20, 25, 30, 40 and 50 hidden layers following recommendations from MATLAB 2019a that the number of hidden layers should over 20. To illustrate, the 30 hidden layers model produces the maximum ANN performance when this research compares R-values and MSEs between 20, 25, 30, 40 and 50 hidden layers models (Table 2).

Table 2 presents the result of the five ANN testing models when the number of hidden layers changes to 20, 25, 30, 40 and 50. There is a tradeoff that the optimal number of hidden layers is 30 because when this research reduces the number of hidden layers to 25 or 20 hidden layers, the MSEs of these models increase slightly and the R-values also decrease slightly. Also, when this research increases the number of hidden layers to 40 hidden layers or 50 hidden layers, the performance of the models does not improve extensively or even worsens compared with the 30 hidden layers model, but the time consumed increases dramatically. The 50 hidden layers model needs more than 10 min on average for each run. As a result, in this research 30

Table 2 ANN testing models for the number of hidden layers

Hidden layers	Time consumed in minutes	Min MSE	Max MSE	Min R per cent	Max R per cent
20	02:51	0.36443	0.36586	84.29	84.30
25	04:11	0.36465	0.38566	83.43	84.29
30	03:17	0.35460	0.36338	84.28	84.50
40	08:28	0.36330	0.37996	83.53	84.43
50	10:42	0.36495	0.37615	83.28	84.42

hidden layers are the optimal number that can maximise the ANN model performance and maintain acceptable time consumption. After selecting the number of hidden layers, the ANN model is ready to train and build.

The research also uses the Levenberg–Marquardt (LMA) training algorithm, which typically requires more computer memory but is less time-consuming. The training process will automatically stop when generalisation stops improving, as indicated by an increase in the MSE of the validation samples (MATLAB 2019). The validating stage uses 15% of the total dataset to measure network generalisation results and also to halt training when generalisation stops improving (MATLAB 2019). Finally, the testing stage uses 15% of the total dataset for an independent measurement of network performance during and after training (MATLAB 2019).

3.2 *Data Processing*

In this research, the Orbis database were used to collect 39,162 firms in Vietnam in the period 2015–2018. The interquartile rule for outliers that multiplies the interquartile range (IQR) by number 1.5 (Tabachnick and Fidell 2007). Missing values are another significant problem in credit rating research in which the dataset contains incomplete data or is missing data altogether. According to Angelini, di Tollo and Roli (2008) and Langkamp et al. (2010), if the missing values are higher than 10%, these observations should be removed. In this research, if a company had missing variables of over 10% in a specific year, this yearly data point was deleted. This recommendation was also applied to the independent variables. If an independent variable had a missing value of more than 10% of all the independent variables, the researcher considered removing this independent variable. However, to remove an independent variable, the researcher needs to evaluate it based on its importance according to the literature review (Altman 1968; Hillegeist et al. 2004; Wu et al. 2010; Giordani et al. 2014; Ohlson 1980; Zmijewski 1984; Shumway 2001; Psillaki et al. 2010; Wu et al. 2010; Duan et al. 2012).

Highly correlated data is defined as an absolute correlation value in the range from 0.7 to 1.0 ($0.7 < |correlation| < 0.01$) (Pearson 2002). If a pair of highly correlated variables was found, this research considered eliminating the weaker impact variable based on the result of the forward elimination method following the recommendations of Blanchet et al. (2008). In addition, the research collected data from the Orbis database and World Bank database to ensure the accuracy of the dataset. As a result of these comprehensive controls, reliability and consistency are not considered to be serious threats in this study, following the recommendation of Dias's (2013) study.

Non-normalness is also a significant problem in credit rating studies which can lead to outliers, skewness and kurtosis issues that significantly affect model estimation and model performance (Jones et al. 2017; Sigrist and Hirnschall 2019). However, ANNs can deal with non-normally distributed data and non-linear data, so the research followed the recommendations of Sigrist and Hirnschall (2019) to keep the data and variables without transformation in order to maintain the accurate meaning of the data. The next section describes the detailed variable selection process.

The variable selection process is a highly significant step in credit rating model development. In this research, the dataset is big data and multi-dimensional that after cleaning the data, the number of variables available included 30 predictor variables. However, not all the variables were useful in the final model of credit rating for Vietnamese firms. To illustrate, Sanche and Lonergan (2006) state that using an extensive range of IVs causes over-fitting issues and even decreases the accuracy or efficiency of a credit rating model. Zhang (2000), Guyon and Elisseeff (2003) and Dias (2013) also suggest that selecting only a small and relevant set of IVs and removing the insignificant variables will increase the effectiveness of the credit rating model, including: (1) making the model easy to understand; (2) decreasing computational capacity and time consumption; (3) reducing data requirements and saving data costs; and (4) maintaining or improving the accuracy of the model. Ideally, the research aimed to build a highly accurate credit rating model (with over 80% AR) that economically uses a relevant, significant and small set of predictor variables. In fact, Farhadieh (2011) states that in financial research, on average only 8–12 variables are relevant out of 50–60 available variables in the final model (16–20% of total available variables). That is the reason why the variable selection process is one of the essential steps in building an adequate credit rating model for firms in Vietnam.

Partitioning of data is an essential task for validating data-mining models (Farhadieh 2011; Jones et al. 2017). The large dataset collected included 39,162 public and non-public firms in Vietnam in many business sectors that had Multi-Objective Rating Evaluation (MORE) credit scores provided by ModeFinance from the Orbis database which include both small, medium and large firms in Vietnam. This research chose only the small, medium and large companies that satisfied the definition of Vietnamese SME in Decree 56/2009/ND-CP of 2009. The primary dataset collected is big enough to partition the dataset into three components, training, validating and testing. As a recommendation from MATLAB R2019a, the training data consists of 70% of the dataset that is employed for learning and fitting the models. The validation data consisting of 15% is employed to observe and tune the model weights throughout estimation and is also used for model assessment. The remaining 15% of the dataset is testing data which are used as a further measurement of predictive accuracy. In MATLAB R2019a, the program can randomly choose 70% training data, 15% validating data and 15% testing data. By using a random selection technique, it is possible to have a different result and a different model every time the analysis is re-run. This research was re-run at least five times for each test and the best models were kept each time the analysis was re-run.

The Data Analysis

The total number of yearly data points in the dataset 2015–2018 is 77,228, which is 29.9% (2015), 33.9% (2016), 35.1% (2017) and 1% (2018). The data was collected from the Orbis database in December 2018, and there were a limited number of observations in 2018. The research uses the dataset 2015–2018 as the primary dataset

Table 3 Data distribution 2015–2018

		Frequency	Percentage	Valid percentage	Cumulative percentage
Validity	2015	23,105	29.9	29.9	29.9
	2016	26,199	33.9	33.9	63.8
	2017	27,134	35.1	35.1	99.0
	2018	790	1.0	1.0	100.0
	Total	77,228	100.0	100.0	

for Research Questions 1 and 2 to find the new predicting model for the credit rating of Vietnamese firms after the implementation of Circular 200 because this dataset reflects the current situation of Vietnamese firms (Table 3).

3.3 Dependent Variables

The dependent variables (DV) which is the ScoreG10 represents the DV credit rating in 10 classes, including AAA, AA, A, BBB, BB, B, CCC, CC, C, and D (Table 4). The ScoreG10 is used as the DV for credit rating purposes. The distribution of the DV ScoreG10 is slightly negatively skewed, –0.005. The top three highest frequency values are 6 occurrences of B with 34.2% of the total dataset; 5 occurrences of BB with 30.0% and 7 occurrences of CCC with 14.9%. On the other hand, the three

Table 4 The dependent variable has the following abbreviations (adapted from Multi-Objective Rating Evaluation (MORE) model—ModeFinance (2016))

Credit rating	Class	Code	Description
Healthy companies	AAA	1	The company is solid to meet its financial commitments
	AA	2	The company is keen to meet its financial commitments
	A	3	The company has good liquidity
Balanced companies	BBB	4	The company's capital structure and economic equilibrium are considered adequate
	BB	5	The company's performances are adequate in its sector and the country
Vulnerable companies	B	6	The company presents weak signals concerning its fundamentals
	CCC	7	The company shows a risky disequilibrium in its capital structure and financial reports
Risky companies	CC	8	The company shows signals of high vulnerability
	C	9	The company shows considerable pathological situations
	D	10	The company does not have enough capacity to meet its financial obligations

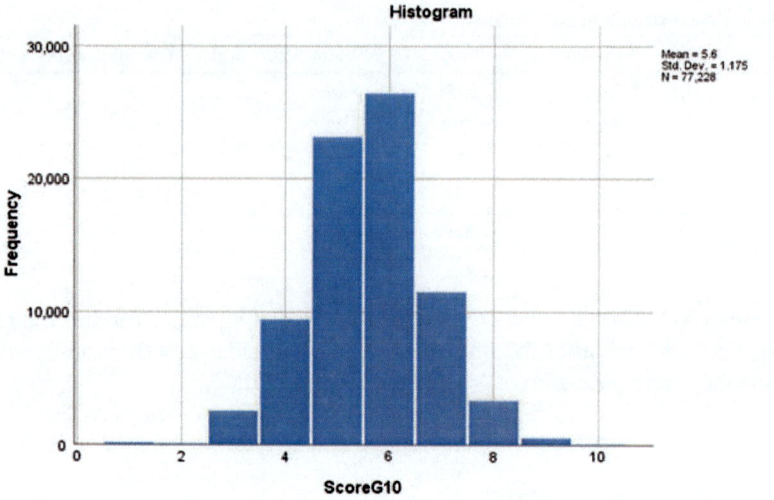

Fig. 2 ScoreG10 frequency histogram of dataset 2015–2018

lowest frequency values are 10 occurrences of D with 0.1% of the total dataset; 2 occurrences of AA with 0.1% and 1 occurrence of A with 0.2% The ScoreG10 frequency histogram of dataset 2015–2018 (Fig. 2) shows that ScoreG10 is almost generally distributed with a mean of 5.61 and standard error 0.004.

3.4 Independent Variables

The IVs include 30 variables that are categorised into four main groups, financial indicators, market indicators, macroeconomic indicators and firm characteristic indicators (Fig. 3).

3.5 Descriptive Statistics

The formula and explanation of the IVs are provided in Table 5.

In general, after deleting missing data, the dataset 2015–2018 presents some problems with highly skewed data distribution and extreme values (outliers) (Table 6). To illustrate, there are highly positively skewed IVs (skewness > 1) including ORSE (skewness of 55.97), ORTA (skewness of 16.47), CR (skewness of 6.91), EBITTA (skewness of 5.89), NITA (skewness of 5.47) and SIZELNNE (skewness of 1.34) while highly negatively skewed distributions include ROE (skewness of –5.82), PM (skewness of –2.67) and GPDG (skewness of –1.49). There are also some

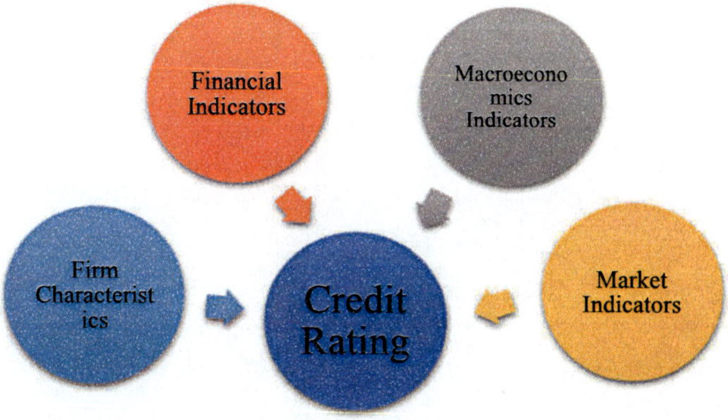

Fig. 3 The proposed research model

Table 5 The independent variable formula

Code	Explanation	Formula
EBITTA	Earnings Before Interest and Tax/Total Assets	= EBIT/TA
NITA	Return of Assets (ROA)	= NI/TA
ORTA	Operating Revenue/Total Assets	= OR/TA
PM	Profit Margin	= EBT/OR
ROE	Return of Equity using Earnings Before Tax (ROE using EBT)	= EBT/Total Equity
ORLG10	Size of the company measures by OR	= Log10 of OR
CR	Current Ratio	= Current Fixed Assets/Current Liability Current Fixed Assets = Stocks + Debtors + Other Current Assets
SolvencyR	Solvency Ratio using Assets Based	= Shareholder Funds/Total Assets Shareholder Funds = Capital + Other Shareholder Funds
ORSE	Operating Revenue/Share Equity	= OR/SE
GDPG	Annual GDP growth of Vietnam	
INF	Annual inflation rate of Vietnam	
RIR	Annual real interest rate of Vietnam	
SIZELNNE	Size of the firm measured by number of employees	Ln of Number of Employees

Table 6 The descriptive statistics for dataset 2015–2018

		ScoreG10	InactiveN	EBITTA	NITA	ORTA	PM	ROE	ORLG10	CR	SolvencyR	ORSE	GDPG	INF	RIR	SIZELNNE
N	Valid	77,228	77,228	77,228	77,228	77,223	73,740	75,361	74,725	74,199	77,052	77,150	77,228	77,228	77,228	69,401
	Missing	0	0	0	0	5	3488	1867	2503	3029	176	78	0	0	0	7827
Mean		5.60	1.00	0.0058	0.0044	1.4398	−0.0001	0.0198	3.1805	3.1315	0.3838	9.4689	0.0651	0.0263	0.0520	3.4066
Std. Error of Mean		0.0040	0.000	0.0001	0.0001	0.0090	0.0003	0.0009	0.0028	0.02996	0.0010	0.4288	0.0001	0.0000	0.0001	0.0044
Median		6.0000	1.00	0.0030	0.0024	0.8962	0.0033	0.0110	3.2068	1.2100	0.3154	2.8489	0.0623	0.0324	0.0579	2.9957
Mode		6	1	0.0000	0.0000	0.0000	0.0006	0.0000	0.0000	1.0300	1.0000	0.0000	0.0901	0.0352	0.0286	2.3026
Std. Deviation		1.1750	0.027	0.0180	0.0159	2.5028	0.0886	0.2351	0.7662	8.1603	0.2892	119.0961	0.0213	0.0115	0.0184	1.1489
Variance		1.3810	0.001	0.0000	0.0000	6.2640	0.0080	0.0550	0.5870	66.5910	0.0840	14,183.8860	0.0000	0.0000	0.0000	1.3200
Skewness		−0.0050	37.433	5.8920	5.4680	16 4720	−2.6650	−5.8160	−0.3180	6.9140	0.5570	55.9710	−0.0290	−0.8440	−0.2510	1.3400
Kurtosis		0.3780	1399.237	133.544	148.77	613.4580	44.8440	473.711	1.271	56.3510	−0.3170	6541.6460	−1.4920	−1.2410	−1.5610	2.4360

slightly skewed variables (−1 < skewness < 1), SolvencyR (skewness of 0.56), INF (skewness of −0.84), ORLG10 (skewness of −0.32) and RIR (skewness of 0.25). Kurtosis of the IVs represents heavy-tailed distributions except for SolvencyR. ORSE, ORTA, ROE, NITA and EBITTA show highly heavy-tailed distributions. As a result, the explanatory variables of the dataset are non-normally distributed.

3.6 Correlation Matrix

Table 7 presents the correlation matrix of the dataset. Multicollinearity is the dataset's issue, which occurs when the IVs are very highly correlated (Tabachnick and Fidell 2007 and Dias 2013). Multicollinearity issues can be found by investigating the correlation matrix of the IVs when their correlations are in excess of 0.9. Table 7 shows that most of the IVs are not highly correlated with their correlations less than 0.9, including ORTA, PM, ROE, ORLG10, CR, solvency ratio, ORSE and SizeLnne.

However, there is a very high correlation between EBITTA and NITA with a correlation of 0.984 (Tabachnick and Fidell 2007 and Dias 2013) of Vietnamese firms in the period of 2015–2018. This relationship illustrates the close relationship between the EBITTA and NITA formulas. The only difference between EBITTA and NITA is the earning before tax and interest and the earnings after tax and interest. Although the EBITTA and NITA correlate very highly, they have essential IVs based on the influential research. Based on those reasons, the research continues to include EBITTA and NITA (ROA) for further forward elimination tests using ANNs. The results of the ANN tests will help the research to decide which IVs should be kept or eliminated in the final model.

In addition, there are some high correlations between GDPG and INF (correlation of 0.891), GDPG and RIR (correlation of −0.989) and INF and RIR (correlation of −0.824) (Tabachnick and Fidell 2007 and Dias 2013). These high correlations illustrate the relationships of the macroeconomic indicators for Vietnam in the period 2015–2018. To explain, the GDP growth in Vietnam in the period 2015–2018 has a close positive relationship with the inflation rate and a clearly negative relationship with RIR. In other words, when GDP growth is high, the inflation rate will also be high and the RIR will be low. These relationships can be explained by the principles of macroeconomics following Mankiw (2014) and Gottheil (2013), which means that as the RIR decreases, people and firms can borrow more money to spend or invest. As a result, a country's GDP grows rapidly and inflation also increases with the rise of aggregate demand. Based on their meaning in the literature, the research continues to include GDPG, INF and RIR for further forward elimination tests using ANNs that will be discussed in the next section.

Table 7 The correlation of the independent variables for Vietnamese firms in the period 2015–2018

Correlation	EBITTA	NITA	ORTA	PM	ROE	ORLG10	CR	SolvencyR	ORSE	GDPG	INF	RIR
EBITTA												
NITA	**0.984**											
ORTA	0.096	0.087										
PM	0.352	0.362	0.010									
ROE	0.307	0.314	0.038	0.093								
ORLG10	0.307	0.291	0.347	0.209	0.081							
CR	−0.119	−0.118	−0.051	−0.080	−0.033	−0.185						
SolvencyR	−0.305	−0.307	−0.185	−0.214	−0.134	−0.354	0.347					
ORSE	0.031	0.032	0.181	0.026	0.109	0.103	−0.012	−0.107				
GDPG	0.021	0.021	−0.029	0.007	0.006	0.066	−0.013	−0.030	−0.002			
INF	0.014	0.015	−0.024	0.008	0.004	0.056	−0.007	−0.025	−0.004	**0.891**		
RIR	0.001	0.000	0.029	0.002	−0.001	−0.055	0.013	0.033	0.001	**−0.989**	**−0.824**	
SIZELNNE	0.159	0.143	−0.096	0.040	0.034	0.283	−0.044	0.045	−0.032	0.042	0.031	−0.028

4 Discussion

For the dataset 2015–2018, this research creates 33 testing models that are divided into three main groups for training, validating and testing purposes (Table 8). For dataset 2008–2014, the data partition includes 14,878 observations in the training dataset, 3,188 observations in the validating dataset and 3,188 observations in the testing dataset, while for dataset 2015–2018, the data partition includes 54,060 observations in the training dataset, 11,584 observations in the validating dataset and 11,584 observations in the testing dataset (Fig. 4).

NITA: Return of assets (ROA) = net income/total assets

Testing model number 2 starts with the input including EBITTA and NITA, while the output is also the ScoreG10. The MSEs for the training dataset, the validation dataset and the testing dataset are in the range from 0.10395 to 0.10482, while the R-values of these datasets are in the range from 49.09% to 49.78%. Comparing the results between testing models number 1 and 2, this research concludes that the ANN performance does not improve when the input includes both EBITTA and NITA. Consequently, this research builds testing model number 14, where the input includes only NITA, to further test for the effect of NITA on the credit rating model. The result of testing model number 14 shows that the MSEs are in the range from 0.10100 to 0.10448, while the R-values are in the range from 49.81% to 50.37%. Comparing the results between testing models number 1 and 14, it is clear that when the input is NITA individually, it is more significant to the credit rating model.

ORTA: Operating revenue/total assets

In testing model number 3 which the input includes EBITTA, NITA and ORTA. The MSEs for the training dataset, the validation dataset and the testing dataset are in the range from 0.10217 to 0.10336, while the R-values of these datasets are in the range from 50.69% to 51.61%. Comparing the results between testing models number 2 and 3, this research concludes that the ANN performance improved slightly. Consequently, to build further testing for the effect of ORTA on the credit rating model, this research builds testing model number 15, where the input includes only NITA and ORTA. The result for testing model number 15 shows that the MSEs are in the range from 0.10584 to 0.10769, while the R-values are in the range from 46.90% to 48.20%. Comparing the results between testing models number 14 and 15, it is clear that ORTA is insignificant to the credit rating model and ORTA is eliminated from the final model for dataset 2015–2018.

PM: Profit margin = earnings before tax/operating revenues

Testing model number 4 starts with the input including EBITTA, NITA, ORTA and PM. The MSEs are in the range from 0.10087 to 0.10228, while the R-values are in the range from 51.47% to 51.78%. Comparing the results between testing models number 3 and 4, this research concludes that the ANN performance also improves slightly when PM is used as an input. As a result, this research builds testing model

Table 8 ANN tests for dataset 2015–2018

Test	Input 1	Input 2	Input 3	Input 4	Input 5	Input 6	Input 7	Input 8	Input 9	Input 10	Input 11	Input 12	Input 13	Result 1	Result 2	Result 3	Result 4	Result 5	Result 6
No	EBITTA	NITA	ORTA	PM	ROE	ORLG10	CURRENT	SOLVENCY	ORSE	GDPG	INF	RIR	SIZELNNE	Training MSE	Validation MSE	Testing MSE	Training R	Validation R	Testing R
1	X													0.10306	0.10570	0.10454	0.4993	0.4920	0.4886
2	X	X												0.10408	0.10482	0.10395	0.4978	0.4910	0.4909
3	X	X	X											0.10217	0.10336	0.10316	0.5069	0.5381	0.5161
4	X	X	X	X										0.10087	0.10228	0.10092	0.5178	0.5162	0.5147
5	X	X	X	X	X									0.63795	0.63379	0.63866	0.7114	0.6998	0.6963
6	X	X	X	X	X	X								0.46527	0.47881	0.48698	0.7966	0.7902	0.7888
7	X	X	X	X	X	X	X							0.40956	0.40738	0.43744	0.8221	0.8194	0.8023
8	X	X	X	X	X	X	X	X						0.34332	0.35857	0.35731	0.8516	0.8478	0.8450
9	X	X	X	X	X	X	X	X	X					0.38223	0.39455	0.39390	0.8332	0.8306	0.8265
10	X	X	X	X	X	X	X	X	X	X				0.34311	0.35951	0.35900	0.8519	0.8457	0.8440
11	X	X	X	X	X	X	X	X	X	X	X			0.34203	0.34836	0.37203	0.8523	0.8480	0.8416
12	X	X	X	X	X	X	X	X	X	X	X	X		0.34283	0.35373	0.36122	0.8534	0.8431	0.8414
13	X	X	X	X	X	X	X	X	X	X	X	X	X	0.36014	0.33698	0.37102	0.8461	0.8425	0.8366
14		X												0.10448	0.10257	0.10100	0.4981	0.5037	0.5019
15		X	X											0.10584	0.10769	0.10703	0.4820	0.4719	0.4690
16		X		X										0.10395	0.10497	0.10414	0.4956	0.4855	0.4936
17		X			X									0.56852	0.56281	0.56623	0.7445	0.7512	0.7435
18		X			X	X								0.57234	0.57489	0.59398	0.7467	0.7366	0.7285
19		X			X		X							0.55042	0.53747	0.58557	0.7513	0.7582	0.7532
20		X			X			X						0.41260	0.41950	0.41036	0.8226	0.8162	0.8247
21		X			X	X		X						0.41740	0.42222	0.42740	0.8200	0.8234	0.8173
22		X			X		X	X						0.37306	0.38881	0.38266	0.8400	0.8295	0.8350
23	X	X			X		X	X						0.37286	0.39055	0.38310	0.8406	0.8284	0.8328

(continued)

Table 8 (continued)

Test	Input 1	Input 2	Input 3	Input 4	Input 5	Input 6	Input 7	Input 8	Input 9	Input 10	Input 11	Input 12	Input 13	Result 1	Result 2	Result 3	Result 4	Result 5	Result 6
24		X			X	X	X	X						0.37360	0.37593	0.39256	0.8392	0.8394	0.8280
25		X			X		X	X		X				0.36682	0.37910	0.38104	0.8426	0.8351	0.8361
26		X			X		X	X		X	X			0.38079	0.39367	0.37933	0.8351	0.8318	0.8376
27		X			X		X	X		X		X		0.37873	0.37400	0.38791	0.8367	0.8402	0.8322
28		X			X		X	X		X			X	0.37457	0.37575	0.38041	0.8402	0.8391	0.8390
29					X									–	–	–	–	–	–
30							X							0.96791	0.97849	0.97097	0.5404	0.5333	0.5406
31								X						0.73513	0.73835	0.73518	0.6823	0.6830	0.6815
32										X				0.13624	0.13876	0.13635	0.9657	0.1061	0.1020
33													X	0.13833	0.13577	0.14214	0.6899	0.5250	0.4720

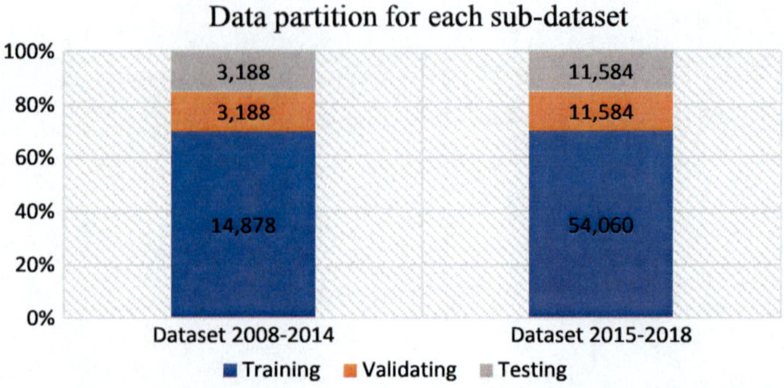

Fig. 4 Data partition for each sub-dataset

number 16, where the input includes only NITA and PM, to further test for the effect of PM on the credit rating model. The result for testing model number 16 indicates that the MSEs are in the range from 0.10395 to 0.10497, while the R-values are in the range from 48.55% to 49.56%. Comparing the results between testing models number 14 and 16, it is clear that PM is insignificant to the credit rating model and PM is also eliminated from the final model for dataset 2015–2018.

ROE: Return of equity using earnings before tax = EBT/total equity

In testing model number 5 which the input includes EBITTA, NITA, ORTA, PM and ROE. The MSEs are in the range from 0.63379 to 0.63866, while the R-values of these datasets are in the range from 69.63% to 71.14%. Comparing the results between testing models numbers 4 and 5, this research concludes that the ANN performance improves significantly when the R-values increase by almost 32.2% on average, but the MSEs also increase by almost 524.4% on average. Consequently, this research builds testing model number 17, where the input includes only NITA and ROE, for further testing of the effect of ROE on the credit rating model. The result for testing model number 17 shows that the MSEs are in the range from 0.56281 to 0.56852, while the R-values of these datasets are in the range from 74.35% to 75.12%. Comparing the results between testing models number 14 and number 17, it is clear that when the inputs are NITA and ROE, they could be the significant predictors because the R-values increase by almost 49.14% on average.

ORLG10: Size of company measures by log 10 of operating revenues

The building process continues with testing model number 6, in which the input includes EBITTA, NITA, ORTA, PM, ROE and ORLG10. The MSEs are in the range from 0.46527 to 0.48698, while the R-values are in the range from 78.88% to 79.66%. Comparing the results for testing models numbers 5 and 6, this research concludes that the ANN performance also improves significantly when the R-values increase by almost 11.98% on average, but the MSEs also decrease by almost 33.78% on average. Consequently, this research builds testing number 18, where the input includes NITA, ROE and ORLG10, for further testing of the effect of ORLG10 on the credit rating model. The result for testing model number 18 shows that the MSEs are in the range from 0.57238 to 0.59398, while the R-values of these datasets are in the range from 72.95% to 74.67%. Comparing the results between testing models number 17 and 18, there is no significant improvement in R-values or MSEs. Consequently, ORLG10 is not a significant factor in the final credit rating model.

CURRENT: Current ratio = current fixed assets/current liability

The next step of the building process continues with testing model number 7, in which the input includes EBITTA, NITA, ORTA, PM, ROE, ORLG10 and the current ratio. The MSEs are in the range from 0.40738 to 0.43744, while the R-values are in the range from 80.23% to 81.94%. Comparing the results for testing models numbers 6 and 7, this research concludes that the ANN performance improves slightly when the R-values increase by almost 3.2% on average, but the MSEs also decrease by almost 11.32% on average. Consequently, this research builds testing model number 19, where the input includes NITA, ROE and the current ratio, for further testing of the effect of the current ratio on the credit rating model. The result for testing model number 19 indicates that the MSEs are in the range from 0.53747 to 0.58557, while the R-values are in the range from 75.13% to 75.82%. Comparing the results for testing models number 17 and 19, although there is only a slight improvement in the model performance, the current ratio could be considered a significant factor but the current ratio needs further testing in testing model number 22 with more significant IVs.

SOLVENCY: Solvency ratio (assets based) = shareholder funds/total assets

For testing model number 8, the input includes EBITTA, NITA, ORTA, PM, ROE, ORLG10, current ratio and solvency ratio. The MSEs are in the range from 0.34332 to 0.35857, while the R-values are in the range from 84.50% to 85.16%. Comparing the results for testing models numbers 7 and 8, it is clear that the ANN performance improves significantly because the R-values increase by almost 3.6% on average, but the MSEs also decrease by almost 22% on average. Thus, this research builds testing model number 20, where the input includes NITA, ROE and the solvency ratio, for further testing of the effect of the solvency ratio. The result for testing model number 20 indicates that the MSEs are in the range from 0.41036 to 0.41950, while the R-values of these datasets are in the range from 81.62% to 82.47%. Comparing the results for testing models number 17 and 20, it is clear that the ANN performance increases significantly when MSEs decrease by 35.52% on average and R-values increase by 9.88% on average, so the solvency ratio is a significant factor in the final credit rating model for the dataset 2015–2018.

ORSE: Operating revenue/share equity

In testing model number 9 which the input includes EBITTA, NITA, ORTA, PM, ROE, ORLG10, current ratio, solvency ratio and ORSE, while the output is unchanged as the ScoreG10. The MSEs of the training dataset, the validation dataset and the testing dataset are in the range from 0.38223 to 0.39455, while the R-values of these datasets are in the range from 82.65% to 83.32%. Comparing the results for testing models numbers 8 and 9, it is clear that the ANN performance does not improve. Consequently, the credit rating model for dataset 2015–2018 excludes ORSE as an insignificant predictor.

GDPG: Gross domestic product growth of the year

Next, testing model number 10 includes the inputs EBITTA, NITA, ORTA, PM, ROE, ORLG10, current ratio, solvency ratio, ORSE and GDPG. As a result, the MSEs are in the range from 0.34311 to 0.35951, while the R-values are in the range from 84.40% to 85.19%. Comparing the results for testing models numbers 9 and 10, this research concludes that the ANN performance improves slightly when the R-values increase by almost 3.6% on average, but the MSEs also decrease by almost 22% on average. Consequently, this research builds testing model number 25, with input including NITA, ROE, current ratio, solvency ratio and GDPG, to further test for the effect of the GDPG on the credit rating model. The result for testing model number 25 shows that the MSEs are in the range from 0.36682 to 0.38104, while the R-values of these datasets are in the range from 83.51% to 84.26%. Comparing the results for testing models number 22 and 25, although there is only a slight improvement in the model performance, the GDPG could be considered a significant factor because, at the end of the building process, it is much harder to improve the ANN performance.

INF: Annual inflation rate

The next step of the building process continues with testing model number 11 in which the input includes EBITTA, NITA, ORTA, PM, ROE, ORLG10, current ratio, solvency ratio, ORSE, GDPG and INF, while the output is also the ScoreG10. The MSEs are in the range from 0.34203 to 0.37203, while the R-values of these datasets are in the range from 84.16% to 85.23%. Comparing the results for testing models numbers 10 and 11, it is clear that the ANN performance does not improve. However, to ensure the result, this research builds testing model 26 including the inputs NITA, ROE, current ratio, solvency ratio, GDPG and INF to further test for the INF effect. As a result of testing model 26, the MSEs are in the range of 0.38079 to 0.39367, and the R-values are in the range of 83.18% to 83.76%. It means that there is no significant improvement in the ANN performance between testing models 25 and 26. The INF is an insignificant predictor.

RIR: Annual real interest rate

Testing model number 12 includes the inputs EBITTA, NITA, ORTA, PM, ROE, ORLG10, current ratio, solvency ratio, ORSE, GDPG and RIR. As a result, the MSEs are in the range from 0.34283 to 0.36122, while the R-values are in the range from 84.14% to 85.34%. Comparing the results for testing models numbers 11 and 12, this research concludes that the ANN performance does not improve. However, to ensure the result for testing model 12, this research builds testing model number 27, with input including NITA, ROE, current ratio, solvency ratio, GDPG and RIR, to further test for the effect of RIR. The result for testing model number 27 shows that the MSEs are in the range from 0.37400 to 0.37873, while the R-values of these datasets are in the range from 83.22% to 84.02%. Comparing the results for testing models numbers 25 and 27, there is no significant improvement in the ANN performance, so that the RIR is also an insignificant predictor. In addition, GDPG and RIR are highly correlated, with a correlation value −0.989. Consequently, the credit rating model for dataset 2015–2018 excludes RIR from the final model.

SIZELNNE: Size of firm measured by Ln number of employees

The next step of the building process continues with testing model number 13 in which the input includes EBITTA, NITA, ORTA, PM, ROE, ORLG10, current ratio, solvency ratio, ORSE, GDPG, INF, RIR and SIZELNNE, while the output is also the ScoreG10. The MSEs are in the range from 0.33698 to 0.37102, while the R-values are in the range from 83.66% to 84.61%. Comparing the results for testing models numbers 12 and 13, it is also clear that the ANN performance does not improve. However, to ensure the result, this research builds testing model 28 including the inputs NITA, ROE, current ratio, solvency ratio, GDPG and SIZELNNE to further test for the SIZELNNE effect. As a result of testing model 28, the MSEs are in the range of 0.37457 to 0.38041, while the R-values are in the range of 83.90% to 84.2%. This means that there is no significant improvement in the ANN performance between testing models 25 and 28. SIZELNNE is an insignificant predictor and is excluded from the credit rating model for the dataset 2015–2018.

To conclude, comparing testing model 13 (with all 13 independent predictors) and testing model 25 (with only five significant independent predictors), it is clear that testing model 25 has the same predicting power as the testing model 13. This result indicates that for the dataset 2015–2018, there are five significant IVs: NITA, ROE, CURRENT, SOLVENCY and GDPG. These significant factors support by the influential literature (Altman 1968; Ohlson 1980; Zmijewski 1984; Shumway 2001; Altman and Sabato 2007; Psillaki et al. 2010; Wu et al. 2010; Duan et al. 2012; Jones et al. 2017; Pham et al. 2018a, b). On the other hand, the eight insignificant IVs are EBITTA, ORTA, PM, ORLG10, ORSE, INF, RIR and SIZELNNE.

Neural network testing models for ranking of significant factors

After identifying the significant predictors, including NITA, ROE, CURRENT, SOLVENCY, GDPD and SIZELNNE, this research also builds testing models to further test for the effect of each significant predictor individually in order to find the ranking of each significant factor. These testing models are testing model 14 and testing models 29 to 33 in dataset 2015–2018.

Based on the results of these testing models, it is clear that the solvency ratio has the highest R-values in the range from 68.15% to 68.3%. The second most significant factor is SIZELNNE, with R-values from 47.2% to 68.99%. Then, the third most significant predictor is the current ratio, with R-values in the range from 53.33% to 54.04%. Next, the fourth most significant predictor is NITA, with R-values in the range from 49.81% to 50.37%. Then, GPDG is the fifth most significant predictor, with R-values fluctuating from 10.20% to 96.57%. When this research builds testing model number 32 for GDPG individually as the only input, the R-values fluctuate significantly, so this result seems to be highly risky for the predicting model. Finally,

Table 9 Accuracy rates of three credit rating model of firms in Vietnam in the period 2015–2018

Model	DV	IVs	AR (%)
ANN	ScoreG10	NITA, ROE, CR, SolvencyR and GPDG	**61.11**
OLR	ScoreG10	NITA, ORTA, PM, ROE, CR, SolvencyR, ORSE, GDPG and SIZELNNE	**55.43**
MDA	ScoreG10	NITA, ORTA, PM, ROE, ORLG10, CR, SolvencyR, ORSE, GDPG and SIZELNNE	**53.10**

ROE is the least significant predictor and this research cannot run an individual test for ROE using the ANN model via MATLAB 2019a.

Predicting the power of credit rating models using ANN

Three different models are built for comparison purposes, including Artificial Neural Networks (ANN), Multiple Discriminant Analysis (MDA). To compare the predicting power of the three different models, this research applies data partition with a training dataset of 70% and a testing dataset of 30%. Then the training dataset is used to build the model and the testing dataset is used to test the Accurate Rate (AR) of each model. The AR is the proportion of correct prediction between the predicted credit rating and the actual credit rating of each model (Table 9). The AR of the ANN model is 61.11%, which is the highest AR compared to the OLR and MDA models. The AR for OLR is 55.43%, which is higher than the MDA model's AR of 53.1%. All of the accurate rates is the range between 0.5 to 0.7 which is inferior and acceptable when the dependent variable has a high number of classes (ten classes) (Chi and Meng 2018). The more classes for the dependent variable the lower accurate rate of the classification results (Chi and Meng 2018). Based on these percentages, the ANN model is arguably the most robust predicting model, then the second most robust is the OLR model and the third most robust is the MDA model. This conclusion is supported by the literature (Gentry et al. 1985; Keasey and Watson 1987; Aziz et al. 1988; Platt and Platt 1990; Ooghe et al. 1995; Mossman et al. 1998; Charitou and Trigeorgis 2000; Atiya 2001; Becchetti and Sierra 2003a; Huang et al. 2004; Altman and Sabato 2007).

Table 8 presents the dependent variable, the independent variables and the AR for each credit rating model, including the ANN, OLR and MDA. Zhang (2000), Guyon and Elisseeff (2003) and Dias (2013) suggest that selecting only a small and relevant set of independent variables and removing the insignificant variables will increase the effectiveness of a credit rating model including: (1) making the model easier to understand; (2) decreasing computational capacity and time consumption; (3) reducing data requirements and saving data costs; and (4) maintaining or improving the accuracy of the credit rating model. Ideally, research builds a highly accurate credit rating model that economically uses a relevant, significant and small set of predictor variables. In this research, it is clear that the ANN model is arguably the most robust predicting model with an AR of 61.11% with five main inputs: NITA, ROE, current ratio, solvency ratio and GPDG. The next best credit rating model is the OLR model with an AR of 55.43% in which the nine main inputs are NITA, ORTA, PM, ROE, current ratio, solvency ratio, ORSE, GDPG and SIZELNNE. The final credit rating model is the MDA model with an AR of 53.1% in which the ten inputs are NITA, ORTA, PM, ROE, ORLG10, current ratio, solvency ratio, ORSE, GDPG and SIZELNNE.

Based on my research analysis and conclusion, ANN, OLR and MDA models are three robust statistical methodologies, but use the same inputs including NITA, ROE, current ratio, solvency ratio and GPDG for the period 2015–2018. This research find evidence that OLR and MDA models use more parameters to build the credit rating model than the ANN model. However, the ARs of OLR and MDA models are lower than the AR of the ANN model. In other words, the ANN model requires fewer input data but has stronger predicting power. It is the confirmation that the ANN model is the most robust methodology/technique to build the credit rating model for firms in Vietnam in the period 2015–2018.

5 Conclusion

This research answers the question of what are the factors that significantly affect the credit rating model for firms in Vietnam using ANN. The significant predictors of the ANN model for the dataset 2015–2018 include NITA, ROE, the current ratio, the solvency ratio and GDPG. These significant predictors support by the influential literature (Altman 1968; Ohlson 1980; Zmijewski 1984; Shumway 2001; Altman and Sabato 2007; Psillaki et al. 2010; Wu et al. 2010; Duan et al. 2012; Jones et al. 2017; Pham et al. 2018a, b). Consequently, NITA, ROE, current ratio, solvency ratio, GDPG and SIZELNNE are robust IVs in the credit rating model. In addition, based on the effects of these predictors to the credit rating model, this research can rank these predictors in order as solvency ratio, SIZELNNE, current ratio, NITA, GPDG, and ROE. In addition, the accurate rate of the ANN model is 61.11%, which is the highest accuracy rate compare to the OLR of 55.43% and MDA of 53.1%. This result indicates that ANN model is the most robust model which use less inputs with higher accurate rate.

The results of this study can use in different functions in the financial markets; firstly, they are a technology tool to alleviate asymmetric information problems between borrowers and lenders (Kumar and Bhattacharya 2006; Benmelech and Dlugosz 2009). Secondly, a reliable credit rating system can help investors with low-cost and useful information which can minimise the level of investment risk significantly. From a macroeconomic view, credit ratings can play a role as a regulatory tool in the financial markets. As a result, these ratings will motivate and strengthen the economic growth of a country (Kumar and Bhattacharya 2006). Credit ratings also help to minimise asymmetric information problems between companies and investors in capital markets, which can address the problem of collective action between dispersed investors (Kumar and Bhattacharya 2006; Benmelech and Dlugosz 2009).

The statistical model constructed in this study can help Vietnamese commercial banks calculate a specific figure as the credit rating of a borrower, so that the bank may offer an appropriate credit policy and interest rate level for each client following the principle of 'high risk, high return'. This model will provide more detailed credit rating information to Vietnamese commercial banks so that it will help to limit subjective errors by credit officers in evaluating customers. In fact, the internal credit rating of Vietnamese commercial banks depends more on qualitative factors that are rated by credit officers, such as the management level of the firm or the relationship between the firm and the commercial banks, than financial factors of the firms (Pham, Truong and Bui 2018). These qualitative evaluations are subjective because they depend on the experience of the credit officers. Besides, this study will also help Vietnamese commercial banks' risk administrators to verify the degree of influence of each indicator (the independent variables) on the credit rating of the firm (the Dependent Variable). The risk officers can then judge clearly which elements have the most significant impact on the repayment capacity of the borrower and will make more accurate decisions about granting credit. These results can be used as reference information for granting credit decisions.

The research faces several limitations if the research findings are applied to the real world. One of the considerations of the implementation of credit rating models is the fair value accounting (FVA) application. FVA requires certain assets and liabilities to be (re)valued at fair value to reflect changes in economic value. The case of Vietnam is a sophisticated transitioning setting with state intervention and the personal interests of officeholders can influence regulatory processes. Given the several challenges identified, this study suggests that the contemporary business environment in Vietnam is less favourable for adopting FVA, at least until these weaknesses are effectively tackled. Although this study does not investigate the independence or professionalism of auditors in Vietnam, these concerns have fuelled debate on FVA in the literature, with Enron and Arthur Andersen being notorious international cases (Benston 2006). This is an area that is worthy of future research (Nguyen 2019). If the Vietnamese accounting system adopts FVA, it could lead to a change in the value of certain assets and liabilities of firms. Consequently, FVA could also affect the financial ratios of firms. If this happened, the credit rating models can be affected significantly.

References

Altman, E.I.: Financial ratios, discriminant analysis and the prediction of corporate bankruptcy. J. Financ. **23**(4), 589–609 (1968)

Altman, E.I., Iwanicz-Drozdowska, M., Laitinen, E.K., Suvas, A.: Financial distress prediction in an international context: a review and empirical analysis of Altman's Z-score model. J. Int. Financ. Manag. Acc. **28**(2), 131–171 (2017)

Altman, E.I., Sabato, G.: Modelling credit risk for SMEs: evidence from the U.S. market. Abacus **43**(3), 332–357 (2007)

Angelini, E., di Tollo, G., Roli, A.: A neural network approach for credit risk evaluation. Q. Rev. Econ. Finance **48**(4), 733–755 (2008)

Argenti, J.: Corporate collapse : the causes and symptoms (1976), London ; New York : McGraw-Hill, viewed 13 September 2019. https://trove.nla.gov.au/version/45431348

Atiya, A.F.: Bankruptcy prediction for credit risk using neural networks: a survey and new results. IEEE Trans. Neural Netw. **12**(4), 929–935 (2001)

Aziz, A., Emanuel, D.C., Lawson, G.H.: Bankruptcy prediction—an investigation of cash flow based models [1]. J. Manage. Stud. **25**(5), 419–437 (1988)

Becchetti, L., Sierra, J.: Bankruptcy risk and productive efficiency in manufacturing firms. J. Bank. Finance **27**(11), 2099–2120 (2003)

Benmelech, E., Dlugosz, J.: 'The Credit Rating Crisis', NBER Macroeconomics Annual (2009)

Benston, G.J.: Fair-value accounting: a cautionary tale from enron. J. Account. Public Policy **25**(4), 465–484 (2006)

Blanchet, F.G., Legendre, P., Borcard, D.: Forward selection of explanatory variables. Ecology **89**(9), 2623–2632 (2008)

Campbell, J.Y., Hilscher, J., Szilagyi, J.: In search of distress risk. J. Financ. **63**(6), 2899–2939 (2008)

Chai, N., Wu, B., Yang, W., Shi, B.: A multicriteria approach for modeling small enterprise credit rating: evidence from China. Emerg. Mark. Financ. Trade **55**(11), 2523–2543 (2019)

Charitou, A., Trigeorgis, L.: Option-based bankruptcy prediction. SSRN Electron. J. (2000)

Chi, G., Meng, B.: Debt rating model based on default identification: empirical evidence from Chinese small industrial enterprises. Manag. Decis. (2018)

Chung, H.M., Gray, P.: Data mining. J. Manag. Inf. Syst. **16**(1), 11–16 (1999)

Cornée, S.: The relevance of soft information for predicting small business credit default: evidence from a social bank. J. Small Bus. Manage. **57**(3), 699–719 (2019)

Craven, M.W., Shavlik, J.W.: Using neural networks for data mining. Futur. Gener. Comput. Syst. **13**(2–3), 211–229 (1997)

Desai, V.S., Crook, J.N., Overstreet, G.A.: A comparison of neural networks and linear scoring models in the credit union environment. Eur. J. Oper. Res. **95**(1), 24–37 (1996)

Dias, R.: The Rise and Fall of Credit Default Swaps: An empirical investigation of global banks and non-bank financial institutions. Swinburne University of Technology, Ph.D. (2013)

Doumpos, M., Lemonakis, C., Niklis, D., Zopounidis, C.: Analytical Techniques in the Assessment of Credit Risk: An Overview of Methodologies and Applications (2019), Springer International Publishing, Cham, viewed 7 September 2019, https://doi.org/10.1007/978-3-319-99411-6

Duan, J.-C., Sun, J., Wang, T.: Multiperiod corporate default prediction—a forward intensity approach. J. Econ. **170**(1), 191–209 (2012)

Farhadieh, F.: A Statistical Framework for Quantifying Adaptive Behavioural Risk for the Banking Industry. Swinburne University of Technology, Ph.D. (2011)

Gentry, J.A., Newbold, P., Whitford, D.T.: Classifying bankrupt firms with funds flow components. J. Account. Res. **23**(1), 146–160 (1985)

Giordani, P., Jacobson, T., von Schedvin, E., Villani, M.: Taking the twists into account: predicting firm bankruptcy risk with splines of financial ratios. J. Financ. Quant. Anal. **49**(4), 1071–1099 (2014)

Goonatilake, S., Treleaven, P.C.: Intelligent Systems for Finance and Business. Wiley (1995)

Gottheil, F.M.:Principles of macroeconomics, Nelson Education (2013)

GSO, V.: General Statistics Office Of Vietnam (2020), viewed 16 February 2020. http://www.gso.gov.vn/default_en.aspx?tabid=622&ItemID=19463

Guyon, I., Elisseeff, A.: An introduction of variable and feature selection. J. Mach. Learn. Res. Spec. Issue Variable Feature Sel. **3**, 1157–1182 (2003)

Hillegeist, S.A., Keating, E.K., Cram, D.P., Lundstedt, K.G.: Assessing the probability of bankruptcy. Rev. Acc. Stud. **9**(1), 5–34 (2004)

Huang, Z., Chen, H., Hsu, C.-J., Chen, W.-H., Wu, S.: Credit rating analysis with support vector machines and neural networks: a market comparative study. Decis. Support Syst. **37**(4), 543–558 (2004)

Jacobson, T., Lindé, J., Roszbach, K.: Firm default and aggregate fluctuations. J. Eur. Econ. Assoc. **11**(4), 945–972 (2013)

Jiang, C., Wang, Z., Wang, R., Ding, Y.: Loan default prediction by combining soft information extracted from descriptive text in online peer-to-peer lending. Ann. Oper. Res. **266**(1–2), 511–529 (2018)

Jiménez, G., Saurina, J.: Collateral, type of lender and relationship banking as determinants of credit risk. J. Bank. Finance **28**(9), 2191–2212 (2004)
Jones, S., Johnstone, D., Wilson, R.: Predicting corporate bankruptcy: an evaluation of alternative statistical frameworks: an evaluation of alternative statistical frameworks. J. Bus. Financ. Acc. **44**(1–2), 3–34 (2017)
Keasey, K., Watson, R.: Non-financial symptoms and the prediction of small company failure: a test of argenti's hypotheses. J. Bus. Financ. Acc. **14**(3), 335–354 (1987)
Kumar, K., Bhattacharya, S.: Artificial neural network vs linear discriminant analysis in credit ratings forecast. Review of Accounting and Finance (2006), viewed 12 September 2019. https://doi.org/10.1108/14757700610686426/full/html
Langkamp, D.L., Lehman, A., Lemeshow, S.: Techniques for handling missing data in secondary analyses of large surveys. Acad. Pediatr. **10**(3), 205–210 (2010)
Le, B.T.: Factors affecting credit risk management system of the Vietnamese commercial banks (2018), University of Economics Ho Chi Minh City, viewed http://digital.lib.ueh.edu.vn/handle/UEH/57976
Lee, T.-S., Chen, I.-F.: A two-stage hybrid credit scoring model using artificial neural networks and multivariate adaptive regression splines. Expert Syst. Appl. **28**(4), 743–752 (2005)
Lukason, O., Hoffman, R.C.: Firm bankruptcy probability and causes: an integrated study. Int. J. Bus. Manag. **9**(11), 80–91 (2014)
Malhotra, R., Malhotra, D.K.: Evaluating consumer loans using neural networks. Omega **31**(2), 83–96 (2003)
Mankiw, N.G.: Brief principles of macroeconomics. Cengage Learning (2020)
Mankiw, N.G.: Brief principles of macroeconomics. Cengage Learning (2014)
Matlab: Choose Neural Network Input-Output Processing Functions—MATLAB & Simulink (2019), viewed 16 September 2019. https://www.mathworks.com/help/deeplearning/ug/choose-neural-network-input-output-processing-functions.html
Mossman, C.E., Bell, G.G., Swartz, L.M., Turtle, H.: An empirical comparison of bankruptcy models. Financ. Rev. **33**(2), 35–54 (1998)
Muñoz-Izquierdo, N., Laitinen, E.K., Camacho-Miñano, M., Pascual-Ezama, D.: Does audit report information improve financial distress prediction over Altman's traditional Z-Score model?. J. Int. Financ. Manag. Account. (2019), viewed 26 September 2019. https://doi.org/10.1111/jifm.12110
Nguyen, L.-U.: The (un)suitability of fair-value accounting in emerging economies: the case of Vietnam. J. Account. Organ. Change (2019), viewed 18 October 2019. https://doi.org/10.1108/JAOC-03-2018-0032/full/html
Ohlson, J.A.: Financial ratios and the probabilistic prediction of bankruptcy. J. Account. Res. **18**(1), 109–131 (1980)
Ooghe, H., Joos, P., De Bourdeaudhuij, C.: Financial distress models in Belgium: the results of a decade of empirical research (1995), 30, viewed 13 September 2019. https://repository.vlerick.com/handle/20.500.12127/143
Parker, L.E.:Notes on multilayer, feedforward neural networks. CS494/594: Projects Mach. Learn. (2006)
Pearson, R.: Outliers in process modeling and identification. Control Syst. Technol. IEEE Trans. **10**, 55–63 (2002)
Pham, Q.H., Truong, T.L.N., Bui, D.T.P.: An alternate internal credit rating system for construction and timber industries using artificial neural network. Econ. Financ. Appl., Springer, Cham, pp. 752–791 (2018a), viewed 29 April 2020. https://doi.org/10.1007/978-3-319-73150-6_59
Pham, V.N.B., Do, T.T., Vo, H.D.: Financial distress and bankruptcy prediction: an appropriate model for listed firms in Vietnam. Econ. Syst. **42**(4), 616–624 (2018b)
Platt, H.D., Platt, M.B.: Development of a class of stable predictive variables: the case of bankruptcy prediction. J. Bus. Financ. Acc. **17**(1), 31–51 (1990)
Psillaki, M., Tsolas, I.E., Margaritis, D.: Evaluation of credit risk based on firm performance. Eur. J. Oper. Res. **201**(3), 873–881 (2010)

Rumelhart, D.E.: Parallel distributed processing: Explorations in the microstructure of cognition. Learni. Int. Representations Error Propag. **1**, 318–362 (1986)

Sanche, R., Lonergan, K.: Variable reduction for predictive modeling with clustering (2006)

Saunders, A., Cornett, M.M.: Financial institutions Management—A risk management approach, 6th edn, The McGraw-Hill (2007), viewed http://www.bulentsenver.com/FIN5477/Financial_Institutions_Management_AntonySaunders_TextBook.pdf

Schechtman, R., Gaglianone, W.P.: Macro stress testing of credit risk focused on the tails. J. Financ. Stability **8**(3), 174–192 (2012). Elsevier

Scott, J.H.: Bankruptcy, secured debt, and optimal capital structure. J. Financ. **32**(1), 1–19 (1977)

Shumway, T.: Forecasting bankruptcy more accurately: a simple hazard model. J. Bus. **74**(1), 101–124 (2001)

Sigrist, F., Hirnschall, C.: Grabit: Gradient tree-boosted Tobit models for default prediction. J. Bank. Finance **102**, 177–192 (2019)

Tabachnick, B., Fidell, L.S.: Multivariate analysis of variance and covariance. Using Multivariate Stat. **3**, 402–407 (2007)

Titman, S., Wessels, R.: The determinants of capital structure choice. J. Financ. **43**(1), 1–19 (1988)

Tsai, M.-C., Lin, S.-P., Cheng, C.-C., Lin, Y.-P.: The consumer loan default predicting model—an application of DEA–DA and neural network. Expert Syst. Appl. **36**(9), 11682–11690 (2009)

Washington, G.K.: Effects of macroeconomic variables on credit risk in the Kenyan banking system. Int. J. Bus. Commerce **3**(9), 1 (2014)

Wu, Y., Gaunt, C., Gray, S.: A comparison of alternative bankruptcy prediction models. J. Contemp. Account. Econ. **6**(1), 34–45 (2010)

Zavgren, C.V.: Assessing the vulnerability to failure of american industrial firms: a logistic analysis. J. Bus. Financ. Acc. **12**(1), 19–45 (1985)

Zhang, G.P.: 'Neural networks for classification: a survey', IEEE transactions on systems, man, and cybernetics. Part C (Appl. Rev.) **30**(4), 451–462 (2000)

Zmijewski, M.E.: Methodological Issues related to the estimation of financial distress prediction models. J. Account. Res. **22**, 59–82 (1984)

Incentives for R&D in Northern Italy Revisited

Chon Van Le

Abstract This paper employs the latest analytical technique of the Sharp regression discontinuity (RD) design to re-evaluate the R&D subsidy program implemented in northern Italy. We find that the program did not exert an impact on firms' investment, regardless of their sizes. Our results are in contrast to those of Bronzini and Iachini (2014) implying that the subsidy generated additional investment for small firms. The general nonexistent impact raises serious doubts as to the efficiency of the program and its procedures of scoring and approving grants.

1 Introduction

Under the growing trend of evidence-based policy making, impact evaluation has attracted a lot of attention among social scientists, program managers and policy makers. However, the analysis of the causal effect of a program (hereinafter called a treatment) on outcomes of interest has encountered several issues. A fundamental one is that at any point in time, a unit either participates in a program or does not participate, so we observe only the value of the potential outcome under the condition actually assigned, while the potential outcome under the other condition is missing. Since the unit cannot be observed simultaneously in two different states, the treatment effect at the individual level is essentially unaccountable.

This counterfactual problem can be statistically resolved with the randomized assignment method, also known as randomized controlled trials (RCTs), in which everyone is given an equal opportunity to participate in the program. But this does not happen in many cases for either ethical or practical reasons. A large number of programs use a score (or an index) to decide that those units whose score exceeds a threshold or cutoff are eligible to get the treatment and that those units whose score is below the cutoff are not. Because we never observe both outcomes for the same value of the score, it is impossible to construct two regression curves so that their vertical

C. Van Le (✉)
International University, Vietnam National University, Ho Chi Minh City, Vietnam
e-mail: lvchon@hcmiu.edu.vn

N. Ngoc Thach et al. (eds.), *Financial Econometrics: Bayesian Analysis, Quantum Uncertainty, and Related Topics*, Studies in Systems, Decision and Control 427,
https://doi.org/10.1007/978-3-030-98689-6_53

difference can be calculated as an estimate of the treatment effect. Nevertheless, we nearly observe both curves at the cutoff. Based on this idea and a few other conditions, the regression discontinuity (RD) design was originally proposed by Thistlethwaite and Campbell (1960) and formalized by Hahn et al. (2001).

There are two types of RD design, i.e., Sharp RD design where units assigned to the treatment condition actually receive it and others do not, and Fuzzy RD design where the treatment assignment is not perfectly complied. Their techniques have evolved and diverged considerably from each other. This paper is to focus on the sharp RD design and then apply it to re-evaluate the study of the research and development (R&D) subsidy program conducted in northern Italy in Bronzini and Iachini (2014). We find that the subsidy did not induce firms to make additional investments, regardless of their sizes. It differs from Bronzini and Iachini's (2014) results indicating that small subsidized firms increased their investments. The overall insignificant impact calls into question the efficacy of this public program.

This paper is structured as follows. Section 2 reviews the sharp RD design's features, local polynomial point estimation and inference, and validation test based on the continuity of the score density around the cutoff. Section 3 reexamines the study by Bronzini and Iachini (2014) and presents new findings. Conclusions follow in Sect. 4.

2 The Sharp Regression Discontinuity Design

In the RD design, each unit receives a score X_i and there is a known cutoff, denoted c. A treatment is assigned to those units with $X_i \geq c$ and not assigned to those units with $X_i < c$. The treatment assignment to unit i is defined as $T_i = 1(X_i \geq c)$, where $1(\cdot)$ is the indicator function. So the conditional probability of treatment assignment given the score changes discontinuously at the cutoff. In the Sharp RD design, the treatment condition assigned is the same as the treatment condition actually received for all units. This implies that the conditional probability of receiving treatment given the score, $P(T_i = 1|X_i = x)$, changes precisely from zero to one at the cutoff.

Following Imbens and Lemieux (2008), we assume that each unit has two potential outcomes: $Y_i(0)$ is the outcome given exposure to the control, and $Y_i(1)$ is the outcome given exposure to the treatment. But only one of them is observed and the other remains latent. The observed outcome is written as

$$Y_i = (1 - T_i) \cdot Y_i(0) + T_i \cdot Y_i(1) = \begin{cases} Y_i(0) \text{ if } X_i < c, \\ Y_i(1) \text{ if } X_i \geq c, \end{cases} \qquad i = 1, \ldots, n.$$

If we consider the data set $(Y_i, X_i)_{i=1}^n$ as a random sample and the potential outcomes $(Y_i(0), Y_i(1))_{i=1}^n$ as random variables, then we can estimate the expected values of the potential outcomes given the score, $\mathrm{E}[Y_i(0)|X_i = x]$ and $\mathrm{E}[Y_i(1)|X_i = x]$. It should be noticed that the conditional expectation function (or regression function) $\mathrm{E}[Y_i(0)|X_i = x]$ is observed only for units located to the left of the cutoff with

$X_i < c$, and $\mathrm{E}[Y_i(1)|X_i = x]$ is observed only for units located to the right of the cutoff with $X_i \geq c$.

The average treatment effect at a particular score is represented by $\mathrm{E}[Y_i(1)|X_i = x] - \mathrm{E}[Y_i(0)|X_i = x]$. This difference cannot be directly estimated because we never observe the pair $Y_i(0)$ and $Y_i(1)$ together at any value of x, which is referred to as the fundamental problem of causal inference. However, extrapolation towards the cutoff point allows us to compare treatment and control units at c. The Sharp RD treatment effect is thus defined at c as

$$\tau_{\mathrm{SRD}} = \mathrm{E}[Y_i(1) - Y_i(0)|X_i = c].$$

Hahn et al. (2001) argued that under certain conditions, if the regression functions $\mathrm{E}[Y_i(0)|X_i = x]$ and $\mathrm{E}[Y_i(1)|X_i = x]$ are assumed to be continuous at $x = c$, then the average causal effect of the treatment in a Sharp RD design is

$$\tau_{\mathrm{SRD}} = \mathrm{E}[Y_i(1) - Y_i(0)|X_i = c] = \lim_{x \downarrow c} \mathrm{E}[Y_i|X_i = x] - \lim_{x \uparrow c} \mathrm{E}[Y_i|X_i = x]. \quad (1)$$

Equation (1) states that the difference between the limits of the treated and control average observed outcomes as the score approaches the cutoff is the average treatment effect at the cutoff. Units just below the cutoff are used to approximate the average outcome that units just above the cutoff would have had if they are exposed to the control condition instead of the treatment. Therefore, the continuity assumption which focuses on observations located in a small neighborhood around the cutoff provides a valid justification for estimating the Sharp RD effect.

Local Polynomial Point Estimation and Inference

Estimation and inference about the treatment effect in Eq. (1) in chapter "On The Skill of Influential Predictions" involves approximating the unknown regression functions $\mathrm{E}[Y_i(0)|X_i = x]$ and $\mathrm{E}[Y_i(1)|X_i = x]$ by a polynomial function of the score. Early empirical work tried to approximate these functions globally using higher-order polynomials, usually of fourth or fifth order. However, a global polynomial approach does not produce a good point estimation and inference as the discussion above implies that the treatment effect is local in nature. A point estimator that is heavily affected by observations far from the boundary tends to give a poor estimate at the boundary point. Gelman and Imbens (2019) argued that the point estimates and standard errors are highly sensitive to the order of the polynomial. Moreover, higher-order polynomials may lead to overfitting and produce too narrow confidence intervals; conventional inference for the treatment effect can hence be misleading.

Modern empirical work approximates the regression functions only near the cutoff and uses lower-order polynomials, which are more robust to boundary and overfitting problems. These polynomials are less sensitive to outliers and other extreme properties of the data generating process that may emerge far from the cutoff. Consequently, the choice of the bandwidth h that determines the size of the neighborhood around the cutoff is very crucial in the RD analysis.

Within this bandwidth, the units closer to the cutoff should receive a bigger weight than those further away. A unit's score can be transformed into $z_i = \frac{X_i - c}{h}$, which is the "standardized" distance between the unit's score and the cutoff. The weights are determined by a kernel function $K(\cdot)$. The triangular kernel function, $K(z) = (1 - |z|)1(|z| \leq 1)$, assigns zero weight to all units with score outside the interval $[c - h, c + h]$, and positive weights to all units with score within this interval. The weight is maximum at the cutoff and decreases linearly and symmetrically as the score gets farther from the cutoff in both directions. The Epanechnikov kernel function, $K(z) = (1 - z^2)1(|z| \leq 1)$, gives quadratic descending weights to units inside the interval and zero weight to the rest. Although point estimation and inference results are generally robust to either kernel function, Cattaneo et al. (2019) suggested the triangular kernel. It has a point estimation mean squared error (MSE) optimality property when used in conjunction with a bandwidth that optimizes the MSE.

The recommended order of the local polynomial approximation is $p = 1$ (Cattaneo and Titiunik 2021). The local linear point estimator in general tends to provide a good trade-off between simplicity, precision, and variability in the RD analysis. The local linear approximation takes the form:

$$\hat{Y}_i = \begin{cases} \hat{\mu}_- + \hat{\mu}_{-,1}(X_i - c) & \text{if } X_i < c, \\ \hat{\mu}_+ + \hat{\mu}_{+,1}(X_i - c) & \text{if } X_i \geq c, \end{cases} \qquad i = 1, \ldots, n. \tag{2}$$

After we run a weighted least squares regression of this model, the intercepts $\hat{\mu}_-$ and $\hat{\mu}_+$ are estimates of the points $\mathrm{E}[Y_i(0)|X_i = c]$ and $\mathrm{E}[Y_i(1)|X_i = c]$, respectively. The estimate of the Sharp RD treatment effect is $\hat{\tau}_{\mathrm{SRD}} = \hat{\mu}_+ - \hat{\mu}_-$.

The properties of local polynomial estimation and inference procedures depend substantially on the width of the neighborhood around the cutoff. A narrower neighborhood is likely to decrease the misspecification error of the local linear approximation, but at the same time to increase the variance of the estimated coefficients due to a smaller number of observations available for estimation, and vice versa. Selecting bandwidth thus entails a "bias-variance trade-off." Since the MSE of the local polynomial RD point estimator is the sum of its squared bias and its variance, the most popular bandwidth selection method tries to optimize the bias-variance trade-off by minimizing the MSE of $\hat{\tau}_{\mathrm{SRD}}$.

Given the polynomial order and kernel function chosen, this method[1] minimizes with respect to h the approximate MSE of the RD treatment effect

$$\mathsf{MSE}\left(\hat{\tau}_{\mathrm{SRD}}\right) = \mathsf{Bias}^2\left(\hat{\tau}_{\mathrm{SRD}}\right) + \mathsf{Variance}\left(\hat{\tau}_{\mathrm{SRD}}\right) = \mathfrak{B}^2 + \mathfrak{V} = h^{2(p+1)}\mathscr{B}^2 + \frac{1}{nh}\mathscr{V},$$

where $\mathscr{B}$ and $\mathscr{V}$ are the bias and variance of $\hat{\tau}_{\mathrm{SRD}}$ not adjusted for sample size, polynomial order, and bandwidth. The MSE-optimal bandwidth is

[1] See Cattaneo et al. (2019) for more detail.

$$h_{\mathrm{MSE}} = \left(\frac{\mathscr{V}}{2(p+1)\mathscr{B}^2 n} \right)^{\frac{1}{2p+3}}. \tag{3}$$

Researchers may choose different bandwidths on each side of the cutoff when the bias and/or variance of the control and treatment groups near the cutoff differ considerably. The two distinct bandwidths h_- and h_+ can be selected by optimizing the approximate MSEs of $\hat{\mu}_-$ and $\hat{\mu}_+$ separately. The MSE-optimal bandwidths are

$$h_{\mathrm{MSE},-} = \left(\frac{\mathscr{V}_-}{2(p+1)\mathscr{B}_-^2 n_-} \right)^{\frac{1}{2p+3}},$$

$$h_{\mathrm{MSE},+} = \left(\frac{\mathscr{V}_+}{2(p+1)\mathscr{B}_+^2 n_+} \right)^{\frac{1}{2p+3}}.$$

In some applications, the estimated bias may be close to zero, the small denominator in (3) can deliver unappropriate bandwidths. The alternative formula is

$$h_{\mathrm{MSE}} = \left(\frac{\mathscr{V}}{[2(p+1)\mathscr{B}^2 + \mathscr{R}]n} \right)^{\frac{1}{2p+3}},$$

where $\mathscr{R}$ is a "regularization" term.

While the local polynomial approximation with an MSE-optimal bandwidth generally carries a bias, conventional least squares inference methods assume that the local polynomial regression model is correctly specified. Therefore, the usual t-statistic and confidence interval should be modified to take into account the misspecification error. The local polynomial RD point estimator $\hat{\tau}_{\mathrm{SRD}}$ is asymptotically normal

$$\frac{\hat{\tau}_{\mathrm{SRD}} - \tau_{\mathrm{SRD}} - \mathfrak{B}}{\sqrt{\mathfrak{V}}} \overset{a}{\sim} N(0, 1).$$

Calonico et al. (2014) proposed a confidence interval that estimates and removes the bias term $\hat{\mathfrak{B}}$ from the RD point estimator and adjusts the variance, denoted by $\hat{\mathfrak{V}}_{\mathrm{bc}}$, to incorporate the additional variability introduced in the bias estimation step. When the MSE-optimal bandwidth is used to estimate the treatment effect, the 95% robust bias corrected confidence interval is

$$\mathrm{CI}_{\mathrm{rbc}} = \left[\hat{\tau}_{\mathrm{SRD}} - \hat{\mathfrak{B}} \pm 1.96 \cdot \sqrt{\hat{\mathfrak{V}}_{\mathrm{bc}}} \right].$$

Because the MSE-optimal bandwidth is not designed to obtain good distributional approximations, an alternative approach is to use different bandwidths for point estimation and inference. The bandwidth for inference is chosen to minimize the asymptotic coverage error rate of the confidence interval $\mathrm{CI}_{\mathrm{rbc}}$. However, $\mathrm{CI}_{\mathrm{rbc}}$ is valid even when h_{MSE} is employed. We can use this bandwidth for both point estimation and inference (Cattaneo et al. 2019).

Validation Tests of the RD Design

The validity of the RD design depends essentially on the plausibility of the RD assumptions. Although the continuity assumptions are not directly testable, there are a number of empirical methods that can provide indirect evidence about the plausibility of the assumptions. The validation tests are based on (i) the null treatment effect on pre-intervention covariates and placebo outcomes, (ii) the continuity of the density function of the score about the cutoff, (iii) the null treatment effect at artificial (or placebo) cutoffs away from the real cutoff, and (iv) the sensitivity of the empirical conclusions to the exclusion of observations near the cutoff and to bandwidth choices.

In this paper, we are going to use the second type of validation test. It examines the null hypothesis that the density of the score is continuous at the cutoff. The test was introduced by McCrary (2008). The density is estimated separately for units below and above the cutoff by a local polynomial density estimator developed by Cattaneo et al. (2020). Failing to reject the null indicates that there is no statistical evidence of manipulation at the cutoff, thus the validity of the RD design is supported.

3 Grants for R&D in Northern Italy

In 2003, the government of Emilia-Romagna launched the "Regional Program for Industrial Research, Innovation and Technological Transfer." The program was to provide grants supporting the R&D expenditure by firms which had main offices and intended to implement projects in the region. The maximum grant for a project was €250,000 with the typical duration from 12 to 24 months. Eligible firms' projects were assessed by a panel of independent experts who gave scores in terms of (i) science and technology, (ii) financial and economic results, (iii) management, and (iv) regional impact. Projects that obtained a total score greater than or equal to 75 out of the maximum 100 points received the grants. After two rounds of applications and evaluations, about €93 million was transferred to subsidized firms.

Since firms cannot internalize all the benefits of their R&D activity, the equilibrium private investment is normally lower than the socially optimal level. In addition, uncertainty and information asymmetry tend to increase the cost of external funding for intangible assets. Public incentives that are able to increase private R&D spending by improving its expected profitabilty to compensate for positive externalities and alleviate liquidity constraints will certainly improve social welfare.

Bronzini and Iachini (2014) applied a Sharp RD design to subsidized and unsubsidized industrial firms with scores near the cutoff, i.e., 75 points, to investigate whether public financial aid in northern Italy really worked. They used two different sample windows: (i) the wide one included 171 firms with scores from 52 to 80, representing 50% of the whole sample, and (ii) the narrow one included 115 firms with scores from 66 to 78, representing 35%. The ranges were selected to roughly balance the numbers of firms above and below the cutoff. The outcome variables were scaled by pre-program sales, capital, or assets which help to avoid potential endogeneity issues.

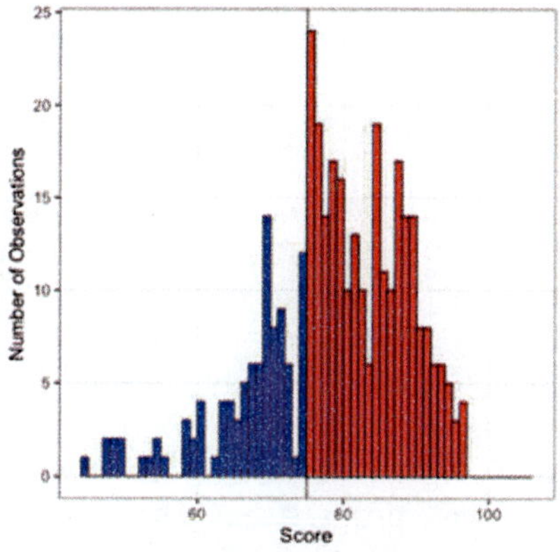

(a) Histogram (full sample)

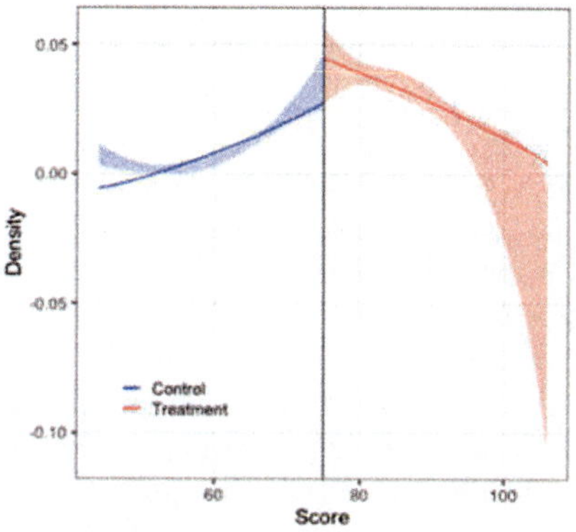

(b) Estimated Density (full sample)

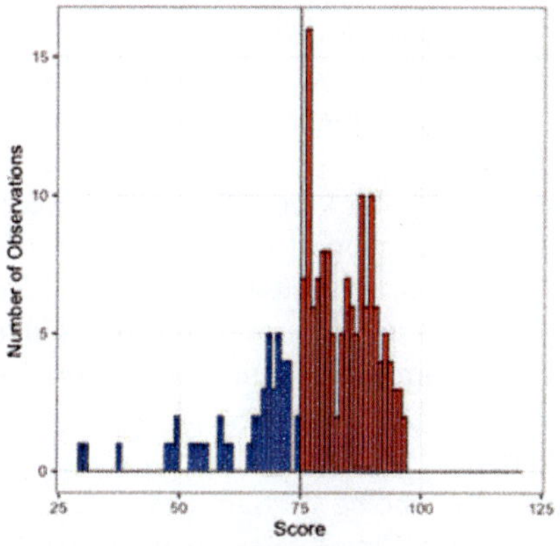

(c) Histogram (large firms)

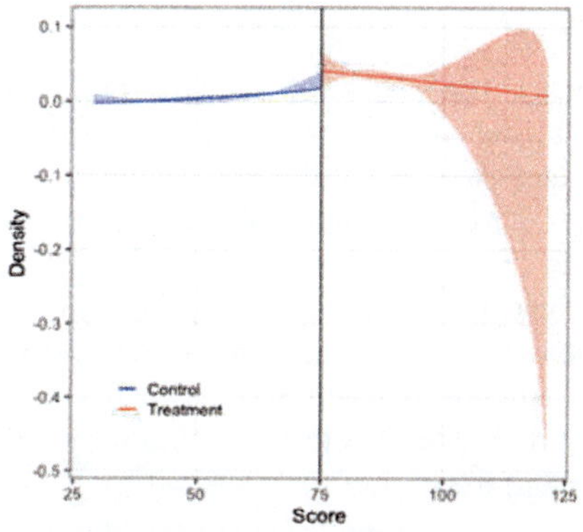

(d) Estimated Density (large firms)

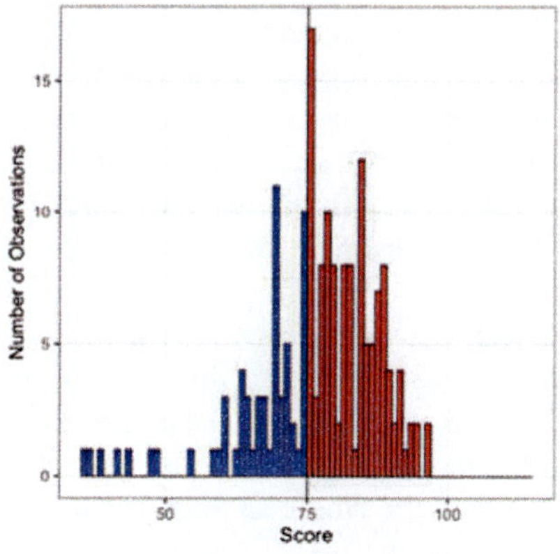

(e) Histogram (small firms)

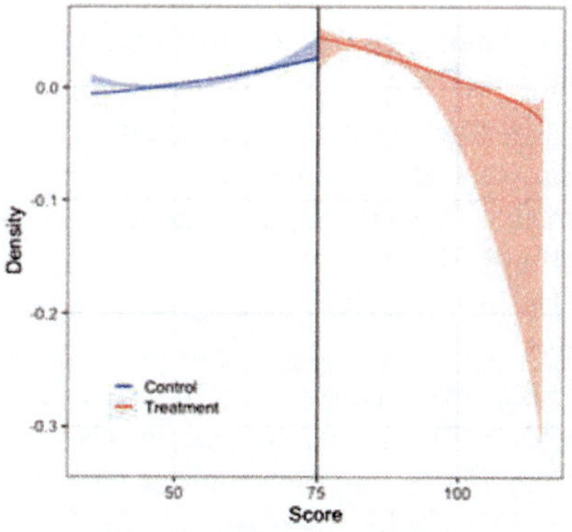

(f) Estimated Density (small firms)

Fig. 1 Histograms and estimated densities of the score

Estimated local quadratic approximation indicated that the program did not stimulate additional investment. However, there was substantial heterogeneity among firms of different sizes. While smaller enterprises increased their investment by generally the grant received, larger firms did not.

Since Bronzini and Iachini's (2014) procedures are not the same as those outlined in Sect. 2, in this paper we would like to estimate the RD treatment effect again using the same data set. First of all, we need to test the null hypothesis that the density of the score is continuous at the cutoff. The density is estimated separately for units below and above the cutoff by a local polynomial density estimator, together with confidence bands, which were proposed by Cattaneo et al. (2020). Figure 1 shows histograms and density estimates with shaded 95% confidence intervals for the whole sample, and the two subsamples of large and small enterprises. It should be noted that the confidence intervals in the three right subfigures may not be centered at the point estimates of the densities because bias correction is employed in the construction of confidence intervals, but not in point estimation. The density estimates for control and treatment units at the cutoff are very close to each other, implying that there appear no breaks in continuity of the density functions. Accordingly, there is no "manipulation" of the density at the cutoff or the validity of the RD design is statistically supported.

Now we use a polynomial order $p = 1$ and a triangular kernel function to select a common MSE-optimal bandwidth for the local linear point estimator $\hat{\tau}_{\text{SRD}}$. The local linear approximation in Eq. (2) is estimated with weighted least-squares and represented by red solid lines in Figs. 2 and 3. These two figures also contain blue dots which are local sample means calculated for non-overlapping evenly-spaced bins (intervals). The number of bins is selected so that the binned means have an asymptotic variability that "mimics" the overall variability of the raw data. The method is referred to as a mimicking variance (MV) choice (Cattaneo et al. 2019).

In Fig. 2, subfigure (a) reveals that subsidized firms' total investment was nearly the same as that of unsubsidized firms. Subfigures (b) and (c) imply that treated enterprises' tangible investment was lower, while their intangible investment was higher than those of control enterprises. However, the Bayes factors in the first column of Table 1 are less than 1, signifying that the data support the model setting the treatment effect equal to zero. There seem no differences at the cutoff, even when we scale total investment by capital and total assets in the pre-program year. It means that the program failed to expand R&D investment of firms located in Emilia-Romagna.

The first two left subfigures in Fig. 3 indicate that large firms which received grants from the local government were likely to make less total investment and tangible investment scaled by sales than those firms which did not. But the Bayes factors in the second column of Table 1 do not confirm this. The overall impact of the subsidy among the group of large enterprises and the whole sample was null. It is in line with Bronzini and Iachini's (2014) findings.

The ineffectiveness of the program also extends to small firms. This is in sharp contrast to Bronzini and Iachini's (2014) endorsed argument that small firms increased their investments considerably, by roughly the amount of the subsidy received, since the grant helped mitigate their particular difficulties in accessing capital markets due

Fig. 2 RD plots of full sample

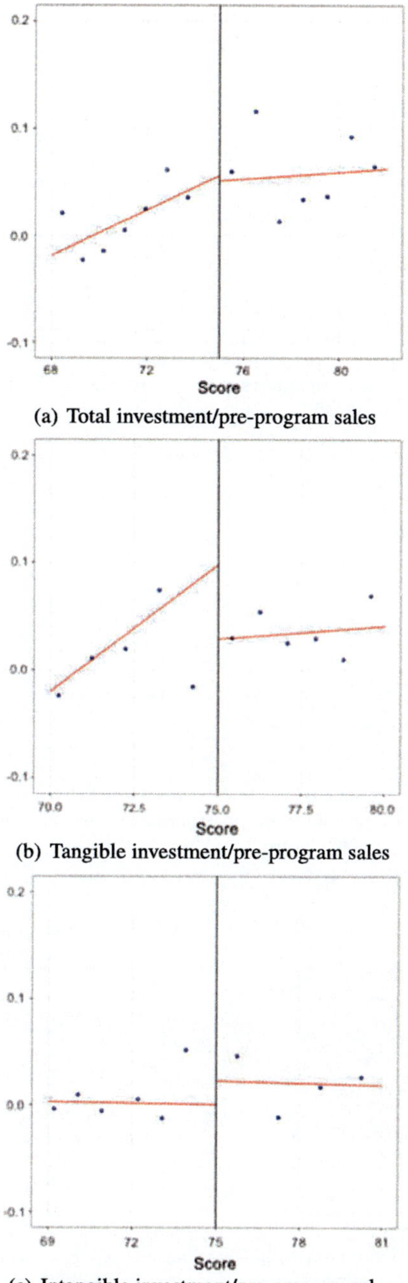

(a) Total investment/pre-program sales

(b) Tangible investment/pre-program sales

(c) Intangible investment/pre-program sales

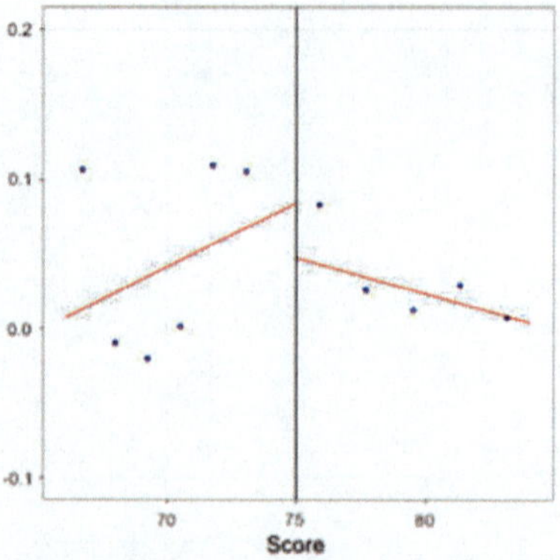

(a) Total investment/pre-program sales (large firms)

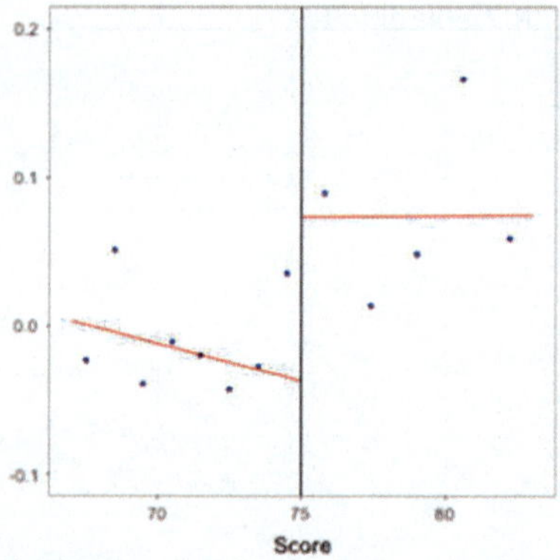

(b) Total investment/pre-program sales (small firms)

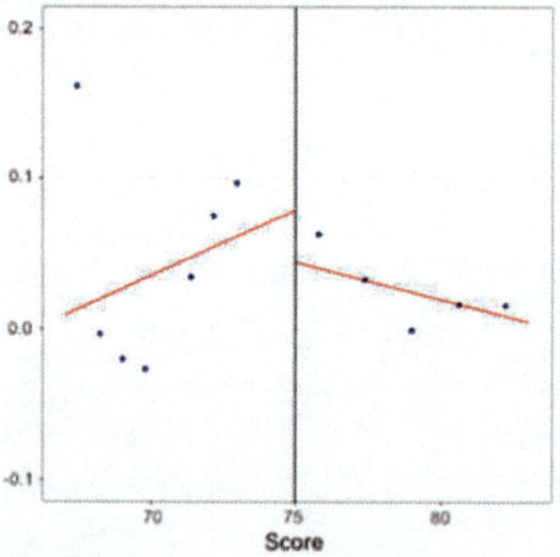

(c) Tangible investment/pre-program sales (large firms)

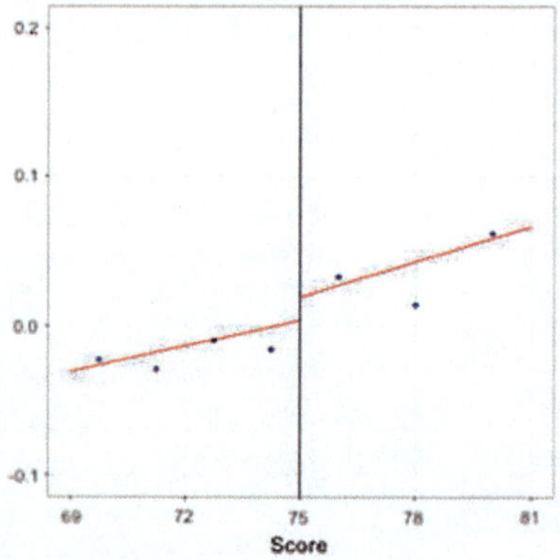

(d) Tangible investment/pre-program sales (small firms)

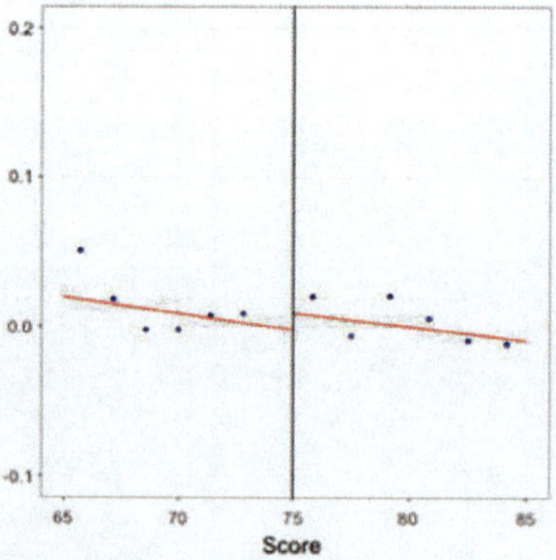

(e) Intangible investment/pre-program sales (large firms)

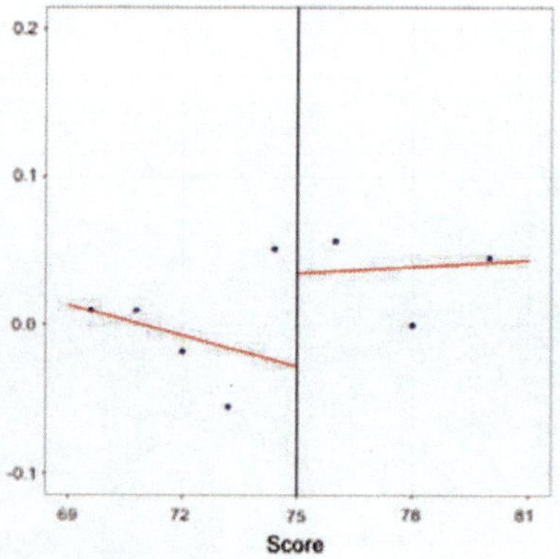

(f) Intangible investment/pre-program sales (small firms)

Fig. 3 RD plots of large and small firms

Table 1 RD program effect on investment

	Full sample	Large firms	Small firms
$\frac{\text{Total Investment}}{\text{pre-program sales}}$	–0.017 (0.11)	–0.080 (0.27)	0.094 (0.57)
$\frac{\text{Tangible Investment}}{\text{pre-program sales}}$	–0.052 (0.30)	–0.089 (0.31)	0.016 (0.11)
$\frac{\text{Intangible Investment}}{\text{pre-program sales}}$	0.029 (0.12)	–0.006 (0.13)	0.065 (0.22)
$\frac{\text{Total Investment}}{\text{pre-program capital}}$	0.879 (0.36)	–0.926 (0.34)	1.895 (2.14)
$\frac{\text{Total Investment}}{\text{pre-program assets}}$	–0.001 (0.09)	–0.103 (0.73)	0.095 (0.33)

Notes: Bayes factors are in parentheses. They are the ratios of the posterior probabilities of the model including the treatment effect and the model excluding it, assuming that the two models have equal prior probability

to strong information asymmetries and insufficient collateral. Although the three right subfigures in Fig. 3 indicate that small treated firms seemed to invest more than control ones, the last column of Table 1 shows that the data provide evidence in favor of the model having zero treatment effect. An exception is total investment scaled by pre-program capital, but according to Jeffreys (1961), it is "barely worth mentioning." Therefore, our empirical results call into question the efficacy of the program or more specifically the procedures of using scores to evaluate R&D projects proposed by enterprises and approving amounts of grant given to subsidized ones.

4 Conclusion

The regression discontinuity (RD) design has emerged as a promising method to study causal treatment effects in non-randomized assignment. It has been popularly used for policy evaluation where everyone does not have an equal opportunity to participate in a program. Its analytical technique has recently evolved substantially. In this paper, we use the latest set of practices to reexamine the subsidy program conducted by the government of Emilia-Romagna in northern Italy in 2003. We find that the program did not have an impact on firms' investment, regardless of firm sizes. Our results are contrary to those of Bronzini and Iachini's (2014) stating that the grants triggered substantial additional investment for small firms. The general negligible treatment effect raises serious doubts as to the efficiency of the program and its relevant procedures.

Acknowledgements The author is very grateful to anonymous referees for their valuable comments.

Conflict of Interest

The author declares that he has no conflict of interest.

References

Bronzini, R., Iachini, E.: Are incentives for R&D effective? Evidence from a regression discontinuity approach. Am. Econ. J. Econ. Policy **6**(4), 100–134 (2014)

Calonico, S., Cattaneo, M.D., Titiunik, R.: Robust nonparametric confidence intervals for regression-discontinuity designs. Econometrica **82**(6), 2295–2326 (2014)

Cattaneo, M.D., Idrobo, N., Titiunik, R.: A Practical Introduction to Regression Discontinuity Designs. Cambridge University Press (2019)

Cattaneo, M.D., Jansson, M., Ma, X.: Simple local polynomial density estimators. J. Am. Stat. Assoc. **115**, 1449–1455 (2020)

Cattaneo, M.D., Titiunik, R.: Regression Discontinuity Designs (2021). https://titiunik.mycpanel.princeton.edu/papers/CattaneoTitiunik2022-ARE.pdf

Gelman, A., Imbens, G.: Why high-order polynomials should not be used in regression discontinuity designs. J. Bus. Econ. Stat. **37**(3), 447–456 (2019)

Hahn, J., Todd, P., Van der Klaauw, W.: Identification and estimation of treatment effects with a regression-discontinuity design. Econometrica **69**(1), 201–209 (2001)

Imbens, G.W., Lemieux, T.: Regression discontinuity designs: a guide to practice. J. Econ. **142**(2), 615–635 (2008)

Jeffreys, H.: The Theory of Probability. Oxford University Press (1961)

McCrary, J.: Manipulation of the running variable in the regression discontinuity design: a density test. J. Econ. **142**(2), 698–714 (2008)

Thistlethwaite, D.L., Campbell, D.T.: Regression-discontinuity analysis: an alternative to the ex-post facto experiment. J. Edu. Psychol. **51**(6), 309–317 (1960)

Study on the Relationship Between the Value of Fixed Assets and Financial Leverage in Vietnam Businesses

Loan Nguyen Thi Duc and Huy Than Quang

Abstract The study was conducted to evaluate the relationship between fixed asset value and financial leverage in Vietnamese enterprises, specifically the construction industry. By using data of 125 construction industry enterprises in Vietnam market from 2015 to 2020, along with the BMA approach, the study shows that there is a positive correlation between the financial fixed assets, fixed asset investment expenditure compared to the financial leverage of the enterprise. Two variables representing fixed assets, TANG (fixed asset value/total assets) and LNTA (Logarithmic total assets), affect financial leverage of 0.45 and 0.07, respectively. When an enterprise increases its fixed asset investment by 1%, it means an increase in long-term debt, the financial leverage coefficient of the enterprise also increases to 0.45 and 0.07 respectively, shown by the dependent variable LTD/TA. In addition, when enterprises increase debt and debt structure tends to favor long-term debt more, the return on assets (ROA) will decrease. This is reflected in the negative coefficient in the return on assets (ROA) variable. When businesses increase the ratio of long-term debt to total assets (LTD/TA) by 1%, the corresponding return on assets (ROA) decreases to nearly 0.28. This result also shows the relationship between debt maturity and fixed assets, which is shown by a positive and statistically significant coefficient of the variables TANG (value of fixed assets/total assets), LNTA (Logarithm of total assets). When the enterprise tends to increase 1% for long-term debt (LTD/TD), the fixed assets of the enterprise is TANG, LNTA also increases by 0.69 and 0.10 respectively. This shows that the long-term borrowing of Vietnamese enterprises is mainly to invest in fixed assets. Enterprises in the construction industry need to be cautious in increasing leverage as this will lead to a more significant decrease in return on assets (ROA) than firms in other industries.

Keywords Fixed asset value · Financial leverage · Bayes · BMA

L. N. T. Duc (✉) · H. T. Quang
Ba Ria Vung Tau University, Vung Tau, Vietnam
e-mail: loanntd@bvu.edu.vn

N. Ngoc Thach et al. (eds.), *Financial Econometrics: Bayesian Analysis, Quantum Uncertainty, and Related Topics*, Studies in Systems, Decision and Control 427,
https://doi.org/10.1007/978-3-030-98689-6_54

1 Introduction

Recently, the country's economic development has led to the formation of new businesses as well as the growth of businesses that are present in many different fields, including the development highlights of enterprises in the field of construction and real estate. According to a report by the Ministry of Construction in Vietnam, the outstanding balance of bank credit for real estate investment and business activities in 2019 was more than VND 521,800 billion. This balance then increased continuously through each quarter of 2020 and reached over VND 633,700 billion at the end of the year. This shows that construction and real estate businesses are mainly dependent on credit supply from joint-stock commercial banks. In order to access credit easily and conveniently, construction and real estate businesses will often mortgage the collateral assets that are projects under construction. The use of such financial leverage will have both advantages and disadvantages that have been mentioned a lot in financial theories. If used correctly and effectively, financial leverage will promote strong asset returns.

Based on actual information at the bank, research has shown that when businesses mortgage highly secure assets, commercial banks are more likely to accept credit supply and therefore access capital. Business loans are more favorable. Along with easily increasing debt, companies will increase investment spending on projects, fixed assets and supplement working capital through investment in asset classes. When the value of fixed assets of an enterprise is larger, it is more convenient to use these fixed assets as collateral at banks to increase loan capital. In other words, in a highly asymmetric information environment, the willingness of credit institutions to provide capital depends on their ability to "dispose of collateral" and thus also affects source of credit for investment in fixed assets of enterprises. This is easily recognized intuitively, but there has not been a specific test in the Vietnamese market about this connection.

Previous studies have focused on many aspects of capital structure, financial leverage, and research results contain mixed evidence on this topic. Particularly in Vietnam, the topic of researching the relationship between fixed asset value and financial leverage as well as the impact of fixed assets on debt maturity has not been approached by researchers.

In this study, the author uses Bayesian approach, a new approach to study the relationship between fixed asset value and financial leverage in Vietnamese enterprises, specifically the construction industry, one of the major contributors to GDP and also an industry that has received much attention from the public in Vietnam in recent times. The objective of the study is to provide information, create favorable conditions for company managers in making appropriate financial policies as well as provide more empirical evidence in the construction industry of the Vietnamese market.

2 An Overview of the Theoretical Basis and Related Studies

2.1 Theoretical Basis

Research on the relationship between fixed asset value and financial leverage in Vietnamese enterprises is based on two main theories: (1) The theory of capital structure by Modigliani and Miller (MM model), and (2) The pecking order theory.

2.1.1 Modigliani and Miller's Theory of Capital Structure (MM Model)

The theory of capital structure was developed by Modigliani and Miller (1958). Modern financial theory today begins with the theory of capital structure of Modigliani and Miller (1958). Before that, no other theory of capital structure had been so widely accepted. MM begins his theory with the assumption that the firm's expected cash flows are predetermined. When a business chooses a certain ratio of debt and equity, it means that the business will distribute its cash flows to investors according to the chosen financing option. Investors and businesses are assumed to have equal access to funding, thereby allowing individual investors to leverage their own finances. When investors have a need to use leverage but cannot find a supply, they can create leverage themselves; Conversely, investors can eliminate financial leverage themselves in case they do not want to use leverage but the capital structure of the business has debt. As a result, financial leverage does not affect the market value of the business.

MM's "Postulate I" states that the market value of any firm is independent of the firm's capital structure. Given the assumptions of perfect financial markets and other ideal conditions such as parallel trading of securities, no taxes and no transaction costs, the law of conservation of value exists. A business cannot change the total value of its securities by dividing its cash flows into different streams: the value of the business is determined by its real assets, not by its securities which issued by the enterprise. Thus, capital structure is not related to firm value when investment decisions are completely separate from financing decisions.

In "Proposition II" of MM, it is shown that Debt has an impact on owner's income and this effect has both advantages and disadvantages. If the company is operating profitably and has excess capacity to pay interest, the use of debt will be beneficial. Debt will act as financial leverage to amplify income for owners. But if the enterprise cannot ensure the above conditions, the use of debt will be disadvantageous. Under perfectly competitive market conditions, individual investors can lend and borrow on the same terms and at the same interest rates as businesses, so individual investors can use self-made financial leverage. to generate income exactly equal to the income that investors receive when investing in a leveraged business. That is, individual investors can do what debt businesses do, so investors will not have any signal effect on whether businesses use debt or not. Thus, under perfect competition market conditions, there exists the law of value preservation and securities trading in the form of parallel trading. Debt does not affect the value of the enterprise but has an impact on income. owner and the effect is risky.

In 1963, Modilligani and Miller presented a follow-up study with the elimination of the corporate income tax hypothesis. According to MM, with corporate income tax, the use of debt will increase the value of the business. Since interest expense is a reasonable expense that is deductible when calculating corporate income tax, a part of the income of a debt-using enterprise is transferred to investors according to the equation: $Vg = Vu + TD$: the value of the debt-using firm (Vg) is equal to the value of the non-debt firm (Vg) plus the gain from the use of debt. Where, D is the total amount of debt used, T is the corporate income tax rate, T.D is the profit from the use of debt" (tax shield).

In defense of this theory, a common argument is often made that: "MM's theory does not describe how firms choose financing options in practice. However, this theory shows us how to find out why funding policy has an impact on firm value." This argument is a valid interpretation of most theories of corporate finance. This influenced the early development of pecking order theory and trade-off theory. In fact, the financial leverage of the business has a great impact and influence on the earnings for the company. Selecting an optimal capital structure is one of the jobs of corporate managers. However, in order to have a suitable financial leverage, the ability to access the credit supply is a very important thing for managers. The topic studies the impact of fixed asset value on financial leverage, debt maturity of enterprises is one of the factors affecting credit supply for enterprises.

2.1.2 Pecking Order Theory

Originally developed by Do Myers and Majluf (1984) as an alternative to the trade-off theory of capital structure. The pecking order theory is viewed as the opposite of the capital structure trade-off theory, which assumes that there is no unambiguously optimal capital structure for a firm. According to this theory, starting with asymmetric information between managers and investors, a phrase used to indicate that managers know more about the potential, risks and value of the business than the outside investors. Asymmetric information affects the choice between internal and external financing and between new issuance of debt and equity securities.

This leads to a pecking order, whereby investments are financed first with retained earnings, primarily reinvested profits; followed by issuing new debt; and finally by issuing new equity (Myers (1984); Myers and Majluf (1984)).

Although there have been many studies testing the superiority of pecking order theory over the capital structure trade-off theory in recent years, the reported results are controversial. Fama and Frech (2002), have argued that no theory has been disproved. Furthermore, Myers (2003) argues that the effectiveness of capital structure theory is based on the different conditions of one firm compared to another, similar to Huang and Song (2006).

2.2 Relevant Empirical Studies

Almeida and Campello (2007) found a link between financial constraints, fixed assets and corporate investment. This study has found a correlation between the investment decisions of enterprises and financial constraints through the credit coefficient to determine. These credit ratios indicate that the relationship between investment and cash flow sensitivity increases with increasing fixed assets in financially constrained firms. Furthermore, a company with a lot of fixed assets is less likely to be financially constrained.

Hall and Joergensen (2008) study the relationship between creditor's rights and corporate leverage with data from small and medium sized firms (SMEs) in 14 transition countries. Regression results in the study indicate that long-term debt used as a measure of financial leverage is positively related to the value of fixed assets of the business. The study uses regression models Pooled OLS, TOBIT and GLM all show that unlisted firms in emerging markets have a positive correlation between creditor's rights and financial leverage as well as financial leverage and debt maturity.

Campello and Giambona (2012) also study the assets and capital structure of enterprises by exploring in terms of the value of fixed assets. The study also shows that there is a relationship between fixed assets and capital structure of firms. By distinguishing different assets in the asset portfolio in the balance sheet, including machinery and equipment, transport vehicles, factories, and land. Campello and Giambona (2012) pointed out that asset liquidity is a factor affecting credit supply and financial leverage of enterprises in the context of a tight economy.

Hall and Joergensen (2008) measure the impact of creditor rights and corporate debt through a measure of long-term debt (LTD) used to reflect financial leverage, measures for owners' rights Debt reflected through the independent variable RULELAW is based on expert research in a survey published in "Central and Eastern Europe Economic Review" in 1998. The study used Pooled OLS, TOBIT and GLM regression models. Research results show that unlisted firms in emerging markets have a positive correlation between creditor's rights and financial leverage as well as debt maturity. Most of the independent variables on fixed asset value (TANG) appear in the pioneering research of author Thomas Hall.

Thomas Hall (2012) studied the relationship between fixed assets and financial leverage in two groups of countries with limited land transferability and countries with unrestricted land transferability. Thomas Hall (2012) used a research model with the debt maturity dependent variable measuring long-term debt on total debt and independent variables including: asset value over total assets, profit after tax on total debt. assets, corporate income tax on profit after tax, logarithm of total assets, logarithm of Vietnam's GDP, logarithm of money supply M2 over GDP of Vietnam, economic growth rate of Vietnam. The article used this model to apply to more than ten thousand companies from different markets in 11 countries. Empirical results show that there exists a relationship between fixed assets and financial leverage of the companies in the research data and this relationship becomes even stronger than that of the group of companies in the unaffected country which limited by land transferability.

3 Method and Data

3.1 Data

Secondary data is collected from 125 companies in the construction industry on the Vietnamese stock exchange, the data collection period is from 2015 to 2020. The time period is over 5 years and the latest updated data to 2020 is enough observations to study the impact of fixed assets on financial leverage of companies on the stock market in Vietnam, specifically the construction industry.

3.2 Method

In recent years, scientists are increasingly realizing the disadvantages of the frequency statistical method (Frequentist) because this method leads to many false conclusions in scientific research (Nguyen Van Tuan 2011). Scientific conclusions in frequency statistics are based on data sets without regard to known information (Nguyen Ngoc Thach 2019). In frequency statistics, the parameters of the population are assumed to be constant but unknown. But for time series data, these parameters will change, so the assumption that the parameters are constant is no longer appropriate. Therefore, more broadly, in Bayesian statistics, the parameters are assumed to be random variables and obey a normal distribution (van de Schoot and Depaoli 2014; Bolstad and Curran 2016). Bayesian conclusions based on a priori information combined with the collected data should have higher accuracy. For frequency statistics, a large enough dataset is required to draw conclusions. While for Bayesian statistics, drawing conclusions does not depend on the size of the data (Baldwin and Fellingham 2013; Depaoli and van de Schoot 2016; Doron and Gaudreau 2014) and overcomes the drawback of frequency statistics. With the development of data science and powerful computational software, Bayesian statistics are increasingly being used by scientists worldwide in their scientific research (Kruschke 2011). Frequency statistics will no longer be used due to its fundamental shortcomings (Nguyen Van Tuan 2011).

From the conditional probability expressed by the formula: p(A/B) = p(A,B)/p(B).

We have Bayes' theorem: p(A/B) = p(B/A)*p(A)/p(B).

Inside, p(A|B): Posterior probability (Posterior), need to find the probability that hypothesis A is true given the collected data;

p(B|A): Data Availability (Likelihood), probability of data collection under the condition that hypothesis A is true (collected data);

p(A): Priori probability, the probability that hypothesis A we believe to happen (true) prior to data collection;

p(B): Constant, probability of data.

One of the most widely used and less error-prone methods today is the Bayesian Model Averaging (BMA) method. This is a method built on the principles of Bayesian statistics, and each model has a predetermined probability, plus actual data, from which it is possible to know which variables are related to the output. out of research. BMA offers 5 best models, thereby offering many different choices depending on the implementation ability and the actual situation/feasibility/flexibility of the model. For each model, the BMA reports the regression coefficient of each prognostic variable, the coefficient R2 (the coefficient that interprets the percentage of variance of the research output variable); value of BIC (Bayesian Information Criterion—"penalty" coefficient for the model) and post-prob (post prob—probability of model occurrence in 100 repeated trials). A nomogram (math) will be built to specify the prognostic score for convenience in the evaluation process.

Some of the studies, Fried et al. (1991) in medicine, used BMA to build predictive models of cardiovascular disease in the United States; Greenland (1993) applies BMA in predicting infant mortality. In sociology, Raftery (1995, 1999) researched on Bayesian information criterion (BIC—the index used in BMA technique) and applied this technique in prediction problems of crime rate, society. In economics, Ley and Steel (2007), Eicher et al. (2011) use BMA in the problem of determining factors affecting economic growth. Jon et al. (2012) built a real-time economic forecasting model with the BMA approach, based on a large number of financial indicators.

Technically, when making a conclusion about an unknown quantity Δ with k statistical models M1, M2,…, Mk, based on the data set D, the posterior probability of Δ is determined by the formula:

$$P(\Delta/D) = \sum_{i=1}^{k} P(M_i, D) * P(M_i, D) \tag{1}$$

where P(Δ|Mi,D) is the posterior probability of Δ given the model Mi, and P(Mi |D) is the posterior probability of the model Mi given the data D. Note that, k can be very large.

Equation (1) shows that the posterior probability of Δ is the weighted average of the posterior probability of Δ in each model.

The posterior probability of Mi is given by the formula:

$$P(M_i|D) = \frac{P(D|M_i)P(M_i)}{\sum_{j=1}^{k} P(D|M_j)P(M_j)} \tag{2}$$

where P(Mi) is the a priori probability of the model Mi and P(D|Mi) is determined by formula (3):

$$P(D|M_i) = \int P(D|\mu_i, M_i)P(D|\mu_i, M_i)d\mu_i \tag{3}$$

where μ_i is the coefficient of the model M_i and P(D|μ_i,Mi) is the rationality function of the coefficient μ_I in the model M_i with D.

The posterior mean and the posterior variance are inferred from the following formula (1):

$$E(\Delta|D) = \sum_{i=1}^{k} \widehat{\Delta} P(M_i|D)$$

$$Var(\Delta|D) = \sum_{i=1}^{k} \mathrm{Var}(\Delta|\mathrm{D}, \mathrm{M_i}) + \widehat{\Delta}_i^2 \mathrm{P}(\mathrm{M_i}|\mathrm{D}) - \mathrm{E}(\Delta|\mathrm{D})^2$$

where: $\Delta_t = \mathrm{E}(\widehat{\Delta}|\mathrm{D},\mathrm{M_i})$.

In practice, the process of implementing calculations according to BMA technique arises some following problems:

- The sum function in formula (1) can contain a lot of terms, making it difficult to calculate;
- The integral in formula (3) is difficult to calculate in many cases, especially when there are many dimensions.

BMA application package version 3.18.6 (2015) developed by Raftery, Hoeting, Volinsky, Painter and Yeung on R statistical language has solved the above problems. Specifically, for the first problem, the sum function in formula (1) will be reduced based on the Occam's window method of Madigan and Raftery (1994). First, the models with a very small posterior probability compared to the model with the largest posterior probability will be discarded, leaving the models in the following set.:

$$\mathcal{A} = \left\{M_j, Max_i[P(M_i|D)] \leq C.P\left(M_j|D\right)\right\}$$

where C is a constant optional for the purposes of the analyst.

Next, we remove models with many variables but the posterior probability is smaller than the model with fewer variables.:

$$\mathcal{B} = \left\{M_l, \exists M_j \in A, M_j \subset M_l, \mathrm{P}\left(M_j|D\right) > \mathrm{P}(M_l|D)\right\}$$

Set $\mathfrak{R} = \mathrm{A} \backslash \mathrm{B}$, then the sum function in (1) is reduced as follows:

$$P(\Delta|D) = \sum_{i \in R} \mathrm{P}(\Delta|M_i, D)\mathrm{P}(M_i, D) \tag{4}$$

The function bic.glm in the above BMA application package used the algorithm of Furnival and Wilson (1974) to quickly search for models in the above $\mathfrak{R}$ set.

For the second problem, according to Raftery (1995), the integral in Eq. (3) can be approximated by the Bayesian Information Criterion (BIC) index:

$$P(D|Mi) \approx e^{-BICi/2}$$

where BIC_i is the index measuring the explanatory ability of the M_i model for data D.

In general, the bic.glm function will look for the best models (with the highest posterior probability) based on Occam's window method, and then take the "average" of these models according to the formula (4) to output the final model.

The author's research paper, based on some previous research results, has found the relationship between financial leverage and fixed asset value of the business. The model below is used based on the pioneering research of author Thomas Hall (2012) on the relationship between fixed asset value and financial leverage of data of more than ten thousand enterprises in other countries:

Model 01: Fixed assets and financial leverage

(LTD/TA)i,t = α + β1TANGi,t + β2CAPi,t + β3ROAi,t + β4TAXi,t + β5LNTAi,t + β6GDPi,t + β7M2/GDPi,t + β8GROWTHi,t + ε

Financial leverage is measured through the dependent variable LTD/TA. Financial leverage is a powerful tool for investors and business owners. There are very few businesses that can say they have no debt. Meanwhile, most companies have to take out a loan at a certain point in time to buy equipment, build new offices, or pay employees. For investors, whether they dare to invest or not depends on whether the institution's debt level is sustainable. To evaluate this, financial leverage coefficients should be considered. Based on previous studies on financial leverage, in this article long-term debt is used as a measure of leverage because long-term debt is more stable and long-term than short-term debt term. Short-term liabilities are volatile, so they do not accurately reflect the debt situation of the business.

Another empirical study is examining debt maturity and fixed asset value through the study of Ratan and Zingales (1995), Wald (1999), Aggarwal and Jamdee (2003) for some G-7 countries as well as Boot et al. (2001), Fan et al. (2008) evaluated developed and developing countries worldwide. The study adopts the following model:

Model 02: Debt maturity and fixed asset value

(LTD/TD)i,t = α + β1TANGi,t + β2CAPi,t + β3ROAi,t + β4TAXi,t + β5LNTAi,t + β6GDPi,t + β7M2/GDPi,t + β8GROWTHi,t + ε

Debt maturity of the enterprise is also one of the factors to evaluate the efficiency and suitability in the use of debt of the enterprise, the measurement model through the dependent variable LTD/TD. This ratio indicates that the debt structure of the enterprise is more inclined to use long-term debt to finance operating and investment expenses or to use short-term debt (Table 1).

Table 1 Description of variables used in the models

Variable name		Measure	Data sources	Expectation sign
Dependent variable	LTD/TA: Long-tem debt/total assets	Long-term debt/total assets	Balance sheets of listed companies	N/A
	LTD/TD: Long-term debt/total debt	Long-term debt/total debt	Balance sheets of listed companies	N/A
Explanatory variable	TANG: Tangible assets: net fixed assets/total assets	Value of fixed assets/total assets	Balance sheets of listed companies	+
	CAP: Capital Expenditure/total assets	Fixed asset investment expenditure/total assets	Balance sheets and cash flows of listed companies	+
	ROA: Return on assets	Profit after tax/total assets	Income statement and balance sheet of listed companies	+
	TAX: Effective tax rate: taxes/(Earning after taxes)	Corporate income tax/profit after tax	Table of business results of listed companies	–
	LNTA: Natural log of total assets	Logarithm of total assets	Balance sheets of listed companies	+
	GDP: Natura log per capital output	Logarithm of Vietnam's GDP	World Bank	N/A
	M2/GPD: Broad money as a portion of GDP	Logarithm of Money Supply M2/GDP of Vietnam	World Bank	N/A
	GROWTH:Annual GDP growth rate	Vietnam's economic growth rate	World Bank	N/A

4 Result

4.1 Financial Leverage (LTD/TA)

The results from Table 2 show that the proposed BMA regression method has 5 models. In which, the first model has four variables TANG, ROA, LNTA, GDP that explain 49.6% and the probability of occurrence of this model is 33.9%. Next, the second model has three variables TANG, ROA, and LNTA that explain 48.9% and the model's probability is 26.1%. The third model has four variables TANG, ROA,

Table 2 Summary of BMA regression results of dependent variable LTD/TA

	p! = 0	EV	SD	Model 1	Model 2	Model 3	Model 4	Model 5
Intercept	100	−0.04	1.39	0.82	−0.81	−0.77	−0.71	5.15
TANG	100	0.45	0.02	0.45	0.45	0.45	0.45	0.45
CAPX	0.00	0.00	0.00					
ROA	95.80	−0.27	0.10	−0.28	−0.28	−0.28	−0.28	−0.29
TAX	0.00	0.00	0.00					
LNTA	100	0.07	0.01	0.07	0.07	0.07	0.07	0.07
GDP	38.80	−0.05	0.10	−0.11				−0.41
M2/GDP	22.90	−0.01	0.06			−0.08		0.23
GROWTH	12.80	0.00	0.01				−0.02	
nVar				4.00	3.00	4.00	4.00	5.00
r2				0.50	0.49	0.50	0.49	0.50
BIC				−333.11	−332.59	−332.07	−331.17	−328.12
Post prob				0.34	0.26	0.20	0.13	0.03

LNTA, M2/GDP that explain 49.5% and the model's probability is 20.1%. The fourth model has four variables TANG, ROA, LNTA, and GROWTH that explain 49.4% and the model's probability is 12.8%. The fifth model with five variables TANG, ROA, LNTA, GDP, M2/GDP explains 49.7% and the model's probability is 2.8%.

Comparing the results of all five models above, we see that the first model is the most optimal model because the probability of occurrence of this model is 33.9%, the highest of the five proposed models, the level The explanation of this model is 49.6% and there are four statistically significant variables: TANG, ROA, LNTA, GDP.

In the following, the author will use a chart to visually represent the occurrence of variables in the five proposed models. In Fig. 1, this histogram is a way for it to

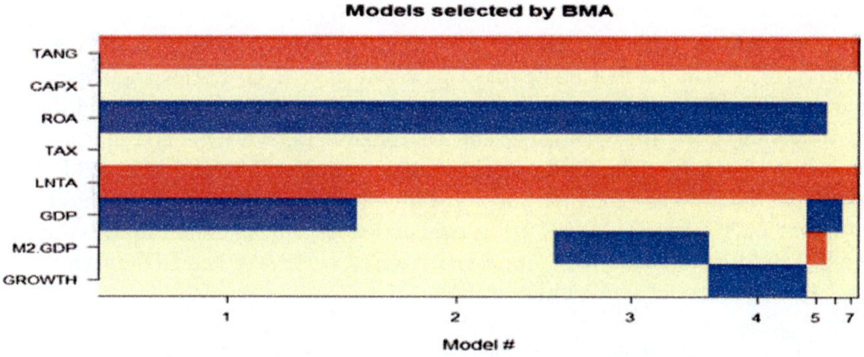

Fig. 1 Graph showing selected variables appearing in the LTD/TA model

show the probability of that variable appearing in a multivariable linear regression model.

On the chart in Fig. 1, there are 2 colors showing the sign of the regression coefficient, the color "green" represents the negative sign (−) and the color "Red" represents the positive sign (+). Through the chart in Fig. 1, it shows that the variable TANG and the variable LNTA affect the dependent variable LTD/TA with the highest consistency (occurring 100% in the model). The second most important factor is the ROA variable and the third is GDP. The remaining factors such as variables M2/GDP, GROWTH have an influence on the dependent variable LTD/TA, but these factors are not as consistent as the factors mentioned above.

So in the end, the author concludes that the most optimal model will be chosen as Model 1:

$$\mathbf{LTD/TA = 0.82 + 0.45\ TANG - 0.28\ ROA + 0.07\ LNTA - 0.11GDP}$$

The independent variables in the model explain 49.6% for the LTD/TA dependent variable and the model's probability is 33.9%.

Financial leverage of the enterprise is positively correlated for the independent variables TANG, LNTA with 99% confidence. When an enterprise increases investment in fixed assets, it means an increase in long-term debt, the financial leverage ratio of the enterprise also increases, as shown by the dependent variable LTD/TA. Two variables representing fixed assets, TANG (fixed asset value/total assets) and LNTA (Logarithmic total assets), affect financial leverage of 0.45 and 0.07, respectively. The author found a negative correlation in return on assets (ROA) and this result is similar to the study of Hall (2002). When businesses increase the ratio of long-term debt to total assets (LTD/TA) by 1%, the ratio of return on assets (ROA) decreases to 0.28.

4.2 Debt Maturity (LTD/TD)

The results from Table 3 show that the proposed BMA regression method has 5 models. In which, the first model has two variables TANG, LNTA explain 53.3% and the probability of occurrence of this model is 28.5%. Next, the second model has three variables TANG, LNTA, and GDP that explain 53.7% and the model's probability is 16.7%. The third model has three variables TANG, ROA, and LNTA that explain 53.7% and the model's probability is 14%. The fourth model has three variables TANG, LNTA, M2/GDP that explain 53.6% and the model's probability is 9.6%. The fifth model has four variables TANG, ROA, LNTA, and GDP that explain 54.2% and the model's probability is 9.4%.

Comparing the results of all five models above, we see that the first model is the most optimal model because the probability of occurrence of this model is 28.5%, the highest of the five proposed models, the level The explanation of this model is 53.3% and there are two statistically significant variables, TANG and LNTA.

Table 3 Summary of regression results of dependent variable LTD/TD

	p! = 0	EV	SD	Model 1	Model 2	Model 3	Model 4	Model 5
Intercept	100	−0.24	2.35	−1.17	0.79	−1.17	−1.13	0.83
TANG	100	0.69	0.03	0.69	0.69	0.69	0.69	0.69
CAPX	1.50	0.00	0.02					
ROA	33.30	−0.08	0.14			−0.25		−0.25
TAX	1.60	0.00	0.00					
LNTA	100	0.10	0.01	0.10	0.10	0.10	0.10	0.10
GDP	30.00	−0.06	0.16		−0.13			−0.14
M2/GDP	18.40	0.00	0.11				−0.09	
GROWTH	9.80	0.00	0.01					
nVar				2.00	3.00	3.00	3.00	4.00
r2				0.53	0.54	0.54	0.54	0.54
BIC				−385.31	−384.25	−383.90	−383.13	−383.10
Post prob				0.29	0.17	0.14	0.10	0.09

In the following, the author will use a chart to visually represent the occurrence of variables in the five proposed models. This graph is a way for it to show the probability of that variable appearing in a multivariable linear regression model.

On the chart in Fig. 2, there are 2 colors showing the sign of the regression coefficient, the color "green" represents the negative sign (−) and the color "Red" represents the positive sign (+). Through the chart in Fig. 2, it shows that the variable TANG and the variable LNTA affect the dependent variable LTD/TD with the highest consistency (occurring 100% in the model). The second most important factor is the ROA variable and the third is GDP. The remaining factors such as variable M2/GDP, GROWTH have an influence on the dependent variable LTD/TD, but these factors are not as consistent as the factors mentioned above.

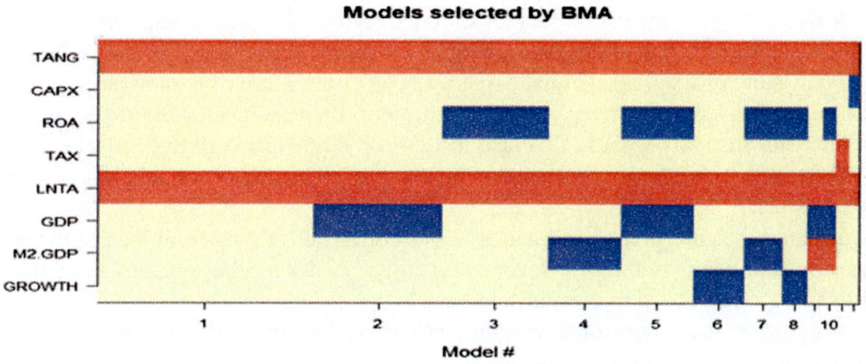

Fig. 2 Graph showing selected variables appearing in the LTD/TD model

So in the end, the author concludes that the most optimal model will be chosen as Model 1:

LTD/TD = −1.17 + 0.69 TANG + 0.10 LNTA

The independent variables in the model explain 53.3% of the dependent variable LTD/TD and the probability of the model appearing is 28.5%.

The regression model explains the same for debt maturity, when enterprises tend to increase by 1% for long-term debt (LTD/TD), fixed assets of the enterprise are TANG, LNTA also increase respectively is 0.69 and 0.10. This shows that the long-term borrowing of Vietnamese enterprises is mainly to invest in fixed assets.

5 Conclude

The study was conducted to evaluate the relationship between fixed asset value and financial leverage in Vietnamese enterprises, specifically the construction industry. By using data of 125 construction industry enterprises in Vietnam market from 2015 to 2020, along with the BMA approach, the study shows that there is a positive correlation between asset value and financial value. fixed assets, fixed asset investment expenditure compared to the financial leverage of the enterprise. Two variables representing fixed assets, TANG (fixed asset value/total assets) and LNTA (Logarithmic total assets), affect financial leverage of 0.45 and 0.07, respectively. When an enterprise increases its fixed asset investment by 1%, it means an increase in long-term debt, the financial leverage coefficient of the enterprise also increases to 0.45 and 0.07 respectively, shown by the dependent variable LTD/TA. In addition, when the enterprise increases debt and the debt structure tends to favor long-term debt more, the return on assets (ROA) will be reduced, which is reflected in the negative coefficient in the ratio variable. return on assets (ROA). When businesses increase the ratio of long-term debt to total assets (LTD/TA) by 1%, the corresponding return on assets (ROA) decreases to 0.28.

This result also shows the relationship between debt maturity and fixed assets, which is shown by a positive and statistically significant coefficient of the variables TANG (value of fixed assets/total assets), LNTA (Logarithm of total assets). When the enterprise tends to increase 1% for long-term debt (LTD/TD), the fixed assets of the enterprise is TANG, LNTA also increases by 0.69 and 0.10 respectively. This shows that the long-term borrowing of Vietnamese enterprises is mainly to invest in fixed assets.

Vietnamese enterprises use fixed assets as collateral for long-term loans and this has affected their own financial leverage. It will be easier for the company to borrow if its fixed assets are of high value and have better liquidity.

Enterprises in the construction industry need to be cautious in increasing leverage as this will lead to a more significant decrease in return on assets (ROA) than firms in other industries.

References

Almeida, H., Campello, M.: Financial constraints, asset tangibility and corporate investment. Rev. Financ. Stud. **20**(5), 1429–1460 (2007)

Andrei, S., Vishny, R.W.: Liquidation values and debt capacity: a market equilibrium approach. J. Financ. **XLVII**(4) (1992)

Baldwin, A., Fellingham, W: Bayesian methods for the analysis of small sample multilevel data with a complex variance structure. Psychol. Methods 18(2), 151–164 (2013). https://doi.org/10.1037/a0030642

Booth, L., Aivazian, V., Demirgüç-Kunt, A., Maksimovic, V.: Capital structures in developing countries. J. Finance **56**, 87–130 (2001)

Campello, M., Giambona, E.: Real assets and capital structure. J. Financ. Quant. Analy., Working paper Cornell University, New York, NY (2012)

Campello, M., Giambona, E., Graham, J.R., Harvey, C.R.: Access to liquidity and corporate investment in Europe during the financial crisis. Rev. Finance **16**(2), 323–346 (2012)

Decree of the Government of Vietnam: Registration of security interests, No. 102/2017ND-CP dated September 1, 2017 (2017)

Decree of the Government of Vietnam, 2006. "Secured Transactions", No. 163/2006/ND-CP dated December 29, 2006

Dietrich, D.: Asset tangibility and capital allocation. J. Corp. Finan. **13**(2007), 995–1007 (2007)

Doron, J., Gaudreau, P.: A point-by-point analysis of performance in a fencing match: psychological processes associated with winning and losing streaks. J. Sport Exerc. Psychol. **36**(1), 3–13 (2014)

Eicher, T., Papageorgiou, C., Raftery, A.: Default priors and predictive performance in Bayesian model averaging, with application to growth determinants. J. Appl. Econometrics **26**, 30–55 (2011)

Fama, E.F., French, K.R.: Testing trade-off and pecking order predictions about dividends and debt. Rev Financ. Stud. **15**(1), 1–33 (2002)

Greenland, S.: Methods for epidemiologic analyses of multiple exposures—a review and comparative study of maximum-likelihood, preliminary testing, and empirical Bayes regression. Stat. Med. **12**, 717–736 (1993)

Hall, T., Jörgensen, F.: Legal rights matter: evidence from panel data on creditor protection and debt. In: Institutional Approach to Global Corporate Governance: Business Systems and Beyond. Emerald Group Publishing Limited (2008)

Huang, G., Song, F.M.: The determinants of capital structure: evidence from China. China Econ. Rev. **17**(1), 14–36 (2006)

Jon, F., Simon, G., Jonathan, H.W., Egon, Z.: Credit spreads as predictors of real-time economic activity: a Bayesian model-averaging approach. In: Finance and Economics Discussion Series: 2012-77. Federal Reserve Board (2012)

Kruschke, K.: Bayesian assessment of null values via parameter estimation and model comparison. Perspect. Psychol. Sci. **6**(3), 299–312 (2011)

Ley, E., Steel, M.: Jointness in Bayesian variable selection with applications to growth regressions. J Macroecon. **29**, 476–493 (2007)

LienVietPostBank: Regulations on guaranteeing credit extension, No. 6469/2018/QD-LienVietPostBank dated 11 July 2018 (2018)

Lu-Andrews, R., Yu-Thompson, Y.: CEO inside debt, asset tangibility, and investment. Int. J. Manag. Finance (2015)

Madigan, D., Raftery, A.E.: Model selection and accounting for model uncertainty in graphical models using Occam's window. J. Am. Stat. Assoc. **89**, 1535–1546 (1994)

Ministry of Finance of Vietnam: "Circular promulgating standards for appraisal of prices No. 08, 09 and 10", No. 126/2015/TT-BTC dated August 20, 2015 (2015)

Modigliani, F., Miller, M.H.: The cost of capital, corporation finance and the theory of investment. Am. Econ. Rev. **48**(3), 261–297 (1958)

Myers, S.C.: The capital structure puzzle. J. Finance **34**(3), 575–592 (1984)

Myers, S.C.: Financing of corporations. In: Constantinides, G., Harris, M., Stulz, R. (eds.) Handbook of the Economics of Finance: Corporate Finance, vol. 1A, pp. 215–253. Elsevier, Amsterdam (2003)

Myers, S.C., Majluf, N.S.: Corporate financing and investment decisions when firms have information that investors do not have. J. Financ. Econ. 13, 187–221 (1984)

Raftery, A.E.: Bayesian model selection in social research. Sociol. Methodol. **25**, 111–163 (1995)

Raftery, A.E.: Bayes factors and BIC: Comment on a critique of the Bayesian information criterion for model selection. Sociol. Methods Res. **27**, 411–427 (1999)

Raghuram, G., Ra, J., Lui, G., Zingales: What do we know about capital structure? Some evidence from international data. J. Financ. **L**(5) (1995)

Rajan, R.G., Zingales, L.: What do we know about capital structure? Some evidence from international data. J. Finance **50**(5), 1421–1460 (1995)

Ran, L.-A., Yin, Y.-T.: CEO inside debt, asset tangibility, and investment. Int. J. Manager. Financ. **11**(4), 451–479 (2015)

Shleifer, A., Vishny, R.W.: Liquidation values and debt capacity: A market equilibrium approach. J. Finance **47**(4), 1343–1366 (1992)

Thomas, W.H.: The collateral chanel: evidence on leverage and assettangibility. J. Corporate Financ., 570–583 (2012)

Van de Schoot, R.: 25 years of Bayes in psychology. Paper presented at the 7th Mplus Users' Meeting, Utrecht, The Netherlands (2016)

Van de Schoot, R., Depaoli, S.: Bayesian analyses: where to start and what to report. Europ. Health Psychol. **16**(2), 75–84 (2014)

Van de Schoot, R., Kaplan, D., Denissen, J., Asendorpf, J.B., Neyer, F.J., Van Aken, M.A.G.: A gentle introduction to Bayesian analysis: applications to developmental research. Child Dev. **85**(3), 842–860 (2014). https://doi.org/10.1111/cdev.12169

Van Tuan, N.: Introduction to the Bayesian method. J. Med. News **63**, 26–34 (2011)

Vig, V.: Access to collateral and corporate debt structure: evidence from a natural experiment. J. Financ. **68**(3), 881–928 (2013)

Factors Affecting Price to Income Rate of Non-financial Company Listed on the Stock Exchange of Vietnam

Loan Nguyen Thi Duc and Huy Than Quang

Abstract The article studies the factors affecting the price-to-earnings ratio, based on collected data of 106 non-financial companies listed on the Vietnam stock market in the period 2015–2020. used Lasso, Ridge, Elactic Net regression estimation methods. The research results are that there are two optimal estimators selected based on the criteria of the smallest MSE (Mean Squared Error) and the highest R2 (Adjusted R Square), which are Lasso and Elactic Net estimators. Research results, applied according to Lasso and Elactic Net models, are that there is a statistically significant correlation between all explanatory variables on the price-to-earnings (P/E) ratio with the explanatory level of 36.2%. In which, the debt-to-equity ratio (DE); Earnings growth (EG) has a negative effect on P/E. The remaining factors are dividend payout ratio (DP); company size (SIZE); return on equity (ROE) and share price to book ratio (MB) both have a positive effect. The research results show that the new approach is more generalized by the Ridge, Elactic Net and Lasso estimation methods. All three methods aim to reduce the coefficients and improve the explanatory level for the model. However, in this study, the coefficients (β) of the explanatory variable did not differ too much between the estimation methods and the level of explanation R2 was also not significantly different. At the same time, the author recommends that investors should focus on factors that have a positive influence on P/E before investing to bring high efficiency in their investment portfolio.

Keywords Influential factors · Price-to-earnings ratio · Invest effects

1 Introduction

In the past 10 years, Vietnam's stock market has made great progress and has attracted a lot of capital flows from foreign investors. According to the State Securities Commission, Vietnam is one of the few countries to keep a positive growth rate

L. N. T. Duc (✉) · H. T. Quang
Ba Ria Vung Tau University, Vũng Tàu, Vietnam
e-mail: loanntd@bvu.edu.vn

N. Ngoc Thach et al. (eds.), *Financial Econometrics: Bayesian Analysis, Quantum Uncertainty, and Related Topics*, Studies in Systems, Decision and Control 427,
https://doi.org/10.1007/978-3-030-98689-6_55

of 2.91% and is forecast to recover at 6.5–6.8% in 2021. This is the main driving force to strengthen confidence. Investors' confidence and attract investment capital flows into the stock market. In order for investors to be able to make decisions and choose which stocks to invest in, in which, investors always focus and pay attention to indicators such as: Price/Sales per share (P/S).), Price-to-Book ratio (P/B), Price to Earning ratio (P/E). P/E is one of the indexes that is always interested by investors, which is the index that investors often apply to stock prices because P/E is easy to use and calculate. The purpose is to value the shares of listed companies.

High or low P/E is one of the important factors that investors today always pay attention to and evaluate before investing in shares of a listed company. Typically, a high P/E represents an investor's expectation of higher earnings growth in the future, and investors are often willing to pay a premium price for the shares of those companies. that business. According to the author's research, stocks of companies with high P/E ratios are currently interested and appreciated by investors, including shares of Vietnam Dairy Products Joint Stock Company (Vietnam Dairy Products Joint Stock Company). VNM), Mobile World Joint Stock Company (MWG), Nhon Trach 2 Petroleum Power Joint Stock Company (NT2), etc. Among them, there are also many stocks of companies with high P/E. But investors underestimated FLC Faros Construction Joint Stock Company (ROS). Therefore, a high or low P/E is always risky because each stock of an active company always has a risk of how it will perform in the future, so it is difficult to determine a price. P/E at what price is the best. Therefore, it is very important for investors to understand P/E as well as the factors that determine P/E.

In this study, the author will focus on researching "Factors affecting the ratio of price to income of non-financial companies listed on Vietnam stock exchange", the article is applied. Lasso, Rigde, Etaclic Net regression methods. With a new methodical approach to assess what factors will affect the price-to-earnings (P/E) ratio in the current period, it also helps securities investors to better understand P/E.

2 Related Studies

Johan and Filip (2007) with the article "A quantitative study of the P/E ratio on the Swedish market". The study was carried out on all companies listed on the Swedish stock exchange from September 1999 to 2007. As a result, dividend payout ratio and interest rates have a negative impact on P/E but interest rates and M/B (market-to-book ratio) have a positive impact on P/E.

Afza and Tahir (2012) research on "Factors affecting P/E ratio on Pakistani stock market". The study was conducted based on data of 25 companies listed on the Pakistani stock market in the period from 2005 to 2009. As a result, factors such as dividend payout ratio (DP), revenue growth (Tobin's Q), and market price volatility (VMP) have a positive impact on the P/E ratio. The size factor of the company (Size) has a negative impact on P/E.

Faezinia (2012) researched on "Factors affecting P/E on the Tehran stock market". The study was conducted based on the data of 120 companies listed on the Tehran Stock Exchange of Iran, from 2005 to 2011. The author's research model, including 8 factors: Interest rate (r), Inflation, Income growth (GR), Risk (β), Return on equity (ROE), Debt-to-equity ratio Equity (Lev), Dividend Ratio (DP), Firm Size (Size). The research results are that the factors have a negative impact on P/E, including: Interest rate (r), Dividend payout ratio (DP). The remaining factors have a positive impact on P/E.

Ramadan (2014) studies on "Factors affecting P/E ratios in companies listed on Amman stock market". The author's research model with P/E ratio is the dependent variable and the explanatory variables, including: Interest rate (r), Income growth (GR), Tobin's Q, Debt to equity ratio (Lev), Dividend payout ratio (DP), Size (business size). This model is applied to all non-financial companies listed in the period 1999–2013 in Jordan.

Research results using OLS regression show that the factors DP, Tobin's Q and Egrowth have the same impact on P/E. Otherwise, the Size factor has the opposite effect.

Dutta et al. (2017) study on "Determining factors to P/E ratio: Empirical research in manufacturing enterprises listed on Dhaka stock exchange" of Bangladesh, for the period from 2011 to 2015, based on data collected from 45 manufacturing enterprises.

Research model, including 8 factors affecting P/E, including: Dividend Growth (DG), Dividend Payout Ratio (DPR), Dividend Yield (DY), Earnings Growth (EG), Leverage (Lev), Net Asset Value Per Share (NAVPS), Return on Assets (ROA), Return on Equity (ROE), Size (SZ), Tobin Q (China).

Research results show that Dividend Payout Ratio (DPR), Size (SZ) have a positive impact on P/E. Leverage Ratio (Lev) and Net Asset Value Per Share (NAVPS) have the opposite effect.

Fesokh and Haddad (2019), study on "Determinants of Price to Income Ratio in Manufacturing Industry in Jordan". The study was conducted based on data collected from audited financial statements of 44 manufacturing companies listed on the Jordan Stock Exchange, during the period 2000–2013. Research model with subvariable depends on P/E and influencing factors, including: market capitalization to total assets of the company (Tobin's Q), leverage (LEV), dividend growth (DG), liquidity (LIQ), company size (Size).

Research results using linear regression (OLS) method show that company size (Size), dividend growth (DG) have a negative influence on P/E. The remaining factors have a positive effect.

Itemgenova and Sikveland (2020) study on "Factors determining the P/E index in the Norwegian aquaculture industry". The study was carried out with eight out of ten seafood companies listed on the Norwegian stock exchange OSE. In which, there are two Scottish Salmon companies, PLC and Hofseth BioCare ASA, excluded from the data because the financial statements of the two companies only show the period 2013–2016 compared to the present.

Research model, using regression Random effects and Fixed effects with dependent variable is P/E and explanatory variables, including: dividend payout ratio (DP), return on equity (ROE), price-to-book ratio (PB), leverage ratio (LEV), salmon price change (RFISH), the change in the NOK's exchange rate against the dollar (NOK/USD), the change in the exchange rate of NOK against EUR (NOK/EUR), change in stock price (LAGGEDPB).

Research results show that regression Random effects is more effective than Fixed effects. The results have 3 factors affecting the P/E index with statistical significance at 1%, including: dividend payout ratio (DP), return on equity (ROE), price change shares (LAGGEDPB). In which, return on equity (ROE) has a negative impact on P/E. The remaining factors have the same effect.

3 Data and Model

3.1 Data

The study is based on the data that the author has collected from the audited financial statements of 106 non-financial companies listed on the Vietnamese stock market in the period 2015–2020.

3.2 Model

Based on related studies, the author proposes the following research model (see Table 1):

$$P/E_{i,t} = \beta o + \beta 1(DP)_{i,t} + \beta 2(DE)_{i,t} + \beta 3(EG)_{i,t} + \beta 4(ROE)_{i,t} + \beta 5(SIZE)_{i,t} + \beta 6(M/B)_{i,t} + u_{i,t}$$

4 Method

4.1 Linear Regression (Multiple Regression Model)

Conventional regression analysis provides a general model with the output being the mean of the dependent variable given the input values of the independent variables of the form:

$$Yi = \beta_o + \beta_1 X_{1i} + \cdots + \beta_p X_{pi} + ui$$

Table 1 Description of variables used in the models

Variable name		Variable name	Measure	Data sources
Dependent variable	P/E	Price-to-earnings ratio	Market price at the time of transaction/Net profit of a stock	Dutta et al. (2017), Itemgenova and Sikveland (2020)
Explanatory variable	DP	Dividend Payout Ratio	Net income – preferred dividends – retained earnings (DPS)/Common shares outstanding (ESP)	Dutta et al. (2017), Itemgenova and Sikveland (2020)
	DE	Debt to equity ratio	Average value of total debt in the year/Average value of equity in the year	Faezinia (2012), Johan and Filip (2007)
	EG	Earnings growth rate	Profit in year t/Profit in year (t−1)	Ramadan (2014), Dutta et al. (2017)
	ROE	Return on equity ratio	Profit after tax/equity	Dutta et al. (2017), Itemgenova and Sikveland (2020)
	SIZE	Enterprise size	Logarithm (total assets each year)	Ramadan (2014), Dutta et al. (2017), Fesokh and Haddad (2019)
	MB	Market-to-book price ratio	Market price of shares/book value of shares	Johan and Filip (2007)

where:

- Y: Dependent variable (explained variable);
- X_j ($j = 1,\ldots,p$): Independent variables (explanatory variables);
- u: Random error with expectation equal to 0.
- The content of linear regression (OLS) estimation can be summarized as follows:

From random sample $\{(Y_i, X_{1i}, \ldots\ldots, X_{pi})\}_{i=1}^{n}$, need to find an estimate $\widehat{\beta} = (\widehat{\beta_0},\ldots,\widehat{\beta_1},\ldots,\widehat{\beta_p})$ so that:

$$\min_{\beta\in\mathbb{R}^p}\left\{\sum_{i=1}^{n}\left(Y_i-\beta_o-\beta_1X_{1i}-\ldots-\beta_pX_{pi}\right)^2\right\}$$

Gauss–Markov (1822) proved and confirmed that the OLS method gives the estimation results that are linear, unbiased and efficient if the following four conditions are satisfied: (1) There is no correlation between random errors course; (2) The conditional expectation of random error is zero; (3) The conditional variance of is constant; (4) There is no perfect multicollinearity between the independent variables.

In OLS estimation, probability values (P-values) are used to confirm the regression coefficient βj, thereby explaining the influence of independent variables X1, X2, ..., Xp on the dependent variable Y. According to Casella and Berger (2002), Cohen (1994), the OLS estimation method has limitations and errors in determining the independent variables that explain X1, X2, ..., Xp on the dependent variable Y., is only at least one variable, or all variables, or only a subset of the variables needed. Tibshirani (1996), said that the OLS estimation method gives non-unique results, it is not possible to determine the degree of influence of each independent variable Xj on the dependent variable Y in the case p > n − 1. Besides, the results of the OLS estimation are usually unbiased and large variance estimates. In these cases, the independent variables have a tight linear correlation, and the number of independent variables in the model is quite large. Because of the large variance, the accuracy of the OLS estimates is not high, hence, the statistical inference results are not reliable.

To solve the limitations of the OLS model, the LASSO (The Least Absolute Shrinkage and Selection Operator) estimation method was introduced by Tibshirani in 1996, the Ridge method (Hoerl and Kennard 1970), Elactic net (Zou and Hastie 2015) in turn was born with the aim of simplifying the effect analysis and improving the predictive power of the OLS regression model.

4.2 Regression Lasso (The Least Absolute Shrinkage and Selection Operator)

LASSO "stands for Least Absolute Reduction and Selection Operator". First introduced by Tibshirani (1996). Lasso regression is a variant of linear regression, like linear regression, Lasso regression tries to fit the data by minimizing RSS (Residual Sum of Squares), with the addition of the expression $\lambda \sum_{J=1}^{P} \|\beta\|_j$ into the model:

$$\sum_{i=n}^{1}\left(y^i - \beta_0 - \sum_{j=1}^{P} \beta_j x_{ij}\right)^2 + \lambda \sum_{J=1}^{P} \beta_J^2 = RSS + \lambda \sum_{J=1}^{P} \|\beta\|_j$$

where, expression $\lambda \sum_{J=1}^{P} \|\beta\|_j$ called norm L1

- With Coefficient β will be taken as absolute value.
- Lamda coefficient (λ), also known as the tuning parameter, or the Penatly parameter, or the Shrinkage parameter, is a number that is always positive, the value at which the linear equation will be refined. adjusted so that the error of the model is minimized, that is, the value of λ in the model that reaches the minimum MSE (Mean Square Error) will be selected. The minimum model MSE (Mean Square Error) has the following form:

$$\min_{\beta^1 \ldots \beta^p} \frac{1}{n} \sum_{i=n}^{1} \left(y^i - \beta_0 - \sum_{j=1}^{p} \beta_j x_{ij} \right)^2 + \lambda \sum_{J=1}^{P} \beta_J^2 = RSS + \lambda \sum_{J=1}^{P} \|\beta\|_j$$

Adjustable parameter λ, rated strength control L1. λ is basically the amount of shrinkage or shrinkage of the model factor:

- When $\lambda = 0$, no parameters are removed. The estimate found is equal to the linear regression (OLS) estimate.
- When λ is non-zero, the coefficients will be reduced to smaller than the linear regression coefficient (OLS).
- When λ increases, more and more coefficients are set to zero and discarded (theoretically, when $\lambda = \infty$, all coefficients are discarded).
- When λ increases, the deviation increases.
- When λ decreases, the variance increases.

From this, we can see that as λ is larger, the variance decreases and the deviation increases. This begs the question: how much bias are we willing to accept to reduce the variance? or what is the optimal value for λ?

There are two possible ways to solve this problem. A more traditional approach would be to choose λ such that information criteria, such as AIC or BIC, are minimal. A more machine learning-like approach is to perform cross-validation and choose the value of λ to minimize the sum of squared residuals (SSR) cross-validation. The first approach emphasizes the fit of the model to the data, while the second approach focuses more on its predictive performance.

Method 1: Minimize information criteria

This approach is to estimate the model with various values for λ and choose one that minimizes the Akaike or Bayes information criterion:

$$AIC_{Ridge} = nlog\left(e^{'}e\right) + 2d \int Lasso,$$

$$BIC_{Ridge} = nlog\left(e^{'}e\right) + 2d \int Lasso\log(n)$$

Method 2: Minimize cross-validation residuals

The cross-validation procedure is one of the methods used to find the best estimator for λ. In the cross-validation procedure, the data set is randomly divided into K subsets, denoted C1, ..., Ck. For each value of λ, the cross-validation procedure is performed as follows: Consider Ci as the test set, (K–1) the remaining set as the training set. Applying the LASSO estimation method on this training set to obtain the model, then using the test set Ci to predict the value of the dependent variable and record the value of MSE_i (Mean Squared Error). Take turns with i = 1, 2, …, K. Let CVM λ be the average of the MSE_i.

$$CVM_\lambda = \frac{\sum_{i=1}^{K} MSE_i}{K}$$

Thus, each λ corresponds to a CVM λ. When is variable over a given set, the best value of λ is the one corresponding to the smallest CVM λ (Hastie et al. 2015).

4.3 Regression Ridge (Shrinkage Regression)

Ridge regression was introduced by Hoerl and Kennard (1970). Ridge regression, like Lasso regression, will minimize RSS (Residual Sum of Squares). By adding the expression $\lambda \sum_{J=1}^{P} \beta_J^2$ into following model:

$$\sum_{i=n}^{1}\left(y^i - \beta_0 - \sum_{j=1}^{p} \beta_j x_{ij}\right)^2 + \lambda \sum_{J=1}^{P} \beta_J^2 = RSS + \lambda \sum_{J=1}^{P} \beta_J^2$$

where: $\lambda \sum_{J=1}^{P} \beta_J^2$, will be called the norm L2

- The coefficient will be the square, not the absolute value like Lasso.
- Lamda coefficient (λ), called the tuning parameter as well as Lasso regression, is a positive number, the value at which the linear equation will be refined so that the error of the model is minimized. maximum, that is, the value of λ in the model that reaches the minimum MSE (Mean Square Error) will be selected. The minimum model MSE (Mean Square Error) has the following form:

$$\underset{\beta^1 \ldots\ldots \beta^p}{min} \frac{1}{n} \sum_{i=n}^{1}\left(y^i - \beta_0 - \sum_{j=1}^{p} \beta_j x_{ij}\right)^2 + \lambda \sum_{J=1}^{P} \beta_J^2 = RSS + \lambda \sum_{J=1}^{P} \beta_J^2$$

- Adjustable parameter λ, control rated strength L2.
- When λ = 0, the result is the same as for conventional linear regression (OLS).
- When λ is non-zero, the coefficients are reduced to smaller than linear regression (OLS).
- When λ increases to infinity, the coefficients become close to zero, not zero like Lasso regression.

In the Ridge model, the selection of λ is similar to the Lasso model, there will be two ways to choose by two methods: Information Criteria Minimization (AIC or BIC, is the smallest) and Cross-validated Residual Reduction Minimization.

4.4 *Elactic Net (Elactic Net Regression)*

Elastic net was introduced by Zou and Hastie (2015). This is a regression model combining the two Lasso and Rigde models. Combining these two models will help us choose the optimal model and solve some limitations of the two Lasso and Ridge models. The general formula is as follows:

$$L_{enet}\left(\hat{\beta}\right) = \frac{\sum_{i-1}^{n}\left(y_i - x_i^j\hat{\beta}\right)^2}{2n} + \lambda\left(\frac{1-\alpha}{2}\sum_{j=1}^{m}\hat{\beta}_j^2 + \alpha\sum_{j=1}^{m}\left|\hat{\beta}_j\right|\right)$$

where α is the mixing parameter between Ridge (α = 0) and Lasso (α = 1). For α = 0.5, is the 50% coordination between the Lasso regression and the 50% Ridge regression.

5 Result

The authors uses R software to perform calculations with the following steps:

- By default, the glmnet () function that performs regression on R software allows automatic selection of λ values. However, here the author has chosen to implement the function on a set of values ranging from $=10^{10}$ to $=10^{-2}$, basically covering all scenarios from the empty model containing only intercept, to the least squares model.
- The multiplier α is deployed on the data set. If α = 0 will be the Ridge model, α = 0.5 will be the Elactic Net model, α = 1 will be the Lasso model.
- Find the most optimal λ by cross-validation based on the appropriate norm for each model, L1 (Lasso model) and L2 (Ridge model), where the MSE (Mean Squared Error) value is the smallest, λ optimal will be selected.
- Calculate the MSE of the models corresponding to the Lasso, Ridge, and Elactic Net estimation methods on the data set.
- Compare the regression results between the models based on the criteria of MSE (Mean Squared Error) and R^2 (Adjusted R Square) to choose the most optimal model.

5.1 Ridge Regression Results

Ridge regression results, with the most optimal λ (λ = 0.163) selected by the cross-validation method (CV) presented in Table 2, shows that all variables have an impact on the dependent variable P/E. In which, variables DE, EG have negative influence on P/E. The remaining variables all have the same effect.

Below is the parameter λ selected by the cross-validation method (CV), in order to choose the most optimal λ (λ = 0.163) with the minimum MSE shown in the following Fig. 1.

Table 2 Estimated results by Ridge method with most optimal

Independent variables	Estimated coefficient (λ = 0.163)
(Intercept)	−48.312253
DP	5.999769
DE	−1.327371
EG	−2.243916
ROE	20.640949
SIZE	1.47832
MB	1.750764

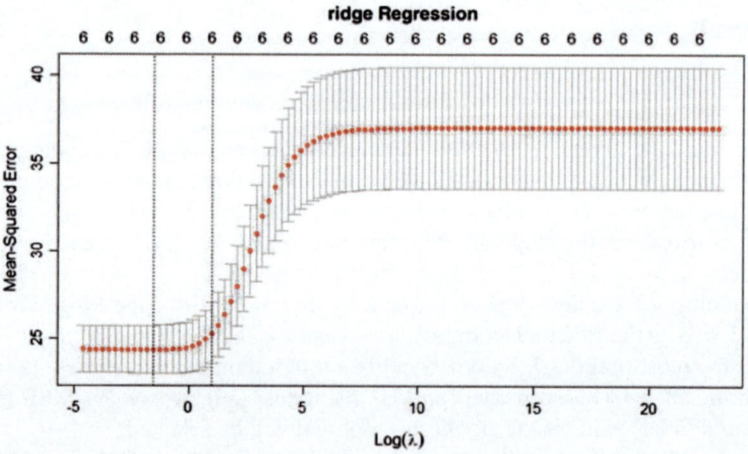

Fig. 1 Variation of MSE according to optimal (λ = 0.163) Ridge regression

Table 3 Estimation results by Elactic Net method with most optimal

Independent variables	Estimated coefficient (λ = 0.01)
(Intercept)	−50.0161
DP	6.034889
DE	−1.35097
EG	−2.2659
ROE	21.27332
SIZE	1.499328
MB	1.806394

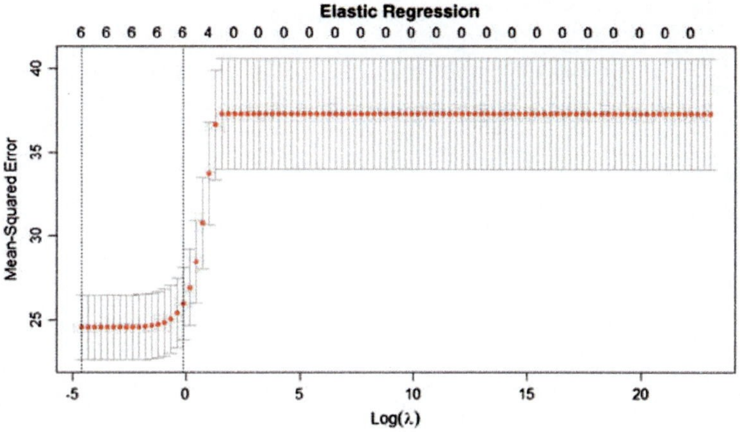

Fig. 2 Variation of MSE according to optimal (λ = 0.01) Elactic Net regression

5.2 *Elactic Net Regression Results*

The results of Elactic Net regression, with being selected the most optimally (λ = 0.01) by the cross-validation method (CV) presented in Table 3, show that all variables have an impact on the dependent variable P/E. In which, variables DE, EG have negative influence on P/E. The remaining variables all have the same effect.

Below is the parameter λ selected by the cross-validation method (CV) to choose the most optimal λ (λ = 0.01) with the minimum MSE shown in the following Fig. 2.

5.3 *Lasso Regression Results*

The results of Lasso regression, with λ being selected the most optimally (λ = 0.01) by cross-validation (CV) method presented in Table 4, show that all variables have

Table 4 Estimation results by Lasso method with the most optimal

Independent variables	Estimated coefficient ($\lambda = 0.01$)
(Intercept)	−50.0161
DP	6.034889
DE	−1.35097
EG	−2.2659
ROE	21.27332
SIZE	1.499328
MB	1.806394

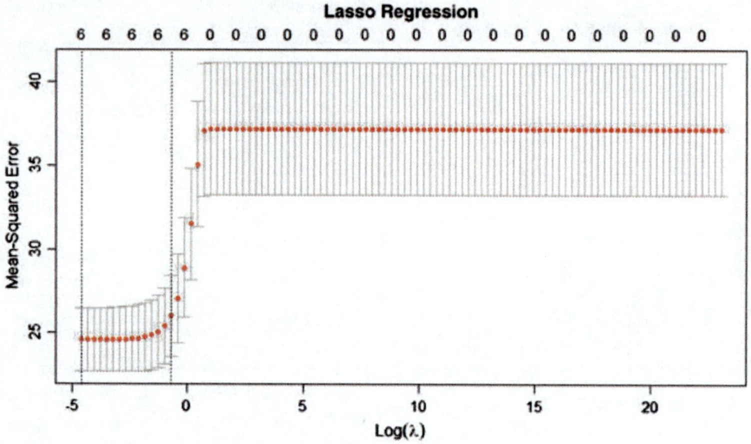

Fig. 3 Variation of MSE according to optimal ($\lambda = 0.01$) Lasso regression

an impact on the dependent variable P/E. In which, variables DE, EG have negative influence on P/E. The remaining variables all have the same effect.

Below is the parameter λ selected by the cross-validation method (CV) to choose the most optimal λ ($\lambda = 0.01$) where the MSE price must be the smallest as shown in the following Fig. 3.

5.4 *Compare the Regression Results Lasso, Ridge, Elactic Net*

To compare and choose which model results will be the most optimal, the author will base on two criteria: MSE (Mean Squared Error) and R^2 (Adjusted R Square). The model with the smallest MSE and the highest R^2 will be the most optimal model selected. The results of Table 5 show that there are both models with the smallest

Table 5 Comparison of MSE and R^2 from LASSO, rigde, Elactic Net

Model criteria	Ridge ($\lambda = 0.163$)	Elactic Net ($\lambda = 0.01$)	Lasso ($\lambda = 0.01$)
R^2	0.331	0.362	0.362
MSE	24.70	23.567	23.567

Table 6 Summary of Lasso and Elactic Net regression results

Independent variables	Lasso	Elactic Net
	Estimated coefficient ($\lambda = 0.01$)	Estimated coefficient ($\lambda = 0.01$)
(Intercept)	−50.0161	-50.0161
DP	6.034889	6.034889
DE	−1.35097	−1.35097
EG	−2.2659	−2.2659
ROE	21.27332	21.27332
SIZE	1.499328	1.499328
MB	1.806394	1.806394

MSE and the highest R^2, which are the Lasso regression model and the Elactic Net model, with MSE = 23,567 and $R^2 = 0.362$.

Thus, choosing the most optimal model based on MSE and R^2 criteria, the result is that both optimal models are selected as Lasso and Elactic Net models. Regression summary results are in Table 6, the estimated coefficients of both models are the same and all variables affect the dependent variable P/E. In which, the variables DE, EG have the opposite effect on the P/E coefficient. The remaining variables all have the same effect.

6 Conclusions and Recommendations

6.1 Conclude

The article studies the factors affecting the price-to-earnings ratio, based on collected data of 106 non-financial companies listed on the Vietnamese stock market in the period 2015–2020. The new approach is to apply Lasso, Ridge, and Elactic Net regression estimation methods. Research results have two selected optimal estimates, Lasso and Elactic Net estimates, both estimates are selected based on the criteria of the smallest MSE (Mean Squared Error) and the highest R^2 with MSE of 23,567 and R^2 is 0.362. At the same time, the regression result is that the coefficients of the two models themselves are the same. Therefore, the interpretation of the effects on the dependent variable is the same and there is no difference.

The content of the study is presented in general about the new asymptote by Ridge, Elactic Net and Lasso estimation methods. All three methods aim to reduce the coefficients and improve the explanatory level for the model. However, in this study, the coefficients (β) of the explanatory variable did not differ too much between the estimation methods and the level of explanation R^2 was also not significantly different.

Research results when applying Lasso and Elactic Net models have a statistically significant correlation between all explanatory variables on the ratio of price to income (P/E) with the explanatory level of 36.2%. In which, the debt-to-equity ratio (DE); Earnings growth (EG) has a negative effect on P/E. The remaining factors are dividend payout ratio (DP); Company size (SIZE); return on equity (ROE) and stock price to book ratio (MB) both have a positive effect.

The author's research results are similar to the studies of Faezinia (2012), Ramadan (2014) when finding a positive correlation between factors of profit growth (EG), stock price to book ratio (MB) and dividend payout ratio (DP) to P/E.

Faezinia (2012), also found a positive correlation between the factors of firm size (SIZE), return on equity (ROE) and P/E. Meanwhile, Itemgenova and Sikveland (2020) found a negative relationship between return on equity (ROE) and P/E.

Johan and Filip (2007), Faezinia (2012) found a negative correlation between the debt-to-equity (DE) ratio and P/E. In the author's research, this relationship is also found.

6.2 Recommendations

Based on the above research results, the author makes some recommendations as follows:

Regarding return on equity (ROE), there is a positive effect on P/E. Therefore, investors should look for companies with high ROE to bring investment efficiency. ROE reflects operating companies with high or low return on equity. If ROE is high, it proves that the company is operating effectively compared to its equity invested, and when it is profitable, the company can pay regular dividends to shareholders, and at the same time can keep it for future purposes. Expanding reinvestment in the company's business activities. On the contrary, if ROE is low, it proves that the company is operating inefficiently and cannot accumulate resources to reinvest. When it needs to expand its business, the company will take out a loan, which will increase credit risk and affect its impact. to the sustainable performance of the company.

Regarding company size (SIZE), there is a positive influence on P/E. This index shows that when a company operates efficiently, it will retain a part of its profits to reinvest and expand its business activities, diversify production activities to increase profits, on the other hand when the increase in the size of the enterprise also shows a positive signal in the business as well as the financial potential of that enterprise. Therefore, investors should focus and consider companies with large scale in the market, as well as with stable business activities before deciding to invest, to bring high efficiency.

Regarding dividend payout ratio (DP), there is a positive influence on P/E. This index shows that companies with stable and effective performance will have a high and regular dividend payout ratio. Therefore, investors should focus on companies with high dividend payout ratio because these companies often operate effectively and stably in the future.

Regarding the ratio of stock price to book price (MB), there is a positive influence on P/E. The higher this index, the more proof that the business is operating effectively and is of interest to investors. On the contrary, a low index indicates that the company is not operating effectively and investors will be afraid, which will cause the stock price to drop. Therefore, investors need to pay attention and consider this index carefully before investing, to bring efficiency.

Regarding the debt-to-equity (DE) ratio, there is a negative effect on P/E. This ratio shows that an increase in debt to equity will decrease the P/E ratio. When a company borrows a lot and uses it for improper purposes, there will always be potential financial risk and insolvency or bankruptcy. Therefore, investors should invest in companies with little debt and good finance to avoid risks.

In terms of profit growth (EG), this index represents the profit of the current year compared to the previous year to evaluate the growth of the company. Research results show that this index negatively affects P/E. In conclusion, the companies that the author collects data in the period 2015–2020, have unstable and sustainable activities, increase or decrease abnormally. Therefore, it is necessary to collect more data and expand the research time to be able to evaluate the company's growth more effectively and accurately.

References

Afza, T., Tahir, S.: Determinants of price-earnings ratio: the case of chemical sector of Pakistan. Int. J. Acad. Res. Bus. Soc. Sci. **2**(8) (2012). ISSN 2222-6990

Casella, G., Berger, R.L.: Statistical Inference. Pacific Grove, CA: Thomson Learning. (2002)

Cohen, J.: The earth is round (p<. 05). American Psychological Association, **49** (12), 997–1003. (1994)

Dutta, K.D., Saha, M., Das, D.C.: Determinants of P/E ratio: an empirical study on listed manufacturing companies in DSE **8**(4), 167–173 (Apr 2018). 5 (2017)

Faezinia, V.: The quantitative study of effective factors on price-earning ratio in capital market of Iran. IJCRB **3**(10) (Feb 2012)

Fesokh, A.A., Haddad, F.S.: Determinants of the price earnings ratio in the manufacturing industries in Jordan. Jordan J. Bus. Adm. **15**(2), 227–235 (2019)

Gordon, M., Shapiro, E.: Capital equipment analysis: the required rate of profit. Manag. Sci. **3**, 102–112 3 (1956)

Gauss, C.F.: Least squares (1809)

Hastie, T., Tibshirani, R., Wainright, M.: Statistical Learning with Sparsity: The Lasso and Generalizations (CRC, Boca Raton, FL). (2015)

Hoerl, A.E., Kennard, R.W.: Ridge regression: applications to nonorthogonal problems. Technometrics, **12**(1), 69-82. (1970)

Itemgenova, A., Sikveland, M.: The determinants of the price-earnings ratio in the Norwegian aquaculture industry. J. Commod. Markets **17**, 100089 (2020)
Johan, K.K., Filip, L.: A quantative study of the P/E ratio on the Swedish market. Master thesis (2007)
Ramadan, I.Z.: Confidence index determinants of the Amman stock exchange listed companies **10**(1) (2014); 2015. 91–104. 6
Tibshirani, R.: Regression shrinkage and selection via the lasso. J R Stat Soci: Series B (Methodological), **58**(1), 267-288. (1996)
Zou, H., Hastie, T.: Regularization and variable selection via the elastic net. J Royal statistical society: series B (statistical methodology), **67**(2), 301–320. (2005)